Deterministic, Stochastic, and Deep Learning Methods for Computational Electromagnetics

Wei Cai

Deterministic, Stochastic, and Deep Learning Methods for Computational Electromagnetics

Second Edition

Science Press
Beijing

Springer

Wei Cai
Department of Mathematics
Southern Methodist University
Dallas, TX, USA

ISBN 978-981-96-0099-1 ISBN 978-981-96-0100-4 (eBook)
https://doi.org/10.1007/978-981-96-0100-4

Jointly published with Science Press
The print edition is not for sale in China (Mainland). Customers from China (Mainland) please order the print book from: Science Press.

Mathematics Subject Classification: 65N30, 65N38, 65N75, 78-10, 65N06

This Springer imprint is published by the registered company Springer Nature Singapore Pte Ltd.
The registered company address is: 152 Beach Road, #21-01/04 Gateway East, Singapore 189721, Singapore

To my wife, Xiaoyan Duan (段晓艳) and my children, Angela Cai (蔡安琪) and Richard Cai (蔡敏琪)

Preface to the Second Edition

Since the publication of the first edition of this book over 11 years ago under the title "*Computational Methods for Electromagnetic Phenomena: Electrostatics in Solvation, Scattering, and Electron Transport*" [1], many new developments in numerical methods have occurred, especially in stochastic and deep learning approaches. To update it with the new results in these and other areas, this book is expanded with work from the research community as well as the author. The current edition, with a new title, includes numerical methods using stochastic and deep neural network learning as well as deterministic approaches.

The intended application area of the computational methods is still electromagnetics, and as in the first edition, the term "electromagnetics" is interpreted in a broader scope than the conventional one, which usually only covers classic electromagnetic (EM) processes described by Maxwell's equations such as EM wave scattering and propagation. Thus, in this book, electromagnetic processes in quantum systems governed by the Schrödinger equation such as electron transport in nanoelectronics and electronic structures of molecules and nanosystems, and their computations are also considered. As in the first edition, the book is divided into three parts: Part I, *Electrostatics*; Part II, *Electromagnetic Scattering*; and Part III, *Electron Transport and Electronic Structure.* Five new chapters have been included in the second edition, giving the book a total of 21 chapters. Corrections and minor changes and literature updates have also been made to other chapters.

The following summarizes the material in the new five chapters, while the synopsis of other 16 chapters can be found in the Preface to the First Edition. Chapter 4 discusses stochastic and probabilistic methods for solving electrostatic problems based on the Feynman–Kac formula, and hybrid boundary integral and random walk-on-spheres methods for solving mixed boundary value problems of the Laplace equation. A concise review of basic stochastic calculus, including Itô's formula and Dynkin's formula as well as stochastic differential equations (SDEs), is included for completeness, and their applications in SDE-based deep neural networks for PDEs are presented in Chap. 6. Chapter 5 discusses stochastic spectral methods for uncertainty quantification and their application in assessing the effect of geometric randomness on the capacitance of rough microchip interconnects. In Chap. 6, several deep neural network learning methods for solving PDEs are discussed, including Deep-Ritz, PINN, and WAN learning methods, as well as SDE-based learning algorithms by leveraging the connection between classic elliptic and parabolic PDEs and forward and backward SDEs. Then, methods to mitigate spectral biases of DNN learning toward lower frequencies are discussed, including phase-shift DNN (PhaseDNN) and multiscale DNN (MscaleDNN). Chapter 10 presents a newly developed fast multipole method for wave source interactions in layered media in three-dimensional space governed by Green's function of the Helmholtz equation. Finally, Chap. 17 discusses density functional theory for electronic structures of quantum systems and its computational methods, including plane wave and finite element methods as well as a linear scaling $O(N)$ density matrix minimization method using the discontinuous Galerkin discretization for computing ground state of quantum systems.

Acknowledgements

The second edition is based on author's lectures given at Southern Methodist University, and includes research work since the first edition by the author and his collaborators, particularly, Bo Wang, Wenzhong Zhang, Pei Elton Hsu, Chanhao Yan, Cuiyan Ding, Yijing Zhou, Ziqi Liu, Zhiqin Xu, Xiaoguang Li, Lizuo Liu, Qing He, Tiao Lu, Haiyan Jiang, and Hengliang Zhu. The author has benefited greatly from many discussions with Weinan E and George Karniadakis on deep learning and uncertainty quantification for scientific computing. The author thanks the following researchers for their help and input during the research, which led to some of the results covered in the book, including Minhyung Cho, Jack Douglas, Qiang Du, Tom Hagstrom, Jiequn Han, Jan Hesthaven, Shi Jin, Lin Lin, Chun Liu, Jianguo Liu, Qing Huo Liu, Jianfeng Lu, Michael Mascagni, Vassilis Papanicoloau, Karl K. Sabelfeld, Chiwang Shu, Tao Tang, Johannes Tausch, Lilian Wang, Dongbin Xiu, Jinchao Xu, Zhenli Xu, Haizhao Yang, George Yin, Lexing Ying, Haomin Zhou, and Tao Zhou. Moreover, the Preface to the First Edition credited many other collaborators of the author for contributing to the results covered in this book.

The author is also thankful for the support of the Betty Clements Chair in Applied Mathematics at Southern Methodist University for providing resources and time, which made the writing of this book possible. Appreciation is given to the US NSF for partially supporting some of the research presented in this book. The author thanks Science Press editor Liping Wang and Springer editors Daniel Wang and Prasanth Anandan for their support and efforts in publishing this book.

Notations

- The symbol Ω denotes random sample space in the probability space triplet $(\Omega, \mathcal{F}, P)$ when discussing stochastic methods in Chaps. 4–6 or a physical domain $\Omega \in \mathbb{R}^d$, $d = 2, 3$ in other chapters.
- The symbol E denotes the mathematical expectation in Chaps. 4–6 or an energy level of a physical system in Part III.

Dallas, Texas
July 2024

Wei Cai
cai@smu.edu

Reference

1. W. Cai. Computational Methods for Electromagnetic Phenomena: electrostatics in solvation, scattering, and electron transport. Cambridge University Press; 2013.

Preface to the First Edition[1]

工欲善其事，必先利其器.
—Analects

Electromagnetic (EM) processes play an important role in many scientific and engineering applications such as the electrostatic forces in biomolecular solvation, radar wave scattering, the interaction of light with electrons in metallic materials, and current flows in nano-electronics, among many others. These are the kinds of electromagnetic phenomena, from atomistic to continuum scales, discussed in this book.

While the focus of the book is on a wide selection of various numerical methods for modeling electromagnetic phenomena, as listed under the entry "numerical methods" in the book index, attention is also given to the underlying physics of the problems under study. As computational research has become strongly influenced by the interaction from many different areas such as biology, physics, chemistry, and engineering, etc., a multifaceted and balanced approach addressing the interconnection among mathematical algorithms and physical principles and applications is needed to prepare graduate students in applied mathematics, sciences, and engineering, to whom this book is aimed, for innovative advanced computational research.

This book arises from courses and lectures the author gave in various universities: the UNC Charlotte and the UC Santa Barbara in the USA, and Peking University, Fudan University, and Shanghai Jiao Tong University in China, to graduate students in applied mathematics and engineering. While attempts are made to include the most important numerical methods, the materials presented are undoubtedly affected by the author's own research experience and knowledge. The principle of selecting the materials is guided by Confucius's teaching above—"For a man to succeed in his endeavors, he must first sharpen his tools". So, emphasis is on the practical

[1] Published by Cambridge University Press, 2013 [1]. Chapter numbers in parentheses refer to the second edition.

and algorithmic aspects of methods ready for applications, instead of detailed and rigorous mathematical elucidation.

The book is divided into three major parts according to three broadly defined though interconnected areas: electrostatics in biomolecules, EM scattering and guiding in microwave and optical systems, and electron transport in semiconductor and plasma media. The first two areas are based on atomistic and continuum EM theory, while the last one is based on Schrödinger quantum and also Maxwell's EM theories. Part I starts with a chapter on the statistical molecular theory of dielectric constants for material polarization in response to an electric field, an important quantity for molecular dynamics simulation of biomolecules and understanding optical properties of materials addressed in the book. Then, the Poisson–Boltzmann (PB) theory for solvation is given in Chapter 2, together with analytical approximation methods such as the generalized Born method for solvation energy and image methods for reaction fields in simple geometries. Chapter 3 contains various numerical methods for solving the linearized PB equations including the boundary integral equation methods, the finite element methods, and the immersed interface methods. Chapter 4(7) presents three methods to handle the long-range electrostatic interactions—a key computational task in molecular dynamics algorithms: the particle-mesh Ewald, the fast multipole method, and a reaction field-based hybrid method.

Part II contains a large collection of numerical techniques for solving the continuum Maxwell's equations for scattering and propagation in time and frequency domains. This part starts with Chapter 5(8) on Maxwell's equations with physical and artificial boundary conditions; the former includes dielectric interface conditions and Leontovich impedance boundary conditions for conductors with a perfect electric conductor (PEC) as a limiting case, and the latter includes local absorbing boundary conditions and uniaxial perfectly matched layer (PML) boundary conditions. Chapter 6(9) discusses the dyadic Green's functions in layered media for Maxwell's equations in the frequency domain and an algorithm for fast computation. High-order surface integral methods for electromagnetic scattering form the subject of Chapter 7(11), which includes the Galerkin method using mixed vector–scalar potentials and the Nyström collocation method for both the hypersingular integral equations and the mixed vector–scalar potential integral equations, and combined integral equations for the removal of resonance in cavities. Finally, the high-order surface current basis for the Galerkin integral equation methods is discussed. Chapter 8(12) on edge elements begins with Nédélec's original construction of the $H(\text{curl})$ conforming basis, and then presents hierarchical high-order elements in 2-D rectangles and 3-D cubes and simplexes in both 2-D and 3-D spaces. Next, time-domain methods, including the discontinuous Galerkin (DG) methods with a high-order hierarchical basis and the finite difference Yee scheme, are given in Chapter 9(13). Numerical methods for periodic structures and surface plasmons in metallic systems are covered in Chapter 10(14), including plane-wave-based methods and transmission spectra calculations for photonics band structures, finite element methods, and volume integral equation (VIE) methods for Maxwell's equations. For the surface plasmons, the DG methods for dispersive media using auxiliary differential equations (ADEs) are given for Debye and Drude media. The final chapter—Chapter 11(15) of Part II contains numerical methods for Schrödinger equations for dielectric optical waveguides and quantum dots: a generalized DG method for the paraxial approximation in optical waveguides, and a VIE method for Schrödinger equations in quantum dots embedded in layered semiconductor materials.

Part III starts with Chapter 12(16) on the electron quantum transport models in semiconductors, which also includes the Fermi–Dirac distribution for electron gas within the Gibbs ensemble theory, density operators, and kinetic descriptions for quantum systems. The quantum transport topics discussed in this chapter include the Wigner transport model in phase space for electrons, the Landauer transmission formula for quantum transport, and the non-equilibrium Green's function (NEGF) method. Then, the non-equilibrium Green's function method in Chapter 13(18) contains the treatment of quantum boundary conditions and finite difference and finite element methods for the NEGF; the latter allows the calculation of the transmission coefficients in the Landauer current formula for general nano-devices. Chapter 14(19) includes numerical methods for the quantum kinetic Wigner equations with the upwinding finite difference and an adaptive cell average spectral element method. Chapter 15(20) first presents the semi-classical Boltzmann and continuum hydrodynamic models for multi-species transport, including electron transport, and then follows with the numerical methods for solving the hydrodynamic equations by Godunov methods and WENO and central differencing methods. In the final chapter of the book, Chapter 16(21), we first present the kinetic Vlasov–Fokker–Planck (VFP) model and the continuum magneto-hydrodynamic (MHD) transport model for electrons in plasma media. Then, several numerical methods are discussed including the VFP scheme in phase space, and the particle-in-cell and constrained transport methods for the MHD model, where the divergence-free condition for the magnetic field is specifically enforced.

In making this book a reality, I credit my education and ways of doing research to my teachers Prof. Zhongci Shi at the University of Science and Technology of China (USTC), who exposed me to the power of non-conforming finite element methods and reminded me that computational research must not be devoid of real science and engineering relevance, and Prof. David Gottlieb (my doctoral thesis advisor) at Brown University, who taught me that simplicity is the beauty in sciences. Also, my scientific research has benefited greatly from encouragements and interactions from the late Prof. Steven Orszag over many years. I have learnt much from interactions with my colleague physicist Prof. Raphael Tsu (a co-inventor of the resonant tunneling diode and a pioneer in quantum superlattices), whose sharp physics insight has always been an inspiration and pleasure during many of our discussions. My former colleague Prof. Boris Rozovsky has provided much encouragement, spurring me to undertake the challenge of writing this book, which started in 2004 during one of my many research collaboration visits with Prof. Pingwen Zhang at Peking University through the Beijing International Center for Mathematical Research. This book would not be possible without the joint research work undertaken in the past few decades with my colleagues Pingwen Zhang and Shaozhong Deng, and my former students and postdoctoral researchers Tiejun Yu, Yijun Yu, Yuchun Lin, Tiao Lu, Xia Ji, Haiyan Jiang, Min hyung Cho, Kai Fan, Sihong Shao, Zhenli Xu, and Jianguo Xin. Special thanks are given for the many useful discussions with my friends and other colleagues, which have contributed to my understanding of various topics in the book, including Achi Brandt, Alexandre Chorin, Weinan E., George Karniadakis, Chiwang Shu, Leslie Greengard, Jan Hesthaven, Tom Hagstrom, Eitan Tadmor, Shiyi Chen, Roger Temam, Weng Cho Chew, Jian-ming Jin, Dian Zhou, Xuan Zeng, Jinchao Xu, Jianguo Liu, Shi Jin, Houde Han, Jing Shi, Ann Gelb, Gang Bao, Jingfang Huang, Bob Eisenberg, Chun Liu, Xianjun Xing, Benzhuo Lu, Tao Tang, Jie Shen, Huazhong Tang, Tsinghua Her, Andrij Baumketner, Donald Jacobs, Guowei Wei, Vasily Astratov, and Greg Gbur. I would like to thank Dr. Shaozhong Deng for his careful reading of the manuscript; many improvements in the presentation of the book have resulted from his suggestions. The author is also grateful for the professional help and great effort of Ms. Irene Pizzie during the copy-editing of the book.

Finally, special acknowledgements are given to the continual support of the Advanced Scientific Computing Research, Office of Science at the Department of Energy (under program managers Sandy Landsberg and Karen Pao) and the Army Research Office (under program manager Joseph Myers) over the years, and to NSF and NIH for allowing me to undertake the research that is behind many results contained in this book.

Charlotte, North Carolina, USA — Wei Cai
September 2012

Reference

1. W. Cai. *Computational Methods for Electromagnetic Phenomena: electrostatics in solvation, scattering, and electron transport*. Cambridge University Press; 2013.

Contents

Part I Electrostatics

1 Dielectric Constant and Fluctuation Formulae for Molecular Dynamics ... 3
1.1 Electrostatics of Charges and Dipoles ... 3
1.2 Polarization **P** and Displacement Flux **D** ... 5
1.2.1 Bound Charges Induced by Polarization ... 6
1.2.2 Electric Field $\mathbf{E}_{\rm pol}(\mathbf{r})$ of a Polarization Density $\mathbf{P}(\mathbf{r})$ ... 8
1.2.3 Singular Integral Expressions of $\mathbf{E}_{pol}(\mathbf{r})$ Inside Dielectrics ... 10
1.3 Clausius–Mossotti and Onsager Formulae for Dielectric Constant ... 11
1.3.1 Clausius–Mossotti Formula for Non-polar Dielectrics ... 11
1.3.2 Onsager Dielectric Theory for Dipolar Liquids ... 13
1.4 Statistical Molecular Theory and Dielectric Fluctuation Formulae ... 18
1.4.1 Statistical Methods for Polarization Density Change $\Delta\mathbf{P}$ ... 20
1.4.2 Classical Electrostatics for Polarization Density Change $\Delta\mathbf{P}$ ... 22
1.4.3 Fluctuation Formulae for Dielectric Constant ϵ ... 24
1.5 Appendices ... 26
1.5.1 Appendix 1: Average Field of a Charge in a Dielectric Sphere ... 26
1.5.2 Appendix 2: Electric Field Due to a Uniformly Polarized Sphere ... 26
1.6 Summary ... 28
References ... 28

2 Poisson–Boltzmann Electrostatics and Analytical Approximations 31
2.1 Poisson–Boltzmann (PB) Model for Electrostatic Solvation 31
2.1.1 Debye–Hückel Poisson–Boltzmann Theory 33
2.1.2 Helmholtz Double Layer and Ion-Size Effect 36
2.1.3 Electrostatic Solvation Energy 40
2.2 Generalized Born (GB) Approximations of Solvation Energy 42
2.2.1 Still's Generalized Born Formulism 42
2.2.2 Integral Expression for Born Radii 43
2.2.3 FFT-Based Algorithm for the Born Radii 45
2.3 Method of Images for Reaction Fields 51
2.3.1 Methods of Images for Simple Geometries 52
2.3.2 Image Methods for Dielectric Spheres 54
2.3.3 Image Methods for Dielectric Spheres in Ionic Solvent 61
2.3.4 Image Methods for Multi-layered Media 63
2.4 Summary 68
References 68

3 Numerical Methods for Poisson–Boltzmann Equations 73
3.1 Boundary Element Methods (BEMs) 73
3.1.1 Cauchy Principal Value (CPV) and Hadamard Finite Part (p.f.) 74
3.1.2 Surface Integral Equations for the PB Equation 79
3.1.3 Computations of CPV and Hadamard p.f. and Collocation BEMs 86
3.2 Finite Element Methods (FEMs) 99
3.3 Immersed Interface Methods (IIMs) 102
3.4 Summary 105
References 106

4 Random Walk Stochastic Methods for PDE Boundary Value Problems 109
4.1 Brownian Motion, Itô Integral, and Itô's Formula 109
4.1.1 Probability Space, Random Variable, and Conditional Expectation 109
4.1.2 Brownian Motion, Itô Integral of Stochastic Processes, and Itô's Formula 113
4.2 Feynman–Kac Formula for Mixed BVPs 117
4.2.1 Dynkin's Formula and Feynman–Kac Formula for Dirichlet BVPs of Elliptic PDEs 117
4.2.2 Local Time $L(t)$ and Feynman–Kac Formulae for Neumann and Robin BVPs of the Schrödinger Equation 121

4.3 Pardoux–Peng's Nonlinear Feynman–Kac Formula for Quasilinear Parabolic PDEs 123
4.4 Walk-On-Spheres (WOS) and Local Time $L(t)$ for Reflecting Brownian Motion 125
4.4.1 WOS Method and Dirichlet BVP 125
4.4.2 Reflecting Brownian Motion and Its Local Time $L(t)$ 127
4.5 Electrical Impedance Tomography (EIT) 129
4.5.1 Feynman–Kac Formula for Mixed BVPs 129
4.5.2 Local EIT Solution with the WOS Method 130
4.6 Boundary Integral Equation and WOS (BIE-WOS) Method for the Dirichlet BVP of the Laplace Equation 132
4.7 Iterative BIE-WOS Method for the Mixed BVP of the Laplace Equation 134
4.8 Summary 136
References 136

5 Stochastic Spectral Methods and Uncertainty Quantification of Microchip Interconnects 139
5.1 Deterministic Spectral Methods 139
5.1.1 Fourier, Chebyshev, Legendre, and Hermite Spectral Approximations 139
5.1.2 Spectral Methods for BVPs of Differential Equations 145
5.2 Stochastic Spectral Methods for Stochastic Differential and Integral Equations 146
5.2.1 Wiener Polynomial Chaos (PC) and Generalized PC (gPC) Expansions 147
5.2.2 Random Variable Dimension Reduction 149
5.2.3 Spectral Galerkin Methods 150
5.2.4 Spectral Collocation Methods and Sparse Grids 153
5.3 Uncertainty Quantification of Rough Microchip Interconnects 155
5.4 Summary 159
References 159

6 Deep Neural Network Learning for PDE Solutions 161
6.1 Deep Neural Network (DNN) 161
6.2 Optimization with Stochastic Gradient Descent Method 163
6.3 DeepRitz Method for the Poisson–Boltzmann Equation of Biomolecular Solvation 164
6.4 Physics-Informed Neural Network (PINN) 167
6.5 Galerkin Weak Form Learning 167

6.6 Stochastic Differential Equation (SDE) Based DNN Methods 168
6.6.1 DeepBSDE for Quasilinear Parabolic Equations via a Terminal Condition Loss 169
6.6.2 Backward SDE Based Algorithms for Quasilinear Parabolic Equations with a Pathwise Loss 169
6.6.3 Martingale Based Learning Algorithm for BVPs 171
6.6.4 Diffusion Monte Carlo DNN Eigensolver 174
6.7 DNNs for Oscillatory Problems and Their Spectral Bias 175
6.7.1 Phase-Shift DNN (PhaseDNN) 175
6.7.2 Multiscale DNN (MscaleDNN) 177
6.7.3 Reduction in the Spectral Bias of DNNs 179
6.8 Summary 183
References 185

7 Fast Algorithms for Long-Range Interactions in Free Space 187
7.1 Ewald Sums for Charges and Dipoles 187
7.2 Particle–Mesh Ewald (PME) Methods 194
7.3 Fast Multipole Method for N-Particle Electrostatic Interactions 197
7.3.1 Multipole Expansions (MEs) 197
7.3.2 A Recursion for Local Expansions (LEs) ($0\to L$-Level) 201
7.3.3 A Recursion for the Multipole Expansion ($L \to 0$-Level) 203
7.3.4 A Pseudo-Code for FMM 204
7.3.5 Conversion Operators for Electrostatic FMM in $\mathbb{R}^3$ 205
7.4 Helmholtz FMM of Wideband of Frequencies for N-Current Source Interactions 207
7.5 Reaction Field Hybrid Model for Electrostatics 210
7.6 Summary 215
References 216

Part II Electromagnetic Scattering

8 Maxwell's Equations, Potentials, and Physical/Artificial Boundary Conditions 221
8.1 Time-Dependent Maxwell's Equations 221
8.1.1 Magnetization **M** and Magnetic Field **H** 222
8.2 Vector and Scalar Potentials 225
8.2.1 Electric and Magnetic Potentials for Time-Harmonic Fields 226

8.3 Physical Boundary Conditions for **E** and **H** ... 228
8.3.1 Interface Conditions Between Dielectric Media ... 228
8.3.2 Leontovich Impedance Boundary Conditions for Conductors ... 230
8.3.3 Sommerfeld and Silver–Müller Radiation Conditions ... 233
8.4 Absorbing Boundary Conditions for **E** and **H** ... 235
8.4.1 One-Way Wave Engquist–Majda Boundary Conditions ... 236
8.4.2 High-Order Local Non-reflecting Bayliss–Turkel Conditions ... 238
8.4.3 Uniaxial Perfectly Matched Layer (UPML) ... 242
8.5 Summary ... 248
References ... 249

9 Dyadic Green's Functions in Layered Media ... 251
9.1 Singular Charge and Current Sources ... 251
9.1.1 Singular Charge Sources ... 251
9.1.2 Singular Hertz Dipole Current Sources ... 253
9.2 Dyadic Green's Functions $\overline{\mathbf{G}}_E(\mathbf{r}|\mathbf{r}')$ and $\overline{\mathbf{G}}_H(\mathbf{r}|\mathbf{r}')$ ... 255
9.2.1 Dyadic Green's Functions for Homogeneous Media ... 256
9.2.2 Dyadic Green's Functions for Layered Media ... 257
9.2.3 Hankel Transform for Radially Symmetric Functions ... 257
9.2.4 Transverse Versus Longitudinal Field Components ... 259
9.2.5 Longitudinal Components of Green's Functions ... 260
9.3 Dyadic Green's Functions for Vector Potentials $\overline{\mathbf{G}}_A(\mathbf{r}|\mathbf{r}')$ ... 265
9.3.1 Sommerfeld Potentials ... 266
9.3.2 Transverse Potentials ... 268
9.4 Fast Computation of Dyadic Green's Functions ... 269
9.5 Summary ... 274
9.6 Appendix: Formula for $\widetilde{G}_1$, $\widetilde{G}_2$, and $\widetilde{G}_3$, etc. ... 275
References ... 276

10 Fast Multipole Methods for Wave Interactions in Layered Media ... 279
10.1 Rederiving ME and LE for the Free Space Helmholtz Green's Function ... 279
10.1.1 ME for the 2-D Free Space Helmholtz Green's Function ... 280
10.1.2 Rederiving h (Multipole) and j (Local)-Expansions of the 3-D Free Space Helmholtz Green's Function, and Translation Operators ... 282

10.2 FMM for the 3-D Helmholtz Equation in Layered Media 287
10.2.1 Helmholtz Green's Function in 3-D Layered Media 287
10.2.2 ME and LE for Reaction Components 290
10.2.3 Polarization Sources and MEs, LEs, and Their Translations 291
10.2.4 A FMM Algorithm for Wave Sources in Layered Media 296
10.3 Summary 299
References 299

11 High-Order Methods for Surface Electromagnetic Integral Equations 301
11.1 Electric and Magnetic Field Surface Integral Equations in Layered Media 301
11.1.1 Integral Representations 301
11.1.2 Singular and Hyper-Singular Surface Integral Equations 307
11.2 Resonance and Combined Integral Equations 315
11.3 Nyström Collocation Methods for Maxwell's Equations 318
11.3.1 Surface Differential Operators 318
11.3.2 Locally Corrected Nyström Method for Hyper-Singular EFIE 320
11.3.3 Nyström Method for Mixed Potential EFIE 324
11.4 Galerkin Methods and High-Order RWG Current Basis 325
11.4.1 Galerkin Method Using Vector–Scalar Potentials 325
11.4.2 Functional Space for Surface Current **J**(**r**) 326
11.4.3 Basis Functions Over Triangular–Triangular Patches 328
11.4.4 Basis Functions over Triangular–Quadrilateral Patches 332
11.5 Summary 338
References 339

12 High-Order Hierarchical Nédélec Edge Elements 341
12.1 Nédélec Edge Elements in H(Curl) 341
12.1.1 Finite Element Method for **E** or **H** Wave Equations 342
12.1.2 Reference Elements and Piola Transformations 344
12.1.3 Nédélec Finite Element Basis in H(Curl) 345
12.2 Hierarchical Nédélec Basis Functions 355
12.2.1 Construction on 2-D Quadrilaterals 356
12.2.2 Construction on 2-D Triangles 357
12.2.3 Construction on 3-D Cubes 360
12.2.4 Construction on 3-D Tetrahedra 361
12.3 Summary 365
References 366

13 Time-Domain Methods: Discontinuous Galerkin Method and Yee Scheme 369
13.1 Weak Formulation of Maxwell's Equations 369
13.2 Discontinuous Galerkin (DG) Discretization 370
13.3 Numerical Flux $\mathbf{h}(\mathbf{u}^-, \mathbf{u}^+)$ 372
13.4 Orthonormal Hierarchical Basis for DG Methods 375
13.4.1 Orthonormal Hierarchical Basis on Quadrilaterals or Hexahedra 375
13.4.2 Orthonormal Hierarchical Basis on Triangles or Tetrahedra 376
13.5 Explicit Formulae of Basis Functions 378
13.6 Computation of Whispering Gallery Modes (WGMs) with DG Methods 380
13.6.1 WGMs in Dielectric Cylinders 380
13.6.2 Optical Energy Transfer in Coupled Micro-Cylinders 381
13.7 Finite Difference Yee Scheme 383
13.8 Summary 388
References 388

14 Scattering in Periodic Structures and Surface Plasmons 391
14.1 Bloch Theory and Band Gap for Periodic Structures 391
14.1.1 Bloch Theory for 1-D Periodic Helmholtz Equations 392
14.1.2 Bloch Wave Expansions 394
14.1.3 Band Gaps of Photonic Structures 395
14.1.4 Plane Wave Method for Band Gap Calculations 396
14.1.5 Rayleigh–Bloch Waves and Band Gaps by Transmission Spectra 398
14.2 Finite Element Methods for Periodic Structures 402
14.2.1 Nédélec Edge Element for Eigen-Mode Problems 402
14.2.2 Time-Domain Finite Element Methods for Periodic Array Antennas 407
14.3 Physics of Surface Plasmon Waves 411
14.3.1 Propagating Plasmons on Planar Surfaces 411
14.3.2 Localized Surface Plasmons 414
14.4 Desingularized Volume Integral Equation (VIE) for Maxwell's Equations 416
14.5 Extraordinary Optical Transmission (EOT) in Thin Metallic Films 420
14.6 Discontinuous Galerkin Method for Resonant Plasmon Couplings 421

14.7 Appendix: Auxiliary Differential Equation (ADE) DG Methods for Dispersive Maxwell's Equations ... 424
14.7.1 Debye Material ... 424
14.7.2 Drude Material ... 430
14.8 Summary ... 431
References ... 432

15 Schrödinger Equations for Waveguides and Quantum Dots ... 435
15.1 Generalized DG (GDG) Methods for Schrödinger Equations ... 435
15.1.1 One-Dimensional Schrödinger Equations ... 436
15.1.2 Two-Dimensional Schrödinger Equations ... 439
15.2 GDG Beam Propagation Methods (BPMs) for Optical Waveguides ... 441
15.2.1 Guided Modes in Optical Waveguides ... 441
15.2.2 Discontinuities in Envelopes of Guided Modes ... 446
15.2.3 GDG-BPM for Electric Fields ... 449
15.2.4 GDG-BPM for Magnetic Fields ... 452
15.3 Volume Integral Equations for Quantum Dots ... 454
15.3.1 One-Particle Schrödinger Equation for Electrons ... 454
15.3.2 VIE for Electrons in Quantum Dots ... 456
15.3.3 Derivation of the VIE for Quantum Dots Embedded in Layered Media ... 458
15.4 Summary ... 463
References ... 463

Part III Electron Transport and Electronic Structure

16 Quantum Electron Transport in Semiconductors ... 467
16.1 Ensemble Theory for Quantum Systems ... 468
16.1.1 Thermal Equilibrium of a Quantum System ... 468
16.1.2 Microcanonical Ensembles ... 469
16.1.3 Canonical Ensembles ... 471
16.1.4 Grand Canonical Ensembles ... 474
16.1.5 Bose–Einstein and Fermi–Dirac Distributions ... 475
16.2 Density Operator $\hat{\rho}$ for Quantum Systems ... 480
16.2.1 One-Particle Density Matrix $\rho(x, x')$ ... 483
16.3 Wigner Transport Equations and Wigner–Moyal Expansions ... 484
16.4 Quantum Wave Transmission and Landauer Current Formula ... 491
16.4.1 Transmission Coefficient $T(E)$ ... 492
16.4.2 Current Formula Through Barriers Via $T(E)$ ... 494

16.5 Non-Equilibrium Green's Function (NEGF) and Transport Current 499
16.5.1 Quantum Devices with One Contact 501
16.5.2 Quantum Devices with Two Contacts 505
16.5.3 Green's Function and Transport Current Formula 506
16.6 Summary 507
References 507

17 Density Functional Theory Computation of Electronic Structure 509
17.1 Kohn–Sham (KS) DFT: From a $3N$-D Linear Eigenvalue Problem to a 3-D Nonlinear One 509
17.2 De-Singularizing $V_{ext}(\mathbf{r})$ in the KS Equations with an Effective Pseudo-Potential 515
17.3 Self-Consistent Solution of the KS Nonlinear Eigenvalue Problem 518
17.3.1 Plane Wave Solutions for the KS Equations 519
17.3.2 Finite Element Methods for the KS Equations 521
17.4 A Linear Scaling Discontinuous Galerkin Density Matrix Minimization (DG-DMM) Method 522
17.4.1 Enforcing Idempotency of Density Matrix $\rho(\mathbf{r}, \mathbf{r}')$ 523
17.4.2 Support Functions in Discontinuous Galerkin Space 525
17.4.3 DG-DMM Method 525
17.5 Summary 529
References 529

18 Non-equilibrium Green's Function (NEGF) Methods for Transport 533
18.1 NEGFs for 1-D Devices 533
18.1.1 1-D Device Boundary Conditions for Green's Functions 533
18.1.2 Finite Difference Methods for 1-D Device NEGFs 536
18.1.3 Finite Element Methods for 1-D Device NEGFs 537
18.2 NEGFs for 2-D Devices 539
18.2.1 2-D Device Boundary Conditions for Green's Functions 540
18.2.2 Finite Difference Methods for 2-D Device NEGFs 541
18.2.3 Finite Element Methods for 2-D Device NEGFs 544
18.3 NEGF Simulation of a 29 nm Double Gate MOSFET 546
18.4 Derivation of Green's Function in 2-D Strip-Shaped Contacts 548
18.5 Summary 549
References 549

19 Numerical Methods for Wigner Quantum Transport 551
19.1 Wigner Equations for Quantum Transport 551
19.1.1 Truncation of Phase Spaces and Charge Conservation 552
19.1.2 Frensley Inflow Boundary Conditions 553
19.2 Adaptive Spectral Element Method (SEM) 554
19.2.1 Cell Averages in k-Space 555
19.2.2 Chebyshev Collocation Methods in x-Space 558
19.2.3 Time Discretization 559
19.2.4 Adaptive Meshes for Wigner Distributions 561
19.3 Upwinding Finite Difference Scheme 561
19.3.1 Selections of L_{coh}, N_{coh}, L_k, and N_k 563
19.3.2 Self-consistent Algorithm Through the Poisson Equation 563
19.3.3 Currents in RTD by NEGF and Wigner Equations 564
19.4 Calculation of Oscillatory Integrals $O_n(z)$ 566
19.5 Summary 567
References 567

20 Hydrodynamic Electron Transport and Finite Difference Methods 569
20.1 Semi-classical and Hydrodynamic Models 569
20.1.1 Semi-classical Boltzmann Equations 569
20.1.2 Hydrodynamic Equations 571
20.2 High-Resolution Finite Difference Methods of Godunov Type 578
20.3 Weighted Essentially Non-Oscillatory (WENO) Finite Difference Methods 583
20.4 Central Differencing Schemes with Staggered Grids 587
20.5 Summary 592
References 592

21 Transport Models in Plasma Media and Numerical Methods 595
21.1 Kinetic and Macroscopic Magneto-Hydrodynamic (MHD) Theories 596
21.1.1 Vlasov–Fokker–Planck Equations 596
21.1.2 MHD Equations for Plasma as a Conducting Fluid 598
21.2 Vlasov–Fokker–Planck (VFP) Schemes 604
21.3 Particle-In-Cell (PIC) Schemes 607
21.4 $\nabla \cdot \mathbf{B} = 0$ Constrained Transport Methods for MHD Equations 608
21.5 Summary 612
References 613

Index 615

Part I
Electrostatics

Chapter 1
Dielectric Constant and Fluctuation Formulae for Molecular Dynamics

The dielectric constant ϵ of a material describes the collective response of its constituent molecules to electric fields, which is of fundamental importance in the study of electromagnetic phenomena in materials. In this chapter, we will present the statistical molecular theory for the dielectric constant. First, a brief review of the classical electrostatic theory of charges and dipoles is given. Then we present the classical Clausius–Mossotti theory for non-polar dielectrics, i.e., materials that do not have molecular dipole moments in the absence of external fields, and the Onsager theory for dipolar dielectrics, specifically for dipolar liquids. Finally, we discuss the statistical molecular theory for the dielectric constant and dielectric formula in terms of dipole moment fluctuations; the latter can be obtained over molecular trajectories in molecular dynamics simulations of dielectric materials.

1.1 Electrostatics of Charges and Dipoles

In this section, we review the basics of electrostatics of charges and dipoles. The Coulombic force of a point charge q at $\mathbf{r}'$ exerting on a test charge Q at $\mathbf{r}$ is given by

$$\mathbf{F}(\mathbf{r}) = \frac{1}{4\pi\epsilon_0} \frac{Qq(\mathbf{r} - \mathbf{r}')}{|\mathbf{r} - \mathbf{r}'|^3}, \tag{1.1}$$

with the vacuum dielectric constant $\epsilon_0 = 8.854 \times 10^{-12}\ \mathrm{C}^2/(\mathrm{N} \cdot \mathrm{m}^2)$ set in the SI base units for force (N), distance (m), and charge (C). The force can also be expressed in terms of the electric field $\mathbf{E}(\mathbf{r})$ generated by the source charge q as

$$\mathbf{F}(\mathbf{r}) = Q\mathbf{E}(\mathbf{r}), \tag{1.2}$$

W. Cai, *Deterministic, Stochastic, and Deep Learning Methods for Computational Electromagnetics*, https://doi.org/10.1007/978-981-96-0100-4_1

and the electric field $\mathbf{E}(\mathbf{r})$ can be written in terms of a scalar electrostatic potential $\Phi(\mathbf{r})$ as

$$\mathbf{E}(\mathbf{r}) = -\nabla\Phi(\mathbf{r}), \tag{1.3}$$

where

$$\Phi(\mathbf{r}) = \frac{1}{4\pi\epsilon_0}\frac{q}{|\mathbf{r}-\mathbf{r}'|}. \tag{1.4}$$

Meanwhile, the potential energy W for the test charge Q in the electrostatic potential field $\Phi(\mathbf{r})$ is simply

$$W = Q\Phi(\mathbf{r}). \tag{1.5}$$

It can be shown easily that the potential $\Phi(\mathbf{r})$ satisfies the following Poisson equation with a Dirac δ source:

$$-\nabla^2\Phi(\mathbf{r}) = q\frac{\delta(\mathbf{r}-\mathbf{r}')}{\epsilon_0}. \tag{1.6}$$

Molecules of many materials (such as dipolar liquids) possess permanent dipole moments (defined below) due to different mass centers for the positive nuclear charges and the negative electron charges. The dipole moments will experience change under external fields, i.e., polarization, which plays a fundamental role in the study of the electrical and optical properties of materials. Moreover, under the influences of an external field, even non-polar molecules can obtain induced dipole moments because of the displacement of the mass centers of the positive and the negative charges.

The electric dipole moment for a pair of opposite charges of magnitude q is defined as the magnitude of the charges times the distance d between the charges. The potential of such a dipole is given by

$$\Phi(\mathbf{r}) = \frac{1}{4\pi\epsilon_0}\left(\frac{q}{|\mathbf{r}-\frac{\mathbf{d}}{2}|} - \frac{q}{|\mathbf{r}+\frac{\mathbf{d}}{2}|}\right), \tag{1.7}$$

where $\mathbf{d}$ is the directional vector pointing from the negative charge to the positive charge. We define the dipole moment vector $\mathbf{p}$ as

$$\mathbf{p} = q\mathbf{d} \;\; (\mathrm{C}\cdot\mathrm{m}). \tag{1.8}$$

In a far-field region, i.e., $|\mathbf{r}| \gg d$, we have

$$\Phi(\mathbf{r}) \approx \frac{1}{4\pi\epsilon_0}\frac{qd\cos\theta}{r^2} = \frac{1}{4\pi\epsilon_0}\frac{\mathbf{p}\cdot\mathbf{r}}{r^3}, \tag{1.9}$$

where θ is the angle between $\mathbf{p}$ and $\mathbf{r}$. The electric field due to the dipole $\mathbf{p}$ is then

$$\mathbf{E}(\mathbf{r}) = -\nabla\Phi(\mathbf{r}) = \frac{1}{4\pi\epsilon_0}\left(\frac{3\mathbf{p}\cdot\mathbf{r}}{r^5}\mathbf{r} - \frac{\mathbf{p}}{r^3}\right), \tag{1.10}$$

which also defines a dipole–dipole tensor $\overline{\mathbf{T}}(\mathbf{r})$ for $\mathbf{r} = (r_1, r_2, r_3)$:

$$T_{\alpha\beta}(\mathbf{r}) = \frac{1}{4\pi\epsilon_0}\frac{1}{r^3}\left(\frac{3r_\alpha r_\beta}{r^2} - \delta_{\alpha\beta}\right), \qquad \alpha, \beta = 1, 2, 3, \tag{1.11}$$

where $\delta_{\alpha\beta}$ is the Kronecker delta.

An ideal point dipole can be obtained by letting the separation distance d tend to zero while holding the product $qd = \mu$ unchanged. The ideal point dipole can be represented by a dipole moment *density* through the Dirac δ function with an orientation along the unit direction $\hat{\mathbf{d}} = \mathbf{d}/|\mathbf{d}|$:

$$\widetilde{\mathbf{p}} = \mu\delta(\mathbf{r})\hat{\mathbf{d}}, \tag{1.12}$$

the spatial integration of which will give the total dipole moment $\mu\hat{\mathbf{d}}$.

For an individual atom or molecule of a dipole moment $\mathbf{p}$, we assume the following linear relation between the induced dipole moment $\Delta\mathbf{p}(\mathbf{r})$ (i.e., $\mathbf{p} \to \mathbf{p} + \Delta\mathbf{p}(\mathbf{r})$) and an external field $\mathbf{E}(\mathbf{r})$:

$$\Delta\mathbf{p}(\mathbf{r}) = \alpha\mathbf{E}(\mathbf{r}), \tag{1.13}$$

where α (cm^3) is the atomic or molecular polarizability. For example, $\alpha/4\pi\epsilon_0 = 0.667$ for H, 0.205 for He, and 24.1 for Na (in units of $10^{-30}\mathrm{m}^3$), respectively [1].

For later use, we consider the potential energy W of a general finite-sized dipole moment $\mathbf{p} = q_1\mathbf{r}_1 + q_2\mathbf{r}_2$, $q_1 + q_2 = 0$, under an external electric field $\mathbf{E}^{\mathrm{ext}} = -\nabla\Phi$. Using (1.5), we have

$$\begin{aligned} W &= q_1\Phi(\mathbf{r}_1) + q_2\Phi(\mathbf{r}_2) \simeq q_1[\Phi(\mathbf{0}) + \nabla\Phi(\mathbf{0})\cdot\mathbf{r}_1] + q_2[\Phi(\mathbf{0}) + \nabla\Phi(\mathbf{0})\cdot\mathbf{r}_2] \\ &= \mathbf{p}\cdot\nabla\Phi(\mathbf{0}). \end{aligned}$$

Therefore, we obtain

$$W = -\mathbf{p}\cdot\mathbf{E}^{\mathrm{ext}}. \tag{1.14}$$

1.2 Polarization P and Displacement Flux D

The collective response of the constituent molecules of a material to an external electric field can be described by a phenomenological quantity, the susceptibility χ of a dielectric material, which measures the displacement (translation or rotation) of permanent dipole moments in polar molecules or the creation of induced dipole moments in non-polar molecules. This process is the so-called *polarization process*. The susceptibility χ and the dielectric constant ϵ are macroscopic quantities; the former links Maxwell's total electric field $\mathbf{E}(\mathbf{r})$ inside the material and the polarization density $\mathbf{P}(\mathbf{r})$ per unit volume by

$$\mathbf{P}(\mathbf{r}) = \epsilon_0 \chi \mathbf{E}(\mathbf{r}), \tag{1.15}$$

where both $\mathbf{E}(\mathbf{r})$ and $\mathbf{P}(\mathbf{r})$ are averaged quantities over a scale larger than the molecular size but smaller than the overall macroscopic scale of the dielectric material under investigation. Specifically, over a physical region V_c around a location $\mathbf{r}$, we define the polarization density function $\mathbf{P}(\mathbf{r})$ through

$$\mathbf{P}(\mathbf{r})|V_c| = \sum_{i \in V_c} \mathbf{p}_i, \tag{1.16}$$

where $|V_c|$ represents the volume of the region V_c, and $\mathbf{p}_i$ is the dipole moment of the ith molecule inside V_c. Therefore, a complete understanding of the dielectric constant can be traced back to the response of each individual molecule under the influence of the external electric field.

1.2.1 Bound Charges Induced by Polarization

For a given polarization density $\mathbf{P}(\mathbf{r})$ within a volume V, and by the linear superposition principle and the far-field approximation of (1.9), the potential $\Phi(\mathbf{r})$ for $\mathbf{r} \notin V$ is

$$\Phi(\mathbf{r}) = \frac{1}{4\pi\epsilon_0} \int_V \frac{\mathbf{P}(\mathbf{r}') \cdot (\mathbf{r} - \mathbf{r}')}{|\mathbf{r} - \mathbf{r}'|^3} \mathrm{d}\mathbf{r}', \tag{1.17}$$

which can be rewritten as

$$\begin{aligned}
\Phi(\mathbf{r}) &= \frac{1}{4\pi\epsilon_0} \int_V \mathbf{P}(\mathbf{r}') \cdot \nabla' \left(\frac{1}{|\mathbf{r} - \mathbf{r}'|} \right) \mathrm{d}\mathbf{r}' \\
&= \frac{1}{4\pi\epsilon_0} \left[\int_V \nabla' \cdot \frac{\mathbf{P}(\mathbf{r}')}{|\mathbf{r} - \mathbf{r}'|} \mathrm{d}\mathbf{r}' - \int_V \frac{1}{|\mathbf{r} - \mathbf{r}'|} \nabla' \cdot \mathbf{P}(\mathbf{r}') \mathrm{d}\mathbf{r}' \right] \\
&= \frac{1}{4\pi\epsilon_0} \left[\int_S \frac{\mathbf{P}(\mathbf{r}') \cdot \mathbf{n}}{|\mathbf{r} - \mathbf{r}'|} \mathrm{d}s' - \int_V \frac{1}{|\mathbf{r} - \mathbf{r}'|} \nabla' \cdot \mathbf{P}(\mathbf{r}') \mathrm{d}\mathbf{r}' \right],
\end{aligned} \tag{1.18}$$

where ∇' denotes differentiation with respect to $\mathbf{r}'$, $S = \partial V$ is the surface of the volume V, and $\mathbf{n}$ represents the outward unit normal vector to the surface. From the last equation, we can conclude that the potential $\Phi(\mathbf{r})$ due to the polarization density $\mathbf{P}(\mathbf{r})$ can be identified as those created by a surface bound charge σ_b on the surface S,

$$\sigma_b(\mathbf{r}) = \mathbf{P}(\mathbf{r}) \cdot \mathbf{n}, \tag{1.19}$$

and a volume bound charge ρ_b inside the volume V,

$$\rho_b(\mathbf{r}) = -\nabla \cdot \mathbf{P}(\mathbf{r}). \tag{1.20}$$

Namely, we have

$$\Phi(\mathbf{r}) = \frac{1}{4\pi\epsilon_0} \int_S \frac{\sigma_b(\mathbf{r}')}{|\mathbf{r} - \mathbf{r}'|} \mathrm{d}s' + \frac{1}{4\pi\epsilon_0} \int_V \frac{\rho_b(\mathbf{r}')}{|\mathbf{r} - \mathbf{r}'|} \mathrm{d}\mathbf{r}'. \tag{1.21}$$

It should be noted that the volume bound charge ρ_b and the surface bound charge σ_b cancel each other to reflect the overall charge neutrality of the dielectric, namely,

$$\int_V \rho_b(\mathbf{r}')\mathrm{d}\mathbf{r}' = -\int_V \nabla' \cdot \mathbf{P}(\mathbf{r}')\mathrm{d}\mathbf{r}' = -\int_S \mathbf{P}(\mathbf{r}') \cdot \mathbf{n}\mathrm{d}s' = -\int_S \sigma_b(\mathbf{r}')\mathrm{d}s'. \tag{1.22}$$

Accounting for the volume bound charge, the Gauss law for the electric field now becomes

$$\nabla \cdot \epsilon_0 \mathbf{E}(\mathbf{r}) = q(\mathbf{r}) + \rho_b(\mathbf{r}), \tag{1.23}$$

where $q(\mathbf{r})$ is the free charge inside the material (in contrast to the charge $\rho_b(\mathbf{r})$ induced by polarization, which is bound to nuclear sites). Using the definition of the volume bound polarization charge $\rho_b(\mathbf{r})$ (1.20), we have

$$\nabla \cdot (\epsilon_0 \mathbf{E}(\mathbf{r}) + \mathbf{P}(\mathbf{r})) = q(\mathbf{r}). \tag{1.24}$$

Introducing a displacement flux $\mathbf{D}(\mathbf{r})$ to account for the polarization effect due to the "displacement" of the dipoles in the dielectric material,

$$\mathbf{D}(\mathbf{r}) = \epsilon_0 \mathbf{E}(\mathbf{r}) + \mathbf{P}(\mathbf{r}) = \epsilon_0(1 + \chi)\mathbf{E}(\mathbf{r}) = \epsilon \mathbf{E}(\mathbf{r}), \tag{1.25}$$

where the material dielectric constant ϵ is defined by

$$\epsilon = \epsilon_0(1 + \chi), \tag{1.26}$$

we arrive at the Gauss law inside the dielectric material:

$$\nabla \cdot \mathbf{D}(\mathbf{r}) = q(\mathbf{r}). \tag{1.27}$$

For convenience, we also define a relative dielectric constant ϵ_r with respect to that of the vacuum ϵ_0 by

$$\epsilon_r = \frac{\epsilon}{\epsilon_0} = 1 + \chi. \tag{1.28}$$

1.2.2 *Electric Field* $\mathbf{E}_{\text{pol}}(\mathbf{r})$ *of a Polarization Density* $\mathbf{P}(\mathbf{r})$

The macroscopic electric polarization field from the potential (1.17) is

$$\mathbf{E}_{\text{pol}}(\mathbf{r}) = -\nabla\Phi(\mathbf{r}) = -\frac{1}{4\pi\varepsilon_0}\nabla\int_V \frac{\mathbf{P}(\mathbf{r}')\cdot(\mathbf{r}-\mathbf{r}')}{|\mathbf{r}-\mathbf{r}'|^3}d\mathbf{r}', \qquad \mathbf{r}\notin V. \tag{1.29}$$

For a field point $\mathbf{r} \notin V$, the integrand in (1.29) is a smooth function; thus, we can move the gradient operator inside the integral to obtain

$$\mathbf{E}_{\text{pol}}(\mathbf{r}) = \int_V \overline{\mathbf{T}}(\mathbf{r}-\mathbf{r}')\mathbf{P}(\mathbf{r}')d\mathbf{r}', \qquad \mathbf{r}\notin V, \tag{1.30}$$

where the dipole–dipole tensor $\overline{\mathbf{T}}$ is defined in (1.11).

For a field point $\mathbf{r}$ inside V, however, the integrand for the potential $\Phi(\mathbf{r})$ in (1.29) will be singular, and more so for the dipole tensor in the expression (1.30) for the electric field $\mathbf{E}_{\text{pol}}(\mathbf{r})$. Nonetheless, we will show in the following that (1.29) still holds even for $\mathbf{r} \in V$; however, (1.30) will be interpreted as a Cauchy principal value integral with an additional term in (1.43).

For a field point $\mathbf{r} \in V$, the polarization field $\mathbf{E}_{\text{pol}}(\mathbf{r})$ from the molecular polarization $\mathbf{p}_i$ can be split into two parts as follows:

$$\mathbf{E}_{\text{pol}}(\mathbf{r}) = \mathbf{E}_{\text{out}}(\mathbf{r}) + \mathbf{E}_{\text{in}}(\mathbf{r}), \tag{1.31}$$

where $\mathbf{E}_{\text{out}}(\mathbf{r})$ and $\mathbf{E}_{\text{in}}(\mathbf{r})$ are the field generated by the dipoles $\mathbf{p}_i$ outside and inside, respectively, a sphere Ω_R of radius R centered at $\mathbf{r}$. Using a far-field approximation similar to (1.17), we have

$$\mathbf{E}_{\text{out}}(\mathbf{r}) = -\nabla\Phi_{\text{out}}(\mathbf{r}), \tag{1.32}$$

where

$$\Phi_{\text{out}}(\mathbf{r}) = \frac{1}{4\pi\epsilon_0}\int_{V/\Omega_R} \frac{\mathbf{P}(\mathbf{r}')\cdot(\mathbf{r}-\mathbf{r}')}{|\mathbf{r}-\mathbf{r}'|^3}d\mathbf{r}'. \tag{1.33}$$

Meanwhile, to be consistent with the fact that Maxwell's electric field is an averaging quantity, $\mathbf{E}_{\text{in}}(\mathbf{r})$ is defined as the average field generated by all dipoles $\mathbf{p}_i$ inside the sphere Ω_R. Each of these dipoles $\mathbf{p}_i$, represented by two point charges $q_i\delta(\mathbf{r}-\mathbf{r}_i-\mathbf{d}_i/2)$ and $-q_i\delta(\mathbf{r}-\mathbf{r}_i+\mathbf{d}_i/2)$, with $\mathbf{p}_i = q_i\mathbf{d}_i$, will generate a microscopic electric field $\mathbf{e}_i(\mathbf{r})$, given by

$$\mathbf{e}_i(\mathbf{r}) = \frac{1}{4\pi\epsilon_0}\left(\frac{q_i(\mathbf{r}-\mathbf{r}_i-\mathbf{d}_i/2)}{|\mathbf{r}-\mathbf{r}_i-\mathbf{d}_i/2|^3} - \frac{q_i(\mathbf{r}-\mathbf{r}_i+\mathbf{d}_i/2)}{|\mathbf{r}-\mathbf{r}_i+\mathbf{d}_i/2|^3}\right), \tag{1.34}$$

which will contribute to the macroscopic electric field $\mathbf{E}_{\text{in}}(\mathbf{r})$ through its average value over the region Ω_R. Using the result (1.140) in Appendix 1 (Sect. 1.5.1), we can compute this average field quantity as

$$\mathbf{E}_{\text{in}}(\mathbf{r}) = \sum_i \bar{\mathbf{e}}_i(\mathbf{r}) = -\frac{1}{4\pi\epsilon_0}\frac{1}{R^3}\sum_i \mathbf{p}_i = -\frac{1}{3\epsilon_0}\mathbf{P}(\mathbf{r}), \tag{1.35}$$

where the last equality follows from

$$\sum_i \mathbf{p}_i = |\Omega_R|\mathbf{P}(\mathbf{r}) = \frac{4}{3}\pi R^3\mathbf{P}(\mathbf{r}). \tag{1.36}$$

Therefore, from (1.31), for a field point $\mathbf{r} \in V$, the polarization field $\mathbf{E}_{\text{pol}}(\mathbf{r})$ can be expressed as

$$\mathbf{E}_{\text{pol}}(\mathbf{r}) = -\frac{1}{4\pi\epsilon_0}\nabla\int_{V/\Omega_R}\frac{\mathbf{P}(\mathbf{r}')\cdot(\mathbf{r}-\mathbf{r}')}{|\mathbf{r}-\mathbf{r}'|^3}d\mathbf{r}' - \frac{1}{3\epsilon_0}\mathbf{P}(\mathbf{r}). \tag{1.37}$$

On the other hand, if we assume that the polarization density $\mathbf{P}(\mathbf{r})$ is uniform inside the sphere Ω_R, then, from (1.146) in Appendix 2 (Sect. 1.5.2), we know that a uniform electric field is created inside the sphere given by

$$-\frac{1}{3\epsilon_0}\mathbf{P}(\mathbf{r}) = -\frac{1}{4\pi\epsilon_0}\nabla\int_{\Omega_R}\frac{\mathbf{P}(\mathbf{r}')\cdot(\mathbf{r}-\mathbf{r}')}{|\mathbf{r}-\mathbf{r}'|^3}d\mathbf{r}'. \tag{1.38}$$

As a result, we have

$$\begin{aligned}\mathbf{E}_{\text{pol}}(\mathbf{r}) &= -\frac{1}{4\pi\epsilon_0}\nabla\int_{V/\Omega_R}\frac{\mathbf{P}(\mathbf{r}')\cdot(\mathbf{r}-\mathbf{r}')}{|\mathbf{r}-\mathbf{r}'|^3}d\mathbf{r}' - \frac{1}{4\pi\epsilon_0}\nabla\int_{\Omega_R}\frac{\mathbf{P}(\mathbf{r}')\cdot(\mathbf{r}-\mathbf{r}')}{|\mathbf{r}-\mathbf{r}'|^3}d\mathbf{r}' \\ &= -\nabla\Phi(\mathbf{r}),\end{aligned} \tag{1.39}$$

where

$$\Phi(\mathbf{r}) = \frac{1}{4\pi\epsilon_0} \int_V \frac{\mathbf{P}(\mathbf{r}') \cdot (\mathbf{r} - \mathbf{r}')}{|\mathbf{r} - \mathbf{r}'|^3} d\mathbf{r}'. \tag{1.40}$$

Therefore, for a material with a *locally uniform polarization density* (i.e., no interior material interfaces), the macroscopic electric polarization field $\mathbf{E}_{\text{pol}}(\mathbf{r})$ can be computed by (1.39), namely, as the negative of the gradient of the potential (1.40) in the whole space $\mathbb{R}^3$.

1.2.3 Singular Integral Expressions of $\mathbf{E}_{\text{pol}}(\mathbf{r})$ *Inside Dielectrics*

For the electric field inside a dielectric with polarization density $\mathbf{P}$, we can extend (1.30) to the case of $\mathbf{r} \in V$ using Cauchy principal integrals. First, we define a de-singularized dipole tensor with $\delta > 0$ by

$$\overline{\mathbf{T}}_\delta(\mathbf{r} - \mathbf{r}') = \begin{cases} 0, & \text{if } |\mathbf{r} - \mathbf{r}'| \le \delta, \\ \overline{\mathbf{T}}(\mathbf{r} - \mathbf{r}'), & \text{if } |\mathbf{r} - \mathbf{r}'| > \delta. \end{cases} \tag{1.41}$$

Then, we have from (1.37) that

$$\begin{aligned} \mathbf{E}_{\text{pol}}(\mathbf{r}) &= -\frac{1}{4\pi\epsilon_0} \nabla \int_{V/\Omega_\delta} \frac{\mathbf{P}(\mathbf{r}') \cdot (\mathbf{r} - \mathbf{r}')}{|\mathbf{r} - \mathbf{r}'|^3} d\mathbf{r}' - \frac{1}{3\epsilon_0} \mathbf{P}(\mathbf{r}) \\ &= \int_V \overline{\mathbf{T}}_\delta(\mathbf{r} - \mathbf{r}') \mathbf{P}(\mathbf{r}') d\mathbf{r}' - \frac{1}{3\epsilon_0} \mathbf{P}(\mathbf{r}). \end{aligned} \tag{1.42}$$

Now, letting $\delta \to 0$, the first term becomes the Cauchy principal value (p.v.) of the singular integral, and we thus have

$$\mathbf{E}_{\text{pol}}(\mathbf{r}) = \text{p.v.} \int_V \overline{\mathbf{T}}(\mathbf{r} - \mathbf{r}') \mathbf{P}(\mathbf{r}') d\mathbf{r}' - \frac{1}{3\epsilon_0} \mathbf{P}(\mathbf{r}). \tag{1.43}$$

More discussion on the Cauchy principal value of singular integrals can be found in Sect. 3.1.1.

1.3 Clausius–Mossotti and Onsager Formulae for Dielectric Constant

1.3.1 Clausius–Mossotti Formula for Non-polar Dielectrics

In this section, we derive a relation between the polarization density $\mathbf{P}(\mathbf{r})$ and Maxwell's electric field $\mathbf{E}(\mathbf{r})$ inside dielectrics. First, from the definition of the polarization density (1.16), the polarization is the combined polarization of all individual molecules, which will be under the effect of a local field $\mathbf{E}_{\text{local}}(\mathbf{r})$ at molecule sites. This local field $\mathbf{E}_{\text{local}}(\mathbf{r})$ is the Lorentz field and should be distinguished from the macroscopic Maxwell's field $\mathbf{E}(\mathbf{r})$. Thus, the polarization density $\mathbf{P}(\mathbf{r})$ can be expressed in terms of the polarization of all individual molecules in a linear sum as

$$\mathbf{P}(\mathbf{r}) = \sum_i N_i \alpha_i (\mathbf{E}_{\text{local}}(\mathbf{r}))_i, \tag{1.44}$$

where α_i is the polarizability of the ith-type particle (molecule) and N_i is the number density of the ith-type particle (per unit volume), respectively.

From (1.10), the microscopic local Lorentz field on each individual molecule is defined as [2]

$$\begin{aligned}\mathbf{E}_{\text{local}}(\mathbf{r}) &= \mathbf{E}'(\mathbf{r}) + \frac{1}{4\pi\epsilon_0}\sum_{j\in V}\left[\frac{3\mathbf{p}_j\cdot(\mathbf{r}-\mathbf{r}_j)}{|\mathbf{r}-\mathbf{r}_j|^5}(\mathbf{r}-\mathbf{r}_j) - \frac{\mathbf{p}_j}{|\mathbf{r}-\mathbf{r}_j|^3}\right]\\ &= \mathbf{E}'(\mathbf{r}) - \frac{1}{4\pi\epsilon_0}\sum_{j\in V}\nabla\frac{\mathbf{p}_j\cdot(\mathbf{r}-\mathbf{r}_j)}{|\mathbf{r}-\mathbf{r}_j|^3},\end{aligned} \tag{1.45}$$

where $\mathbf{E}'(\mathbf{r})$ denotes the external electric field when the dielectric material is absent, and $\mathbf{p}_j$ is the dipole moment of the molecule at $\mathbf{r}_j$.

We split the summation in (1.45) into two groups of dipoles: those dipoles inside a sphere Ω_R around $\mathbf{r}$ and the rest in the exterior of the sphere $\Omega_R^c = V\backslash\Omega_R$. Then, we have

$$\begin{aligned}\mathbf{E}_{\text{local}}(\mathbf{r}) &= \mathbf{E}'(\mathbf{r}) - \frac{1}{4\pi\epsilon_0}\nabla\left(\sum_{j\in\Omega_R} + \sum_{j\in\Omega_R^c}\right)\frac{\mathbf{p}_j\cdot(\mathbf{r}-\mathbf{r}_j)}{|\mathbf{r}-\mathbf{r}_j|^3}\\ &= \mathbf{E}'(\mathbf{r}) - \frac{1}{4\pi\epsilon_0}\left[\nabla\sum_{j\in\Omega_R}\frac{\mathbf{p}_j\cdot(\mathbf{r}-\mathbf{r}_j)}{|\mathbf{r}-\mathbf{r}_j|^3} + \nabla\int_{\Omega_R^c}\frac{\mathbf{P}(\mathbf{r}')\cdot(\mathbf{r}-\mathbf{r}')}{|\mathbf{r}-\mathbf{r}'|^3}d\mathbf{r}'\right],\end{aligned} \tag{1.46}$$

where a far-field approximation has been used in replacing $\mathbf{p}_j$ by $\mathbf{P}(\mathbf{r}')$ for the summation over Ω_R^c. It should be noted that a more appropriate expression for the field of the dipole $\mathbf{p}_j$ inside Ω_R should be (1.34); however, it would not affect the discussion and conclusion below.

Meanwhile, Maxwell's electric field $\mathbf{E}(\mathbf{r})$, being the sum of the external field $\mathbf{E}'(\mathbf{r})$ and the field generated by the polarizations $\mathbf{E}_{\text{pol}}(\mathbf{r})$ in (1.37), can be expressed as

$$\mathbf{E}(\mathbf{r}) = \mathbf{E}'(\mathbf{r}) - \frac{1}{4\pi\epsilon_0}\nabla \int_{\Omega_R^c} \frac{\mathbf{P}(\mathbf{r}')\cdot(\mathbf{r}-\mathbf{r}')}{|\mathbf{r}-\mathbf{r}'|^3} d\mathbf{r}' - \frac{1}{3\epsilon_0}\mathbf{P}(\mathbf{r}). \tag{1.47}$$

Subtracting (1.47) from (1.46), we have

$$\mathbf{E}_{\text{local}}(\mathbf{r}) - \mathbf{E}(\mathbf{r}) = -\frac{1}{4\pi\epsilon_0}\nabla \sum_{j\in\Omega_R} \frac{\mathbf{p}_j\cdot(\mathbf{r}-\mathbf{r}_j)}{|\mathbf{r}-\mathbf{r}_j|^3} + \frac{1}{3\epsilon_0}\mathbf{P}(\mathbf{r}).$$

The summation over $j \in \Omega_R$ will vanish if we assume a cubic lattice for the material [3, p. 168] and a constant polarization $\mathbf{p}_j$ inside Ω_R. Therefore, we have

$$\mathbf{E}_{\text{local}}(\mathbf{r}) = \mathbf{E}(\mathbf{r}) + \frac{1}{3\epsilon_0}\mathbf{P}(\mathbf{r}), \tag{1.48}$$

which is defined as the *Lorentz local field* [4, p. 388] acting on an individual molecule. From (1.146) and (1.144), it can be seen that the field correction $\mathbf{P}(\mathbf{r})/(3\epsilon_0)$ over Maxwell's field $\mathbf{E}(\mathbf{r})$ in the Lorentz local field $\mathbf{E}_{\text{local}}(\mathbf{r})$ is caused by the surface bound polarization charge on the boundary of the region Ω_R (zero volume bound charge due to the assumption of a uniform polarization inside).

Next, using the ansatz of a linear relation (1.15) between the polarization and Maxwell's field, (1.48) becomes

$$\mathbf{E}_{\text{local}}(\mathbf{r}) = \mathbf{E}(\mathbf{r}) + \frac{\chi}{3}\mathbf{E}(\mathbf{r}) = \left(1+\frac{\chi}{3}\right)\mathbf{E}(\mathbf{r}). \tag{1.49}$$

Substituting (1.49) into the right-hand side of (1.44) and using (1.15) for its left-hand side, we have

$$\epsilon_0\chi\mathbf{E}(\mathbf{r}) = \sum_i N_i\alpha_i\left(1+\frac{\chi}{3}\right)\mathbf{E}(\mathbf{r}),$$

and, after eliminating the field $\mathbf{E}(\mathbf{r})$, we get

$$\frac{\chi}{3+\chi}\epsilon_0 = \frac{1}{3}\sum_i N_i\alpha_i, \tag{1.50}$$

which gives the well-known Clausius–Mossotti formula by using (1.28) [3, 5, 6]:

$$\frac{\epsilon_r - 1}{\epsilon_r + 2}\epsilon_0 = \frac{1}{3}\sum_i N_i\alpha_i. \tag{1.51}$$

1.3.2 Onsager Dielectric Theory for Dipolar Liquids

In the Onsager dielectric theory for a material with permanent molecular dipole moments, the polarization of the material is considered to come from two different sources, i.e.,

$$\mathbf{P} = \mathbf{P}_\alpha + \mathbf{P}_\mu, \tag{1.52}$$

where $\mathbf{P}_\alpha$ is the induced polarization from the translation of the atom's electron/nuclear charges and $\mathbf{P}_\mu$ is the dipole polarization by the orientation change of permanent dipoles, respectively. If α_i is the atomistic polarizability of the ith-type particle, then

$$\mathbf{P}_\alpha = \left(\sum_i N_i \alpha_i\right) \mathbf{E}_{\text{local}} \tag{1.53}$$

and

$$\mathbf{P}_\mu = \sum_i N_i \overline{\boldsymbol{\mu}}_i, \tag{1.54}$$

where $\overline{\boldsymbol{\mu}}_i$ is the permanent dipole vector obtained through Gibbs-averaging over all orientations under the effect of a directing field $\mathbf{E}_d$; the latter is only part of the local field $\mathbf{E}_{\text{local}}(\mathbf{r})$ acting on the molecules.

In deriving the Onsager theory, we will take a different approach from the procedure used in deriving (1.48), where a virtual sphere Ω_R of uniform polarization is used. Here, in order to find $\mathbf{E}_{\text{local}}$ and $\mathbf{E}_d$, we enclose each molecule in a spherical cavity Ω_R (a real physical cavity) without any other molecules inside (i.e., assuming no rotation hindrance from neighboring molecules). The electric field inside the spherical cavity acting on the molecule now comes from two sources, the first being the reaction field of the surrounding dielectric continuum outside the cavity on the dipole moment of the molecule, and the second being the field generated by the presence of the cavity, termed the cavity field. For both fields, a self-consistent argument will be needed as the dipole is polarizable, namely the reaction field or the directional field will change the dipole moment, which in turn will affect the reaction field or the directional field, respectively.

In the following, we will follow the discussion from [3]. First, we will find the average direction for the molecular permanent dipole. We recall from (1.14) that the energy of a dipole $\boldsymbol{\mu}$ under $\mathbf{E}_d$ is

$$W = -\boldsymbol{\mu} \cdot \mathbf{E}_d = -\mu E_d \cos\theta, \tag{1.55}$$

where the angle θ is the inclination angle measured from the direction of $\mathbf{E}_d$, which is taken to be the x-axis.

The relative probability of the orientations of the dipole $\boldsymbol{\mu}$ follows Boltzmann's distribution. If $\mathbf{E}_d = 0$, all directions θ have the same probability, so

$$p(\theta)\mathrm{d}\theta = \frac{2\pi r(\sin\theta) r \mathrm{d}\theta}{4\pi r^2} = \frac{\sin\theta}{2}\mathrm{d}\theta, \tag{1.56}$$

where the numerator in the first fraction is the surface area of a ring patch of inclination angle between θ and $\theta + \mathrm{d}\theta$ on the surface of the sphere of radius r. If $\mathbf{E}_d \neq 0$, then we should have

$$p(\theta)\mathrm{d}\theta = \frac{\mathrm{e}^{-\frac{W}{k_B T}}}{Z}\frac{\sin\theta}{2}\mathrm{d}\theta, \tag{1.57}$$

where k_B is the Boltzmann constant, T is the temperature, and

$$Z = \int_0^{\pi} \mathrm{e}^{\mu E_d \cos\theta / k_B T}\frac{\sin\theta}{2}\mathrm{d}\theta. \tag{1.58}$$

In order to find the average of the dipole moments under the directing field $\mathbf{E}_d$, we consider the average energy $\langle W \rangle$ of the dipole, which requires the average of $\cos\theta$:

$$\begin{aligned} \langle \cos\theta \rangle \equiv \overline{\cos\theta} &= \frac{\int_0^{\pi} \cos\theta \mathrm{e}^{\mu E_d \cos\theta / k_B T}\frac{1}{2}\sin\theta \mathrm{d}\theta}{Z} \\ &= \frac{1}{a}\frac{\int_{-a}^{a} x\mathrm{e}^x\,\mathrm{d}x}{\int_{-a}^{a}\mathrm{e}^x\,\mathrm{d}x} = \coth a - \frac{1}{a} = L(a), \end{aligned} \tag{1.59}$$

where $a = \mu E_d/(k_B T)$, coth is the hyperbolic cotangent function, and $L(a)$ is the so-called Langevin function. For $a \ll 1$, we have

$$L(a) = \frac{a}{3} - \frac{a^3}{45} + \frac{2a^5}{945} + \cdots. \tag{1.60}$$

Therefore, we obtain

$$\langle \cos\theta \rangle = \frac{a}{3} = \frac{\mu E_d}{3k_B T}, \tag{1.61}$$

and the average energy $\langle W \rangle$ can be identified with an averaged dipole moment $\overline{\boldsymbol{\mu}} \parallel \mathbf{E}_d$ (by a symmetric argument) with a magnitude

$$\overline{\mu} = \mu\langle\cos\theta\rangle = \frac{\mu^2}{3k_B T}E_d. \tag{1.62}$$

Thus, from (1.52)–(1.54), we have

$$\mathbf{P} = \epsilon_0\chi\mathbf{E} = \sum_i N_i\left(\alpha_i\mathbf{E}_{\text{local}} + \frac{\mu_i^2}{3k_B T}\mathbf{E}_d\right). \tag{1.63}$$

Calculation of $\mathbf{E}_{\text{local}}$ and $\mathbf{E}_d$

- **Reaction field of non-polarizable dipoles**

A dipole $\boldsymbol{\mu}$ inside a spherical cavity Ω_R of radius R of molecular scale will induce a reaction field from the surrounding dielectric material with a relative dielectric constant ϵ_r outside the cavity, and the potential can be written as [7]

$$\phi_1(\mathbf{r}) = \sum_{n=0}^{\infty} \frac{B_n}{r^{n+1}} P_n(\cos\theta), \qquad \text{if } |\mathbf{r}| \geq R, \tag{1.64}$$

$$\phi_2(\mathbf{r}) = \frac{1}{4\pi\epsilon_0} \frac{\mu}{r^2} \cos\theta + \sum_{n=0}^{\infty} C_n r^n P_n(\cos\theta), \qquad \text{if } |\mathbf{r}| \leq R, \tag{1.65}$$

where the series sum in $\phi_2(\mathbf{r})$ is the reaction potential, $P_n(x)$ is the Legendre polynomial of order n, and B_n and C_n are expansion coefficients to be determined by the following boundary conditions:

$$\phi_1 \to 0 \quad \text{as} \quad |\mathbf{r}| \to \infty, \tag{1.66}$$

$$\phi_1 = \phi_2|_{r=R}, \qquad \epsilon_1 \frac{\partial \phi_1}{\partial \mathbf{n}} = \epsilon_2 \left. \frac{\partial \phi_2}{\partial \mathbf{n}} \right|_{r=R}, \tag{1.67}$$

where $\epsilon_1 = \epsilon_r \epsilon_0$ and $\epsilon_2 = \epsilon_0$.

Using the boundary conditions, we can show that all coefficients are zero except

$$B_1 = \frac{3}{2\epsilon_r + 1} \frac{\mu}{4\pi\epsilon_0}, \qquad C_1 = -\frac{1}{4\pi\epsilon_0} \frac{2(\epsilon_r - 1)}{2\epsilon_r + 1} \frac{\mu}{R^3}.$$

Therefore, the reaction field at the center of Ω_R is

$$\mathbf{E}_{\text{rf}} = -\nabla C_1 x = \frac{1}{4\pi\epsilon_0} \frac{1}{R^3} \frac{2(\epsilon_r - 1)}{2\epsilon_r + 1} \boldsymbol{\mu} = f\boldsymbol{\mu}, \tag{1.68}$$

where $\boldsymbol{\mu}$ is assumed to be in the x-coordinate direction, and

$$f = \frac{1}{4\pi\epsilon_0} \frac{1}{R^3} \frac{2(\epsilon_r - 1)}{2\epsilon_r + 1}. \tag{1.69}$$

- **Reaction field of polarizable dipoles**

A polarizable dipole $\boldsymbol{\mu}$ will produce a reaction field $\mathbf{E}_{\text{rf}}$ inside the cavity Ω_R which is parallel to the dipole itself, as shown in (1.68) above; as a result, the dipole moment $\boldsymbol{\mu}$ will be increased to $\boldsymbol{\mu} + \alpha\mathbf{E}_{\text{rf}}$. To be self-consistent, the reaction field $\mathbf{E}_{\text{rf}}$ should then be related to the total dipole moment, i.e.,

$$\mathbf{E}_{\text{rf}} = f(\boldsymbol{\mu} + \alpha\mathbf{E}_{\text{rf}}). \tag{1.70}$$

Therefore

$$\mathbf{E}_{\mathrm{rf}} = \frac{f}{1 - f\alpha}\boldsymbol{\mu}. \tag{1.71}$$

It is noted that the reaction field to a dipole $\boldsymbol{\mu}$, being parallel to $\boldsymbol{\mu}$, will not change the orientation of the dipole.

- **Cavity field $\mathbf{E}_c$ as part of the directing field $\mathbf{E}_d$**

From (1.71), the reaction field from a permanent dipole of any orientation is parallel to the dipole itself, so it will not contribute to the rotation of that dipole. Therefore, there is no need to include as part of $\mathbf{E}_d$ the reaction field of the Gibbs-averaged permanent dipole moment $\overline{\boldsymbol{\mu}}$.

A cavity in a dielectric system will modify a given field $\mathbf{E} = (E, 0, 0)$ far away from the cavity, however, resulting in a cavity field $\mathbf{E}_c$ parallel to the external field $\mathbf{E}$ inside the cavity. Thus, the cavity field $\mathbf{E}_c$ is not parallel to the permanent dipole, and it will contribute to the overall directing field $\mathbf{E}_d$ affecting the dipole.

Let us first find the cavity field $\mathbf{E}_c$. Assume that the external field is along the x-direction for illustration, namely $\mathbf{E} = (E, 0, 0)$. Then, the potential outside and inside the cavity is given as [7]

$$\phi_1(\mathbf{r}) = \sum_{n=0}^{\infty} \frac{B_n}{r^{n+1}} P_n(\cos\theta) - Er\cos\theta, \qquad \text{if } |\mathbf{r}| \geq R, \tag{1.72}$$

$$\phi_2(\mathbf{r}) = \sum_{n=0}^{\infty} C_n r^n P_n(\cos\theta), \qquad \text{if } |\mathbf{r}| \leq R. \tag{1.73}$$

By using the interface continuity conditions (1.66) and (1.67), we obtain

$$B_1 = \frac{\epsilon_2 - \epsilon_1}{\epsilon_2 + 2\epsilon_1} R^3 E, \quad C_1 = -\frac{3\epsilon_1}{2\epsilon_1 + \epsilon_2} E, \quad B_n = C_n = 0, \quad n \neq 1,$$

where ϵ_1 and ϵ_2 are the dielectric constants outside and inside the cavity, respectively. Therefore, we have

$$\phi_2(\mathbf{r}) = -\frac{3\epsilon_1}{2\epsilon_1 + \epsilon_2} Ex. \tag{1.74}$$

For $\epsilon_1 = \epsilon_r\epsilon_0$ and $\epsilon_2 = \epsilon_0$, the field inside the cavity, $\mathbf{E}_c$, will be

$$\mathbf{E}_c = \frac{3\epsilon_r}{2\epsilon_r + 1}\mathbf{E}. \tag{1.75}$$

The cavity field $\mathbf{E}_c$ will be part of the directing field $\mathbf{E}_d$ producing a rotation polarization of the dipole inside the cavity; however, there are other fields there which will contribute to $\mathbf{E}_d$. Due to the polarizability of the molecule inside the cavity, the directing field $\mathbf{E}_d$ will induce a dipole moment $\alpha\mathbf{E}_d$, which in turn will give a reaction field $f\alpha\mathbf{E}_d$ inside the cavity. Together with the cavity field $\mathbf{E}_c$, we then have the total directing field $\mathbf{E}_d$ in a self-consistent manner as

$$\mathbf{E}_d = \mathbf{E}_c + f\alpha\mathbf{E}_d. \tag{1.76}$$

It should be noted that a more accurate reaction field factor, namely $f/(1 - f\alpha)$ from (1.71), could be used here. As a result, we get

$$\mathbf{E}_d = \frac{1}{1 - f\alpha}\mathbf{E}_c. \tag{1.77}$$

Finally, combining (1.75) and (1.76) will give

$$\mathbf{E}_d = \frac{1}{1 - f\alpha}\mathbf{E}_c = \frac{1}{1 - f\alpha}\frac{3\epsilon_r}{2\epsilon_r + 1}\mathbf{E}. \tag{1.78}$$

- **Local field $\mathbf{E}_{\text{local}}$**

The local field acting on a molecule with a permanent dipole moment will now consist of two parts: the reaction field from the permanent dipole $\overline{\boldsymbol{\mu}}$ and the directing field $\mathbf{E}_d$. Namely, we have

$$\mathbf{E}_{\text{local}} = \mathbf{E}_d + \overline{\mathbf{E}}_{\text{rf}}. \tag{1.79}$$

The reaction field $\overline{\mathbf{E}}_{\text{rf}}$ due to the Gibbs-averaged permanent dipole moment $\overline{\boldsymbol{\mu}}$ is obtained from (1.71) and (1.62):

$$\overline{\mathbf{E}}_{\text{rf}} = \frac{f}{1 - f\alpha}\overline{\boldsymbol{\mu}} = \frac{f}{1 - f\alpha}\frac{\mu^2}{3k_BT}\mathbf{E}_d. \tag{1.80}$$

Thus, it follows from (1.78)–(1.80) that

$$\begin{aligned}\mathbf{E}_{\text{local}} &= \left(1 + \frac{f}{1 - f\alpha}\frac{\mu^2}{3k_BT}\right)\mathbf{E}_d \\ &= \left(1 + \frac{f}{1 - f\alpha}\frac{\mu^2}{3k_BT}\right)\frac{1}{1 - f\alpha}\frac{3\epsilon_r}{2\epsilon_r + 1}\mathbf{E}. \end{aligned}\tag{1.81}$$

Now substituting (1.81) and (1.78) into (1.63) and eliminating $\mathbf{E}$, and using the identity $\chi = \epsilon_r - 1$, we arrive at the following Onsager equation [8]:

$$\frac{(\epsilon_r - 1)(2\epsilon_r + 1)\epsilon_0}{3\epsilon_r} = \sum_i N_i \frac{1}{1 - f\alpha_i}\left(\alpha_i + \frac{1}{3k_BT}\frac{\mu_i^2}{1 - f\alpha_i}\right), \tag{1.82}$$

where f also depends on ϵ_r and is given by

$$f = \frac{1}{4\pi\epsilon_0}\frac{1}{R^3}\frac{2(\epsilon_r - 1)}{2\epsilon_r + 1}.$$

1.4 Statistical Molecular Theory and Dielectric Fluctuation Formulae

In this section, we will derive fluctuation formulae for computing dielectric constants using dipole moment fluctuations obtained through molecular dynamics simulations of dipolar liquids, which allow realistic modeling of the material systems. We consider a classical system V of molecules specified by a microstate Γ of discrete dipole moments,

$$\Gamma = \left\{ (\mathbf{r}_i, \boldsymbol{\mu}_i),\ i = 1, 2, \ldots, N \right\}, \tag{1.83}$$

where $\mathbf{r}_i$ and $\boldsymbol{\mu}_i$ are the position and the point dipole moment of the ith molecule, respectively. Also, we assume that each individual point dipole $\boldsymbol{\mu}_i$, in the absence of all other dipoles $\boldsymbol{\mu}_j$, $j \neq i$, has a polarizability matrix $\boldsymbol{\alpha}_i$ as defined in (1.13), namely a change of dipole moment $\Delta\boldsymbol{\mu}_i$ will be induced on the individual dipole $\boldsymbol{\mu}_i$ under an external field $\mathbf{E}^{\text{ext}}$:

$$\Delta\boldsymbol{\mu}_i = \boldsymbol{\alpha}_i \mathbf{E}^{\text{ext}}, \tag{1.84}$$

where in general $\boldsymbol{\alpha}_i$ should be a matrix quantity, i.e., $\boldsymbol{\alpha}_i = (\alpha_{i\alpha\beta})_{3\times3}$.

The polarization density of (1.16) can be rewritten to include all the point dipole moments:

$$\mathbf{P}(\mathbf{r}) = \sum_i \boldsymbol{\mu}_i \delta(\mathbf{r} - \mathbf{r}_i), \tag{1.85}$$

and the total dipole moment $\mathbf{M}$ of the polarization density is then given by

$$\mathbf{M} = \int_V \mathbf{P}(\mathbf{r}) \mathrm{d}\mathbf{r} = \sum_i \boldsymbol{\mu}_i. \tag{1.86}$$

Our objective is to relate the dielectric constant ϵ of (1.26) to *the fluctuation in the total dipole moment* $\mathbf{M}$. We could achieve this goal by first finding the polarization of the system in response to the external field macroscopically with the classical electrostatics and microscopically with statistical Gibbs averages, respectively. Then, we combine these two results to yield a relation between the dielectric constant and the dipole moment fluctuations.

The dipole moment at $\mathbf{r}_i$ with the zero external field $\mathbf{E}' = \mathbf{0}$ is $\boldsymbol{\mu}_i$. Once an external field $\mathbf{E}'$ is applied, there will be a change in the dipole moment, $\boldsymbol{\mu}_i \to \boldsymbol{\mu}_i + \Delta\boldsymbol{\mu}_i$. We will need to calculate the change of all dipole moments $\Delta\boldsymbol{\mu}_i$. The key to this computation is to realize that a change in a given dipole moment will induce an electric field, which will in turn polarize further the rest of the dipoles [9]. So, to find all $\Delta\boldsymbol{\mu}_i$, a self-consistent approach is required.

Using the local (atomic) polarizability matrix $\boldsymbol{\alpha}_i$ in (1.84) and considering the total contribution of polarization field from all other dipoles $\Delta\boldsymbol{\mu}_j$, the induced change in the dipole moment at the location $\mathbf{r}_i$ will be

$$\Delta\boldsymbol{\mu}_i = \boldsymbol{\alpha}_i \left[\mathbf{E}' + \sum_{j \neq i} \mathbf{T}(\mathbf{r}_i - \mathbf{r}_j)\Delta\boldsymbol{\mu}_j \right], \tag{1.87}$$

i.e.,

$$\Delta\mu_{i\alpha} = \alpha_{i\alpha\beta} \left[E'_\beta + \sum_{j \neq i} \overline{T}_{\beta\gamma}(\mathbf{r}_i - \mathbf{r}_j)\Delta\mu_{j\gamma} \right],$$

and the change in the total dipole moment of the system is

$$\Delta\mathbf{M} = \sum_i \Delta\boldsymbol{\mu}_i. \tag{1.88}$$

We rewrite (1.87) as

$$\sum_j B_{ij\alpha\beta}\Delta\mu_{j\beta} = E'_\alpha, \tag{1.89}$$

where

$$B = \begin{bmatrix} \boldsymbol{\alpha}_1^{-1} & \overline{\mathbf{T}}(\mathbf{r}_1 - \mathbf{r}_2) & \cdots \\ \overline{\mathbf{T}}(\mathbf{r}_2 - \mathbf{r}_1) & \boldsymbol{\alpha}_2^{-1} & \cdots \\ \vdots & \vdots & \ddots \end{bmatrix}, \tag{1.90}$$

$$\Delta\mu_{i\alpha} = \sum_j B^{-1}_{ij\alpha\beta} E'_\beta. \tag{1.91}$$

Then, the induced polarization density should increase by the amount

$$(\Delta\mathbf{P})_\alpha = \left(\mathbf{P}' - \mathbf{P}\right)_\alpha = \sum_i \Delta\mu_{i\alpha}\delta(\mathbf{r} - \mathbf{r}_i) = \sum_{ij} B^{-1}_{ij\alpha\beta}\delta(\mathbf{r} - \mathbf{r}_i)E'_\beta. \tag{1.92}$$

Let us define a local polarizability of the whole system as

$$\mathbf{a}(\mathbf{r}) = \sum_{ij} B_{ij}^{-1}\delta(\mathbf{r} - \mathbf{r}_j), \tag{1.93}$$

and the total system polarizability as

$$\mathbf{A} = \int_V \mathbf{a}(\mathbf{r})\mathrm{d}\mathbf{r} = \sum_{ij} B_{ij}^{-1}. \tag{1.94}$$

Thus, the change in the total dipole moment $\Delta\mathbf{M}$ from (1.92) is simply

$$\Delta \mathbf{M} = \mathbf{A}\mathbf{E}', \tag{1.95}$$

where

$$\Delta \mathbf{M} = \int_V \Delta \mathbf{P} d\mathbf{r}. \tag{1.96}$$

1.4.1 *Statistical Methods for Polarization Density Change* $\Delta \mathbf{P}$

The statistical mechanics theory of dielectric constant introduced by [10, 11] is based on a linear response of the molecule's dipole moments to an external field. In this statistical theory, the average polarization density at a thermal equilibrium of temperature T is defined through the Gibbs average (refer to Sect. 16.1.3 for discussion on the Gibbs ensemble averages) as follows:

$$\langle \mathbf{P}(\mathbf{r}) \rangle = \frac{\int \mathbf{P}(\mathbf{r}) \mathrm{e}^{-H/k_B T} \mathrm{d}^N \Gamma}{\int \mathrm{e}^{-H/k_B T} \mathrm{d}^N \Gamma}, \tag{1.97}$$

where $\mathrm{d}^N \Gamma = \mathrm{d}\mathbf{r}_1 \ldots \mathrm{d}\mathbf{r}_N \mathrm{d}\boldsymbol{\mu}_1 \ldots \mathrm{d}\boldsymbol{\mu}_N$, T is the temperature, and H, the Hamiltonian of the N-molecular system in the absence of any external field, is given by

$$H = \frac{1}{2} \sum_i m \left(\dot{\mathbf{r}}_i\right)^2 + \sum_{i<j} U(\mathbf{r}_i - \mathbf{r}_j), \tag{1.98}$$

where m is the mass of the molecule and $U(\mathbf{r}_i - \mathbf{r}_j)$ is the binary molecular force potential, which could include bond (valance, angle, and torque) and non-bond (electrostatic and van der Waals) forces [12].

Let $\mathbf{E}'$ be a uniform external field. In the linear response theory, the polarization of the dipoles under the external field is assumed to be

$$P'_\alpha(\mathbf{r}) = P_\alpha(\mathbf{r}) + a_{\alpha\beta}(\mathbf{r}) E'_\beta, \tag{1.99}$$

while the Hamiltonian of the polarized system is changed from H to

$$H' = H - \mathbf{M} \cdot \mathbf{E}' - \frac{1}{2} \Delta \mathbf{M} \cdot \mathbf{E}', \tag{1.100}$$

where $\mathbf{M}$ is the zero-field total dipole moment, and the third term is the self-energy [11, p. 169] due to the induced dipole moment change $\Delta \mathbf{M}$ which is given in (1.95). Therefore, we have

$$H' = H - M_\alpha E'_\alpha - \frac{1}{2} E'_\alpha A_{\alpha\beta} E'_\beta. \tag{1.101}$$

Similar to (1.97), we calculate the Gibbs average of the polarization density $\mathbf{P}'$ under the external field $\mathbf{E}'$ by

$$\langle P'_\alpha(\mathbf{r})\rangle_{E'} = \frac{\displaystyle\int [P_\alpha(\mathbf{r}) + a_{\alpha\beta}(\mathbf{r})E'_\beta]\mathrm{e}^{-H'/k_BT}\mathrm{d}^N\Gamma}{\displaystyle\int \mathrm{e}^{-H'/k_BT}\mathrm{d}^N\Gamma} \equiv \frac{F}{G}, \tag{1.102}$$

where H' is given in (1.101).

For small $E' \ll 1$, by a Taylor expansion

$$\langle P'_\alpha(\mathbf{r})\rangle_{E'} \approx \langle P_\alpha(\mathbf{r})\rangle + \frac{\partial}{\partial E'_\beta}\langle P'_\alpha(\mathbf{r})\rangle\bigg|_{E'=0} E'_\beta, \tag{1.103}$$

and

$$\frac{\partial\langle P'_\alpha(\mathbf{r})\rangle_{E'}}{\partial E'_\beta} = \frac{F'G - FG'}{G^2}, \tag{1.104}$$

we have

$$\begin{aligned} F' = \frac{\partial F(E')}{\partial E'_\beta}\bigg|_{E'=0} &= \int a_{\alpha\beta}\mathrm{e}^{-H/k_BT}\mathrm{d}^N\Gamma \\ &- \frac{1}{k_BT}\int P_\alpha \mathrm{e}^{-H/k_BT}\cdot(-M_\beta - A_{\alpha\beta}E'_\beta)\mathrm{d}^N\Gamma \\ &= \int a_{\alpha\beta}\mathrm{e}^{-H/k_BT}\mathrm{d}^N\Gamma + \frac{1}{k_BT}\int P_\alpha M_\beta \mathrm{e}^{-H/k_BT}\mathrm{d}^N\Gamma, \end{aligned} \tag{1.105}$$

$$G' = \frac{1}{k_BT}\int M_\beta \mathrm{e}^{-H/k_BT}\mathrm{d}^N\Gamma. \tag{1.106}$$

Therefore, we get

$$\frac{\partial\langle P'_\alpha(\mathbf{r})\rangle_{E'}}{\partial E'_\beta}\bigg|_{E'_\beta=0} = \langle a_{\alpha\beta}\rangle + \frac{1}{k_BT}\left[\langle P_\alpha M_\beta\rangle - \langle P_\alpha\rangle\langle M_\beta\rangle\right]. \tag{1.107}$$

So

$$\begin{aligned} \langle P'_\alpha(\mathbf{r})\rangle_{E'} &\approx \langle P_\alpha(\mathbf{r})\rangle \\ &+ \left\{\frac{1}{k_BT}[\langle P_\alpha M_\beta\rangle - \langle P_\alpha\rangle\langle M_\beta\rangle] + \langle a_{\alpha\beta}\rangle\right\} E'_\beta, \end{aligned} \tag{1.108}$$

which yields a local polarization formula in terms of the dipole moment averages,

$$\Delta P_\alpha = h_{\alpha\beta}(\mathbf{r}) E'_\beta, \tag{1.109}$$

where

$$h_{\alpha\beta}(\mathbf{r}) = \frac{1}{k_B T}[\langle P_\alpha M_\beta \rangle - \langle P_\alpha \rangle \langle M_\beta \rangle] + \langle a_{\alpha\beta}(\mathbf{r}) \rangle.$$

To obtain a relation for the total polarization of the system, we integrate (1.109) over the whole system sample V to arrive at

$$\Delta M_\alpha = H_{\alpha\beta} E'_\beta, \tag{1.110}$$

where

$$H_{\alpha\beta} = \int_V h_{\alpha\beta}(\mathbf{r}) \mathrm{d}\mathbf{r} = \frac{1}{k_B T} \left[\langle M_\alpha M_\beta \rangle - \langle M_\alpha \rangle \langle M_\beta \rangle + \langle A_{\alpha\beta} \rangle \right]. \tag{1.111}$$

1.4.2 Classical Electrostatics for Polarization Density Change $\Delta\mathbf{P}$

In the classical linear theory of polarization, Maxwell's field is defined in (1.47) as the superposition of the external field and the polarization field $\mathbf{E}_{\mathrm{pol}}(\mathbf{r})$, namely

$$\mathbf{E}(\mathbf{r}) = \mathbf{E}^{\mathrm{ext}}(\mathbf{r}) + \mathbf{E}_{\mathrm{pol}}(\mathbf{r}), \tag{1.112}$$

where $\mathbf{E}^{\mathrm{ext}}(\mathbf{r})$ is the external field and $\mathbf{E}_{\mathrm{pol}}(\mathbf{r})$ is the polarization field produced by the polarization density $\mathbf{P}(\mathbf{r})$ in (1.39),

$$\mathbf{E}_{\mathrm{pol}}(\mathbf{r}) = -\nabla \Phi(\mathbf{r}), \tag{1.113}$$

with

$$\Phi(\mathbf{r}) = \frac{1}{4\pi\epsilon_0} \int_V \frac{\mathbf{P}(\mathbf{r}') \cdot (\mathbf{r} - \mathbf{r}')}{|\mathbf{r} - \mathbf{r}'|^3} \mathrm{d}\mathbf{r}'. \tag{1.114}$$

Equivalently, from (1.43), we have the singular integral representation for the polarization field

$$\mathbf{E}_{\mathrm{pol}}(\mathbf{r}) = \mathrm{p.v.} \int_V \overline{\mathbf{T}}(\mathbf{r} - \mathbf{r}') \mathbf{P}(\mathbf{r}') \mathrm{d}\mathbf{r}' - \frac{1}{3\epsilon_0} \mathbf{P}(\mathbf{r}). \tag{1.115}$$

Therefore, Eq. (1.112) becomes

$$\mathbf{E}(\mathbf{r}) = \mathbf{E}^{\text{ext}}(\mathbf{r}) + \text{p.v.} \int\limits_V \overline{\mathbf{T}}(\mathbf{r} - \mathbf{r}')\mathbf{P}(\mathbf{r}')\text{d}\mathbf{r}' - \frac{1}{3\epsilon_0}\mathbf{P}(\mathbf{r}), \tag{1.116}$$

which can be simplified by using (1.15) to

$$\frac{1}{\lambda}\mathbf{P}(\mathbf{r}) = \mathbf{E}^{\text{ext}}(\mathbf{r}) + \text{p.v.} \int\limits_V \overline{\mathbf{T}}(\mathbf{r} - \mathbf{r}')\mathbf{P}(\mathbf{r}')\text{d}\mathbf{r}', \tag{1.117}$$

where

$$\frac{1}{\lambda} = \frac{1}{\epsilon_0}\left(\frac{1}{\chi} + \frac{1}{3}\right).$$

To obtain $\mathbf{P}(\mathbf{r})$, we need to solve the above integral equation, which is in fact a continuous analog of (1.87). There are different ways to solve this integral equation depending on how system V is arranged (whether confinement by a vacuum [11] and the surrounding dielectrics [13], or the geometry of the system, for example layered or spherical [9, 14]). Let us assume *a periodic system with a truncated dipole interaction*. We can then use Fourier series to solve the integral equation (1.116) [13] when the Fourier series of $\mathbf{P}(\mathbf{r})$ is defined as

$$\hat{\mathbf{P}}(\mathbf{k}) = \frac{1}{|V|} \int\limits_V \mathbf{P}(\mathbf{r})\text{e}^{-\text{i}\mathbf{k}\cdot\mathbf{r}}\text{d}\mathbf{r}. \tag{1.118}$$

Then, as $\delta \to 0$, (1.117) in the Fourier space becomes

$$\hat{\mathbf{P}}(\mathbf{k}) = \lambda\left[\hat{\mathbf{E}}^{\text{ext}}(\mathbf{k}) + \hat{\mathbf{T}}_\delta(\mathbf{k})\hat{\mathbf{P}}(\mathbf{k})\right]. \tag{1.119}$$

Setting $\mathbf{k} = \mathbf{0}$, we get

$$\hat{\mathbf{P}}(\mathbf{0}) = \lambda\hat{\mathbf{E}}^{\text{ext}}(\mathbf{0}). \tag{1.120}$$

Due to anti-symmetry, we can show $\hat{\mathbf{T}}_\delta(\mathbf{0}) = \mathbf{0}$ [13], and

$$\hat{\mathbf{E}}^{\text{ext}}(\mathbf{0}) = \frac{1}{|V|} \int\limits_V \mathbf{E}^{\text{ext}}(\mathbf{r})\text{d}\mathbf{r} = \mathbf{E}^{\text{ext}}, \quad \text{for a constant field.} \tag{1.121}$$

Now, varying the external field by an increment $\mathbf{E}'$,

$$\mathbf{E}^{\text{ext}} \to \mathbf{E}^{\text{ext}} + \mathbf{E}',$$

we have

$$\mathbf{P} \to \mathbf{P} + \Delta\mathbf{P}.$$

The variation of (1.120) yields

$$\Delta\hat{\mathbf{P}}(\mathbf{0}) = \lambda\hat{\mathbf{E}}'(\mathbf{0}). \tag{1.122}$$

However, by definition,

$$\Delta\hat{\mathbf{P}}(\mathbf{0}) = \frac{1}{|V|}\int_V \Delta\mathbf{P}d\mathbf{r} = \frac{1}{|V|}\Delta\mathbf{M}. \tag{1.123}$$

Therefore, Eq. (1.122) gives for a constant external field $\mathbf{E}'$

$$\Delta\mathbf{M} = \lambda|V|\mathbf{E}'. \tag{1.124}$$

Assuming a z-directed uniform external field

$$\mathbf{E}' = (0, 0, E'_z),$$

we have

$$\Delta M_z = \lambda|V|E'_z. \tag{1.125}$$

1.4.3 Fluctuation Formulae for Dielectric Constant ϵ

Assuming a z-directed uniform external field and forming the ratio between (1.110) and (1.125), we have

$$\lambda = \frac{H_{zz}}{|V|}, \tag{1.126}$$

namely

$$\frac{3(\epsilon_r - 1)\epsilon_0}{\epsilon_r + 2} = \frac{H_{zz}}{|V|}, \tag{1.127}$$

where, assuming the zero average of the total polarization, i.e., $\langle A_{\alpha\beta}\rangle = 0$,

$$H_{zz} = \frac{1}{k_B T}\left[\langle M_z M_z\rangle - \langle M_z\rangle^2\right]. \tag{1.128}$$

For a general external field

$$\langle M_z\rangle = \frac{1}{3}\langle\mathbf{M}\rangle, \quad \langle M_z^2\rangle = \frac{1}{3}\langle\mathbf{M}^2\rangle, \tag{1.129}$$

we have the Clausius–Mossotti-type fluctuation formula [9, 13] for the dielectric constant:

$$\frac{3(\epsilon_r - 1)}{\epsilon_r + 2} = \frac{1}{3\epsilon_0|V|k_B T}\left[\langle\mathbf{M}^2\rangle - \langle\mathbf{M}\rangle^2\right]. \tag{1.130}$$

The Gibbs-averaged quantities $\langle\rangle$ in (1.130) can be obtained by molecular dynamics simulation of the system in various ensembles (constant temperature in this case) [15].

To close the discussion on fluctuation formulae, it should be mentioned that the Formula (1.130) is for a periodic system and other fluctuation formulae for different configurations are derived for planar layers [14] and liquids encapsulated in spherical cavities [16–18].

Formulae for the dielectric constants of liquids encapsulated in spherical cavities have been derived previously [14, 16–18]. In the model devised by Berendsen [16], a central sphere of radius R_c with a permittivity ϵ_r is enclosed in a spherical layer of thickness τ with permittivity ϵ', and then the larger spherical region of radius $R_c + \tau$ is embedded in a dielectric continuum with permittivity ϵ_{ext}. This model generalizes the cavity model to cases where a transitional dielectric layer may exist. The simple cavity model is recovered by setting ϵ' to either ϵ_r or ϵ_{ext}. The dielectric constant in this model for $\langle \mathbf{M} \rangle = 0$ is given by

$$\epsilon_r = \frac{1 + \frac{B}{A} 2\epsilon' \left[(2\epsilon_{\text{ext}} + \epsilon') + \left(\frac{R}{R_c + \tau} \right)^3 (\epsilon_{\text{ext}} - \epsilon') \right]}{1 - \frac{B}{A} \left[(2\epsilon_{\text{ext}} + \epsilon') + 2 \left(\frac{R}{R_c + \tau} \right)^3 (\epsilon_{\text{ext}} - \epsilon') \right]}, \tag{1.131}$$

where

$$A = (2\epsilon_{\text{ext}} + \epsilon')(2\epsilon' + 1) - 2 \left(\frac{R}{R_c + \tau} \right)^3 (\epsilon_{\text{ext}} - \epsilon')(1 - \epsilon'), \tag{1.132}$$

$$B = \frac{1}{\epsilon_0} \frac{\langle \mathbf{M}^2 \rangle}{3k_B T |V(R)|}. \tag{1.133}$$

Here, B describes the fluctuation of the total dipole moment $\mathbf{M}(R)$ of a spherical sample V with radius R, and $|V(R)|$ is the volume of the sample. Equation (1.131) is reduced to a simpler expression in [14] for the dielectric cavity if ϵ' is set to ϵ_{ext}:

$$\epsilon_r = \frac{1 + B \frac{2\epsilon_{\text{ext}}}{2\epsilon_{\text{ext}} + 1}}{1 - B \frac{1}{2\epsilon_{\text{ext}} + 1}}. \tag{1.134}$$

This latter expression can be further reduced to the known Kirkwood–Fröhlich expression if ϵ_{ext} is set to ϵ_r [10, 11, 19]

$$\frac{(\epsilon_r - 1)(2\epsilon_r + 1)}{3\epsilon_r} = B, \tag{1.135}$$

and to the Clausius–Mossotti-type formulae in (1.130) if ϵ_{ext} is set to 1.

1.5 Appendices

1.5.1 Appendix 1: Average Field of a Charge in a Dielectric Sphere

For a source charge q located at $\mathbf{r}' \in \Omega_R$, the electric field $\mathbf{e}$ at a distance $\mathbf{r}$ from $\mathbf{r}'$ is given by

$$\mathbf{e}(\mathbf{r}) = \frac{q}{4\pi\epsilon_0}\frac{\mathbf{r}}{|\mathbf{r}|^3}. \tag{1.136}$$

The average of the field over the sphere Ω_R is defined as

$$\langle\mathbf{e}\rangle = \frac{1}{|\Omega_R|}\int_{\Omega_R}\mathbf{e}(\mathbf{r})\,\mathrm{d}\mathbf{r} = \frac{1}{\frac{4}{3}\pi R^3}\int_{\Omega_R}\frac{q}{4\pi\epsilon_0}\frac{\mathbf{r}}{|\mathbf{r}|^3}\,\mathrm{d}\mathbf{r} = -\int_{\Omega_R}\frac{\rho\,\mathrm{d}(-\mathbf{r})}{4\pi\epsilon_0}\frac{-\mathbf{r}}{|-\mathbf{r}|^3}. \tag{1.137}$$

With the constant $\rho = q/\left(\frac{4}{3}\pi R^3\right)$, the term $\frac{\rho\,\mathrm{d}(-\mathbf{r})}{4\pi\epsilon_0}\frac{-\mathbf{r}}{|-\mathbf{r}|^3}$ can be viewed as the electric field at $\mathbf{r}'$ due to a charge element $\rho\,\mathrm{d}(-\mathbf{r})$. As a result, $\langle\mathbf{e}\rangle$ is exactly the electric field at $\mathbf{r}'$ from a uniformly ρ-charged sphere [1, p. 156]. Now, using Gauss's theorem and the symmetry of the field inside a uniformly charged sphere, we can calculate the amount of charge inside the sphere $|\mathbf{r}| \le r'$ and obtain

$$4\pi r'^2\epsilon_0 e = \int_{|\mathbf{r}|=r'}\epsilon_0\mathbf{e}\cdot\mathbf{n}\;\;\mathrm{d}s = \rho\frac{4}{3}\pi r'^3. \tag{1.138}$$

Therefore, we have

$$e = \frac{1}{4\pi r'^2\epsilon_0}\rho\frac{4}{3}\pi r'^3 = \frac{q}{4\pi\epsilon_0 R^3}r', \tag{1.139}$$

which implies that

$$\langle\mathbf{e}\rangle = -\frac{q}{4\pi\epsilon_0}\frac{1}{R^3}\mathbf{r}'. \tag{1.140}$$

1.5.2 Appendix 2: Electric Field Due to a Uniformly Polarized Sphere

Assume a constant polarization density $\mathbf{P}(\mathbf{r}) = \mathbf{P} = (0, 0, p)$ inside the sphere Ω_R, namely, in the spherical polar coordinates (ρ, ϕ, θ),

$$\begin{aligned}\rho_b &= -\nabla\cdot\mathbf{P} = 0,\\ \sigma_b &= \mathbf{P}\cdot\mathbf{n} = p\cos\theta,\end{aligned} \tag{1.141}$$

for which the potential $\Phi(\mathbf{r})$ can be shown to be

$$\Phi(\mathbf{r}) = \begin{cases} \sum_{n=0}^{\infty} A_n r^n P_n(\cos\theta), & \text{if } |\mathbf{r}| \le R, \\ \sum_{n=0}^{\infty} \frac{B_n}{r^{n+1}} P_n(\cos\theta), & \text{if } |\mathbf{r}| \ge R. \end{cases} \tag{1.142}$$

Note that the potential satisfies the following interface conditions (refer to (8.65)):

$$\Phi(R^+, \theta) - \Phi(R^-, \theta) = 0,$$
$$\epsilon_0 \frac{\partial \Phi}{\partial r}(R^+, \theta) - \epsilon_0 \frac{\partial \Phi}{\partial r}(R^-, \theta) = -\sigma_b.$$

We can show that

$$\Phi(\mathbf{r}) = \begin{cases} \frac{p}{3\epsilon_0} r\cos\theta, & \text{if } |\mathbf{r}| \le R, \\ \frac{p}{3\epsilon_0}\frac{R^3}{r^2}\cos\theta, & \text{if } |\mathbf{r}| \ge R. \end{cases} \tag{1.143}$$

On the other hand, from (1.17) the potential from the polarized sphere is

$$\begin{aligned} \Phi(\mathbf{r}) &= \frac{1}{4\pi\epsilon_0}\int_{\Omega_R} \frac{\mathbf{P}(\mathbf{r}')\cdot(\mathbf{r}-\mathbf{r}')}{|\mathbf{r}-\mathbf{r}'|^3}\,\mathrm{d}\mathbf{r}' = \frac{1}{4\pi\epsilon_0}\int_{\Omega_R} \mathbf{P}(\mathbf{r}')\cdot\nabla'\frac{1}{|\mathbf{r}-\mathbf{r}'|}\,\mathrm{d}\mathbf{r}' \\ &= \frac{1}{4\pi\epsilon_0}\int_{\Omega_R} \nabla'\cdot\frac{\mathbf{P}(\mathbf{r}')}{|\mathbf{r}-\mathbf{r}'|}\,\mathrm{d}\mathbf{r}' = \frac{1}{4\pi\epsilon_0}\int_{\partial\Omega_R} \frac{\mathbf{P}(\mathbf{r}')\cdot\mathbf{n}}{|\mathbf{r}-\mathbf{r}'|}\,\mathrm{d}s'. \end{aligned} \tag{1.144}$$

The electric field is

$$E_z = -\frac{\partial \Phi}{\partial z} = -\frac{1}{4\pi\epsilon_0}\int_{\partial\Omega_R} \frac{(z-z')\mathbf{P}(\mathbf{r}')\cdot\mathbf{n}}{|\mathbf{r}-\mathbf{r}'|^3}\,\mathrm{d}s'.$$

Consequently, we have

$$E_z|_{r=0} = -\frac{1}{4\pi\epsilon_0}\int_0^{\pi}\mathrm{d}\theta\int_0^{2\pi}\mathrm{d}\phi\frac{R\cos\theta\, p\cos\theta}{R^3}R^2\sin\theta = -\frac{1}{3\epsilon_0}p. \tag{1.145}$$

Note that

$$E_x = E_y = 0.$$

Therefore, finally we have

$$\mathbf{E} = -\nabla\Phi(\mathbf{r}) = -\frac{1}{4\pi\epsilon_0}\nabla\int_{\Omega_R}\frac{\mathbf{P}(\mathbf{r}')\cdot(\mathbf{r}-\mathbf{r}')}{|\mathbf{r}-\mathbf{r}'|^3}\,\mathrm{d}\mathbf{r}' = -\frac{1}{3\epsilon_0}\mathbf{P}. \tag{1.146}$$

1.6 Summary

The simple Clausius–Mossotti and Onsager formulae of dielectric constants in terms of atomistic/molecular polarizability are based on a continuum environment for individual atoms/molecules and the effect of the reaction field on the latter; however, more detailed molecular-level interactions, such as van der Waals forces, are not accounted for. On the other hand, the statistical molecular theory allows the incorporation of specific molecular interactions in its formulation; together with molecular dynamics simulations with appropriate force fields representing those interactions, a general way of computing dielectric properties of materials at thermal equilibrium is made possible.

References

1. D.J. Griffiths, *Introduction to Electrodynamics*, 3rd edn. (Benjamin Cummings, San Francisco, CA, 1999)
2. L. Kantorovich, *Quantum Theory of the Solid State: An Introduction (Fundamental Theories of Physics)* (Kluwer Academic Publishers, Norwell, MA, 2004)
3. C.J.F. Böttcher, *Theory of Electric Polarization, Vol. 1: Dielectrics in Static Fields*, 2nd edn. (Elsevier Sciences, Amsterdam, 1973)
4. C. Kittel, *Introduction to Solid State Physics*, 8th edn. (Wiley, New York, 2004)
5. O.F. Mossotti, Discussione analitica sul'influenza che l'azione di un mezzo dielettrico ha sulla distribuzione dell'electricitá alla superficie di piú corpi elettrici disseminati in eso. Mem. di Mathem. e Fis. Mod. **24**, 49–74 (1850)
6. R. Clausius, *Die mechanische Wärmetheorie II* (Braunschweig, Friedrich Vieweg und Sohn, 1879), pp.62–97
7. J.D. Jackson, *Classical Electrodynamics*, 3rd edn. (Wiley, Hoboken, NJ, 2001)
8. L. Onsager, Electric moments of molecules in liquids. J. Am. Chem. Soc. **58**, 1486–1493 (1936)
9. H.A. Stern, S.E. Feller, Calculation of the dielectric permittivity profile for a nonuniform system: application to a lipid bilayer simulation. J. Chem. Phys. **118**, 3401–3412 (2003)
10. J.G. Kirkwood, The dielectric polarization of polar liquids. J. Chem. Phys. **7**, 911–919 (1939)
11. H. Fröhlich, *Theory of Dielectrics: Dielectric Constant and Dielectric Loss*, 2nd edn. (Clarendon Press, Oxford, UK, 1958)
12. A.R. Leach, *Molecular Modelling: Principles and Applications*, 2nd edn. (Prentice-Hall, Englewood Cliffs, NJ, 2001)
13. M. Neumann, Dipole moment fluctuation formulas in computer simulations of polar systems. Mol. Phys. **50**, 841–858 (1983)
14. V. Ballenegger, J.-P. Hansen, Dielectric permittivity profiles of confined polar fluids. J. Chem. Phys. **122**, 114711 (2005)
15. D. Frenkel, B. Smit, *Understanding Molecular Simulation*, 2nd edn. (Academic Press, New York, 2001)
16. H.J.C. Berendsen, *Molecular dynamics and Monte Carlo calculations on water, 1972* (Orsay, France, Report of a C.E.C.A.M. Workshop, 1972), pp. 29–39

17. D.J. Adams, I.-R. McDonald, Thermodynamic and dielectric properties of polar lattices. Mol. Phys. **32**, 931–947 (1976)
18. J.G. Powles, R.F. Fowler, W.A.B. Evans, The dielectric constant of a polar liquid by the simulation of microscopic drops. Chem. Phys. Lett. **107**, 280–283 (1984)
19. H. Fröhlich, General theory of the static dielectric constant. Trans. Faraday Soc. **44**, 238–243 (1948)

Chapter 2
Poisson–Boltzmann Electrostatics and Analytical Approximations

In this chapter, first, we introduce the Poisson–Boltzmann (PB) equation, which is based on the Debye–Hückel potential of mean force (PMF) approximation for electrostatic interactions for biomolecules in ionic solvents, and second, we introduce the concept of electrostatic solvation energy. Several analytical approximation methods for solving electrostatic solvation problems are discussed. First, the generalized Born approximation is described for the electrostatic solvation energy using Born radii for atoms embedded in molecules. A fast Fourier transform (FFT)-based algorithm for calculating the Born radii is given. Then we present various image approximations to electrostatic reaction fields in the Poisson and Poisson–Boltzmann electrostatic models in the presence of dielectric or perfectly conducting materials with boundaries such as single or multiple planes and spherical and cylindrical geometries.

2.1 Poisson–Boltzmann (PB) Model for Electrostatic Solvation

The electrostatic force is one of the most important forces in ion–ion, ion–solvent, and solute molecule–solvent interactions [1, 2] for understanding the structure and stability of biomolecules in an aqueous environment. Such interactions are defined as a solvation process of ions or solutes by solvent molecules, such as the formation of hydration shells around ions by water molecules. The classical electrostatic theory can be applied to a system comprising solute macromolecules and surrounding ionic solvent environments [3, 4]. In this classical continuum approach, the solute is described as a region with a low dielectric constant ϵ_i, typically $\epsilon_i = 1 \sim 4$, and partial charges q_j are assigned to atomic locations $\mathbf{r}_j$, leading to an interior charge density

W. Cai, *Deterministic, Stochastic, and Deep Learning Methods for Computational Electromagnetics*, https://doi.org/10.1007/978-981-96-0100-4_2

$$\rho(\mathbf{r}) = \sum_{j=1}^{N} q_j \delta(\mathbf{r} - \mathbf{r}_j), \tag{2.1}$$

where the partial charges q_j can be computed using quantum mechanics calculations [5] or molecular mechanics force fields [6], and the atomic locations $\mathbf{r}_j$ are taken as the nuclear centers of the atoms inside the solute. The solute boundary Γ is defined by the molecular surface (see Fig. 2.1, which was produced using visual molecular dynamics (VMD) software [7]), employing either the van der Waals (vdW) surface (composed of the sum of overlapping vdW spheres), or the solvent-accessible surface (SAS) (generated by rolling a small sphere on the vdW surface) [8]. The solvent, occupying the exterior of the solute, is assigned a higher dielectric constant $\epsilon_o \approx 80$, and, in general, is an ionic liquid with the mobile ionic charge number density $n_i(\mathbf{r})$ for the ions of the type i.

The macroscopic potential $\Phi(\mathbf{r})$, due to the embedded charges in the solute and the mobile ionic charges in the solvent, is then governed by the Poisson equation

$$-\nabla \cdot \epsilon(\mathbf{r}) \nabla \Phi(\mathbf{r}) = \rho(\mathbf{r}) + \rho_{\text{ion}}(\mathbf{r}), \tag{2.2}$$

where the total ion charge density $\rho_{\text{ion}}(\mathbf{r}) = \sum_i e z_i n_i(\mathbf{r})$, z_i is the charge of individual i-type ions, and $n_i(\mathbf{r})$ is the number density of i-type ions at the position $\mathbf{r}$. Due to the discontinuities of dielectric constants inside and outside the solute, two interface conditions on Γ for the continuities of the potential $\Phi(\mathbf{r})$ and the normal displacement flux are required, i.e.,

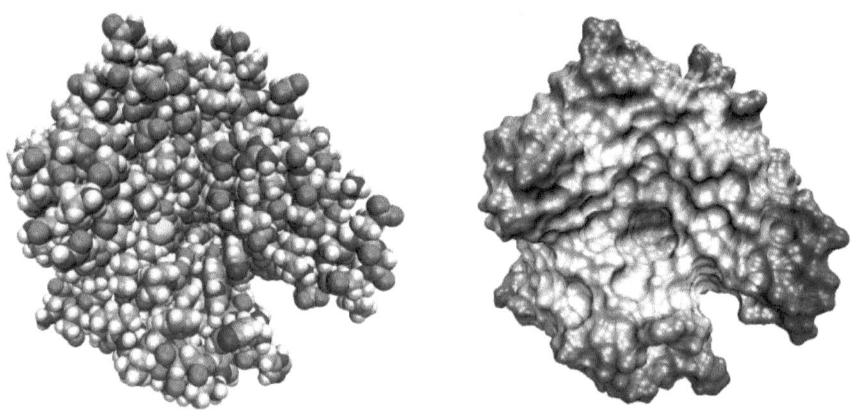

Fig. 2.1 Molecular surfaces of carbonic anhydrase-II. (Left) van der Waals surface and (right) solvent-accessible surface (SAS). This image was made with VMD software. VMD is developed with NIH support by the Theoretical and Computational Biophysics group at the Beckman Institute, UIUC

$$\Phi(\mathbf{r}^-) = \Phi(\mathbf{r}^+), \quad \epsilon_i \frac{\partial \Phi(\mathbf{r}^-)}{\partial \mathbf{n}} = \epsilon_o \frac{\partial \Phi(\mathbf{r}^+)}{\partial \mathbf{n}}, \tag{2.3}$$

for $\mathbf{r} \in \Gamma$, where $\mathbf{r}^-$ and $\mathbf{r}^+$ are, respectively, the inner and the outer limits at the position $\mathbf{r}$, and $\mathbf{n}$ is the outward unit normal to the surface of the solute.

2.1.1 Debye–Hückel Poisson–Boltzmann Theory

To determine the electrostatic potential $\Phi(\mathbf{r})$ by (2.2) and (2.3), we need to know the i-type ion density distribution $n_i(\mathbf{r})$, which will be derived based on the Debye–Hückel theory for homogeneous electrolytes of various ions. In the Debye–Hückel theory of ionic solvent of electrolytes, to compute the electric potential of an overall neutral system, an *ion-cloud* model is adopted [2]. In this model, a single ion is selected as the reference point of the system, called a j-ion (due to the homogeneity of the electrolytes, which j-ion is selected is not essential), and the rest of the ions in the solvent will be modeled as a *cloud* made of *continuous charge density distribution*. The resulting potential is denoted as $\Phi_j(\mathbf{r})$, whereas the j-ion is represented by a spherical cavity of radius a with a total charge at its center ez_j (the ions are considered non-polarizable by other ions or external charges). Despite the fact that all types of ions are of finite size, in the ion-cloud model [9] all ions except the j-ion will be represented by a continuous charge number distribution $n_i^j(\mathbf{r})$ per unit volume, centering around the j-ion. Therefore, $\Phi_j(\mathbf{r})$ satisfies the following Poisson equation [10]:

$$-\epsilon\nabla^2\Phi_j(\mathbf{r}) = \begin{cases} ez_j\delta(\mathbf{r}), & \text{if } r < a, \\ e\sum\limits_i z_i n_i^j(\mathbf{r}), & \text{if } r > a, \end{cases} \tag{2.4}$$

where $n_i^j(\mathbf{r})$ denotes the number concentration per unit volume of the i-type ions around the selected j-ion. The distribution of the i-type ion around the j-ion is a function of the distance r only due to the homogeneity of the system, given in terms of a radial distribution function $g_{ij}(r)$:

$$n_i^j(\mathbf{r}) = n_i^0 g_{ij}(\mathbf{r}), \tag{2.5}$$

where n_i^0 is the number density of the i-type ions in the bulk, and $g_{ij}(r) \to 1$ as $r \to \infty$.

The function $g_{ij}(r)$ is the radial distribution between two types of ions, where $4\pi r^2 g_{ij}(r)dr$ gives the probability of finding an i-type ion in the shell of $[r, r + dr]$ surrounding the j-ion. Note that $g_{ij}(r)$ can be expressed in terms of a Gibbs average over all other ions and charge configurations, i.e., by a Boltzmann factor weighted integration over all other ion and charge positions in phase space [10, Sect. 17 4], i.e.,

$$g_{ij}(\mathbf{r}) = g_{ij}(\mathbf{r}_i, \mathbf{r}_j) = N(N-1)\frac{\displaystyle\int \mathrm{e}^{-U/k_BT}\mathrm{d}\Gamma'}{Z}, \tag{2.6}$$

where $\mathrm{d}\Gamma' = \mathrm{d}r_1 \ldots \mathrm{d}r_N/\mathrm{d}r_i\mathrm{d}r_j$, $U(\mathbf{r}_1, \ldots, \mathbf{r}_N)$ is the interaction potential among all N particles (ions, solvent molecules, etc.), and Z is the normalizing factor (partition function):

$$Z = \int \mathrm{e}^{-U/k_BT}\mathrm{d}r_1 \ldots \mathrm{d}r_N. \tag{2.7}$$

To see the physical meaning of $g_{ij}(\mathbf{r}_i, \mathbf{r}_j)$, we rewrite (2.6) in the following form:

$$g_{ij}(\mathbf{r}_i, \mathbf{r}_j) = \mathrm{e}^{-w(\mathbf{r}_i,\mathbf{r}_j)/k_BT} = \mathrm{e}^{-w_{ij}/k_BT}. \tag{2.8}$$

Then, we have

$$-w_{ij} = k_BT \ln \int \mathrm{e}^{-U/k_BT}\mathrm{d}\Gamma' + C, \tag{2.9}$$

where C is a constant. Differentiating w_{ij} with respect to the position of one of the two particles, say r_i, we have

$$-\nabla_i w_{ij} = \frac{\displaystyle\int \mathrm{e}^{-U/k_BT}(-\nabla_i U)\mathrm{d}\Gamma'}{\displaystyle\int \mathrm{e}^{-U/k_BT}\mathrm{d}\Gamma'} \equiv \frac{\displaystyle\int \mathrm{e}^{-U/k_BT}\mathbf{f}_i\mathrm{d}\Gamma'}{\displaystyle\int \mathrm{e}^{-U/k_BT}\mathrm{d}\Gamma'} \equiv \langle \mathbf{f}_i \rangle_{\mathrm{mean}}, \tag{2.10}$$

where $\mathbf{f}_i = -\nabla_i U$ is the force acting on the particle i for any given configuration of other $N-1$ particles. Therefore, $-\nabla_i w_{ij}$ gives the mean force $\langle \mathbf{f}_i \rangle_{\mathrm{mean}}$ on the particle i from all possible configurations of the other $N-1$ particles. In this sense, w_{ij} is called the potential of mean force (PMF).

In the Debye–Hückel theory [11], an important assumption is made that the mean force on the i-type ions is just the electrostatic force, namely, in terms of the electric potentials, the PMF

$$w_{ij} = ez_i\Phi_j(\mathbf{r}). \tag{2.11}$$

Therefore, non-electrostatic potentials such as short-range van der Waals potentials are ignored in this theory.

Combining (2.11) and (2.8) into (2.5), we have the distribution of the i-type ions as follows:

$$n_i(\mathbf{r}) = n_i^0 \exp\left(-\frac{ez_i\Phi_j}{k_BT}\right). \tag{2.12}$$

Substituting (2.12) into (2.2), we obtain a nonlinear Poisson–Boltzmann (PB) equation for the electrostatic potential $\Phi(\mathbf{r})$ (after dropping the subscript j) for the solute–solvent system:

$$-\nabla \cdot \epsilon(\mathbf{r})\nabla\Phi(\mathbf{r}) = \rho(\mathbf{r}) + \sum_i ez_i n_i^0 \exp\left(-\frac{ez_i\Phi}{k_B T}\right). \tag{2.13}$$

In the Debye–Hückel theory of electrolytes, a linearization of (2.13) is made [10, p. 325] to give

$$-\nabla \cdot \epsilon(\mathbf{r})\nabla\Phi(\mathbf{r}) = \rho(\mathbf{r}) + \sum_i ez_i n_i^0 \left(1 - \frac{ez_i\Phi}{k_B T}\right), \tag{2.14}$$

provided that

$$\frac{ez_i\Phi}{k_B T} \ll 1. \tag{2.15}$$

Equation (2.14) can be further simplified due to the neutrality of the solution $\sum_i ez_i n_i^0 = 0$ to

$$\nabla^2\Phi(\mathbf{r}) - \kappa^2\Phi(\mathbf{r}) = -\frac{1}{\epsilon}\rho(\mathbf{r}), \tag{2.16}$$

where the Debye–Hückel inverse length κ is given by

$$\kappa^2 = \frac{e}{\epsilon k_B T}\sum_i en_i^0 z_i^2 = \frac{2e}{\epsilon k_B T} I, \tag{2.17}$$

and the ionic strength I is defined as

$$I = \frac{1}{2}\sum_i en_i^0 z_i^2. \tag{2.18}$$

Although in some biological systems the assumption of small Φ in (2.15) is not justified, the linearized Poisson–Boltzmann equation has been widely used in biomolecular applications. Various work has been carried out on its mathematical analysis [12], numerical solutions [13], and dynamic simulations [14].

To simplify the notation in the rest part of this chapter, we will write the Poisson equation in the solute and the PB equation in the ionic solvent in a unified form as follows:

$$-\nabla \cdot \epsilon(\mathbf{r})\nabla\Phi(\mathbf{r}) + \lambda^2\Phi(\mathbf{r}) = \rho, \quad \mathbf{r} \in \Omega_\mathrm{i} \text{ or } \mathbf{r} \in \Omega_\mathrm{o}, \tag{2.19}$$

where the dielectric constant and the ionic density are assumed to be constants inside Ω_i and Ω_o:

$$\epsilon(\mathbf{r}) = \begin{cases} \epsilon_\mathrm{i}, & \text{if } \mathbf{r} \in \Omega_\mathrm{i}, \\ \epsilon_\mathrm{o}, & \text{if } \mathbf{r} \in \Omega_\mathrm{o}, \end{cases} \qquad \lambda^2 = \begin{cases} \epsilon_\mathrm{i}\kappa_\mathrm{i}^2, & \text{if } \mathbf{r} \in \Omega_\mathrm{i}, \\ \epsilon_\mathrm{o}\kappa_\mathrm{o}^2, & \text{if } \mathbf{r} \in \Omega_\mathrm{o}. \end{cases} \tag{2.20}$$

Here $\kappa_{\rm i} = 0$ as the solute interior is modeled by the Poisson equation. The potential $\Phi(\mathbf{r})$ satisfies the interface condition (2.3) and a decaying condition at infinity, namely

$$\lim_{r\to\infty} \Phi(\mathbf{r}) = 0. \tag{2.21}$$

2.1.2 *Helmholtz Double Layer and Ion-Size Effect*

In the derivation of the Debye–Hückel theory, which leads to the PB model of electrostatic solvation of biomolecules, we have ignored the finite size of various i-types of ions and also ion correlations beyond those associated with the formation of ion clouds around an opposite j-type ion in (2.4). It has been found [15] that the PB model overestimates the ion density near charged surfaces such as DNA and amino acids. Near a charged surface, ions of opposite signs will be attracted to the surface, whereas ions of the same sign will be repelled to form a so-called Helmholtz double layer, first studied by Helmholtz [16]. The width of the Helmholtz layer is about the same as the radius of the attracted ions, possibly including the hydration shells of water. Later, it was found that some of the dehydrated ions (usually of the same sign as the charges on the surface) or polar water molecules could be adsorbed to the charged surface due to the van der Waals force to form another layer, which is defined as the inner Helmholtz plane (IHP), the former being known as the outer Helmholtz plane (OHP) [17]. The Helmholtz double layer of two oppositely charged planes (the charged surface and the OHP) defines a potential drop from the charged surface to the OHP. In this model of electrostatic potential, the thermal motion of the ions is not considered, contrary to Gouy and Chapman [9, 18], who introduced a diffused "double layer" due to the thermal motion of the ions. In their diffused layer, the ions obey the Boltzmann distribution, resulting in excess ions of opposite signs near the charged surface and reduced ions of the same sign. However, the Helmholtz double layer is ignored in their model. Later, Stern [19] proposed combining the Helmholtz double layer and the Gouy and Chapman diffused double layer, the latter starting from the OHP. In practice, due to the long-range correlations and finite-size effects of the ions, the structure of ionic solutions near a charged surface is more complex (such as non-monotonic charge profile and layering) than is predicted by the PB theory [20, 21].

Figure 2.2 shows the overall picture of the Stern model for a simple planar surface with negative charges [17]. A consequence of the Stern model is that the OHP depends on the type and the amount of ions or water molecules adsorbed on the charged surface. Let the OHP be at $x = d$ (where d is the width of the IHP plus $a/2$, with a being the diameter of hydrolated ions of positive sign). Then the potential will satisfy a 1-D PB equation, and the ion distribution has the following form based on (2.12):

$$n_i(x) = n_i^0 \exp\left(-\frac{z_i e[\phi(x) - \phi_\infty]}{k_B T}\right), \tag{2.22}$$

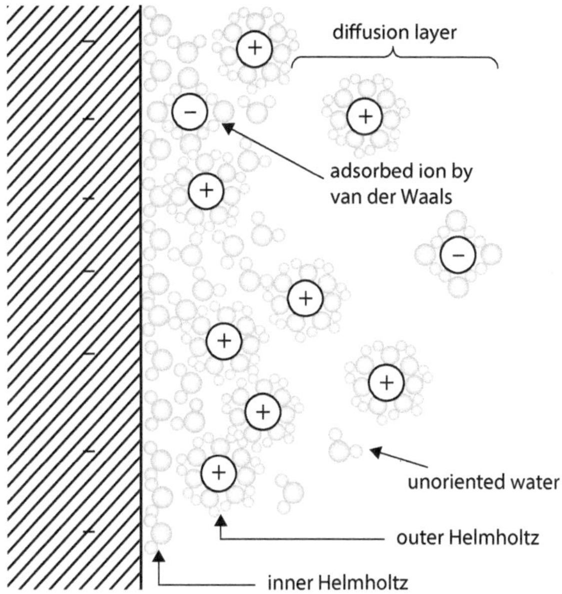

Fig. 2.2 The Stern model indicating the Helmholtz double layer: inner Helmholtz plane; outer Helmholtz plane

where the potential ϕ_∞ at infinity is introduced to reflect the fact that when $x \to \infty$, $n_i(x) \to n_i^0$. Also, the linearized version of the 1-D PB equation takes the form

$$\frac{\mathrm{d}^2}{\mathrm{d}x^2}\phi(x) - \kappa^2\left[\phi(x) - \phi_\infty\right] = 0, \tag{2.23}$$

whose solution is given by

$$\phi(x) - \phi_\infty = (\phi_{\text{OHP}} - \phi_\infty)\,\mathrm{e}^{-\kappa(x-d)}. \tag{2.24}$$

We can calculate the double layer width $l_{\text{double-layer}}$ by considering the potential drop from the OHP by a factor of e^{-1}, i.e., $\phi(x) - \phi_\infty = (\phi_{\text{OHP}} - \phi_\infty)\,\mathrm{e}^{-1}$, giving

$$l_{\text{double-layer}} = d + \kappa^{-1}. \tag{2.25}$$

The existence and the size of the double layer demonstrate the need to include the ion size in a theory for ionic solvents, especially in the presence of a charged surface associated with biomolecules. It is clear that the packing density of the ions near the charged surface within the Helmholtz layer will generate a saturation limit for the ion density near the charged surface, while, on the contrary, the PB model is known to produce unbounded ion density as the surface charge increases. To remedy this overestimation of the ion density, attempts are made to introduce the ion-size effect into electrolyte theory within the easy-to-use PB framework. For this purpose,

Borukhov introduced a mean field free energy for the electrolyte, which explicitly includes the ion size, and a modified PB model is derived from its Euler–Lagrange equations, presented below based on [15, 22].

Let us consider the free energy of a $1 : z$ asymmetric electrolyte of two ion species. Both ion species and the solvent molecules have the same size, a^3. The grand canonical Gibbs free energy functional for the electrolyte can be expressed in terms of the electric potential $\Phi(\mathbf{r})$ and ion concentrations $n_-(\mathbf{r})$ and $n_+(\mathbf{r})$ as

$$F = U - TS - V, \tag{2.26}$$

where the electrostatic energy is defined as

$$U = \int \left(-\frac{\epsilon}{2} |\nabla \Phi|^2 + en_+ \Phi - zen_- \Phi \right) \mathrm{d}\mathbf{r}, \tag{2.27}$$

the entropy in terms of the ion concentrations and solvent density is given by

$$\begin{aligned} -TS = \frac{k_B T}{a^3} \int & [n_+ a^3 \ln(n_+ a^3) + n_- a^3 \ln(n_- a^3) \\ & + (1 - n_+ a^3 - n_- a^3) \ln(1 - n_+ a^3 - n_- a^3)] \mathrm{d}\mathbf{r}, \end{aligned} \tag{2.28}$$

and the chemical potential is defined as

$$V = \int (\mu_+ n_+ + \mu_- n_-) \, \mathrm{d}\mathbf{r}, \tag{2.29}$$

respectively.

The entropy function contains the solvent density $(1/a^3)(1 - n_+ a^3 - n_- a^3) \ln(1 - n_+ a^3 - n_- a^3)$ to include the steric effect of the ion exclusion volume by requiring the density to remain positive.

The extreme conditions at equilibrium,

$$\frac{\delta F}{\delta n_+} = 0, \quad \frac{\delta F}{\delta n_-} = 0, \tag{2.30}$$

imply that

$$e\Phi - \mu_+ + \frac{k_B T}{a^3} \left[a^3 \ln(n_+ a^3) - a^3 \ln(1 - n_+ a^3 - n_- a^3) \right] = 0, \tag{2.31}$$

$$-ze\Phi - \mu_- + \frac{k_B T}{a^3} \left[a^3 \ln(n_- a^3) - a^3 \ln(1 - n_+ a^3 - n_- a^3) \right] = 0. \tag{2.32}$$

Subtracting (2.32) from (2.31) gives

$$n_+ = n_- \mathrm{e}^{\beta(\mu_+ - \mu_-) - \beta(1+z)e\Phi}, \tag{2.33}$$

where $\beta = 1/(k_B T)$.

In the region away from the molecules where Φ is small, the ion density will approach its bulk density n_+^0 and n_-^0 and the neutrality implies that $n_+^0 = zn_-^0 \equiv zn^0$. Therefore, Eq. (2.33) with $\Phi = 0$ implies that

$$e^{\beta(\mu_+ - \mu_-)} = z, \tag{2.34}$$

and

$$n_+ = zn_- e^{-\beta(1+z)e\Phi}. \tag{2.35}$$

Rewrite (2.31) as

$$\frac{n_+ a^3}{1 - n_+ a^3 - n_- a^3} = e^{\beta(\mu_+ - e\Phi)}, \tag{2.36}$$

and use (2.35) to eliminate n_+ in (2.32), which can be then rewritten as

$$\frac{n_- a^3}{1 - zn_- e^{-\beta(1+z)e\Phi} a^3 - n_- a^3} = e^{\beta(\mu_- + ze\Phi)} = \frac{1}{e^{-\beta\mu_-} e^{-z\beta e\Phi}}. \tag{2.37}$$

Solving n_- in (2.37), we have

$$n_- = \frac{1}{a^3} \frac{1}{1 + e^{-\beta\mu_-} e^{-z\beta e\Phi} + ze^{-\beta(1+z)e\Phi}}. \tag{2.38}$$

Meanwhile, note that $n_- \to n^0$ as $\Phi \to 0$. From (2.38), we obtain

$$e^{-\beta\mu_-} = \frac{1 - v_0}{a^3 n^0}, \tag{2.39}$$

where v_0 is the ion volume fraction: $v_0 = za^3 n^0$.

Plugging (2.39) into (2.38), we have

$$n_- = \frac{n^0 e^{\beta ze\Phi}}{1 - v_0 + v_0(e^{\beta ze\Phi} + ze^{-\beta e\Phi})/(1+z)}, \tag{2.40}$$

and with (2.33)

$$n_+ = \frac{zn^0 e^{-\beta e\Phi}}{1 - v_0 + v_0(e^{\beta ze\Phi} + ze^{-\beta e\Phi})/(1+z)}. \tag{2.41}$$

Finally, we have an ion-size modified Poisson–Boltzmann equation from (2.2):

$$\nabla^2\Phi = \frac{zen^0}{\epsilon} \frac{e^{z\beta e\Phi} - e^{-\beta e\Phi}}{1 - v_0 + v_0(e^{\beta ze\Phi} + ze^{-\beta e\Phi})/(1+z)}. \tag{2.42}$$

Other modified PB theories for ions with different sizes can be found in [22–24].

2.1.3 Electrostatic Solvation Energy

The potential energy of a solute–solvent system comes from short-range forces such as van der Waals forces and long-range electrostatic forces. The solvation of a solute inside a solvent involves the competition between these forces. The free energy required during the solvation, denoted as ΔG_{sol}, corresponds to the energy associated with the transfer of a solute molecule from a vacuum to a solvent environment (the symbol Δ indicates that only potential energy change is relevant as a reference value for potential energy is always implicitly implied). The solvation energy is most conveniently decomposed into two components, ΔG_{pol} and ΔG_{np}, which are referred to as the polar and non-polar solvation energy, respectively [25]. The non-polar part is associated with the step of an insertion process into the solvent, where an empty space is created to form a cavity to accommodate the solute atoms whose charges are nullified at this step. The polar solvation energy ΔG_{pol} results from the electrostatic interaction in the form of solvent polarization and redistribution of the mobile ion charges [10].

The solvation energy was computed by Born [26] for an ion of charge q, modeled as a spherical cavity of radius a containing both the bare ion and the first water molecules around the ion (the so-called solvation shell), which is surrounded by a dielectric continuum solvent of dielectric constant ϵ_{o}. The solvation energy can be obtained by introducing a coupling parameter λ between the ion and the solvent, where $\lambda = 0$ indicates no electrostatic coupling and $\lambda = 1$ restores the full Coulombic coupling, respectively. The electrostatic potential energy $U^{(el)}(\lambda)$ of the ion and the solvent outside the cavity is given by the reaction field $\Phi_{\text{rf}}(0, \lambda)$, where λ indicates the λ-reduction of the Coulomb potential [10, Eq. (18.20)]:

$$U^{(el)}(\lambda) = q\Phi_{\text{rf}}(0, \lambda) = q\lambda\Phi_{\text{rf}}(0, 1) = \frac{1}{4\pi}\frac{q^2}{a}\lambda\left(\frac{1}{\epsilon_{\text{o}}} - \frac{1}{\epsilon_{\text{i}}}\right), \tag{2.43}$$

where $\Phi_{\text{rf}}(0, 1)$ is given by (2.115). Here $U^{(el)}(\lambda)\mathrm{d}\lambda$ equals the work needed for the increase of the coupling parameter by $\mathrm{d}\lambda$. As usual, the infinite self-energy created by the ion's Coulomb field at the ion itself is ignored. Therefore, the total work required to effect the complete coupling of the ion to the solvent can be additively computed as

$$W = \int_0^1 \lambda q\Phi_{\text{rf}}(0, 1)\mathrm{d}\lambda. \tag{2.44}$$

Substituting (2.43) into (2.44), we obtain the well-known Born formula for the electrostatic solvation energy:

$$\Delta G_{\text{pol}}^{\text{Born}} = \frac{1}{4\pi}\frac{1}{2}\frac{q^2}{a}\left(\frac{1}{\epsilon_{\text{o}}} - \frac{1}{\epsilon_{\text{i}}}\right). \tag{2.45}$$

As seen in (2.44), the effect of the coupling parameter λ can be viewed equivalently as a scaling factor of the ion charge. Therefore, this coupling process can also be considered as a *charging process*, first proposed by Onsager [27] and Kirkwood [28], where $\lambda = 0$ is the uncharged state and $\lambda = 1$ is the fully charged state. The work in (2.44) can now be identified as the charging energy.

Thus, for a general solute molecule, we can treat the electrostatic free energy ΔG_{pol} as the work needed to charge the solute atomic charge from zero to its full charge value in the ionic solvent environment [4, 29, 30]:

$$\Delta G_{\text{pol}} = W_{\text{ch}}, \tag{2.46}$$

where the charging energy W_{ch} is defined similarly as in (2.44):

$$\Delta G_{\text{pol}} = W_{\text{ch}} \equiv \int_0^1 \mathrm{d}\lambda \int \mathrm{d}\mathbf{r}\rho(\mathbf{r})\Phi_{\text{rf}}(\mathbf{r}, \lambda). \tag{2.47}$$

In order to find the reaction field Φ_{rf}, (2.13) will be solved twice and the dielectric constant $\epsilon(\mathbf{r})$ is described with a two-constant model, i.e., $\epsilon(\mathbf{r}) = \epsilon_{\text{i}}$ inside the solute and $\epsilon(\mathbf{r}) = \epsilon_{\text{ext}}$ for the exterior of the solute. Firstly, Eq. (2.13) is solved with $\epsilon_{\text{ext}} = \epsilon_{\text{o}}$ to produce a potential $\Phi(\mathbf{r})$ in the solvent environment; secondly, it is solved with $\epsilon_{\text{ext}} = \epsilon_{\text{i}}$ to produce a potential in a reference environment [31], denoted as $\Phi_{\text{ref}}(\mathbf{r})$. The difference of these two potentials then gives the reaction field $\Phi_{\text{rf}} = \Phi - \Phi_{\text{ref}}$. For the linearized PB electrostatic model, $\Phi_{\text{rf}}(\mathbf{r}, \lambda) = \lambda\Phi_{\text{rf}}(\mathbf{r})$, and the electrostatic free energy ΔG_{pol} is given by

$$\Delta G_{\text{pol}} = \frac{1}{2}\int_{\mathbb{R}^3} \Phi_{\text{rf}}(\mathbf{r})\rho(\mathbf{r})\mathrm{d}\mathbf{r}, \tag{2.48}$$

which is the electrostatic energy needed to transfer the solute from the reference environment to the solvent dielectric [29–32].

The exact solution of the reaction field $\Phi_{\text{rf}}(\mathbf{r})$ from the PB equation (2.13) is unknown due to the nonlinearity of the equation and the complex geometric shape of the solute biomolecule. Approximate solutions are given later in this chapter.

From a different point of view, the electrostatic free energy of the system can be defined using an energy density in a variational principle for an energy functional W_{ed} for the nonlinear PB electrostatic model for the solute–solvent system [29]:

$$W_{\text{ed}} = \int \mathrm{d}\mathbf{r}\left[\rho(\mathbf{r})\Phi(\mathbf{r}) - \int_0^{\Phi(\mathbf{r})} S(\phi)\mathrm{d}\phi - \frac{\epsilon(\mathbf{r})}{2}|\nabla\Phi(\mathbf{r})|^2\right], \tag{2.49}$$

where

$$S(\phi) = -\sum_i q_i n_i^0 \exp\left(-\frac{q_i \phi}{k_B T}\right).$$

By using Gauss's law, W_{ed} can be shown [30] to be equivalent to

$$W_{\text{ed}} = \int \mathrm{d}\mathbf{r} \left[\frac{1}{2}\rho(\mathbf{r})\Phi(\mathbf{r}) + \frac{1}{2}S(\Phi(\mathbf{r}))\Phi(\mathbf{r}) - \int_0^{\Phi(\mathbf{r})} S(\phi)\mathrm{d}\phi\right], \tag{2.50}$$

which, for the linearized PB equation (2.16) where S becomes a linear function, can be further simplified to

$$W_{\text{ed}} = \frac{1}{2}\int \mathrm{d}\mathbf{r}\rho(\mathbf{r})\Phi(\mathbf{r}). \tag{2.51}$$

This result is consistent with (2.47) for the linearized PB model except for the included additional self-energy term from the Coulomb potential generated by the density $\rho(\mathbf{r})$ in the reference environment, usually taken to be the vacuum.

2.2 Generalized Born (GB) Approximations of Solvation Energy

Using (2.48) to find the solvation energy is not a trivial task as finding the reaction field involves large costs in solving the PB equations in 3-D spatial regions with complex solute molecular shapes. In Chap. 3, several numerical methods will be discussed. Meanwhile, in this section we present an analytical method, a generalized Born (GB) approximation method, to find the solvation energy without solving the reaction potential explicitly. The GB theory [33] generalizes the Born formula (2.45) by extending the concept of the ion radius to any atom embedded inside a solute. The resulting Born radius for an atom in some sense measures how deeply the atom is buried inside the molecule. Here, we present a brief overview of the methodology of the GB methods. Interested readers may refer to the works of Bashford and Case [31], Feig and Brooks [14, 34, 35] for more systematic reviews and discussions.

2.2.1 *Still's Generalized Born Formulism*

Still et al. [33] proposed the GB method to approximate the solvation free energy by an analytical pairwise sum over the atoms in the molecule:

$$\Delta G_{\text{pol}} = \frac{1}{4\pi}\frac{1}{2}\left(\frac{1}{\epsilon_{\text{o}}} - \frac{1}{\epsilon_{\text{i}}}\right)\sum_{i,j}\frac{q_i q_j}{f_{ij}}, \tag{2.52}$$

and

$$f_{ij} = \sqrt{r_{ij}^2 + R_i R_j \exp\left(-\frac{g r_{ij}^2}{R_i R_j}\right)}, \tag{2.53}$$

where g is a positive constant and R_i is the so-called effective Born radius of atom i, defined through its self-energy ΔG^i_{pol} using the Born solvation energy formula (2.45):

$$R_i = \frac{1}{2}\left(\frac{1}{\epsilon_\text{o}} - \frac{1}{\epsilon_\text{i}}\right)\frac{1}{4\pi}\frac{q_i^2}{\Delta G^i_{\text{pol}}}. \tag{2.54}$$

The self-energy ΔG^i_{pol} of atom i in (2.54) can be computed directly from solving the Poisson equation by setting atomic charges of all atoms to zero except that of atom i itself [36], namely the work needed to charge only atom i in the presence of the solvent following the argument of (2.44). In practice, the self-energy ΔG^i_{pol} will be approximated.

Equation (2.52) is an interpolation between two extreme cases, for which (2.52) is exact, of the inter-particle distance: the Born limit (2.45) [26] at small distances ($r_{ij} = 0$) and the Coulomb limit at large distances. The ionic effects can be incorporated by substituting $\mathrm{e}^{-\kappa r_{ij}}/\epsilon_\text{o}$ for $1/\epsilon_\text{o}$ in the formulation, which also satisfies the limit conditions at the Debye–Hückel level [37]. Here, κ is the inverse Debye–Hückel screening parameter in the linearized PB equation (2.16) and (2.17). The inverse of the parameter g in (2.53) can be a value from 0.1 to 0.5; most commonly 0.25 is used due to historical reasons [33]. The zero limit of $g = 0$ was suggested by Grycuk [38], resulting in a simpler function, $f_{ij} = \sqrt{r_{ij}^2 + R_i R_j}$. Other variations of Still's pairwise formula (2.53) were also used, such as $f_{ij} = r_{ij} + 0.5(R_i + R_j)\exp[-2r_{ij}/(R_i + R_j)]$, which also has a better performance than the original one for the spherical case [39].

2.2.2 *Integral Expression for Born Radii*

In the GB models, a Coulomb field approximation (CFA) [31] was used. The basic assumption of the CFA is that the electric displacement flux $\mathbf{D}$ due to charge i at $\mathbf{r} = 0$ remains in Coulombic form, i.e.,

$$\mathbf{D}_i \approx \frac{1}{4\pi}\frac{q_i \mathbf{r}}{r^3}, \tag{2.55}$$

even for dielectrics outside the solute molecule varying from ϵ_i to ϵ_o during the solvation process. Thus, the work of assembling the charge i at its location within the molecule is

$$W_i = \frac{1}{2}\int_{\mathbb{R}^3} (\mathbf{E}\cdot\mathbf{D})\mathrm{d}\mathbf{r} \approx \frac{1}{(4\pi)^2}\frac{1}{2}\int_{\mathbb{R}^3} \frac{q_i^2}{\epsilon(\mathbf{r})r^4}\mathrm{d}\mathbf{r}, \tag{2.56}$$

where the origin is set at the charge location, and a linear response $\mathbf{D} = \epsilon(\mathbf{r})\mathbf{E}$ is used to define the relation of the electric field $\mathbf{E}$ and the displacement $\mathbf{D}$. The formula for the work in (2.56) using the electric field $\mathbf{E}$ and the displacement vector $\mathbf{D}$ is equivalent to the electrostatic solvation energy (2.48) after excluding the infinite self-energy terms associated with point charges [40, Sect. 4.7]. The electrostatic free energy of solvation is then obtained by taking the difference of the work done W_i between the solvent environment $\epsilon_{\text{ext}} = \epsilon_{\text{o}}$ and the reference environment $\epsilon_{\text{ext}} = \epsilon_{\text{i}}$ of the exterior domain of the molecule:

$$\begin{aligned}\Delta G^i_{\text{pol}} &= \frac{1}{(4\pi)^2}\left(\frac{1}{2}\int_{\mathbb{R}^3} \frac{q_i^2}{\epsilon(\mathbf{r})r^4}\mathrm{d}\mathbf{r} - \frac{1}{2}\int_{\mathbb{R}^3} \frac{q_i^2}{\epsilon_{\text{i}} r^4}\mathrm{d}\mathbf{r}\right) \\ &= \frac{1}{(4\pi)^2}\frac{q_i^2}{2}\left(\frac{1}{\epsilon_{\text{o}}} - \frac{1}{\epsilon_{\text{i}}}\right)\int_{\Omega_{\text{o}}} \frac{1}{r^4}\mathrm{d}\mathbf{r}.\end{aligned} \tag{2.57}$$

Comparing (2.57) with the Born solvation energy form (2.45), we can define a generalized Born radius R_i as

$$\frac{1}{R_i} = \frac{1}{4\pi}\int_{\Omega_{\text{o}}} \frac{1}{r^4}\mathrm{d}\mathbf{r}. \tag{2.58}$$

In the following we introduce a method to calculate the generalized Born radius where the singularity of the kernel around the atom site in (2.58) is replaced by a smoother function. We rewrite (2.58) in the following form:

$$\frac{1}{R_i} = \frac{1}{4\pi}\int_{\mathbb{R}^3} G(\mathbf{r}-\mathbf{r}_i)\mathrm{d}\mathbf{r} - \frac{1}{4\pi}\int_{\Omega_{\text{i}}} G(\mathbf{r}-\mathbf{r}_i)\mathrm{d}\mathbf{r}, \tag{2.59}$$

where we assume that the atomic excluded sphere S_i, embedded inside the molecule, has a common radius $a_i = a$ for every atom i. Note that G is a smoothed version of the function $1/r^4$ inside the excluded sphere S_i, i.e.,

$$G(\mathbf{r}) = \begin{cases} W_a^n(r), & \text{if } r \le a, \\ 1/r^4, & \text{if } r > a, \end{cases} \tag{2.60}$$

where the smoother $W_a^n(r)$ produces an nth-order continuity of $G(\mathbf{r})$ at $r = a$. For example,

$$W_a^1(r) = -\frac{2}{a^6}r^2 + \frac{3}{a^4}, \tag{2.61}$$

$$W_a^2(r) = \frac{3}{a^8}r^4 - \frac{8}{a^6}r^2 + \frac{6}{a^4}, \tag{2.62}$$

$$W_a^3(r) = -\frac{4}{a^{10}}r^6 + \frac{15}{a^8}r^4 - \frac{20}{a^6}r^2 + \frac{10}{a^4}. \tag{2.63}$$

Note that a larger n will lead to a faster decay in the spectrum of $G(\mathbf{r})$ in the Fourier frequency domain, and such a fast decay makes an efficient method for calculating the Born radius with the FFT possible. The first integration on the right-hand side of (2.59) can be calculated analytically as

$$\frac{1}{4\pi}\int_{\mathbb{R}^3} G(\mathbf{r}-\mathbf{r}_i)d\mathbf{r} = \frac{1}{4\pi}\int_{\mathbb{R}^3\backslash S_i} \frac{1}{|\mathbf{r}-\mathbf{r}_i|^4}d\mathbf{r} + \frac{1}{4\pi}\int_{S_i} G(\mathbf{r}-\mathbf{r}_i)d\mathbf{r}. \tag{2.64}$$

In (2.64), the first integral on the right-hand side equals to $1/a$ while the second term is the integral of the smoother $W_a^n(r)$ inside S_i and equals $\frac{1}{4\pi}\int_{S_i} W_a^n(r)d\mathbf{r} = 3/(5a)$, $29/(35a)$, and $65/(63a)$ when $n = 1, 2$, and 3, respectively.

The radius a can be chosen arbitrarily, and if it is taken as an atomic radius, for instance the van der Waals radius, the sphere S_i is completely inside Ω_i. On the other hand, if the sphere S_i is not completely inside Ω_i, Eq. (2.59) can be rewritten as

$$\frac{1}{R_i} = \frac{1}{4\pi}\int_{\mathbb{R}^3} G(\mathbf{r}-\mathbf{r}_i)d\mathbf{r} - \frac{1}{4\pi}\int_{\Omega_i} G(\mathbf{r}-\mathbf{r}_i)d\mathbf{r}$$
$$+ \frac{1}{4\pi}\int_{A_i}\left[\frac{1}{|\mathbf{r}-\mathbf{r}_i|^4} - G(\mathbf{r}-\mathbf{r}_i)\right]d\mathbf{r}, \tag{2.65}$$

in which A_i is the portion of S_i outside Ω_i. Since the center of S_i is inside Ω_i, the integral over the region A_i is not singular, and it can be calculated by a numerical quadrature or by an approximate analytical formula [41].

2.2.3 FFT-Based Algorithm for the Born Radii

The FFT can be used for the evaluation of the second integral on the right-hand side of (2.59) or (2.65), which takes on the form

$$\Phi(\mathbf{r}) = \int_{\Omega_i} G(\mathbf{r}-\mathbf{r}')d\mathbf{r}'. \tag{2.66}$$

Once $\Phi(\mathbf{r})$ is calculated on grid lattice points, the value $\Phi(\mathbf{r}_i)$ corresponding to the ith off-grid lattice atom can be obtained by a simple interpolation from the nearby data on the lattice sites surrounding the atom. In order to use the FFT, we define an indicator function for the molecular volume domain Ω_i as follows:

$$f(\mathbf{r}) = \begin{cases} 1, \text{ if } \mathbf{r} \in \Omega_\mathrm{i}, \\ 0, \text{ if } \mathbf{r} \notin \Omega_\mathrm{i}. \end{cases} \tag{2.67}$$

Then, the integral in (2.66) can be extended to the full space as

$$\Phi(\mathbf{r}) = \int_{\mathbb{R}^3} G(\mathbf{r} - \mathbf{r}')f(\mathbf{r}')\mathrm{d}\mathbf{r}' = G * f(\mathbf{r}), \tag{2.68}$$

which is a convolution suitable for evaluation using the Fourier transform

$$\hat{f}(\boldsymbol{\xi}) = \frac{1}{(2\pi)^{3/2}} \int_{\mathbb{R}^3} f(\mathbf{r})\mathrm{e}^{-i\mathbf{r}\cdot\boldsymbol{\xi}}\, \mathrm{d}\mathbf{r}$$

and the discrete fast Fourier transform.

The FFT-based method will give $\Phi(\mathbf{r}_{ijk})$ on the grid lattice sites $\mathbf{r}_{ijk} = (x_i, y_j, z_k)$, $0 \le i, j, k \le N$, at a cost of $O(N^3 \log N)$. Then, $\Phi(\mathbf{r}_\alpha)$ for the αth off-grid lattice site atom can be obtained by an interpolation from $\Phi(\mathbf{r}_{ijk})$ at a cost of $O(M)$, for instance with a linear interpolation for M atom sites. Here, N is independent of M and only depends on the shape of the molecule Ω_i, i.e., the lattice should be fine enough to resolve the boundary of the molecule within a prescribed accuracy. Therefore, the total complexity of the FFT-based method is $O(N^3 \log N + M)$.

Using the FFT to compute (2.68) **with the smoother kernel** G

The FFT algorithm has an $O(N\log N)$ complexity of evaluating the following two transforms between data $\{f(x_j) : -N/2 \le j \le N/2 - 1\}$ and discrete Fourier coefficients $\{\hat{f}_k : -N/2 \le k \le N/2 - 1\}$:

$$f(x_j) = \sum_{k=-N/2}^{N/2-1} \hat{f}_k \mathrm{e}^{ikx_j}, \quad \text{for } x_j = j\frac{2\pi}{N}, \quad -\frac{N}{2} \le j \le \frac{N}{2} - 1, \tag{2.69}$$

$$\hat{f}_k = \frac{1}{N} \sum_{j=-N/2}^{N/2-1} f(x_j)\mathrm{e}^{-ikx_j}, \quad \frac{N}{2} \le k \le \frac{N}{2} - 1. \tag{2.70}$$

To illustrate the idea of using the FFT for (2.66), let us consider the 1-D analog of (2.66) for the evaluation of

$$\Phi(x) = \int_{-b}^{b} G(x - x')\mathrm{d}x', \tag{2.71}$$

for $x \in V = (-b, b)$ and

$$G(x-x') = \begin{cases} W_a^n(|x-x'|), & \text{if } |x-x'| \le a, \\ \frac{1}{|x-x'|^4}, & \text{if } |x-x'| > a. \end{cases}$$

If $f(x)$ is the indicator function for the domain V as defined in (2.67), then the 1-D convolution corresponding to (2.68) is given by

$$\Phi(x) = \int_{-\infty}^{\infty} G(x-x') f(x') \mathrm{d}x' = G * f(x). \tag{2.72}$$

Applying the Fourier transform, we have

$$\hat{\Phi}(\xi) = \hat{G}(\xi) \hat{f}(\xi), \tag{2.73}$$

with

$$\hat{G}(\xi) = \mathcal{F}\{G(x)\} = \frac{1}{\sqrt{2\pi}} \int_{-\infty}^{\infty} G(x) \mathrm{e}^{-i\xi x} \mathrm{d}x, \tag{2.74}$$

$$\hat{f}(\xi) = \mathcal{F}\{f(x)\} = \frac{1}{\sqrt{2\pi}} \int_{-\infty}^{\infty} f(x) \mathrm{e}^{-i\xi x} \mathrm{d}x, \tag{2.75}$$

and using the inverse Fourier transform we have

$$\Phi(x) = \mathcal{F}^{-1}\left\{\hat{G}(\xi)\hat{f}(\xi)\right\} = \frac{1}{\sqrt{2\pi}} \int_{-\infty}^{\infty} \hat{G}(\xi)\hat{f}(\xi) \mathrm{e}^{+i\xi x}\, \mathrm{d}\xi. \tag{2.76}$$

Due to the fact that $f(x)$ is discontinuous at $x = \pm b$ and $G(x)$ is C^n-continuous at $x = \pm a$ with the smoother W_a^n, the decay conditions of $\hat{f}(\xi)$ and $\hat{G}(\xi)$ are

$$\hat{f}(\xi) = O\left(\frac{1}{\xi}\right), \quad |\xi| \to +\infty, \tag{2.77}$$

$$\hat{G}(\xi) = O\left(\frac{1}{\xi^{n+1}}\right), \quad |\xi| \to +\infty. \tag{2.78}$$

Let ε be an error tolerance of the whole algorithm, against which we truncate the integral over $\xi \in (-\infty, +\infty)$, i.e.,

$$\Phi(x) = \frac{1}{\sqrt{2\pi}} \int_{-\infty}^{\infty} \mathrm{e}^{i\xi x} \hat{f}(\xi)\hat{G}(\xi)\mathrm{d}\xi \approx \frac{1}{\sqrt{2\pi}} \int_{-\Omega\pi}^{\Omega\pi} \mathrm{e}^{i\xi x} \hat{f}(\xi)\hat{G}(\xi)\mathrm{d}\xi, \tag{2.79}$$

with the truncation parameter Ω defined as follows based on the decay conditions (2.77) and (2.78):

$$\frac{1}{(\pi\Omega)^{n+1}} = \varepsilon. \tag{2.80}$$

An N-point rectangular quadrature rule for the integral in (2.79) yields

$$\Phi(x) \approx \frac{\sqrt{2\pi}\Omega}{N} \sum_{k=-N/2}^{N/2-1} e^{i\xi_k x} \hat{f}(\xi_k)\hat{G}(\xi_k), \tag{2.81}$$

where $\xi_k = k(2\pi\Omega/N)$ and N will be selected based on the Shannon sampling rate of $T = \pi/L$ [42, 43] for a plane wave $e^{i\xi L}$ of wave number L in the ξ variable:

$$N = \frac{2\pi\Omega}{T} = 2\Omega L. \tag{2.82}$$

In principle, the selection of N should also depend on the oscillatory behavior of the spectral functions $\hat{f}(\xi)$ and $\hat{G}(\xi)$. In the case that a larger N is needed to resolve the oscillations in $\hat{f}(\xi)$ and $\hat{G}(\xi)$, we can achieve that by increasing the size of L.

Calculation of $\Phi(x_j)$, $x_j = j(2L/N) \in [-L, L]$, $-N/2 \le j \le N/2 - 1$

Next, we calculate the value of $\Phi(x_j)$ at N points inside the interval $[-L, L]$. Again, the size of N will be based on the Shannon sampling rate $T = 1/\Omega$ for function $e^{\pm i\Omega\pi x}$ in the x variable, which gives

$$N = \frac{2L}{T} = 2\Omega L. \tag{2.83}$$

Let

$$x_j = j\frac{2L}{N} \in [-L, L], \quad -\frac{N}{2} \le j \le \frac{N}{2} - 1. \tag{2.84}$$

Then

$$\Phi(x_j) \approx \frac{\sqrt{2\pi}\Omega}{N} \sum_{k=-N/2}^{N/2-1} \hat{f}(\xi_k)\hat{G}(\xi_k)e^{ik\frac{2\pi\Omega}{N}j\frac{2L}{N}} = \frac{\sqrt{2\pi}\Omega}{N} \sum_{k=-N/2}^{N/2-1} \hat{f}(\xi_k)\hat{G}(\xi_k)e^{ikj\frac{2\pi}{N}}, \tag{2.85}$$

which can be evaluated by one FFT at a cost of $O(N \log N)$.

Calculation of $\hat{f}(\xi_k)$, $\xi_k = k(2\pi\Omega/N)$, $|k| \le N/2$

As $\sup(f) \subset [-L, L]$, we have

$$\hat{f}(\xi) = \frac{1}{\sqrt{2\pi}} \int_{-\infty}^{\infty} f(x)e^{-i\xi x}dx = \frac{1}{\sqrt{2\pi}} \int_{-L}^{L} f(x)e^{-i\xi x}dx, \tag{2.86}$$

which is approximated by an N-point rectangular quadrature rule

$$\hat{f}(\xi) \approx \frac{2L}{N\sqrt{2\pi}} \sum_{j=-N/2}^{N/2-1} f(x_j) \mathrm{e}^{-i\xi x_j}, \tag{2.87}$$

where $x_j = j(2L/N)$. As $N = 2\Omega L$, we have for $-N/2 \le k \le N/2$

$$\hat{f}(\xi_k) = \frac{2L}{N\sqrt{2\pi}} \sum_{j=-N/2}^{N/2-1} f(x_j) \mathrm{e}^{-ik\frac{2\pi\Omega}{N} j \frac{2L}{N}} = \frac{2L}{N\sqrt{2\pi}} \sum_{j=-N/2}^{N/2-1} f(x_j) \mathrm{e}^{-ikj\frac{2\pi}{N}}, \tag{2.88}$$

to be evaluated by one FFT at a cost of $O(N \log N)$.

Calculation of $\Phi(x_m, y_n, z_l)$ in the 3-D case

The 3-D Fourier transform $\hat{G}(\boldsymbol{\xi})$ can be found analytically, which is defined by

$$\hat{G}(\boldsymbol{\xi}) = \frac{1}{(2\pi)^{3/2}} \int_{\mathbb{R}^3} G(\mathbf{r}) \mathrm{e}^{i\boldsymbol{\xi}\cdot\mathbf{r}} \mathbf{dr}, \tag{2.89}$$

and is a spherically symmetric function of $\boldsymbol{\xi}$ due to the spherical symmetry of $G(\mathbf{r})$ in the spatial domain. Therefore, the Fourier transform at a radial distance ρ (by letting $\boldsymbol{\xi} = (0, 0, \rho)$) is

$$\hat{G}(\boldsymbol{\xi}) = \frac{1}{(2\pi)^{3/2}} \int_0^{2\pi} \int_0^{\pi} \int_0^{\infty} G(r) \mathrm{e}^{ir\rho\cos\psi} r^2 \sin\psi \mathrm{d}r \mathrm{d}\psi \mathrm{d}\theta, \tag{2.90}$$

where $\rho = |\boldsymbol{\xi}|$, and (r, θ, ψ) are the spatial spherical coordinates with $x = r\cos\theta\sin\psi$, $y = r\sin\theta\sin\psi$, and $z = r\cos\psi$. Integration in ψ and substitution of the piecewise definition of $G(r)$ yield

$$\begin{aligned}
\hat{G}(\boldsymbol{\xi}) &= \frac{\sqrt{2}}{\pi} \int_0^{\infty} G(r) r \frac{\sin(r\rho)}{\rho} \mathrm{d}r \\
&= \frac{\sqrt{2}}{\pi} \left[\int_0^{a} W_a^n(r) r \frac{\sin(r\rho)}{\rho} \mathrm{d}r + \int_a^{\infty} \frac{\sin(r\rho)}{r^3\rho} \mathrm{d}r \right] \\
&= \frac{\sqrt{2}}{\pi} (I_n + II).
\end{aligned} \tag{2.91}$$

The second integral II can be integrated to give

$$\int_a^\infty \frac{\sin(r\rho)}{r^3\rho}\mathrm{d}r = \frac{1}{a}\left[-\frac{\pi\delta}{4} + {}_1F_2\left(-\frac{1}{2};\frac{1}{2},\frac{3}{2};-\frac{1}{4}\delta^2\right)\right], \tag{2.92}$$

where $\delta = a\rho$, and ${}_1F_2(\alpha;\beta,\gamma;x)$ is the hypergeometric function

$${}_1F_2(\alpha;\beta,\gamma;x) = \sum_{m=0}^{\infty}\frac{(\alpha)_m}{(\beta)_m(\gamma)_m}\frac{x^m}{m!},$$

with $(\alpha)_m = \alpha(\alpha+1)\dots(\alpha+m-1)$ as the rising factorial.

For the first integral in (2.91), we have, for $n = 1, 2$, and 3,

$$I_1 = \frac{1}{a\delta^5}\left[\delta(\delta^2+12)\cos\delta + 3(\delta^2-4)\sin\delta\right], \tag{2.93}$$

$$I_2 = \frac{1}{a\delta^7}\left[\delta(\delta^4-12\delta^2+360)\cos\delta + 3(\delta^4+44\delta^2-120)\sin\delta\right] \tag{2.94}$$

$$I_3 = \frac{1}{a\delta^9}\left[\delta(\delta^6-12\delta^4-1560\delta^2+20160)\cos\delta \right.$$
$$\left. + 3(\delta^6-20\delta^4+2760\delta^2-6720)\sin\delta\right]. \tag{2.95}$$

It should be noted that the form for I_n as defined is not usable at $\delta = 0$ due to the denominator in I_n having higher-order infinitesimals than the numerator. Therefore, $I_n(0)$ should be calculated by an extrapolation; for example, for a first-order accurate extrapolation we can use $I_n(0) \simeq 2I_n(a/2) - I_n(a)$.

Algorithm

Let the molecule be contained in a rectangular box of size $[-L_x, L_x] \times [-L_y, L_y] \times [-L_z, L_z]$. If the smallest box that contains the molecule is $[-a, a] \times [-b, b] \times [-c, c]$, then, due to the periodicity of the FFT, the computational box $[-L_x, L_x] \times [-L_y, L_y] \times [-L_z, L_z]$ should be chosen such that $L_x \geq 2a$, $L_y \geq 2b$, and $L_z \geq 2c$ to avoid the overlap of the images of f and G.

The following steps form the flow of the algorithm in the 3-D case [41].

- Step 1: For an nth-order smoother $W_a^n(r)$ in (2.60) and an error tolerance $\varepsilon > 0$, choose the truncation parameter Ω by

$$\Omega = \frac{1}{\pi\varepsilon^{1/(n+1)}} \tag{2.96}$$

and set

$$N_x = 2\Omega L_x,\quad N_y = 2\Omega L_y,\quad N_z = 2\Omega L_z, \tag{2.97}$$

and

$$N = N_x N_y N_z.$$

- Step 2: Compute $\hat{f}(\xi_i, \eta_j, \chi_k), (\xi_i, \eta_j, \chi_k) = (i\frac{2\pi\Omega}{N_x}, j\frac{2\pi\Omega}{N_y}, k\frac{2\pi\Omega}{N_z}), -N_x/2 \leq i \leq N_x/2 - 1, -N_y/2 \leq j \leq N_y/2 - 1, -N_z/2 \leq k \leq N_z/2 - 1$, using one 3-D FFT for the following sums at a cost of $O(N \log N)$:

$$\hat{f}(\xi_i, \eta_j, \chi_k) = \frac{8L_x L_y L_z}{(2\pi)^{3/2} N} \sum_{m=-N_x/2}^{N_x/2-1} \sum_{n=-N_y/2}^{N_y/2-1} \sum_{l=-N_z/2}^{N_z/2-1} f(x_m, y_n, z_l) \cdot \exp\left(-i2\pi\left(\frac{mi}{N_x} + \frac{nj}{N_y} + \frac{lk}{N_z}\right)\right). \tag{2.98}$$

- Step 3: Compute $\Phi(x_m, y_n, z_l), (x_m, y_n, z_l) = (m\frac{2L_x}{N_x}, n\frac{2L_y}{N_y}, l\frac{2L_z}{N_z}) \in [-L_x, L_x] \times [-L_y, L_y] \times [-L_z, L_z], -N_x/2 \leq m \leq N_x/2 - 1, -N_y/2 \leq n \leq N_y/2 - 1, -N_z/2 \leq l \leq N_z/2 - 1$, using one 3-D FFT for the following sums at a cost of $O(N \log N)$:

$$\Phi(x_m, y_n, z_l) = \frac{(2\pi)^{3/2}\Omega^3}{N} \sum_{i=-N_x/2}^{N_x/2-1} \sum_{j=-N_y/2}^{N_y/2-1} \sum_{k=-N_z/2}^{N_z/2-1} \hat{f}(\xi_i, \eta_j, \chi_k)\hat{G}(\xi_i, \eta_j, \chi_k) \cdot \exp\left(i2\pi\left(\frac{mi}{N_x} + \frac{nj}{N_y} + \frac{lk}{N_z}\right)\right). \tag{2.99}$$

In the 3-D case, the function $f(x, y, z)$ is the indicator function of the solute molecule. Therefore, N_x, N_y, and N_z should be large enough such that the boundary of the solute molecule is well resolved on the $N_x N_y N_z$-lattice grid to ensure a prescribed accuracy in the Fourier transform.

2.3 Method of Images for Reaction Fields

Image methods with virtual charges were first used by Lord Kelvin in the nineteenth century to represent the polarization field of dielectric materials [44]. In the Poisson electrostatic model, the electrostatic potential $\Phi(\mathbf{r})$ of a source charge q located at $\mathbf{r}_s$ inside a region Ω satisfies the Poisson equation

$$\nabla \cdot (\epsilon(\mathbf{r})\nabla\Phi(\mathbf{r})) = -q\delta(\mathbf{r} - \mathbf{r}_s), \quad \mathbf{r} \in \Omega, \tag{2.100}$$

together with a homogeneous boundary condition on the boundary $\partial\Omega$ if Ω^c (the exterior of Ω) is a perfect conductor,

$$\Phi(\mathbf{r}) = 0, \qquad \mathbf{r} \in \partial\Omega, \tag{2.101}$$

or a continuity condition if both Ω and Ω^c are dielectric materials:

$$\Phi(\mathbf{r}^+) = \Phi(\mathbf{r}^-),$$
$$\epsilon_o \frac{\partial \Phi(\mathbf{r}^+)}{\partial \mathbf{n}} = \epsilon_i \frac{\partial \Phi(\mathbf{r}^-)}{\partial \mathbf{n}}, \tag{2.102}$$

where $\mathbf{n}$ is the external normal to the boundary, ϵ_i is the dielectric constant inside Ω and ϵ_o the dielectric constant outside Ω, and $+$ and $-$ denote the limit taken from outside and inside Ω, respectively. Meanwhile, the potential $\Phi(\mathbf{r})$ is assumed to decay to zero when $\mathbf{r} \to \infty$.

In general, the solution to (2.100) has to be solved with numerical methods as discussed in Chap. 3, and analytical forms of the solution are only available for simple geometries. Still, the potential $\Phi(\mathbf{r})$ can be decomposed into two parts: one part for the potential due to the source charge q at $\mathbf{r}_s$ and the other part for the reaction field $\Phi_{rf}(\mathbf{r})$ that reflects the polarization of the material in Ω^c, namely

$$\Phi(\mathbf{r}) = \frac{q}{4\pi\epsilon_i |\mathbf{r} - \mathbf{r}_s|} + \Phi_{rf}(\mathbf{r}). \tag{2.103}$$

2.3.1 *Methods of Images for Simple Geometries*

For selected geometries, methods of images can provide simple and analytical solutions to the reaction field $\Phi_{rf}(\mathbf{r})$ in (2.103). The following are some classical image solutions to simple geometries of conducting bodies and dielectrics. More details on other types of dielectric shapes can be found in Smythe [45].

- **Potential of a point charge in the presence of a conducting sphere**

For a charge q outside a conducting sphere of radius a, the potential in $\Omega = f\{\mathbf{r}: |\mathbf{r}| > a\}$ outside the sphere is given as the sum of the primary potential from the charge at $\mathbf{r}_s = (r_s, \theta_s, \phi_s)$ in the spherical coordinates and the potential of an image charge $-q$ at the Kelvin image location [44] inside the sphere, which is the inversion point

$$\mathbf{r}_k = \left(\frac{a^2}{r_s}, \theta_s, \phi_s \right) \tag{2.104}$$

with respect to the sphere

$$\Phi(\mathbf{r}) = \frac{q}{4\pi\epsilon_i |\mathbf{r} - \mathbf{r}_s|} - \frac{q}{4\pi\epsilon_i |\mathbf{r} - \mathbf{r}_k|}, \qquad \mathbf{r} \in \Omega. \tag{2.105}$$

- **Potential of a point charge in the presence of a conducting 3-D half-space**

Consider a point charge located at $\mathbf{r}_s = (0, 0, d)$ along the z-axis above a conducting plane ($z = 0$) at zero potential. So, we have $\Omega = \{\mathbf{r} = (x, y, z): z \geq 0\}$, and the homogeneous boundary condition (2.101) is assumed at $z = 0$. The solution to (2.100) is given by (2.103). The reaction field in this case can be represented by an image charge $q' = -q$ located below the conducting plane $z = 0$ at the mirror image location $\mathbf{r}_{\rm im} = (0, 0, -d)$, and the effect of this image potential (the reaction potential) is to satisfy the required zero potential boundary condition (2.101). So, we have

$$\Phi_{\rm rf}(\mathbf{r}) = -\frac{q}{4\pi\epsilon_i|\mathbf{r} - \mathbf{r}_{\rm im}|}. \tag{2.106}$$

- **Potential of a point charge in the presence of a dielectric 3-D half-space**

Here we have a point charge located at $\mathbf{r}_{\rm s} = (0, 0, d) \in \Omega = \{\mathbf{r} = (x, y, z) : z \geq 0\}$ above a dielectric half space, and the potential will have to satisfy the continuity condition (2.102) at the interface $z = 0$. Again, the potential in the upper half space is given by (2.103), where the reaction field can be represented by an image charge. Specifically, for $z > 0$ we have

$$\Phi_{\rm rf}(\mathbf{r}) = \frac{q'}{4\pi\epsilon_i|\mathbf{r} - \mathbf{r}_{\rm im}|}, \tag{2.107}$$

where

$$\mathbf{r}_{\rm im} = (0, 0, -d), \quad q' = -\left(\frac{\epsilon_{\rm o} - \epsilon_i}{\epsilon_{\rm o} + \epsilon_i}\right) q\ .$$

The potential in the lower half-space can also be represented by another image located at the source point in the upper half-space, consistent with the non-singular feature of the potential for $z < 0$:

$$\Phi(\mathbf{r}) = \frac{q''}{4\pi\epsilon_{\rm o}|\mathbf{r} - \mathbf{r}_{\rm s}|}, \tag{2.108}$$

where

$$q'' = \left(\frac{2\epsilon_{\rm o}}{\epsilon_{\rm o} + \epsilon_i}\right) q.$$

- **Potential of a line charge and a dielectric cylinder**

In this case, the potential problem is a 2-D one for the cross section of the cylinder. Let $\Omega = \{\mathbf{r} = (r, \theta) : r > a\}$ denote the exterior of the cylinder of radius a which contains a line source charge q at the location $\mathbf{r}_{\rm s} = (r_{\rm s}, \theta_{\rm s} = 0)$, $r_{\rm s} > a$, in polar coordinates.

The potential in Ω is the superposition of the potential from the source charge and two images at 0 and $\mathbf{r}_{\rm im} = (a^2/r_{\rm s}, 0)$, respectively [45], i.e.,

$$\Phi(\mathbf{r}) = -\frac{q}{2\pi\epsilon_i}\ln(|\mathbf{r} - \mathbf{r}_{\rm s}|) + \frac{q}{2\pi\epsilon_i}\ln(|\mathbf{r}|) - \frac{q'}{2\pi\epsilon_i}\ln(|\mathbf{r} - \mathbf{r}_{\rm im}|),\ |\mathbf{r}| > a, \tag{2.109}$$

where

$$q' = -\left(\frac{\epsilon_o - \epsilon_i}{\epsilon_o + \epsilon_i}\right) q.$$

Meanwhile, the potential inside the cylinder is given by a second image q'' at the location $\mathbf{r}_s$:

$$\Phi(\mathbf{r}) = -\frac{q''}{2\pi\epsilon_o} \ln(|\mathbf{r} - \mathbf{r}_s|), \ |\mathbf{r}| < a, \tag{2.110}$$

where

$$q'' = \left(\frac{2\epsilon_o}{\epsilon_o + \epsilon_i}\right) q.$$

- **Potential of a line charge and a conducting cylinder**

As a limiting case of (2.109), we can find the potential outside the cylinder centered at the origin due to a line charge placed at $(r_s, \theta_s = 0)$, $r_s > a$, as

$$\Phi(\mathbf{r}) = -\frac{q}{2\pi\epsilon_i} \ln(|\mathbf{r} - \mathbf{r}_s|) + \frac{q}{2\pi\epsilon_i} \ln(|\mathbf{r} - \mathbf{r}_{im}|), \ |\mathbf{r}| > a, \tag{2.111}$$

where $\mathbf{r}_{im} = (a^2/r_s, 0)$.

2.3.2 *Image Methods for Dielectric Spheres*

In this section, we will present discrete image approximations to the reaction field for a dielectric sphere Ω $\{= \mathbf{r}\text{: } |\mathbf{r}| < a\}$ with a dielectric constant ϵ_i; the dielectric constant outside the sphere is assumed to be ϵ_o. The starting point is the classical Kirkwood series expansion [46] for the reaction fields. Extension to the Poisson–Boltzmann equation will be discussed in Sect. 2.3.3 where the medium outside the sphere has a mobile ion density.

Friedman's One Image Approximation

For a charge q inside a dielectric sphere Ω (see Fig. 2.3), the potential is the solution of (2.100) and (2.102) for Ω, given by (2.103). The reaction field $\Phi_{rf}(\mathbf{r})$ at an observation point $\mathbf{r} = (r, \theta, \phi)$ inside the sphere can be expressed in terms of the Legendre polynomials of $\cos\theta$ [46]:

$$\begin{aligned}\Phi_{rf}(\mathbf{r}) &= \frac{q(\epsilon_i - \epsilon_o)}{4\pi\epsilon_i a} \sum_{n=0}^{\infty} \left(\frac{n+1}{\epsilon_i n + \epsilon_o(n+1)}\right) \left(\frac{r r_s}{a^2}\right)^n P_n(\cos\theta) \\ &= \sum_{n=0}^{\infty} A_n r^n P_n(\cos\theta),\end{aligned} \tag{2.112}$$

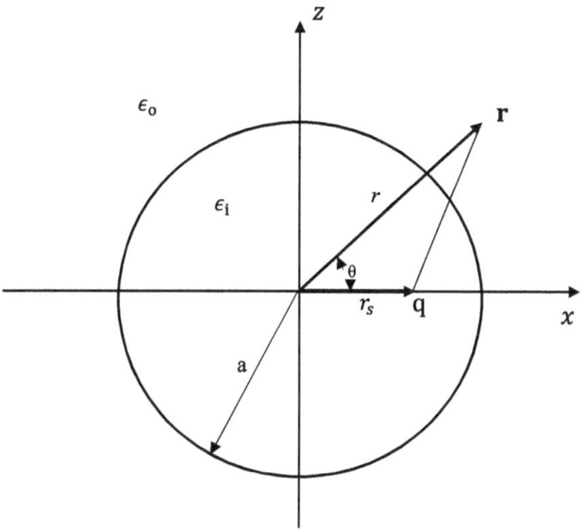

Fig. 2.3 A point charge and a dielectric sphere. The point charge is inside the sphere ($r_s < a$)

where $P_n(x)$, $n = 0, 1, 2, \ldots$, are the Legendre polynomials defined in (5.9),

$$A_n = \frac{q}{4\pi\epsilon_i} \cdot \frac{r_s^n}{a^{2n+1}} \cdot \gamma \cdot \left(1 + \frac{1+\gamma}{1-\gamma+2n}\right), \quad n \geq 0, \tag{2.113}$$

and

$$\gamma = \frac{\epsilon_i - \epsilon_o}{\epsilon_i + \epsilon_o}. \tag{2.114}$$

In particular, the first coefficient A_0 is used in the Born solvation energy in (2.45):

$$\Phi_{rf}(\mathbf{0}) = A_0 = \frac{q}{4\pi\epsilon_i}\frac{1}{a}\left(\frac{1}{\epsilon_o} - \frac{1}{\epsilon_i}\right). \tag{2.115}$$

The well-known Kirkwood expansion (2.112) converges fast when **r** is away from the boundary of the sphere. However, in the case that the point charge is close to the boundary, the convergence rate by the Kirkwood series expansion is slow due to $rr_s/a^2 \approx 1$, requiring a great number of terms in the expansion to achieve reasonable accuracy for the reaction field.

Before we proceed to derive the image approximation, we quote a useful identity, namely the harmonic expansion of the potential at **r** from a point charge q at $\mathbf{r}_s$ in a homogeneous dielectric:

$$\frac{q}{4\pi\epsilon|\mathbf{r} - \mathbf{r}_s|} = \frac{q}{4\pi\epsilon r}\sum_{n=0}^{\infty}\left(\frac{r_s}{r}\right)^n P_n(\cos\theta), \quad \text{if } r > r_s, \tag{2.116}$$

or

$$\frac{q}{4\pi\epsilon|\mathbf{r}-\mathbf{r}_\mathrm{s}|}=\frac{q}{4\pi\epsilon r_\mathrm{s}}\sum_{n=0}^{\infty}\left(\frac{r}{r_\mathrm{s}}\right)^n P_n(\cos\theta),\quad \text{if } r<r_\mathrm{s}. \tag{2.117}$$

Following Deng et al. [47], by expanding the term $(n+1)/(\epsilon_i n+\epsilon_\mathrm{o}(n+1))$ in (2.112) in terms of $\epsilon_i/((\epsilon_i+\epsilon_\mathrm{o})(n+1))<1$, we obtain

$$\frac{n+1}{\epsilon_i n+\epsilon_\mathrm{o}(n+1)}=\frac{1}{\epsilon_i+\epsilon_\mathrm{o}}\left[1+\frac{\epsilon_i}{\epsilon_i+\epsilon_\mathrm{o}}\frac{1}{n+1}+\left(\frac{\epsilon_i}{\epsilon_i+\epsilon_\mathrm{o}}\right)^2\left(\frac{1}{n+1}\right)^2+\cdots\right], \tag{2.118}$$

allowing us to write the reaction field given in (2.112) as

$$\Phi_{\mathrm{rf}}(\mathbf{r})=R^{(0)}(\mathbf{r})+R^{(1)}(\mathbf{r})+R^{(2)}(\mathbf{r})+\ldots, \tag{2.119}$$

where for $k=0,1,2,\ldots$ we have

$$R^{(k)}(\mathbf{r})=\frac{(\epsilon_i-\epsilon_\mathrm{o})\epsilon_i^k}{4\pi\epsilon_i(\epsilon_i+\epsilon_\mathrm{o})^{k+1}}\frac{a}{r_\mathrm{s}}\frac{q}{(a^2/r_\mathrm{s})}\sum_{n=0}^{\infty}\left(\frac{1}{n+1}\right)^k\left(\frac{r}{a^2/r_\mathrm{s}}\right)^n P_n(\cos\theta). \tag{2.120}$$

In particular, the first term in (2.119) is

$$R^{(0)}(\mathbf{r})=\frac{\epsilon_i-\epsilon_\mathrm{o}}{4\pi\epsilon_i(\epsilon_i+\epsilon_\mathrm{o})}\frac{a}{r_\mathrm{s}}\frac{q}{(a^2/r_\mathrm{s})}\sum_{n=0}^{\infty}\left(\frac{r}{a^2/r_\mathrm{s}}\right)^n P_n(\cos\theta), \tag{2.121}$$

which is the Legendre polynomial expansion of the Coulomb potential at the point **r** inside the sphere due to a point charge of strength q_F outside the sphere at the Kelvin image location $\mathbf{r}_\mathrm{k}$ in (2.104), namely

$$R^{(0)}(\mathbf{r})=\frac{q_\mathrm{F}}{4\pi\epsilon_i|\mathbf{r}-\mathbf{r}_\mathrm{k}|}, \tag{2.122}$$

where

$$q_\mathrm{F}=-\left(\frac{\epsilon_\mathrm{o}-\epsilon_i}{\epsilon_\mathrm{o}+\epsilon_i}\right)\frac{a}{r_\mathrm{s}}q.$$

The Friedman image approximation to the reaction field [48] is thus defined as

$$\Phi_{\mathrm{rf}}(\mathbf{r})\approx\Phi_{\mathrm{F}}(\mathbf{r})=R^{(0)}(\mathbf{r}). \tag{2.123}$$

The Friedman image approximation has been applied in many areas, including molecular dynamics or Monte Carlo simulations [49–51]. It is clear from (2.120) that the approximation holds even when the source charge r_s approaches the boundary of the sphere.

Multiple Image Approximations with Controllable Accuracy

The source charge location $\mathbf{r}_s$ can be inside or outside the sphere $\Omega = \{\mathbf{r}\colon |\mathbf{r}| \le a\}$, and, to illustrate the procedure used to derive the multiple image approximation to the electrostatic potential in the whole space, we will consider the case where a source charge q at $\mathbf{r}_s = (r_s, \theta_s, \phi_s) \in \Omega$ is inside the sphere, as shown in Fig. 2.3. First, let us recall that the Kelvin image point

$$\mathbf{r}_k = (a^2/r_s, \theta_s, \phi_s). \tag{2.124}$$

We will show that the potentials inside and outside the sphere can be approximated by those of a point charge at the Kelvin image location and a line image charge with a power law distribution density along a ray extending from the Kelvin image point $\mathbf{r}_k$ to infinity (for the reaction field inside the sphere) or along a line segment between the origin and the Kelvin image point (for the reaction field outside the sphere if the source charge is outside the sphere). This representation has been discovered independently by several authors. The result was first obtained by Neumann [52], and then by Finkelsteĭn [53], and was rediscovered in the 1990s independently by Lindell [54] and Norris [55].

To obtain the potential inside the sphere due to the polarization, we plug the expansion coefficients A_n in (2.113) into (2.112) to obtain the reaction field:

$$\begin{aligned}
\Phi_{\mathrm{rf}}(\mathbf{r}) &= \sum_{n=0}^{\infty} A_n r^n P_n(\cos\theta) \\
&= \sum_{n=0}^{\infty} \frac{q}{4\pi\epsilon_i} \cdot \frac{r_s^n}{a^{2n+1}} \cdot \gamma \left(1 + \frac{1+\gamma}{1-\gamma+2n}\right) r^n P_n(\cos\theta) \\
&= \frac{\gamma q}{4\pi\epsilon_i r_k} \frac{a}{r_s} \sum_{n=0}^{\infty} \left(\frac{r}{r_k}\right)^n P_n(\cos\theta) \\
&\quad + \frac{q}{4\pi\epsilon_i} \frac{\gamma(1+\gamma)}{2} \sum_{n=0}^{\infty} \frac{r_s^n}{a^{2n+1}} \cdot \frac{2}{1-\gamma+2n} \cdot r^n P_n(\cos\theta) \\
&= S_1 + S_2.
\end{aligned}$$

The first part, S_1, becomes exactly the expansion obtained from (2.117) by putting $r_s = r_k$ and $\epsilon = \epsilon_i$ for a point charge of magnitude

$$q_k = \gamma \frac{a}{r_s} q, \tag{2.125}$$

outside the sphere at the inversion point $\mathbf{r}_k$. For the second part, we use the following simple integral identity for $\alpha = (1-\gamma)/2 + n > 0$:

$$\frac{1}{\alpha r_{\mathrm{k}}^{\alpha}} = \int_{r_{\mathrm{k}}}^{\infty} \frac{1}{x^{\alpha+1}} \mathrm{d}x, \quad \text{for} \quad \alpha > 0, \tag{2.126}$$

which expresses the fraction $2/(1-\gamma+2n) = 1/\alpha$ as an integral. As a result, we have

$$\begin{aligned} S_2 &= \frac{q}{4\pi\epsilon_i} \frac{\gamma(1+\gamma)}{2} \sum_{n=0}^{\infty} \left[\frac{a^{-\gamma}}{r_{\mathrm{s}}^{\frac{1-\gamma}{2}}} \int_{r_{\mathrm{k}}}^{\infty} \frac{1}{x^{\frac{1-\gamma}{2}+n+1}} \mathrm{d}x \right] r^n P_n(\cos\theta) \\ &= \int_{r_{\mathrm{k}}}^{\infty} \left[\frac{q}{4\pi\epsilon_i x} \frac{1}{a} \frac{\gamma(1+\gamma)}{2} \left(\frac{x}{r_{\mathrm{k}}}\right)^{-\frac{1-\gamma}{2}} \sum_{n=0}^{\infty} \left(\frac{r}{x}\right)^n P_n(\cos\theta) \right] \mathrm{d}x \\ &= \int_{r_{\mathrm{k}}}^{\infty} \left[\frac{q'(x)}{4\pi\epsilon_i x} \sum_{n=0}^{\infty} \left(\frac{r}{x}\right)^n P_n(\cos\theta) \right] \mathrm{d}x, \end{aligned} \tag{2.127}$$

where

$$q'(x) = \frac{q}{a} \frac{\gamma(1+\gamma)}{2} \left(\frac{x}{r_{\mathrm{k}}}\right)^{-\frac{1-\gamma}{2}}, \quad r_{\mathrm{k}} \le x. \tag{2.128}$$

Putting $r_{\mathrm{s}} = x$ and $\epsilon = \epsilon_i$ in (2.117), we can see that the inside of the above integral in (2.127) represents the potential generated by charge $q'(x)$ at $\mathbf{x}$.

This result shows that the reaction field $\Phi_{\mathrm{rf}}(\mathbf{r})$ inside the sphere is

$$\Phi_{\mathrm{rf}}(\mathbf{r}) = \frac{q_{\mathrm{k}}}{4\pi\epsilon_i |\mathbf{r} - \mathbf{r}_{\mathrm{k}}|} + \int_{r_{\mathrm{k}}}^{\infty} \frac{q'(x)}{4\pi\epsilon_i |\mathbf{r} - \mathbf{x}|} \mathrm{d}x, \quad \mathbf{r} \in \Omega, \tag{2.129}$$

where

$$\mathbf{x} = \frac{x}{r_{\mathrm{s}}} \mathbf{r}_{\mathrm{s}}, \quad q_{\mathrm{k}} = \gamma \frac{a}{r_{\mathrm{s}}} q, \quad r_{\mathrm{k}} = \frac{a^2}{r_{\mathrm{s}}}, \tag{2.130}$$

and

$$q'(x) = \frac{\epsilon_i(\epsilon_i - \epsilon_{\mathrm{o}})}{(\epsilon_i + \epsilon_{\mathrm{o}})^2} \frac{q}{a} \left(\frac{r_{\mathrm{k}}}{x}\right)^{-\frac{1-\gamma}{2}}, \quad r_{\mathrm{k}} \le x. \tag{2.131}$$

The potential outside the sphere can also be represented by a point charge and a line image charge. Similar results can be obtained for the case when the source charge q is outside the dielectric sphere.

The following lists the results for all cases. The first subscript on the images indicates the location r_s of the source charge q and the second indicates the field location.

1. Internal source ($r_s < a$)
   - Interior Field: Position and magnitude of point image charge $r_k(\geq a)$, $q'_{ii} = \gamma\frac{a}{r_s}q$, and distributed image line charges $q''_{ii}(x) = \frac{q}{a}\frac{\gamma(1+\gamma)}{2}\left(\frac{x}{r_k}\right)^{-\frac{1-\gamma}{2}}$, $r_k \leq x$.
   - Exterior field: Position and magnitude of point image charge $r_s(\leq a)$, $q'_{io} = (1+\gamma)q$, and distributed image line charges $q''_{io}(x) = \frac{q}{r_s}\frac{\gamma(1+\gamma)}{2}\left(\frac{x}{r_s}\right)^{-\frac{1+\gamma}{2}}$, $0 \leq x \leq r_s$.
2. External source ($r_s > a$)
   - Interior Field: Position and magnitude of point image charge $r_s(\geq a)$, $q'_{oi} = (1-\gamma)q$, and distributed image line charges $q''_{oi}(x) = \frac{q}{r_s}\frac{\gamma(1-\gamma)}{2}\left(\frac{x}{r_s}\right)^{-\frac{1-\gamma}{2}}$, $r_s \leq x$.
   - Exterior field: Position and magnitude of point image charge $r_k(\leq a)$, $q'_{oo} = -\gamma\frac{a}{r_s}q$, and distributed image line charges $q''_{oo}(x) = \frac{q}{a}\frac{\gamma(1-\gamma)}{2}\left(\frac{x}{r_k}\right)^{-\frac{1+\gamma}{2}}$, $0 \leq x \leq r_k$.

Note: a is the radius of the sphere; r_s is the radial position of the source charge q; $r_k = a^2/r_s$ is the inversion point, and $-1 < \gamma < 1$. The potential from the source point charge q at $\mathbf{r}_s$ will be added to the potential from the images for interior field points when q is inside the sphere, and for exterior field points when q is outside the sphere.

Next, we derive discrete image charge approximation of the potentials based on the line image representation (2.129). In order to achieve this, we discretize the following line integral by an appropriate numerical quadrature

$$I = \int_{r_k}^{\infty} \frac{1}{|\mathbf{r}-\mathbf{x}|}\left(\frac{x}{r_k}\right)^{-\frac{1-\gamma}{2}} \mathrm{d}x. \tag{2.132}$$

Firstly, by introducing the change of variables $r_k/x = ((1-s)/2)^\tau$ with $\tau > 0$, we have

$$I = \tau \cdot 2^{\frac{\gamma-1}{2}\tau} \int_{-1}^{1} (1-s)^\alpha \cdot h(\mathbf{r}, s; \tau)\mathrm{d}s, \tag{2.133}$$

where $\alpha = (1-\gamma)\tau/2 - 1$ and

$$h(\mathbf{r}, s; \tau) = \frac{2^\tau r_k}{|(1-s)^\tau \mathbf{r} - 2^\tau \mathbf{r}_k|}. \tag{2.134}$$

Next, we employ a numerical quadrature to approximate the integral I in (2.133). Note that $s = -1$ corresponds to the Kelvin image location $x = r_k$. Also we have $\alpha > -1$ since $-1 < \gamma < 1$ and $\tau > 0$. Therefore, we can choose the Gauss–Radau quadrature based on Jacobi polynomials. The Jacobi polynomials $J_n^{(\alpha,\beta)}(s)$ on the interval $[-1, 1]$ are orthogonal polynomials under the Jacobi weight $w^{\alpha,\beta}(s) = (1-s)^\alpha (1+s)^\beta$, i.e.,

$$\int_{-1}^{1} (1-s)^\alpha (1+s)^\beta J_j^{(\alpha,\beta)}(s)\, J_k^{(\alpha,\beta)}(s)\, \mathrm{d}s = \gamma_j^{\alpha,\beta} \delta_{jk},$$

where $\alpha > -1$ and $\beta > -1$ and $\gamma_j^{\alpha,\beta}$ can be found in (5.5) [56].

Let s_m and $\omega_m, m = 0, 1, 2, \ldots, M$, be the Jacobi–Gauss–Radau points and weights on the interval $[-1, 1]$ with $s_0 = -1, s_M < 1$, and $\alpha = (1-\gamma)\tau/2 - 1$, $\beta = 0$. Then, the numerical quadrature for approximating I in (2.133) is

$$I \approx \tau \cdot 2^{\frac{\gamma-1}{2}\tau} \sum_{m=0}^{M} \omega_m h(\mathbf{r}, s_m; \tau), \tag{2.135}$$

which yields

$$\int_{r_k}^{\infty} \frac{q'(x)}{4\pi\epsilon_i\, |\mathbf{r}-\mathbf{x}|} \mathrm{d}x \approx \sum_{m=0}^{M} \frac{q_m}{4\pi\epsilon_i |\mathbf{r}-\mathbf{x}_m|}, \tag{2.136}$$

where for $m = 0, 1, 2, \ldots, M$

$$q_m = 2^{\frac{\gamma-1}{2}\tau - 1} \gamma(1+\gamma)\tau\omega_m q \cdot \frac{x_m}{a}, \tag{2.137}$$

and

$$\mathbf{x}_m = \mathbf{r}_k \cdot \left(\frac{2}{1-s_m}\right)^{\tau}. \tag{2.138}$$

Finally, we have an approximation of the total potential inside the sphere in terms of the potentials of the source charge at $\mathbf{r}_s$ and those of $M+1$ image charges [57]:

$$\Phi(\mathbf{r}) \approx \frac{q}{4\pi\epsilon_i\, |\mathbf{r}-\mathbf{r}_s|} + \frac{q_k}{4\pi\epsilon_i\, |\mathbf{r}-\mathbf{r}_k|} + \sum_{m=0}^{M} \frac{q_m}{4\pi\epsilon_i |\mathbf{r}-\mathbf{x}_m|}. \tag{2.139}$$

Due to the use of the Jacobi–Gauss–Radau points, the first image location $\mathbf{x}_0$ coincides with the Kelvin image location $\mathbf{r}_k$, i.e., $\mathbf{x}_0 = \mathbf{r}_k$.

The following summarizes the magnitudes and locations of discrete images $(q_m, \mathbf{x}_m)$, $m = 0, 1, 2, \ldots, M$ for potentials inside (internal field) and outside (external field) the sphere for both cases of inside and outside source charge q. Again, the first subscript on the images indicates the location r_s of the source charge q and the second indicates the field location.

1. Internal source ($r_s < a$)
   - Interior Field: Position and magnitude of discrete image charges $x_m^{ii} = r_k \cdot \left(\frac{2}{1-s_m}\right)^\tau$, $q_m^{ii} = 2^{\frac{\gamma-1}{2}\tau-1}\gamma(1+\gamma)\tau\omega_m q \cdot \frac{x_m^{ii}}{a}$.
   - Exterior field: Position and magnitude of discrete image charges $x_m^{io} = r_s \cdot \left(\frac{1-s_m}{2}\right)^\tau$, $q_m^{io} = 2^{\frac{\gamma-1}{2}\tau-1}\gamma(1+\gamma)\tau\omega_m q$.
2. External source ($r_s > a$)
   - Interior Field: Position and magnitude of discrete image charges $x_m^{oi} = r_s \cdot \left(\frac{2}{1-s_m}\right)^\tau$, $q_m^{oi} = 2^{\frac{\gamma-1}{2}\tau-1}\gamma(1-\gamma)\tau\omega_m q \cdot \frac{x_m^{oi}}{r_s}$.
   - Exterior field: Position and magnitude of discrete image charges $x_m^{oo} = r_k \cdot \left(\frac{1-s_m}{2}\right)^\tau$, $q_m^{oo} = 2^{\frac{\gamma-1}{2}\tau-1}\gamma(1-\gamma)\tau\omega_m q \cdot \frac{a}{r_s}$.

2.3.3 Image Methods for Dielectric Spheres in Ionic Solvent

In this section, we consider the case when there is a mobile ion density in the solvent outside the sphere. As discussed earlier in the Debye–Hückel theory, the mobile ion concentration in the ionic solvent is given by a Boltzmann distribution in the mean field approximation. For a solvent of weak ionic strength, the linearized Poisson–Boltzmann equation (2.19) can be used outside the sphere. The reaction field inside the sphere is defined by

$$\Phi_{\mathrm{rf}}(\mathbf{r}) = \sum_{n=0}^{\infty} A_n r^n P_n(\cos\theta). \tag{2.140}$$

The expansion coefficients A_n are found from the boundary conditions in (2.102) to be

$$A_n = \frac{q}{4\pi\epsilon_i a}\frac{1}{r_k^n}\frac{\epsilon(n+1)k_n(u)+uk_n'(u)}{\epsilon n k_n(u)-uk_n'(u)} = \frac{q}{4\pi\epsilon_i a}\frac{1}{r_k^n}\frac{\epsilon(n+1)\mathcal{S}_n(u)+1}{\epsilon n\mathcal{S}_n(u)-1}, \tag{2.141}$$

where $u = \lambda a$, $\epsilon = \epsilon_i/\epsilon_o$, and $\mathcal{S}_n(u) = k_n(u)/(uk_n'(u))$. Here, $k_n(r)$ is the modified spherical Hankel function of order n defined by

$$k_n(r) = \frac{\pi e^{-r}}{2r}\sum_{k=0}^{n}\frac{(n+k)!}{k!(n-k)!}\frac{1}{(2r)^k}. \tag{2.142}$$

For an ionic solvent, the modified spherical Hankel function has the following asymptotic expansion in terms of $u = \lambda a < 1$:

$$k_n(u) = \pi \frac{(2n)!}{n!} \frac{1}{(2u)^{n+1}} + O\left(\frac{1}{u^{n-1}}\right), \quad \text{for } n \geq 1, \tag{2.143}$$

and

$$k_n(u) = \pi \frac{(2n)!}{n!} \frac{1}{(2u)^{n+1}} - \frac{\pi}{2} + O\left(\frac{1}{u^{n-1}}\right), \quad \text{for } n = 0. \tag{2.144}$$

We then have, for $n \geq 0$,

$$\begin{aligned} \mathcal{S}_n(u) &= -\frac{1}{n+1+u} + O\left(\frac{1}{u^2}\right), \quad u \to \infty, \\ \mathcal{S}_n(u) &= -\frac{1}{n+1+u} + O\left(u^2\right), \quad u \to 0. \end{aligned} \tag{2.145}$$

Plugging (2.145) into (2.141) leads to an approximation of the expansion coefficient A_n as

$$A_n \approx \frac{q}{4\pi\epsilon_i a} \frac{1}{r_{\mathrm{k}}^n} \left(\gamma + \frac{\delta}{n+\sigma}\right), \tag{2.146}$$

where

$$\sigma = \frac{1+u}{1+\epsilon}, \qquad \delta = \gamma(1-\sigma) - \frac{u}{1+\epsilon}.$$

As σ and r_{k} are both positive real constants, by the integral identity (2.126) with $\alpha = n + \sigma$, we can rewrite the approximation of the reaction field (2.140) as

$$\begin{aligned} \Phi_{\mathrm{rf}}(\mathbf{r}) \approx{}& \frac{q\gamma a}{4\pi\epsilon_i r_s} \frac{1}{r_{\mathrm{k}}} \sum_{n=0}^{\infty} \left(\frac{r}{r_{\mathrm{k}}}\right)^n P_n(\cos\theta) \\ &+ \int_{r_{\mathrm{k}}}^{\infty} \frac{q\delta}{4\pi\epsilon_i a} \left(\frac{x}{r_{\mathrm{k}}}\right)^{-\sigma} \frac{1}{x} \sum_{n=0}^{\infty} \left(\frac{r}{x}\right)^n P_n(\cos\theta)\mathrm{d}x. \end{aligned} \tag{2.147}$$

Using the expansion of (2.117), we obtain the following line image approximation for the reaction field [58]:

$$\Phi_{\mathrm{rf}}(\mathbf{r}) \approx \frac{q_{\mathrm{k}}}{4\pi\epsilon_i |\mathbf{r} - \mathbf{r}_{\mathrm{k}}|} + \int_{r_{\mathrm{k}}}^{\infty} \frac{q'(x)}{4\pi\epsilon_i |\mathbf{r} - \mathbf{x}|} \mathrm{d}x, \tag{2.148}$$

where q_{k} is defined in (2.125), and the line image charge is now

$$q'(x) = q\frac{\delta}{a}\left(\frac{x}{r_{\mathrm{k}}}\right)^{-\sigma}, \quad \text{for } \ r_{\mathrm{k}} \leq x. \tag{2.149}$$

To obtain multiple discrete image charge approximations to the reaction field, the line image charge introduced in (2.148) can be similarly discretized as in (2.136).

2.3.4 *Image Methods for Multi-layered Media*

Series Solution for a Layered Medium for PB Equations

Consider a three-layer model representing the dielectric environment of a cell membrane immersed in ionic fluids [59], as shown in Fig. 2.4. The cell membrane is treated as an infinite layer (region II), separated from the bulk solvents (regions I and III) by two parallel planar boundaries. A point charge q_s is located at $\mathbf{x}_s$ inside region II. As the membrane is made of hydrocarbons, it can be described by a dielectric continuum with a dielectric constant ϵ_m. The electric potential Φ_m in the membrane layer, i.e., in the intermediate layer, satisfies a Poisson equation (2.100) with $\epsilon = \epsilon_m$. The potentials in regions I and III are denoted by $\Phi_{\text{up}}(\mathbf{x}, \mathbf{x}_s)$ and $\Phi_{\text{down}}(\mathbf{x}, \mathbf{x}_s)$, respectively.

The ionic solvents in areas I and III are characterized by a dielectric permittivity ϵ_s and parameters λ_1 and λ_2. The potentials in these two regions are governed by the linearized Poisson–Boltzmann equation (2.16).

The potential inside the membrane layer can be decomposed as in (2.103). By expanding both sides of the Poisson equation (2.100) with orthonormal functions and noting the vanishing of the Coulombic potential at infinity, the potential excluding a prefactor $\frac{1}{4\pi}$ from the source can be expanded as follows:

$$\frac{q_s}{\epsilon_m|\mathbf{x}-\mathbf{x}_s|} = \int_0^\infty\int_0^\infty d\alpha d\beta \cos\alpha(x-x_s)\cos\beta(y-y_s)\frac{2q_s}{\epsilon_m\pi u}e^{-u|z-z_s|}, \quad (2.150)$$

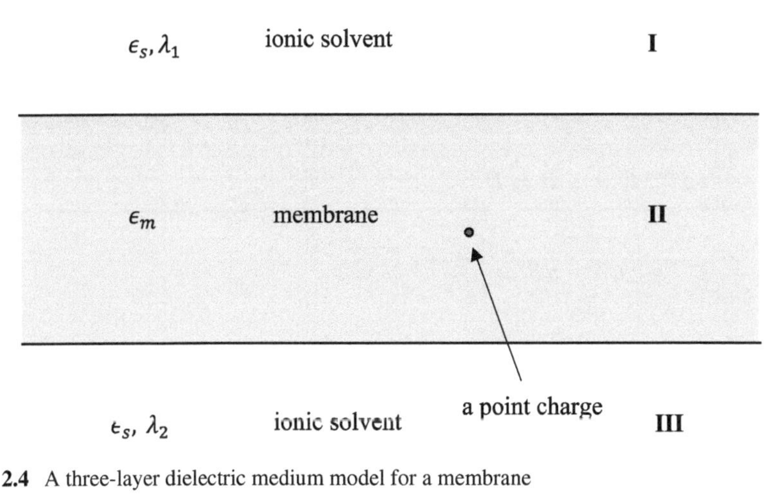

Fig. 2.4 A three-layer dielectric medium model for a membrane

where $u = \sqrt{\alpha^2 + \beta^2}$. Equation (2.150) is a Sommerfeld-type identity well known in electromagnetic scattering theory [60, 61].

As Φ_{rf} satisfies the Laplace equation and Φ_{up} and Φ_{down} satisfy the linearized Poisson–Boltzmann equation, these three potentials excluding a pre-factor $\frac{1}{4\pi}$ can also be expanded, respectively, as follows:

$$\Phi_{\mathrm{rf}}(\mathbf{x}, \mathbf{x}_{\mathrm{s}}) = \int_0^\infty \int_0^\infty \mathrm{d}\alpha \mathrm{d}\beta \cos\alpha(x - x_{\mathrm{s}}) \cos\beta(y - y_{\mathrm{s}}) \cdot [A(\alpha, \beta)\mathrm{e}^{uz} + B(\alpha, \beta)\mathrm{e}^{-uz}], \tag{2.151}$$

$$\Phi_{\mathrm{up}}(\mathbf{x}, \mathbf{x}_{\mathrm{s}}) = \int_0^\infty \int_0^\infty \mathrm{d}\alpha \mathrm{d}\beta \cos\alpha(x - x_{\mathrm{s}}) \cos\beta(y - y_{\mathrm{s}}) C(\alpha, \beta)\mathrm{e}^{-\sqrt{u^2+\lambda_1^2}z}, \tag{2.152}$$

$$\Phi_{\mathrm{down}}(\mathbf{x}, \mathbf{x}_{\mathrm{s}}) = \int_0^\infty \int_0^\infty \mathrm{d}\alpha \mathrm{d}\beta \cos\alpha(x - x_{\mathrm{s}}) \cos\beta(y - y_{\mathrm{s}}) D(\alpha, \beta)\mathrm{e}^{\sqrt{u^2+\lambda_2^2}z}, \tag{2.153}$$

where $\lambda_i^2 = \epsilon_s \kappa_i^2, i = 1, 2$, and for simplicity of notation, the factor of $1/4\pi$ has been dropped for the potentials.

Suppose the planar surfaces are located at $z = 0$ and $z = l$, with l being the thickness of the membrane. Two interface conditions for the continuities of the potential and the normal displacements at each interface are given by

$$\Phi_m = \Phi_{\mathrm{down}}, \quad \epsilon_m \frac{\partial \Phi_m}{\partial z} = \epsilon_s \frac{\partial \Phi_{\mathrm{down}}}{\partial z}, \quad \text{for } z = 0, \tag{2.154}$$

$$\Phi_m = \Phi_{\mathrm{up}}, \quad \epsilon_m \frac{\partial \Phi_m}{\partial z} = \epsilon_s \frac{\partial \Phi_{\mathrm{up}}}{\partial z}, \quad \text{for } z = l. \tag{2.155}$$

Substituting (2.151)–(2.153) into the boundary conditions (2.154) and (2.155) yields a linear system for the coefficients A, B, C, and D:

$$\begin{cases} \dfrac{2q}{\epsilon_m \pi u}\mathrm{e}^{-uz_{\mathrm{s}}} + A + B = D, \\[2ex] \dfrac{2q}{\epsilon_m \pi}\mathrm{e}^{-uz_{\mathrm{s}}} + uA - uB = \dfrac{\epsilon_s}{\epsilon_m}\sqrt{u^2 + \lambda_2^2}D, \\[2ex] \dfrac{2q}{\epsilon_m \pi u}\mathrm{e}^{-u(l-z_{\mathrm{s}})} + A\mathrm{e}^{ul} + B\mathrm{e}^{-ul} = C\mathrm{e}^{-\sqrt{u^2+\lambda_1^2}l}, \\[2ex] -\dfrac{2q}{\epsilon_m \pi}\mathrm{e}^{-u(l-z_{\mathrm{s}})} + uA\mathrm{e}^{ul} - uB\mathrm{e}^{-ul} = -\dfrac{\epsilon_s}{\epsilon_m}\sqrt{u^2 + \lambda_1^2}C\mathrm{e}^{-\sqrt{u^2+\lambda_1^2}l}. \end{cases} \tag{2.156}$$

For convenience, let $\epsilon = \epsilon_s/\epsilon_m$, $\tau_i = \epsilon\sqrt{u^2+\lambda_i^2}$, $i = 1, 2$. Solving the linear system (2.156) leads to the coefficients in the reaction potential Φ_{rf}:

$$A(\alpha,\beta) = \frac{2q_\mathrm{s}}{\epsilon_m \pi u}\frac{\mathrm{e}^{-u(l-z_\mathrm{s})}(u-\tau_1)(u+\tau_2)+\mathrm{e}^{-u(l+z_\mathrm{s})}(u-\tau_1)(u-\tau_2)}{(u+\tau_1)(u+\tau_2)\mathrm{e}^{ul}-(u-\tau_1)(u-\tau_2)\mathrm{e}^{-ul}}, \quad (2.157)$$

$$B(\alpha,\beta) = \frac{2q_\mathrm{s}}{\epsilon_m \pi u}\frac{\mathrm{e}^{u(l-z_\mathrm{s})}(u+\tau_1)(u-\tau_2)+\mathrm{e}^{-u(l-z_\mathrm{s})}(u-\tau_1)(u-\tau_2)}{(u+\tau_1)(u+\tau_2)\mathrm{e}^{ul}-(u-\tau_1)(u-\tau_2)\mathrm{e}^{-ul}}. \quad (2.158)$$

Method of Images in a Layered Non-ionic Solution

First, we consider the case of pure water where no mobile charges are present in the solvents (pure water). Then, the parameters $\lambda_1 = \lambda_2 = 0$. Here, we will illustrate the mathematical equivalence between (2.151) and an image representation. Recall that $\tau_1 = \tau_2 = \epsilon u$ in this case. Equation (2.151) can be rewritten in four parts as follows:

$$\Phi_{\mathrm{rf}}(\mathbf{x},\mathbf{x}_\mathrm{s}) = \int_0^\infty\int_0^\infty \mathrm{d}\alpha \mathrm{d}\beta \cos\alpha(x-x_\mathrm{s})\cos\beta(y-y_\mathrm{s})[I+II+III+IV], \quad (2.159)$$

where

$$I = \frac{2q}{\epsilon_m\pi u}\frac{\mathrm{e}^{-u(2l-z_\mathrm{s}-z)}\sigma}{1-\mathrm{e}^{-2ul}\sigma^2}, \qquad II = \frac{2q}{\epsilon_m\pi u}\frac{\mathrm{e}^{-u(2l+z_\mathrm{s}-z)}\sigma^2}{1-\mathrm{e}^{-2ul}\sigma^2},$$

$$III = \frac{2q}{\epsilon_m\pi u}\frac{\mathrm{e}^{u(-z_\mathrm{s}-z)}\sigma}{1-\mathrm{e}^{-2ul}\sigma^2}, \qquad IV = \frac{2q}{\epsilon_m\pi u}\frac{\mathrm{e}^{-u(2l-z_\mathrm{s}+z)}\sigma^2}{1-\mathrm{e}^{-2ul}\sigma^2},$$

and

$$\sigma = \frac{1-\epsilon}{1+\epsilon}.$$

As $\left|\mathrm{e}^{-2ul}\sigma^2\right| < 1$, we have the following geometric progression:

$$\frac{1}{1-\mathrm{e}^{-2ul}\sigma^2} = \sum_{k=0}^{+\infty}\mathrm{e}^{-2kul}\sigma^{2k} = \sum_{k=0}^{-\infty}\mathrm{e}^{2kul}\sigma^{-2k}.$$

Then, Eq. (2.159) becomes

$$\begin{aligned}
\Phi_{\mathrm{rf}}(\mathbf{x}, \mathbf{x}_{\mathrm{s}}) &= \sum_{k=0}^{+\infty} \int_0^\infty \int_0^\infty \mathrm{d}\alpha \mathrm{d}\beta \cos \alpha(x - x_{\mathrm{s}}) \cos \beta(y - y_{\mathrm{s}}) \frac{2q}{\epsilon_m \pi u} \\
&\times \left(\mathrm{e}^{-u(2kl+2l-z_{\mathrm{s}}-z)} \sigma^{2k+1} + \mathrm{e}^{-u(2kl+2l+z_{\mathrm{s}}-z)} \sigma^{2k+2}\right) \\
&+ \sum_{k=0}^{-\infty} \int_0^\infty \int_0^\infty \mathrm{d}\alpha \mathrm{d}\beta \cos \alpha(x - x_{\mathrm{s}}) \cos \beta(y - y_{\mathrm{s}}) \frac{2q}{\epsilon_m \pi u} \\
&\times \left(\mathrm{e}^{u(2kl-z_{\mathrm{s}}-z)} \sigma^{-2k+1} + \mathrm{e}^{-u(-2kl+2l-z_{\mathrm{s}}+z)} \sigma^{-2k+2}\right) \\
&= \sum_{k=0}^{+\infty} \frac{Q_{2k+1}}{\epsilon_m |\mathbf{x} - \mathbf{x}_{2k+1}|} + \frac{Q_{2k+2}}{\epsilon_m |\mathbf{x} - \mathbf{x}_{2k+2}|} \\
&+ \sum_{k=0}^{-\infty} \frac{Q_{2k-1}}{\epsilon_m |\mathbf{x} - \mathbf{x}_{2k-1}|} + \frac{Q_{2k-2}}{\epsilon_m |\mathbf{x} - \mathbf{x}_{2k-2}|} \\
&= \sum_{k=-\infty,\ k\neq 0}^{+\infty} \frac{Q_k}{\epsilon_m |\mathbf{x} - \mathbf{x}_k|},
\end{aligned} \tag{2.160}$$

where the second to last equation holds from the Sommerfeld identity (2.150). Hence, we have the following image representation [59, 62] of the reaction field:

$$\Phi_{\mathrm{rf}}(\mathbf{x}, \mathbf{x}_s) = \sum_{k=-\infty,\ k\neq 0}^{+\infty} \frac{Q_k}{\epsilon_m |\mathbf{x} - \mathbf{x}_k|}, \tag{2.161}$$

with

$$\mathbf{x}_k = \left(x_{\mathrm{s}}, y_{\mathrm{s}}, (-1)^k \left(z_{\mathrm{s}} - \frac{l}{2}\right) + \left(k + \frac{1}{2}\right) l\right), \quad Q_k = q_{\mathrm{s}} \left(\frac{\epsilon_m - \epsilon_s}{\epsilon_m + \epsilon_s}\right)^{|k|}. \tag{2.162}$$

Method of images in layered ionic solutions

In the following the locations and the magnitudes of the image charges will be produced through a Prony-type approximation to the Fourier transform of the reaction potential and the Sommerfeld-type identity (2.150). In the Prony approximation [63, 64], a sum of exponentials is used to approximate the Fourier transform of the exact reaction field potential. Rewrite the exact reaction field (2.151) in the following form:

$$\Phi_{\mathrm{rf}}(\mathbf{x}, \mathbf{x}_{\mathrm{s}}) = \int_0^\infty \int_0^\infty \mathrm{d}\alpha \mathrm{d}\beta \cos \alpha(x - x_{\mathrm{s}}) \cos \beta(y - y_{\mathrm{s}}) \frac{2q_{\mathrm{s}}}{\epsilon_m \pi u} [\hat{A} \mathrm{e}^{uz} + \hat{B} \mathrm{e}^{-uz}], \tag{2.163}$$

where

$$\hat{A}(u) = \frac{\mathrm{e}^{-u(l-z_s)}(u-\tau_1)(u+\tau_2) + \mathrm{e}^{-u(l+z_s)}(u-\tau_1)(u-\tau_2)}{(u+\tau_1)(u+\tau_2)\mathrm{e}^{ul} - (u-\tau_1)(u-\tau_2)\mathrm{e}^{-ul}},$$
$$\hat{B}(u) = \frac{\mathrm{e}^{u(l-z_s)}(u+\tau_1)(u-\tau_2) + \mathrm{e}^{-u(l-z_s)}(u-\tau_1)(u-\tau_2)}{(u+\tau_1)(u+\tau_2)\mathrm{e}^{ul} - (u-\tau_1)(u-\tau_2)\mathrm{e}^{-ul}}. \tag{2.164}$$

By comparing (2.163) with the Sommerfeld identity (2.150), and using the symmetry of the rectangular system, a natural approach can be obtained to approximate the functions $\hat{A}(u)$ and $\hat{B}(u)$ by a sum of K exponential functions as

$$\hat{A}^{\mathrm{app}}(u) = \sum_{k=1}^{K} \left(a_{2k-1}\mathrm{e}^{-(z_{2k-1}-c_k)u} + a_{2k}\mathrm{e}^{-(z_{2k}+c_k)u}\right),$$
$$\hat{B}^{\mathrm{app}}(u) = \sum_{k=-1}^{-K} \left(a_{2k+1}\mathrm{e}^{(z_{2k+1}+c_k)u} + a_{2k}\mathrm{e}^{(z_{2k}-c_k)u}\right), \tag{2.165}$$

where $z_k = (-1)^k(z_s - l/2) + (k+1/2)l$ are the locations of image charges in the case $\lambda_1 = \lambda_2 = 0$, $a_{\pm k}$ and $c_{\pm k}$ are unknown variations of the kth pair of image charges. Imposing the following conditions:

$$z_{2k-1} - c_k > l, \quad z_{2k} + c_k > l, \qquad \text{for} \quad k > 0, \tag{2.166}$$

$$z_{2k+1} + c_k < 0, \quad z_{2k} - c_k < 0, \qquad \text{for} \quad k < 0, \tag{2.167}$$

and then applying the Sommerfeld-type identity (2.150) yields [59]

$$\begin{aligned}
\Phi_{\mathrm{rf}}(\mathbf{x}, \mathbf{x}_s) &= \int_0^\infty \int_0^\infty \mathrm{d}\alpha \mathrm{d}\beta \cos\alpha(x-x_s)\cos\beta(y-y_s)\frac{2q_s}{\epsilon_m \pi u} \\
&\quad \times \left[\sum_{k=1}^{K}(a_{2k-1}\mathrm{e}^{-(z_{2k-1}-c_k-z)u} + a_{2k}\mathrm{e}^{-(z_{2k}+c_k-z)u})\right. \\
&\quad \left. + \sum_{k=-1}^{-K}(a_{2k+1}\mathrm{e}^{-(z-z_{2k+1}-c_k)u} + a_{2k}\mathrm{e}^{-(z-z_{2k}+c_k)u})\right] \\
&= \sum_{k=1}^{K} \frac{q_s}{\epsilon_m}\left[\frac{a_{2k-1}}{|\mathbf{x}-\mathbf{x}_{2k-1}|} + \frac{a_{2k}}{|\mathbf{x}-\mathbf{x}_{2k}|}\right] \\
&\quad + \sum_{k=-1}^{-K} \frac{q_s}{\epsilon_m}\left[\frac{a_{2k+1}}{|\mathbf{x}-\mathbf{x}_{2k+1}|} + \frac{a_{2k}}{|\mathbf{x}-\mathbf{x}_{2k}|}\right],
\end{aligned} \tag{2.168}$$

with

$$\mathbf{x}_{2k-1} = (x_s, y_s, z_{2k-1} - c_k), \qquad \mathbf{x}_{2k} = (x_s, y_s, z_{2k} + c_k), \qquad k > 0, \tag{2.169}$$

$$\mathbf{x}_{2k+1} = (x_s, y_s, z_{2k+1} + c_k), \qquad \mathbf{x}_{2k} = (x_s, y_s, z_{2k} - c_k), \qquad k < 0. \tag{2.170}$$

Once a_k and c_k are found, the image charges are given by the Sommerfeld-type identity (2.150). As the functions $\hat{A}(u)$ and $\hat{B}(u)$ are damped exponentials, such an approximation is reasonable and effective. The parameters a_k and c_k can be calculated by solving a simple minimization problem of the L_2 errors, sampling u at selected $u_j, j = 1, 2, \ldots, J$.

2.4 Summary

The Poisson–Boltzmann theory of electrostatic interactions in ionic solutions is based on the Debye–Hückel mean field theory of the ions where the ions' discreteness and long-range correlation effects beyond the Debye screen effect are not included. As a result, the PB theory may lose accuracy near charged surfaces such as a protein's surface and along an ion channel wall. A size modified PB equation is introduced to address the steric effect of ions of finite size near surfaces. Meanwhile, the Born solvation model and the generalized Born (GB) formula provide analytical ways to compute the electrostatic solvation energy, for molecules of general shape with the GB formula. We have provided an FFT-based method to compute the Born radius required in the GB formula. For simple geometries, including dielectric spheres and layered media, the image approximation of the reaction field for the PB electrostatics allows a quick calculation of the solvation energy. More discussion on analytical approximations to PB electrostatic theory can be found in the review of Xu and Cai [65].

References

1. S.R. Milner, The virial of a mixture of ions. Phil. Mag. **6**, 551–578 (1912)
2. J.O'M. Bockris, A.K.N. Reddy, M. Gamboa-Aldeco. *Modern Electrochemisty 2A: Fundamentals of Electrodics*, 2nd edn. (Kluwer Academic Publishers, Norwell, MA, 2000)
3. B. Honig, A. Nicholls, Classical electrostatics in biology and chemistry. Science **268**, 1144–1149 (1995)
4. F. Fogolari, A. Brigo, H. Molinari, The Poisson-Boltzmann equation for biomolecular electrostatics: a tool for structural biology. J. Mol. Recognit. **15**, 377–392 (2002)
5. M.E. Davis, J.A. McCammon, Electrostatics in biomolecular structure and dynamics. Chem. Rev. **90**, 509–521 (1990)
6. A.R. Leach, *Molecular Modelling: Principles and Applications*, 2nd edn. (Prentice-Hall, Englewood Cliffs, NJ, 2001)

7. W. Humphrey, A. Dalke, K. Schulten, VMD—Visual Molecular Dynamics. J. Mol. Graph. **14**, 33–38 (1996)
8. S. Lindskog, Structure and mechanism of carbonic anhydrase. Pharmacol. Therapeut. **74**, 1–20 (1997)
9. G. Gouy, About the electric charge on the surface of an electrolyte. J. Phys. (Fr.) **9**, 457–468 (1910)
10. T.L. Hill, *An Introduction to Statistical Thermodynamics* (Dover Publications, New York, 1987)
11. P. Debye, E.Hückel, The theory of electrolytes. I. Lowering of freezing point and related phenomena. Physik. Z. **24**, 185–206 (1923)
12. B. Li, Minimization of electrostatic free energy and the Poisson-Boltzmann equation for molecular solvation with implicit solvent. SIAM J. Math. Anal. **40**, 2536–2566 (2009)
13. B. Lu, Y.C. Zhou, M.J. Holst, J.A. McCammon, Recent progress in numerical methods for the Poisson-Boltzmann equation in biophysical applications. Commun. Comput. Phys. **3**, 973–1009 (2008)
14. M. Feig, C.L. Brooks III., Recent advances in the development and application of implicit solvent models in biomolecule simulations. Curr. Opin. Struc. Biol. **14**, 217–224 (2004)
15. I. Borukhov, D. Andelman, H. Orland, Steric effects in electrolytes: a modified Poisson-Boltzmann equation. Phys. Rev. Lett. **79**, 435–438 (1997)
16. H. Helmholtz, Ueber einige Gesetze der Verteilung elektrischer Ströme in körperlichen Leitern mit Anwendung auf diethierisch-elektrischen Versuche. Pogg. Ann. Physik. Chemie. **89**, 211–233 (1853)
17. C.H. Hamann, A. Hamnett, W. Vielstich, *Electrochemistry* (Wiley-VCH, Hoboken, NJ, 2007)
18. D.L. Chapman, A contribution to the theory of electrocapillarity. Phil. Mag. **25**, 475–481 (1913)
19. O. Stern, Zur theorie der elektrolytischen Doppelschicht. Z. Elektrochem. **30**, 508–516 (1924)
20. D. Boda, W.R. Fawcett, D. Henderson, S. Sokołowski, Monte Carlo, density functional theory, and Poisson-Boltzmann theory study of the structure of an electrolyte near an electrode. J. Chem. Phys. **116**, 7170–7176 (2002)
21. D. Henderson, D. Boda, Insights from theory and simulation on the electrical double layer. Phys. Chem. Chem. Phys. **11**, 3822–3830 (2009)
22. B. Lu, Y.C. Zhou, Poisson-Nernst-Planck equations for simulating biomolecular diffusion-reaction processes II: size effects on ionic distributions and diffusion-reaction rates. Biophys. J. **100**, 2475–2485 (2011)
23. V.B. Chu, Y. Bai, J. Lipfert, D. Herschlag, S. Doniach, Evaluation of ion binding to DNA duplexes using a size-modified Poisson-Boltzmann theory. Biophys. J. **93**, 3202–3209 (2007)
24. B. Eisenberg, Y. Hyon, C. Liu, Energy variational analysis of ions in water and channels: field theory for primitive models of complex ionic fluids. J. Chem. Phys. **133**, 104104 (2010)
25. B. Roux, Implicit solvent models, in *Computational Biochemistry and Biophysics*. ed. by O.M. Becker, A.D. MacKerell Jr., B. Roux, M. Watanabe (Marcel Dekker, New York, 2001), pp. 133–151
26. M. Born, Volumes and heats of hydration of ions. Z. Phys. **1**, 45–48 (1920)
27. L. Onsager, Theories of concentrated electrolytes. Chem. Rev. **13**, 73–89 (1933)
28. J.G. Kirkwood, Statistical mechanics of fluid mixtures. J. Chem. Phys. **3**, 300–313 (1935)
29. K.A. Sharp, B. Honig, Calculating total electrostatic energies with the nonlinear Poisson-Boltzmann equation. J. Phys. Chem. **94**, 7684–7692 (1990)
30. H.-X. Zhou, Macromolecular electrostatic energy within the nonlinear Poisson-Boltzmann equation. J. Chem. Phys. **100**, 3152–3162 (1994)
31. D. Bashford, D.A. Case, Generalized Born models of macromolecular solvation effects. Ann. Rev. Phys. Chem. **51**, 129–152 (2000)
32. G. Sigalov, P. Scheffel, A. Onufriev, Incorporating variable dielectric environments into the generalized Born model. J. Chem. Phys. **122**, 094511 (2005)
33. W.C. Still, A. Tempczyk, R.C. Hawley, T. Hendrickson, Semianalytical treatment of solvation for molecular mechanics and dynamics. J. Am. Chem. Soc. **112**, 6127–6129 (1990)
34. A. Onufriev, A.W. Ralph, C.S. David, Implicit solvent models in molecular dynamics simulations: a brief overview. Ann. Rep. Comput. Chem. **4**, 125–137 (2008)

35. A. Onufriev, Continuum electrostatics solvent modeling with the generalized Born model, in *Modeling Solvent Environments*. ed. by M. Feig (Wiley, New York, 2010), pp. 127–165
36. A. Onufriev, D.A. Case, D. Bashford, Effective Born radii in the generalized Born approximation: the importance of being perfect. J. Comput. Chem. **23**, 1297–1304 (2002)
37. J. Srinivasan, M.W. Trevathan, P. Beroza, D.A. Case, Application of a pairwise generalized Born model to proteins and nucleic acids: inclusion of salt effects. Theor. Chem. Acc. **101**, 426–434 (1999)
38. T. Grycuk, Deficiency of the Coulomb-field approximation in the generalized Born model: an improved formula for Born radii evaluation. J. Chem. Phys. **119**, 4817–4826 (2003)
39. M.S. Lee, F.R. Salsbury Jr., M.A. Olson, An efficient hybrid explicit/implicit solvent method for biomolecular simulations. J. Comput. Chem. **25**, 1967–1978 (2004)
40. J.D. Jackson, *Classical Electrodynamics*, 3rd edn. (Wiley, Hoboken, NJ, 2001)
41. W. Cai, Z. Xu, A. Baumketner, A new FFT-based algorithm to compute Born radii in the generalized Born theory of biomolecule solvation. J. Comput. Phys. **227**, 10162–10177 (2008)
42. C. Shannon, W. Weaver, *The Mathematical Theory of Communication* (University of Illinois Press, Urbana and Chicago, IL, 1963)
43. I. Daubechies, *Ten Lectures on Wavelets* (SIAM, Philadelphia, PA, 1992)
44. W. (L.K.) Thomson, *Reprint of Papers on Electrostatics and Magnetism*, 2nd edn. (Macmillan, London, 1884)
45. W.R. Smythe, *Static and Dynamic Electricity*, 3rd edn. (Hemisphere Publishing, New York, 1989)
46. J.G. Kirkwood, Theory of solutions of molecules containing widely separated charges with special applications to zwitterions. J. Chem. Phys. **2**, 351–361 (1934)
47. S. Deng, W. Cai, D. Jacobs, A comparable study of image approximations to the reaction field. Comput. Phys. Commun. **177**, 689–699 (2007)
48. H.L. Friedman, Image approximation to the reaction field. Mol. Phys. **29**, 1533–1543 (1975)
49. J.A.C. Rullmann, P.T.V. Duijnen, Analysis of discrete and continuum dielectric models; application to the calculation of protonation energies in solution. Mol. Phys. **61**, 293–311 (1987)
50. A. Wallqvist, On the implementation of Friedman boundary conditions in liquid water simulations. Mol. Simul. **10**, 13–17 (1993)
51. L. Wang, J. Hermans, Reaction field molecular dynamics simulation with Friedman's image charge method. J. Phys. Chem. **99**, 12001–12007 (1995)
52. C. Neumann, *Hydrodynamische Untersuchen nebst einem Anhang uber die Probleme der Elecktrostatik und der magnetischen Induktion* (Teubner, Leipzig, 1883), pp. 279–282
53. A. V. Finkelsteǐn, Electrostatic interactions of charged groups in an aqueous medium and their effect on the formation of polypeptide chain secondary structure. Mol. Biol. **11**, 811–819 (1977)
54. I.V. Lindell, Electrostatic image theory for the dielectric sphere. Radio Sci. **27**, 1–8 (1992)
55. W.T. Norris. Charge images in a dielectric sphere. IEE Proc.—Sci. Meas. Technol. **142**, 142–150 (1995)
56. W. Gautschi, Algorithm 726: ORTHPOL–a package of routines for generating orthogonal polynomials and Gauss-type quadrature rules. ACM Trans. Math. Softw. **20**, 21–62 (1994)
57. W. Cai, S. Deng, D. Jacobs, Extending the fast multipole method to charges inside or outside a dielectric sphere. J. Comput. Phys. **223**, 846–864 (2007)
58. Z. Xu, S. Deng, W. Cai, Image charge approximations of reaction fields in solvents with arbitrary ionic strength. J. Comput. Phys. **228**, 2092–2099 (2009)
59. H. Lin, Z. Xu, H. Tang, W. Cai, Image approximations to electrostatic potentials in layered electrolytes/dielectrics and an ion-channel model. J. Sci. Comput. **53**, 249–67 (2012)
60. A. Sommerfeld, *Partial Differential Equations in Physics* (Academic Press, New York, 1949)
61. W.C. Chew, *Waves and Fields in Inhomogeneous Media* (Van Nostrand Reinhold, New York, 1990)
62. P.-K. Yang, S.-H. Liaw, C. Lim, Representing an infinite solvent system with a rectangular finite system using image charges. J. Phys. Chem. B **106**, 2973–2982 (2002)

63. R. Prony, Essai éxperimental et analytique: sur les lois de la dilatabilité de fluides élastique et sur celles de la force expansive de la vapeur de l'alkool, à différentes températures. *J. l'École Polytech. (Paris)* **1**, 24–76 (1795)
64. L. Weiss, R.N. McDonough, Prony's method, z-transforms, and Padé approximation. SIAM Rev. **5**, 145–149 (1963)
65. Z.L. Xu, W. Cai, Fast analytical methods for macroscopic electrostatic models in biomolecular simulations. SIAM Rev. **53**, 683–720 (2011)

Chapter 3
Numerical Methods for Poisson–Boltzmann Equations

Numerical methods for solving the PB equations are indispensable for finding accurate solutions for molecular solvation energies; three methods, including the boundary element method, the finite element method, and the finite difference immersed interface method (IIM), will be discussed in this chapter.

Achieving accuracy and speed in solving the PB equations requires the capabilities to handle complicated molecular surfaces and singular source charges inside the solute molecules. The integral equation method transforms the infinite domain problem to the molecular surface, thus removing the issue of the interior singular source charges. For both finite element and finite difference methods, singularity subtraction techniques can be used to remove the effect of the singular point sources. Meanwhile, to treat the molecular surfaces, the finite difference IIM uses a special difference formula near the surfaces. On the other hand, finite element methods, based on an unstructured mesh, allow highly accurate resolution of the complicated molecular surfaces and fast multigrid solutions of the resulting matrix equations.

3.1 Boundary Element Methods (BEMs)

Let us consider a second-order elliptic equation in $\mathbb{R}^3$:

$$\mathcal{L}u = f(\mathbf{r}), \tag{3.1}$$

where

$$\mathcal{L} = \Delta + k^2. \tag{3.2}$$

W. Cai, *Deterministic, Stochastic, and Deep Learning Methods for Computational Electromagnetics*, https://doi.org/10.1007/978-981-96-0100-4_3

The fundamental solution to (3.1) is given by

$$G_k(\mathbf{r}) = \frac{1}{4\pi}\frac{\mathrm{e}^{-\mathrm{i}kr}}{r}, \tag{3.3}$$

namely

$$\mathcal{L}G_k(\mathbf{r}) = -\delta(\mathbf{r}). \tag{3.4}$$

For a real k, we have the Helmholtz equation for wave propagation, while for a pure imaginary $k = -\mathrm{i}\kappa$, we have the linearized PB equation (2.16):

$$\mathcal{L}u = \Delta u - \kappa^2 u = f(\mathbf{r}). \tag{3.5}$$

Equations (3.1) and (3.5) require a boundary condition at infinity, that is, a Sommerfeld radiation condition for the Helmholtz equation or a vanishing condition for the PB equation, i.e.,

$$\begin{cases} \text{(Helmholtz equation)} & \left(\frac{\partial}{\partial r} + \mathrm{i}k\right) u(r) = O\left(\frac{1}{r^2}\right), r \to \infty \\ & \text{Sommerfeld radiation condition} \\ \text{(PB equation)} & \lim_{r\to\infty} u(r) = 0. \end{cases} \tag{3.6}$$

These boundary conditions will ensure the uniqueness of the solution of both equations [1].

3.1.1 *Cauchy Principal Value (CPV) and Hadamard Finite Part (p.f.)*

A surface charge or dipole density generates electrostatic potentials in the whole space, which can be expressed as linear superpositions of the potential of a single point charge in (1.4) or a dipole moment in (1.9), resulting in the so-called single- or double-layer potentials. These potentials also appear naturally in the direct integral equation for (3.1) based on Green's second identity (3.43) in Sect. 3.1.2.

Let us consider the following three types of singular potentials defined on the interface Γ between the interior region Ω_i and its complement exterior region Ω_o, $\Omega_\mathrm{i} \cup \Omega_\mathrm{o} = \mathbb{R}^3$. In this section, $G(\mathbf{r}, \mathbf{r}') = G_k(\mathbf{r} - \mathbf{r}')$.

Definition 3.1 A single-layer potential $u(\mathbf{r})$ is defined by

$$u(\mathbf{r}) = S(q) \equiv \int_\Gamma G(\mathbf{r}, \mathbf{r}')q(\mathbf{r}')\mathrm{d}s', \mathbf{r} \notin \Gamma, \tag{3.7}$$

where $q(\mathbf{r})$ is the charge density on Γ. Then, $u(\mathbf{r})$ satisfies the elliptic equation (3.1) with $f = 0$ in both Ω_i and Ω_o, and also the boundary condition (3.6) at infinity.

Definition 3.2 A double-layer potential $u(\mathbf{r})$ is defined by

$$u(\mathbf{r}) = D(\phi) \equiv \int_{\Gamma} \frac{\partial G(\mathbf{r}, \mathbf{r}')}{\partial \mathbf{n}'} \phi(\mathbf{r}')\mathrm{d}s', \mathbf{r} \notin \Gamma, \tag{3.8}$$

where $\phi(\mathbf{r})$ is the dipole surface density and $\mathbf{n}'$ is the outward normal on the boundary Γ. Again, $u(\mathbf{r})$ satisfies the elliptic equation (3.1) with $f = 0$ in both Ω_{i} and Ω_{o}, and also the boundary condition (3.6) at infinity.

Definition 3.3 A hyper-singular potential is defined by

$$u(\mathbf{r}) = N(\sigma) \equiv \frac{\partial}{\partial \mathbf{n}} \int_{\Gamma} \frac{\partial G(\mathbf{r}, \mathbf{r}')}{\partial \mathbf{n}'} \sigma(\mathbf{r}')\mathrm{d}s'$$
$$= \int_{\Gamma} \frac{\partial^2 G(\mathbf{r}, \mathbf{r}')}{\partial \mathbf{n} \partial \mathbf{n}'} \sigma(\mathbf{r}')\mathrm{d}s', \mathbf{r} \notin \Gamma, \tag{3.9}$$

where $\sigma(\mathbf{r})$ is the dipole density. Again, $u(\mathbf{r})$ satisfies the elliptic equation (3.1) with $f = 0$ in both Ω_{i} and Ω_{o}, and also the condition (3.6) at infinity.

The functions $u(\mathbf{r})$ given by the singular potentials above are well defined for all $\mathbf{r} \in \mathbb{R}^3 \backslash \Gamma$ and satisfy the Helmholtz (or PB) equation there. However, to find the limiting values of $u(\mathbf{r})$ or $\partial u / \partial \mathbf{n}(\mathbf{r})$ at the defining surface Γ, care is needed as the surface integrals over Γ involve unbounded functions when $\mathbf{r} \in \Gamma$. Specifically, the concept of Cauchy principal values of singular integrals will be needed.

The Cauchy principal value (CPV) of a singular function $f(\mathbf{r})$ over Γ is defined as

$$\text{p.v.} \int_{\Gamma} f(\mathbf{r}')\mathrm{d}s' = \lim_{\varepsilon \to 0} \int_{\Gamma \backslash S_\varepsilon} f(\mathbf{r}')\mathrm{d}s' \tag{3.10}$$

if the limit on the right-hand side exists, where the surface patch S_ε of area size ε contains the singularity of the integrand.

Even though for $\mathbf{r} \in \Gamma$ the surface integrals in (3.7)–(3.9) themselves could be divergent, the limits of $u(\mathbf{r})$ as $\mathbf{r}$ approaches the surface Γ are expected to remain finite and well defined for smooth Γ, as in electrostatics $u(\mathbf{r})$ and its derivatives represent a physical potential and an electric field, respectively, generated by finite charges or dipole densities. The concept of the Cauchy principal value or, for that matter, the Hadamard finite part comes into play due to the specific way of taking the limit of the potential toward the boundary. We illustrate this concept by considering the well-known Hilbert transform of a function $\phi(x)$ [2], $x \in \mathbb{R} = (-\infty, \infty)$:

$$H(\phi)(z) = \frac{1}{2\pi \mathrm{i}} \int_{-\infty}^{\infty} \frac{\phi(x)}{x - z} \mathrm{d}x, z \notin \mathbb{R}. \tag{3.11}$$

To compute the limit of $z = x + \mathrm{i}y \in \mathbb{C} \to x_0$, i.e., $y \to 0^+$, we can deform the real line contour to include a semi-circle $S_\varepsilon(x_0)$ of radius ε centered at x_0 in the lower half-plane. Then, we have

$$
\begin{aligned}
&\lim_{z\to x_0} H(\phi)(z) \\
&= \frac{1}{2\pi\mathrm{i}} \lim_{z\to x_0} \left[\int_{-\infty}^{x_0-\varepsilon} + \int_{x_0+\varepsilon}^{\infty} \right] \frac{\phi(x)}{x-z}\mathrm{d}x + \frac{1}{2\pi\mathrm{i}} \lim_{z\to x_0} \int_{S_\varepsilon(x_0)} \frac{\phi(z')}{z'-z}\mathrm{d}z' \\
&= \frac{1}{2\pi\mathrm{i}} \left[\int_{-\infty}^{x_0-\varepsilon} + \int_{x_0+\varepsilon}^{\infty} \right] \frac{\phi(x)}{x-x_0}\mathrm{d}x + \frac{1}{2\pi\mathrm{i}} \int_{S_\varepsilon(x_0)} \frac{\phi(z')}{z'-x_0}\mathrm{d}z' \\
&= \frac{1}{2\pi\mathrm{i}} \left[\int_{-\infty}^{x_0-\varepsilon} + \int_{x_0+\varepsilon}^{\infty} \right] \frac{\phi(x)}{x-x_0}\mathrm{d}x + \frac{\phi(x_0)}{2} + O(\varepsilon).
\end{aligned}
\tag{3.12}
$$

Now let $\varepsilon \to 0$. Assuming the two integrals in the square bracket of the last equation above have a limit, we have

$$
\lim_{z\to x_0} H(\phi)(z) = \frac{1}{2\pi\mathrm{i}} \,\text{p.v.} \int_{-\infty}^{\infty} \frac{\phi(x)}{x-x_0}\mathrm{d}x + \frac{\phi(x_0)}{2}, \tag{3.13}
$$

where the first term is the Cauchy principal value of the form (3.10), i.e.,

$$
\text{p.v.} \int_{-\infty}^{\infty} \frac{\phi(x)}{x-x_0}\mathrm{d}x = \lim_{\varepsilon\to 0} \left[\int_{-\infty}^{x_0-\varepsilon} + \int_{x_0+\varepsilon}^{\infty} \right] \frac{\phi(x)}{x-x_0}\mathrm{d}x. \tag{3.14}
$$

In fact, for a smooth function $\phi(x)$, the Cauchy principal value of (3.14) can be calculated as follows. For example, for $\phi(x) = 1$, we have

$$
\text{p.v.} \int_{a}^{b} \frac{1}{x-c}\mathrm{d}x = \lim_{\varepsilon\to 0} \left[\int_{a}^{c-\varepsilon} + \int_{c+\varepsilon}^{b} \right] \frac{1}{x-c}\mathrm{d}x = \ln\left(\frac{b-c}{c-a}\right), \tag{3.15}
$$

and for a Hölder continuous function $\phi(x)$ with index $\lambda > 0$, we have

$$
\text{p.v.} \int_a^b \frac{\phi(x)}{x-c} \mathrm{d}x = \int_a^b \frac{\phi(x)-\phi(c)}{x-c} \mathrm{d}x + \phi(c)\ \text{p.v.} \int_a^b \frac{1}{x-c} \mathrm{d}x
$$

$$
= \int_a^b \frac{\phi(x)-\phi(c)}{x-c} \mathrm{d}x + \phi(c) \ln\left(\frac{b-c}{c-a}\right), \tag{3.16}
$$

where the first integral in the last equation is a well-defined regular integral.

For the hyper-singular integral as in (3.9), the above procedure may fail, as the limit of the integral over $\Gamma \backslash S_\varepsilon(\mathbf{r}_0)$ may diverge as $\varepsilon \to 0$. In many cases that integral takes the form

$$
\int_{\Gamma \backslash S_\varepsilon(\mathbf{r}_0)} \frac{\partial^2 G(\mathbf{r}_0, \mathbf{r}')}{\partial \mathbf{n} \partial \mathbf{n}'} \sigma(\mathbf{r}') \mathrm{d}s' = C + \Lambda(\mathbf{r}_0; \varepsilon, \sigma), \ \ \mathbf{r}_0 \in \Gamma, \tag{3.17}
$$

where $\Lambda(\mathbf{r}_0; \varepsilon, \sigma) = O(1/\varepsilon^\lambda)$, $\lambda > 0$, and C is some finite value. Therefore, Eq. (3.17) will become infinite as $\varepsilon \to 0$. But this does not necessarily imply that $\lim_{\mathbf{r}\to\mathbf{r}_0} u(\mathbf{r})$ in (3.9) will be unbounded, as the limit is usually well defined and actually could be computed directly by using Gauss's theory on the surface patch $S_\varepsilon(\mathbf{r}_0)$ of finite size ε to reduce the singularity of the surface integral (3.9) (as in Sect. 11.1.2), and then taking the limit $\mathbf{r} \to \mathbf{r}_0$. The fact is that the surface integral over $S_\varepsilon(\mathbf{r}_0)$ will in general produce a similar term $O(1/\varepsilon^\lambda)$ but of opposite sign, thus canceling the ε-divergent term in (3.17). Therefore, only the finite value C in (3.17) is of significance as far as the limit of the potential toward the surface is concerned. In fact, the Hadamard finite part (p.f.) [3] takes this finite value C in (3.17), by subtracting the divergent term $O(1/\varepsilon^\lambda)$, i.e.,

$$
\text{p.f.} \int_\Gamma \frac{\partial^2 G(\mathbf{r}_0, \mathbf{r}')}{\partial \mathbf{n} \partial \mathbf{n}'} \sigma(\mathbf{r}') \mathrm{d}s' = \lim_{\varepsilon \to 0} \left[\int_{\Gamma \backslash S_\varepsilon(\mathbf{r}_0)} \frac{\partial^2 G(\mathbf{r}_0, \mathbf{r}')}{\partial \mathbf{n} \partial \mathbf{n}'} \sigma(\mathbf{r}') \mathrm{d}s' - \Lambda(\mathbf{r}_0; \varepsilon, \sigma) \right]. \tag{3.18}
$$

The exact form of the $O(1/\varepsilon^\lambda)$ term $\Lambda(\mathbf{r}_0; \varepsilon, \sigma)$ depends on the type of the kernel $\partial^2 G/\partial \mathbf{n} \partial \mathbf{n}'$ in question and the density function σ [4, Sect. 3.2]. For example, the Hadamard finite part of the following 1-D hyper-singular integral is defined in Fox [5] as

$$
\text{p.f.} \frac{1}{\pi} \int_{-1}^{1} \frac{\sigma(t)}{(t-x)^2} \mathrm{d}t = \frac{1}{\pi} \lim_{\varepsilon \to 0} \left[\int_{-1}^{x-\varepsilon} \frac{\sigma(t)}{(t-x)^2} \mathrm{d}t + \int_{x+\varepsilon}^{1} \frac{\sigma(t)}{(t-x)^2} \mathrm{d}t - \Lambda(x; \varepsilon, \sigma) \right], \tag{3.19}
$$

where

$$
\Lambda(x; \varepsilon, \sigma) = \frac{\sigma(x-\varepsilon) + \sigma(x+\varepsilon)}{\varepsilon}. \tag{3.20}
$$

Moreover, for a curve Γ in $\mathbb{R}^2$ of smoothness $C^{2+\alpha}$, $0 < \alpha < 1$, for $G(\mathbf{r}_0, \mathbf{r}) = -(1/2\pi)\ln|\mathbf{r} - \mathbf{r}_0|$, in the Hadamard finite part of the hyper-singular integral (3.18) it can be shown [4, p. 107] that

$$\Lambda(\mathbf{r}_0; \varepsilon, \sigma) = -\frac{\sigma(\mathbf{r}_0)}{\pi\varepsilon} + O(\varepsilon^\alpha). \tag{3.21}$$

With the concept of the Cauchy principal value and the Hadamard finite part, we are ready to study the limiting property toward the surface Γ for functions $u(\mathbf{r})$ defined by the three types of layer potential operators (3.7)–(3.9). Firstly, let us define the interior and the exterior trace on Γ, for $\mathbf{r} \in \Gamma$,

$$u_{\mathrm{i}}(\mathbf{r}) = \lim_{\mathbf{x}\in\Omega_{\mathrm{i}}\to\mathbf{r}} u(\mathbf{x}), \tag{3.22}$$

$$u_{\mathrm{o}}(\mathbf{r}) = \lim_{\mathbf{x}\in\Omega_{\mathrm{o}}\to\mathbf{r}} u(\mathbf{x}), \tag{3.23}$$

and denote the jump of the traces on Γ as

$$[u] = u_{\mathrm{o}} - u_{\mathrm{i}}. \tag{3.24}$$

Moreover, we define the limiting value of the normal derivative of u as

$$\left(\frac{\partial u}{\partial \mathbf{n}}\right)_{\mathrm{i}}(\mathbf{r}) = \lim_{\mathbf{x}\in\Omega_{\mathrm{i}}\to\mathbf{r}} \frac{\partial u}{\partial \mathbf{n}}(\mathbf{x}), \tag{3.25}$$

$$\left(\frac{\partial u}{\partial \mathbf{n}}\right)_{\mathrm{o}}(\mathbf{r}) = \lim_{\mathbf{x}\in\Omega_{\mathrm{o}}\to\mathbf{r}} \frac{\partial u}{\partial \mathbf{n}}(\mathbf{x}), \tag{3.26}$$

where $\mathbf{n}$ is the outer normal of the boundary $\partial\Omega_{\mathrm{i}} = \Gamma$.

We also define the difference of the normal derivatives as

$$\left[\frac{\partial u}{\partial \mathbf{n}}\right] = \left(\frac{\partial u}{\partial \mathbf{n}}\right)_{\mathrm{o}} - \left(\frac{\partial u}{\partial \mathbf{n}}\right)_{\mathrm{i}}. \tag{3.27}$$

Theorem 3.1 ([1, Theorem 3.1.2]) *(a) The single-layer potential $u(\mathbf{r}) = S(q)$ is a continuous function across Γ, and*

$$u(\mathbf{r}) = p.v. \int_\Gamma G(\mathbf{r}, \mathbf{r}')q(\mathbf{r}')ds' \quad \textit{for} \quad \mathbf{r} \in \Gamma, \tag{3.28}$$

while $\partial u/\partial \mathbf{n}$ has a jump, namely for $\mathbf{r} \in \Gamma$ we have

$$\left(\frac{\partial u}{\partial \mathbf{n}}\right)_{\mathrm{i}}(\mathbf{r}) = \frac{q(\mathbf{r})}{2} + p.v. \int_\Gamma \frac{\partial G(\mathbf{r}, \mathbf{r}')}{\partial \mathbf{n}} q(\mathbf{r}')ds', \tag{3.29}$$

$$\left(\frac{\partial u}{\partial \mathbf{n}}\right)_{\mathrm{o}}(\mathbf{r}) = -\frac{q(\mathbf{r})}{2} + p.v. \int_{\Gamma} \frac{\partial G(\mathbf{r}, \mathbf{r}')}{\partial \mathbf{n}} q(\mathbf{r}')\mathrm{d}s', \tag{3.30}$$

which implies

$$\left[\frac{\partial u}{\partial \mathbf{n}}\right] = -q(\mathbf{r}) \textit{ for } \mathbf{r} \in \Gamma. \tag{3.31}$$

(b) The double-layer potential $u(\mathbf{r}) = D(\phi)$ *is discontinuous across* Γ, *and for* $\mathbf{r} \in \Gamma$ *we have*

$$u_{\mathrm{i}}(\mathbf{r}) = -\frac{\phi(\mathbf{r})}{2} + p.v. \int_{\Gamma} \frac{\partial G(\mathbf{r}, \mathbf{r}')}{\partial \mathbf{n}'} \phi(\mathbf{r}')\mathrm{d}s', \tag{3.32}$$

$$u_{\mathrm{o}}(\mathbf{r}) = \frac{\phi(\mathbf{r})}{2} + p.v. \int_{\Gamma} \frac{\partial G(\mathbf{r}, \mathbf{r}')}{\partial \mathbf{n}'} \phi(\mathbf{r}')\mathrm{d}s', \tag{3.33}$$

and

$$[u] = \phi(\mathbf{r}). \tag{3.34}$$

Meanwhile, the normal derivative of the double-layer potential is continuous across Γ, *and for* $\mathbf{r} \in \Gamma$ *we have*

$$\frac{\partial u}{\partial \mathbf{n}}(\mathbf{r}) = p.f. \int_{\Gamma} \frac{\partial^2 G(\mathbf{r}, \mathbf{r}')}{\partial \mathbf{n} \partial \mathbf{n}'} \phi(\mathbf{r}')\mathrm{d}s'. \tag{3.35}$$

Remark 3.1 **(Physical meaning of the jump conditions)** (1) In interpreting the meaning of q in the single-layer potential (3.7) and (3.31), we can consider the displacement flux $\mathbf{D}$ given in terms of u, i.e., $\mathbf{D} = -\nabla u$. Then the normal component of $\mathbf{D}$ should satisfy the physical boundary condition $[\mathbf{D} \cdot \mathbf{n}] = -[\partial u/\partial \mathbf{n}] - q$ (see (8.65)) as there is a charge distribution q on Γ, which is consistent with the conclusion in (3.31). (2) Similarly, for the double-layer potential generated by a dipole density $\phi(\mathbf{r})$ on Γ, the electric potential will have a drop (jump) (3.34) from the negative to positive charges inside a dipole on the surface. However, as the dipole is charge neutral, no net surface charge exists on the surface. Therefore, the electric field (the normal derivative of the potential in (3.35)) will be continuous, consistent with (8.65) with zero surface charges.

3.1.2 Surface Integral Equations for the PB Equation

In this section, we will derive the surface integral equations for the solution of the PB equation in (2.19) using fundamental solutions and Green's second identity.

For simplicity of notation, we will write the equations in $\Omega_{\rm i}$ and $\Omega_{\rm o}$ in a uniform format as

$$\mathcal{L}u(\mathbf{r}) = -\rho(\mathbf{r}), \tag{3.36}$$

where $\mathcal{L} = \Delta + k^2$ is defined in (3.2). For the PB equation (3.5) of main concern here, we have $k = -{\rm i}\kappa$ and

$$\kappa^2 = \begin{cases} \kappa_{\rm i}^2 = \lambda_{\rm i}^2/\epsilon_{\rm i}, & \text{if } \mathbf{r} \in \Omega_{\rm i}, \\ \kappa_{\rm o}^2 = \lambda_{\rm o}^2/\epsilon_{\rm o}, & \text{if } \mathbf{r} \in \Omega_{\rm o}, \end{cases} \tag{3.37}$$

and

$$\rho(\mathbf{r}) = \begin{cases} \dfrac{1}{\epsilon_{\rm i}}\displaystyle\sum_{j=1}^{N} q_j \delta(\mathbf{r} - \mathbf{r}_j), & \text{if } \mathbf{r} \in \Omega_{\rm i}, \\ 0, & \text{if } \mathbf{r} \in \Omega_{\rm o}. \end{cases} \tag{3.38}$$

Theorem 3.2 *Let $u(\mathbf{r})$ satisfy the elliptic equation in (3.36)–(3.38) for $\mathbf{r} \in \Omega_{\rm i} \cup \Omega_{\rm o}$ with $k = 0$ inside $\Omega_{\rm i}$ and also either the Sommerfeld radiation condition at infinity for the case of real k, $|\partial u/\partial r + {\rm i}ku| = O\left(1/r^2\right)$, or the vanishing condition (2.21) at infinity for the case of a pure imaginary $k = -{\rm i}\kappa$. Then for $\mathbf{r} \in \Omega_{\rm i}$, we have the representation for the solution of the Poisson equation*

$$\begin{aligned} u(\mathbf{r}) = &\int_\Gamma \left[G_0(\mathbf{r}, \mathbf{r}') \left(\frac{\partial u(\mathbf{r}')}{\partial \mathbf{n}'} \right)_{\rm i} - \frac{\partial G_0(\mathbf{r}, \mathbf{r}')}{\partial \mathbf{n}'} u_{\rm i}(\mathbf{r}') \right] {\rm d}s' \\ &+ \frac{1}{\epsilon_{\rm i}} \sum_{j=1}^{N} q_j G_0(\mathbf{r}, \mathbf{r}_j), \end{aligned} \tag{3.39}$$

where $\mathbf{n}'$ is the outer normal of the boundary $\Gamma = \partial\Omega_{\rm i}$, and for $\mathbf{r} \in \Omega_{\rm o}$ we have the representation for the solution of the PB equation (3.5):

$$u(\mathbf{r}) = -\int_\Gamma \left[G_{-{\rm i}\kappa}(\mathbf{r}, \mathbf{r}') \left(\frac{\partial u(\mathbf{r}')}{\partial \mathbf{n}'} \right)_{\rm o} - \frac{\partial}{\partial \mathbf{n}'} G_{-{\rm i}\kappa}(\mathbf{r}, \mathbf{r}') u_o(\mathbf{r}') \right] {\rm d}s'. \tag{3.40}$$

For the Helmholtz equation, replace $G_{-{\rm i}\kappa}(\mathbf{r}, \mathbf{r}')$ by $G_k(\mathbf{r}, \mathbf{r}')$. Moreover, for $\mathbf{r} \in \Gamma$ we have

$$\begin{aligned} \frac{1}{2} u_{\rm i}(\mathbf{r}) = p.v. &\int_\Gamma \left[G_0(\mathbf{r}, \mathbf{r}') \left(\frac{\partial u(\mathbf{r}')}{\partial \mathbf{n}'} \right)_{\rm i} - \frac{\partial}{\partial \mathbf{n}'} G_0(\mathbf{r}, \mathbf{r}') u_{\rm i}(\mathbf{r}') \right] {\rm d}s' \\ &+ \frac{1}{\epsilon_{\rm i}} \sum_{j=1}^{N} q_j G_0(\mathbf{r}, \mathbf{r}_j) \end{aligned} \tag{3.41}$$

and

$$\frac{1}{2}u_{\mathrm{o}}(\mathbf{r}) = p.v. \int_{\Gamma} \left[-G_{-\mathrm{i}\kappa}(\mathbf{r}, \mathbf{r}') \left(\frac{\partial u(\mathbf{r}')}{\partial \mathbf{n}'} \right)_{\mathrm{o}} + \frac{\partial}{\partial \mathbf{n}'} G_{-\mathrm{i}\kappa}(\mathbf{r}, \mathbf{r}') u_{\mathrm{o}}(\mathbf{r}') \right] \mathrm{d}s'. \quad (3.42)$$

Again for the Helmholtz equation, replace $G_{-\mathrm{i}\kappa}(\mathbf{r}, \mathbf{r}')$ by $G_k(\mathbf{r}, \mathbf{r}')$.

Proof We follow the idea of the proof of Theorem 3.1.1 in Nédélec [1]. For any domain Ω and a point $\mathbf{r}' \notin \Omega$, the fundamental solution $G_k(\mathbf{r}, \mathbf{r}')$ in the domain satisfies the homogeneous elliptic equation

$$\mathcal{L}G_k(\mathbf{r}, \mathbf{r}') = 0,$$

while the solution $u(\mathbf{r})$ satisfies

$$\mathcal{L}u = -\rho.$$

Then, we have

$$\int_{\Omega} (G_k \mathcal{L}u - u \mathcal{L}G_k) \mathrm{d}\mathbf{r} = -\int_{\Omega} \rho G_k \mathrm{d}\mathbf{r}.$$

By using Green's second identity, we have

$$-\int_{\Omega} \rho G_k \mathrm{d}\mathbf{r} = \int_{\partial\Omega} \left(\frac{\partial u}{\partial \mathbf{n}} G_k - \frac{\partial G_k}{\partial \mathbf{n}} u \right) \mathrm{d}s, \quad (3.43)$$

where $\mathbf{n}$ is the outer normal on $\partial\Omega$.

We apply (3.43) to $(B_R \cap \Omega_{\mathrm{o}}) \setminus B_\varepsilon$, where $B_\varepsilon = B_\varepsilon(\mathbf{r}')$ is a ball of radius ε centered at $\mathbf{r}'$ and $B_R = B_R(\mathbf{0})$ is a ball of radius R enclosing Ω_{i}, to get

$$\begin{aligned} -\int_{(B_R \cap \Omega_{\mathrm{o}}) \setminus B_\varepsilon} \rho G \mathrm{d}\mathbf{r} &= \left(\int_{S_\varepsilon} + \int_{S_R} \right) \left[\frac{\partial u}{\partial \mathbf{n}}(\mathbf{r}) G(\mathbf{r}, \mathbf{r}') - \frac{\partial}{\partial \mathbf{n}} G(\mathbf{r}, \mathbf{r}') u(\mathbf{r}) \right] \mathrm{d}s \\ &+ \int_{\Gamma} \left[\frac{\partial u}{\partial \mathbf{n}_{\mathrm{i}}}(\mathbf{r}) G(\mathbf{r}, \mathbf{r}') - \frac{\partial}{\partial \mathbf{n}_{\mathrm{i}}} G(\mathbf{r}, \mathbf{r}') u(\mathbf{r}) \right] \mathrm{d}s, \end{aligned} \quad (3.44)$$

where $G(\mathbf{r}, \mathbf{r}') = G_k(\mathbf{r}, \mathbf{r}')$, and $\mathbf{n}_{\mathrm{i}}$ is the normal pointing into Ω_{i}; $\mathbf{n}$ on S_ε also points into B_ε. We will examine the surface integrals as follows.

(a) We firstly examine the first surface integral:

$$\left|\int_{S_\varepsilon} \frac{\partial u}{\partial \mathbf{n}} G(\mathbf{r}, \mathbf{r}')\mathrm{d}s\right| \le \sup_{\mathbf{r}\in B_\varepsilon} \left|\frac{\partial u}{\partial \mathbf{n}}(\mathbf{r})\right| \int_{S_\varepsilon} |G(\mathbf{r}, \mathbf{r}')|\mathrm{d}s$$
$$= \sup_{\mathbf{r}\in B_\varepsilon} \left|\frac{\partial u}{\partial \mathbf{n}}\right| 4\pi\varepsilon^2 \frac{1}{4\pi\varepsilon} = \varepsilon \sup_{\mathbf{r}\in B_\varepsilon} \left|\frac{\partial u}{\partial \mathbf{n}}\right| \to 0 \text{ as } \varepsilon \to 0.$$

Secondly, for a fixed $\mathbf{r}'$ we have

$$\int_{S_\varepsilon} \frac{\partial}{\partial \mathbf{n}} G(\mathbf{r}, \mathbf{r}')u(\mathbf{r})\mathrm{d}s = u(\mathbf{r}') \int_{S_\varepsilon} \frac{\partial}{\partial \mathbf{n}} G(\mathbf{r}, \mathbf{r}')\mathrm{d}s + \int_{S_\varepsilon} \frac{\partial}{\partial \mathbf{n}} G(\mathbf{r}, \mathbf{r}')(u(\mathbf{r}) - u(\mathbf{r}'))\mathrm{d}s. \tag{3.45}$$

To estimate the second term in (3.45), we write $R = |\mathbf{r} - \mathbf{r}'|$ to obtain

$$\left|\frac{\partial}{\partial \mathbf{n}} G(\mathbf{r}, \mathbf{r}')\right| = \left|\frac{\partial G}{\partial r}\right| = \left|\left(-\frac{\mathrm{i}k}{4\pi R} - \frac{1}{4\pi R^2}\right)\mathrm{e}^{-\mathrm{i}kR}\right|$$
$$\le \frac{k}{4\pi R} + \frac{1}{4\pi R^2} = \frac{k}{4\pi\varepsilon} + \frac{1}{4\pi\varepsilon^2},$$

so that

$$\left|\int_{S_\varepsilon} \frac{\partial}{\partial \mathbf{n}} G(\mathbf{r}, \mathbf{r}')(u(\mathbf{r}) - u(\mathbf{r}'))\mathrm{d}s\right| \le \sup_{\mathbf{r},\mathbf{r}'\in B_\varepsilon} |u(\mathbf{r}) - u(\mathbf{r}')| \int_{S_\varepsilon} \left(\frac{k}{4\pi\varepsilon} + \frac{1}{4\pi\varepsilon^2}\right)\mathrm{d}s$$
$$= (\varepsilon k + 1) \sup_{\mathbf{r},\mathbf{r}'\in B_\varepsilon} |u(\mathbf{r}) - u(\mathbf{r}')| \to 0 \text{ as } \varepsilon \to 0.$$

As for the first term on the right-hand side of (3.45), we have

$$\int_{S_\varepsilon} \frac{\partial}{\partial \mathbf{n}} G(\mathbf{r}, \mathbf{r}')\mathrm{d}s = -\int_{S_\varepsilon} \frac{\partial}{\partial r} G(\mathbf{r}, \mathbf{r}')\mathrm{d}s = (1 + \mathrm{i}k\varepsilon)\mathrm{e}^{-\mathrm{i}k\varepsilon} \to 1 \text{ as } \varepsilon \to 0.$$

Thus

$$\int_{S_\varepsilon} \left[\frac{\partial u}{\partial \mathbf{n}}(\mathbf{r})G(\mathbf{r}, \mathbf{r}') - \frac{\partial}{\partial \mathbf{n}} G(\mathbf{r}, \mathbf{r}')u(\mathbf{r})\right]\mathrm{d}s \to -u(\mathbf{r}'). \tag{3.46}$$

(b) Next, we look at the second surface integral in (3.44) in the case of real k (Helmholtz equation):

$$
\begin{aligned}
&\int_{S_R} \left[\frac{\partial u}{\partial \mathbf{n}}(\mathbf{r}) G(\mathbf{r}, \mathbf{r}') - \frac{\partial G(\mathbf{r}, \mathbf{r}')}{\partial \mathbf{n}} u(\mathbf{r}) \right] \mathrm{d}s \\
&\quad = \int_{S_R} \left[\left(\frac{\partial u}{\partial \mathbf{n}} + \mathrm{i}ku \right) G(\mathbf{r}, \mathbf{r}') - \left(\frac{\partial G(\mathbf{r}, \mathbf{r}')}{\partial \mathbf{n}} + \mathrm{i}kG(\mathbf{r}, \mathbf{r}') \right) u(\mathbf{r}) \right] \mathrm{d}s \\
&\quad = E_1 + E_2.
\end{aligned}
$$

Using the Sommerfeld radiation condition (for the PB equation, there is no need to insert the term iku above), we have

$$
|E_1| \leq \frac{C}{R^2} \int_{S_R} \frac{1}{4\pi R} \mathrm{d}s = \frac{C}{R} \to 0 \qquad \text{as} R \to +\infty;
$$

also

$$
\left| \frac{\partial G}{\partial r} + \mathrm{i}kG \right| \leq \frac{1}{4\pi R^2},
$$

as the maximum $\sup_{\mathbf{r} \in S_R} |u(\mathbf{r})|$ decays as $O(1/R)$ [1], and

$$
|E_2| \leq \frac{1}{4\pi R^2} \int_{S_R} |u(\mathbf{r})| \mathrm{d}s \leq \sup_{\mathbf{r} \in S_R} |u(\mathbf{r})| \to 0 \qquad \text{as } R \to +\infty.
$$

Therefore, we have

$$
\int_{S_R} \left[\frac{\partial u}{\partial \mathbf{n}}(\mathbf{r}) G(\mathbf{r}, \mathbf{r}') - \frac{\partial G(\mathbf{r}, \mathbf{r}')}{\partial \mathbf{n}} u(\mathbf{r}) \right] \mathrm{d}s \to 0 \text{ as } R \to \infty. \tag{3.47}
$$

Finally, Eq. (3.44) together with (3.46) and (3.47) implies (3.40) after using the definition of ρ from (3.38) and $G(\mathbf{r}, \mathbf{r}') = G_{-\mathrm{i}\kappa}(\mathbf{r}, \mathbf{r}')$ in Ω_o and replacing $\mathbf{r}'$ by $\mathbf{r}$. Equation (3.39) can be proven similarly.

Moreover, Eqs. (3.41) and (3.42) result from the limiting results of the singular boundary operator in (3.28)–(3.35). □

Remark 3.2 As an alternative and direct way to obtain (3.41) for $\mathbf{r} \in \Gamma$, we can select a local patch γ_ε of size ε on Γ, which can be just the intersection of Γ with a ball $V_\varepsilon(\mathbf{r})$ centered at $\mathbf{r}$, namely $\gamma_\varepsilon = V_\varepsilon(\mathbf{r}) \cap \Gamma$. Then we perturb the part γ_ε of the boundary Γ around $\mathbf{r}$ to obtain a larger domain $\Omega_\mathrm{i} \cup V_\varepsilon$, within which $\mathbf{r}$ will be an interior point. We denote $S_\varepsilon = \partial(V_\varepsilon \backslash \Omega_\mathrm{i})$, namely the part of the spherical surface of $V_\varepsilon(\mathbf{r})$ outside Ω_i. Then we apply the representation result (3.39) to the expanded domain $\Omega_\mathrm{i} \cup V_\varepsilon$ and, as $\varepsilon \to 0$, the boundary integration over S_ε will modify the coefficient of $u(\mathbf{r})$ on the left-hand side of (3.39) to be $1/2$ if $\mathbf{r} \subset \Gamma$ is a smooth point (with a uniquely defined tangential plane) while the surface integral over $\Gamma \backslash \gamma_\varepsilon$ becomes the Cauchy principal value (CPV) on the right-hand side, as indicated in

(3.41). Equation (3.42) can be similarly derived, but from the side of the domain Ω_o. More discussions on the contribution of the surface S_ε are given in Sect. 3.1.3, where methods of direct computation of the CPV and the Hadamard finite part are discussed.

Remark 3.3 (Hyper-singular integral equations) By applying a differential operation $\partial/\partial\mathbf{n}$ to (3.39) and (3.40), and then allowing $\mathbf{r}$ to approach the boundary Γ, with the limit properties in (3.29), (3.30), and (3.35), we obtain the following hyper-singular equations for $\mathbf{r} \in \Gamma$:

$$\frac{1}{2}\frac{\partial}{\partial\mathbf{n}}u_i(\mathbf{r}) = \text{p.v.}\int_\Gamma \frac{\partial}{\partial\mathbf{n}}G_0(\mathbf{r},\mathbf{r}')\left(\frac{\partial u(\mathbf{r}')}{\partial\mathbf{n}'}\right)_i \mathrm{d}s' - \text{p.f.}\int_\Gamma \frac{\partial^2}{\partial\mathbf{n}\partial\mathbf{n}'}G_0(\mathbf{r},\mathbf{r}')u_i(\mathbf{r}')\mathrm{d}s' + \frac{1}{\epsilon_i}\sum_{j=1}^N q_j\frac{\partial}{\partial\mathbf{n}}G_0(\mathbf{r},\mathbf{r}_j) \tag{3.48}$$

and

$$\frac{1}{2}\frac{\partial}{\partial\mathbf{n}}u_o(\mathbf{r}) = -\text{p.v.}\int_\Gamma \frac{\partial}{\partial\mathbf{n}}G_{-i\kappa}(\mathbf{r},\mathbf{r}')\left(\frac{\partial u(\mathbf{r}')}{\partial\mathbf{n}'}\right)_o \mathrm{d}s' + \text{p.f.}\int_\Gamma \frac{\partial^2}{\partial\mathbf{n}\partial\mathbf{n}'}G_{-i\kappa}(\mathbf{r},\mathbf{r}')u_o(\mathbf{r}')\mathrm{d}s'. \tag{3.49}$$

Equations (3.41) and (3.42) will be completed with the continuity equations for the potential and the displacement fluxes for $\mathbf{r} \in \Gamma$:

$$u_i(\mathbf{r}) = u_o(\mathbf{r}), \tag{3.50}$$

$$\epsilon_i\left(\frac{\partial u(\mathbf{r})}{\partial\mathbf{n}}\right)_i = \epsilon_o\left(\frac{\partial u(\mathbf{r})}{\partial\mathbf{n}}\right)_o. \tag{3.51}$$

Introducing the density variables

$$q(\mathbf{r}) = \left(\frac{\partial u(\mathbf{r})}{\partial\mathbf{n}}\right)_i, \tag{3.52}$$

$$\sigma(\mathbf{r}) = u_i(\mathbf{r}), \tag{3.53}$$

then from (3.51) we have

$$\left(\frac{\partial u(\mathbf{r})}{\partial\mathbf{n}}\right)_o = \gamma q(\mathbf{r}), \tag{3.54}$$

where $\gamma = \epsilon_i/\epsilon_o$.

In terms of the density variables, Eqs. (3.41) and (3.42) can be expressed as

$$\frac{1}{2}\sigma(\mathbf{r}) = \text{p.v.}\int_{\Gamma}\left[G_0(\mathbf{r},\mathbf{r}')q(\mathbf{r}') - \frac{\partial}{\partial \mathbf{n}'}G_0(\mathbf{r},\mathbf{r}')\sigma(\mathbf{r}')\right]\mathrm{d}s' + \frac{1}{\epsilon_\mathrm{i}}\sum_{j=1}^{N} q_j G_0(\mathbf{r},\mathbf{r}_j) \tag{3.55}$$

and

$$\frac{1}{2}\sigma(\mathbf{r}) = \text{p.v.}\int_{\Gamma}\left[-G_{-\mathrm{i}\kappa}(\mathbf{r},\mathbf{r}')\gamma q(\mathbf{r}') + \frac{\partial}{\partial \mathbf{n}'}G_{-\mathrm{i}\kappa}(\mathbf{r},\mathbf{r}')\sigma(\mathbf{r}')\right]\mathrm{d}s', \tag{3.56}$$

or by using the single- and the double-layer potential operators as follows:

$$\frac{1}{2}\sigma = S_0(q) - D_0(\sigma) + \frac{1}{\epsilon_\mathrm{i}}\sum_{j=1}^{N} q_j G_0(\mathbf{r},\mathbf{r}_j) \tag{3.57}$$

and

$$\frac{1}{2}\sigma = -\gamma S_{-\mathrm{i}\kappa}(q) + D_{-\mathrm{i}\kappa}(\sigma). \tag{3.58}$$

It turns out that (3.57) and (3.58) are in fact ill-conditioned and not suitable for numerical solutions. To obtain a better conditioned integral equation, we can combine (3.48) with (3.49), and (3.55) with (3.56), respectively, to arrive at two new integral equations for the densities [6, 7]:

$$\begin{aligned}\left(\frac{1}{2}+\frac{1}{2\gamma}\right)\sigma = \text{p.v.}\int_{\Gamma} & [G_0(\mathbf{r},\mathbf{r}') - G_{-\mathrm{i}\kappa}(\mathbf{r},\mathbf{r}')]q(\mathbf{r}') \\ & - \left[\frac{\partial}{\partial \mathbf{n}'}G_0(\mathbf{r},\mathbf{r}') - \frac{1}{\gamma}\frac{\partial}{\partial \mathbf{n}'}G_{-\mathrm{i}\kappa}(\mathbf{r},\mathbf{r}')\right]\sigma(\mathbf{r}')\mathrm{d}s' \\ & + \frac{1}{\epsilon_\mathrm{i}}\sum_{j=1}^{N} q_j G_0(\mathbf{r},\mathbf{r}_j), \ \mathbf{r}\in\Gamma,\end{aligned} \tag{3.59}$$

and

$$\begin{aligned}\left(\frac{1}{2}+\frac{\gamma}{2}\right)q = \text{p.v.}\int_{\Gamma} & \left[\frac{\partial}{\partial \mathbf{n}}G_0(\mathbf{r},\mathbf{r}') - \gamma\frac{\partial}{\partial \mathbf{n}}G_{-\mathrm{i}\kappa}(\mathbf{r},\mathbf{r}')\right]q(\mathbf{r}')\mathrm{d}s' \\ - \text{p.f.}\int_{\Gamma} & \left[\frac{\partial^2}{\partial \mathbf{n}\partial \mathbf{n}'}G_0(\mathbf{r},\mathbf{r}') - \frac{\partial^2}{\partial \mathbf{n}\partial \mathbf{n}'}G_{-\mathrm{i}\kappa}(\mathbf{r},\mathbf{r}')\right]\sigma(\mathbf{r}')\mathrm{d}s' \\ + \frac{1}{\epsilon_\mathrm{i}} & \sum_{j=1}^{N} q_j \frac{\partial}{\partial \mathbf{n}}G_0(\mathbf{r},\mathbf{r}_j), \quad \mathbf{r}\in\Gamma.\end{aligned} \tag{3.60}$$

Due to the difference between the two hyper-singular integrands used in (3.60), the Hadamard finite part (p.f.) of the singular integral is just the normal Cauchy principal value (CPV). This type of cancelation of hyper-singularity through a difference was first used in sect. 23 of [8] and also in [9].

3.1.3 Computations of CPV and Hadamard p.f. and Collocation BEMs

The surface singular integral equations (3.41) and (3.42) or the surface hyper-singular integral equations (3.48) and (3.49) are derived using the potential representation formula through a limiting process as mentioned in Remarks 3.2 and 3.3, respectively. The CPV and the Hadamard finite part integrals are introduced as a result, and how we compute these is important for a successful implementation. In this section, two methods will be presented on how to handle these singular integrals in collocation boundary element methods (BEMs) for the surface integral equations. The first is a regularization method using simple solutions of the electrostatics problems (Laplace equations) [10–12]. The second is a direct calculation with regular Gauss quadratures [13–15].

Collocation BEMs

In a collocation method of the surface integral equations, the surface is usually decomposed into patches $\{S_i\}_{i=1}^N$ of triangular and quadrilateral shapes, each of which is parameterized by an analytical mapping

$$\mathbf{r} = \chi(\mathbf{u}) = (\chi_1(\mathbf{u}), \chi_2(\mathbf{u}), \chi_3(\mathbf{u})), \mathbf{r} \in S_i, \tag{3.61}$$

with variables $\mathbf{u} = (u_1, u_2)$ in a reference triangle

$$T_0 = \{(u_1, u_2) : 0 \le u_1, u_2, u_1 + u_2 \le 1\} \tag{3.62}$$

or square

$$\Omega_0 = \{(u_1, u_2) : 0 \le u_1, u_2 \le 1\}, \tag{3.63}$$

and shape functions $N_l(\mathbf{r})$ are constructed on the reference domains and mapped into the patches to represent the physical density (charge or dipole). The density $u(\mathbf{r})$ will be approximated by the following shape functions:

$$u(\mathbf{r}) = \sum_l u(\mathbf{r}_l) N_l(\mathbf{r}), \mathbf{r} \in S_i, \tag{3.64}$$

where $u(\mathbf{r}_l)$ are the nodal values at $\mathbf{r}_l$ on each patch S_i. For a piecewise constant approximation, the center of each patch can be used as the only node, and for a piecewise linear approximation, the three (or four) vertices of the triangular (quadrilateral) patch can be used as the nodes. The shape functions for the linear approximation on the reference triangle or square can be found in Sects. 11.4.3 and 11.4.4.

Regularization of Singular Integrals

A regularization of the singular equations (3.41) and (3.42) and the hyper-singular equations (3.48) and (3.49) can be achieved using the following identities of Green's function $G_0(\mathbf{r}, \mathbf{r}')$ [16, 17], and the collocation method can then be applied to the resulting weakly singular integral equations (of only $O(1/r)$). Other regularization methods using Stokes' theorem have also been studied [11, 18, 19].

Let $\mathbf{n}$ and $\mathbf{n}'$ be the outer normals on $\partial\Omega_i$. We then have the following identities:

- The first identity

$$\int_\Gamma \frac{\partial}{\partial \mathbf{n}'} G_0(\mathbf{r}, \mathbf{r}')\mathrm{d}s' = \begin{cases} -1, & \forall \mathbf{r} \in \Omega_i, \\ 0, & \forall \mathbf{r} \in \Omega_o. \end{cases} \tag{3.65}$$

- The second identity

$$\int_\Gamma \frac{\partial^2}{\partial \mathbf{n}\partial \mathbf{n}'} G_0(\mathbf{r}, \mathbf{r}')\mathrm{d}s' = 0, \forall \mathbf{r} \in \Omega_i \cup \Omega_o. \tag{3.66}$$

- The third identity

$$\int_\Gamma \frac{\partial}{\partial \mathbf{n}} G_0(\mathbf{r}, \mathbf{r}') n_k(\mathbf{r}')\mathrm{d}s' - \int_\Gamma \frac{\partial^2}{\partial \mathbf{n}\partial \mathbf{n}'} G_0(\mathbf{r}, \mathbf{r}')(x_k' - x_k)\mathrm{d}s'$$
$$= \begin{cases} n_k(\mathbf{r}), & \forall \mathbf{r} \in \Omega_i, \\ 0, & \forall \mathbf{r} \in \Omega_o. \end{cases} \tag{3.67}$$

Remark 3.4 The identity (3.65) can be obtained by using the simple solution $u = 1$ in the representation formula (3.39) with no source term for $\mathbf{r} \in \Omega_i$, while for $\mathbf{r} \in \Omega_o$ we can enclose Ω_o with a large ball $B_R(\mathbf{r})$ of radius R centered at $\mathbf{r}$ and then apply the representation formula on $\Omega_o \cap B_R(\mathbf{r})$. The identity (3.66) can be obtained by differentiating (3.65), and the identity (3.67) is obtained by using the simple solution $u(\mathbf{r}) = (\mathbf{r})_k = x_k' - x_k$ in the representation formula (3.39), followed by a differentiation with respect to $\mathbf{n}$.

In order to regularize the CPV boundary integral, we consider the representation for the domain

$$\Omega_\varepsilon = \Omega_i \cup V_\varepsilon(\mathbf{r}), \tag{3.68}$$

which is an enlarged version of Ω_i by the ball of $V_\varepsilon(\mathbf{r})$ (refer to Fig. 3.1 for the case where Ω_ε is formed by excluding $V_\varepsilon(\mathbf{r})$ from Ω_i), and we denote the part of the spherical surface $\partial V_\varepsilon(\mathbf{r})$ outside Ω_i by s_ε and the intersection of $\Gamma = \partial\Omega_i$ with $V_\varepsilon(\mathbf{r})$ as γ_ε, i.e., $\gamma_\varepsilon = \Gamma \cap V_\varepsilon$, respectively. The boundary of Ω_ε is denoted by

$$\Gamma_\varepsilon \equiv \partial\Omega_\varepsilon = \Gamma - \gamma_\varepsilon + s_\varepsilon. \tag{3.69}$$

We apply Green's second formula to the domain Ω_ε and assume all interior charges q_j in (3.38) are zero. Then, as in (3.39), we have for all $\varepsilon > 0$

$$\Pi_1(\mathbf{r}) \equiv \int\limits_{\Gamma_\varepsilon} \left[G_0(\mathbf{r}, \mathbf{r}') \frac{\partial u(\mathbf{r}')}{\partial \mathbf{n}'} - \frac{\partial}{\partial \mathbf{n}'} G_0(\mathbf{r}, \mathbf{r}') u(\mathbf{r}') \right] \mathrm{d}s' = cu(\mathbf{r}), \tag{3.70}$$

where $c = 1$ for $\mathbf{r} \in \Omega_\varepsilon$, and $c = 0$ for $\mathbf{r} \in \Omega_\varepsilon^c$. As $\mathbf{r}$ is not on the surface Γ_ε, a differentiation $\partial/\partial\mathbf{n}$ can be applied and moved inside the integral of (3.70), so we also have

$$\Pi_2(\mathbf{r}) \equiv \int\limits_{\Gamma_\varepsilon} \left[\frac{\partial}{\partial \mathbf{n}} G_0(\mathbf{r}, \mathbf{r}') \frac{\partial u(\mathbf{r}')}{\partial \mathbf{n}'} - \frac{\partial^2}{\partial \mathbf{n} \partial \mathbf{n}'} G_0(\mathbf{r}, \mathbf{r}') u(\mathbf{r}') \right] \mathrm{d}s' = c \frac{\partial u(\mathbf{r})}{\partial \mathbf{n}}. \tag{3.71}$$

By using the identity (3.65) on the surface Γ_ε, Eq. (3.70) can be rewritten, for $c = 1$, as

$$\begin{aligned} cu(\mathbf{r}) &= -u(\mathbf{r}) \int\limits_{\Gamma_\varepsilon} \frac{\partial G_0(\mathbf{r}, \mathbf{r}')}{\partial \mathbf{n}'} \mathrm{d}s' \\ &\quad + \int\limits_{\Gamma_\varepsilon} \left\{ G_0(\mathbf{r}, \mathbf{r}') \frac{\partial u(\mathbf{r}')}{\partial \mathbf{n}'} - \frac{\partial G_0(\mathbf{r}, \mathbf{r}')}{\partial \mathbf{n}'} \left[u(\mathbf{r}') - u(\mathbf{r}) \right] \right\} \mathrm{d}s' \\ &= \left(\int\limits_{\Gamma - \gamma_\varepsilon} + \int\limits_{s_\varepsilon} \right) \left\{ G_0(\mathbf{r}, \mathbf{r}') \frac{\partial u(\mathbf{r}')}{\partial \mathbf{n}'} - \frac{\partial}{\partial \mathbf{n}'} G_0(\mathbf{r}, \mathbf{r}') \left[u(\mathbf{r}') - u(\mathbf{r}) \right] \right\} \mathrm{d}s' + u(\mathbf{r}). \end{aligned} \tag{3.72}$$

The integral over s_ε vanishes as $\varepsilon \to 0$ if the potential u is Hölder continuous, i.e., there are $\alpha > 0$ and $\beta > 0$ such that

$$|u(\mathbf{x}) - u(\mathbf{y})| \leq \beta |\mathbf{x} - \mathbf{y}|^\alpha, \quad \forall \mathbf{x}, \mathbf{y} \in \Gamma. \tag{3.73}$$

Namely, as $\varepsilon \to 0$ the surface area of s_ε is of the order $O(\varepsilon^2)$ and the integrand is of the order of $O(1/\varepsilon^{2-\alpha})$. Therefore, we have the following weakly singular form of the CPV singular integral equation (3.39) with no source charges (i.e. $q_j = 0$) for $c = 1$

$$(c-1)u(\mathbf{r}) = \int\limits_{\Gamma} \left\{ \frac{\partial}{\partial \mathbf{n}'} G_0(\mathbf{r}, \mathbf{r}') \left[u(\mathbf{r}) - u(\mathbf{r}') \right] + G_0(\mathbf{r}, \mathbf{r}') \frac{\partial u(\mathbf{r}')}{\partial \mathbf{n}'} \right\} \mathrm{d}s', \ \mathbf{r} \in \Gamma. \tag{3.74}$$

Next, in order to regularize the hyper-singular integral equation (3.48) with no source charges (i.e., $q_j = 0$), we take the normal derivative $\partial/\partial\mathbf{n}$ of (3.39) over Γ_ε and rewrite the resulting equation as follows:

$$\frac{\partial u(\mathbf{r})}{\partial\mathbf{n}} = \int\limits_{\Gamma-\gamma_\varepsilon} \left[\frac{\partial}{\partial\mathbf{n}}G_0(\mathbf{r},\mathbf{r}')\frac{\partial u(\mathbf{r}')}{\partial\mathbf{n}'} - \frac{\partial^2}{\partial\mathbf{n}\partial\mathbf{n}'}G_0(\mathbf{r},\mathbf{r}')u(\mathbf{r}')\right]\mathrm{d}s' + \int\limits_{s_\varepsilon} \left[\frac{\partial}{\partial\mathbf{n}}G_0(\mathbf{r},\mathbf{r}')\frac{\partial u(\mathbf{r}')}{\partial\mathbf{n}'} - \frac{\partial^2}{\partial\mathbf{n}\partial\mathbf{n}'}G_0(\mathbf{r},\mathbf{r}')u(\mathbf{r}')\right]\mathrm{d}s'. \quad (3.75)$$

Let us examine the two integrals over s_ε. By using the mean value theorem, the first integral over s_ε in (3.75) can be rewritten as

$$\lim_{\varepsilon\to 0}\int\limits_{s_\varepsilon}\frac{\partial}{\partial\mathbf{n}}G_0(\mathbf{r},\mathbf{r}')\frac{\partial u(\mathbf{r}')}{\partial\mathbf{n}'}\mathrm{d}s' = \lim_{\varepsilon\to 0}\left[\int\limits_{s_\varepsilon}\frac{\partial}{\partial\mathbf{n}}G_0(\mathbf{r},\mathbf{r}')n_k(\mathbf{r}')\mathrm{d}s'\right]\frac{\partial u(\mathbf{r})}{\partial x_k}$$
$$= \lim_{\varepsilon\to 0}\left[\int\limits_{\Gamma_\varepsilon}\frac{\partial}{\partial\mathbf{n}}G_0(\mathbf{r},\mathbf{r}')n_k(\mathbf{r}')\mathrm{d}s' - \int\limits_{\Gamma-\gamma_\varepsilon}\frac{\partial}{\partial\mathbf{n}}G_0(\mathbf{r},\mathbf{r}')n_k(\mathbf{r}')\mathrm{d}s'\right]\frac{\partial u(\mathbf{r})}{\partial x_k},$$

where the double index indicates a summation.

Using the identity (3.67) for the integral over the closed surface Γ_ε we have

$$\lim_{\varepsilon\to 0}\int\limits_{s_\varepsilon}\frac{\partial}{\partial\mathbf{n}}G_0(\mathbf{r},\mathbf{r}')\frac{\partial u(\mathbf{r}')}{\partial\mathbf{n}'}\mathrm{d}s'$$
$$= \lim_{\varepsilon\to 0}\left[n_k(\mathbf{r}) + \int\limits_{\Gamma_\varepsilon}\frac{\partial^2}{\partial\mathbf{n}\partial\mathbf{n}'}G_0(\mathbf{r},\mathbf{r}')(x_k' - x_k)\mathrm{d}s'\right]\frac{\partial u(\mathbf{r})}{\partial x_k}$$
$$- \lim_{\varepsilon\to 0}\int_{\Gamma-\gamma_\varepsilon}\frac{\partial}{\partial\mathbf{n}}G_0(\mathbf{r},\mathbf{r}')n_k(\mathbf{r}')\mathrm{d}s'\frac{\partial u(\mathbf{r})}{\partial\mathbf{x}_k}$$
$$= \frac{\partial u(\mathbf{r})}{\partial\mathbf{n}} + \lim_{\varepsilon\to 0}\int\limits_{\Gamma_\varepsilon}\frac{\partial^2}{\partial\mathbf{n}\partial\mathbf{n}'}G_0(\mathbf{r},\mathbf{r}')\left[(x_k' - x_k)\frac{\partial u(\mathbf{r})}{\partial x_k}\right]\mathrm{d}s'$$
$$- \lim_{\varepsilon\to 0}\int_{\Gamma-\gamma_\varepsilon}\frac{\partial}{\partial\mathbf{n}}G_0(\mathbf{r},\mathbf{r}')\frac{\partial u(\mathbf{r})}{\partial x_k}n_k(\mathbf{r}')\mathrm{d}s'. \quad (3.76)$$

On the other hand, the second integral over s_ε in (3.75) can be rewritten as

$$\lim_{\varepsilon\to 0}\int\limits_{s_\varepsilon}\frac{\partial^2}{\partial\mathbf{n}\partial\mathbf{n}'}G_0(\mathbf{r},\mathbf{r}')u(\mathbf{r}')\mathrm{d}s'$$
$$= \lim_{\varepsilon\to 0}\int_{s_\varepsilon}\frac{\partial^2}{\partial\mathbf{n}\partial\mathbf{n}'}G_0(\mathbf{r},\mathbf{r}')\left[u(\mathbf{r}) + (x_k' - x_k)\frac{\partial u(\mathbf{r})}{\partial x_k} + O(\varepsilon^2)\right]\mathrm{d}s'$$

$$
\begin{aligned}
&= \lim_{\varepsilon\to 0}\left(\int\limits_{\Gamma_\varepsilon} - \int\limits_{\Gamma-\gamma_\varepsilon}\right) \frac{\partial^2}{\partial\mathbf{n}\partial\mathbf{n}'} G_0(\mathbf{r},\mathbf{r}')u(\mathbf{r})\mathrm{d}s' \\
&\quad + \lim_{\varepsilon\to 0}\int\limits_{s_\varepsilon} \frac{\partial^2}{\partial\mathbf{n}\partial\mathbf{n}'} G_0(\mathbf{r},\mathbf{r}')\left[(x_k'-x_k)\frac{\partial u(\mathbf{r})}{\partial x_k} + O(\varepsilon^2)\right]\mathrm{d}s' \\
&= -\lim_{\varepsilon\to 0}\int_{\Gamma-\gamma_\varepsilon} \frac{\partial^2}{\partial\mathbf{n}\partial\mathbf{n}'} G_0(\mathbf{r},\mathbf{r}')u(\mathbf{r})\mathrm{d}s' \\
&\quad + \lim_{\varepsilon\to 0}\int\limits_{s_\varepsilon} \frac{\partial^2}{\partial\mathbf{n}\partial\mathbf{n}'} G_0(\mathbf{r},\mathbf{r}')\left[(x_k'-x_k)\frac{\partial u(\mathbf{r})}{\partial x_k}\right]\mathrm{d}s', \qquad (3.77)
\end{aligned}
$$

where the identity (3.66) was used in the last equation of (3.77) over the closed surface Γ_ε.

Substituting (3.76) and (3.77) into (3.75) and using the fact that $\Gamma_\varepsilon - s_\varepsilon = \Gamma - \gamma_\varepsilon$, we arrive at

$$
\begin{aligned}
&\lim_{\varepsilon\to 0}\int\limits_{\Gamma-\gamma_\varepsilon} \frac{\partial^2}{\partial\mathbf{n}\partial\mathbf{n}'} G_0(\mathbf{r},\mathbf{r}')\left[u(\mathbf{r}') - u(\mathbf{r}) - (x_k'-x_k)\frac{\partial u(\mathbf{r})}{\partial x_k}\right]\mathrm{d}s' \\
&\qquad = \lim_{\varepsilon\to 0}\int\limits_{\Gamma-\gamma_\varepsilon} \frac{\partial}{\partial\mathbf{n}} G_0(\mathbf{r},\mathbf{r}')\left[\frac{\partial u(\mathbf{r}')}{\partial\mathbf{n}'} - \frac{\partial u(\mathbf{r})}{\partial\mathbf{n}'}\right]\mathrm{d}s',
\end{aligned}
$$

which results in the weakly singular integral formulation of the hyper-singular integral equation (3.48):

$$
\begin{aligned}
&\int\limits_{\Gamma} \frac{\partial^2}{\partial\mathbf{n}\partial\mathbf{n}'} G_0(\mathbf{r},\mathbf{r}')\left[u(\mathbf{r}') - u(\mathbf{r}) - (x_k'-x_k)\frac{\partial u(\mathbf{r})}{\partial x_k}\right]\mathrm{d}s' \\
&\qquad = \int\limits_{\Gamma} \frac{\partial}{\partial\mathbf{n}} G_0(\mathbf{r},\mathbf{r}')\left[\frac{\partial u(\mathbf{r}')}{\partial\mathbf{n}'} - \frac{\partial u(\mathbf{r})}{\partial\mathbf{n}'}\right]\mathrm{d}s', \qquad (3.78)
\end{aligned}
$$

provided that ∇u is Hölder continuous, i.e., there are $\alpha > 0$ and $\beta > 0$ such that

$$
\left|\frac{\partial}{\partial x_i}u(\mathbf{x}) - \frac{\partial}{\partial y_i}u(\mathbf{y})\right| \le \beta|\mathbf{x}-\mathbf{y}|^\alpha,\ \forall\mathbf{x},\mathbf{y}\in\Gamma, i = 1,2,3. \qquad (3.79)
$$

In the weakly singular integral equations (3.74) and (3.78), the potential u is from the interior domain Ω_i, and similar weakly integral equations can also be obtained for the potential u from the outside domain by using the identities (3.65)–(3.67) for $\mathbf{r} \in \Omega_\mathrm{o}$. Also, the regularization methods can be applied to Green's function $G_k(\mathbf{r},\mathbf{r}')$ by rewriting

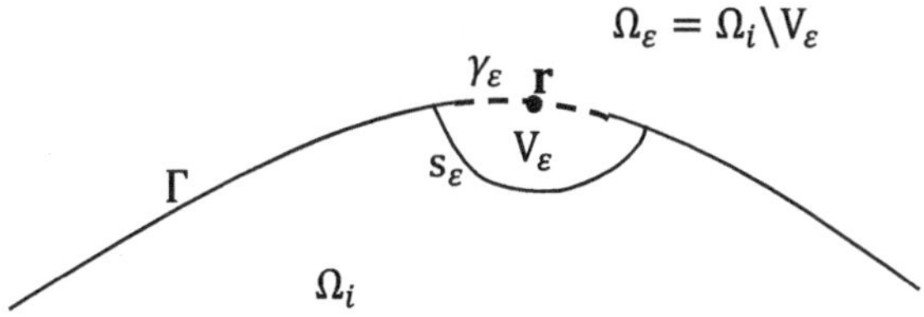

Fig. 3.1 Ω_ε, excluding the singular point on the boundary of $\Gamma = \partial\Omega_\mathrm{i}$

$$G_k(\mathbf{r}, \mathbf{r}') = \left[G_k(\mathbf{r}, \mathbf{r}') - G_0(\mathbf{r}, \mathbf{r}')\right] + G_0(\mathbf{r}, \mathbf{r}') \equiv H(\mathbf{r}, \mathbf{r}') + G_0(\mathbf{r}, \mathbf{r}'),$$

where $H(\mathbf{r}, \mathbf{r}')$ will be a regular smooth kernel and only regularization for the $G_0(\mathbf{r}, \mathbf{r}')$ is required [11].

Direct Computations of CPV and Hadamard p.f. Integrals

In the regularization procedures discussed above, the identities of Green's functions (3.65)–(3.67) are required, which will not be available for Green's functions of non-homogeneous media, such as the layered media discussed in Chap. 2. In the following, methods of direct computation of the singular integrals will be presented which are applicable for more general Green's functions.

For a collocation point $\mathbf{r} \in \Gamma$, let us define a domain excluding a vanishing volume V_ε around $\mathbf{r}$, which is usually taken to be a ball of radius ε centered at $\mathbf{r}$, and denote the part of the surface of V_ε inside Ω_i as s_ε, i.e., $s_\varepsilon = \Omega_\mathrm{i} \cap \partial V_\varepsilon$ and the intersection of Γ and V_ε as γ_ε, i.e., $\gamma_\varepsilon = \Gamma \cap V_\varepsilon$, respectively. The new domain as depicted in Fig. 3.1, which excludes $\mathbf{r}$, is denoted as

$$\Omega_\varepsilon = \Omega_\mathrm{i} \backslash V_\varepsilon. \tag{3.80}$$

It is noted that a different definition is used in (3.68) where $\mathbf{r}$ is included in Ω_ε and s_ε is the part of the spherical surface $\partial V_\varepsilon(\mathbf{r})$ outside Ω_i.

In both (3.70) and (3.71) the potential and its normal derivative are the interior fields from the domain Ω_i where the subscript i is omitted for simplicity.

Equations (3.70) and (3.71) will be the starting point to derive the surface integral equations for the potential and its derivative, and in the process of taking the limit $\varepsilon \to 0$, the CPV and the Hadamard finite part together with free term coefficients will be clarified and computed.

As $G_k(\mathbf{r}, \mathbf{r}')$ and $G_0(\mathbf{r}, \mathbf{r}')$ share the same type of singularity at $\mathbf{r} = \mathbf{r}'$, the direct calculation methods described below will be applicable for both kernels.

- **Computation of CPV**

In this case, as $\mathbf{r} \notin \Omega_\varepsilon$, similar to the way in which (3.72) is derived using Green's second identity, we can show that $\Pi_1(\mathbf{r}) = 0$, where $\Pi_1(\mathbf{r})$ is defined in (3.70). We can also rewrite $\Pi_1(\mathbf{r})$ as

$$\Pi_1(\mathbf{r}) = \int\limits_{\Gamma-\gamma_\varepsilon} \left[G_0(\mathbf{r},\mathbf{r}')\frac{\partial u(\mathbf{r}')}{\partial \mathbf{n}'} - \frac{\partial}{\partial \mathbf{n}'}G_0(\mathbf{r},\mathbf{r}')u(\mathbf{r}') \right] \mathrm{d}s'$$
$$+ \int\limits_{s_\varepsilon} \left[G_0(\mathbf{r},\mathbf{r}')\frac{\partial u(\mathbf{r}')}{\partial \mathbf{n}'} - \frac{\partial}{\partial \mathbf{n}'}G_0(\mathbf{r},\mathbf{r}')\left[u(\mathbf{r}') - u(\mathbf{r})\right] \right] \mathrm{d}s'$$
$$- u(\mathbf{r}) \int\limits_{s_\varepsilon} \frac{\partial}{\partial \mathbf{n}'}G_0(\mathbf{r},\mathbf{r}')\mathrm{d}s'. \quad (3.81)$$

As $\varepsilon \to 0$, the second integral above will vanish, as the surface area of s_ε is on the order of $O(\varepsilon^2)$ and the integrand is on the order of $O(1/\varepsilon)$. The limit of the third integral defines the so-called free term coefficient denoted by

$$c(\mathbf{r}) = \lim_{\varepsilon\to 0} \int_{s_\varepsilon} \frac{\partial}{\partial \mathbf{n}'}G_0(\mathbf{r},\mathbf{r}')\mathrm{d}s', \quad (3.82)$$

where the normal $\mathbf{n}'$ points to the interior of the exclusion volume $V_\varepsilon(\mathbf{r})$.

Now, Eq. (3.81) becomes

$$c(\mathbf{r})u(\mathbf{r}) = \int\limits_{\Gamma} G_0(\mathbf{r},\mathbf{r}')\frac{\partial u(\mathbf{r}')}{\partial \mathbf{n}'}\mathrm{d}s' - \lim_{\varepsilon\to 0} \int\limits_{\Gamma-\gamma_\varepsilon} \frac{\partial}{\partial \mathbf{n}'}G_0(\mathbf{r},\mathbf{r}')u(\mathbf{r}')\mathrm{d}s'. \quad (3.83)$$

If $\mathbf{r}$ is a smooth point on Γ (as a result, Γ will have a unique tangent plane at $\mathbf{r}$), then $c(\mathbf{r}) = 1/2$, and (3.83) is exactly (3.41) with $q_j = 0$. For a point $\mathbf{r}$ at an edge or a corner (i.e., Γ has multiple tangent planes), the free term coefficient $c(\mathbf{r})$ will depend on the geometry of s_ε, which can still be calculated [13].

The collocation method requires that (3.83) is satisfied at specific locations $\mathbf{r}$, which could be the same nodes that define the shape functions on each patch or some other points inside the patch. Let S_p^i, $i = 1, 2, \ldots, m$, be the patches which share the node $\mathbf{r}$ as one of its vertices and let a composite patch S_p be

$$S_p = \bigcup_{i=1}^{m} S_p^i, \quad (3.84)$$

which contains $\mathbf{r}$ at the center. In the limit of (3.83), we only need to show how to compute the limit of the integral over $S_p - \gamma_\varepsilon$, i.e.,

$$\text{p.v.} \int\limits_{S_p} \frac{\partial}{\partial \mathbf{n}'}G_0(\mathbf{r},\mathbf{r}')u(\mathbf{r}')\mathrm{d}s' \equiv \lim_{\varepsilon\to 0} \int\limits_{S_p-\gamma_\varepsilon} \frac{\partial}{\partial \mathbf{n}'}G_0(\mathbf{r},\mathbf{r}')u(\mathbf{r}')\mathrm{d}s'$$
$$= \lim_{\varepsilon\to 0} \sum_{i=1}^{m} \int_{S_p^i-\gamma_\varepsilon^i} \frac{\partial}{\partial \mathbf{n}'}G_0(\mathbf{r},\mathbf{r}')u(\mathbf{r}')\mathrm{d}s', \quad (3.85)$$

where

$$\gamma_\varepsilon^i = \gamma_\varepsilon \bigcap S_p^i, \text{ and } \gamma_\varepsilon = \bigcup_{i=1}^m \gamma_\varepsilon^i. \tag{3.86}$$

For each i, the integral over $S_p^i - \gamma_\varepsilon^i$ will be calculated over the corresponding region $\chi^{-1}(S_p^i - \gamma_\varepsilon^i)$ in the reference domain in the polar coordinates

$$\chi^{-1}(S_p^i - \gamma_\varepsilon^i) = \left\{(\rho, \theta) : \theta_1^i \le \theta \le \theta_2^i, \alpha^i(\theta, \varepsilon) \le \rho \le \overline{\rho}^i(\theta)\right\}, \tag{3.87}$$

where the center of the polar coordinates is $\mathbf{u}^* = (\rho^*, \theta^*)$:

$$\mathbf{r} = \chi(\mathbf{u}^*) = \chi(\rho^*, \theta^*), \mathbf{r}' = \chi(\mathbf{u}) = \chi(\rho, \theta). \tag{3.88}$$

Here we have assumed that the node $\mathbf{r}$ is mapped to one of the vertices of the reference domain, and the end range $\overline{\rho}^i(\theta)$ for ρ is then independent of ε.

As

$$\frac{\partial}{\partial \mathbf{n}'} G_0(\mathbf{r}, \mathbf{r}') = \frac{1}{4\pi} \frac{\partial}{\partial \mathbf{n}'} \frac{1}{|\mathbf{r} - \mathbf{r}'|} = \frac{1}{4\pi} \frac{1}{|\mathbf{r} - \mathbf{r}'|^2} \sum_{k=1}^3 \frac{(x_k' - x_k)}{|\mathbf{r} - \mathbf{r}'|} n_k', \tag{3.89}$$

we consider a typical term in $\partial G_0(\mathbf{r}, \mathbf{r}')/\partial \mathbf{n}'$:

$$T_k(\mathbf{r}, \mathbf{r}') = \frac{1}{|\mathbf{r} - \mathbf{r}'|^2} \frac{(x_k' - x_k)}{|\mathbf{r} - \mathbf{r}'|} n_k', \; k = 1, 2, 3, \tag{3.90}$$

where $\mathbf{n}' = (n_1', n_2', n_3')$, $\mathbf{r} = (x_1, x_2, x_3)$, and $\mathbf{r}' = (x_1', x_2', x_3')$.

Meanwhile, we assume that the potential has been expressed in terms of the shape functions over each patch, so we can set $u(\mathbf{r}')$ to be one of the shape functions $N_l(\mathbf{r}')$ and consider

$$\begin{aligned}
\int\limits_{S_p^i - \gamma_\varepsilon^i} T_k(\mathbf{r}, \mathbf{r}') N_l(\mathbf{r}') \mathrm{d}s' &= \int\limits_{\theta_1^i}^{\theta_2^i} \int\limits_{\alpha^i(\theta,\varepsilon)}^{\overline{\rho}^i(\theta)} T_k\left((\rho^*, \theta^*), (\rho, \theta)\right) N_l(\rho, \theta) J(\rho, \theta) \rho \mathrm{d}\rho \mathrm{d}\theta \\
&= \int\limits_{\theta_1^i}^{\theta_2^i} \mathrm{d}\theta \int\limits_{\alpha^i(\theta,\varepsilon)}^{\overline{\rho}^i(\theta)} F_k(\rho, \theta) \mathrm{d}\rho,
\end{aligned} \tag{3.91}$$

where the Jacobian of the parametric mapping χ is given by

$$J(\rho, \theta) = \left| \frac{\partial \chi}{\partial u_1} \times \frac{\partial \chi}{\partial u_2} \right|, \tag{3.92}$$

and

$$F_k(\rho, \theta) = T_k\left((\rho^*, \theta^*), (\rho, \theta)\right) N_l(\rho, \theta) J(\rho, \theta)\rho. \tag{3.93}$$

Consider the Taylor expansion of the parametric mapping χ_k:

$$\begin{aligned} x_k' - x_k &= \frac{\partial \chi_k}{\partial u_1}(u_1 - u_1^*) + \frac{\partial \chi_k}{\partial u_2}(u_2 - u_2^*) + \sum_{i,j=1}^{2} \frac{1}{2} \frac{\partial^2 \chi_k}{\partial u_i \partial u_j}(u_i - u_i^*)(u_j - u_j^*) \\ &\equiv \rho A_k(\theta) + \rho^2 B_k(\theta) + O(\rho^3), \end{aligned}$$

where, in terms of the polar angle,

$$A_k(\theta) = \frac{\partial \chi_k}{\partial u_1}(\mathbf{u}^*) \cos\theta + \frac{\partial \chi_k}{\partial u_2}(\mathbf{u}^*) \sin\theta, \tag{3.94}$$

$$B_k(\theta) = \frac{\partial^2 \chi_k}{\partial u_1^2}(\mathbf{u}^*) \frac{\cos^2\theta}{2} + \frac{\partial^2 \chi_k}{\partial u_1 \partial u_2}(\mathbf{u}^*) \cos\theta \sin\theta + \frac{\partial^2 \chi_k}{\partial u_2^2}(\mathbf{u}^*) \frac{\sin^2\theta}{2}. \tag{3.95}$$

It can be shown [14] that

$$\frac{x_k' - x_k}{|\mathbf{r} - \mathbf{r}'|} = \frac{A_k(\theta)}{A(\theta)} + O(\rho), \tag{3.96}$$

where

$$A(\theta)^2 = \sum_{k=1}^{3} |A_k(\theta)|^2$$

and

$$|\mathbf{r} - \mathbf{r}'|^2 = \rho^2 \sum_{k=1}^{3} |A_k(\theta)|^2 + 2\rho^3 \sum_{k=1}^{3} A_k(\theta) B_k(\theta) + O(\rho^4). \tag{3.97}$$

So on γ_ε we should have

$$\varepsilon^2 = \rho^2 \sum_{k=1}^{3} |A_k(\theta)|^2 + 2\rho^3 \sum_{k=1}^{3} A_k(\theta) B_k(\theta) + O(\rho^4),$$

which implies that on $\rho = \alpha(\theta, \varepsilon) = \boldsymbol{\chi}^{-1}(\gamma_\varepsilon)$ in polar coordinates

$$\begin{aligned} \alpha(\theta, \varepsilon) &= \varepsilon \frac{1}{A(\theta)} - \varepsilon^2 \frac{\sum_{k=1}^{3} A_k(\theta) B_k(\theta)}{A^4(\theta)} + O(\varepsilon^3) \\ &\equiv \varepsilon \beta(\theta) + \varepsilon^2 \gamma(\theta) + O(\varepsilon^3), \end{aligned} \tag{3.98}$$

where the final equation defines $\beta(\theta)$ and $\gamma(\theta)$.

Therefore, due to the continuity of $N_l(\rho, \theta)$ and the Jacobian $J(\rho, \theta)$, we have

$$F_k(\rho, \theta) = \frac{1}{\rho}\left[f_k(\theta) + O(\rho)\right], \tag{3.99}$$

where

$$f_k(\theta) = \frac{A_k(\theta)}{A^3(\theta)} N_l(\rho^*, \theta^*) J(\rho^*, \theta^*). \tag{3.100}$$

Finally, Eq. (3.85) becomes

$$\begin{aligned}
&\text{p.v.} \int_{S_p} \frac{\partial}{\partial \mathbf{n}'} G_0(\mathbf{r}, \mathbf{r}') u(\mathbf{r}') \mathrm{d}s' \\
&= \frac{1}{4\pi} \sum_{k=1}^{3} \sum_{i=1}^{m} \lim_{\varepsilon \to 0} \int_{\theta_1^i}^{\theta_2^i} \mathrm{d}\theta \int_{\alpha^i(\theta,\varepsilon)}^{\overline{\rho}^i(\theta)} F_k(\rho, \theta) \mathrm{d}\rho \\
&= \frac{1}{4\pi} \sum_{k=1}^{3} \sum_{i=1}^{m} \lim_{\varepsilon \to 0} \int_{\theta_1^i}^{\theta_2^i} \mathrm{d}\theta \int_{\alpha^i(\theta,\varepsilon)}^{\overline{\rho}^i(\theta)} \left[F_k(\rho, \theta) - \frac{f_k(\rho, \theta)}{\rho}\right] \mathrm{d}\rho \\
&\quad + \frac{1}{4\pi} \sum_{k=1}^{3} \sum_{i=1}^{m} \lim_{\varepsilon \to 0} \int_{\theta_1^i}^{\theta_2^i} f_k(\theta) \ln\left(\frac{\overline{\rho}^i(\theta)}{\alpha^i(\theta, \varepsilon)}\right) \mathrm{d}\theta.
\end{aligned}$$

From (3.98), as $\varepsilon \to 0$, we finally obtain the CPV through two regular integrals which can be readily computed by Gauss quadratures:

$$\begin{aligned}
\text{p.v.} \int_{S_p} \frac{\partial}{\partial \mathbf{n}'} G_0(\mathbf{r}, \mathbf{r}') u(\mathbf{r}') \mathrm{d}s' &= \frac{1}{4\pi} \sum_{k=1}^{3} \sum_{i=1}^{m} \int_{\theta_1^i}^{\theta_2^i} \mathrm{d}\theta \int_{0}^{\overline{\rho}^i(\theta)} \left[F_k(\rho, \theta) - \frac{f_k(\rho, \theta)}{\rho}\right] \mathrm{d}\rho \\
&\quad - \frac{1}{4\pi} \sum_{k=1}^{3} \sum_{i=1}^{m} \int_{\theta_1^i}^{\theta_2^i} f_k(\theta) \ln\left(\frac{\beta(\theta)}{\overline{\rho}^i(\theta)}\right) \mathrm{d}\theta,
\end{aligned} \tag{3.101}$$

provided that

$$\lim_{\varepsilon \to 0} (\ln \varepsilon) \sum_{i=1}^{m} \int_{\theta_1^i}^{\theta_2^i} f_k(\theta) \mathrm{d}\theta = 0, \quad \text{for } k = 1, 2, 3. \tag{3.102}$$

Equation (3.101) shows that the CPV can be computed directly with regular 1-D Gauss quadratures applied to the θ and ρ integrations if the solution $u(\mathbf{r})$ satisfies the

Hölder continuity condition (3.73) and the parameterization mapping $\chi(\boldsymbol{u})$ satisfies the smoothness condition (3.102).

- **Computation of Hadamard finite part (p.f.)**

Introducing a variable

$$q(\mathbf{r}) = \frac{\partial u(\mathbf{r})}{\partial \mathbf{n}}, \tag{3.103}$$

we will assume the following regularity conditions on $u(\mathbf{r})$ and $q(\mathbf{r})$:

$$u \in C^{1,\alpha},\ q \in C^{0,\alpha}. \tag{3.104}$$

Thus, we have

$$\begin{aligned} u(\mathbf{r}') &= u(\mathbf{r}) + \nabla u(\mathbf{r}) \cdot (\mathbf{r}' - \mathbf{r}) + O(|\mathbf{r}' - \mathbf{r}|^{1+\alpha}), \\ q(\mathbf{r}') &\equiv \nabla u(\mathbf{r}') \cdot \mathbf{n}' = \nabla u(\mathbf{r}) \cdot \mathbf{n} + O(|\mathbf{r}' - \mathbf{r}|^{\alpha}). \end{aligned} \tag{3.105}$$

The smoothness of the densities is required for the existence of the Hadamard finite part at a collocation point; detailed discussions can be found in [20–23].

Using (3.105), we first rewrite Π_2 of (3.71) as

$$\begin{aligned} \Pi_2(\mathbf{r}) &= \int_{\Gamma_\varepsilon} \left[\frac{\partial}{\partial \mathbf{n}} G_0(\mathbf{r}, \mathbf{r}') q(\mathbf{r}') - \frac{\partial^2}{\partial \mathbf{n} \partial \mathbf{n}'} G_0(\mathbf{r}, \mathbf{r}') u(\mathbf{r}') \right] \mathrm{d}s' \\ &= \int_{\Gamma - \gamma_\varepsilon} \left[\frac{\partial}{\partial \mathbf{n}} G_0(\mathbf{r}, \mathbf{r}') q(\mathbf{r}') - \frac{\partial^2}{\partial \mathbf{n} \partial \mathbf{n}'} G_0(\mathbf{r}, \mathbf{r}') u(\mathbf{r}') \right] \mathrm{d}s' \\ &+ \int_{s_\varepsilon} \left\{ \frac{\partial G_0(\mathbf{r}, \mathbf{r}')}{\partial \mathbf{n}} \left[q(\mathbf{r}') - \nabla u(\mathbf{r}) \cdot \mathbf{n} \right] \right. \\ &\left. - \frac{\partial^2 G_0(\mathbf{r}, \mathbf{r}')}{\partial \mathbf{n} \partial \mathbf{n}'} \left[u(\mathbf{r}') - u(\mathbf{r}) - \nabla u(\mathbf{r}) \cdot (\mathbf{r}' - \mathbf{r}) \right] \right\} \mathrm{d}s' \\ &- \nabla u(\mathbf{r}) \cdot \int_{s_\varepsilon} \left\{ \frac{\partial^2 G_0(\mathbf{r}, \mathbf{r}')}{\partial \mathbf{n} \partial \mathbf{n}'} (\mathbf{r}' - \mathbf{r}) - \frac{\partial G_0(\mathbf{r}, \mathbf{r}')}{\partial \mathbf{n}} \mathbf{n} \right\} \mathrm{d}s' \\ &- u(\mathbf{r}) \int_{s_\varepsilon} \frac{\partial^2 G_0(\mathbf{r}, \mathbf{r}')}{\partial \mathbf{n} \partial \mathbf{n}'} \mathrm{d}s'. \end{aligned} \tag{3.106}$$

The second integral in (3.106) will go to zero as the surface area of s_ε is on the order of $O(\varepsilon^2)$ and the integrand is of at most order of $O(1/\varepsilon)$ as a result of (3.105). The third integral will give a free matrix term $\overline{\mathbf{c}} = (c_{ki}(\mathbf{r}))_{3\times 3}$, where, for $i, k = 1, 2, 3$:

$$\int_{S_\varepsilon} \left[\frac{\partial^2}{\partial x_i \partial \mathbf{n}'} G_0(\mathbf{r}, \mathbf{r}')(x_k' - x_k) - \frac{\partial}{\partial x_i} G_0(\mathbf{r}, \mathbf{r}') n_k \right] \mathrm{d}s' = c_{ki}(\mathbf{r}) + O(\varepsilon). \quad (3.107)$$

The fourth integral can be shown [15] to have the following expansion for $i = 1, 2, 3$:

$$\int_{S_\varepsilon} \frac{\partial^2}{\partial x_i \partial \mathbf{n}'} G_0(\mathbf{r}, \mathbf{r}') \mathrm{d}s' = \frac{b_i(\mathbf{r})}{\varepsilon} + a_i(\mathbf{r}) + O(\varepsilon). \quad (3.108)$$

As $\mathbf{r} \notin \Omega_\varepsilon$, (3.71) becomes $\Pi_2(\mathbf{r}) = 0$. Substituting (3.107) and (3.108) into (3.106), and taking the limit $\varepsilon \to 0$, then the equation $\Pi_2(\mathbf{r}) = 0$ can be rewritten as

$$\begin{aligned} &\mathbf{n}^{\mathrm{T}} \cdot \mathbf{a}(\mathbf{r}) u(\mathbf{r}) + \mathbf{n}^{\mathrm{T}} \overline{\mathbf{c}}(\mathbf{r}) \nabla u(\mathbf{r}) \\ &- \lim_{\varepsilon \to 0} \left[\int_{\Gamma - \gamma_\varepsilon} \left[\frac{\partial}{\partial \mathbf{n}} G_0(\mathbf{r}, \mathbf{r}') q(\mathbf{r}') - \frac{\partial^2}{\partial \mathbf{n} \partial \mathbf{n}'} G_0(\mathbf{r}, \mathbf{r}') u(\mathbf{r}') \right] \mathrm{d}s' - \frac{\mathbf{b}(\mathbf{r}) \cdot n}{\varepsilon} u(\mathbf{r}) \right] \\ &= 0. \end{aligned} \quad (3.109)$$

If $\mathbf{r}$ is a smooth point on Γ (as a result, Γ will have a unique tangent plane at $\mathbf{r}$), then $\mathbf{a}(\mathbf{r}) = 0$, $\overline{\mathbf{c}}(\mathbf{r}) = (1/2)\overline{\mathbf{I}}$, and the second integral in (3.109) will be shown later to be finite, which is exactly the Hadamard finite part. Also, the divergent part is explicitly identified in (3.109), which will be canceled by a similar divergent term with opposite sign from the integration over $\Gamma - \gamma_\varepsilon$. Finally, Eq. (3.83) becomes the hyper-singular integral equation, where the Hadamard finite part concept can be used to represent the limit of the second integral in (3.109), i.e.,

$$\frac{1}{2} \frac{\partial u(\mathbf{r})}{\partial \mathbf{n}} + \text{p.f.} \int_\Gamma \frac{\partial^2}{\partial \mathbf{n} \partial \mathbf{n}'} G_0(\mathbf{r}, \mathbf{r}') u(\mathbf{r}') \mathrm{d}s' - \text{p.v.} \int_\Gamma \frac{\partial}{\partial \mathbf{n}} G_0(\mathbf{r}, \mathbf{r}') \frac{\partial u(\mathbf{r}')}{\partial \mathbf{n}} \mathrm{d}s' = 0. \quad (3.110)$$

Next, as in the case of the CPV, we show how to compute the Hadamard finite part explicitly using regular Gauss quadratures. The evaluation of the limit of the first term in (3.109) involving $\frac{\partial}{\partial \mathbf{n}} G_0(\mathbf{r}, \mathbf{r}')$ is a Cauchy principal value integral. We need only show how to compute the following limit over a patch S_p, and u is taken to be one of the shape functions over S_p in (3.64):

$$\Sigma_i = \lim_{\varepsilon \to 0} \left[\int_{S_p - \gamma_\varepsilon} \frac{\partial^2}{\partial x_i \partial \mathbf{n}'} G_0(\mathbf{r}, \mathbf{r}') N_l(\mathbf{r}') \mathrm{d}s' + \frac{b_i(\mathbf{r})}{\varepsilon} N_l(\mathbf{r}) \right], i = 1, 2, 3. \quad (3.111)$$

Remark 3.5 It should be noted that if $\mathbf{r}$ is at a vertex shared by many patches, the continuity condition of u and q will be hard to satisfy, as a $C^{1,\alpha}$ approximation to u will be required. If such a continuity is not met, the collocation point $\mathbf{r}$ will be taken to be inside one patch; in this case, S_p will just consist of one patch.

For simplicity, we will assume that **r** is an interior collocation point to the single patch S_p. Then

$$\chi^{-1}(S_p - \gamma_\varepsilon^i) = \{(\rho,\theta) : 0 \le \theta \le 2\pi, \alpha(\theta,\varepsilon) \le \rho \le \overline{\rho}(\theta)\}. \tag{3.112}$$

Using the polar coordinates in the parametric space, we have for $i = 1, 2, 3$

$$\begin{aligned}\Sigma_i &= \lim_{\varepsilon\to 0}\left[\int\limits_{S_p-\gamma_\varepsilon} \frac{\partial^2}{\partial x_i \partial \mathbf{n}'} G_0(\mathbf{r},\mathbf{r}')N_l(\mathbf{r}')\mathrm{d}s' + \frac{b_i(\mathbf{r})}{\varepsilon}N_l(\mathbf{r})\right] \\ &= \lim_{\varepsilon\to 0}\left[\int\limits_0^{2\pi}\mathrm{d}\theta \int\limits_{\alpha(\theta,\varepsilon)}^{\overline{\rho}(\theta)} F_i(\rho,\theta)\mathrm{d}\rho + \frac{b_i(\mathbf{r})}{\varepsilon}N_l(\mathbf{r})\right],\end{aligned} \tag{3.113}$$

where

$$F_i(\rho,\theta) = \frac{\partial^2}{\partial x_i \partial \mathbf{n}'} G_0\left((\rho^*,\theta^*),(\rho,\theta)\right) N_l(\rho,\theta) J(\rho,\theta)\rho. \tag{3.114}$$

The hyper-singularity of the integral implies that

$$F_i(\rho,\theta) = \frac{F_{-2}(\theta)}{\rho^2} + \frac{F_{-1}(\theta)}{\rho} + O(1), \tag{3.115}$$

where the coefficients $F_{-1}(\theta)$ and $F_{-2}(\theta)$ are given in [14, Eq. (C19) of app. C].

Then we get

$$\begin{aligned}\Sigma_i &= \lim_{\varepsilon\to 0}\left[\int\limits_0^{2\pi}\mathrm{d}\theta \int\limits_{\alpha(\theta,\varepsilon)}^{\overline{\rho}(\theta)} \left(F_i(\rho,\theta) - \frac{F_{-2}(\theta)}{\rho^2} - \frac{F_{-1}(\theta)}{\rho}\right)\mathrm{d}\rho\right] \\ &+ \lim_{\varepsilon\to 0}\int\limits_0^{2\pi}\mathrm{d}\theta \int\limits_{\alpha(\theta,\varepsilon)}^{\overline{\rho}(\theta)} \frac{F_{-2}(\theta)}{\rho^2}\mathrm{d}\rho \\ &+ \lim_{\varepsilon\to 0}\left[\int\limits_0^{2\pi}\mathrm{d}\theta \int\limits_{\alpha(\theta,\varepsilon)}^{\overline{\rho}(\theta)} \frac{F_{-1}(\theta)}{\rho}\mathrm{d}\rho + \frac{b_i(\mathbf{r})}{\varepsilon}N_l(\mathbf{r})\right].\end{aligned} \tag{3.116}$$

Each of the three limits has been shown to be finite [14], with (3.98) giving the Hadamard finite part over S_p as

$$\Sigma_i = \text{p.f.} \int\limits_{S_p} \frac{\partial^2}{\partial x_i \partial \mathbf{n}'} G_0(\mathbf{r}, \mathbf{r}') N_l(\mathbf{r}') \mathrm{d}s'$$
$$= \int\limits_0^{2\pi} \mathrm{d}\theta \int\limits_0^{\overline{\rho}(\theta)} \left[F_i(\rho, \theta) - \frac{F_{-2}(\theta)}{\rho^2} - \frac{F_{-1}(\theta)}{\rho} \right] \mathrm{d}\rho$$
$$+ \int\limits_0^{2\pi} F_{-1}(\theta) \ln \left| \frac{\overline{\rho}(\theta)}{\beta(\theta)} \right| \mathrm{d}\theta - \int\limits_0^{2\pi} F_{-2}(\theta) \left[\frac{\gamma(\theta)}{\beta^2(\theta)} + \frac{1}{\overline{\rho}(\theta)} \right] \mathrm{d}\theta. \tag{3.117}$$

3.2 Finite Element Methods (FEMs)

The difficulties in solving the PB equations (2.19)–(2.21) numerically are due to the following factors: (1) the singular behavior of the solution at the point charges q_j; (2) accurate approximations of solutions near the molecular surfaces; and (3) the treatment of the infinite domain of the exterior region of the molecules.

For the finite element method, we first truncate the whole space $\mathbb{R}^3$ to a finite computational domain $\Omega \supset \Omega_{\mathrm{i}}$ and consider a numerical boundary condition on $\partial\Omega$:

$$u|_{\partial\Omega} = g. \tag{3.118}$$

Various choices of g can be considered, such as the Coulomb potential of the point charges with the dielectric constant ϵ_0. However, for simplicity we will set $g = 0$. The finite element method is based on a weak formulation of (2.19) with the boundary condition (3.118) [24].

Due to the singular source terms in (3.38), the potential u will have large gradients near the point charges at $\mathbf{r}_j$. To resolve these gradients, a singular function G is defined with screened Coulomb potentials from these charges [25, 26]

$$G(\mathbf{r}) = \sum_{j=1}^{N} \frac{q_j}{4\pi\epsilon_{\mathrm{i}}} \frac{\mathrm{e}^{-\lambda_{\mathrm{i}}|\mathbf{r}-\mathbf{r}_j|}}{|\mathbf{r}-\mathbf{r}_j|}, \tag{3.119}$$

which satisfies the following inhomogeneous PB equation:

$$-\nabla \cdot \epsilon_{\mathrm{i}} \nabla G + \lambda_{\mathrm{i}}^2 G = \rho, \quad \mathbf{r} \in \mathbb{R}^3. \tag{3.120}$$

Now we consider the function w:

$$w = u - G - \widetilde{g}, \tag{3.121}$$

where $\widetilde{g}$ is any locally supported function near $\partial\Omega$ such that the support has no intersection with $\Gamma = \partial\Omega_i$, i.e.,

$$\sup(\widetilde{g}) \cap \Gamma = \emptyset, \tag{3.122}$$

and also

$$\widetilde{g}(\mathbf{r}) = G(\mathbf{r}) - g, \ \mathbf{r} \in \partial\Omega, \tag{3.123}$$

which produces a homogeneous boundary condition for w, i.e.,

$$w|_{\partial\Omega} = 0. \tag{3.124}$$

Next, by subtracting (3.120) from (2.19), we can see that w satisfies the following inhomogeneous PB equation with a much smoother right-hand side:

$$-\nabla \cdot \epsilon \nabla w + \lambda^2 w = f(\mathbf{r}), \quad \mathbf{r} \in \Omega_i \text{ or } \mathbf{r} \in \Omega_o, \tag{3.125}$$

where

$$f(\mathbf{r}) = \begin{cases} 0, & \text{if } \mathbf{r} \in \Omega_i, \\ \frac{\epsilon_o \lambda_i^2 - \epsilon_i \lambda_o^2}{\epsilon_i} G(\mathbf{r}) + \left(\epsilon_o \nabla^2 - \lambda_o^2\right) \widetilde{g}(\mathbf{r}), & \text{if } \mathbf{r} \in \Omega_o. \end{cases} \tag{3.126}$$

Moreover, a transmission condition for w can be derived from (2.3):

$$w_i - w_o = 0, \tag{3.127}$$

$$\epsilon_i \left(\frac{\partial w}{\partial \mathbf{n}}\right)_i - \epsilon_o \left(\frac{\partial w}{\partial \mathbf{n}}\right)_o = (\epsilon_o - \epsilon_i)\frac{\partial G}{\partial \mathbf{n}} \equiv \sigma_\Gamma. \tag{3.128}$$

Now, by integrating (3.125) over $\Omega_i \cup \Omega_o$ with a test function $v \in H_0(\Omega)$ (subspace of the Sobolev space $H(\Omega)$ of functions with L^2-integrable first derivatives where boundary trace is zero [24]) and using integration by parts and the transmission condition (3.127) and (3.128), we have the following weak form of (3.125).

Find $w \in H_0(\Omega)$, such that

$$\forall v \in H_0(\Omega), \quad a(w, v) = F(v), \tag{3.129}$$

where $a(w, v)$ is a bilinear form

$$a(w, v) = \int_\Omega \left(\epsilon \nabla w \nabla v + \lambda^2 w v\right) d\mathbf{r}, \tag{3.130}$$

and the functional $F(v)$ is

$$F(v) = \int_{\Omega} f v \mathrm{d}\mathbf{r} + \int_{\Gamma} \sigma_{\Gamma} v \mathrm{d}s. \tag{3.131}$$

The finite element is based on a triangulation of the solution domain Ω into elements K which conform to the molecular interface Γ. A polynomial space $P_k(K)$ is used for the element subspace, where the basis functions are usually associated with the nodal values at specific points on K, such as the vertices of K or middle points on the edges of the element K. For a reference tetrahedron $\hat{K}$ with vertices $\{(0,0,0), (1,0,0), (0,1,0), (0,0,1)\} \equiv \{v_i\}_{i=0}^{3}$, the following basis functions for $P_1(K)$ can be used:

$$\begin{aligned}
\hat{\phi}_0(\hat{x}, \hat{y}, \hat{z}) &= 1 - \hat{x} - \hat{y} - \hat{z}, \\
\hat{\phi}_1(\hat{x}, \hat{y}, \hat{z}) &= \hat{x}, \\
\hat{\phi}_2(\hat{x}, \hat{y}, \hat{z}) &= \hat{y}, \\
\hat{\phi}_3(\hat{x}, \hat{y}, \hat{z}) &= \hat{z}.
\end{aligned} \tag{3.132}$$

The basis functions on a general tetrahedron K are defined by a pull-back function $F(\hat{x}, \hat{y}, \hat{z})$:

$$(\hat{x}, \hat{y}, \hat{z}) \in \hat{K} \to (x, y, z) = F(\hat{x}, \hat{y}, \hat{z}) \in K, \tag{3.133}$$

and the basis functions on K take the following form

$$\phi_i(x, y, z) = \hat{\phi}_i \circ F^{-1}(x, y, z) = \phi_i(\hat{x}, \hat{y}, \hat{z}), i = 0, 1, 2, 3. \tag{3.134}$$

A finite element subspace $V_h \subset H(\Omega)$ can be constructed by combining the element subspace where the common nodal values at the shared vertices between adjacent elements will ensure the global continuity of the function in V_h:

$$V_h = \{v_h : v_h|_K \in P_k(K)\}. \tag{3.135}$$

Let $\{\phi_i(x, y, z)\}|_{i=1}^{N_K}$ be the basis functions indexed by the N_K vertices of the triangulation. Then

$$V_h = \left\{ v_h : v_h = \sum_{i=1}^{N_K} v_i \phi_i(x, y, z) \right\} \subset H(\Omega), \tag{3.136}$$

where the nodal unknown is $v_i = v_h(\alpha_i)$ at the node α_i. Correspondingly, a subspace for $H_0(\Omega)$ can be defined by setting the nodal values at the boundary nodes as zero. Let N_{int} denote the total number of the interior nodes and N_{b} the total number of the boundary nodes as listed in the summation. Then $N_K = N_{\text{int}} + N_{\text{b}}$, and

$$V_{0h} = \left\{ v_h : v_h = \sum_{i=1}^{N_{\text{int}}} v_i \phi_i(x, y, z) \right\} \subset H_0(\Omega). \tag{3.137}$$

The finite element solution to the weak form (3.129) is as follows.

Find $w_h \in V_{0h}$, such that

$$\forall v_h \in V_{0h}, \quad a(w_h, v_h) = F(v_h). \tag{3.138}$$

By taking $v_h = \phi_j(x, y, z)$, $j = 1, 2, \ldots, N_{\text{int}}$, we arrive at the following linear system for the unknowns $\mathbf{w} = (w_1, \ldots, w_{N_{\text{int}}})^{\mathrm{T}}$:

$$A\mathbf{w} = \mathbf{f}, \tag{3.139}$$

where the stiffness matrix A is defined through

$$a_{ij} = a(\phi_i, \phi_j), 1 \leq i, j \leq N_{\text{int}}, \tag{3.140}$$

and the source $\mathbf{f}$ is defined through

$$f_i = F(\phi_i), 1 \leq i \leq N_{\text{int}}. \tag{3.141}$$

The matrix equation is a sparse linear system, and for large N_{int} a multigrid iterative method is usually used to solve (3.139) [26].

3.3 Immersed Interface Methods (IIMs)

Because of the dissimilarity of the dielectric constants of the molecule and its solvent, the derivative of the potential is discontinuous as shown in (3.128), which poses difficulties in finite difference discretization of the PB equation. The finite difference IIM for (3.125) [27] incorporates the jump condition on the molecular surface Γ into the construction of the difference formula near the molecular boundary. Other types of modified finite difference methods, such as the matched interface and boundary (MIB) method, have been applied to treat the molecular interfaces in the solutions of PB equations [28].

Let us consider the PB equation in (3.125) and write the transmission condition (3.127) and (3.128) in a more general notation:

$$w^+ - w^- = u, \; \epsilon^+ \left(\frac{\partial w}{\partial \mathbf{n}} \right)^+ - \epsilon^- \left(\frac{\partial w}{\partial \mathbf{n}} \right)^- = v, \tag{3.142}$$

where + indicates the limit taken from Ω_{o} and – the limit from Ω_{i}, and $\mathbf{n}$ is the outer normal on $\partial\Omega_{\mathrm{i}} = \Gamma$.

A uniform mesh will be used to discretize (3.125), and the mesh point is indexed by (i, j, k) with a spacing h in all three directions. The interface is described by a level set function φ [29, 30]

$$\Gamma = \{(x, y, z) : \varphi(x, y, z) = 0\}, \tag{3.143}$$

and the interior and the exterior of the molecule are described by

$$\Omega_{\rm i} = \{(x, y, z) : \varphi(x, y, z) < 0\}, \Omega_{\rm o} = \{(x, y, z) : \varphi(x, y, z) > 0\}, \tag{3.144}$$

respectively.

By using the level set function φ, a mesh point (i, j, k) is identified as an irregular point if

$$\varphi_{ijk}^{\min}\varphi_{ijk}^{\max} < 0, \tag{3.145}$$

where

$$\varphi_{ijk}^{\min} = \min\{\varphi(i \pm 1, j, k), \varphi(i, j \pm 1, k), \varphi(i, j, k \pm 1)\},$$

$$\varphi_{ijk}^{\max} = \max\{\varphi(i \pm 1, j, k), \varphi(i, j \pm 1, k), \varphi(i, j, k \pm 1)\}.$$

The rest of the mesh points will be considered as regular points where the following difference equation for (3.125) is used [27, 31]:

$$\begin{aligned}
&\epsilon\left(i - \frac{1}{2}, j, k\right)[w(i-1, j, k) - w(i, j, k)] \\
&+ \epsilon\left(i + \frac{1}{2}, j, k\right)[(w(i+1, j, k) - w(i, j, k)] \\
&+ \epsilon\left(i, j - \frac{1}{2}, k\right)[w(i, j-1, k) - w(i, j, k)] \\
&+ \epsilon\left(i, j + \frac{1}{2}, k\right)[(w(i, j+1, k) - w(i, j, k)] \\
&+ \epsilon\left(i, j, k - \frac{1}{2}\right)[w(i, j, k-1) - w(i, j, k)] \\
&+ \epsilon\left(i, j, k + \frac{1}{2}\right)[(w(i, j, k+1) - w(i, j, k)] \\
&- \lambda^2(i, j, k)w(i, j, k) = -\frac{f(i, j, k)}{h}.
\end{aligned} \tag{3.146}$$

For irregular mesh points identified by (3.145), a different difference formula is used in the IIM in the following form:

$$\sum_{m}^{n_s} \gamma_m w(i + i_m, j + j_m, k + k_m) = f(i, j, k) + C(i, j, k), \tag{3.147}$$

where n_s is the number of the mesh points around (i, j, k) used in the difference formula, and the coefficients γ_m are to be determined by a requirement on the truncation error at X^*,

$$T(i, j, k) = \sum_{m}^{n_s} \gamma_m w(i + i_m, j + j_m, k + k_m) - f(i, j, k) - C(i, j, k) = O(h), \tag{3.148}$$

where X^* is the projection of the mesh point (i, j, k) onto the interface Γ.

To satisfy the requirement of (3.148), we expand all terms in $T(i, j, k)$ by a Taylor expansion at X^* and use the jump conditions for the partial derivatives of u and v to obtain [31]

$$\begin{aligned}
w^+ &= w^- + u, \\
w_\xi^+ &= \frac{\epsilon^-}{\epsilon^+} w_\xi^- + \frac{v}{\epsilon^+}, \\
w_\eta^+ &= w_\eta^- + u_\eta, \\
w_\tau^+ &= w_\tau^- + u_\tau, \\
w_{\eta\tau}^+ &= w_{\eta\tau}^- + (w_\xi^- - w_\xi^+)\chi_{\eta\tau} + u_{\eta\tau}, \\
w_{\eta\eta}^+ &= w_{\eta\eta}^- + (w_\xi^- - w_\xi^+)\chi_{\eta\eta} + u_{\eta\eta}, \\
w_{\tau\tau}^+ &= w_{\tau\tau}^- + (w_\xi^- - w_\xi^+)\chi_{\tau\tau} + u_{\tau\tau}, \\
w_{\xi\eta}^+ &= \frac{\epsilon^-}{\epsilon^+} w_{\xi\eta}^- + \left(w_\eta^+ - \frac{\epsilon^-}{\epsilon^+} w_\eta^-\right)\chi_{\eta\eta} \\
&\quad + \left(w_\tau^+ - \frac{\epsilon^-}{\epsilon^+} w_\tau^-\right)\chi_{\eta\tau} + \frac{v_\eta}{\epsilon^+}, \\
w_{\xi\tau}^+ &= \frac{\epsilon^-}{\epsilon^+} w_{\xi\tau}^- + \left(w_\eta^+ - \frac{\epsilon^-}{\epsilon^+} w_\eta^-\right)\chi_{\eta\tau} \\
&\quad + \left(w_\tau^+ - \frac{\epsilon^-}{\epsilon^+} w_\tau^-\right)\chi_{\tau\tau} + \frac{v_\tau}{\epsilon^+},
\end{aligned}$$

$$\begin{aligned}
w_{\xi\xi}^+ &= \frac{\epsilon^-}{\epsilon^+} w_{\xi\xi}^- + \left(\frac{\epsilon^-}{\epsilon^+} - 1\right) w_{\eta\eta}^- + \left(\frac{\epsilon^-}{\epsilon^+} - 1\right) w_{\tau\tau}^- \\
&\quad + (w_\xi^+ - w_\xi^-)(\chi_{\eta\eta} + \chi_{\tau\tau}) - u_{\eta\eta} - u_{\tau\tau} + \frac{1}{\epsilon^+}([\lambda^2]w^- - (\lambda^2)^+ [w]),
\end{aligned} \tag{3.149}$$

where $[w] = w^+ - w^-$.

Then it can be shown that the truncation error is given by

$$T(i,j,k) = a_1 w^- + a_2 w_\xi^- + a_3 w_\eta^- + a_4 w_\tau^- + a_5 w_{\eta\tau}^- + a_6 w_{\eta\eta}^- + a_7 w_{\tau\tau}^-$$
$$+ a_8 w_{\xi\eta}^- + a_9 w_{\xi\tau}^- + a_{10} w_{\xi\xi}^- + (\hat{T} - C(i,j,k)) + O(h), \qquad (3.150)$$

where all a_i and $\hat{T}$ are functions of the 27 coefficients γ_m. Therefore, in order to satisfy (3.148), we require all coefficients a_i to be zero and set $C(i,j,k) = \hat{T}$, which results in an over-determined system for γ_m:

$$B\gamma = \mathbf{b}. \qquad (3.151)$$

To determine the 27 unknown coefficients, the following minimization problem is solved for the optimal choice:

$$\min_{\gamma_m} \frac{1}{2} \sum_m (\gamma_m - d_m)^2 \qquad (3.152)$$
$$\text{subject to } B\gamma = \mathbf{b},$$
$$\gamma_m < 0,\ (i_m,\ j_m,\ k_m) = (0,0,0),$$
$$\gamma_m > 0,\ (i_m,\ j_m,\ k_m) \neq (0,0,0),$$

where the reference coefficients are

$$d_m = \frac{\epsilon_{i+\frac{i_m}{2},j+\frac{j_m}{2},k+\frac{k_m}{2}}}{h^2},\ \text{if } i_m^2 + j_m^2 + k_m^2 = 1,$$
$$d_m = 0,\quad \text{otherwise, and } d_0 = -\frac{1}{h^2}\sum_{m\neq 0} \epsilon_{i+\frac{i_m}{2},j+\frac{j_m}{2},k+\frac{k_m}{2}}. \qquad (3.153)$$

3.4 Summary

Boundary element methods based on singular integrals are popular methods, especially coupled with fast multipole methods to be discussed in the next chapter, for molecules of general shapes. The singular integrals in the BEMs in the form of Cauchy principal values and the Hadamard finite part of Green's functions can be treated with subtraction and direct evaluation techniques, where the latter technique is also applicable to more general Green's functions. Mesh-based methods such as finite element and finite difference methods produce sparse linear systems for which fast solvers such as multigrid methods [32] can be used.

References

1. J.C. Nédélec, *Acoustic and Electromagnetic Equations: Integral Representations for Harmonic Problems* (Springer, Berlin, 2001)
2. N.I. Muskhelishvili, *Singular Integral Equations: Boundary Problems of Function Theory and Their Application to Mathematical Physics* (Noordhoff International, Leiden, 1953)
3. J. Hadamard, *Lectures on Cauchy Problems in Linear Partial Differential Equations* (Dover Publications, New York, 2003)
4. G.C. Hsiao, W.L. Wendland, *Boundary Integral Equations* (Springer, Berlin, 2008)
5. C. Fox, A generalization of the Cauchy principal value. Can. J. Math. **9**, 110–117 (1957)
6. A.H. Juffer, E.F.F. Botta, B.A.M. van Keulen, A. van der Ploeg, H.J.C. Berendsen, The electric potential of a macromolecule in a solvent: a fundamental approach. J. Comput. Phys. **97**, 144–171 (1991)
7. B. Lu, X. Cheng, J. Huang, J.A. McCammon, Order N algorithm for computation of electrostatic interactions in biomolecular systems. Proc. Natl. Acad. Sci. USA **103**, 19314–19319 (2006)
8. C. Müller, *Foundations of the Mathematical Theory of Electromagnetic Waves* (Springer, New York, 1969)
9. V. Rokhlin, Solution of acoustic scattering problem by means of second kind of integral equations. Wave Motion **5**, 257–272 (1983)
10. J. Giroire, J.C. Nédélec, Numerical solution of an exterior Neumann problem using a double layer potential. Math. Comp. **32**, 973–990 (1978)
11. Y. Liu, F.J. Rizzo, A weakly singular form of the hypersingular boundary integral equation applied to 3-D acoustic wave equations. Comput. Meth. Appl. Mech. Eng. **96**, 271–287 (1992)
12. Y. Liu, *Fast Multipole Boundary Element Method: Theory and Applications in Engineering* (Cambridge University Press, Cambridge, UK, 2009)
13. M. Guiggiani, A. Gigante, A general algorithm for multidimensional Cauchy principal value integrals in the boundary element method. ASME J. Appl. Mech. **57**, 906–915 (1990)
14. M. Guiggiani, G. Krishnasamy, T.J. Rudolphi, F.J. Rizzo, A general algorithm for the numerical solution of hypersingular boundary integral equations. ASME J. Appl. Mech. **59**, 604–614 (1992)
15. M. Guiggiani, Formulation and numerical treatment of boundary integral equations with hypersingular kernels, in *Singular Integrals in Boundary Element Methods*, ed. by V. Sladek, J. Sladek (Computational Mechanics Publications, Southampton Chap. 3, 1998)
16. Y. Liu, T.J. Rudolphi, Some identities for fundamental solutions and their applications to weakly-singular boundary element formulations. Eng. Anal. Bound. Elem. **8**, 301–311 (1991)
17. Y. Liu, T.J. Rudolphi, New identities for fundamental solutions and their applications to nonsingular boundary element formulations. Comput. Mech. **24**, 286–292 (1999)
18. G. Krishnasamy, L.W. Schmerr, T.J. Rudolphi, F.J. Rizzo, Hypersingular boundary integral equations: some applications in acoustic and elastic wave scattering. ASME J. Appl. Mech. **57**, 404–414 (1990)
19. T.J. Rudolphi, The use of simple solutions in the regularization of hypersingular boundary integral equations. Math. Comput. Model. **15**, 269–278 (1991)
20. N.M. Gunter, *Potential Theory and its Applications to Basic Problems of Mathematical Physics* (Fredrick Ungar, New York, 1967)
21. P.A. Martin, F.J. Rizzo, On boundary integral equations for crack problems. Proc. R. Soc. Lond. A **421**, 341–355 (1989)
22. P.A. Martin, F.J. Rizzo, Hypersingular integrals: how smooth must the density be? Int. J. Numer. Meth. Eng. **39**, 687–704 (1996)
23. G. Krishnasamy, F.J. Rizzo, T.J. Rudolphi, Continuity requirements for density functions in the boundary integral equation method. Comput. Mech. **9**, 267–284 (1992)
24. P.G. Ciarlet, *The Finite Element Method for Elliptic Problems* (North-Holland Publishing, Amsterdam, 1978)

25. I.-L. Chern, J.-G. Liu, W.-C. Wang, Accurate evaluation of electrostatics for macromolecules in solution. Meth. Appl. Anal. **10**, 309–328 (2003)
26. L. Chen, M. Holst, J. Xu, The finite element approximation of the nonlinear Poisson-Boltzmann equation. SIAM J. Numer. Anal. **45**, 2298–2320 (2007)
27. J. Wang, Q. Cai, Z.-L. Li, H.-K. Zhao, R. Luo, Achieving energy conservation in Poisson-Boltzmann molecular dynamics: accuracy and precision with finite-difference algorithms. Chem. Phys. Lett. **468**, 112–118 (2009)
28. S. Yu, W. Geng, G.W. Wei, Treatment of geometric singularities in implicit solvent models. J. Chem. Phys. **126**, 244108 (2007)
29. J.A. Sethian, *Level Set Methods and Fast Marching Methods: Evolving Interfaces in Computational Geometry, Fluid Mechanics, Computer Vision, and Materials Science*, 2nd edn. (Cambridge University Press, Cambridge, UK, 1999)
30. S.J. Osher, R.P. Fedkiw, *Level Set Methods and Dynamic Implicit Surfaces* (Springer, Berlin, 2002)
31. Z. Li, K. Ito, *The Immersed Interface Method: Numerical Solutions of PDEs Involving Interfaces and Irregular Domains* (SIAM, Philadelphia, PA, 2006)
32. A. Brandt, *Guide to Multigrid Development*, Lecture Notes in Mathematics, vol. 960 (Springer, Berlin, 1982)

Chapter 4
Random Walk Stochastic Methods for PDE Boundary Value Problems

The Feynman–Kac formula provides a fundamental relation between the solution of an elliptic partial differential equation (PDE) and a stochastic process, providing a probabilistic representation of the PDE's solution at any given location and a Monte Carlo solution [1–4]. In this chapter, we first briefly present Itô integrals of stochastic processes and their martingale properties, discuss the local time of reflecting Brownian motions, and derive the Feynman–Kac formulae for mixed boundary value problems (BVPs) using the Itô formula for functions of stochastic processes. Then, the walk-on-spheres (WOS) method is introduced for the computation of the local time. Additionally, the Pardoux–Peng theory of the nonlinear Feynman–Kac formula for quasilinear parabolic equations is presented, which will be key for designing stochastic process-based deep learning algorithms for solving PDEs in Chap. 6. Next, as an application of the Feynman–Kac formula for a local solution, we solve the forward problem in electrical impedance tomography. Furthermore, to find the global solution of BVPs using the Feynman–Kac formula, a hybrid method, BIE-WOS, that combines local boundary integral equation (BIE) and the WOS Feynman–Kac approach, is introduced for the Dirichlet BVP of the Laplace equation. Finally, an iterative BIE-WOS method is presented for solving mixed BVPs of the Laplace equation without using reflecting Brownian motions.

4.1 Brownian Motion, Itô Integral, and Itô's Formula

4.1.1 Probability Space, Random Variable, and Conditional Expectation

A probability space consists of a triplet $(\Omega, \mathcal{F}, P)$. The sample space $\Omega = \{\omega_i\}$ contains outcomes ω_i of some random experiments, such as heads or tails in the toss-up of some coins. The σ-algebra $\mathcal{F}$ consists of subsets (termed events) of Ω, satisfying

W. Cai, *Deterministic, Stochastic, and Deep Learning Methods for Computational Electromagnetics*, https://doi.org/10.1007/978-981-96-0100-4_4

the following conditions including the closedness conditions under countable set union and complement operations, i.e.,

1. $\Omega \in \mathcal{F}$;
2. if $A \in \mathcal{F}$, then $A^c \in \mathcal{F}$ where $A^c = \Omega \backslash A$ is the complement of A in Ω;
3. if $A_1, A_2, \cdots \in \mathcal{F}$, then $\bigcup_{i=1}^{\infty} A_i \in \mathcal{F}$.

Finally, a probability measure P is a mapping from $\mathcal{F}$ into the interval [0, 1] and $P(A), \forall A \in \mathcal{F}$ denotes the probability of the occurrence of an event A. P satisfies the following conditions:

1. $P(\emptyset) = 0$, $P(\Omega) = 1$;
2. if $A_1, A_2, \cdots \in \mathcal{F}$ are pairwise disjoint, i.e., $A_i \cap A_j = \emptyset$ if $i \neq j$, then

$$P\left(\bigcup_{i=1}^{\infty} A_i\right) = \sum_{i=1}^{\infty} P(A_i). \tag{4.1}$$

- Conditional probability.
  For two events A and B, if we are given the condition that event B has already occurred, we can define the conditional probability of an event A occurring as

$$P(A|B) = \frac{P(A \cap B)}{P(B)}. \tag{4.2}$$

- Random variable (r.v.). $X(\omega)$ is a $\mathcal{F}$-measurable real-valued function into the real line $\mathbb{R}$

$$X : \omega \in \Omega \to X(\omega) \in \mathbb{R}, \tag{4.3}$$

  and, for each $B \in \mathbb{B}(\mathbb{R})$—Borel set of $\mathbb{R}$ (the smallest σ-algebra containing all intervals $(a, b]$, $a < b \in \mathbb{R}$),

$$X^{-1}(B) \in \mathcal{F}.$$

- Probability distribution function and probability distribution density function (pdf) of a random variable. We define the probability distribution function as

$$F(x) = P\{X \leq x\} \equiv P\{\omega \in \Omega, X(\omega) \leq x\}, \tag{4.4}$$

  and the probability distribution density function $f(x)$ as the derivative of $F(x)$

$$f(x) = F'(x), \tag{4.5}$$

  where $f(x)$ could contain a Dirac delta function $\delta(x - x_i)$ (i.e., a point mass at x_i) at jump discontinuities of $F(x)$ as in the case of a discrete random variable X taking only a finite number of values in $\mathbb{R}$. Therefore, in general, we have

$$P\{X \leq x\} = \int_{-\infty}^{x} f(u)du, \tag{4.6}$$

and, for any $a < b \in \mathbb{R}$,

$$P\{\omega \in \Omega, a < X(\omega) \leq b\} = P\{a < X \leq b\} = \int_{a}^{b} f(u)du. \tag{4.7}$$

For an infinitesimal interval $(x, x + dx]$, we use the shorthand $X^{-1}((x, x + dx]) = d\omega$ and $P(d\omega) = f(x)dx$.

- Expectation $E(X)$ and variance $\text{Var}(X)$ of an r.v. X.

$$E(X) = \int_{\Omega} X(\omega)P(d\omega) = \int_{-\infty}^{\infty} xf(x)dx, \tag{4.8}$$

$$\begin{aligned} \text{Var}(X) &= E((X(\omega) - E(X))^2) = \int_{\Omega} (X(\omega) - E(X))^2 \, P(d\omega) \\ &= \int_{-\infty}^{\infty} (x - E(X))^2 \, f(x)dx. \end{aligned} \tag{4.9}$$

We can also define a conditional expectation when X is restricted to $B \in \mathcal{F}$ as

$$E(X|B) = \int_{\Omega} X(\omega)I_B(\omega)P(d\omega|B) = \int_{B} X(\omega)P(d\omega|B). \tag{4.10}$$

4.1.1.1 Conditional Expectation $E(X|G)$

The conditional expectation $E(X|G)$ given any σ-algebra $G \subset \mathcal{F}$ is defined as a G-measurable random variable, which will be the best approximation to X by all G-measurable functions, defined by the following orthogonality condition,

$$E((X - E(X|G))\,\xi) = 0, \tag{4.11}$$

where ξ is any G-measurable function. Therefore, if X itself is G-measurable, then we have

$$E(X|G) = X. \tag{4.12}$$

By setting $\xi(\omega) = I_B(\omega)$ for any $B \in G$, condition (4.11) implies that we have

$$\int_B X(\omega)P(d\omega) = \int_B E(X|G)P(d\omega). \tag{4.13}$$

In many situations, such as the study of stochastic processes in Sect. 4.1.2, we would like to understand one r.v. X given the information about a second r.v. Y, namely, the σ-algebra generated by Y,

$$G = \sigma_Y = \sigma\{Y^{-1}(B), \forall B \in \mathbb{B}(\mathbb{R})\}, \tag{4.14}$$

the smallest σ-algebra containing all sets $Y^{-1}(B)$. In this case, we denote $E(X|G) \equiv E(X|Y)$, which will be the best approximation to X by any random function of the form $\xi = g(Y)$.

To better understand $E(X|Y) = g(Y)$, we consider the cases of two discrete r.v.'s X and Y, say, $X \in \{x_1, x_2, \ldots\}$, $Y \in \{y_1, y_2, \ldots\}$. Then, the sample space Ω has the following disjoint decomposition,

$$\Omega = \bigcup_i B_i, \;\; B_i = \{Y = y_i\}. \tag{4.15}$$

We now show that $g(Y)$ is a random function from $\Omega \to \mathbb{R}$, which takes a constant value on each B_i. Thus, $E(X|Y)$ is a piecewise constant function approximation to the r.v. X. To see this, we first show that the best constant approximation to any random variable is its expectation by considering the following quadratic function

$$I(c) = E((X-c)^2),$$

whose minimum should occur at $I'(c) = 0$, namely, $2E(X-c) = 0$, i.e.,

$$c = E(X). \tag{4.16}$$

Next, using the partition (4.15), we evaluate the L^2 norm between X and $g(Y)$,

$$\begin{aligned}
E((X-g(Y))^2) &= \int_\Omega (X(\omega) - g(Y))^2 \, P(d\omega) \\
&= \sum_i \int_{B_i} (X(\omega) - g(Y))^2 \, P(d\omega) \\
&\overset{c_i = g(y_i)}{=\!=\!=} \sum_i P(B_i) \int_{B_i} (X(\omega) - c_i)^2 \, \frac{P(d\omega)}{P(B_i)} \\
&= \sum_i P(B_i) \int_{B_i} (X(\omega) - c_i)^2 \, \frac{P(d\omega)}{P(B_i)} \\
&= \sum_i P(B_i) \int_{B_i} (X(\omega) - c_i)^2 \, P(d\omega|B_i).
\end{aligned}$$

Now, minimizing the sum in the last line is equivalent to minimizing each individual positive term. Therefore, from (4.16), we must have

$$c_i = E(X|B_i) = E(X|Y = y_i),$$

i.e., we have

$$g(Y(\omega)) = g(y_i) = c_i = E(X|Y = y_i), \quad \omega \in B_i,$$

or

$$g(Y(\omega)) = E(X|Y), \quad \omega \in \Omega.$$

Namely, the best approximation to X among all $g(Y)$'s should be

$$g(Y) = E(X|Y).$$

The following lists some properties of the conditional expectation to be used later, and their proofs can be found in Sect. 2.4 of [5].

1. If $G = \{\emptyset, \Omega\}$, then
$$E(X|G) = E(X); \tag{4.17}$$
2. If X is G measurable, then
$$E(XY|G) = XE(Y|G); \tag{4.18}$$
3. (Tower property) If $G_1 \subset G_2$, then
$$E(E(X|G_2)|G_1) = E(X|G_1), \tag{4.19}$$
and, in particular, by taking $G_1 = \{\emptyset, \Omega\}$, we have for any G,
$$E(E(X|G)) = E(X). \tag{4.20}$$

4.1.2 Brownian Motion, Itô Integral of Stochastic Processes, and Itô's Formula

4.1.2.1 Stochastic Process and Brownian Motion

A stochastic process $X(t, \omega) = X_t(\omega) = X(t)$ is a series of t-indexed random variables. For example, Brownian motion arises as a mathematical idealization of the motion of pollen under bombardment by surrounding water molecules, which can be viewed as a continuous-time random walk in space. Wiener provided the mathematical proof of the existence of Brownian motion $B(t, \omega) = B_t(\omega)$ approximately 1920, satisfying the following conditions:

1. For each $\omega \in \Omega$, $B(t)$ has a continuous path and $B(0) = 0$.
2. (Independent increment) For any $0 < u \leq s < t$, $B(t) - B(s)$ and $B(u) - B(0) = B(u)$ are independent.
3. (Normal increment) For $s < t$, $B(t) - B(s) \sim N(0, t - s)$ where $N(\mu, \sigma^2)$ denotes the r.v. of a normal distribution with mean μ and variance σ^2.

4.1.2.2 Itô Integral of Stochastic Processes

A stochastic process $X_t(\omega)$ is associated with a filtration $\mathcal{F}_t$, which contains the information of the collection of random variable $X_s(\omega)$ up to time t. Here, we let $\mathcal{F}_t$ be the natural filtration generated by the Brownian motion, i.e. $\mathcal{F}_t$ is the smallest σ-algebra which includes all sets of $\{B_s^{-1}(\Gamma), s \leq t, \Gamma$ is any Borel set of $\mathbb{R}\}$,

$$\mathcal{F}_t = \sigma \left\{ \bigcup_{s \leq t} \{B_s^{-1}(\Gamma), \Gamma \in \mathbb{B}(\mathbb{R})\} \right\}. \tag{4.21}$$

For a $\mathcal{F}_t$-adapted process, $X(t, \omega)$, i.e., $X(t)$ is $\mathcal{F}_t$ measurable for all $t > 0$, satisfying with a probability one that $\int_0^T X^2(s)ds < \infty, T > 0$, its Itô integral over $[0, t]$ is a stochastic process $Y(t)$ defined through the following limit over partitions $0 = t_0 < t_1 < \cdots < t_n = t$, $\delta_n = \max_{0 \leq i \leq n-1} |t_{i+1} - t_i| \to 0$,

$$Y(t) = \int_0^t X(s) dB(s) \triangleq \lim_{\delta_n \to 0} \sum_{i=0}^{n-1} X(t_i) \left(B(t_{i+1}) - B(t_i) \right). \tag{4.22}$$

The following theorem summarizes some key properties of the Itô integral

Theorem 4.1 *An Itô integral has the following properties:*

1. *(Zero mean)* $E\left(\int_0^t X(s) dB(s)\right) = 0.$
2. *(Second moment with isometry property)* $E\left(\int_0^t X(s) dB(s)\right)^2 = \int_0^t EX^2(s) ds.$
3. *(Martingale property)* $Y(t)$ *is a Martingale, namely, for* $s \leq t$,

$$E\left(Y(t) | \mathcal{F}_s\right) = Y(s). \tag{4.23}$$

Specifically, we have

$$E\left(Y(t) | \mathcal{F}_0\right) = Y(0) = 0. \tag{4.24}$$

Proof Properties 1 and 2 can be shown first for the finite sum in the definition of the Itô integral, then pass the limit inside the expectation. For instance, by applying the expectation to the finite term sum in (4.22) and using the independence of $X(t_i)$ and Brownian motion increment at time t_i, $B(t_{i+1}) - B(t_i)$, we see that $E\left(X(t_i)\left(B(t_{i+1}) - B(t_i)\right)\right) = EX(t_i)E\left(B(t_{i+1}) - B(t_i)\right) = 0$, thus the mean of the finite sum is zero. A similar approach can be applied to the second moment of the finite term sum.

To show the martingale property of $Y(t)$, we first show that

$$E\left(\int_s^t X(u)dB(u)\middle|\mathcal{F}_s\right) = 0, \tag{4.25}$$

which can be computed by the limit of the Itô integral over $s = t_0 < t_1 < \cdots < t_n = t$

$$\begin{aligned}
E\left(\int_s^t X(u)dB(u)\middle|\mathcal{F}_s\right) &\sim E\left(\sum_{i=0}^{n-1} X(t_i)\left(B(t_{i+1}) - B(t_i)\right)\middle|\mathcal{F}_s\right)\\
&= \sum_{i=0}^{n-1} E\Big(E\left(X(t_i)\left(B(t_{i+1}) - B(t_i)\right)|\mathcal{F}_{t_i}\right)|\mathcal{F}_s\Big)\\
&= \sum_{i=0}^{n-1} E\Big(X(t_i)E\left(B(t_{i+1}) - B(t_i)|\mathcal{F}_{t_i}\right)|\mathcal{F}_s\Big)\\
&= \sum_{i=0}^{n-1} E\Big(X(t_i)E\left(B(t_{i+1}) - B(t_i)\right)|\mathcal{F}_s\Big) = 0,
\end{aligned}$$

where in the second equation we use the fact that $\mathcal{F}_s \subset \mathcal{F}_{t_i}$ and the tower property of conditional expectation (4.19), in the third equation, the fact that $X(t_i)$ is $\mathcal{F}_{t_i}$-measurable, and in the fourth equation, that $B(t_{i+1}) - B(t_i)$ is independent of $\mathcal{F}_{t_i}$.

Next, we have

$$\begin{aligned}
E\left(Y(t)|\mathcal{F}_s\right) &= E\left(\int_0^s X(u)dB(u) + \int_s^t X(u)dB(u)|\mathcal{F}_s\right)\\
&= E\left(\int_0^s X(u)dB(u)|\mathcal{F}_s\right) + E\left(\int_s^t X(u)dB(u)|\mathcal{F}_s\right)\\
&= E\left(\int_0^s X(u)dB(u)|\mathcal{F}_s\right) = \int_0^s X(u)dB(u) = Y(s),
\end{aligned}$$

where in the second but last step, we have used the fact that $\int_0^s X(u)dB(u)$ is $\mathcal{F}_s$-measurable and the conditional expectation property (4.12). □

4.1.2.3 Itô's Formula for Functions of Stochastic Processes

An Itô process $X_t(\omega) \equiv X(t, \omega)$ is defined by the following stochastic differential

$$dX(t) = \mu(t)dt + \sigma(t)dB(t), \tag{4.26}$$

which is to be understood in the sense of an Itô integral due to the nondifferentiability of Brownian path $B(t)$,

$$X(t) = X(0) + \int_0^t \mu(s)ds + \int_0^t \sigma(s)dB(s), \tag{4.27}$$

where the first integral is a Riemann integral and the second integral is an Itô integral.

Next, we can define a new stochastic process $Y(t, \omega)$ with any $C^{1,2}$ function $f(t, x)$ as follows

$$Y_t(\omega) \equiv Y(t, \omega) = f(t, X(t, \omega)), \tag{4.28}$$

for which the differential for $Y(t, \omega)$ is given by the following Itô's formula [6, 7]

$$\begin{aligned} dY_t = df(t, X_t) &= \left(\frac{\partial}{\partial t} + \mu(t)\frac{\partial}{\partial x} + \frac{\sigma^2(t)}{2}\frac{\partial^2}{\partial x^2}\right) f(t, X_t)dt \\ &+ \frac{\partial}{\partial x} f(t, X_t)\sigma(t)dB(t). \end{aligned} \tag{4.29}$$

By defining a generator $\mathcal{L}$ for the Itô process $X_t(\omega)$ as

$$\mathcal{L} = \mu(t)\frac{\partial}{\partial x} + \frac{\sigma^2(t)}{2}\frac{\partial^2}{\partial x^2}, \tag{4.30}$$

the Itô's formula (4.29) can be written in the following compact form

$$df(t, X_t) = \left(\frac{\partial}{\partial t} + \mathcal{L}\right) f(t, X_t)dt + \frac{\partial}{\partial x} f(t, X_t)\sigma(t)dB(t). \tag{4.31}$$

A d-dimensional Ito process $X \in \mathbb{R}^d$ driven by a d-dimensional independent Brownian motion $B(t) = (B_1(t), \ldots, B_d(t))^\intercal \in \mathbb{R}^d$ is defined by the following differential

$$dX(t) = \mu(t)dt + \sigma(t)dB(t), \tag{4.32}$$

with a drift vector $\mu(t) \in \mathbb{R}^d$ and a $d \times d$ diffusion matrix σ, and its generator will be defined by

$$\mathcal{L} = \mu \cdot \nabla + \frac{1}{2} \sum_{i,j=1}^{d} a_{i,j} \frac{\partial^2}{\partial x_i \partial x_j} = \mu \cdot \nabla + \frac{1}{2} \mathrm{Tr}(A \nabla \nabla^{\mathsf{T}}), \tag{4.33}$$

where the $d \times d$ diffusion coefficient matrix $A = (a_{i,j}) = \sigma \sigma^{\mathsf{T}}$.

Then, the Itô's formula for $Y(t, \omega) = f(t, X_1, X_2, \ldots, X_d)$ with $f \in C^{1,2}$ will take the same form as (4.31) with the generator $\mathcal{L}$ given in (4.33) and $\frac{\partial}{\partial x}$ replaced by ∇.

4.2 Feynman–Kac Formula for Mixed BVPs

4.2.1 Dynkin's Formula and Feynman–Kac Formula for Dirichlet BVPs of Elliptic PDEs

4.2.1.1 Dynkin's Formula

Let $X(t) \in \mathbb{R}^d$ be an Itô diffusion defined by the following stochastic differential equation

$$dX(t) = \mu(t, X(t))dt + \sigma(t, X(t))dB(t), \tag{4.34}$$

with a generator $\mathcal{L}$ given by (4.33) with spatial-temporal dependent coefficients. For any given $C^{1,2}$ function $f(t, x)$, we define the following process

$$M_f(t) = f(t, X(t)) - \int_0^t \left(\mathcal{L}_u f + \frac{\partial f}{\partial u} \right) (u, X(u))du. \tag{4.35}$$

Theorem 4.2 *$M_f(t)$ is a martingale.*

Proof By using the Itô's formula (4.31), we have

$$\begin{aligned} df(t, X(t)) = &\left(\mathcal{L}_t f(t, X(t)) + \frac{\partial f}{\partial t}(t, X(t)) \right) dt \\ &+ \nabla f(t, X(t))\sigma(t, X(t))dB(t), \end{aligned}$$

and, by integrating on both sides, we obtain

$$\begin{aligned} &f(t, X(t)) - \int_0^t \left(\mathcal{L}_u f + \frac{\partial f}{\partial u} \right) (u, X(u))du \\ &= f(0, X(0)) + \int_0^t \nabla f(u, X(u))\sigma(u, X(u))dB(u), \end{aligned}$$

thus,

$$M_f(t) = f(0, X(0)) + \int_0^t \nabla f(u, X(u))\sigma(u, X(u))dB(u),$$

which is a martingale from (4.23). □

Theorem 4.3 (Dynkin's formula) *For any $C^{1,2}$ function $f(t, \boldsymbol{x})$, we have following identity*

$$Ef(t, X(t)) = f(0, X(0)) + E\int_0^t \left(\mathcal{L}_u f + \frac{\partial f}{\partial u}\right)(u, X(u))du. \tag{4.36}$$

Proof As $M_f(t)$ is a martingale, we have for $0 \le s \le t$

$$E(M_f(t)|\mathcal{F}_s) = M_f(s),$$

and, by taking the expectation on both sides, we have

$$E(E(M_f(t)|\mathcal{F}_s)) = E(M_f(s)).$$

Now, using the double expectation property (4.20), we have

$$E(M_f(t)) = E(M_f(s)),$$

thus, we have a constant mean for the martingale, i.e.,

$$E(M_f(t)) = E(M_f(0)),$$

namely,

$$Ef(t, X(t)) - E\int_0^t \left(\mathcal{L}_u f + \frac{\partial f}{\partial u}\right)(u, X(u))du = f(0, X(0)),$$

thus, the Dynkin's formula (4.36) holds. □

Theorem 4.4 (Dynkin's formula with a stopping time [5]) *Let a r. v. τ be a stopping time, i.e., $\{\omega, \tau(\omega) \le t\} \in \mathcal{F}_t, t > 0$, then, the following identity holds*

$$Ef(\tau, X(\tau)) = f(0, X(0)) + E\int_0^\tau \left(\mathcal{L}_u f + \frac{\partial f}{\partial u}\right)(u, X(u))du. \tag{4.37}$$

4.2.1.2 Feynman–Kac Formula for BVPs with Dirichlet Data

Consider the BVPs of the following elliptic operator $\mathcal{L}$ of the form (4.33) with $\mu = \mu(\mathbf{x}), \sigma = \sigma(\mathbf{x})$,

$$\begin{cases} \mathcal{L}u = -g, \mathbf{x} \in D \\ u|_{\partial D} = \phi \end{cases}, \tag{4.38}$$

or

$$\begin{cases} \mathcal{L}u + qu = -g, \mathbf{x} \in D \\ u|_{\partial D} = \phi \end{cases}. \tag{4.39}$$

A probabilistic representation of solution u can be derived using the Dynkin's formula for the exit time of Itô diffusion.

First, we define the exit time of Itô diffusion of a domain of an open set D as

$$\tau_D(\omega) = \inf\{t, X(t) \notin D\}. \tag{4.40}$$

Theorem 4.5 *τ_D is a stopping time, namely, the set $\{\tau_D \le t\} \in \mathcal{F}_t, \forall t > 0$.*

Proof Consider a collection of increasing nested closed sets $K_m \subset K_n$, $m \le n$ such that $D = \bigcup_m K_m$. To show that τ_D is a stopping time, we can use the following identity [8]

$$\{\omega|\tau_D(\omega) \le t\} = \bigcap_m \bigcup_{s<t} \{\omega|X_s(\omega) \notin K_m\}, \tag{4.41}$$

together with the fact that $\{\omega \mid X_s(\omega) \notin K_m\} \in \mathcal{F}_s \subset \mathcal{F}_t$, and here, s will be taken to be the rational number for countable sets.

To show that (4.41) holds, we will prove the mutual inclusion of the left- and right-hand sides of (4.41).

On the one hand, let $\omega \in \{\omega \mid \tau_D(\omega) \le t\}$, then for each m, there exists an $s < t$ such that

$$X_s(\omega) \notin K_m, X_s(\omega) \in D,$$

thus

$$\omega \in \bigcup_{s<t} \{\omega|X_s(\omega) \notin K_m\},$$

therefore,

$$\omega \in \bigcap_m \bigcup_{s<t} \{\omega|X_s(\omega) \notin K_m\}.$$

On the other hand, let us assume that

$$\omega \in \bigcap_m \bigcup_{s<t} \{\omega|X_s(\omega) \notin K_m\}. \tag{4.42}$$

We will show that $\omega \in \{\omega \mid \tau_D(\omega) \leq t\}$. If not, we will have $\tau_D(\omega) \geq t + \varepsilon > t$ for some positive ε, which will result in the assumption (4.42) being invalid as shown below, thus this is a contradiction.

Now, with $\tau_D(\omega) \geq t + \varepsilon$, we can show that there exists an m that $X_{[0,t)}(\omega) \in K_m$ (see Fig. 4.1). If no such K_m exists, there will be a sequence of time $t_n \in [0, t)$ such that $\lim_{n\to\infty} \text{dist}(X(t_n) - \partial D) = 0$. We can find a convergent subsequence $\{t_{n'}\} \subset \{t_n\}$ such that $t_{n'} \to t^* \in [0, t]$, which implies that $\tau_D(\omega) \leq t^* \leq t$, contradicting $\tau_D(\omega) \geq t + \varepsilon$.

Thus, there exists a K_m such that $X_{[0,t)}(\omega) \in K_m$, then,

$$\omega \in \bigcap_{s<t} \{\omega \mid X_s(\omega) \in K_m\}$$

$\iff \omega \notin \left(\bigcap_{s<t}\{\omega \mid X_s(\omega) \in K_m\}\right)^c = \bigcup_{s<t}\{\omega \mid X_s(\omega) \notin K_m\}$, which means that $\omega \notin \bigcap_m \bigcup_{s<t}\{\omega|X_s(\omega) \notin K_m\}$, a contradiction to the assumption (4.42). Therefore, we must have

$$\omega \in \{\omega \mid \tau_D(\omega) \leq t\}.$$

□

Next, we present the Feynman–Kac formulae for the probabilistic solution of BVPs (4.38) and (4.39).

Theorem 4.6 *Assuming that the generator $\mathcal{L}$ is uniformly elliptic with uniform Lipschitz continuous coefficients and the boundary of D belongs to C^2, then, the solution to (4.38) is given by*

$$u(\boldsymbol{x}) = E\phi(X(\tau_D) + E\int_0^{\tau_D} g(X(u))du, \ \forall \boldsymbol{x} \in D, \tag{4.43}$$

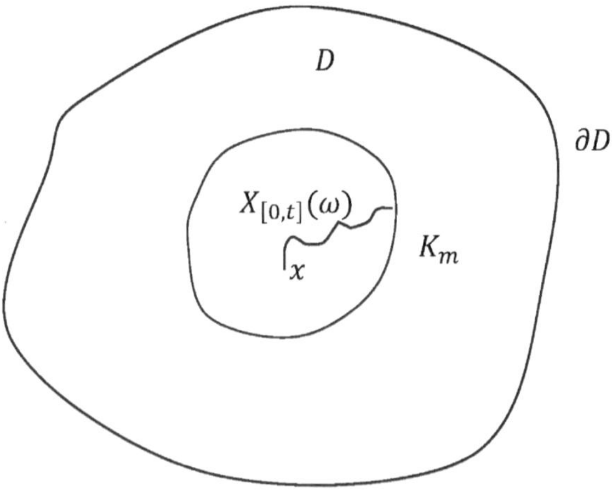

Fig. 4.1 A Brownian path $X_t(\omega)$, $X_0 = x$ within a set K_m

where the Itô diffusion $X(t)$ is the solution to (4.34) with an initial condition $X(0) = \boldsymbol{x}$.

Proof To prove, we set $f(t, \mathbf{x}) = u(\mathbf{x})$ in (4.37), due to the fact $\frac{\partial f}{\partial t} = 0$ and the PDE and the boundary condition in (4.37), we arrive at (4.43). □

For PDEs with the zero-th order term in (4.39), we have the following probabilistic representation of their solutions [9].

Theorem 4.7 (Feynman–Kac formula) *With similar conditions for the differential operator $\mathcal{L}$ as above and for positive Holder continuous coefficient q in $\overline{D}$, the solution to (4.39) is given by*

$$u(\boldsymbol{x}) = E\left(e_q(\tau_D)\phi(X(\tau_D))\right) + E\int_0^{\tau_D} e_q(u)g(X(u))du, \ \forall \boldsymbol{x} \in D, \tag{4.44}$$

where the gauge function $e_q(t)$ is defined as

$$e_q(t) = e^{\int_0^t q(X(s))ds}. \tag{4.45}$$

4.2.2 *Local Time $L(t)$ and Feynman–Kac Formulae for Neumann and Robin BVPs of the Schrödinger Equation*

4.2.2.1 Local Time $L(t)$ of Reflecting Brownian Motion

To formulate probabilistic solutions for Neumann and Robin BVPs of the Schrödinger equation, a reflecting Brownian motion (RBM), which undergoes specular reflections upon hitting a domain boundary, and its local time, as a measure of the frequency of such reflections will be needed. We introduce the boundary local time $L(t)$ for an RBM through a Skorokhod problem.

Skorokhod problem Assume that D is a bounded domain in $\mathbb{R}^d$ with a C^2 boundary. Let $f(t)$ be a continuous path in $\mathbb{R}^d$ with $f(0) \in \bar{D}$. A pair $(\xi(t), L(t))$ is a solution to the Skorokhod problem $S(f; D)$ if the following conditions are satisfied:

1. $\xi(t)$ is a path in $\bar{D}$;
2. $L(t)$ is a nondecreasing function that increases only when $\xi \in \partial D$, namely,

$$L(t) = \int_0^t I_{\partial D}(\xi(s))L(ds), \tag{4.46}$$

3. The Skorokhod equation holds,

$$S(f;D): \qquad \xi(t) = f(t) - \int_0^t \mathbf{n}(\xi(s))L(ds), \tag{4.47}$$

where $\mathbf{n}(\mathbf{x})$ is the outward unit normal vector at $\mathbf{x} \in \partial D$.

In the case where $f(t) = B_t$, the corresponding $\xi(t)$ will be the RBM X_t. As the name suggests, an RBM behaves like a Brownian motion as long as its path remains inside the domain D, but it will be reflected back inwardly along the normal direction of the boundary when the path attempts to pass through the boundary. The fact that X_t is a diffusion process can be proven by using martingale formulation and showing that X_t is the solution to the corresponding martingale problem with the Neumann boundary condition [10].

The boundary local time $L(t)$ for an RBM measures the amount of time the RBM spends near the boundary and at the same time the frequency that the RBM hits the boundary. Numerical methods for simulating the RBM and computing its local time will be given in Sect. 4.4.2 [11–13].

4.2.2.2 Feynman–Kac Formulae for Neumann and Robin BVPs of the Schrödinger Equation

We will consider the solution $u(\mathbf{x})$ of the following mixed BVPs of the Schrödinger equation,

$$\begin{aligned} \frac{1}{2}\Delta u + qu &= -f(\mathbf{x}), \quad && \mathbf{x} \in D, \\ u &= \phi_1(\mathbf{x}), && \text{on } \Lambda_1, \\ \frac{\partial u}{\partial \mathbf{n}} &= \phi_2(\mathbf{x}), && \text{on } \Lambda_2, \\ \frac{\partial u}{\partial \mathbf{n}} - cu &= \phi_3(\mathbf{x}), && \text{on } \Lambda_3, \end{aligned} \tag{4.48}$$

where $\mathbf{n}$ is the outward normal on the boundary ∂D, and $\partial D = \Lambda_1 \cup \Lambda_2 \cup \Lambda_3$ is assumed to be C^2, and the boundary data ϕ_i are assumed to be C^0. A probabilistic representation of $u(\mathbf{x})$ for the mixed boundary value problem of the Schrödinger equation is given by the following Feynman–Kac formula [9, 10, 14],

$$\begin{aligned} u(\mathbf{x}) = E\left(\hat{e}_c(\tau_{\Lambda_1})e_q(\tau_{\Lambda_1})\phi_1(X_{\tau_{\Lambda_1}})\right) + E\int_0^\infty \hat{e}_c(t)e_q(t)\phi_{2,3}(X_t)dL(t) \\ + E\int_0^\infty \hat{e}_{-c}(t)e_q(t)f(X_t)dt, \quad \mathbf{x} \in D, \end{aligned} \tag{4.49}$$

where X_t is a partially reflecting Brownian motion, $X_0 = \mathbf{x}$, whose path will terminate once it hits the Dirichlet boundary Λ_1, and $L(t)$ is the corresponding local time, and the Feynman–Kac functional

$$\hat{e}_c(t) := e^{\int_0^t c(X_s)dL(s)}. \tag{4.50}$$

The Feynman–Kac formula (4.49) can be extended for the mixed BVP of general linear elliptic operators [15] while the reflection of the diffusion process will be oblique on the boundary with respect to the direction $A\mathbf{n}$ and the Skorokhod equation (4.47) will also be modified, accordingly [16].

4.3 Pardoux–Peng's Nonlinear Feynman–Kac Formula for Quasilinear Parabolic PDEs

To develop deep learning algorithms for solving nonlinear parabolic equations in high dimensions, in Chap. 6. Here, we present the probabilistic representation for the terminal value problem solution $u(t, \mathbf{x})$, $t \in [0, T]$, $\mathbf{x} \in \mathbb{R}^d$ for the following d-dimensional quasilinear parabolic PDE,

$$\partial_t u + \mathcal{L}(u) = \phi(\mathbf{x}), \quad \mathbf{x} \in \mathbb{R}^d \tag{4.51}$$

with a terminal condition at $t = T$,

$$u(T, \mathbf{x}) = g(\mathbf{x}), \tag{4.52}$$

where $\mathcal{L}$ is given by (4.33) but with $\sigma = \sigma(t, \mathbf{x}, u)$, $\mu = \mu(t, \mathbf{x}, u, \nabla u)$, and $\phi = \phi(t, \mathbf{x}, u, \nabla u)$, which are functions with ranges in spaces with dimensions $d \times d$, d and 1, respectively. Our problem is to find the value $u(0, \mathbf{x}_0)$ at $t = 0$ for $\mathbf{x}_0 \in \mathbb{R}^d$. Therefore, in some sense, the problem is equivalent to a time reverse problem for a time reversed version of (4.51) with an initial data at $t = 0$.

According to the Pardoux–Peng theory [17], under certain regularity conditions, a forward–backward SDE (FBSDE) reformulation gives a nonlinear implicit Feynman–Kac formula for the solution of the parabolic PDE (4.51).

Let $B_t = (B_1(t), \ldots, B_d(t))$ where each $B_j(t)$ is a Brownian motion, and $\{\mathcal{F}_t : 0 \le t \le T\}$ its natural filtration on the time interval $[0, T]$, similarly defined as in (4.21). Then, we have the following stochastic equations for processes X_t, Y_t and Z_t in d, 1 and d dimensions that are adaptive to the filtration $\{\mathcal{F}_t : 0 \le t \le T\}$, respectively,

$$\begin{aligned} dX_t &= \mu(t, X_t, Y_t, Z_t)dt + \sigma(t, X_t, Y_t)dB_t, \\ X_0 &= \mathbf{x}_0, \end{aligned} \tag{4.53}$$

$$\begin{aligned} dY_t &= \phi(t, X_t, Y_t, Z_t)dt + Z_t^T \sigma(t, X_t, Y_t)dB_t, \\ Y_T &= g(X_T). \end{aligned} \tag{4.54}$$

If μ and σ do not explicitly depend on Y_t or Z_t, the FBSDEs are *decoupled*. It was proven [17] that processes defined by

$$Y_t = u(t, X_t), \quad Z_t = \nabla u(t, X_t) \tag{4.55}$$

in fact satisfy the above equations (4.53) and (4.54). This can be seen formally by using the Ito's formula (4.31) and the forward SDE of X_t (4.53), i.e.,

$$\begin{aligned} dY_t = du(t, X_t) &= (\partial_t u + \mathcal{L}(u))\, dt + \nabla^T u \sigma dB_t \\ &= (\partial_t u + \mathcal{L}(u))\, dt + Z_t^T \sigma dB_t, \end{aligned} \tag{4.56}$$

which implies the PDE (4.51) by comparing (4.56) with the backward SDE (4.54) for Y_t.

- Connection with the classic Feynman–Kac formula [18].

There is a connection between the FBSDE representation of parabolic equations and the classic Feynman–Kac formula. In the parabolic PDE (4.51), if ϕ has a linear dependence on u, i.e.,

$$\phi(t, \mathbf{x}, u, \nabla u) = c(t, \mathbf{x})u(t, \mathbf{x}) + f(t, \mathbf{x}), \tag{4.57}$$

then, the backward SDE (4.54) has an explicit solution

$$\begin{aligned} Y_t = e^{-\int_t^T c(s, X_s)ds} g(X_T) &- \int_t^T e^{-\int_t^s c(\tau, X_\tau)d\tau} f(s, X_s)ds \\ &- \int_t^T e^{-\int_t^s c(\tau, X_\tau)d\tau} Z_s^T \sigma(s, X_s, Y_s)dB_s. \end{aligned} \tag{4.58}$$

By taking the conditional expectation on both sides and using the martingale property of Itô integrals (4.24), we arrive at

$$Y_t = E\left(e^{-\int_t^T c(s, X_s)ds} g(X_T) - \int_t^T e^{-\int_t^s c(\tau, X_\tau)d\tau} f(s, X_s)ds \,\middle|\, \mathcal{F}_t \right). \tag{4.59}$$

For $(t, \mathbf{x}) \in [0, T] \times \mathbb{R}^d$, using $X_t = \mathbf{x}$ as the initial condition of the forward SDE (4.53) on the time interval $[t, T]$, the traditional Feynman–Kac formula for the initial value problem of a parabolic equation [6] is then recovered,

$$u(t, \mathbf{x}) = E\left(e^{-\int_t^T c(s,X_s)ds} g(X_T) - \int_t^T e^{-\int_t^s c(\tau,X_\tau)d\tau} f(s, X_s)ds \,\middle|\, X_t = \mathbf{x} \right). \quad (4.60)$$

For a general parabolic equation with a nonlinear function $\phi(t, \mathbf{x}, u, \nabla u)$, we have

$$Y_t = E\left(g(X_T) - \int_t^T \phi(s, X_s, Y_s, Z_s)ds \,\middle|\, \mathcal{F}_t \right),$$

and for given $(t, \mathbf{x}) \in [0, T] \times \mathbb{R}^d$, the following nonlinear integral equation for $u(t, \mathbf{x})$ is obtained

$$u(t, \mathbf{x}) = E\left(g(X_T) - \int_t^T \phi(s, X_s, u(s, X_s), \nabla u(s, X_s))ds \,\middle|\, X_t = \mathbf{x} \right). \quad (4.61)$$

4.4 Walk-On-Spheres (WOS) and Local Time $L(t)$ for Reflecting Brownian Motion

In this section, and throughout the rest of the chapter, we will discuss the walk-on-spheres method for solving mixed boundary value problems (BVPs) of the Laplace equation.

4.4.1 WOS Method and Dirichlet BVP

The random walk-on-spheres method was first proposed by Müller [19] to simulate the exit time of a Brownian motion for the Dirichlet BVP problem of the Laplacian operator with the probabilistic representation of the solution (4.43). It was further developed in [3, 20, 21], for Neumann and Robin BVPs [11–13, 22], as well as for the fractional Laplacian operator [4].

To illustrate how the WOS method can be used to implement (4.43) for the Dirichlet problem (4.38), formula (4.43) can be rewritten in terms of a harmonic measure of boundary hitting probability μ_D^x with a density $p(\mathbf{x}, \mathbf{y}), \mathbf{y} \in \partial D$ for each $\mathbf{x} \in D$ defined by

$$\mu_D^x(F) = P^x \left\{ X_{\tau_D} \in F, X_0 = \mathbf{x} \right\} = \int_F p(\mathbf{x}, \mathbf{y}) ds_y, \quad F \subset \partial D, \tag{4.62}$$

thus, (4.43) becomes

$$u(\mathbf{x}) = E^x(\phi(X_{\tau_D})) = \int_{\partial D} \phi(\mathbf{y}) d\mu_D^x = \int_{\partial D} \phi(\mathbf{y}) p(\mathbf{x}, \mathbf{y}) ds_y. \tag{4.63}$$

Using the integral representation (3.39) of the solution of the Laplace equation via a homogeneous Green's function $G_{\rm h}(\mathbf{x}, \mathbf{y})$ of the domain, it is clear that the harmonic measure density is in fact given by [23],

$$p(\mathbf{x}, \mathbf{y}) = -\frac{\partial G_{\rm h}(\mathbf{x}, \mathbf{y})}{\partial \mathbf{n}_y}, \tag{4.64}$$

where the homogeneous Green's function of the domain $G_{\rm h}(\mathbf{x}, \mathbf{y})$ satisfies for each $\mathbf{y} \in D$

$$\begin{cases} -\Delta G_{\rm h}(\mathbf{x}, \mathbf{y}) = \delta(\mathbf{x} - \mathbf{y}), & \mathbf{x} \in D, \\ G_{\rm h}(\mathbf{x}, \mathbf{y}) = 0, & \mathbf{x} \in \partial D. \end{cases} \tag{4.65}$$

Thus, if the starting point $\mathbf{x}$ of a Brownian motion is at the center of a ball, the probability of the Brownian motion exiting a portion of the boundary of the ball will be proportional to the portion's area. It is known that all sample paths of a Brownian motion starting in the domain intersects the boundary ∂D almost surely [19]. Sampling a Brownian path with a series of balls within the domain, regardless of how the path navigates in the interior of the ball, can significantly reduce the path sampling time. To be more specific, given a starting point $\mathbf{x}$ inside the domain D, we simply draw a ball of the largest possible radius fully contained in D, and then the next location of the Brownian path on the surface of the ball can be sampled, using a uniform distribution on the sphere, for example, at $\mathbf{x}_1$. We then treat $\mathbf{x}_1$ as the new starting point, draw a second largest possible ball fully contained in D, and jump from $\mathbf{x}_1$ to $\mathbf{x}_2$ on the surface of the second ball as before. Repeat this procedure until the path hits an absorption ϵ-shell of the domain [3, 20]. When this happens, we decide that the path has hit the boundary ∂D (see Fig. 4.2 for an illustration).

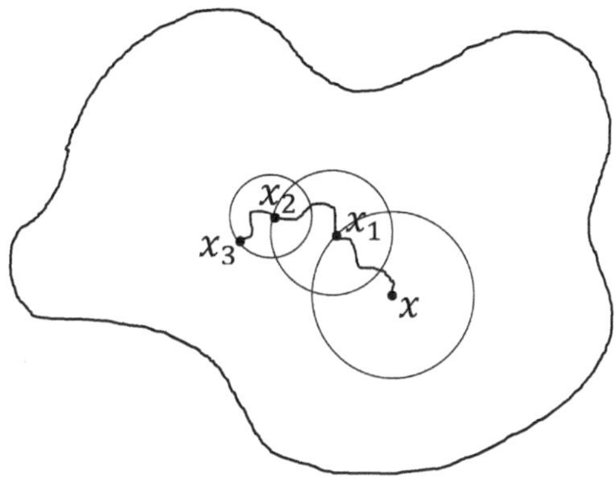

Fig. 4.2 The walk-on-spheres (WOS) method

Next we define an estimator of (4.63) by

$$u(\mathbf{x}) \approx \frac{1}{N}\sum_{i=1}^{N}\phi(\mathbf{x}_i), \tag{4.66}$$

where N is the number of Brownian paths sampled and $\mathbf{x}_i$ is the first hitting point of the ith path on the boundary. Using a jump size (radius of the ball) δ on each step for the WOS, an $O(1/\delta^2)$ (δ is chosen independently of ϵ) number of steps will be needed for a Brownian path to reach the boundary [24].

4.4.2 Reflecting Brownian Motion and Its Local Time $L(t)$

In this section, we will apply the WOS method to sample the path of the RBM and compute its local time $L(t)$, which is required for the Neumann and Robin BVPs.

Levy proposed [25] that the local time can be computed through the following limiting process

$$L(t) \equiv \lim_{\epsilon \to 0} \frac{\int_0^t I_{D_\epsilon}(X_s)ds}{2\epsilon}, \tag{4.67}$$

where D_ϵ is the ϵ-shell region of width ϵ containing ∂D and $D_\epsilon \subset D$ as shown in Fig. 4.3 (left). This limit exists both in L^2 and P^x-$a.s.$ for any $x \in \overline{D}$ [25].

A reflecting Brownian motion path is simulated by the WOS method inside D. Once it enters the D_ϵ-region, we continue the WOS in D_ϵ but with a fixed radius Δx (in a fraction of ϵ e.g. 1/4–1/6). To simulate the path of the RBM, at some point in time, the Brownian path will run out of the domain. For this to happen, the radius of WOS increases to $2\Delta x$ when the path is close to the boundary at a distance less than Δx. In this way, the Brownian path will have a chance to escape the domain, and when that happens, we then project it back to the nearest point on the boundary along

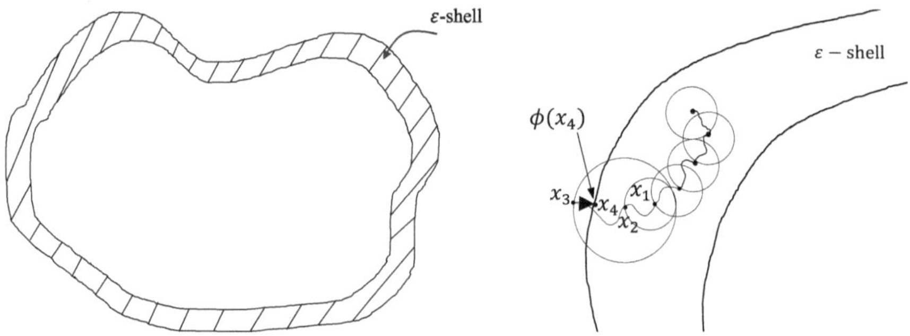

Fig. 4.3 (Left) A D_ϵ-region for a bounded domain in $\mathbb{R}^3$, (Right) The WOS inside the D_ϵ-region

the normal of the boundary. Afterwards, the Brownian motion path will continue, completing a reflection from the boundary. After the reflection, the radius of WOS changes back to Δx for the next step of the WOS. Figure 4.3 (right) illustrates the movement of the RBM in the ϵ-shell region D_ϵ.

To compute the local time, we first recall the fact [12] that the elapsed time Δt for a random walk step on average is proportional to the square of the step size, in fact,

$$<\Delta t> = \frac{(\Delta x)^2}{3}. \tag{4.68}$$

We keep track of a vector n_i of the WOS steps as well as each entry (t_{in}) and exit time (t_{out}) of the path into the D_ϵ region, and then allocate the local time increment for each possible collision with the boundary within this time period. Namely, to compute δn_{t_j} corresponding to the collision points X_{t_j}, $j = 1, \ldots, m_j, m_j \geq 0$, we record the increment of the WOS steps $n_{j,out} - n_{j,in}$ between one entry into and one exit out of the D_ϵ region by a path, and the number of collisions m_j at the boundary during this sojourn into the D_ϵ region (see Fig. 4.4). Then, using the occupation time definitions in (4.67) and (4.68), we set

$$dL(t_j) = \frac{\delta n_{t_j} \Delta t}{2\epsilon} = \delta n_{t_j} \frac{(\Delta x)^2}{6\epsilon}, \; j = 1, \ldots, m_j, \tag{4.69}$$

where

$$\delta n_{t_j} = \begin{cases} \frac{n_{j,out} - n_{j,in}}{m_j}, & \text{if } m_j > 0 \\ 0 \quad, & \text{if } m_j = 0 \end{cases}. \tag{4.70}$$

This definition will give a zero local time increment if there is no collision during the time period $[t_{\text{in}}, t_{\text{out}}]$, i.e. $m_j = 0$, which is consistent with the condition (4.46) in the Skorokhod problem, allowing an increment in $L(t)$ only when the path hits the boundary.

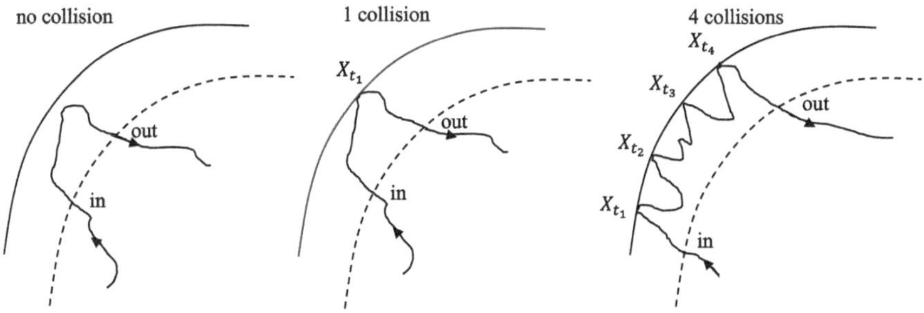

Fig. 4.4 Three situations of the path walking in the D_ϵ-region. (left) Number of hits, $m = 0$. (middle) Number of hits, $m = 1$ (right) Number of hits, $m = 4$

4.5 Electrical Impedance Tomography (EIT)

In this section, we will apply the WOS method and the probabilistic solution of BVP problems in (4.49) as a path integral Monte Carlo method for the mixed boundary value problem in the forward problem in the EIT.

4.5.1 *Feynman–Kac Formula for Mixed BVPs*

First, the time interval in the Feynman–Kac formula (4.49) is truncated to $[0, T]$ and the number of WOS steps for each sample path is set to be N_{path}, the WOS method (4.49) for the mixed BVPs of the Laplace equation (i.e., $f = 0, q = 0$) will be

$$\tilde{u}_{\text{mix}}(\mathbf{x}) = E\left(\int_0^T \hat{e}_c(t)\phi_{2,3}(X_t)dL(t)\right) + E\left(\hat{e}_c(\tau_{\Gamma_1})\phi_1(X_{\tau_{\Gamma_1}})\right), \tag{4.71}$$

or

$$\tilde{u}_{\text{mix}}(\mathbf{x}) = E\left(\sum_{j'=0}^{N_{\text{path}}} \hat{e}_c(t_j)\phi_{2,3}(X_{t_j})dL(t_j)\right) + E\left(\hat{e}_c(\tau_{\Gamma_1})\phi_1(X_{\tau_{\Gamma_1}})\right). \tag{4.72}$$

Equivalently,

$$\tilde{u}_{\text{mix}}(\mathbf{x}) = E\left(\sum_{j:X_{t_j}\in\Gamma_2\cup\Gamma_3}^{N_{\text{path}}} \hat{e}_c(t_j)\phi_{2,3}(X_{t_j})dL(t_j)\right) + E\left(\hat{e}_c(\tau_{\Gamma_1})\phi_1(X_{\tau_{\Gamma_1}})\right), \tag{4.73}$$

where j' denotes each WOS step of the path, and j denotes the steps where the reflecting boundary is hit by the path. The approximation of the Feynman–Kac functional $\hat{e}_c(t)$ is computed by

$$\hat{e}_c(t) \approx e^{\sum_{t_k \leq t} c(X_{t_k}) dL(t_k)}, \quad t = t_j \text{ or } \tau_{\Gamma_1}, \tag{4.74}$$

where $t_k : X_{t_k} \in \Gamma_3$, i.e. k denotes the steps where the path hits the Robin boundary before time t.

Therefore, finally, with the local time increment (4.69), we have the approximation to the probabilistic solution for the Laplace equation (4.48) with $q = 0$, $f = 0$,

$$\tilde{u}_{\text{mix}}(\mathbf{x}) = E\left(\sum_{j'=0}^{N_{\text{path}}} e^{\sum_{k=0}^{j} c(X_{t_k})\delta n_{t_k} \frac{(\Delta x_{j'})^2}{6\epsilon}} \phi_{2,3}(X_{t_j})\delta n_{t_j} \frac{(\Delta x_{j'})^2}{6\epsilon}\right)$$
$$+ E\left(e^{\sum_{t_k < \tau_{\Gamma_1}} c(X_{t_k})\delta n_{t_k} \frac{(\Delta x_{j'})^2}{6\epsilon}} \phi_1(X_{\tau_{\Gamma_1}})\right). \tag{4.75}$$

4.5.2 Local EIT Solution with the WOS Method

In this section, we will apply the WOS method to solve the forward problem in inverse electrical impedance tomography [11]. As only solutions for electrodes are needed in the forward problem, the WOS method provides a local solution approach instead of costly global methods such as finite element or whole domain-boundary element methods.

As shown in Fig. 4.5, the domain of the object for the EIT problem is denoted as Ω, embedded within which there is an anomaly $\Omega_0 \subset \Omega$. The domain Ω is assumed to have a smooth boundary with a limited number of electrodes E_l, $l = 1, \ldots, L$ attached

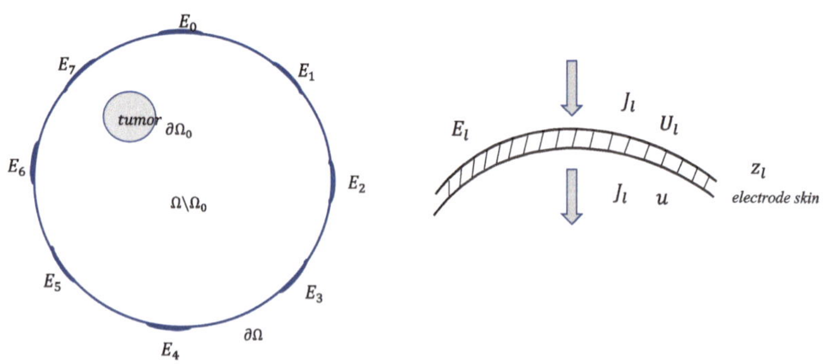

Fig. 4.5 A complete electrode model: (left) domain $\Omega\backslash\Omega_0$, (right) The contact impedance z_l

to $\partial\Omega$. The scalar electric potential u for the complete electrode model [26] satisfies the following boundary value problem

$$\nabla \cdot \gamma \nabla u = 0, \qquad \text{in } \Omega \backslash \Omega_0, \tag{4.76a}$$

$$\frac{1}{|E_l|} \int_{E_l} \gamma \frac{\partial u}{\partial \mathbf{n}} dS = J_l, \quad \text{on } E_l, \quad l = 1, 2, \ldots, L, \tag{4.76b}$$

$$\gamma \frac{\partial u}{\partial \mathbf{n}} = 0, \qquad \text{on } \partial\Omega \backslash \cup_{l=1}^{L} E_l, \tag{4.76c}$$

$$u + z_l \gamma \frac{\partial u}{\partial \mathbf{n}} = U_l, \quad \text{on } E_l, \quad l = 1, 2, \ldots, L, \tag{4.76d}$$

$$\sum_{l=1}^{L} |E_l| J_l = 0, \tag{4.76e}$$

$$u = u_0 \qquad \text{on } \partial\Omega_0, \tag{4.76f}$$

where γ is the conductivity inside $\Omega \backslash \Omega_0$. J_l is the current density sent to the lth electrode with surface impedance z_l and U_l is the corresponding measured voltage. $|E_l|$ is the area of the lth electrode and $\mathbf{n}$ denotes the outer unit normal vector on $\partial\Omega$.

The six equations (4.76a)–(4.76f) constitute the elliptic PDE and the boundary conditions of the domain $\Omega \backslash \Omega_0$. In the inverse EIT problem, we deduce that the interior of the object, such as the presence or absence of a tumor and its location, through the current–voltage pairs (J_l, U_l) on the electrodes. Thus, solving the forward problem is equivalent to obtaining the voltage-to-current map given a known internal configuration, namely, solving a mixed BVP consisting of (4.76a), (4.76c), (4.76d), and (4.76f). The WOS method of (4.75) can produce accurate potentials on each electrode and Fig. 4.6 shows the computed potential within a 1% relative error on the electrode E_3 with or without the tumor [11].

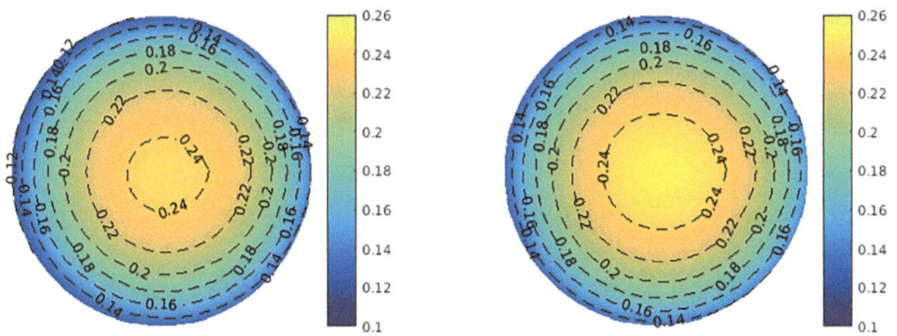

Fig. 4.6 The electric potential on the electrode E_3 by the WOS method (left) no tumor, (right) with tumor at (0.183,0.213,0.586)

4.6 Boundary Integral Equation and WOS (BIE-WOS) Method for the Dirichlet BVP of the Laplace Equation

The Feynman–Kac formula provides only a single location solution for BVPs with an intrinsic parallelism in the sampling of random paths. In this and subsequent sections, we will develop a hybrid method, which combines a deterministic local integral equation and the WOS method for finding the global solution of the Laplace equation, i.e., Eq. (4.38) with $\mathcal{L} = \frac{1}{2}\Delta,\ g = 0$. The resulting method, the BIE-WOS method [27], will retain both the parallelism and the capability of handling complex geometry of a Monte Carlo method.

For the exterior Dirichlet boundary value problem, with the integral representation of the solution of the Laplace equation in (3.39), we only need to find the normal derivative of the solution on the boundary (Neumann data) from the given boundary value of the solution (Dirichlet data), i.e., the DtN map (Dirichlet to Neumann map).

First, we superimpose a hemisphere over any selected portion of the domain boundary $\partial\Omega$; see Fig. 4.7. The intersection of the domain boundary and the hemisphere is denoted by S and the surface of the hemisphere outside the domain Ω is denoted by Γ. The region bounded by S and Γ is denoted by Ω_S. Let $G_{\mathrm{h}}(\mathbf{x}, \mathbf{y})$ be the homogeneous Green's function of the ball in (4.65), defined with the Kelvin image method (2.105). Then, the integral representation of (3.39) can be applied to the boundary of the domain Ω_S to yield the following identity

$$
\begin{aligned}
u(\mathbf{x}) = &- \int\limits_{\Gamma} \frac{\partial G_{\mathrm{h}}(\mathbf{x}, \mathbf{y})}{\partial \mathbf{n}_{\mathbf{y}}} u(\mathbf{y}) ds_y \\
&+ \int\limits_{S} \left[-\frac{\partial G_{\mathrm{h}}(\mathbf{x}, \mathbf{y})}{\partial \mathbf{n}_{\mathbf{y}}} u(\mathbf{y}) + G_{\mathrm{h}}(\mathbf{x}, \mathbf{y}) \frac{\partial u(\mathbf{y})}{\partial \mathbf{n}_{\mathbf{y}}} \right] ds_y, \quad \mathbf{x} \in \Omega_S.
\end{aligned} \tag{4.77}
$$

It should be noted that the integral over Γ involves only the normal derivative of the Green's function since G_{h} vanishes on Γ by definition, and as a result, on Γ only the solution $u(\mathbf{y})$ is needed. However, both the solution $u(\mathbf{y})$ and its normal derivative $\frac{\partial u(\mathbf{y})}{\partial \mathbf{n}}$ appear in the integral over S. The solution $u(\mathbf{y})$ over Γ can be computed with the Feynman–Kac formula (4.63) using the WOS method with the BC in (4.38), i.e.,

$$
u(\mathbf{y}) = \int\limits_{\partial\Omega} \phi(\mathbf{z}) p(\mathbf{y}, \mathbf{z}) ds_z, \tag{4.78}
$$

where the transition probability $p(\mathbf{y}, \mathbf{z})$ defined in (4.64) can be simulated by the WOS method.

Then, the Neumann data over S can be solved from the following integral equation:

$$
K\left[\frac{\partial u}{\partial \mathbf{n}}\right](\mathbf{x}) = b(\mathbf{x}), \quad \mathbf{x} \in S, \tag{4.79}
$$

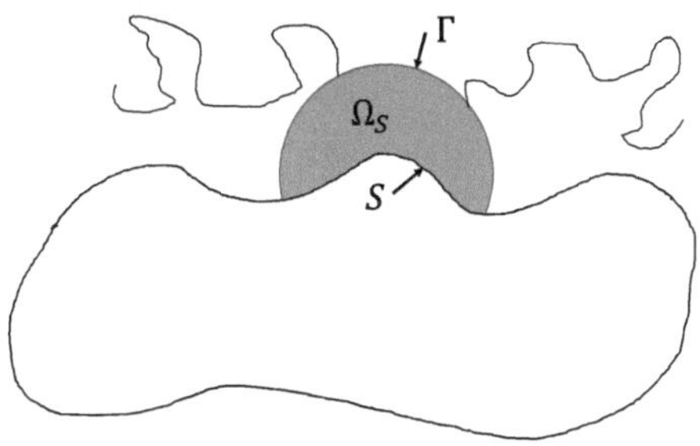

Fig. 4.7 The BIE–WOS method for finding Neumann data over a patch S on a curved boundary

where

$$K\left[\frac{\partial u}{\partial \mathbf{n}}\right] \equiv \int\limits_S G_{\mathrm{h}}(\mathbf{x},\mathbf{y})\frac{\partial u(\mathbf{y})}{\partial \mathbf{n_y}}ds_y, \tag{4.80}$$

and

$$b(\mathbf{x}) \equiv \left[\frac{u(\mathbf{x})}{2} + \text{p.v.}\int\limits_S \frac{\partial G_{\mathrm{h}}(\mathbf{x},\mathbf{y})}{\partial \mathbf{n_y}}u(\mathbf{y})ds_y\right] + \int\limits_\Gamma \frac{\partial G_{\mathrm{h}}(\mathbf{x},\mathbf{y})}{\partial \mathbf{n_y}}u(\mathbf{y})ds_y. \tag{4.81}$$

Here, p.v. represents the Cauchy principal value of the double layer potential defined in (3.10).

The integral equation of the first kind (4.79) is ill-conditioned and may cause numerical difficulties, especially when the linear algebraic system arising from discretization is large. To avoid the ill-conditioning issue, a well-conditioned integral equation of the second kind can be obtained by taking the normal derivatives of both sides of (4.77), resulting in the following identity

$$\begin{aligned}\frac{\partial}{\partial \mathbf{n_x}}u(\mathbf{x}) = &-\int\limits_\Gamma \frac{\partial^2 G_{\mathrm{h}}(\mathbf{x},\mathbf{y})}{\partial \mathbf{n_x}\partial \mathbf{n_y}}u(\mathbf{y})ds_y\\ &+\int\limits_S \left[-\frac{\partial^2 G_{\mathrm{h}}(\mathbf{x},\mathbf{y})}{\partial \mathbf{n_x}\partial \mathbf{n_y}}u(\mathbf{y}) + \frac{\partial G_{\mathrm{h}}(\mathbf{x},\mathbf{y})}{\partial \mathbf{n_x}}\frac{\partial u(\mathbf{y})}{\partial \mathbf{n_y}}\right]ds_y, \quad \mathbf{x}\in\Omega_S. \end{aligned}\tag{4.82}$$

If we let $\mathbf{x}$ approach the boundary S, we then obtain the following second-kind integral equation

$$\left(\frac{1}{2}I - D\right)\left[\frac{\partial u}{\partial \mathbf{n}}\right](\mathbf{x}) = b(\mathbf{x}), \quad \mathbf{x}\in S, \tag{4.83}$$

where the integral operator of the double layer potential is

$$D\left[\frac{\partial u}{\partial \mathbf{n}}\right](\mathbf{x}) \equiv \int_S \frac{\partial G_{\mathrm{h}}(\mathbf{x}, \mathbf{y})}{\partial \mathbf{n}_{\mathbf{x}}} \frac{\partial u(\mathbf{y})}{\partial \mathbf{n}_{\mathbf{y}}} ds_y, \tag{4.84}$$

and the new right-hand side

$$b(\mathbf{x}) \equiv -\int_\Gamma \frac{\partial^2 G_{\mathrm{h}}(\mathbf{x}, \mathbf{y})}{\partial \mathbf{n}_{\mathbf{x}} \partial \mathbf{n}_{\mathbf{y}}} u(\mathbf{y}) ds_y - \text{p.f.} \int_S \frac{\partial^2 G_{\mathrm{h}}(\mathbf{x}, \mathbf{y})}{\partial \mathbf{n}_{\mathbf{x}} \partial \mathbf{n}_{\mathbf{y}}} u(\mathbf{y}) ds_y, \quad \mathbf{x} \in S. \tag{4.85}$$

Here, p.f. the Hadamard finite part limit for the hyper-singular integral defined in (3.18), which will be handled by a regularization technique.

BIE–WOS Algorithm. The BIE–WOS method for obtaining Neumann data over a patch S consists of the following two steps:

- Step 1: Apply the Feynman–Kac formula (4.78) using the WOS sampling technique to compute the solution $u(\mathbf{y}_{i,j})$ at the Gauss quadrature points $\mathbf{y}_{i,j} \in \Gamma$, and compute the right-hand side function $b(\mathbf{x})$ in (4.81) or (4.85) by appropriate Gauss quadrature.
- Step 2: Solve the BIE (4.79) or (4.83) with a collocation method to obtain the Neumann data $\frac{\partial u}{\partial \mathbf{n}}$ over S.

4.7 Iterative BIE-WOS Method for the Mixed BVP of the Laplace Equation

In this section, we extend the BIE-WOS to mixed Neumann and Robin BVPs of the Laplace equation of (4.48) with $f = g = 0$ using an iterative approach [28]. The Dirichlet data is available only on part of the boundary of the domain, the BIE-WOS previously developed for a pure Dirichlet problem cannot be applied directly. The iterative approach will start by initializing Dirichlet data on both the Neumann and Robin parts of the boundary with a zero Dirichlet data $u^{(0)} = 0$, the BIE-WOS is applied to solve for an updated value for the Dirichlet data $u^{(1)}$ on the non-Dirichlet part of the boundary, this procedure can continue until convergence of the Dirichlet data on the whole boundary within some prescribed error tolerance.

The iterative BIE-WOS method for the Laplace equation over the domain boundary consists of boundary preparation, Dirichlet data initialization, and solving a local integral equation over each patch within each iteration. This iterative method converges quickly to mixed-type BVP problems [28].

- **Boundary patch preparation**: the boundary $\partial\Omega$ will be covered with M overlapping patches S_j, $j = 1, \ldots, M$, obtained by the intersection of half-spheres Ω_{S_j},i.e., $S_j = \Omega_{S_j} \cap \partial\Omega$.
- **Initialization of Dirichlet data** on $\partial\Omega = \Lambda_1 \cup \Lambda_2 \cup \Lambda_3$, $u^{(0)}|_{\partial\Omega}$: $u^{(0)}|_{\Lambda_1} = \phi_1, u^{(0)}|_{\Lambda_2} = 0, u^{(0)}|_{\Lambda_3} = 0$.

- For $n = 1, 2, \ldots,$ given $u^{(n-1)}|_{\partial\Omega}$, for $n < n_{\max}$, or $\max_{\partial\Omega} |u^{(n)} - u^{(n-1)}| > \varepsilon$, with a prescribed number of iterations $n_{\max}$ and an error tolerance ε.
Step 1: At each Gaussian points $\mathbf{y}_i$ over each half-spherical surface Γ_j, a WOS solution is obtained for the Laplace equation with a complete Dirichlet data $u^{(n-1)}$ on $\partial\Omega$, using a killed Brownian motion WOS as in (4.78), i.e.,

$$u^{(n)}(\mathbf{y}_i)|_{\Gamma_j} = \int_{\partial\Omega} p(\mathbf{y}_i, \mathbf{x}')u^{(n-1)}(\mathbf{x}')ds_{x'}. \tag{4.86}$$

Step 2: Over each patch S_j $j = 1, \ldots, M$, the following four local BIEs will be solved.

Case 1: If S_j is given a Dirichlet data, i.e. $u|_{S_j} = \phi_1$, solve the following integral equation over S_j for the Neumann data on S_j, denoted as $\frac{\partial u^{(n)}(\mathbf{x})}{\partial \mathbf{n}_\mathbf{x}}|_{S_j}$,

$$-\int_{S_j} G_{\rm h}(\mathbf{x}, \mathbf{y})\frac{\partial u(\mathbf{y})}{\partial \mathbf{n_y}}ds_y = f_j(\mathbf{x}) - \frac{1}{2}\phi_1(\mathbf{x}) - \int_{S_j} \frac{\partial G_{\rm h}(\mathbf{x}, \mathbf{y})}{\partial \mathbf{n_y}}\phi_1(\mathbf{y})ds_y, \mathbf{x} \in S_j, \tag{4.87}$$

where

$$f_j(\mathbf{x}) = -\int_{\Gamma_j} \frac{\partial G_{\rm h}(\mathbf{x}, \mathbf{y})}{\partial \mathbf{n_y}}u(\mathbf{y})ds_y, \tag{4.88}$$

where the integral can be evaluated with a Gaussian quadrature over Γ_j with the value of u at the nodes computed by (4.86).

Case 2: If S_j is given a Neumann data, i.e. $\frac{\partial u(\mathbf{x})}{\partial \mathbf{n}_\mathbf{x}}|_{S_j} = \phi_2$, solve the following integral equation over S_j for the Dirichlet data on S_j, denoted as $u^{(n)}|_{S_j}$,

$$\frac{1}{2}u(\mathbf{x}) + \int_{S_j} \frac{\partial G_{\rm h}(\mathbf{x}, \mathbf{y})}{\partial \mathbf{n_y}}u(\mathbf{y})ds_y = f_j(\mathbf{x}) + \int_{S_j} G_{\rm h}(\mathbf{x}, \mathbf{y})\phi_2(\mathbf{y})ds_y, \quad \mathbf{x} \in S_j. \tag{4.89}$$

Case 3: If S_j is given Robin data, i.e., $\frac{\partial u(\mathbf{x})}{\partial \mathbf{n}_\mathbf{x}} - cu(\mathbf{x})|_{S_j} = \phi_3$, solve the following integral equation over S_j for the Dirichlet data on S_j, denoted as $u^{(n)}|_{S_j}$,

$$\begin{aligned}
&\frac{1}{2}u(\mathbf{x}) + \int_{S_j} \left[\frac{\partial G_{\rm h}(\mathbf{x}, \mathbf{y})}{\partial \mathbf{n_y}} - cG_{\rm h}(\mathbf{x}, \mathbf{y})\right] u(\mathbf{y})ds_y, \\
&= f_j(\mathbf{x}) + \int_{S_j} G_{\rm h}(\mathbf{x}, \mathbf{y})\phi_3(\mathbf{y})ds_y, \quad \mathbf{x} \in S_j.
\end{aligned} \tag{4.90}$$

Case 4: S_j is given a mixed type of boundary condition that includes Dirichlet, Neumann, and Robin type. An integral equation involving the combination of (4.87), (4.89), and (4.90) will be derived and solved.

At the end of **Step 2**, the updated Dirichlet data $u^{(n)}|_{\partial\Omega}$ will be obtained from (4.89) and (4.90). If $n < n_{\max}$, or $\max_{\partial\Omega} |u^{(n)} - u^{(n-1)}| > \varepsilon$, then return to **Step 1**, otherwise, stop.

4.8 Summary

Random walk-on-spheres stochastic methods based on the Feynman–Kac formula provide effective methods for local solutions of the Laplace equation with mixed boundary conditions where they are needed. Using local integral equations and an iterative approach, the BIE-WOS method provides a massively parallel algorithm for the global solution of mixed problems of the Laplace equation with the Feynman–Kac formula involving only killed Brownian motions.

References

1. C. Graham, D. Talay, *Stochastic Simulation and Monte Carlo Methods: Mathematical Foundations of Stochastic Simulation* (Springer Science & Business Media, 2013)
2. K.K. Sabelfeld, *Monte Carlo Methods in Boundary Value Problems* (Springer, 1991)
3. J.A. Given, C.-O. Hwang, M. Mascagni, First-and last-passage Monte Carlo algorithms for the charge density distribution on a conducting surface. Phys. Rev. E **66**, 056704 (2002)
4. A.E. Kyprianou, A. Osojnik, T. Shardlow, Unbiased 'walk-on-spheres' Monte Carlo methods for the fractional Laplacian. IMA J. Numer. Anal. **38**, 1550–1578 (2017)
5. S.N. Cohen, R.J. Elliott, *Stochastic Calculus and Applications* (Birkhäuser, New York, 2015)
6. F.C. Klebaner. *Introduction to Stochastic Calculus with Applications* (World Scientific Publishing Company, 2012)
7. W.E, T. Li, E. Vanden-Eijnden, *Applied Stochastic Analysis* (American Mathematical Society, 2021)
8. B. Oksendal. *Stochastic Differential Equations: An Introduction with Applications* (Springer Science & Business Media, 2013)
9. M. Freidli,. *Functional Integration and Partial Differential Equations* (Princeton University Press, 1985)
10. E.P. Hsu, Reflecting Brownian motion, boundary local time and the Neumann problem. Dissertation Abstracts International Part B: Science and Engineering [Diss. Abst. Int. PT. B: Sci. Eng.] **45**(6) (1984)
11. C.Y. Ding, Y.J. Zhou, W. Cai, X. Zeng, C.H. Yan, A path integral Monte Carlo (PIMC) method based on Feynman-Kac formula for electrical impedance tomography. J. Comput. Phys. **476**, 111862 (2023)
12. Y. Zhou, W. Cai, E.P. Hsu, Computation of local time of reflecting Brownian motion and probabilistic representation of the Neumann Problem. Commun. Math. Sci. **15**, 237–259 (2017)
13. Y. Zhou, W. Cai, Numerical solution of the Robin problem of Laplace equations with a Feynman-Kac formula and reflecting Brownian motions. J. Sci. Comput. **69**(1), 107–121 (2016)

14. V.G. Papanicolaou, The probabilistic solution of the third boundary value problem for second order elliptic equations. Probab. Theory Relat. Fields **87**, 27–77 (1990)
15. J. Duan, J. Peng, The probabilistic solution of a system of semilinear elliptic PDEs under the third boundary conditions. Potential Anal. **4**, 1571–97 (2023)
16. Z.Q. Chen, On reflecting diffusion processes and Skorokhod decompositions. Prob. Theory Related Fields **94**, 281–315 (1993)
17. E. Pardoux, S. Peng, Backward stochastic differential equations and quasilinear parabolic partial differential equations, in *Stochastic Partial Differential Equations and Their Applications* (Springer, Berlin, Heidelberg, 1992)
18. W.Z. Zhang, W. Cai, FBSDE based neural network algorithms for high-dimensional quasilinear parabolic PDEs. J. Comput. Phys. **470**, 111557 (2022)
19. M.E. Müller, Some continuous Monte Carlo methods for the Dirichlet problem. Ann. Math. Stat. **27**(3), 569–589 (1956)
20. M. Mascagni, C.O. Hwang, ϵ-Shell error analysis for "Walk On Spheres" algorithms. Math. Comput. Simul. **63**(2), 93–104 (2003)
21. J.A. Given, J.B. Hubbard, J.F. Douglas, A first-passage algorithm for the hydrodynamic friction and diffusion-limited reaction rate of macromolecules. J. Chem. Phys. **106**(9), 3761–3771 (1997)
22. S. Maire, M. Simon, A partially reflecting random walk on spheres algorithm for electrical impedance tomography. J. Comput. Phys. **303**, 413–430 (2015)
23. K.L. Chung, *Green, Brown, and Probability* (World scientific, Singapore, 1995)
24. I. Binder, M. Braverman, The rate of convergence of the walk of sphere algorithm. Geomet. Funct. Anal. **22**, 558–587 (2012)
25. P. Lévy, *Processus Stochastiques et Mouvement Brownien* (Gauthier-Villars, Paris, 1948)
26. M. Cheney, D. Isaacson, J.C. Newell, Electrical impedance tomography. SIAM Rev. **41**(1), 85–101 (1999)
27. C.H. Yan, W. Cai, X. Zeng, A parallel method for solving Poisson equations with Dirichlet data using local boundary integral equations and random walks. SIAM J. Sci. Comput. **35**(4), B868–B889 (2013)
28. C.Y. Ding, C.H. Yan, X. Zeng, W. Cai, A parallel iterative probabilistic method for mixed problems of Laplace equations with the Feynman-Kac formula of killed Brownian motions. SIAM Sci. Comput. **44**, A3413–A3435 (2022)

Chapter 5
Stochastic Spectral Methods and Uncertainty Quantification of Microchip Interconnects

Stochastic effects are ubiquitous in many scientific and engineering applications, and quantifying the resulting uncertainty can provide important guidance in the design of robust engineering systems, ensuring that desired performance criteria are met. In this chapter, we will present stochastic spectral methods, including the Galerkin and collocation methods, using Wiener polynomial chaos and generalized polynomial chaos for the solution of partial differential or integral equations with stochastic effects. An application of the stochastic collocation method is presented for solving a stochastic integral equation for capacitance extraction of microchip interconnects, where stochastic geometric variance arises from manufacturing processes.

5.1 Deterministic Spectral Methods

Spectral methods have been applied to solve PDEs and study fluid flows and electromagnetics [1] since their invention by Gottlieb and Orszag [2]. The methods are based on series expansions of eigenfunctions of some Sturm–Liouville self-adjoint (often singular) operators. The most popular cases are Fourier basis for periodic problems, Chebyshev and Legendre polynomials for non-periodic problems, and Hermite polynomials for problems involving unbounded domains.

5.1.1 Fourier, Chebyshev, Legendre, and Hermite Spectral Approximations

A function $f(x)$ defined on an interval $[a, b]$ can be approximated by a series expansion using the orthonormal eigenfunctions $\phi_n(x)$ of some Sturm–Liouville differential operator

W. Cai, *Deterministic, Stochastic, and Deep Learning Methods for Computational Electromagnetics*, https://doi.org/10.1007/978-981-96-0100-4_5

$$f(x) = \sum_{n=0}^{\infty} a_n \phi_n(x), \quad x \in [a, b]. \tag{5.1}$$

The basis function $\phi_n(x)$ is defined for the following cases [3].

- Fourier basis for $2\pi-$periodic functions.

$$[a, b] = [0, 2\pi], \ \ \phi_n(x) = \frac{1}{\sqrt{2\pi}} e^{inx}. \tag{5.2}$$

For non-periodic problems, the family of Jacobi polynomials $J_n^{(\alpha,\beta)}(x)$ can be used, which are orthogonal polynomial over $[-1, 1]$ with respect to the weight

$$\omega^{\alpha,\beta}(x) = (1-x)^{\alpha}(1+x)^{\beta}, \tag{5.3}$$

and,

$$\int_{-1}^{1} \omega^{\alpha,\beta}(x) J_m^{(\alpha,\beta)}(x) J_n^{(\alpha,\beta)}(x) dx = \gamma_n^{\alpha,\beta} \delta_{m,n}, \tag{5.4}$$

where

$$\gamma_n^{\alpha,\beta} = \frac{2^{\alpha+\beta+1}\Gamma(n+\alpha+1)\Gamma(n+\beta+1)}{(2n+\alpha+\beta+1)n!\Gamma(n+\alpha+\beta+1)}. \tag{5.5}$$

The Jacobi polynomials are the solutions of the following singular Sturm–Liouville eigenvalue problem

$$-(1-x)^{-\alpha}(1+x)^{-\beta}\partial_x\left((1-x)^{\alpha+1}(1+x)^{\beta+1}\partial_x J_n^{(\alpha,\beta)}(x)\right) = \lambda_n^{\alpha,\beta} J_n^{(\alpha,\beta)}(x), \tag{5.6}$$

where the eigenvalue

$$\lambda_n^{\alpha,\beta} = n(n+\alpha+\beta+1). \tag{5.7}$$

Two special cases are Chebyshev $T_n(x) = J_n^{(-\frac{1}{2},-\frac{1}{2})}(x)$ and Legendre polynomial $L_n(x) = J_n^{(0,0)}(x)$.

- Chebyshev polynomials for non-periodic functions.

$$[a, b] = [-1, 1], \ \ \phi_n(x) = T_n(x) = \cos(n\cos^{-1}(x)), \text{(Chebyshev polynomial)}, \tag{5.8}$$

with a three-term recursive relation

$$\begin{aligned} T_0(x) &= 1, \quad T_1(x) = x, \\ T_{n+1}(x) &= 2xT_n(x) - T_{n-1}(x), \ n \geq 1. \end{aligned}$$

- Legendre polynomials for non-periodic functions.

$$[a, b] = [-1, 1], \;\; \phi_n(x) = P_n(x), \text{(Legendre polynomial)}, \tag{5.9}$$

with a three-term recursive relation

$$P_0(x) = 1, \quad P_1(x) = x, \tag{5.10}$$
$$(n+1)P_{n+1}(x) = (2n+1)\,xP_n(x) - nP_{n-1}(x), \; n \geq 1. \tag{5.11}$$

- Hermite polynomials for functions over $(-\infty, \infty)$,

$$[a, b] = (-\infty, \infty), \;\; \phi_n(x) = H_n(x), \text{(Hermite polynomial)}, \tag{5.12}$$

with a three-term recursive relation

$$H_0(x) = 1, \quad H_1(x) = x, \tag{5.13}$$
$$H_{n+1}(x) = xH_n(x) - nH_{n-1}(x), n \geq 1, \tag{5.14}$$

$$\int_{-1}^{1} \frac{1}{\sqrt{2\pi}} e^{-\frac{x^2}{2}} H_m(x) H_m(x) dx = \gamma_n^{(H)} \delta_{m,n}, \tag{5.15}$$

where

$$\gamma_n^{(H)} = n!. \tag{5.16}$$

In summary, for all cases, we have the orthonormal conditions for the basis functions under a weighted inner product,

$$<\phi_n(x), \phi_m(x)>_\omega \equiv \int_a^b \omega(x)\phi_n(x)\phi_m(x) dx = \gamma_n \delta_{n,m}, \tag{5.17}$$

and the weight function

$$\omega(x) = \begin{cases} 1 & \text{Fourier base on } [0, 2\pi] \\ \frac{1}{\sqrt{1-x^2}} & \text{Chebyshev polynomial on } [-1, 1] \\ 1 & \text{Legendre polynomial on } [-1, 1] \\ \frac{1}{\sqrt{2\pi}} e^{-\frac{x^2}{2}} & \text{Hermite polynomial } (-\infty, \infty) \end{cases} . \tag{5.18}$$

and the normalization constant $\gamma_n = 1$ for the Fourier basis, $\gamma_0 = \pi$, $\gamma_n = \pi/2, n \geq 1$ for the Chebyshev polynomials, $\gamma_n = 2/(2n+1)$ for the Legendre polynomials, and γ_n is given in (5.16) for the Hermite polynomials.

Galerkin approximation. Let the subspace

$$X_N = \text{span}\{\phi_n(x)\}_{n=0}^{N}, \tag{5.19}$$

and a Galerkin approximation $f_N(x) \in X_N$ to $f(x)$ is defined by the $L^2_\omega-$ orthogonal projection

$$<f_N(x) - f(x), \phi_i(x)>_\omega \; = \; 0, \quad \forall \phi_i(x) \in X_N, \tag{5.20}$$

i.e.,

$$\int_a^b \omega(x) f_N(x)\phi_i(x)dx = \int_a^b \omega(x) f(x)\phi_i(x)dx, \quad 0 \le i \le N. \tag{5.21}$$

Set

$$f_N(x) = \sum_{n=0}^{N} a_n \phi_n(x). \tag{5.22}$$

Due to the orthogonality of the basis function, we have the "Fourier" coefficient

$$a_n = \frac{1}{\gamma_n} \int_a^b \omega(x) f(x)\phi_n(x)dx. \tag{5.23}$$

The Galerkin approximation to a smooth function has spectral convergence (refer to Eqs. (2.48), (3.264), and (7.127) [4]),

$$||f(x) - f_N(x)||_\omega \le \frac{C}{N^{\delta k}}, \text{ for } f(x) \in C^k[a, b], \tag{5.24}$$

where $\delta = 1$ except for the Hermite polynomials, for which $\delta = 1/2$, and the constant C depends on a weighted norm of $f^{(k)}(x)$.

Collocation approximation. A collocation $I_N f(x) \in X_N$ of the function $f(x)$ is defined so that at the collocation points $\{x_i\}_{i=0}^{N}$, we have

$$I_N f(x_i) = f(x_i), \quad i = 0, \ldots, N. \tag{5.25}$$

The interpolant $I_N f(x)$ can be expressed in terms of the Lagrangian polynomial

$$\begin{aligned} l_i(x) &= \frac{(x - x_0)\cdots(x - x_{i-1})(x - x_{i+1})\cdots(x - x_N)}{(x_i - x_0)\cdots(x_i - x_{i-1})(x_i - x_{i+1})\cdots(x_i - x_N)} \in X_N, \\ l_i(x_j) &= \delta_{ij}, \end{aligned} \tag{5.26}$$

and

$$I_N f(x) = \sum_{i=0}^{N} f(x_i) l_i(x). \tag{5.27}$$

The interpolant $I_N f(x)$ can be used to define a quadrature rule for integration,

$$\int_a^b \omega(x) f(x) dx \simeq \sum_{j=0}^{N} \omega_j f(x_j), \tag{5.28}$$

and the quadrature rule weights

$$\omega_i = \int_a^b \omega(x) l_i(x) dx, \quad 0 \le i \le N. \tag{5.29}$$

Meanwhile, as $I_N f(x) \in X_N$, we can also have

$$I_N f(x) = \sum_{n=0}^{N} \widehat{a}_n \phi_n(x), \tag{5.30}$$

where $\widehat{a}_n$ is named the discrete "Fourier"-expansion coefficient, whose explicit formulae are given below [4].

- **Fourier case**. The collocation points are $x_i = i\frac{2\pi}{N} \in [0, 2\pi],\ 0 \le i \le N-1$, and

$$l_i(x) = \frac{1}{N} \sin\big(N\frac{x - x_i}{2}\big) \cot\big(\frac{x - x_i}{2}\big), \tag{5.31}$$

and

$$\widehat{a}_n = \frac{1}{N}\sum_{j=0}^{N} f(x_j) e^{-inx_j} \sim \frac{1}{2\pi}\int_0^{2\pi} f(x) e^{-inx} dx. \tag{5.32}$$

- **Chebyshev case**.

(a) Chebyshev–Gauss points—the collocation points are the zeros of the $T_{N+1}(x)$, i.e.,

$$x_j = -\cos\frac{(2j+1)\pi}{2N+2}, \quad \omega_j = \frac{\pi}{N+1}, \ 0 \le j \le N, \tag{5.33}$$

and the quadrature rule (5.28) is exact for polynomials of order $2N + 1$.

(b) Chebyshev–Gauss–Labotto (CGL) points:

$$x_j = -\cos\frac{j\pi}{N}, \quad \omega_j = \frac{\pi}{\bar{c}_j N}, \ 0 \le j \le N, \tag{5.34}$$

and $\overline{c}_0 = \overline{c}_N = 2, \overline{c}_j = 1,\ 1 \leq j \leq N-1$, and the quadrature rule (5.28) is exact for polynomials of order $2N-1$.

(c) Chebyshev–Gauss–Radau (CGR) points:

$$x_j = -\cos\frac{2j\pi}{2N+1},\ \omega_0 = \frac{\pi}{2N+1},\ \omega_j = \frac{2\pi}{2N+1},\ 1 \leq j \leq N, \tag{5.35}$$

and the quadrature rule (5.28) is exact for polynomials of order $2N$.

- **Legendre case**.

(a) Legendre–Gauss points—the collocation points are the zeros of the $P_{N+1}(x)$, and the quadrature weights are

$$\omega_j = \frac{2}{(1-x_j^2)\left[P'_{N+1}(x_j)\right]^2},\ 0 \leq j \leq N, \tag{5.36}$$

and the quadrature rule (5.28) is exact for polynomials of order $2N+1$.

(b) Legendre–Gauss–Labotto (LGL) points are the zeros of $(1-x^2)P'_N(x)$, $x_0 = -1, x_N = 1$ and the quadrature weights are

$$\omega_j = \frac{2}{N(N+1)}\frac{1}{\left[P_N(x_j)\right]^2},\ 0 \leq j \leq N, \tag{5.37}$$

and $\overline{c}_0 = \overline{c}_N = 2, \overline{c}_j = 1, 1 \leq j \leq N-1$, and the quadrature rule (5.28) is exact for polynomials of order $2N-1$.

(c) Legendre–Gauss–Radau (LGR) points are the zeros of $P_N(x) + P_{N+1}(x)$, $x_0 = -1$,

$$\omega_j = \frac{1}{(N+1)^2}\frac{(1-x_j)}{\left[P_N(x_j)\right]^2},\ 0 \leq j \leq N, \tag{5.38}$$

and the quadrature rule (5.28) is exact for polynomials of order $2N$.

For both the Chebyshev ($\alpha = \beta = -\frac{1}{2}$) and Legendre polynomials ($\alpha = \beta = 0$), the discrete Fourier coefficients are given by Theorem 3.28 [4],

$$\widehat{a}_n = \frac{1}{\delta_n^{\alpha,\beta}}\sum_{j=0}^{N} f(x_j)J_n^{(\alpha,\beta)}(x_j)\omega_j, \tag{5.39}$$

where $\delta_n^{\alpha,\beta} = \gamma_n^{\alpha,\beta}$ for $0 \leq n \leq N-1$ and

$$\delta_N^{\alpha,\beta} = \begin{cases} \gamma_N^{\alpha,\beta} \text{ for Gauss and Gauss–Radau points} \\ (2 + \frac{\alpha+\beta+1}{N})\gamma_N^{\alpha,\beta} \text{ for Gauss–Labotto points} \end{cases}. \tag{5.40}$$

- **Hermite case.**
  Hermite–Gauss points—the collocation points are the zeros of the $H_{N+1}(x)$, and the quadrature weights are (Eq. (7.79) [4])

$$\omega_j = \frac{2^N N!}{(N+1)\left[H_N(x_j)\right]^2}, \; 0 \le j \le N, \tag{5.41}$$

  and the quadrature rule (5.28) is exact for polynomials of order $2N+1$.
  Discrete expansion coefficients

$$\widehat{a}_n = \frac{1}{\sqrt{\pi}2^n n!} \sum_{j=0}^{N} f(x_j) H_n(x_j)\omega_j. \tag{5.42}$$

The collocation points of the Gauss quadrature type above also produce an interpolant for smooth functions with a spectral accuracy (refer to Theorems 3.41–3.43, and 7.17 in [4]), namely

$$||f(x) - I_N f(x)|| \le \frac{C}{N^{\delta k+\mu}} \text{ for } f(x) \in C^k[a,b], \tag{5.43}$$

where $||\cdot||$ is a weighted norm, and $\delta = 1$, $\mu = 0$ except for the Hermite polynomials, for which $\delta = 1/2$, $\mu = -1/6$, and the constant C depends on a weighted norm of $f^{(k)}(x)$.

5.1.2 *Spectral Methods for BVPs of Differential Equations*

Consider the following BVP of a variable coefficient ordinary differential equation (ODE):

$$\mathcal{L}[u](x) \equiv -\alpha u'' + \beta u' + \gamma u = f(x), \quad x \in [-1, 1], \tag{5.44}$$

with boundary conditions (BCs)

$$u(-1) = u(1) = 0. \tag{5.45}$$

Galerkin method. A Galerkin method with minimizing residuals seeks a solution $u_N(x) \in X_{N,0} = \{\phi \in X_N, \phi(-1) = \phi(1) = 0\}$ such that its residual is orthogonal to $X_{N,0}$, i.e.,

$$< -\alpha u_N'' + \beta u_N' + \gamma u_N - f, \psi_i(x) >_\omega \; = \; 0, \; \forall \psi_i(x) \in X_{N,0}, \tag{5.46}$$

namely,

$$\int_{-1}^{1} \left(-\alpha u_N'' + \beta u_N' + \gamma u_N\right) \psi_i(x)\omega(x)dx = \int_{-1}^{1} f(x)\psi_i(x)\omega(x)dx. \tag{5.47}$$

Galerkin–Petrov method. The difference from the Galerkin method above is that the test function $\psi_i(x)$ can come from another finite dimension space, i.e., $\psi_i(x) \in Y_N$.

Collocation method. In this case $\psi_i(x) = \delta(x - x_i)$, namely, the collocation solution $u_N(x) \in X_N$ satisfies the ODE (5.44) at the collocation points

$$-\alpha u_N'' + \beta u_N' + \gamma u_N = f|_{x=x_i},\ 1 \le i \le N-1. \tag{5.48}$$

in addition to the BCs $u_N(-1) = u_N(1) = 0$.

Both the Galerkin and collocation methods have a spectral convergence to smooth solutions of the differential equations [4, Chaps. 4 and 7].

- Connection between the Galerkin and collocation methods

Considering the Legendre–Galerkin method (5.47) with a numerical quadrature with the LGL formula (5.37), we have for $\forall \psi_i(x) \in X_{N,0}$

$$\sum_{j=0}^{N} \omega_j \left(-\alpha u_N''(x_j) + \beta u_N'(x_j) + \gamma u_N(x_j)\right) \psi_i(x_j) = \sum_{j=0}^{N} \omega_j f(x_j)\psi_i(x_j), \tag{5.49}$$

and, taking $\psi_i(x) = l_i(x) \in X_{N,0}$, $1 \le i \le N-1$, using the fact that $l_i(x_j) = \delta_{ij}$, we obtain

$$-\alpha u_N''(x_i) + \beta u_N'(x_i) + \gamma u_N(x_i) = f(x_i),\ 1 \le i \le N-1, \tag{5.50}$$

which is exactly the collocation method in (5.48). Therefore, a spectral collocation method for the differential equation is equivalent to a spectral Galerkin method with a numerical quadrature for the integrals in (5.47).

5.2 Stochastic Spectral Methods for Stochastic Differential and Integral Equations

In this section, we apply spectral methods to the random variables in stochastic differential or integral equation of the following general operator form

$$\mathcal{L}(\mathbf{x}, \theta)[u] = f(\mathbf{x}, \theta), \quad \mathbf{x} \in D, \tag{5.51}$$

where $\theta \in \Omega$ indicates the randomness of some relevant stochastic effects, which could manifest as the physical parameters of the PDE coefficients, represented by a random field, or the stochastic geometry variations in domain D from a manufacturing process. Two typical cases are given below.

- Stochastic elliptic differential operator

$$\mathcal{L}(\mathbf{x}, \theta)[u] = -\nabla \left(a(\mathbf{x}, \theta)\nabla u\right) + b(\mathbf{x}, \theta)u, \quad \mathbf{x} \in D, \tag{5.52}$$

  and stochastic BCs

$$u(\mathbf{x}, \theta) = h(\mathbf{x}, \theta), \quad \mathbf{x} \in \partial D. \tag{5.53}$$

- Integral operator over a stochastic boundary

$$\mathcal{L}(\mathbf{x}, \theta)[u] = \int_{\partial D(\theta)} K(\mathbf{x}, \mathbf{y})u(\mathbf{y})d\mathbf{y}, \tag{5.54}$$

  where the boundary of the domain $\partial D(\theta)$ contains stochastic variance, which can be written as a random perturbation from some deterministic nominal surface S_0, i.e.,

$$\partial D(\theta) = S_0 + \Sigma(\mathbf{x}, \theta), \quad \mathbf{x} \in S_0, \tag{5.55}$$

  where $\Sigma(\mathbf{x}, \theta)$ is a random field.

Due to the randomness of the coefficients $a(\mathbf{x}, \theta), b(\mathbf{x}, \theta)$, or the boundary of the domain $D(\theta)$, the solution to (5.51) will also be a random field $u(\mathbf{x}, \theta), x \in D$, and stochastic spectral methods based on the Wiener polynomial chaos (PC) or its generalization can be used for its approximation.

5.2.1 Wiener Polynomial Chaos (PC) and Generalized PC (gPC) Expansions

Wiener polynomial chaos (PC) Given an infinite set of i.i.d Gaussian r.v.'s $\xi = \{\xi_i, i \in I = \{1, 2, \ldots\}\}$, the concept of a pth-order polynomial chaos $\widehat{\Gamma}_p$ was proposed by Wiener [5] as

$$\widehat{\Gamma}_p = \widehat{\Gamma}_p(\{\xi_i\}_{i\in I}) = \left\{ \begin{matrix} \xi_{i_1}^{n_1}\xi_{i_2}^{n_2}\cdots\xi_{i_k}^{n_k}, 0 \le n_1 + \cdots n_k \le p \\ \text{distinct } \{i_1, \ldots, i_k\} \subset I \end{matrix} \right\}, \tag{5.56}$$

and, there exists an orthogonal "homogeneous PC" $\Gamma_i \subset \widehat{\Gamma}_i$, $E[\Gamma_i\Gamma_j] = \delta_{ij}$ (indicating the orthogonality of elements in the two spaces),

$$\widehat{\Gamma}_p = \Gamma_0 \oplus \Gamma_1 \oplus \cdots \oplus \Gamma_p. \tag{5.57}$$

As the probability density function for a normal r.v. ξ is

$$\rho(x) = \frac{1}{\sqrt{2\pi}} e^{-\frac{x^2}{2}}, \tag{5.58}$$

which is exactly the weight in the definition of Hermite polynomial $H_i(x)$ in (5.15). The orthogonality condition (5.15) is equivalent to the orthogonality of two r.v.'s $H_i(\xi)$, $H_j(\xi)$,

$$E[H_i(\xi)H_j(\xi)] = \delta_{ij}. \tag{5.59}$$

Therefore, the homogeneous polynomial chaos Γ_p is simply

$$\Gamma_p = \text{span}\left\{ \begin{array}{l} \psi_j(\boldsymbol{\xi}) \equiv H_{n_1}(\xi_{i_1})H_{n_2}(\xi_{i_2})\cdots H_{n_k}(\xi_{i_k}), \\ n_1 + \cdots n_k = p, \text{distinct}\{i_1, \ldots, i_k\} \subset I \end{array} \right\}. \tag{5.60}$$

It can be shown that $\{\Gamma_i\}_{i=0}^{\infty}$ forms a complete orthonormal basis for the space of square integrable r.v.'s over a probability space $(\Omega, \mathcal{F}, P)$, $\Theta = \{\mu(\theta), E[\mu^2] < \infty\}$. Namely, $\forall \mu(\theta) \in \Theta$, we have the following PC expansion,

$$\begin{aligned} \mu(\theta) &= \sum_{p=0}^{\infty} \sum_{\substack{\{i_1,\ldots,i_k\}\subset I \\ \text{distinct}}}^{n_1+\cdots n_k = p} a_{i_1,\ldots,i_k}^{n_1,\ldots,n_k} H_{n_1}(\xi_{i_1})H_{n_2}(\xi_{i_2})\cdots H_{n_k}(\xi_{i_k}) \\ &= \sum_{i=1}^{\infty} a_i \psi_i(\boldsymbol{\xi}), \end{aligned} \tag{5.61}$$

where a single index i under a given order mapping

$$a_i \longleftrightarrow a_{i_1,\ldots,i_k}^{n_1,\ldots,n_k}, \tag{5.62}$$

is used to order all the basis functions, and the coefficients can be obtained by the orthogonality of the basis function

$$a_{i_1,\ldots,i_k}^{n_1,\ldots,n_k} = E[\mu(\theta)H_{n_1}(\xi_{i_1})H_{n_2}(\xi_{i_2})\cdots H_{n_k}(\xi_{i_k})], \tag{5.63}$$

and a truncation of the series for $p \le P$ will give the Pth-order PC expansion.

Remark 5.1 In cases where the r.v. $\mu(\theta) = \mu(\xi_1(\theta), \cdots \xi_d(\theta))$, $\{\xi_i, 1 \le i \le d\}$ are i.i.d. Gaussian r.v.'s, then the polynomial chaos space Γ_p of (5.60) and $\widehat{\Gamma}_p$ of (5.57) are all finite dimensions, and $N_p = \dim\widehat{\Gamma}_p = \binom{d+p}{d}$, $\dim\Gamma_p = N_p - N_{p-1} = \binom{d+p-1}{d-1}$ while $I = \{1, 2, \ldots, d\}$ in their definitions.

Generalized polynomial chaos (gPC). The original Wiener Hermite PC was later generalized [6] to cases where the probability distribution function $\rho_i(\xi_i)$, $\xi_i \in \Xi_\mathbf{i} \subset \mathbb{R}$, can be different for each random variable used in the expansion (5.61). The exact same procedure described above can be carried out, and the polynomial $\phi_n^{(i)}(\xi_i)$ will be the corresponding orthogonal polynomials with respect to the weight function $\rho_i(\xi_i)$, i.e.,

$$\int_{\Xi_\mathrm{i}} \phi_m^{(i)}(x)\phi_n^{(i)}(x)\rho_i(x)dx = \delta_{mn}, \tag{5.64}$$

and the gPC expansion [7] will then be

$$\begin{aligned}\mu(\theta) &= \sum_{p=0}^{\infty} \sum_{\substack{\{i_1,\ldots,i_k\}\subset I \\ \text{distinct}}}^{n_1+\cdots n_k=p} a_{i_1,\ldots,i_k}^{n_1,\ldots,n_k} \phi_{n_1}^{(i_1)}(\xi_{i_1})\phi_{n_2}^{(i_2)}(\xi_{i_2})\cdots\phi_{n_k}^{(i_k)}(\xi_{i_k}) \\ &= \sum_{i=1}^{\infty} a_i \psi_i(\boldsymbol{\xi}),\end{aligned} \tag{5.65}$$

where the expansion coefficient $a_{i_1,\ldots,i_k}^{n_1,\ldots,n_k}$ can be computed similarly to that in (5.63) due to the orthogonality of the polynomials in each variable, and a truncation of the series for $p \le P$ gives a convergent Pth-order gPC expansion as $P \to \infty$ [8].

A class of Asky orthogonal polynomials [6], as a subfamily of hypergeometric functions, can be used for various distributions, including the Laguerre polynomials for the Gamma distribution over $[0, \infty)$, the Jacobi polynomials for the Beta distribution over $[a, b]$, the Legendre polynomials for the uniform distribution over $[a, b]$, the Charlier polynomials over a discrete set $\{0, 1, 2, \ldots\}$ for the Poisson distribution, and the Krawtchouk polynomials for the binomial distribution $\{0, 1, \ldots, N\}$, etc.

5.2.2 *Random Variable Dimension Reduction*

Before we apply the PC or gPC expansions to construct a spectral method for the stochastic operator equation (5.51), we can reduce the infinite number of random variables in the random field $a(\mathbf{x}, \theta)$ in (5.52) and $\Sigma(\mathbf{x}, \theta)$ in (5.55) to a finite number. This reduction can be achieved by using a truncated Karhunen–Lòve expansion for the random field with a known correlation function.

Karhunen–Loève expansion: Given a second-order process $X_t(\theta) = X(t, \theta)$, where the index $t \in K$ can be a time index ($K = [0, T]$) or a spatial domain $K = D \subset \mathbb{R}^d$, $E[X_t^2] < \infty$, then, we have the following eigenfunction expansion

$$X(t, \theta) = \sum_{i=1}^{\infty} \xi_n(\theta)e_n(t), \tag{5.66}$$

where the basis functions $\{e_n(t)\}_{n=1}^{\infty}$ form a complete orthonormal eigenfunction set for the correlation matrix $R(s,t)$ of the process $X_t(\theta)$, i.e.,

$$\int_K R(s,t)e_n(s)ds = \lambda_n e_n(t),\ n = 1, 2, \ldots,\ t \in K, \tag{5.67}$$

and the correlation matrix $R(s,t)$ is defined as

$$R(s,t) = E[X_s X_t], \tag{5.68}$$

and, meanwhile,

$$\xi_n(\theta) = \int_K X(t,\theta)e_n(t)dt. \tag{5.69}$$

5.2.3 Spectral Galerkin Methods

The approximation provided by a finite truncation of the PC expansion (5.61) and gPC chaos expansion (5.65) allows for seeking a solution for the stochastic operator equation (5.51) in the following form

$$u(\mathbf{x},\boldsymbol{\theta}) = \sum_{j=1}^{N_p} u_j(\mathbf{x})\psi_j(\boldsymbol{\xi}), \quad \forall \mathbf{x} \in D, \tag{5.70}$$

where N_p is defined in Remark 5.1 and the d-independent random variable

$$\boldsymbol{\xi} = \{\xi_i\}_{i=1}^{d} \in \Xi = \prod_{i=1}^{d} \Xi_\mathbf{i}, \quad \xi_i \in \Xi_\mathbf{i} \subset \mathbb{R}, \tag{5.71}$$

with probability density functions (PDFs) $\{\rho_i\}_{i=1}^{d}$ are assumed to model the stochasticity at random parameters and data in the stochastic equations. The state space $\Xi_\mathbf{i}$ of each r.v. ξ_i is assumed to be the whole $\mathbb{R}$ or a subset of $\mathbb{R}$ where its PDF will be zero outside the subset. Now, in (5.51)–(5.53), we have

$$\mathcal{L}(\mathbf{x},\boldsymbol{\theta}) = \mathcal{L}(\mathbf{x},\xi_1(\theta),\ldots\xi_d(\theta)), \quad \mathbf{x} \in \mathbb{R}, \tag{5.72}$$

and

$$f(\mathbf{x},\boldsymbol{\theta}) = f(\mathbf{x},\xi_1(\theta),\ldots\xi_d(\theta)),$$

$$a(\mathbf{x},\boldsymbol{\theta}) = a(\mathbf{x},\xi_1(\theta),\ldots\xi_d(\theta)), \tag{5.73}$$

$$b(\mathbf{x},\boldsymbol{\theta}) = b(\mathbf{x},\xi_1(\theta),\ldots\xi_d(\theta)), \tag{5.74}$$

$$h(\mathbf{x}, \boldsymbol{\theta}) = h(\mathbf{x}, \xi_1(\theta), \ldots \xi_d(\theta)),$$

where for each r.v. $\xi_i(\theta)$: $\theta \in \Omega_i \to \mathbb{R}$ (the same symbol θ is used for all sample spaces Ω_i when no confusion exists) and $\boldsymbol{\theta} = \{\theta_1, \ldots, \theta_d\} \in \Omega = \prod_{i=1}^{d} \Omega_i$, and, in (5.55),

$$\Sigma(\mathbf{x}, \boldsymbol{\theta}) = \Sigma(\mathbf{x}, \xi_1(\theta), \ldots \xi_d(\theta)). \tag{5.75}$$

The stochastic spectral Galerkin method [9] searches for a solution $u(\mathbf{x}, \theta)$ of the form (5.51) such that

$$\int_{\Omega} \mathcal{L}(\mathbf{x}, \boldsymbol{\xi})[u]\psi_i(\boldsymbol{\xi})P(d\omega) = \int_{\Omega} f(\mathbf{x}, \boldsymbol{\xi})\psi_i(\boldsymbol{\xi})P(d\omega), \; i = 1, \ldots, N_p, \tag{5.76}$$

i.e.,

$$E[\mathcal{L}(\mathbf{x}, \boldsymbol{\xi})[u]\psi_i(\boldsymbol{\xi})] = E[f(\mathbf{x}, \boldsymbol{\xi})\psi_i(\boldsymbol{\xi})]. \tag{5.77}$$

In terms of the probability density function $\rho(\mathbf{z})$ for the d-variate random vector ξ

$$\rho(\mathbf{z}) = \prod_{k=1}^{d} \rho_k(z_k), \quad \mathbf{z} \in \Xi, \tag{5.78}$$

where $\rho_k(z_k)dz_k = P(\theta \in \Omega_k \mid \xi_k(\theta) \in (z_k \cdot z_k + dz_k]\}$, we have for each $\mathbf{x} \in D$

$$\int_{\Xi} \mathcal{L}(\mathbf{x}, \mathbf{z})[u]\psi_i(\mathbf{z})\rho(\mathbf{z})d\mathbf{z} = \int_{\Xi} f(\mathbf{x}, \mathbf{z})\psi_i(\mathbf{z})\rho(\mathbf{z})d\mathbf{z}, \; i = 1, \ldots, N_p, \tag{5.79}$$

which exactly mimics the spectral Galerkin method (5.21) or (5.47) using orthogonal polynomial $\psi_i(\mathbf{z})$ over $\mathbf{z} \in \Xi$ under the weight $\rho(\mathbf{z})$. Therefore, if $\rho_i(z)$ are all Gaussian PDFs, then $\psi_i(\mathbf{z})$ will be the tensor product of Hermite polynomials, i.e.,

$$\psi_i(\mathbf{z}) = H_{n_1}(z_{i_1})H_{n_2}(z_{i_2}) \cdots H_{n_k}(z_{i_k}), \tag{5.80}$$

assuming the index ordering (5.62).

Equation (5.79) gives a system of N_p (coupled) deterministic partial differential equations over domain D for unknown functions $\{u_j(\mathbf{x})\}_{j=1}^{N_p}$, which will be discretized by traditional numerical methods such as finite element, finite difference or spectral methods.

To see the resulting system, let us consider the case of the stochastic elliptic equation (5.52) and (5.53) while the case of the stochastic integral equation will be considered in Sect. 5.3. The system (5.79) now gives

$$\int_{\Xi} \big(-\nabla(a(\mathbf{x},\xi)\nabla u(\mathbf{x},\xi)) + b(\mathbf{x},\xi)u(\mathbf{x},\xi)\big)\psi_i(\mathbf{z})\rho(\mathbf{z})d\mathbf{z}$$
$$= \int_{\Xi} f(\mathbf{x},\mathbf{z})\psi_i(\mathbf{z})\rho(\mathbf{z})dz$$
$$\equiv f_i(\mathbf{x}), \quad \mathbf{x} \in D, \tag{5.81}$$

which becomes by using the PC or gPC expansion for the solution $u(\mathbf{x}, \xi)$,

$$-\sum_{j=1}^{N_p} \nabla \left(\int_{\Xi} a(\mathbf{x},\mathbf{z})\psi_j(\mathbf{z})\psi_i(\mathbf{z})\rho(\mathbf{z})d\mathbf{z} \right) \nabla u_j(\mathbf{x})$$
$$+ \left(\int_{\Xi} b(\mathbf{x},\mathbf{z})\psi_j(\mathbf{z})\psi_i(\mathbf{z})\rho(\mathbf{z})d\mathbf{z} \right) u_j(\mathbf{x}) = f_i(\mathbf{x}), \tag{5.82}$$

where $f_i(\mathbf{x})$ is the PC expansion coefficient for $f(\mathbf{x}, \boldsymbol{\xi})$ in (5.61) or (5.65). Namely,

$$\sum_{j=1}^{N_p} \big(-\nabla a_{ij}(\mathbf{x})\nabla u_j(\mathbf{x}) + b_{ij}(\mathbf{x})u_j(\mathbf{x})\big) = f_i(\mathbf{x}), \quad i = 1, \ldots, N_p, \tag{5.83}$$

where

$$a_{ij}(\mathbf{x}) = \int_{\Xi} a(\mathbf{x},\mathbf{z})\psi_j(\mathbf{z})\psi_i(\mathbf{z})\rho(\mathbf{z})d\mathbf{z}, \tag{5.84}$$

$$b_{ij}(\mathbf{x}) = \int_{\Xi} b(\mathbf{x},\mathbf{z})\psi_j(\mathbf{z})\psi_i(\mathbf{z})\rho(\mathbf{z})d\mathbf{z}. \tag{5.85}$$

Meanwhile, the boundary condition for (5.83) is simply

$$u_i(\mathbf{x}) = h_i(\mathbf{x}), \quad \mathbf{x} \in \partial D, \tag{5.86}$$

where $h_i(\mathbf{x})$ are the PC or gPC expansion coefficients for $h(\mathbf{x}, \theta)$.

Clearly, the stochastic Galerkin method produces a coupled N_p deterministic PDE system, which needs to be solved together; thus, this approach is computationally expensive for higher-order PC and gPC approximations to the SPDE problems (5.52) and (5.53).

5.2.4 Spectral Collocation Methods and Sparse Grids

An alternative approach for solving the stochastic equation (5.51) is using a collocation approach [10] as in (5.25) or (5.48), which is simply taking the test function $\psi_i(\mathbf{z}) = \delta(\mathbf{z} - \mathbf{z}_i)$ where $\{\mathbf{z}_i \in \Xi\}_{i=1}^{N_c}$ are N_c−collocation points in the state space of the random vector $\boldsymbol{\xi}(\theta)$. Now, the Eq. (5.81) will become simply

$$\mathcal{L}(\mathbf{x}, \mathbf{z}_i)[u] = f(\mathbf{x}, \mathbf{z}_i), \quad i = 1, 2, \ldots N_c, \ \mathbf{x} \in D. \tag{5.87}$$

For the stochastic elliptic equations (5.52) and (5.53), we will simply have the following N_c−uncoupled deterministic PDEs, for $i = 1, 2, \ldots N_c$,

$$-\nabla\big(a(\mathbf{x}, \mathbf{z}_i)\nabla u(\mathbf{x}, \mathbf{z}_i)\big) + b(\mathbf{x}, \mathbf{z}_i)u(\mathbf{x}, \mathbf{z}_i) = f(\mathbf{x}, \mathbf{z}_i), \qquad \mathbf{x} \in D, \tag{5.88}$$

$$u(\mathbf{x}, \mathbf{z}_i) = h(\mathbf{x}, \mathbf{z}_i), \qquad \mathbf{x} \in \partial D. \tag{5.89}$$

The stochastic collocation method is very similar to Monte Carlo (MC) sampling strategy, but the state space of the random vector $\boldsymbol{\xi}$ is sampled at some prescribed collocation points (e.g., the zeros of some orthogonal polynomials), instead of distribution-guided samples in the traditional MC method. To minimize the number of collocation points while still maintaining the overall accuracy of the collocation method, a sparse grid method is usually used.

Sparse grids for interpolation and integration. For simplicity of argument, we assume that all the r.v. values are in the range $[-1, 1]$. We first start with 1-d m_i interpolation points of level-i,

$$\Theta_1^i = \{z_1^{(i)}, \ldots, z_{m_i}^{(i)}\} = \{z_j^{(i)}, 1 \le j \le m_i\} \tag{5.90}$$

where the collocation points $z_j^{(i)}$ will be the zeros of some orthogonal polynomials for best approximations, say, $H_{m_i}(z_j) = 0$, $1 \le j \le m_i$ if the random variable has a Gaussian distribution. However, for efficient implementation, nested collocation points between different levels are often used, instead. The simplest nested collocation points are those from an equally spaced composite trapezoidal rule. Another popular choice is the Clenshaw–Curtis or Fejér quadrature points over [−1,1] [11], which are defined as the extrema of Chebyshev polynomials (Fejér points ignore the end points −1, 1),

$$z_j^{(i)} = -\cos\Big(\frac{(j-1)\pi}{m_i - 1}\Big), \ m_i = 2^{i-1} + 1, 1 \le j \le m_i, i > 1, \tag{5.91}$$

and by default, we set $m_1 = 1$, $z_1^{(1)} = 0$.

A corresponding level i interpolation $I^i(f)$ for function $f(z)$ using m_i values can be defined as in (5.27)

$$I^i(f)(z) = \sum_{j=1}^{m_i} f(z_j^{(i)}) l_j^{(i)}(z), \tag{5.92}$$

where the basis function $l_j^{(i)}(z)$ can be the Lagrangian function defined in (5.26) or other continuous piecewise polynomial functions. Correspondingly, the one dimensional level-i quadrature formula can be defined as

$$\int_{\Xi_i} f(z)\rho_i(z)dz = \sum_{j=1}^{m_i} \omega_j^i f(z_j), \tag{5.93}$$

where $\Xi_i = \text{supp}\rho_i(z)$.

A tensor product extension for an $\mathbf{i} = (i_1, \ldots, i_d)$-level interpolation for a d-dimensional function $f(z_1, \ldots z_d)$ is then

$$\begin{aligned} I_d^{\text{tensor}}(f)(z_1, \ldots, z_d) &\equiv (I^{i_1} \otimes \cdots \otimes I^{i_d})(f) \\ &= \sum_{j_1=1}^{m_{i_1}} \cdots \sum_{j_d=1}^{m_{i_d}} f(z_{j_1}^{(i_1)}, \ldots z_{j_d}^{(i_d)}) l_{j_1}^{(i_1)}(z_1) \cdots l_{j_d}^{(i_d)}(z_d), \end{aligned} \tag{5.94}$$

where I^{i_k}, $1 \le k \le d$ is the i_k-level interpolation operator for the z_k variable. Equation (5.94) involves $m_{i_1} \cdots m_{i_d} = m^d$ number of terms if $m_{i_1} = \cdots = m_{i_d} = m$, which grows exponentially in terms of the dimension d, resulting in a problem of the curse of dimensionality (CoD).

To avoid the CoD problem, a k-level Smolyak sparse grid interpolation [12] on a d-dimensional cube $\Xi = [-1, 1]^d$, proposed in [13–15], can be used to define the sparse interpolation operator as follows [16]

$$I_d^{\text{sparse}}(f) \equiv \sum_{k+1 \le i_1 + \cdots + i_d \le d+k} (-1)^{d+k-|\mathbf{i}|_1} \binom{d-1}{d+k-|\mathbf{i}|_1} (I^{i_1} \otimes \cdots \otimes I^{i_d})(f), \tag{5.95}$$

where the d-variate interpolation level index $\mathbf{i} = (i_1, \ldots, i_d)$, $|\mathbf{i}|_1 = i_1 + \cdots + i_d$. (5.95) only requires the values of the function f at a subset Θ_d^k of the tensor product grid

$$\Theta_d^k = \bigcup_{k+1 \le i_1 + \cdots + i_d \le d+k} (\Theta_1^{i_1} \times \cdots \times \Theta_1^{i_d}) \subset [-1, 1]^d. \tag{5.96}$$

It was shown [17] that

$$\dim(\Theta_d^k) \sim \frac{2^k}{k!} d^k \sim 2^k \dim(\pi_d^k), \tag{5.97}$$

which grows algebraically in terms of the dimension d, and here π_d^k denotes the space of d-variable polynomials of total degree at most k. A corresponding quadrature weights can also be defined as

$$\omega^{i_1,\ldots,i_d}_{j_1,\ldots,j_d} = (-1)^{d+k-|\mathbf{i}|_1} \binom{d-1}{d+k-|\mathbf{i}|_1} (\omega^{i_1}_{j_1} \cdots \omega^{i_d}_{j_d}), \tag{5.98}$$

and, any multidimensional integral can then be approximated by the following quadrature,

$$\int_{\Xi} f(\mathbf{z})\rho(\mathbf{z})d\mathbf{z} \sim \sum_{(z^{(i_1)}_{j_1},\ldots,z^{(i_d)}_{j_d}),1\le j_l\le m_{i_l},1\le l\le d} \omega^{i_1,\ldots,i_d}_{j_1,\ldots,j_d} f(z^{(i_1)}_{j_1},\ldots,z^{(i_d)}_{j_d}), \tag{5.99}$$

which can be shown to be exact for the d-variable polynomial space π^k_d [18], and can be used to compute the statistical moments of random function $f(\boldsymbol{\theta})$. For software implementation of sparse grids, refer to [16].

5.3 Uncertainty Quantification of Rough Microchip Interconnects

In this section, we apply the stochastic collocation method for the uncertainty quantification of the effects of the stochastic variance due to manufacturing processes on the electrical properties of interconnects in nanoscale microchips. Two important physical parameters for passive interconnects, the capacitance and inductance, are most needed in establishing equivalent discrete circuit models of interconnects for the simulation of large-scale microchip components. Here, we will consider the capacitance parameter extraction problem for a collection of $m+1$-conductors in 3-D, as shown in Fig. 5.1.

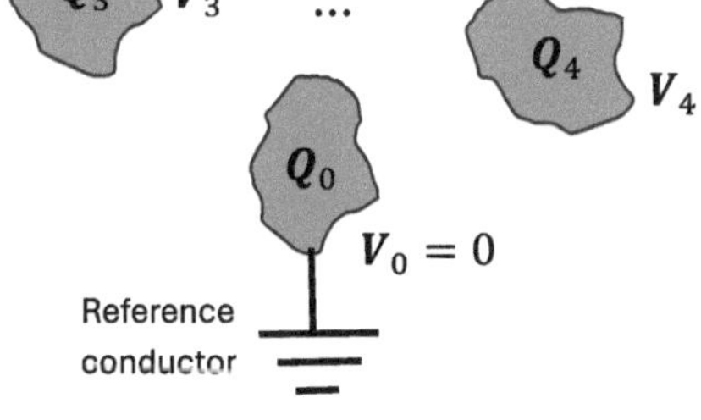

Fig. 5.1 A collection of conductors in space with one reference conductor grounded at zero voltage

First, we define the following physical quantities:

- V_i—the voltage on the ith conductor, the voltage on a selected reference conductor is set to ground zero voltage, $V_0 = 0$.
- Q_i—the total surface charge on the ith conductor
- C_{ij}—the coupling capacitance between ith and jth conductors.

Then, the charge on the ith conductor is the combined contribution from the jth conductor, which is the product of the coupling capacitance C_{ij} times the voltage difference $V_i - V_j$. Namely

$$\begin{aligned} Q_1 &= C_{10}V_1 + C_{12}\left(V_1 - V_2\right) + \cdots C_{1m}\left(V_1 - V_m\right), \\ Q_2 &= C_{21}\left(V_2 - V_1\right) + C_{20}V_2 + \cdots C_{2m}\left(V_2 - V_m\right), \\ &\vdots \\ Q_m &= C_{m1}\left(V_m - V_1\right) + C_{m2}\left(V_m - V_2\right) + \cdots C_{m0}V_m, \end{aligned} \tag{5.100}$$

and the parameter C_{ij} can be inverted from this system by setting the voltage at 1 on one of the conductors and 0 on the others, and computing the corresponding charge distribution by solving the electric potential problem with some numerical methods.

The charge distribution $q(\mathbf{r})$ on all the conductor surfaces and the electric potential $\phi(\mathbf{r})$ satisfies the following integral equation from the principle of superposition,

$$\phi(\mathbf{r}) = \int_{\partial C} q(\mathbf{r}')G(\mathbf{r}, \mathbf{r}')d\mathbf{r}', \quad \mathbf{r} \in \mathbb{R}^3, \tag{5.101}$$

where $\partial C = \cup_{i=0}^{m}\partial c_i$ denotes the collection of surfaces of all conductor c_i, and the free-space Green's function for the Coulomb potential

$$G(\mathbf{r}, \mathbf{r}') = \frac{1}{4\pi\varepsilon_0|\mathbf{r} - \mathbf{r}'|}. \tag{5.102}$$

Therefore, the charge Q_i on the ith conductor ∂c_i is

$$Q_i = \int_{\partial c_i} q(\mathbf{r})ds. \tag{5.103}$$

We assume

$$\partial C(\mathbf{r}) = \partial C_0(\mathbf{r}) + \Sigma(\mathbf{r}, \theta), \tag{5.104}$$

where the random field $\Sigma(\mathbf{r}, \theta)$ indicates the stochastic variance of the conductor surface.

We first mesh a nominal surface (see an example for a 2×2-bit bus interconnect in Fig. 5.2) into M-panels

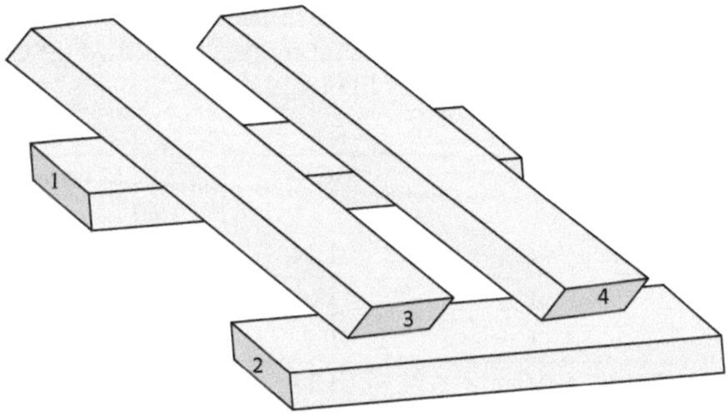

Fig. 5.2 A 2 × 2-bit bus interconnect with a perfect surface

$$\partial C_0(\mathbf{r}) = \bigcup_{i=0}^{M} \mathbf{a}_i, \tag{5.105}$$

and over each panel, along its normal direction, a random displacement $\sigma_i(\theta)$ will be assigned at the center s_i of the panel

$$\partial C(\mathbf{r}) = \bigcup_{i=0}^{M} (\mathbf{a}_i + \sigma_i(\theta)). \tag{5.106}$$

The Nyström collocation method described in Sect. 11.3, applied at the center of each panel, can be used to solve the integral equation for a given realization of a random conductor surface to give a matrix equation

$$P\mathbf{q} = \mathbf{v}, \tag{5.107}$$

where $\mathbf{q} = (q_1, q_2, \ldots, q_M)^{\perp}, \mathbf{v} = (v_1, v_2, \ldots, v_M)^{\perp}$ are the charges and voltages at the center of all panels, and the discretized integral operator matrix

$$p_{i,j}(\theta) = \frac{1}{4\pi\varepsilon_0} \int_{\mathbf{a}_j} \frac{1}{|\mathbf{a}_i - r_j + (\sigma_i - \sigma_j)|} ds_j. \tag{5.108}$$

To apply the stochastic collocation method to solve (5.101), we first reduce the dimension of the random variable $\sigma = (\sigma_1, \sigma_2, \ldots, \sigma_M)^{\perp}$ into a small set of iid Gaussian random variable $\boldsymbol{\xi} = \{\xi_i\}_{i=1}^{d}$ as in Sect. 5.2.2, and, then, Eq. (5.107) becomes

$$P(\boldsymbol{\xi})\mathbf{q} = \mathbf{v}, \tag{5.109}$$

which can be solved using the collocation method at the collocation point $\mathbf{z}_k \in \Xi$

$$P(\mathbf{z}_k)\mathbf{q}^{(k)} = \mathbf{v}^{(k)}, \tag{5.110}$$

Table 5.1 The mean and variance of the capacitance of a 2-bit bus and a 2 × 2-bit bus with a second-order PC stochastic spectral collocation method (SSCM) and a Monte Carlo result. Taken from [19] with permission by EDAA

$\overline{C} = E[C]$				$\sigma = \sqrt{E[(C - \overline{C})^2]}$	
		MC	SSCM error (%)	MC	SSCM error (%)
2-bit bus	C_{11}	1.14	0.09	0.21	4.1
	C_{22}	1.60	0.14	0.39	4.6
	C_{12}	−0.7	0.16	0.20	4.8
2 × 2 bus	C_{11}	1.17	0.03	0.23	3.4
	C_{12}	−0.7	0.04	0.21	4.0

and, then a PC expansion for the random vector $\mathbf{q}$ can be obtained by either interpolation or a least square fitting for the expansion coefficients.

Table 5.1 shows the mean and variance of the computed capacitance of a second-order PC stochastic spectral collocation method (SSCM) compared with a Monte Carlo result. Moreover, Fig. 5.3 shows the cumulative density function for the capacitance parameter C_{22} computed by the SSCM and Monte Carlo methods as well as a perturbation method [19].

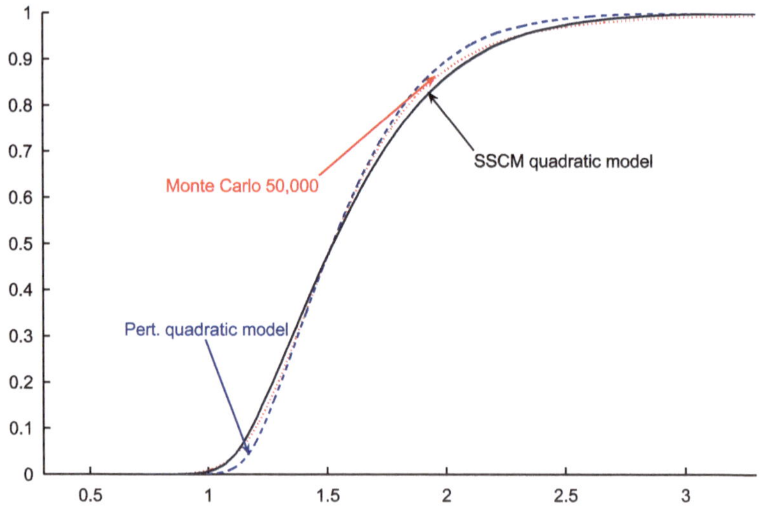

Fig. 5.3 Cumulative density function for C_{22} for a 3-bit bus interconnect: solid line—second-order SSCM, dashes—Monte Carlo method with 50,000 samples, dots—a second-order perturbation method. Taken from [19] with permission by EDAA

5.4 Summary

Stochastic spectral methods, particularly stochastic collocation methods based on the Wiener polynomial chaos and the generalized polynomial chaos, provide efficient numerical techniques for solving stochastic differential and integral equations to estimate the uncertainty of the physical quantities of interest due to stochastic variance from material properties or manufacturing processes. For random variables of modest dimensions, numerical experiments have shown that spectral collocation methods perform more efficiently than traditional Monte Carlo simulations. However, as the random dimension increases, the high dimensionality of the problem becomes a challenge for stochastic spectral methods even with sparse grids.

References

1. C. Canuto, M.Y. Hussaini, A. Quarteroni, T.A. Zang, *Spectral Methods in Fluid Dynamics* (Springer, 1988)
2. D. Gottlieb, S. Orszag, *Numerical Analysis of Spectral Methods: Theory and Applications* (SIAM-CBMS, Philadelphia, PA, 1987)
3. M. Abramowitz, I.A. Stegun, *Handbook of Mathematical Functions with Formulas, Graphs, and Mathematical Tables*, 9th edn. (Dover Publications, New York, 1972)
4. J. Shen, T. Tang, L.L. Wang, *Spectral Methods: Algorithms, Analysis and Applications* (Springer Science & Business Media, 2011)
5. N. Wiener, The homogeneous chaos. Amer. J. Math. **60**, 897–936 (1938)
6. D. Xiu, G.E. Karniadakis, The Wiener-Asky polynomial chaos for stochastic differential equations. SIAM J. Sci. Comput. **24**(2), 619–644 (2002)
7. D. Xiu. *Numerical Methods for Stochastic Computations: A Spectral Method Approach* (Princeton University Press, 2010)
8. O.G. Ernst, A. Mugler, H.J. Starkloff, E. Ullmann, On the convergence of generalized polynomial chaos expansions. ESAIM: Math. Model. Numer. Anal. **46**(2), 317–339 (2012)
9. R. Ghanem, P. Spanos, *Stochastic Finite Elements: A Spectral Approach* (Springer, New York, 1991)
10. D. Xiu, J.S. Hesthaven, High-order collocation methods for differential equations with random inputs. SIAM J. Sci. Comput. **27**(3), 1118–1139 (2005)
11. J. Waldvogel, Fast construction of the Fejér and Clenshaw-Curtis quadrature rules. BIT Num. Math. **46**, 195–202 (2006)
12. S.A. Smolyak, Quadrature and interpolation formulas for tensor products of certain classes of functions. Soviet Math. **4**, 240–243 (1963)
13. G. Wasilkowski, H. Wozniakowski, Explicit cost bounds of algorithms for multivariate tensor product problems. J. Complex. **11**, 1–56 (1995)
14. V. Barthelmann, E. Novak, K. Ritter, High dimensional polynomial interpolation on sparse grids. Adv. Comput. Math. **12**, 273–288 (2000)
15. A. Klimke, B. Wohlmuth. Algorithm 847: Spinterp: piecewise multilinear hierarchical sparse grid interpolation in MATLAB. ACM Trans. Math. Software **31** (2005)
16. R.C. Smith, *Uncertainty Quantification: Theory, Implementation, and Applications* 2nd edn. (Society for Industrial and Applied Mathematics, 2024)
17. E. Novak, K. Ritter, High dimensional integration of smooth functions over cubes. Numer. Math. **75**, 79–97 (1996)

18. E. Novak, K. Ritter, Simple cubature formulas with high polynomial exactness. Constr. Approx. **15**(4), 449–522 (1999)
19. H. Zhu, X. Zeng, W. Cai, J. Xue, D. Zhou, A sparse grid based spectral stochastic collocation method for variations-aware capacitance extraction of interconnects under nanometer process technology, in *2007 Design, Automation and Test in Europe Conference and Exhibition* (IEEE, 2007), pp. 1–6

Chapter 6
Deep Neural Network Learning for PDE Solutions

Deep neural networks (DNNs) employ successive function compositions of affine mappings and nonlinear activation function operations to create approximations for general functions. The coefficients in the affine mappings comprise the neural networks' trainable parameters in machine learning methods. The mathematical foundation for the DNN approximation is based on its universal approximation property, which states that any continuous function can be arbitrarily closely approximated by a sufficiently wide one-hidden-layer DNN with a continuous sigmoidal nonlinear activation function (i.e., a function that saturates as the input variable goes to infinity) [1–3]. Quantitative approximation error estimates of deep ReLU networks for continuous and smooth functions in terms of width and depth have been established [4, 5].

Learning PDE solutions with DNNs is performed differently from the traditional discretization and solving linear system approach in finite element or finite difference methods. The DNN solution is formulated as the minimizer of some objective loss function, which is found by a stochastic gradient descent (SGD) optimization process (i.e., a learning process) over the neural network parameters while taking advantage of the efficient calculation of the derivatives of DNNs through auto-differentiation. Stochasticity is introduced in the SGD search algorithm, enabling it to escape the possibly numerous local minima of non-convex objective loss functions.

6.1 Deep Neural Network (DNN)

A deep neural network is composed of a sequential alternate composition with linear functions and nonlinear activation functions. Given $m, n \geq 1$, let $\Theta(\mathbf{x}) : \mathbb{R}^n \to \mathbb{R}^m$ be a linear function of the form $\Theta(\mathbf{x}) = \mathbf{W}\mathbf{x} + \mathbf{b}$, where $\mathbf{W} = (w_{ij}) \in \mathbb{R}^{m\times n}$, and $\mathbf{b} \in \mathbb{R}^m$ are called weights and biases, respectively. $\sigma(u) : \mathbb{R} \to \mathbb{R}$ denotes a nonlinear

W. Cai, *Deterministic, Stochastic, and Deep Learning Methods for Computational Electromagnetics*, https://doi.org/10.1007/978-981-96-0100-4_6

activation function. By applying $\sigma(u)$ component-wise, we can extend the activation function to $\sigma(\mathbf{u}) : \mathbb{R}^n \to \mathbb{R}^n$. A large depth L gives a DNN, and a $L+1$ layer neural network can be expressed in a compact form as

$$\begin{aligned} f_{\theta}(\mathbf{x}) &= f(\mathbf{x}, \boldsymbol{\theta}) \equiv f_L(\mathbf{x}), \\ f_l(\mathbf{x}) &= [\Theta_l \circ \sigma] \circ f_{l-1}(\mathbf{x}), \quad l = L, L-1, \ldots 1, \end{aligned} \tag{6.1}$$

with $f_0(\mathbf{x}) = \Theta_0(\mathbf{x})$, or equivalently,

$$f_{\theta}(\mathbf{x}) = f(\mathbf{x}, \boldsymbol{\theta}) = \Theta_L \circ \sigma \circ \Theta_{L-1} \circ \sigma \cdots \circ \Theta_1 \circ \sigma \circ \Theta_0(\mathbf{x}). \tag{6.2}$$

The DNN is said to have L hidden layers, the lth layer has a width of n^l neurons, and all weight and bias parameters in the DNN are collectively denoted by $\boldsymbol{\theta}$.

Commonly used activation functions are given below.

- Rectified linear unit-ReLU [6]

$$\mathrm{ReLU}(x) = \max\{0, x\}. \tag{6.3}$$

- LeakyReLU [7]

$$\mathrm{LeakyReLU}(x) = \begin{cases} x & \text{for } x \geq 0 \\ \alpha x & \text{for } x < 0 \end{cases} \text{ with } \alpha \in \mathbb{R}. \tag{6.4}$$

- ReLU^n

$$\mathrm{ReLU}^n(x) = \max\{0, x^n\},\ n \in \mathbb{N}^+. \tag{6.5}$$

- Exponential linear unit—ELU [8]

$$\mathrm{ELU}(x) = \begin{cases} x & \text{for } x \geq 0 \\ \alpha(e^x - 1) & \text{for } x < 0 \end{cases} \text{ with } \alpha \in \mathbb{R}. \tag{6.6}$$

- Scaled exponential linear unit—SELU [9]

$$\mathrm{SELU}(x) = \lambda \mathrm{ELU}(x),\ \lambda \in (0, \infty). \tag{6.7}$$

- Softplus [10]

$$\mathrm{Softplus}(x) = \ln(1 + e^x). \tag{6.8}$$

- Sigmoid

$$\mathrm{Sigmoid}(x) = \frac{1}{1 + e^{-x}}. \tag{6.9}$$

- Tanh

$$\mathrm{Tanh}(x) = \frac{e^x - e^{-x}}{e^x + e^{-x}}. \tag{6.10}$$

6.2 Optimization with Stochastic Gradient Descent Method

A DNN can be used to approximate a target function or the solution of PDEs. Finding the DNN parameter $\boldsymbol{\theta}$ is usually done through an optimization procedure by minimizing a loss function $L(\boldsymbol{\theta})$ with a gradient decent method as follows:

$$\boldsymbol{\theta}_{n+1} = \boldsymbol{\theta}_n - \eta_n \nabla_{\boldsymbol{\theta}} L(\boldsymbol{\theta}_n), \tag{6.11}$$

where $\eta_n < 1$ is the learning rate and ∇_θ calculated by the back-propagation technique of Hinton (1986) in his 2024 Nobel Prize work [11]. As the objective function $L(\theta)$ is generally non-convex and may have many local minima, the stochastic gradient descent (SGD) [12, 13] is often used in deep learning. The stochasticity could be introduced by a realization $\widetilde{L}(\boldsymbol{\theta})$ of the loss function, which is viewed as an expectation of a random variable X through sampling (mini-batch). Namely, if

$$L(\boldsymbol{\theta}) = \int_D l(\mathbf{x}, \boldsymbol{\theta})\pi(\mathbf{x})d\mathbf{x} = E_{X\sim\pi}[l(X, \boldsymbol{\theta})], \tag{6.12}$$

then, a n-point sample in D according to the distribution $\pi(x)$ (i.e., a mini-batch of size n) will give an empirical loss

$$\widetilde{L}(\boldsymbol{\theta}) = \frac{1}{n}\sum_{i=1}^{n} l(\mathbf{x}_i, \boldsymbol{\theta}), \quad \{\mathbf{x}_i\} \sim \pi(\mathbf{x}), \tag{6.13}$$

and, the SGD in a mini-batch implementation is then

$$\boldsymbol{\theta}_{n+1} = \boldsymbol{\theta}_n - \eta_n \nabla_{\boldsymbol{\theta}} \widetilde{L}(\boldsymbol{\theta}_n). \tag{6.14}$$

Other sources of stochasticity can come from stochastic variance in measurement data or a stochastic model of the physical problem or even the definition of the loss function itself (see Sect. 6.6).

A popular efficient optimizing algorithm in deep learning is the ADAM (adaptive moment estimation), an adaptive algorithm that scales component-wise the gradient and moment directions [14]. The ADAM for the empirical loss is given as follows

$$\begin{aligned}
v_{n+1} &= \alpha v_n + (1-\alpha)(\nabla_{\boldsymbol{\theta}} \widetilde{L}(\boldsymbol{\theta}_n))^2, \\
m_{n+1} &= \beta m_n + (1-\beta)\nabla_{\boldsymbol{\theta}} \widetilde{L}(\boldsymbol{\theta}_n), \\
\boldsymbol{\theta}_{n+1} &= \boldsymbol{\theta}_n - \eta_n \frac{m_{n+1}}{\sqrt{v_{n+1}} + \varepsilon},
\end{aligned} \tag{6.15}$$

where $(\nabla_{\boldsymbol{\theta}} \widetilde{L}(\boldsymbol{\theta}_n))^2$ in the equation for the vector v_n denotes the component-wise product of the gradient vector, and $\sqrt{v_{n+1}}$ denotes the square root of each component of the vector v_{n+1} as well as the division in the SGD update equation for θ_{n+1}, ε is a small quantity used to avoid overflow in the division. Finally, parameters $\alpha, \beta \in (0, 1]$ reflect the memory effect of the past gradients and second moment of the gradients.

In a regression problem of learning a target function, the loss function $L(f_{\boldsymbol{\theta}}(\mathbf{x}), \boldsymbol{y})$ uses the difference between a prediction of the DNN and a true label. The empirical loss for a set $S = \{(\mathbf{x}_i, \boldsymbol{y}_i)\}_{i=1}^n$ is then

$$\widetilde{L}(\boldsymbol{\theta}) = \frac{1}{n}\sum_{i=1}^{n} \ell(f_{\boldsymbol{\theta}}(\mathbf{x}_i), \boldsymbol{y}_i), \tag{6.16}$$

where $\ell(f_{\boldsymbol{\theta}}(\mathbf{x}_i), \boldsymbol{y}_i) = |f_{\boldsymbol{\theta}}(\mathbf{x}_i) - \boldsymbol{y}_i|^2$, for instance.

In learning PDE solutions, the empirical loss can be defined without labels using the least square residual of the PDEs as in the physics-informed neural network (PINN) [15, 16] or the variational loss of an energy functional in the DeepRitz method [17] and others, to be discussed in the sections below.

6.3 DeepRitz Method for the Poisson–Boltzmann Equation of Biomolecular Solvation

In this section, we present the DeepRitz method [17] based on the variational formulation of a self-adjoint elliptic PDE. Here, we consider the Poisson–Boltzmann equation (PBE) (2.19) with a right-hand side $f(\mathbf{x})$ within a finite domain $D = (\Omega_i \cup \Omega_o) \cap \{||\mathbf{x}|| \leq R\}$, $R > 0$, together with the interface condition (2.3) on $\Gamma \subset D$, which can be the surface of a solute molecule inside a solvent Ω_o. For simplicity, instead of the decay condition at infinity (2.21), an approximate homogeneous boundary condition on ∂D is used, i.e.,

$$u|_{\partial D} = 0. \tag{6.17}$$

The deep Ritz method seeks a variational solution $u(\mathbf{x})$ of Eqs. (2.3), (2.19) and (6.17) through the following minimization problem

$$u = \arg \min_{v \in H_0^1(D)} J(v), \tag{6.18}$$

where the energy functional is defined as

$$\begin{aligned} J(v) = J(v, f) &= \int_D \frac{1}{2}\left(\epsilon(\mathbf{x})|\nabla v(\mathbf{x})|^2 + \lambda^2(\mathbf{x})v(\mathbf{x})^2\right) d\mathbf{x} - \int_D f(\mathbf{x})v(\mathbf{x})d\mathbf{x} \\ &\triangleq \int_D e(v(\mathbf{x}))\mathrm{d}\mathbf{x}. \end{aligned} \tag{6.19}$$

A DNN $u_{\boldsymbol{\theta}}(\mathbf{x})$ will be used to represent the trial function $v(\mathbf{x})$ above, and the DNN solution u_{θ^*} is defined by

$$\boldsymbol{\theta}^* = \arg\min_{\boldsymbol{\theta}} J(u_{\boldsymbol{\theta}}). \tag{6.20}$$

The minimizer $\boldsymbol{\theta}^*$ is found by the SGD method (6.14) or (6.15). To introduce stochasticity into the SGD, the integral in Eq. (6.19) is computed with a Monte Carlo method with randomly samples at $\mathbf{x}_i, i = 1, \ldots, n$ at each training step, namely,

$$\nabla_{\boldsymbol{\theta}} J(u_{\boldsymbol{\theta}}(\mathbf{x})) \sim \nabla_{\boldsymbol{\theta}} \frac{1}{n} \sum_{i=1}^{n} e(u_{\boldsymbol{\theta}}(\mathbf{x}_i)). \tag{6.21}$$

At convergence, $\boldsymbol{\theta}^{(n)} \to \boldsymbol{\theta}^*$, we obtain a DNN approximate solution $u_{\boldsymbol{\theta}^*}(\mathbf{x})$ to $u(\mathbf{x})$.

Variational Functional for Non-homogeneous Dirichlet Boundary Conditions

To derive the functional (6.19) for the PB equation with a non-homogeneous boundary condition

$$u|_{\partial D} = g, \tag{6.22}$$

we can construct a spatial extension function $\widetilde{g} \in C^2(D)$, such that

$$\widetilde{g}|_{\partial D} = g, \ \frac{\partial \widetilde{g}}{\partial n}|_{\partial D} = 0, \ \sup(\widetilde{g}) \cap \Gamma = \emptyset, \tag{6.23}$$

and consider the function

$$w = u - \widetilde{g}, \tag{6.24}$$

which satisfies the equation (2.19) with a new right-hand side

$$\widetilde{f} = f + \epsilon(\mathbf{x})\Delta\widetilde{g} - \lambda^2(\mathbf{x})\widetilde{g}, \tag{6.25}$$

and a homogeneous boundary condition on ∂D. Therefore, the solution can be found as the minimizer of the following minimization problem

$$w^* = \arg \min_{v \in H_0^1(D)} J(v, \tilde{f}). \tag{6.26}$$

Now consider the set

$$V_g = \{w \in H^1(D) \mid w = \widetilde{g} + v, v \in H_0^1(\Omega)\} = \widetilde{g} \oplus H_0^1(D) \subset H^1(D). \tag{6.27}$$

Using the definitions of V_g and $\widetilde{f}$ in (6.25), we can show that for a piecewise constant $\epsilon(\mathbf{x})$,

$$
\begin{aligned}
J(v;\widetilde{f}) &= \frac{1}{2}\int_D \left(|\epsilon(\mathbf{x})\,(\nabla w - \nabla\widetilde{g})\,|^2 + \lambda^2(\mathbf{x})\,(w-\widetilde{g})^2\right) d\mathbf{x} \\
&\quad - \int_D \left(f + \epsilon(\mathbf{x})\Delta\widetilde{g} - \lambda^2(\mathbf{x})\widetilde{g}\right)(w-\widetilde{g})\, d\mathbf{x} \\
&= J(w;f) - \int_D \epsilon\nabla w\nabla\widetilde{g} d\mathbf{x} - \int_D \epsilon\Delta\widetilde{g} w dx + C(f,\widetilde{g}) \\
&= J(w;f) - \int_{\partial D} \epsilon\frac{\partial\widetilde{g}}{\partial n} w dx + C(f,\widetilde{g}) \\
&= J(w;f) + C(f,\widetilde{g}),
\end{aligned}
$$

where $\frac{\partial\widetilde{g}}{\partial n}|_{\partial D} = 0$ is used to arrive at the last equation and the constant term

$$
\begin{aligned}
C(f,\widetilde{g}) &= \frac{1}{2}\int_D \left(|\epsilon(\mathbf{x})\nabla\widetilde{g}|^2 + \lambda^2(\mathbf{x})\widetilde{g}^2\right) d\mathbf{x} \\
&\quad + \int_D f\widetilde{g} d\mathbf{x} + \int_D \left(\epsilon(\mathbf{x})\Delta\widetilde{g} - \lambda^2(\mathbf{x})\widetilde{g}\right)\widetilde{g} d\mathbf{x}.
\end{aligned}
\tag{6.28}
$$

Therefore, we have [35]

$$
\arg\min_{v\in H_0^1(D)} J(v,\widetilde{f}) = \arg\min_{w=v+\tilde{g}\in V_g} J(w,f). \tag{6.29}
$$

In practice, a penalty term can be used in the objective loss function to directly enforce the boundary condition as follows:

$$
w^* = \arg\min_{w\in H^1(D)} J(w,f) + \beta||w-g||^2. \tag{6.30}
$$

In the computation, the DeepRitz empirical loss function for the DNN u_θ is taken to be

$$
\begin{aligned}
L_{\mathrm{ritz}}(\theta) &= \frac{1}{|S|}\sum_{\mathbf{x}_i\in S}\left(\frac{1}{2}\epsilon(\mathbf{x}_i)|\nabla u_{\boldsymbol{\theta}}(\mathbf{x}_i)|^2 + \frac{1}{2}\lambda^2(\mathbf{x}_i)u_{\boldsymbol{\theta}}(\mathbf{x}_i)^2 - f(\mathbf{x}_i)u_{\boldsymbol{\theta}}(\mathbf{x}_i)\right) \\
&\quad + \beta L_{\mathrm{bdry}}(\boldsymbol{\theta}),
\end{aligned}
\tag{6.31}
$$

where the boundary loss

$$
L_{\mathrm{bdry}}(\boldsymbol{\theta}) = \frac{1}{|\tilde{S}|}\sum_{\mathbf{x}_i\in\tilde{S}}(u_{\boldsymbol{\theta}}(\mathbf{x}_i) - g(\mathbf{x}_i))^2, \tag{6.32}
$$

and S is the sample set in D and $\tilde{S}$ the sample set on ∂D, and β is a penalty parameter.

6.4 Physics-Informed Neural Network (PINN)

An alternative approach for finding a PDE's solution is to minimize the least squares of the PDE's residual for the DNN approximation u_θ, similar to the least squares finite element method (LSFEM) [18, 19]. This approach is used in the PINN [16] and LSNN [20].

Consider the following elliptic BVP problem

$$\mathcal{L}[u] + V(\mathbf{x}, u) = f(\mathbf{x}), \quad \mathbf{x} \in D, \qquad u = g, \quad \mathbf{x} \in \partial D, \tag{6.33}$$

where

$$\mathcal{L} = \frac{1}{2} \sum_{i,j=1}^{3} a_{i,j} \frac{\partial^2}{\partial x_i \partial x_j} + \mu \cdot \nabla. \tag{6.34}$$

The PINN loss function is given by

$$L_{\text{pinn}}(\boldsymbol{\theta}) = L_{\text{pde}}(\boldsymbol{\theta}) + \beta L_{\text{bdry}}(\boldsymbol{\theta}), \tag{6.35}$$

where

$$L_{\text{pde}}(\boldsymbol{\theta}) = \frac{1}{|S|} \sum_{\mathbf{x}_i \in S} \left(\mathcal{L}[u](\mathbf{x}_i) + V(x_i, u(x_i)) - f(\mathbf{x}_i) \right)^2, \tag{6.36}$$

and S is the sample set from D to evaluate the residuals of the PDE from the DNN $u_{\boldsymbol{\theta}}(\mathbf{x})$, and $L_{\text{bdry}}(\boldsymbol{\theta})$ is the boundary loss term in (6.32) with a penalty parameter β. The PINN solution $u_{\boldsymbol{\theta}^*}$ is then found by the following minimization problem

$$\boldsymbol{\theta}^* = \arg\min_{\boldsymbol{\theta}} L_{\text{pinn}}(\boldsymbol{\theta}). \tag{6.37}$$

6.5 Galerkin Weak Form Learning

Moreover, using a weak form of the PDE (6.33), a weak adversarial network (WAN) [21] can be constructed to find a DNN solution.

Galerkin weak form. Find $u \in H_0^1(D)$, such that for every test function $\phi \in H_0^1(D)$

$$<A[u], \phi> \equiv \int_D \left(\frac{1}{2} \sum_{i,j=1}^{d} a_{ij} \partial_j u \partial_i u + \mu \cdot \nabla u \phi + V(\mathbf{x}, u)\phi - f\phi \right) d\mathbf{x} = 0, \tag{6.38}$$

$$u = 0, \quad \mathbf{x} \in \partial D.$$

The weak solution u^* can be characterized by the minimum operator norm of the operator $A[u]$ defined by (6.38), i.e.,

$$\min_{u \in H_0^1} ||A[u]||_{op} = ||A[u^*]||_{op} = 0, \tag{6.39}$$

where the operator norm is defined as

$$||A[u]||_{\text{op}} \triangleq \max\{<A[u], \phi>^2/||\phi||_2^2, \quad \phi \in H_0^1(D), \phi \neq 0\}. \tag{6.40}$$

Therefore, the weak solution u^* can be found by solving the following min–max saddle-point problem: Find $u \in H_0^1(D)$ such that

$$u = \arg \min_{v \in H_0^1(D)} \max_{\phi \in H_0^1(D)} <A[v], \phi>^2/||\phi||_2^2. \tag{6.41}$$

The WAN method approximates the weak solution u via a DNN $u_{\boldsymbol{\theta}}$ while the test function ϕ_η is taken to be an adversarial network with parameter η, learned to challenge u_θ by maximizing $<A[u_{\boldsymbol{\theta}}], \phi_\eta>^2/||\phi_\eta||_2^2$ for each $u_{\boldsymbol{\theta}}$. In [21], the following loss function is used to learn both networks,

$$L_{\text{wan}}(\boldsymbol{\theta}, \boldsymbol{\eta}) = \log <A[u_{\boldsymbol{\theta}}], \phi_\eta>^2 - \log ||\phi_\eta||_2^2 + L_{\text{bdry}}(\boldsymbol{\theta}), \tag{6.42}$$

where a boundary loss $L_{\text{bdry}}(\boldsymbol{\theta})$ of (6.32) is used to enforce the boundary condition.

The WAN solution $u_{\boldsymbol{\theta}^*}$ is obtained through the following saddle-point problem,

$$(\boldsymbol{\theta}^*, \boldsymbol{\eta}^*) = \arg \min_{\boldsymbol{\theta}} \max_{\boldsymbol{\eta}} L_{\text{wan}}(\boldsymbol{\theta}, \boldsymbol{\eta}). \tag{6.43}$$

Moreover, a Friedrich weak form [22] for the (6.33) rewritten in the first system can be used to develop a Friedrich-learning scheme [23], which allows the treatment of solutions of less regularities such as jump discontinuities, again through a saddle-point min–max problem for the DNN solution and the test function networks.

6.6 Stochastic Differential Equation (SDE) Based DNN Methods

When the dimension of the PDEs is high, their solutions will be beyond the reach of traditional finite element or spectral methods due to the curse of dimensionality (CoD). However, DNNs can be used to address the CoD problem. The Pardoux–Peng nonlinear Feynman–Kac theory in Sect. 4.3 and martingale formulations of PDEs can be used to construct the loss function for the optimization problem for training the DNN solution of the PDE by sampling the paths of some stochastic processes, discretized in time by the Euler–Maruyama scheme [24].

6.6.1 DeepBSDE for Quasilinear Parabolic Equations via a Terminal Condition Loss

The DeepBSDE method [25] is based on the Pardoux–Peng theory of the nonlinear Feynman–Kac formula for quasilinear parabolic equations in Sect. 4.3. It trains a network solution for the terminal value problem of (4.51) (4.52) with a loss function using the terminal condition for r.v. Y_T in the BSDE (4.54).

For a given time interval $[0, T]$, we define a partition

$$0 = t_1 < t_2 < \cdots < t_i < t_{i+1} < \cdots < t_N = T, \tag{6.44}$$

and, on each interval $[t_n, t_{n+1}]$, use the Euler–Maruyama scheme for (4.53) (4.54) to calculate $X_{n+1} \in \mathbb{R}^d$ and $Y_{n+1} \in \mathbb{R}$ with realizations of Brownian motion B_n, i.e.,

$$X_{n+1} = X_n + \mu(t_n, X_n, Y_n, Z_n)\Delta t_n + \sigma(t_n, X_n, Y_n)\Delta B_n, \tag{6.45}$$

$$Y_{n+1} = Y_n + \phi(t_n, X_n, Y_n, Z_n)\Delta t_n + Z_{n+1}(t_n)^T \sigma(t_n, X_n, Y_n)\Delta B_n, \tag{6.46}$$

$$Z_{n+1} = f^{(n+1)}_{\theta^{n+1}}(X_{n+1}), \tag{6.47}$$

where $X_0 = \mathbf{x}_0 \in \mathbb{R}^d$ and $\Delta t_n = t_{n+1} - t_n$, $\Delta B_n = B_{n+1} - B_n$ and a fully connected feed-forward network $f^{(n+1)}_{\theta^{n+1}}$ at each time location t_{n+1} to be trained such that

$$f^{(n+1)}_{\theta^{n+1}}(X_{n+1}) \approx \nabla u(t_{n+1} \cdot X_{n+1}). \tag{6.48}$$

The parameters for the subnetworks $f^{(n)}_{\theta^n}(\cdot)$ at $\{t_n\}$ are denoted collectively by $\boldsymbol{\theta} = \{\theta^n\}_{n=1}^N$ and the output Y_N is expected to be an approximation of $u(T, X_N)$. Thus, the loss function using the terminal condition is defined by a Monte Carlo approximation of

$$L_{\text{bsde}}(\boldsymbol{\theta}, Y_0) = E\,\|Y_N - g(X_N)\|^2, \tag{6.49}$$

whose minimizer Y_0^* produces an approximation to $u(0, \mathbf{x}_0)$ of (4.51) and (4.52).

6.6.2 Backward SDE Based Algorithms for Quasilinear Parabolic Equations with a Pathwise Loss

The DeepBSDE can be used to provide the solution of the parabolic equation (4.51) and (4.52) at one single space–time location. An alternative way is to use a loss function composed of pathwise differences between two stochastic processes related to the FBSDE of (4.53) and (4.54) and $u(t, \mathbf{x})$. Both processes are expected to converge to the same continuous diffusion process if the time discretization of the forward and backward SDEs (FBSDEs) and DNN approximations $u_{\boldsymbol{\theta}}(\mathbf{x})$ to the PDE solution $u(\mathbf{x})$ converge. The resulting FBSNN method [26, 27], which provides a DNN approximation $u_\theta(t, \mathbf{x})$, consists of the following steps.

1. Given $X_0 = \mathbf{x}_0$ and let $Y_0 = u_{\boldsymbol{\theta}}(t_0, X_0)$, $Z_0 = \nabla u_\theta(t_0, X_0)$.
2. Generate two sets of discrete M paths. For each time interval $[t_n, t_{n+1}]$, $0 \le n \le N-1$, calculate $X_{n+1} \in \mathbb{R}^d$ and $Y_{n+1} \in \mathbb{R}$ as in (6.45) and (6.46) while

$$Z_{n+1} = \nabla u_{\boldsymbol{\theta}}(t_{n+1}, X_{n+1}). \tag{6.50}$$

   Next, we define a second r.v. using the DNN approximation $u_{\boldsymbol{\theta}}$ for the PDE solution,

$$\tilde{Y}_{n+1} = u_{\boldsymbol{\theta}}(t_{n+1}, X_{n+1}). \tag{6.51}$$

3. For a batch size of M paths identified by $\{\omega_i\}_{i=1}^M$, the loss function is defined by

$$\begin{aligned} L_{\text{fbsnn}}(\boldsymbol{\theta}) = \frac{1}{M} \sum_{\omega_i, i=1}^{M} \Bigg[\frac{1}{N} \sum_{n=1}^{N} \left\| Y_n - \tilde{Y}_n \right\|^2 + \beta_1 \left\| \tilde{Y}_N - g(X_N) \right\|^2 \\ + \beta_2 \left\| Z_N - \nabla g(X_N) \right\|^2 \Bigg], \end{aligned} \tag{6.52}$$

   where β_1 and β_2 are penalty parameters for enforcing the terminal conditions, and the minimizer $\boldsymbol{\theta}^*$ of this loss function will give the DNN approximation $u_{\boldsymbol{\theta}^*}(t, \mathbf{x})$ in the region visited by $X_n \in \mathbb{R}^d$.

Figure 6.1 shows the numerical results of the FBSNN method for the following 100-dimensional Black–Scholes–Barenblatt equation excluding 4 holes with an oscillatory solution, which achieved the half-order convergence of the underlying Euler–Maruyama for discretizing the diffusion process X_t with a mean relative error ranging from $1.67e-03$ to $6.77e-4$ for $N = 48$ to $N = 768$ [27].

$$\begin{aligned} \partial_t u + \frac{1}{2} \text{Tr}[\sigma^2 \nabla \nabla u] = \phi, \quad \mathbf{x} \in \mathbb{R}^d,\ d = 100 \\ u(T, \mathbf{x}) = g(\mathbf{x}), \end{aligned} \tag{6.53}$$

where $T = 1.0$, $\sigma = 0.4$ and $r = 0.05$ and

$$g(\mathbf{x}) = \|\mathbf{x}\|^2 \left(1 + \alpha \sin\left(\beta S_1 - \gamma T\right)\right), \tag{6.54}$$

$$\phi(t, \mathbf{x}, u, \nabla u) = r(u - \nabla u \cdot \mathbf{x}) + \alpha e^{(r+\sigma^2)(T-t)} P(t, \mathbf{x}), \tag{6.55}$$

$$P(t, \mathbf{x}) = (r\beta S_1 S_2 - \gamma S_2 + 2\sigma^2 \beta S_3) \cos(\beta S_1 - \gamma t) - \frac{\sigma^2 \beta^2}{2} S_2^2 \sin(\beta S_1 - \gamma t), \tag{6.56}$$

and $S_j = \sum_{i=1}^d x_i^j$, $j = 1, 2$. The PDE (6.53) has an exact solution

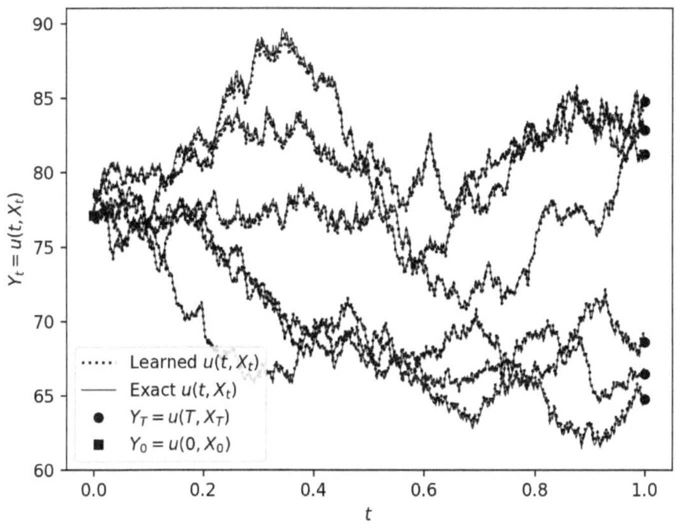

Fig. 6.1 Learned PDE solution (dots) vs exact solution $u(t, x)$ (line) for the Black–Scholes–Barenblatt equation (6.53) ($\alpha = 0.025$, $\beta = 0.25$ and $\gamma = 32$) with $N = 192$ along 6 sample paths in $\mathbb{R}^{100}$ originating from a 0.125-sized cubic neighborhood of $\mathbf{x}_0 = (1, 0.5, 1, 0.5, \ldots, 1, 0.5)^\top \in \mathbb{R}^{100}$

$$u(t, \mathbf{x}) = e^{(r+\sigma^2)(T-t)} \|\mathbf{x}\|^2 \left(1 + \alpha \sin\left(\beta S_1 - \gamma t\right)\right), \tag{6.57}$$

and corresponds to the coupled FBSDEs

$$\begin{aligned} dX_t &= \sigma \operatorname{diag}(X_t) dB_t, \\ X_0 &= \mathbf{x}_0, \\ dY_t &= \left(r(Y_t - Z_t \cdot X_t) + \alpha e^{(r+\sigma^2)(T-t)} P(t, X_t)\right) dt + \sigma Z_t^T \operatorname{diag}(X_t) dB_t, \\ Y_T &= g(X_T). \end{aligned} \tag{6.58}$$

6.6.3 Martingale Based Learning Algorithm for BVPs

Another way to construct a loss function for a neural network approximation $u_\theta(t, \mathbf{x})$ is to use the martingale property in the stochastic representation for the solution of PDE (6.33) [28].

The time independent vector $\mu = \mu(\mathbf{x}) \in \mathbb{R}^d$, matrix $\sigma = \sigma_{d\times d}(\mathbf{x})$ in the definition of $A = (a_{ij}) = \sigma\sigma^\top$ in the elliptic operator $\mathcal{L}$ of (6.34) can be identified with the drift and diffusion coefficients of the following stochastic Itô process $X_t(\omega) \in \mathbb{R}^d$ with $\mathcal{L}$ as its generator

$$dX_t = \mu dt + \sigma \cdot dB_t, \qquad X_0 = \mathbf{x}_0 \in D,$$

where $B_t = (B_t^1, \ldots, B_t^d)^\top \in \mathbb{R}^d$ is a Brownian motion in $\mathbb{R}^d$.

Using the Itô's formula (4.31), it can be shown that the following process M_t defines a martingale with respect to the natural filtration $\mathcal{F}_t$ of B_t,

$$M_t = u(X_t) - u(X_0) - \int_0^t [f(X_s) - V(X_s, u(X_s))]\, ds, \tag{6.59}$$

which implies that for any $A \in \mathcal{F}_s, s < t$,

$$E[M_t - M_s | A] = 0, \tag{6.60}$$

namely,

$$\int_A [M_t - M_s] P(d\omega) = 0. \tag{6.61}$$

Let $u_\theta(\mathbf{x})$ be a neural network approximating the PDE's solution, and for any given time interval $[0, T]$ with a partition (6.44), we generate K-discrete realizations of the Itô process using the Euler–Maruyama scheme from K-realizations of the Brownian motions $B_i^{(k)}, 0 \le k \le K$. Namely, with $\Delta t_i = t_{i+1} - t_i$,

$$\begin{aligned} X_{i+1}^{(k)} &= X_i^{(k)} + \mu(X_i^{(k)})\Delta t_i + \sigma(X_i^{(k)}) \cdot \Delta B_i^{(m)}, \quad i = 0, 1, 2, \ldots, \\ X_0^{(k)} &= \mathbf{x}_0. \end{aligned}$$

Loss function $L_{\text{mart}}(\boldsymbol{\theta})$ for the neural network $u_{\boldsymbol{\theta}}(\mathbf{x})$ can be defined to enforce the martingale property (6.60) as follows:

$$\begin{aligned} L_{\text{mart}}(\boldsymbol{\theta}) &= \frac{1}{N} \sum_{i=0}^{N-1} \left(E[M_{t_{i+1}} - M_{t_i} | A_i]\right)^2 \qquad (6.62) \\ &\simeq \frac{1}{N} \sum_{i=0}^{N-1} \frac{1}{|A_i|^2} \left(\sum_{k=1}^{|A_i|} u_{\boldsymbol{\theta}}(X_{i+1}^{(k)}) - u_{\boldsymbol{\theta}}(X_i^{(k)}) - \left[f(X_i^{(k)}) - V(X_i^{(k)}, u_{\boldsymbol{\theta}}(X_i^{(k)})) \right] \Delta t_i \right)^2, \end{aligned}$$

where the integral in (6.59) over the time interval $[t_i, t_{i+1}]$ is approximated by a left-end one point quadrature, and A_i is a randomly selected subset, as required by the martingale definition of the process M_t (making the martingale approach a natural fit for SGD deep learning), of the K paths. An additional boundary loss term of (6.32) is added to enforce the Dirichlet boundary condition.

The DeepMartNet [28] based on this approach can learn a good approximation $u_{\theta^*}(x)$ to the solutions of high-dimensional PDEs (6.33) by solving the following optimization problem using the SGD training,

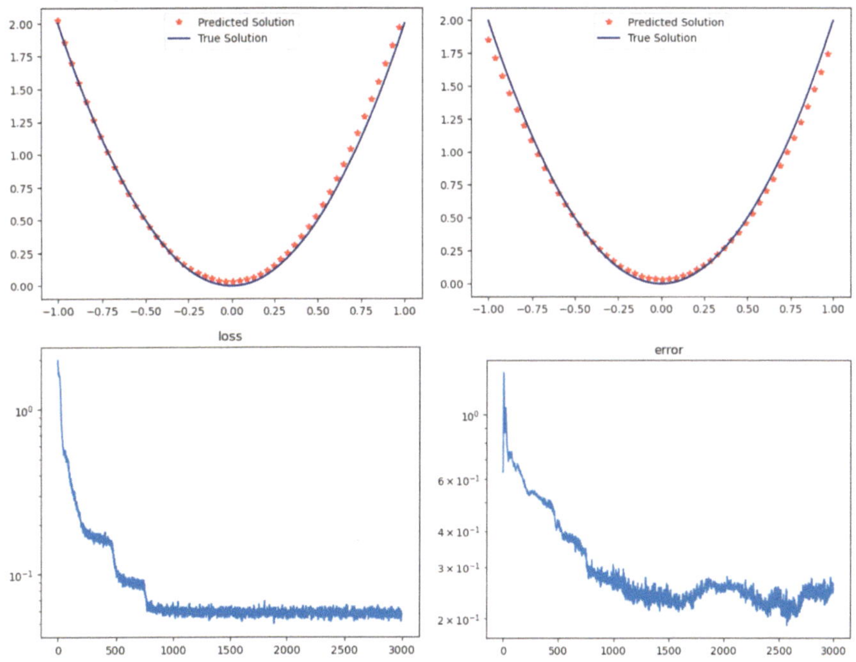

Fig. 6.2 DeepMartNet solution of the nonlinear PBE (6.64) in a 10-dimensional unit ball. Upper left: $u(x\boldsymbol{e}_1)$ where $\boldsymbol{e}_1 = (1, 0, \ldots, 0)$; Upper right: $u(x\boldsymbol{e})$ where $\boldsymbol{e} = (1, 1, \ldots, 1)$; Lower left: loss vs. epoch; right: relative L^2 error vs epoch

$$\boldsymbol{\theta}^* = \arg\min_{\boldsymbol{\theta}}[L_{\text{mart}}(\boldsymbol{\theta}) + L_{\text{bdry}}(\boldsymbol{\theta})]. \tag{6.63}$$

Nonlinear PBE in a d=10-dimensional unit ball. For a 1:1 symmetric two species ionic solvent, the electrostatic potential based on the Debye–Hückel theory is given by the nonlinear PBE in (2.13),

$$\begin{cases} -\Delta u + \sinh u = f & \mathbf{x} \in D, \\ u = g & \mathbf{x} \in \partial D, \end{cases} \tag{6.64}$$

where $D = \{x \in \mathbb{R}^d : \|\mathbf{x}\|_2 \leq L\}$, $L = 1$. We consider the case of a true solution as $u = \alpha \sum_{i=1}^{d} x_i^2$, $\alpha = 2$. Figure 6.2 shows the numerical results of DeepMartNet with a four-layer fully connected DNN with (10, 10, 10, 1) neurons. The activation function for the first hidden layer is Tanh, and that for the second hidden layer is the GELU with the Tanh approximation. $K = 10^6$ paths are used, and the sizes of the mini-batch $|A| = 4000$ paths are sampled in each epoch. The total number of training epochs is 3000. The terminal time for the paths is $T = 0.25$, and the time step is $t = 0.01$. The learning rate starts at 0.01, and decreases by a factor of 0.99 every 100 epochs.

6.6.4 Diffusion Monte Carlo DNN Eigensolver

In an approach similar to the power iteration method for the linear algebra eigenvalue problem [29], a diffusion Monte Carlo (DMC) method was proposed [30] through a fixed point of a semi-group for the eigenvalue problem of the following linear elliptic operator $\mathcal{L}$ of (6.34) (i.e., $\mu = \mu(\mathbf{x})$, $A = (a_{i,j}) = \sigma\sigma^\top, \sigma = \sigma(\mathbf{x})$),

$$\mathcal{L}\Psi(\mathbf{x}) = \lambda\Psi(\mathbf{x}), \quad \mathbf{x} \in \mathbb{R}^d. \tag{6.65}$$

Equation (6.65) can be reformulated as a virtual time dependent backward parabolic PDE with the sought-after eigenfunction as the terminal condition, i.e.,

$$\partial_t u(t, \mathbf{x}) + \mathcal{L}u(t, \mathbf{x}) - \lambda u(t, \mathbf{x}) = 0, \quad u(T, \mathbf{x}) = \Psi(\mathbf{x}), \quad \mathbf{x} \in \mathbb{R}^d. \tag{6.66}$$

Thus, the following fixed-point property holds,

$$P_t^\lambda \Psi \equiv u(T - t, \cdot), \quad P_T^\lambda \Psi = \Psi, \tag{6.67}$$

where the semi-group for the backward-in-time evolutionary system can be formally defined by

$$P_t^\lambda = e^{-(\mathcal{L}-\lambda)t}. \tag{6.68}$$

The loss function is set to be $||P_T^\lambda \Psi - \Psi||^2$, while the evolution of the PDE solution is performed by integrating the following two related SDEs, per the Pardoux–Peng theory in Sect. 4.3, over a partition $\{t_i = i\Delta t, \Delta t = T/N\}_{i=0}^N$ of the time interval $[0, T]$, instead of solving the PDE in $\mathbb{R}^d$,

$$\begin{aligned} X_{n+1} &= X_n + \sigma\Delta B_n, \quad X_0 = \mathbf{x}_0 \\ u_{n+1} &= u_n + (\lambda\Psi_\theta - \mu^T\nabla\Psi_\theta)(X_n)\Delta t + (\nabla\Psi_\theta(X_n))^\top\sigma\Delta B_n, \quad u_0 = \Psi_\theta(\mathbf{x}_0). \end{aligned} \tag{6.69}$$

The original algorithm in [30] uses a separate DNN to approximate the gradient of $\Psi(\mathbf{x})$. The loss function for the DNN $\Psi_\theta(\mathbf{x})$, which approximates the eigenfunction, and the eigenvalue λ is then defined by

$$L_{\mathrm{dmc}}(\theta, \lambda) = E_{X_0\sim\pi_0}[|u_N - \Psi_\theta(X_N)|^2], \tag{6.70}$$

whose minimizer will produce the appoximation to the eigenpair.

6.7 DNNs for Oscillatory Problems and Their Spectral Bias

Many neural networks, such as fully connected and convolution neural networks (CNNs), demonstrate a spectral bias of frequency-dependent convergence behavior [31, 32]. During training, the DNNs tend to approximate the low-frequency components of the targeted functions first and quickly, and more slowly for higher frequency components. This spectral bias degenerates the overall convergence of DNNs for problems with oscillatory solutions. In this section, we will present two approaches to overcome the spectral bias of DNNs for oscillatory problems.

6.7.1 *Phase-Shift DNN (PhaseDNN)*

A PhaseDNN converts high-frequency components of the data to a lower frequency spectrum with frequency shifts, then transforms the learned representations to the original frequency range [33]. A series of sub-DNNs is used to achieve a uniform convergence across many frequencies in approximating high-frequency functions.

We use a one-dimensional case to illustrate how to arrive at the PhaseDNN. Consider a band-limited function within a frequency range $[-K_0, K_0]$, i.e.,

$$\mathrm{Supp}\hat{f}(k) \subset [-K_0, K_0] = [-M\Delta k, M\Delta k], \quad \Delta k = K_0/M. \tag{6.71}$$

We first partition $[-K_0, K_0]$ with uniform sub-intervals,

$$[-K_0, K_0] \subset \bigcup_{j=-M}^{M} A_j, A_j = \left[\omega_j - \frac{\Delta k}{2}, \omega_j + \frac{\Delta k}{2}\right], \ \omega_j = j\Delta k, j = -M, \ldots, M. \tag{6.72}$$

Then, we can decompose the target function $f(x)$ in Fourier space as follows:

$$\hat{f}(k) = \sum_{j=-M}^{M} \chi_{A_j}(k)\hat{f}(k) \triangleq \sum_{j=-M}^{M} \hat{f}_j(k), \tag{6.73}$$

where χ_{A_j} is the indicator function for the interval A_j. The above decomposition gives a corresponding one in the physical x-space as

$$f(x) = \sum_{j=-M}^{M} f_j(x), \tag{6.74}$$

where

$$f_j(x) = \mathcal{F}^{-1}[\hat{f}_j](x). \tag{6.75}$$

The decomposition (6.74) involves $2M + 1$ functions $f_j(x)$, whose frequency spectrum is limited to $[\omega_j - \frac{\Delta k}{2}, \omega_j + \frac{\Delta k}{2}]$. Therefore, a simple phase shift by ω_j could translate its spectrum to $[-\Delta k/2, \Delta k/2]$, allowing it to be learned quickly by a DNN $T_j(x, \boldsymbol{\theta}_j)$. Namely, as $\hat{f}_j(k + \omega_j)$ is supported in $[-\Delta k/2, \Delta k/2]$, its inverse Fourier transform $\mathcal{F}^{-1}\left[\hat{f}_j(k + \omega_j)\right]$, denoted as

$$f_j^{\text{shift}}(x) = \mathcal{F}^{-1}\left[\hat{f}_j(k + \omega_j)\right](x), \tag{6.76}$$

can be learned quickly by a DNN $T_j(x, \boldsymbol{\theta}_j)$.

Meanwhile, in the physical domain, we have

$$f_j^{\text{shift}}(x) = e^{-i\omega_j x} f_j(x). \tag{6.77}$$

Equation (6.77) shows that once $f_j^{\text{shift}}(x)$ is learned, $f_j(x)$ is also learned by simply removing the phase factor, i.e.,

$$f_j(x) \approx e^{i\omega_j x} T_j(x, \boldsymbol{\theta}_j). \tag{6.78}$$

With all $f_j(x)$ for $-M \leq j \leq M$ learned, we have an approximation to $f(x)$ over the whole frequency range $[-M\Delta k, M\Delta k] = [-K_0, K_0]$ as follows

$$f(x) \approx T_{\boldsymbol{\theta}}(x) = \sum_{j=-M}^{M} e^{i\omega_j x} T_j(x, \boldsymbol{\theta}_j), \tag{6.79}$$

where $\boldsymbol{\theta} = \{\boldsymbol{\theta}_j\}_{j=-M}^{M}$ are the parameters of the combined DNN.

For real-valued solutions, the PhaseDNN can be written in the following form,

$$T_{\boldsymbol{\theta}}(x) = \sum_{j=0}^{M} A_j(x) \cos(\omega_j x) + B_j(x) \sin(\omega_j x), \tag{6.80}$$

where A_j,B_j are DNNs with their individual network parameters and the shift frequencies ω_j can also be used as trainable parameters to best fit the target function.

Numerical results in [33] have shown the effectiveness of the PhaseDNN together with the PINN in Sect. 6.4 in approximating highly oscillatory solutions with tens of wavelengths of the following Helmholtz equation (6.81). Meanwhile, a normal fully connected DNN failed to capture the high-frequency oscillation in the solutions.

$$\Delta u(\mathbf{x}) + \left(\lambda^2 + \omega(\mathbf{x})\right) u(\mathbf{x}) = f(\mathbf{x}), \quad \mathbf{x} \in \mathbb{R}^d. \tag{6.81}$$

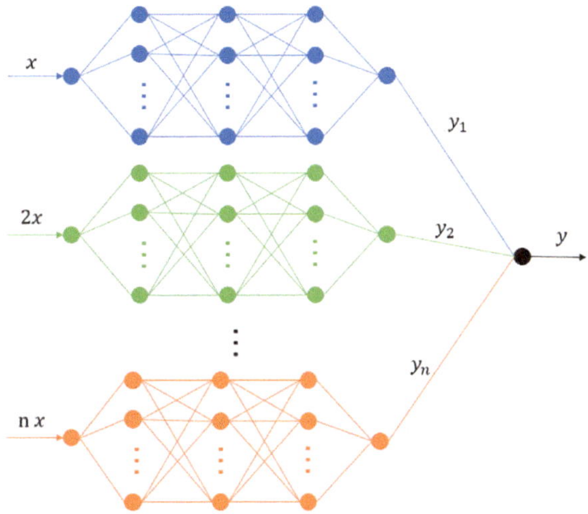

Fig. 6.3 A parallel implementation of the MscaleDNN with scales $\alpha_i = i$. Reprinted from [35], copyright (2020) by Global Science, Ltd.

6.7.2 *Multiscale DNN (MscaleDNN)*

The PhaseDNN will suffer the CoD problem for high-dimensional problems due to the number of shifts required from all coordinate directions. In contrast, the MscaleDNN [34–36] employs a radial scaling in the frequency domain, together with activation functions with localized or fast decaying frequency profiles, such as smooth functions with a compact support or sin and cos functions. Radial scaling again converts the problem of approximating high frequency contents of target functions or PDE solutions to a problem of learning lower frequency functions. Frequency-localized activation functions help frequency separation in the target function to be approximated by individual DNNs in the MscaleDNN.

To illustrate how the MscaleDNN arises, we again consider a band-limited function $f(\mathbf{x})$, $\mathbf{x} \in \mathbb{R}^d$, whose Fourier transform $\widehat{f}(\mathbf{k})$ has a compact support, i.e.,

$$\mathrm{Supp}\widehat{f}(\mathbf{k}) \subset B(K_{\max}) = \{\mathbf{k} \in \mathbb{R}^d, |\mathbf{k}| \leq K_{\max}\}. \tag{6.82}$$

We now partition the domain $B(K_{\max})$ as a union of M concentric annulus with uniform or non-uniform width. For the case of a uniform width K_0,

$$A_i = \{\mathbf{k} \in \mathbb{R}^d, (i-1)K_0 \leq |\mathbf{k}| \leq iK_0\}, \quad K_0 = K_{\max}/M, \quad 1 \leq i \leq M, \tag{6.83}$$

$$B(K_{\max}) = \bigcup_{i=1}^{M} A_i. \tag{6.84}$$

Now, we can decompose the function $\widehat{f}(\mathbf{k})$ in the Fourier domain as before

$$\widehat{f}(\mathbf{k}) = \sum_{i=1}^{M} \chi_{A_i}(\mathbf{k}) \widehat{f}(\mathbf{k}) \triangleq \sum_{i=1}^{M} \widehat{f}_i(\mathbf{k}), \tag{6.85}$$

$$\mathrm{Supp}\, \widehat{f}_i(\mathbf{k}) \subset A_i. \tag{6.86}$$

The decomposition in Fourier space gives a corresponding decomposition in the physical space

$$f(\mathbf{x}) = \sum_{i=1}^{M} f_i(\mathbf{x}), \tag{6.87}$$

where

$$f_i(\mathbf{x}) = \mathcal{F}^{-1}[\widehat{f}_i(\mathbf{k})](\mathbf{x}). \tag{6.88}$$

From (6.86), we can apply simple downward scaling to convert the high frequency region A_i to a low-frequency region. Namely, we define a scaled version of $\widehat{f}_i(\mathbf{k})$ as

$$\widehat{f}_i^{(\mathrm{scale})}(\mathbf{k}) = \widehat{f}_i(\alpha_i \mathbf{k}), \qquad \alpha_i > 1, \tag{6.89}$$

and, correspondingly, in the physical space

$$f_i^{(\mathrm{scale})}(\mathbf{x}) = \frac{1}{\alpha_i^d} f_i\left(\frac{1}{\alpha_i}\mathbf{x}\right), \tag{6.90}$$

or

$$f_i(\mathbf{x}) = \alpha_i^d f_i^{(\mathrm{scale})}(\alpha_i \mathbf{x}). \tag{6.91}$$

The spectrum of the scaled function $\widehat{f}_i^{(\mathrm{scale})}(\mathbf{k})$ will be of low frequency if α_i is sufficiently large, i.e.,

$$\mathrm{Supp}\, \widehat{f}_i^{(\mathrm{scale})}(\mathbf{k}) \subset \left\{ \mathbf{k} \in \mathbb{R}^d, \frac{(i-1)K_0}{\alpha_i} \le |\mathbf{k}| \le \frac{iK_0}{\alpha_i} \right\}. \tag{6.92}$$

With a DNN's learning preference toward low frequency, with iK_0/α_i being small, we can train a DNN $f_{\theta^{n_i}}(\mathbf{x})$ to learn $f_i^{(\mathrm{scale})}(\mathbf{x})$ quickly

$$f_i^{(\mathrm{scale})}(\mathbf{x}) \sim f_{\theta^{n_i}}(\mathbf{x}), \tag{6.93}$$

thus, allowing an approximation to $f_i(\mathbf{x})$

$$f_i(\mathbf{x}) \sim \alpha_i^d f_{\theta^{n_i}}(\alpha_i \mathbf{x}), \tag{6.94}$$

and, to $f(\mathbf{x})$ as well

$$f(\mathbf{x}) \sim \sum_{i=1}^{M} \alpha_i^d f_{\theta^{n_i}}(\alpha_i \mathbf{x}), \tag{6.95}$$

giving the format of the MscaleDNN [35, 36]. The scale factor α_i can also be used as a trainable parameter to best fit the target function. One typical parallel implementation of MscaleDNNs is shown in Fig. 6.3 [35].

Figure 6.4 shows the solution obtained by the MscaleDNN DeepRitz method for the Poisson equation over the domain $[-1, 1]^2$ excluding 4 holes with an oscillatory solution given by $u(\mathbf{x}) = \sin(\mu x_1)\sin(\mu x_2)$, $\mu = 7\pi$. Moreover, Fig. 6.5 shows the ability of the MscaleDNN to handle geometric singularity in solving the Poisson–Boltzmann equation (2.19) for biomolecular solvation applications.

6.7.3 *Reduction in the Spectral Bias of DNNs*

6.7.3.1 Spectral Bias Reduction with the PhaseDNN

We will show that training the PhaseDNN for fitting a function with many frequencies is equivalent to training a series of DNNs for functions of low frequency only, thus avoiding the spectral bias for the PhaseDNN.

Let $T_j(x)$ in (6.79) be written as $T_j(x) = u_j(\eta w_j x + b_j)$, where $\eta > 0$ is assumed to be a small parameter (i.e., small DNN parameters), and $u_j(t)$ is a (complex valued) DNN function. It can be shown [33] that the Fourier transform of (6.79) is

$$\hat{T}(k) = \sum_{j=-M}^{M} \hat{T}_j(k - \omega_j) = \sum_{j=-M}^{M} \frac{1}{\eta |w_j|} \exp\left(- \frac{i b_j (k - \omega_j)}{\eta w_j} \right) \hat{u}_j \left(\frac{k - \omega_j}{\eta w_j} \right), \tag{6.96}$$

and, with the Parseval equality, the loss function in (6.16) for the PhaseDNN becomes

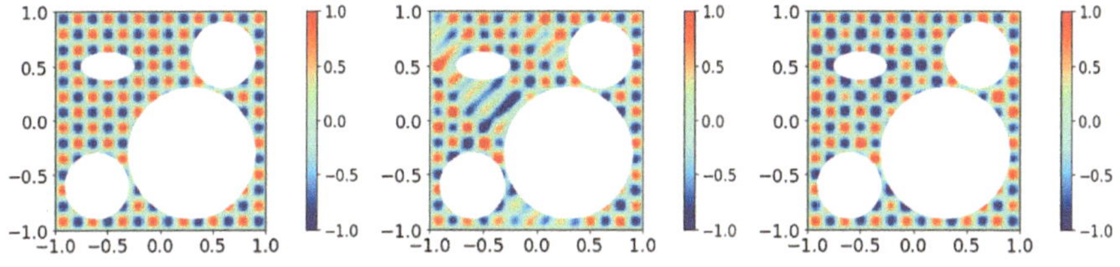

Fig. 6.4 DNN results of a Poisson equations in a multiconnected domain with oscillatory solutions (left) Exact solution, (middle) a normal DNN with sizes 1-1000-1000-1000-1 , (right) MscaleDNN with five subnetworks of size 1-1000-1000-1000-1, and scale coefficients $\{1, 2, 4, 8, 16\}$. The MscaleDNN captures oscillations much better than the normal fully connected network. Reprinted from [35], copyright (2020) by Global Science, Ltd.

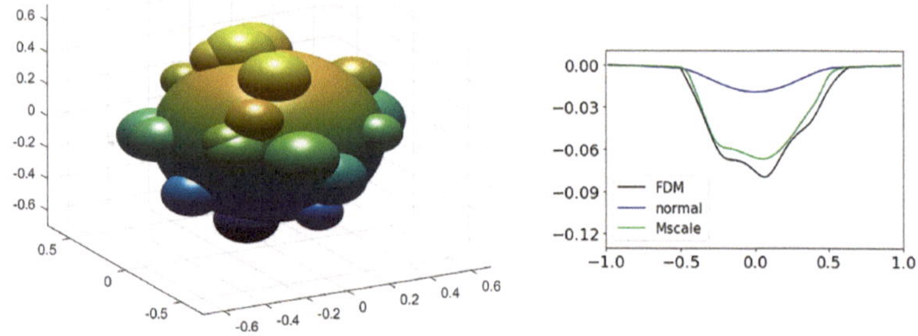

Fig. 6.5 Numerical solutions of the PB equation for a molecule $-\Delta u + \mu^2 u = f(\mathbf{x})$ on line $x_1 = x_3 = 0$, $f(\mathbf{x}) = \frac{e^{\sin \mu x_1 + \sin \mu x_2 + \sin \mu x_3}}{|\mathbf{x}|^2+1}(|\mathbf{x}|^2 - 1)$ with $\mu = 20$, $\epsilon(\mathbf{x}) = 1$ for $\mathbf{x}$ in molecule, $\epsilon(\mathbf{x}) = 80$ for $\mathbf{x}$ in water, $\lambda^2(\mathbf{x}) = 1$.(Left) Biomolecule boundary Γ in (2.3), (right) comparison of solutions between the normal DNN (top line) and MscaleDNN (middle line) against the reference solution (bottom line) from a finite difference method (FDM). Reprinted from [35], copyright (2020) by Global Science, Ltd.

$$L_{\text{phase}}(\boldsymbol{\theta}) = \int_{-\infty}^{\infty} |\hat{f} - \hat{T}|^2 dk = \int_{-\infty}^{\infty} |\hat{f}|^2 dk - 2\text{Re} \int_{-\infty}^{\infty} \hat{T}\bar{\hat{f}} dk + \int_{-\infty}^{\infty} |\hat{T}|^2 dk. \quad (6.97)$$

When η is small, we have the following approximation for the loss function:

$$\begin{aligned} L_{\text{phase}}(\boldsymbol{\theta}) &\approx \int_{-\infty}^{\infty} \sum_{j=-M}^{M} |\hat{f}_j(k)|^2 - 2\text{Re}\hat{T}_j(k-\omega_j)\bar{\hat{f}}_j + |\hat{T}_j(k-\omega_j)|^2 dk \\ &= \sum_{j=-M}^{M} \int_{-\infty}^{\infty} |\hat{f}_j(k) - \hat{T}_j(k-\omega_j)|^2 dk = \sum_{j=-M}^{M} \int_{-\infty}^{\infty} |\mathcal{F}^{-1}[\hat{f}_j(k+\omega_j)] - T_j(x)|^2 dx \\ &= \sum_{j=-M}^{M} \int_{-\infty}^{\infty} |e^{-i\omega_j x}\mathcal{F}^{-1}[\hat{f}_j](x) - T_j(x)|^2 dx, \end{aligned} \quad (6.98)$$

where $e^{-i\omega_j x}\mathcal{F}^{-1}[\hat{f}_j](x)$ is exactly the $f_j^{\text{shift}}(x)$ in (6.76) and (6.77), whose frequency content is in the low-frequency range $[-\Delta k/2, \Delta k/2]$. Hence, the total loss function $L_{\text{phase}}(\theta)$ is approximately the sum of those of the $2M+1$ separate DNNs T_j, which can easily learn the low-frequency function $f_j^{\text{shift}}(x)$, individually. As a result, the PhaseDNN will have a uniform convergence over all frequencies concerned.

6.7.3.2 Spectral Bias Reduction with the MscaleDNN

Using the neural tangent kernel (NTK) framework for DNN training dynamics [37], we derive a diffusion equation model for the error evolution in the frequency domain [38] for the learning algorithm by the MscaleDNN in fitting a target function with the loss function $L_s(\boldsymbol{\theta})$ in (6.16). We will assume a vanishing learning rate and an infinite wide network in a two-layered network with an activation function $\sigma(x) = \sin x$. The MscaleDNN [36] with one hidden layer is given by

$$f(\mathbf{x}, \boldsymbol{\theta}) = \frac{1}{\sqrt{N}} \sum_{j=0}^{s} \alpha_j^d \sum_{k=1}^{q} \sigma(\boldsymbol{\theta}_{jq+k}^{\mathrm{T}} \alpha_j \mathbf{x} + \mathbf{b}_{jq+k}), \quad \mathbf{x} \in D = [-1, 1]^d, \tag{6.99}$$

where $s + 1$ is the number of scales used, q is the number of neurons for each scale, and $m = (s + 1)q$ is the total number of neurons in the hidden layer. This network reduces to a one-scale normal fully connected neural network if $s = 0$. Using the gradient descent method (6.14) with a vanishing learning rate, the training dynamics of the output function $f(\mathbf{x}, \boldsymbol{\theta})$ are then

$$\begin{aligned}
\frac{\mathrm{d}}{\mathrm{d}t} f(\mathbf{x}, \boldsymbol{\theta}) &= \nabla_{\boldsymbol{\theta}} f(\mathbf{x}, \boldsymbol{\theta}) \cdot \dot{\boldsymbol{\theta}} = -\nabla_{\boldsymbol{\theta}} f(\mathbf{x}, \boldsymbol{\theta}) \cdot \nabla_{\boldsymbol{\theta}} L_s(\boldsymbol{\theta}) \\
&= -\nabla_{\boldsymbol{\theta}} f(\mathbf{x}, \boldsymbol{\theta}) \cdot \sum_{i=1}^{n} \nabla_{\boldsymbol{\theta}} f(\mathbf{x}_i, \boldsymbol{\theta})(f(\mathbf{x}_i, \boldsymbol{\theta}) - f(\mathbf{x}_i)) \\
&= -\sum_{i=1}^{n} K_m(\mathbf{x}, \mathbf{x}_i)(f(\mathbf{x}_i, \boldsymbol{\theta}) - f(\mathbf{x}_i)), \quad \mathbf{x} \in D,
\end{aligned} \tag{6.100}$$

and the NTK $K_m(\mathbf{x}, \mathbf{x}')(t)$ at time t and evaluated at $(\mathbf{x}, \mathbf{x}') \in D \times D$ is defined by

$$K_m(\mathbf{x}, \mathbf{x}')(t) = \nabla_{\boldsymbol{\theta}} f(\mathbf{x}, \boldsymbol{\theta}(t)) \cdot \nabla_{\boldsymbol{\theta}} f(\mathbf{x}', \boldsymbol{\theta}(t)). \tag{6.101}$$

Moreover, for the MscaleDNN, we can show that [38]

$$\begin{aligned}
K_s(\mathbf{x}, \mathbf{x}'; \boldsymbol{\theta}) = \sum_{j=0}^{s} \frac{\alpha_j^{2d}(\alpha_j^2 \mathbf{x}^{\mathrm{T}} \mathbf{x}' + 1)}{N} &\sum_{k=1}^{q} \sigma'(\boldsymbol{\theta}_{jq+k}^{\mathrm{T}} \alpha_p \mathbf{x} + \mathbf{b}_{jq+k}) \\
&\cdot \sigma'(\boldsymbol{\theta}_{jq+k}^{\mathrm{T}} \alpha_j \mathbf{x}' + \mathbf{b}_{jq+k}).
\end{aligned} \tag{6.102}$$

Set the activation function $\sigma(x) = \sin(x)$ and assume that all the parameters $\{\theta_{p,k}\}$ in $\boldsymbol{\theta}_p = (\theta_{p,1}, \theta_{p,2}, \ldots, \theta_{p,d})^{\mathrm{T}}$, and $\{\mathbf{b}_p\}$ are independent random variables of the normal distribution. Then, using the law of large numbers and the identity

$$(2\pi)^{-\frac{d+1}{2}} \int_{\mathbb{R}^{d+1}} e^{\mathrm{i}(\boldsymbol{\theta}^{\mathrm{T}} \boldsymbol{x} + yb)} e^{-\frac{|\boldsymbol{\theta}|^2 + b^2}{2}} d\boldsymbol{\theta} db = e^{-\frac{|\boldsymbol{x}|^2 + y^2}{2}}, \quad \forall \boldsymbol{x} \in \mathbb{R}^d, \quad y \in \mathbb{R},$$

we have

$$\lim_{q\to\infty} K_s(\mathbf{x},\mathbf{x}';\boldsymbol{\theta}) = \sum_{j=0}^{s} \frac{\alpha_j^{2d}(\alpha_j^2\mathbf{x}^{\mathrm{T}}\mathbf{x}'+1)}{s+1} E\left(\cos(\boldsymbol{\theta}_1^{\mathrm{T}}\alpha_j\mathbf{x}+\mathbf{b}_1)\cos(\boldsymbol{\theta}_1^{\mathrm{T}}\alpha_j\mathbf{x}'+\mathbf{b}_1)\right)$$
$$= \sum_{j=0}^{s} \frac{\alpha_j^{2d}(\alpha_j^2\mathbf{x}^{\mathrm{T}}\mathbf{x}'+1)}{2(s+1)} \left[e^{-2}e^{-\frac{\alpha_j^2|\mathbf{x}+\mathbf{x}'|^2}{2}} + e^{-\frac{\alpha_j^2|\mathbf{x}-\mathbf{x}'|^2}{2}}\right]. \tag{6.103}$$

Therefore, as the network width approaches infinity and the learning rate approaches zero, gradient descent learning renders the dynamic system (6.100) in the following form (after replacing the discrete sum in (6.100) with an integral) [38],

$$\begin{aligned}&\partial_t(f(\mathbf{x},\boldsymbol{\theta}) - f(\mathbf{x}))\\ &= -\sum_{j=0}^{s} \frac{\alpha_j^{2(d+1)}\mathbf{x}^{\mathrm{T}}}{2(s+1)} \int_D \left[e^{-2}\mathcal{G}_j(\mathbf{x}+\mathbf{x}') + \mathcal{G}_j(\mathbf{x}-\mathbf{x}')\right]\mathbf{x}'(f(\mathbf{x}',\boldsymbol{\theta}) - f(\mathbf{x}'))\mathrm{d}\mathbf{x}'\\ &\quad -\sum_{j=0}^{s} \frac{\alpha_j^{2d}}{2(s+1)} \int_D \left[e^{-2}\mathcal{G}_j(\mathbf{x}+\mathbf{x}') + \mathcal{G}_j(\mathbf{x}-\mathbf{x}')\right](f(\mathbf{x}',\boldsymbol{\theta}) - f(\mathbf{x}'))\mathrm{d}\mathbf{x}',\end{aligned} \tag{6.104}$$

where

$$\mathcal{G}_j(\boldsymbol{x}) := e^{-\alpha_j^2|\boldsymbol{x}|^2/2}, \quad \boldsymbol{x} \in \mathbb{R}^d, \tag{6.105}$$

is a scaled Gaussian function.

By defining a zero extension of the error of the network by

$$u(\mathbf{x},\boldsymbol{\theta}) = \begin{cases} 0, & \mathbf{x} \notin D, \\ f(\mathbf{x},\boldsymbol{\theta}) - f(\mathbf{x}), & \mathbf{x} \in D, \end{cases} \tag{6.106}$$

then, the dynamic system (6.104) can be rewritten for $\mathbf{x} \in D$

$$\begin{aligned}&\partial_t u(\mathbf{x},\boldsymbol{\theta}) I_D(\mathbf{x})\\ &= -\sum_{j=0}^{s} \frac{I_D(\mathbf{x})\alpha_j^{2(d+1)}\mathbf{x}^{\mathrm{T}}}{2(s+1)} \int_{\mathbb{R}^d} \left[e^{-2}\mathcal{G}_j(\mathbf{x}+\mathbf{x}') + \mathcal{G}_j(\mathbf{x}-\mathbf{x}')\right]\mathbf{x}' u(\mathbf{x}',\boldsymbol{\theta})\mathrm{d}\mathbf{x}'\\ &\quad -\sum_{j=0}^{s} \frac{I_\Omega(\mathbf{x})\alpha_j^{2d}}{2(s+1)} \int_{\mathbb{R}^d} \left[e^{-2}\mathcal{G}_j(\mathbf{x}+\mathbf{x}') + \mathcal{G}_j(\mathbf{x}-\mathbf{x}')\right]u(\mathbf{x}',\boldsymbol{\theta})\mathrm{d}\mathbf{x}',\ \mathbf{x} \in \mathbb{R}^d,\end{aligned} \tag{6.107}$$

where an indicator function $I_D(\mathbf{x})$ is used to extend the equation to the whole space, and $\hat{I}_D(\boldsymbol{\xi}) \to \delta(\boldsymbol{\xi})$ if the domain D expands to the whole space.

Therefore, in the limit of a large domain D, after taking the Fourier transform on both sides of (6.107) with respect to $\mathbf{x}$, we arrive at a diffusion-reaction equation in the frequency variable $\boldsymbol{\xi} \in \mathbb{R}^d$,

$$\frac{\partial \hat{u}(\boldsymbol{\xi},\boldsymbol{\theta}(t))}{\partial t} = \nabla_{\boldsymbol{\xi}} \cdot \left[\sum_{j=0}^{s} \frac{\alpha_j^{2(d+1)}\widehat{\mathcal{G}}_j(\boldsymbol{\xi})}{8\pi^2(s+1)}\left(\nabla_{\boldsymbol{\xi}}\hat{u}(\boldsymbol{\xi},\boldsymbol{\theta}(t)) - e^{-2}\nabla_{\boldsymbol{\xi}}\overline{\hat{u}(\boldsymbol{\xi},\boldsymbol{\theta}(t))}\right)\right]$$
$$- \sum_{j=0}^{s} \frac{\alpha_j^{2d}\widehat{\mathcal{G}}_j(\boldsymbol{\xi})}{2(s+1)}[e^{-2}\bar{\hat{u}}(\boldsymbol{\xi},\boldsymbol{\theta}(t)) + \hat{u}(\boldsymbol{\xi},\boldsymbol{\theta}(t))], \quad \boldsymbol{\xi} \in \mathbb{R}^d. \tag{6.108}$$

Defining

$$A_s^{\pm}(\boldsymbol{\xi}) = \frac{1 \pm e^{-2}}{8\pi^2(s+1)}\sum_{j=0}^{s}\alpha_j^{2(d+1)}\widehat{\mathcal{G}}_j(\boldsymbol{\xi}), \quad B_s^{\pm}(\boldsymbol{\xi}) = \frac{1 \pm e^{-2}}{2(s+1)}\sum_{j=0}^{s}\alpha_j^{2d}\widehat{\mathcal{G}}_j(\boldsymbol{\xi}), \tag{6.109}$$

and denoting by $\hat{u}^{\pm}(\boldsymbol{\xi},\boldsymbol{\theta}(t))$ the real and imaginary parts of $\hat{u}(\boldsymbol{\xi},\boldsymbol{\theta}(t))$, we arrive at the following diffusion equations for the error in the frequency domain.

Error Diffusion Model

$$\partial_t \hat{u}^{\pm}(\boldsymbol{\xi},t) = \nabla_{\boldsymbol{\xi}} \cdot \left[A_s^{\mp}(\boldsymbol{\xi})\nabla_{\boldsymbol{\xi}}\hat{u}^{\pm}(\boldsymbol{\xi},t)\right] - B_s^{\pm}(\boldsymbol{\xi})\hat{u}^{\pm}(\boldsymbol{\xi},t), \quad \boldsymbol{\xi} \in \mathbb{R}^d. \tag{6.110}$$

This diffusion model matches well with the error decay of practical training of MscaleDNNs in approximating oscillatory functions [38].

Cases of $\{A_s^{\mp}(\xi), B_s^{\pm}(\xi)\}$ for a one-dimensional case are plotted in Fig. 6.6 for an up to four-scale two-layer MscaleDNN. We can see that both the diffusion coefficients $A_s^{\mp}(\xi)$ and damping factors $B_s^{\pm}(\xi)$ have larger support and larger magnitude in the frequency domain with increasing s, implying faster error decays over larger frequency regions, and thus, providing the basis for the reduced spectral bias capability of the MscaleDNN. Moreover, for $s = 0$, this result confirms the spectral bias of a two-layer fully connected neural network with a sine-activation function.

6.8 Summary

Due to the compositional structure of the DNNs and meshless random domain or path sampling used in their learning, algorithms based on DNNs have the potential to tackle high-dimensional problems without suffering from the curse of dimensionality, where traditional numerical methods cannot handle them at all. DNN methods also avoid the costs and difficulties of meshing complex geometries. Although still in the early stages of development, DNN learning of PDE solutions has found many applications in scientific and engineering computing [39–41]. More efficient DNN learning algorithms and a deeper understanding of their convergence behavior, training, and generalizations will help realize their full potential.

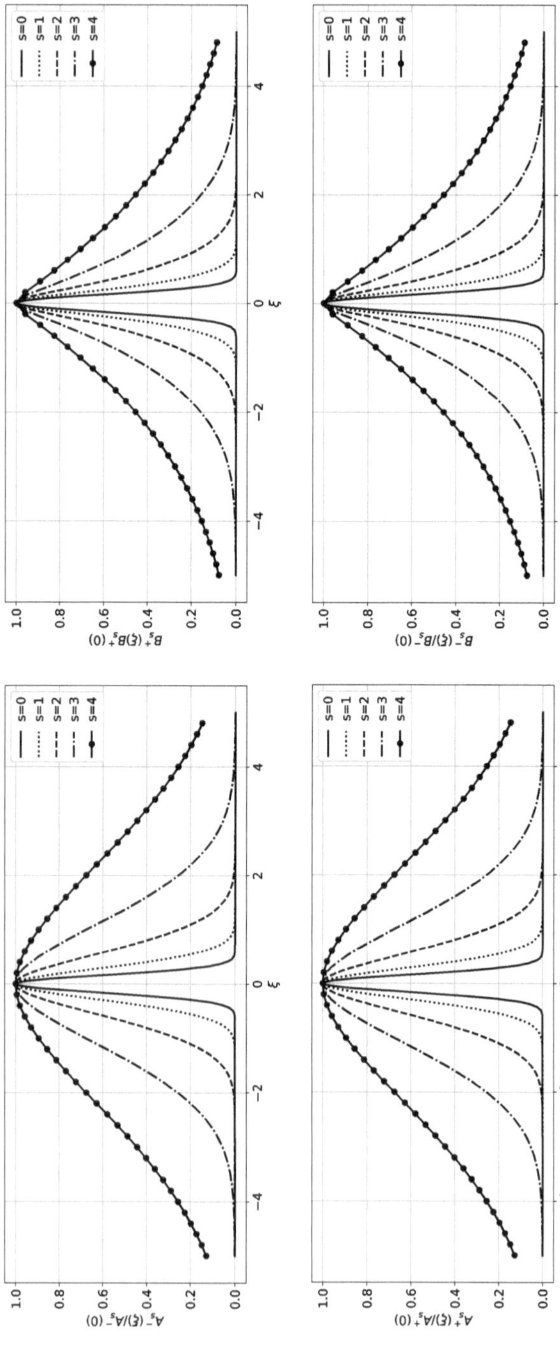

Fig. 6.6 Scaled diffusion and damping coefficients $A_s^-(\xi)$ (left/up), $A_s^+(\xi)$ (left/down), $B_s^+(\xi)$ (right/up) , and $B_s^-(\xi)$ (right/down) with scaling factor $\alpha_j = 2^j$ for scales $s = 0, 1, 2, 3, 4$

References

1. G. Cybenko, Approximation by superpositions of a sigmoidal function. Math. Control, Signals Syst. **2**, 303–314 (1989)
2. K. Hornik, M. Stinchcombe, H. White, Multilayer feedforward networks are universal approximators. Neural Netw. **2**, 359–366 (1989)
3. K. Hornik, Approximation capabilities of multilayer feedforward networks. Neural Netw. **4**, 251–257 (1991)
4. Z. Shen, H. Yang, S. Zhang, Deep network approximation characterized by number of neurons. Commun. Comput. Phys. **28**(5), 1768–811 (2020)
5. J. Lu, Z. Shen, H. Yang, S. Zhang, Deep network approximation for smooth functions. SIAM J. Math. Anal. **53**, 5465–5506 (2021)
6. V. Nair, G.E. Hinton, Rectified linear units improve restricted Boltzmann machines, in *Proceedings of the 27th International Conference on International Conference on Machine Learning, ICML'10*, Madison, WI, USA (Omnipress, 2010), pp. 807–814
7. A.L. Maas, A.Y. Hannun, A.Y. Ng, *Rectifier nonlinearities improve neural network acoustic models, in ICML, Workshop on Deep Learning for Audio, (Speech, and Language Processing* (Atlanta, Georgia, USA, 2013)
8. D. A. Clevert, T. Unterthiner, S. Hochreiter, Fast and accurate deep network learning by exponential linear units (ELUs), in *4th International Conference on Learning Representations, ICLR 2016*, ed. by Y. Bengio, Y. LeCun, San Juan, Puerto Rico, May 2–4, 2016, Conference Track Proceedings (2016)
9. G. Klambauer, T. Unterthiner, A. Mayr, S. Hochreiter. Self normalizing neural networks, *Advances in Neural Information Processing Systems*, ed. by I. Guyon, U. Von Luxburg, S. Bengio, H. Wallach, R. Fergus, S. Vishwanathan, R. Garnett, vol. 30 (Curran Associates, Inc., 2017)
10. X. Glorot, A. Bordes, Y. Bengio, Deep sparse rectifier neural networks, in *Proceedings of the Fourteenth International Conference on Artificial Intelligence and Statistics*, ed. by G. Gordon, D. Dunson, M. Dudí k, Volume 15 of Proceedings of Machine Learning Research, Fort Lauderdale, FL, USA, 11–13 April 2011 (2011), pp. 315–323
11. D.E. Rumelhart, G.E. Hinton, R.J. Williams, Learning representations by back-propagating errors. Nature. **323**(6088), 533–566 (1986)
12. H. Robbins, S. Monro, A stochastic approximation method. Ann. Math. Statist. **22**, 400–407 (1951)
13. L. Bottou, F.E. Curtis, J. Nocedal, Optimization methods for large-scale machine learning. SIAM Rev. **60**(2), 223–311 (2018)
14. D.P. Kingma, J.L. Ba, Adam: a method for stochastic optimization, in *International Conference on Learning Representations* (2015), pp. 1–13
15. M. Dissanayake, N. Phan-Thien, Neural-network-based approximations for solving partial differential equations. Commun. Numer. Methods Eng. **10**, 195–201 (1994)
16. M. Raissi, P. Perdikaris, G.E. Karniadakis, Physics-informed neural networks: a deep learning framework for solving forward and inverse problems involving nonlinear partial differential equations. J. Comput. Phys. **378**, 686–707 (2019)
17. E. Weinan, B. Yu, The deep Ritz method: a deep learning-based numerical algorithm for solving variational problems. Commun. Math. Stat. **6**(1), 1–12 (2018)
18. B.N. Jiang, L.A. Povinelli, Least-squares finite element method for fluid dynamics. Comput. Methods Appl. Mech. Eng. **81**(1), 13–37 (1990)
19. P.B. Bochev, M.D. Gunzburger, Finite element methods of least-squares type. SIAM Rev. **40**(4), 789–837 (1998)
20. Z. Cai, J. Chen, M. Liu, X. Liu, Deep least-squares methods: an unsupervised learning-based numerical method for solving elliptic PDEs. J. Comput. Phys. **420**, 109707 (2020)
21. Y. Zang, G. Bao, X. Ye, H. Zhou, Weak adversarial networks for high-dimensional partial differential equations. J. Comput. Phys. **411**, 109409 (2020)

22. K. Friedrichs, Symmetric positive linear differential equations. Commun. Pure Appl. Mathe. **11**, 333–418 (1958)
23. F. Chen, J. Huang, C. Wang, H. Yang, Friedrichs learning: weak solutions of partial differential equations via deep learning. SIAM Sci. Comput. **45**, A1271-99 (2023)
24. P.E. Kloeden, E. Platen, *Numerical Solution of Stochastic Differential Equations* (Springer Science & Business Media; 2013)
25. J. Han, A. Jentzen, E. Weinan, Solving high-dimensional partial differential equations using deep learning. Proc. Natl. Acad. Sci. **115**, 8505–8510 (2018)
26. M. Raissi. Forward-backward stochastic neural networks: deep learning of high-dimensional partial differential equations (2018), arXiv e-prints arXiv:1804.07010
27. W.Z. Zhang, W. Cai, FBSDE based neural network algorithms for high-dimensional quasilinear parabolic PDEs. J. Comput. Phys. **470**, 111557 (2022)
28. W. Cai, A. He, D. Margolis, Deepmartnet—a martingale based deep neural network learning method for Dirichlet BVP and eigenvalue problems of elliptic PDEs (2023), arXiv preprint arXiv:2311.09456, also see arXiv:2408.14395
29. L.N. Trefethen, D. Bau, *Numerical Linear Algebra* (SIAM, 2022)
30. J. Han, J. Lu, M. Zhou, Solving high-dimensional eigenvalue problems using deep neural networks: a diffusion Monte Carlo like approach. J. Comput. Phys. **423**, 109792 (2020)
31. Z.-Q.J. Xu, Y. Zhang, T. Luo, Y. Xiao, Z. Ma, Frequency principle: Fourier analysis sheds light on deep neural networks. Commun. Comput. Phys. **28**, 1746–1767 (2020)
32. N. Rahaman, A. Baratin, D. Arpit, F. Draxler, M. Lin, F. Hamprecht, Y. Bengio, A. Courville, On the spectral bias of neural networks, in *International Conference on Machine Learning, PMLR* (2019), pp. 5301–5310
33. W. Cai, X. Li, L. Liu, A phase shift deep neural network for high frequency approximation and wave problems. SIAM J. Sci. Comput. **42**, A3285–A3312 (2020)
34. W. Cai, Z.-Q.J. Xu, Multi-scale deep neural networks for solving high dimensional PDEs (2019), arXiv preprint arXiv:1910.11710
35. Z. Liu, W. Cai, Z.-Q.J. Xu, Multi-scale deep neural network (MscaleDNN) for solving Poisson-Boltzmann equation in complex domains. Commun. Comput. Phys. **28**, 1970–2001 (2020)
36. L. Zhang, W. Cai, Z.-Q.J. Xu, A correction and comments on "multi-scale deep neural network (MscaleDNN) for solving Poisson-Boltzmann equation in complex domains CICP", **28**(5), 1970–2001 (2020). Commun. Comput. Phys. **33**, 1509–1513 (2023)
37. A. Jacot, F. Gabriel, C. Hongler, Neural tangent kernel: convergence and generalization in neural networks. Adv. Neural Inform. Process. Syst. **31**, 8580–8589 (2018)
38. B. Wang, H. Yuan, L. Liu, W. Zhang, W. Cai, On spectral bias reduction of multi-scale neural networks for regression problems. Neural Net. **21**, 107179 (2025)
39. E. Weinan, Machine learning and computational mathematics. Commun. Comput. Phys. **28**, 1639–1670 (2020)
40. E. Weinan, J. Han, A. Jentzen, Algorithms for solving high dimensional PDEs: from nonlinear Monte Carlo to machine learning. Nonlinearity **35**, 278 (2021)
41. G.E. Karniadakis, I.G. Kevrekidis, L. Lu, P. Perdikaris, S. Wang, L. Yang, Physics-informed machine learning. Nat. Rev. Phys. **3**(6), 422–440 (2021)

Chapter 7
Fast Algorithms for Long-Range Interactions in Free Space

Fast algorithms are indispensable for computing long-range interactions between electric charges or current sources, which is one of the most important computational efforts in molecular dynamics simulations of biological systems [1], and the simulations of wave interactions. Three methods will be discussed. The first is the well-known Ewald summation [2] for a periodic system of charges and/or dipoles, and its particle–mesh Ewald (PME) implementation. The PME uses charge interpolation onto a regular lattice and the fast Fourier transformation (FFT) to produce an algorithm of $O(N \log N)$ complexity for an N-charge and/or dipole system. The second is the fast multipole method (FMM) [3] for N particles, applicable to both periodic and non-periodic systems, using multipole expansions to reduce the cost of computing the far-field potential of charges (or current sources) and a hierarchical oct-tree structure for an $O(N)$ (electrostatics) or $O(N \log N)$ (current sources) multi-level algorithm. Finally, a hybrid multi-scale method combines the image charge approximation to the reaction field of charges inside a spherical cavity surrounded by a dielectric medium and the FMM for the interaction of all charges and their image charges, resulting in an overall $O(N)$ algorithm.

7.1 Ewald Sums for Charges and Dipoles

In 1921, Ewald proposed a method to compute the potential due to an infinite array of charges from periodic image copies of N charges q_i in a primary box Λ with an overall charge neutrality, i.e.,

$$\sum_{j-1}^{N} q_j = 0. \tag{7.1}$$

W. Cai, *Deterministic, Stochastic, and Deep Learning Methods for Computational Electromagnetics*, https://doi.org/10.1007/978-981-96-0100-4_7

The primary box Λ is defined by three elementary lattice vectors $\{\mathbf{a}_i, i = 1, 2, 3\}$, the so-called Bravais lattice vectors [4]. The three corresponding elementary reciprocal lattice vectors $\{\mathbf{b}_k, k = 1, 2, 3\}$ are also defined such that

$$\mathbf{a}_i \cdot \mathbf{b}_k = 2\pi\delta_{ik}, \tag{7.2}$$

where δ_{ik} is the Kronecker delta. The general reciprocal lattice vector $\mathbf{G}$ is given by

$$\mathbf{G} = m_1\mathbf{b}_1 + m_2\mathbf{b}_2 + m_3\mathbf{b}_3, \quad m_k \in \mathbb{Z},\ k = 1, 2, 3, \tag{7.3}$$

where $\mathbb{Z}$ represents the set of all integers.

The N charge locations $\mathbf{r}_j \in \Lambda$ are identified by a triplet of fractional coordinates (f_{1j}, f_{2j}, f_{3j}):

$$\mathbf{r}_j = f_{1j}\mathbf{a}_1 + f_{2j}\mathbf{a}_2 + f_{3j}\mathbf{a}_3, \tag{7.4}$$

where, for $1 \le j \le N, 1 \le k \le 3$,

$$f_{kj} = \frac{1}{2\pi}\mathbf{r}_j \cdot \mathbf{b}_k. \tag{7.5}$$

The periodic image $\Lambda_\mathbf{n}$ of the box Λ is constructed by $\mathbf{n} = (n_1,n_2,n_3)$-shifts, $n_i \in \mathbb{Z},\ i = 1, 2, 3$, as

$$\Lambda_\mathbf{n} = \Lambda + n_1\mathbf{a}_1 + n_2\mathbf{a}_2 + n_3\mathbf{a}_3 \equiv \Lambda + \mathbf{n}_a, \tag{7.6}$$

and the image copies of the N charges in the primary box Λ are then defined by the same shifts in the corresponding box $\Lambda_\mathbf{n}$:

$$\mathbf{r}_{j\mathbf{n}} = \mathbf{r}_j + \mathbf{n}_a. \tag{7.7}$$

The Ewald summation is used to evaluate the potential $V(\mathbf{r})$ from the N charges at $\mathbf{r}_j$, $1 \le j \le N$, in Λ and their periodic images at $\mathbf{r}_{j\mathbf{n}}$ in $\Lambda_\mathbf{n}$, i.e.,

$$V(\mathbf{r}) = \sum_\mathbf{n}\sum_j \frac{q_j}{4\pi\epsilon_0|\mathbf{r} - \mathbf{r}_{j\mathbf{n}}|}. \tag{7.8}$$

As the series (7.8) is only conditionally convergent and the limit in fact depends on the order of the summation in $\mathbf{n}$, Ewald [2] proposed to split the series into two parts: a direct part $V_{\text{dir}}(\mathbf{r})$ (carried out in the physical space) and a reciprocal part $V_{\text{rec}}(\mathbf{r})$ (carried out in the reciprocal Fourier G-space) (see Fig. 7.1). This splitting is done by placing a diffused charge density $\sigma_{j\mathbf{n}}(\mathbf{r})$ with a total charge $-q_j$ centered at each charge $\mathbf{r}_{j\mathbf{n}}$:

$$\sigma_{j\mathbf{n}}(\mathbf{r}) = -q_j\sigma(|\mathbf{r} - \mathbf{r}_{j\mathbf{n}}|), \tag{7.9}$$

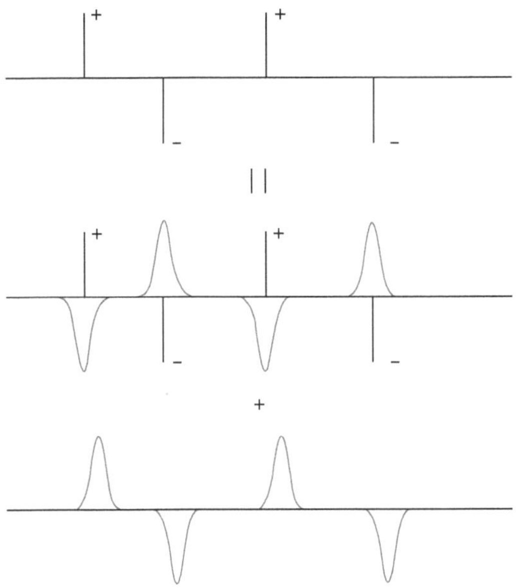

Fig. 7.1 Wald summation with real and reciprocal spaces

where σ is usually taken to be a Gaussian with a variance $1/(\sqrt{2}\alpha)$ (other types of local density shape have also been studied [5])

$$\sigma(u) = \frac{\alpha^3}{\pi^{3/2}} e^{-\alpha^2 u^2}. \tag{7.10}$$

The diffused charge $\sigma_{j\mathbf{n}}(\mathbf{r})$ of opposite sign imposed on each charge q_j at $\mathbf{r}_{j\mathbf{n}}$ effectively screens the latter charge, resulting in a short-ranged interaction potential. The potential of all these screened charges is denoted as $V_{\text{dir}}(\mathbf{r})$. The potential $\phi_\sigma(\mathbf{r})$ due to a single Gaussian charge density $\sigma(|\mathbf{r} - \mathbf{r}_j|)$ satisfies the Poisson equation (7.13) with $-\sigma(|\mathbf{r} - \mathbf{r}_j|)$ as the right-hand side, whose solution is given by the $\text{erf}(x)$ function [6]:

$$\phi_\sigma(\mathbf{r}) = \frac{1}{4\pi\epsilon_0} \frac{\text{erf}(\alpha|\mathbf{r} - \mathbf{r}_j|)}{|\mathbf{r} - \mathbf{r}_j|}, \tag{7.11}$$

where

$$\text{erf}(x) = \frac{2}{\sqrt{\pi}} \int_0^x e^{-u^2} du.$$

Therefore, the direct potential $V_{\text{dir}}(\mathbf{r})$ is simply

$$\begin{aligned} V_{\text{dir}}(\mathbf{r}) &= \sum_{\mathbf{n}}\sum_{j}\frac{q_j}{4\pi\epsilon_0}\left(\frac{1}{|\mathbf{r}-\mathbf{r}_{j\mathbf{n}}|}-\frac{\text{erf}(\alpha|\mathbf{r}-\mathbf{r}_{j\mathbf{n}}|)}{|\mathbf{r}-\mathbf{r}_{j\mathbf{n}}|}\right) \\ &= \sum_{\mathbf{n}}\sum_{j}\frac{q_j}{4\pi\epsilon_0}\frac{\text{erfc}(\alpha|\mathbf{r}-\mathbf{r}_{j\mathbf{n}}|)}{|\mathbf{r}-\mathbf{r}_{j\mathbf{n}}|}, \end{aligned} \tag{7.12}$$

where the complementary error function $\text{erfc}(x) = 1 - \text{erf}(x)$.

The exponential decay of the function $\text{erfc}(x)$ reflects the screening effect of the diffused Gaussian density $\sigma_{j\mathbf{n}}(\mathbf{r})$ superposed on the charge q_j at $\mathbf{r}_{j\mathbf{n}}$. Therefore, for each given field location $\mathbf{r} = \mathbf{r}_i$, only a few terms need to be included in the summation (7.12) by using a simple truncation cut-off (proportional to $1/\alpha$) based on the distance $|\mathbf{r}_{j\mathbf{n}} - \mathbf{r}_i|$. As a result, the computation of $V_{\text{dir}}(\mathbf{r}_i)$ at all field locations $\mathbf{r}_i$, $i = 1, 2, \ldots, N$, will only cost $O(N)$ operations.

Next, we consider the potential $V_{\text{rec}}(\mathbf{r})$ from all the diffused charges $\sigma_{j\mathbf{n}}(\mathbf{r})$, which in fact satisfies the following Poisson equation:

$$\epsilon_0\nabla^2 V_{\text{rec}}(\mathbf{r}) = -\rho(\mathbf{r}), \tag{7.13}$$

where

$$\rho(\mathbf{r}) = \sum_{\mathbf{n}}\sum_{j}\sigma_{j\mathbf{n}}(\mathbf{r}) = \sum_{\mathbf{n}}\sum_{j}q_j\sigma(|\mathbf{r}-\mathbf{r}_{j\mathbf{n}}|). \tag{7.14}$$

From the periodicity of the density $\rho(\mathbf{r})$, we can expand both $V_{\text{rec}}(\mathbf{r})$ and $\rho(\mathbf{r})$ in terms of Fourier series:

$$V_{\text{rec}}(\mathbf{r}) = \sum_{G} V_{\text{rec}}(\mathbf{G})e^{\mathrm{i}\mathbf{G}\cdot\mathbf{r}} \tag{7.15}$$

and

$$\rho(\mathbf{r}) = \sum_{G}\rho(\mathbf{G})e^{\mathrm{i}\mathbf{G}\cdot\mathbf{r}}, \tag{7.16}$$

where

$$\rho(\mathbf{G}) = \frac{1}{|\Lambda|}\int_{\Lambda}\rho(\mathbf{r})e^{-\mathrm{i}\mathbf{G}\cdot\mathbf{r}}d\mathbf{r}. \tag{7.17}$$

The Fourier coefficient $\rho(\mathbf{G})$ in (7.17) can be calculated as follows:

$$\begin{aligned} \rho(\mathbf{G}) &= \frac{1}{|\Lambda|}\int_{\Lambda}\sum_{\mathbf{n}}\sum_{j}q_j\sigma(|\mathbf{r}-\mathbf{r}_j-\mathbf{n}_a|)e^{-\mathrm{i}\mathbf{G}\mathbf{r}}d\mathbf{r} \\ &= \frac{1}{|\Lambda|}\sum_{j}\sum_{\mathbf{n}}\int_{\Lambda}q_j\sigma(|\mathbf{r}-\mathbf{r}_j-\mathbf{n}_a|)e^{-\mathrm{i}\mathbf{G}\cdot\mathbf{r}}d\mathbf{r} \\ &= \frac{1}{|\Lambda|}\sum_{j}\sum_{\mathbf{n}}\int_{\Lambda_{\mathbf{n}}}q_j\sigma(|\mathbf{r}-\mathbf{r}_j|)e^{-\mathrm{i}\mathbf{G}\cdot\mathbf{r}}d\mathbf{r}, \end{aligned}$$

where $\exp(-\mathrm{i}\mathbf{G}\cdot\mathbf{n}_a)=1$ is used in the final equality, and then

$$\begin{aligned}\rho(\mathbf{G}) &= \frac{1}{|\Lambda|}\sum_j\int_{\mathbb{R}^3} q_j\sigma(|\mathbf{r}-\mathbf{r}_j|)\mathrm{e}^{-\mathrm{i}\mathbf{G}\cdot\mathbf{r}}\mathrm{d}\mathbf{r} = \frac{1}{|\Lambda|}\sum_j\int_{\mathbb{R}^3} q_j\sigma(|\mathbf{r}|)\mathrm{e}^{-\mathrm{i}\mathbf{G}\cdot(\mathbf{r}+\mathbf{r}_j)}\mathrm{d}\mathbf{r} \\ &= \frac{1}{|\Lambda|}\sum_j q_j\mathrm{e}^{-\mathrm{i}\mathbf{G}\cdot\mathbf{r}_j}\int_{\mathbb{R}^3}\sigma(|\mathbf{r}|)\mathrm{e}^{-\mathrm{i}\mathbf{G}\cdot\mathbf{r}}\mathrm{d}\mathbf{r} = \frac{S(\mathbf{G})}{|\Lambda|}\int_{\mathbb{R}^3}\sigma(|\mathbf{r}|)\mathrm{e}^{-\mathrm{i}\mathbf{G}\cdot\mathbf{r}}\mathrm{d}\mathbf{r} \\ &= \frac{\alpha^3}{\pi^{3/2}}\frac{1}{|\Lambda|}\mathrm{e}^{-\frac{G^2}{4\alpha^2}}S(\mathbf{G})\int_{\mathbb{R}^3}\mathrm{e}^{-(\alpha\mathbf{r}+\frac{\mathrm{i}}{2\alpha}\mathbf{G})^2}\mathrm{d}\mathbf{r} = \frac{1}{|\Lambda|}S(\mathbf{G})\mathrm{e}^{-\frac{G^2}{4\alpha^2}},\end{aligned} \tag{7.18}$$

where a quantity $S(\mathbf{G})$ (the structure factor [4]) is introduced as follows:

$$S(\mathbf{G}) = \sum_{j=1}^{N} q_j\mathrm{e}^{-\mathrm{i}\mathbf{G}\cdot\mathbf{r}_j}. \tag{7.19}$$

Plugging (7.15) and (7.16) into the Poisson equation (7.13), and equating the Fourier coefficients, we have

$$V_{\mathrm{rec}}(\mathbf{G}) = \frac{1}{\epsilon_0|\Lambda|G^2}S(\mathbf{G})\mathrm{e}^{-\frac{G^2}{4\alpha^2}},\quad \mathbf{G}\neq 0, \tag{7.20}$$

for which we obtain the solution for $V_{\mathrm{rec}}(\mathbf{r})$ in the Fourier space as

$$V_{\mathrm{rec}}(\mathbf{r}) = \frac{1}{\epsilon_0|\Lambda|}\sum_{G\neq 0}\frac{1}{G^2}S(\mathbf{G})\mathrm{e}^{-\frac{G^2}{4\alpha^2}}\mathrm{e}^{\mathrm{i}\mathbf{G}\cdot\mathbf{r}}, \tag{7.21}$$

where the constant term $V_{\mathrm{rec}}(\mathbf{C}=\mathbf{0})$ for the potential function $V_{\mathrm{rec}}(\mathbf{r})$ can be set to be zero.

Finally, the potential at $\mathbf{r}=\mathbf{r}_i$ can be computed as follows:

$$\begin{aligned}V(\mathbf{r}_i) &= V_{\mathrm{dir}}(\mathbf{r}_i)+V_{\mathrm{rec}}(\mathbf{r}_i)-\frac{2\alpha q_i}{4\pi\epsilon_0\sqrt{\pi}}+\frac{1}{\epsilon_0(2\epsilon_{\mathrm{sur}}+1)|\Lambda|}\mathbf{r}_i\sum_j q_j\mathbf{r}_j \\ &= \sum_{\mathbf{n}}{}'\sum_j\frac{1}{4\pi\epsilon_0}\frac{q_j\,\mathrm{erfc}(\alpha|\mathbf{r}_i-\mathbf{r}_{j\mathbf{n}}|)}{|\mathbf{r}_i-\mathbf{r}_{j\mathbf{n}}|}+\frac{1}{\epsilon_0|\Lambda|}\sum_{G\neq 0}\frac{1}{G^2}S(\mathbf{G})\mathrm{e}^{-\frac{G^2}{4\alpha^2}}\mathrm{e}^{\mathrm{i}\mathbf{G}\cdot\mathbf{r}_i} \\ &\quad -\frac{2\alpha q_i}{4\pi\epsilon_0\sqrt{\pi}}+\frac{1}{\epsilon_0(2\epsilon_{\mathrm{sur}}+1)|\Lambda|}\mathbf{r}_i\sum_j q_j\mathbf{r}_j,\end{aligned} \tag{7.22}$$

where the prime on the **n**-summation indicates the exclusion of the self interaction potential (unbounded) at $\mathbf{r}_i=\mathbf{r}_j$ for $\mathbf{n}=\mathbf{0}$. Two correction terms have been introduced in (7.22): the first accounts for the self-interaction of the Gaussian density

$q_j\sigma(|\mathbf{r}-\mathbf{r}_j|)$, and the second is a surface term coming from the exterior surface of the finite crystal system during the summation process. Here, ϵ_{sur} is the dielectric constant assumed for the environment outside the finite system [7, 8].

The potential energy of the primary cell within the periodic charge system is

$$U=\frac{1}{2}\sum_{\mathbf{n}}{}'\sum_{i=1}^{N}\sum_{j=1}^{N}\frac{1}{4\pi\epsilon_0}\frac{q_iq_j}{|\mathbf{r}_i-\mathbf{r}_{j\mathbf{n}}|}, \tag{7.23}$$

which, under the Ewald splitting procedure for the electric potential (7.22), can be computed as

$$\begin{aligned}U=&\frac{1}{2}\sum_i\sum_j\sum_{\mathbf{n}}{}'\frac{1}{4\pi\epsilon_0}\frac{q_iq_j\,\text{erfc}(\alpha|\mathbf{r}_i-\mathbf{r}_{j\mathbf{n}}|)}{|\mathbf{r}_i-\mathbf{r}_{j\mathbf{n}}|}+\frac{1}{2\epsilon_0|\Lambda|}\sum_i q_i\sum_{G\neq 0}\frac{1}{G^2}S(\mathbf{G})\mathrm{e}^{-\frac{G^2}{4\alpha^2}}\mathrm{e}^{\mathrm{i}\mathbf{G}\cdot\mathbf{r}_i}\\&-\frac{\alpha}{4\pi\epsilon_0\sqrt{\pi}}\sum_{i=1}^{N}q_i^2+\frac{1}{2\epsilon_0(2\epsilon_{\text{sur}}+1)|\Lambda|}\left|\sum_{i=1}^{N}q_i\mathbf{r}_i\right|^2,\end{aligned} \tag{7.24}$$

where the second to last term subtracts the self-interaction energy of each charge and, again, the final correction term [8, 9] is the polarization energy of the total dipole moment of the system charges from the surrounding dielectric media outside the regions where the summation over the finite range of $\mathbf{n}$ is taken. The final term is sometimes also identified as an extrinsic potential [7, 10, 11].

The Ewald sum (7.24) for the potential energy was given a strict mathematical derivation in [9] by using a convergence factor in the form of $\mathrm{e}^{-s|\mathbf{n}|}$ in (7.23) and taking the limit $s\to 0$.

Ewald Sums for Systems with Dipoles

In a periodic molecular system, where the interaction of the molecules can be described through their partial charges and dipole moments, the Ewald sums can be extended to compute the interaction of charges and dipoles [8].

The potential of a collection of molecules (or particles) described by partial charges q_j and dipole moments $\boldsymbol{\mu}_j$, $1\le j\le N$, is given at location $\mathbf{r}_i$ by the following Ewald sum:

$$V(\mathbf{r}_i)=V_{\text{dir}}(\mathbf{r}_i)+V_{\text{rec}}(\mathbf{r}_i)+V_{\text{self}}(\mathbf{r}_i)+V_{\text{surf}}(\mathbf{r}_i), \tag{7.25}$$

where the direct part evaluated in the real space is given by

$$V_{\text{dir}}(\mathbf{r}_i)=\sum_{j\neq i}\left(\hat{T}_{ij}q_j+\hat{T}_{ij}^{\alpha}\mu_{j,\alpha}\right), \tag{7.26}$$

and, for simplicity of notation, the screened electric potential $\hat{T}_{ij}$ and its derivatives $\hat{T}_{ij}^{\alpha}$ and $\hat{T}_{ij}^{\alpha\beta}$ are denoted as

$$\hat{T}_{ij} = \frac{1}{4\pi\epsilon_0}\frac{\mathrm{erfc}(\alpha|\mathbf{r}_i - \mathbf{r}_j|)}{|\mathbf{r}_i - \mathbf{r}_j|}, \tag{7.27}$$

$$\hat{T}_{ij}^{\alpha} = \nabla_\alpha \hat{T}_{ij}, \tag{7.28}$$

$$\hat{T}_{ij}^{\alpha\beta} = \nabla_\alpha \nabla_\beta \hat{T}_{ij}. \tag{7.29}$$

The reciprocal potential part evaluated in the Fourier space is given by

$$V_{\mathrm{rec}}(\mathbf{r}_i) = \frac{1}{\epsilon_0|\Lambda|}\sum_{G\neq 0}\frac{1}{G^2}\mathrm{e}^{-\frac{G^2}{4\alpha^2}} S^{q\mu}(\mathbf{G})\mathrm{e}^{\mathrm{i}\mathbf{G}\cdot\mathbf{r}_i}, \tag{7.30}$$

where the structure factor $S^{q\mu}(\mathbf{G})$ for the charge–dipole system is

$$S^{q\mu}(\mathbf{G}) = \sum_{j=1}^{N}\left(q_j - \mathrm{i}\boldsymbol{\mu}_j\cdot\mathbf{G}\right)\mathrm{e}^{-\mathrm{i}\mathbf{G}\cdot\mathbf{r}_j}. \tag{7.31}$$

The self-energy potential,

$$V_{\mathrm{self}}(\mathbf{r}_i) = -\frac{2\alpha q_i}{4\pi\epsilon_0\sqrt{\pi}} - \sum_{j\neq i; i,j\in p}\left(\widetilde{T}_{ij}q_j + \widetilde{T}_{ij}^{\alpha}\mu_{j,\alpha}\right), \tag{7.32}$$

where $i, j \in p$ indicates that sites i and j both belong to the same particle (or molecule), and a modified intra-molecular interaction potential $\widetilde{T}_{ij}$ and its derivatives $\widetilde{T}_{ij}^{\alpha}$ are used:

$$\widetilde{T}_{ij} = \frac{1}{4\pi\epsilon_0}\frac{\mathrm{erf}(\alpha|\mathbf{r}_i - \mathbf{r}_j|)}{|\mathbf{r}_i - \mathbf{r}_j|}, \tag{7.33}$$

$$\widetilde{T}_{ij}^{\alpha} = \nabla_\alpha \widetilde{T}_{ij}, \tag{7.34}$$

$$\widetilde{T}_{ij}^{\alpha\beta} = \nabla_\alpha \nabla_\beta \widetilde{T}_{ij}. \tag{7.35}$$

Finally, the surface potential

$$V_{\mathrm{surf}}(\mathbf{r}_i) = \frac{1}{\epsilon_0(2\epsilon_{\mathrm{sur}}+1)|\Lambda|}\mathbf{r}_i\sum_j\left(q_j\mathbf{r}_j + \boldsymbol{\mu}_j\right). \tag{7.36}$$

Meanwhile, the potential energy of the system of charges and dipoles can also be found by the Ewald sum as

$$U = U_{\text{dir}} + U_{\text{rec}} + U_{\text{self}} + U_{\text{surf}}, \tag{7.37}$$

where the potential energy of the screened charges and dipoles evaluated in the real space is given by

$$U_{\text{dir}} = \frac{1}{2}\sum_{\mathbf{n}}{}' \sum_{i} \sum_{i,j} \left(q_i q_j \hat{T}_{ij} + q_i \hat{T}_{ij}^{\alpha} \mu_{j,\alpha} - \mu_{i,\alpha} \hat{T}_{ij}^{\alpha} q_j - \mu_{i,\alpha} \hat{T}_{ij}^{\alpha,\beta} \mu_{i,\beta} \right); \tag{7.38}$$

the potential energy of the screening Gaussian charges evaluated in the Fourier space is given by

$$U_{\text{rec}} = \frac{1}{2\epsilon_0 |\Lambda|} \left(\sum_{G \neq 0} \frac{1}{G^2} \mathrm{e}^{-\frac{G^2}{4\alpha^2}} |S^{q\mu}(\mathbf{G})|^2 \right); \tag{7.39}$$

the self-energy consists of that of the Gaussian density and the charges and the dipoles inside each individual molecule:

$$\begin{aligned} U_{\text{self}} &= -\frac{1}{4\pi\epsilon_0} \sum_{i=1}^{N} \left(\frac{\alpha}{\sqrt{\pi}} q_i^2 + \frac{2\alpha^3}{3\sqrt{\pi}} \mu_i^2 \right) \\ &\quad - \frac{1}{2} \sum_{p} \sum_{j \neq i; i,j \in p} \left(q_i \widetilde{T}_{ij} q_j + q_i \widetilde{T}_{ij}^{\alpha} \mu_{j,\alpha} - \mu_{i,\alpha} \widetilde{T}_{ij}^{\alpha} q_j - \mu_{i,\alpha} \widetilde{T}_{ij}^{\alpha\beta} \mu_{j,\beta} \right); \end{aligned} \tag{7.40}$$

and, finally, the surface energy due to the exterior ϵ_{sur} dielectric environment is given by

$$U_{\text{surf}} = \frac{1}{2\epsilon_0 (2\epsilon_{\text{sur}} + 1)|\Lambda|} \left[\left| \sum_{i=1}^{N} q_i \mathbf{r}_i \right|^2 + 2 \left(\sum_{i=1}^{N} q_i \mathbf{r}_i \right) \cdot \left(\sum_{i=1}^{N} \boldsymbol{\mu}_i \right) + \left| \sum_{i=1}^{N} \boldsymbol{\mu}_i \right|^2 \right]. \tag{7.41}$$

The Ewald sum can also be extended to a system of quadrupoles [12, 13] and polarizable dipoles [8, 14].

7.2 Particle–Mesh Ewald (PME) Methods

The cost for the original Ewald sum [2] is expensive due to the computation of the structure factors for non-uniform charges in (7.19) (or charges–dipoles in (7.31)). In applying (7.21) in computer simulation, a truncation, proportional to α, on the reciprocal vector $\mathbf{G}$ space will be required. Assuming that the Fourier indices $\mathbf{G} = (m_1, m_2, m_3)$ for $V_{\text{rec}}(\mathbf{r})$ are truncated within the following range:

$$-\frac{M_\alpha}{2} \le m_\alpha \le \frac{M_\alpha}{2}, \; M_\alpha = O(N^{1/3}), \;\; \alpha = 1, 2, 3, \tag{7.42}$$

for which $S(\mathbf{G})$ will be computed, the cost of computing each $S(\mathbf{G})$ in (7.19) will be $O(N)$. So, the total cost will be $O(N^2)$.

Thus, in order to reduce the cost in computing $V_{\text{rec}}(\mathbf{r})$, the key is to reduce the computation of all structure factors $S(\mathbf{G})$ in (7.19), which involves the evaluation of the exponential $\exp(-\mathrm{i}\mathbf{G} \cdot \mathbf{r}_j)$ at irregular locations $\mathbf{r}_j$. To achieve this goal, an auxiliary regular lattice mesh, previously proposed in the particle–particle particle–mesh (PPPM) method [15], is introduced in the PME method [16–18]:

$$\mathbf{r}_\mathbf{k} = \frac{k_1}{K_1}\mathbf{a}_1 + \frac{k_2}{K_2}\mathbf{a}_2 + \frac{k_3}{K_3}\mathbf{a}_3, \;\; 0 \le k_\alpha \le K_\alpha, \;\; \alpha = 1, 2, 3, \tag{7.43}$$

where K_α is the total number of mesh points along the $\mathbf{a}_\alpha$ direction. Associated with this mesh, we define an interpolation function $L_\mathbf{k}(\mathbf{r})$, $\mathbf{k} = (k_1, k_2, k_3)$:

$$L_\mathbf{k}(\mathbf{r}) = L_{k_1}(f_1)L_{k_2}(f_2)L_{k_3}(f_3), \;\; \text{if} \;\; \mathbf{r} = \frac{f_1}{K_1}\mathbf{a}_1 + \frac{f_2}{K_2}\mathbf{a}_2 + \frac{f_3}{K_3}\mathbf{a}_3, \tag{7.44}$$

where L_k can just be the piecewise linear hat function in the finite element method such that $L_k(l) = \delta_{kl}$, and $\text{supp}(L_k) = (k-1, k+1)$, ensuring that

$$L_\mathbf{k}(\mathbf{r}_\mathbf{m}) = \delta_{\mathbf{km}}. \tag{7.45}$$

As in the finite element method, the function $L_\mathbf{k}(\mathbf{r})$ can be used to interpolate a function at $\mathbf{r}$ using its values at $\mathbf{r}_\mathbf{k}$ by the following formula:

$$f(\mathbf{r}) = \sum_\mathbf{k} f(\mathbf{r}_\mathbf{k})L_\mathbf{k}(\mathbf{r}). \tag{7.46}$$

Next, the interpolation formula (7.46) is applied to the exponential function

$$\mathrm{e}^{-\mathrm{i}\mathbf{G}\cdot\mathbf{r}_j} = \sum_{\mathbf{k}\in\Delta} \mathrm{e}^{-\mathrm{i}\mathbf{G}\cdot\mathbf{r}_\mathbf{k}} L_\mathbf{k}(\mathbf{r}_j), \tag{7.47}$$

where the index set Δ is

$$\Delta = \{(k_1, k_2, k_3) : \; 0 \le k_\alpha \le K_\alpha - 1, \alpha = 1, 2, 3\}. \tag{7.48}$$

In obtaining (7.47), we have used the compact support of the interpolant functions and we have also assumed implicitly that no charge q_j falls into the support of the boundary interpolant functions $L_\mathbf{k}$, when $\delta_{k_1K_1} + \delta_{k_2K_2} + \delta_{k_3K_3} \neq 0$. For those charges, we can treat their contributions to $S(\mathbf{G})$ separately without changing the overall complexity of the algorithm.

From (7.19) we have

$$
\begin{aligned}
S(\mathbf{G}) &= \sum_j q_j \sum_{\mathbf{k}\in\Delta} \mathrm{e}^{-\mathrm{i}\mathbf{G}\cdot\mathbf{r_k}} L_{\mathbf{k}}(\mathbf{r}_j) = \sum_{\mathbf{k}\in\Delta} \mathrm{e}^{-\mathrm{i}\mathbf{G}\cdot\mathbf{r_k}} \sum_j q_j L_{\mathbf{k}}(\mathbf{r}_j) \\
&= \sum_{k_1=0}^{K_1-1}\sum_{k_2=0}^{K_2-1}\sum_{k_3=0}^{K_3-1} Q_{\mathbf{k}} \mathrm{e}^{-\mathrm{i}\mathbf{G}\cdot\mathbf{r_k}} = \sum_{k_1=0}^{K_1-1}\sum_{k_2=0}^{K_2-1}\sum_{k_3=0}^{K_3-1} Q_{\mathbf{k}} \mathrm{e}^{-\mathrm{i}2\pi\left(\frac{m_1k_1}{K_1}+\frac{m_2k_2}{K_2}+\frac{m_3k_3}{K_3}\right)},
\end{aligned}
\tag{7.49}
$$

where the definitions of $\mathbf{G}$ in (7.3) and of $\mathbf{r_k}$ in (7.43) and relation (7.2) have been used, along with

$$
Q_{\mathbf{k}} = \sum_j q_j L_{\mathbf{k}}(\mathbf{r}_j), \quad \mathbf{k} \in \Delta. \tag{7.50}
$$

The calculation of each $Q_{\mathbf{k}}$ will involve only a few charges due to the small support of the function $L_{\mathbf{k}}(\mathbf{r}_j)$. Therefore, $\mathbf{Q} = \{Q_{\mathbf{k}} : \mathbf{k} \in \Delta\}$ can be calculated in $O(N)$ operations. From (7.49), $S(\mathbf{G})$ is simply the discrete Fourier transform of the 3-D data array $\mathbf{Q}$ defined on a regular lattice, readily achieved using the FFT at a cost of $O(N \log N)$ if we select $K_\alpha = M_\alpha = O(N^{1/3})$, $\alpha = 1, 2, 3$.

The reciprocal potential $V_{\mathrm{rec}}(\mathbf{r_k})$ on the auxiliary mesh lattices can be approximated by truncating the Fourier series in (7.21), namely

$$
V_{\mathrm{rec}}(\mathbf{r_k}) \simeq \frac{1}{\epsilon_0|\Lambda|} \sum_{m_1=-\frac{M_1}{2}}^{\frac{M_1}{2}-1} \sum_{m_2=-\frac{M_2}{2}}^{\frac{M_2}{2}-1} \sum_{m_3=-\frac{M_3}{2}}^{\frac{M_3}{2}-1} \frac{\mathrm{e}^{-\frac{G^2}{4\alpha^2}} S(\mathbf{G})}{G^2} \mathrm{e}^{\mathrm{i}2\pi\left(\frac{m_1k_1}{M_1}+\frac{m_2k_2}{M_2}+\frac{m_3k_3}{M_3}\right)}, \tag{7.51}
$$

(assuming $K_\alpha = M_\alpha$, $\alpha = 1, 2, 3$), which again can be implemented by an FFT.

Finally, the potential $V_{\mathrm{rec}}(\mathbf{r_i})$ off the lattice is obtained by using the interpolation formula (7.46) from $V_{\mathrm{rec}}(\mathbf{r_k})$.

The accuracy of the PME can be easily controlled by the size of the auxiliary mesh $K = K_1K_2K_3$, and the truncation used in the Fourier series for $V_{\mathrm{rec}}(\mathbf{r})$ above (which depends on the magnitude of α), and also the type of the interpolant function $L_{\mathbf{k}}$ in (7.46). The linear Lagrange interpolation is used in the original paper of Darden et al. [16], and later a smoother cubic spline is used, which allows continuous differentiation of the potential for force calculations that require the derivatives of the potentials [17].

An alternative way for evaluating the reciprocal potential $V_{\mathrm{rec}}(\mathbf{r}_i)$ at particle location $\mathbf{r}_i$ without using FFT is the random batch Ewald method (RB-Ewald) [19, 20]. RB-Ewald employs an importance sampling approach over $\mathbf{G}$ vectors by treating $\mathrm{e}^{-G^2/4\alpha^2}$ in (7.51) as a discrete Gaussian distribution (after a normalization). And, the Metropolis–Hastings method can then be utilized to sample a mini-batch of the $\mathbf{G}$

vectors in Fourier space, and the resulting sum over the sampled **G** vectors at all particle locations can be computed independently from the sums of other mini-batches. This approach leads to an efficient parallel implementation of the Ewald method.

7.3 Fast Multipole Method for N-Particle Electrostatic Interactions

The fast multipole method (FMM) is used to compute N-body interactions from discrete charges, whose potential field is given by Green's functions for the following Poisson equations:

$$\text{(2-D)} \qquad -\Delta\Phi(\mathbf{r}) = \delta(\mathbf{r}), \qquad \Phi(\mathbf{r}) = -\frac{1}{2\pi}\log(|\mathbf{r}|), \tag{7.52}$$

$$\text{(3-D)} \qquad -\Delta\Phi(\mathbf{r}) = \delta(\mathbf{r}), \qquad \Phi(\mathbf{r}) = \frac{1}{4\pi|\mathbf{r}|}. \tag{7.53}$$

The FMM can be illustrated in a 2-D electrostatic problem [3], with a potential given by (7.52). The goal is to evaluate the potential field at M points from N charges $q_1, q_2, \ldots, q_N$ distributed at $\mathbf{r}_1, \mathbf{r}_2, \ldots, \mathbf{r}_N$. The potential at a field point $\mathbf{r}$ due to a charge q_i is given by $(1/(2\pi)$ is omitted for simplicity)

$$\Phi_i(\mathbf{r}) = -q_i \log(|\mathbf{r} - \mathbf{r}_i|). \tag{7.54}$$

From the principle of linear superposition, the total potential at $\mathbf{r}$ is given by

$$\Phi(\mathbf{r}) = \sum_{i=1}^{N} \Phi_i(\mathbf{r}) = -\sum_{i=1}^{N} q_i \log(|\mathbf{r} - \mathbf{r}_i|). \tag{7.55}$$

The calculation of the potential from all N charges for each field point requires $O(N)$ flops (floating point operations). Therefore, for M field points, the cost by direct evaluation of (7.55) is $O(MN)$ operations.

- The goal of the FMM is to reduce the complexity for N-particle electrostatic interactions from $O(N^2)$ to $O(N)$ when $M = N$.

7.3.1 *Multipole Expansions (MEs)*

Key Idea

The key idea of multipole expansion is that it applies to a potential in the far field (i.e., at a distance greater than $2a$ from the source charges inside a sphere of radius a).

We make this observation: far away from a group of source charges, the combined potential from those source charges could be represented by a source at one point, located in the center of the group of the source charges, with multipole source components beyond the single pole of the original Coulomb potential, such as dipole, quadrupole, etc.

Mathematically, this amounts to an expansion of negative powers for the combined potential from all charges at a common point. We illustrate this idea using the 2-D potential (7.54), where the total potential at $\mathbf{r} = (x, y)$ is given by (7.55).

To facilitate the calculation, we introduce the complex number $z = x + \mathrm{i}y$, and then the potential for the 2-D Poisson equation will be just the real part of an analytical complex function away from z_i, i.e.,

$$\Phi_i(\mathbf{r}) = -q_i \log(|\mathbf{r} - \mathbf{r}_i|) = -q_i \mathrm{Re}(\log(z - z_i)), \tag{7.56}$$

where

$$\log z = \log\left(\sqrt{x^2 + y^2}\mathrm{e}^{\mathrm{i}\theta}\right) = \log\left(\sqrt{x^2 + y^2}\right) + \mathrm{i}\theta. \tag{7.57}$$

Furthermore, the total complex potential at z can be computed by the following multipole expansion (ME):

$$\begin{aligned} \Phi(z) &= \sum_{i=1}^{N} \Phi_i(z) = -\sum_{i=1}^{N} q_i \log(z - z_i) \\ &= Q \log z + \sum_{k=1}^{\infty} \frac{a_k}{z^k}, \end{aligned} \tag{7.58}$$

where

$$Q = -\sum_{i=1}^{N} q_i, \quad a_k = -\sum_{i=1}^{N} \frac{q_i z_i^k}{k}. \tag{7.59}$$

The proof of (7.58) is based on a simple Taylor expansion as follows:

$$\log(z - z_0) = \log(z) + \log\left(1 - \frac{z_0}{z}\right) = \log(z) + \sum_{k=1}^{\infty} \frac{1}{k}\left(\frac{z_0}{z}\right)^k. \tag{7.60}$$

As $|z_0/z| < 1$, we have $\log(1 - x) = \sum_{k=1}^{\infty}(x^k/k)$ for $x = |z_0/a|$, $|x| < 1$, and we can truncate the series to the Pth term with an error estimate as follows:

$$\left|\Phi(z) - Q\log(z) - \sum_{k=1}^{P} \frac{a_k}{z^k}\right| \le \left(\frac{A}{C-1}\right)\left(\frac{1}{C}\right)^P = A\left(\frac{1}{2}\right)^P, \tag{7.61}$$

where $C = |z/a| \ge 2$.

For example, for an accuracy at 10^{-4}, we require

$$\left(\frac{1}{2}\right)^P = 10^{-4} \rightarrow P \sim 13,$$

and the following approximation will have the desired accuracy:

$$\Phi(z) \approx Q \log z + \sum_{k=1}^{P} \frac{a_k}{z^k}, \qquad \text{for } |z| \geq 2a. \tag{7.62}$$

- **Cost of the multipole expansion**

We see that

$$\text{the cost of calculating the charge } Q = N,$$

$$\text{the cost of calculating all } a_k = 4NP, \tag{7.63}$$

so the total cost is given by

$$(4P + 1)N \sim O(N). \tag{7.64}$$

Now with

$$\Phi(z_j) = Q \log z_j + \sum_{k=1}^{P} \frac{a_k}{z_j^k}, \tag{7.65}$$

for M field points there will be $(3P + 1)M$ operations. Therefore, the total cost for computing the potential at all M field points is given by

$$(4P + 1)N + (3P + 1)M \sim 7PN = O(N) \text{ if } M = N. \tag{7.66}$$

Multi-level algorithm

As illustrated above, when the field point is far away from the source charges, a substantial cost reduction can be achieved by using Taylor expansions and pre-calculating the multipole expansion coefficients. However, in most application problems, the source charges and the potential field points are intermingled, and sometimes, they are at the same set of physical locations. So, in general, it will be difficult to satisfy the far-field requirement between the potential field locations and all source charge locations. The solution to this problem is to create a hierarchical structure of subdivision of the physical region into nested rectangular/cubic boxes [21], as the far-field concept is applicable to every scale of distance. This means that at each given level of boxes of size a, the multipole expansion idea can be used at that level.

In practice, the tree structure of the nested rectangular boxes in Fig. 7.2 is so formed such that the smallest box contains only 3–4 charges. For an L-level hierarchical

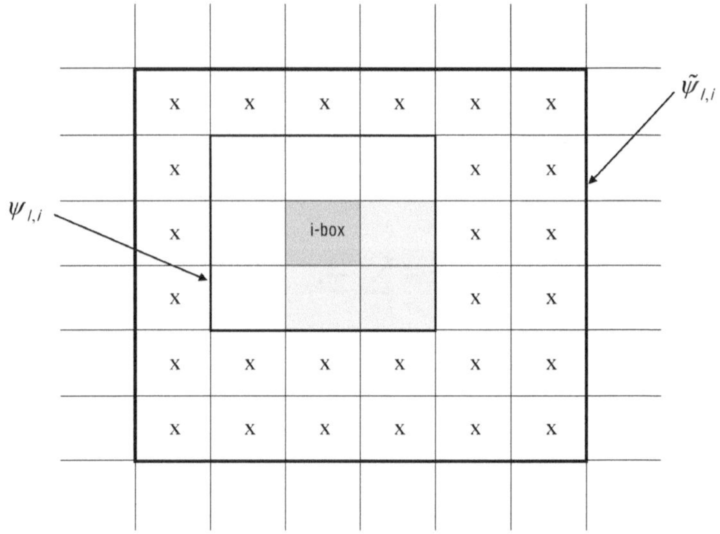

Fig. 7.2 FMM mesh setup, showing i-box (dark shaded box) and its parent (light shaded boxes); the interaction boxes (marked by x), which are bounded by thinner and thicker rectangular boundaries; the thinner rectangular boundary outside which the charges define the local expansion $\psi_{l,i}$ for the i-boxes; the thicker rectangular boundary outside which the charges define $\tilde{\psi}_{l,i}$ for the i-boxes

structure of nested boxes, there are 4^L boxes in 2-D (8^L boxes in 3-D) at the finest level (Lth level).

- **Concept of local expansion for far field**

We consider any i-box at the level l. Our goal is then to calculate the potential at z inside the i-box, which can be decomposed as follows:

$$\Phi(z) = \sum_{q_i \text{ inside i-box \& its neighbors}} \Phi_{q_i}(z) + \text{ potential from all other charges outside.} \tag{7.67}$$

The potential contribution to the i-box from all charges in a j-box centered at z_0 from the second group in (7.67) can be represented by a multipole expansion as (7.58), as each j-box is one box away from the i-box. This procedure can be carried out for all j-boxes, resulting in a multipole expansion at the center of each j-box, which together can give the potential inside the i-box. To avoid using many centers from all the multipole expansions, after realizing the fact that it is the potential inside the same i-box that we are interested in and that all multipole expansions are smooth functions inside the i-box, we can convert all the multipole expansions to a Taylor expansion at $x = 0$, the center of the i-box. This Taylor expansion is defined as a *local expansion.*

Our ultimate goal is to find the local expansion of all i-boxes on the L th level by converting all multipole expansions from charges outside the i-box using the following translations.

- **Multipole to local translation**

Consider a multipole expansion (ME) at z_0:

$$\Phi(z) = Q\log(z - z_0) + \sum_{k=1}^{P} \frac{a_k}{(z - z_0)^k}, \tag{7.68}$$

which diverges to ∞ at z_0. However, if we are only concerned with the region $|z - z_0| > 2a$, then we can convert the ME there to a local (Taylor) expansion at $z = 0$. Namely, by using the Taylor expansion of $f(z) = 1/(z - z_0)^k$,

$$\frac{1}{(z - z_0)^k} = z_0^{-k}\left(\frac{z}{z_0} - 1\right)^{-k} = \sum_{k=0}^{\infty} \frac{f^{(k)}(0)}{k!} z^k, \tag{7.69}$$

the ME (7.68) can be expressed as a local Taylor expansion at 0, i.e.,

$$\Phi(z) = \sum_{l=0}^{\infty} b_l z^l, \tag{7.70}$$

where

$$\begin{aligned} b_0 &= \sum_{k=1}^{P} \frac{a_k}{z_0^k}(-1)^k + a_0\log(-z_0), \\ b_l &= \left[\frac{1}{z_0^l}\sum_{k=1}^{P} \frac{a_k}{z_0^k}\binom{l+k-1}{k-1}(-1)^k\right] - \frac{a_0}{l z_0^l}, \quad l \geq 1. \end{aligned} \tag{7.71}$$

7.3.2 A Recursion for Local Expansions (LEs) (0 → L-Level)

Denote the potential due to all charges *outside the i-box and its eight neighbors* by a function $\psi_{l,i}(z)$, i.e.,

$$\begin{aligned} \psi_{l,i}(z) = {} & \text{local expansion for the potential} \\ & \text{due to all charges outside the i-box and} \\ & \text{the eight neighbors of the i-box,} \end{aligned} \tag{7.72}$$

where $0 \leq l \leq L$ is the level of the tree structure.

To generate a recursive process, we also define another function $\tilde{\psi}_{l,j}(z)$ for the *parent of a j-box on level l* in the same spirit, namely

$$\tilde{\psi}_{l,j}(z) = \text{local expansion for the potential due to all charges outside the parent } \textit{of the j-box and} \text{ the eight neighbors of the parent.} \quad (7.73)$$

Therefore, if the i-box at level l has four children $j = 1, 2, 3, 4$ at level $l + 1$, then it is clear that by identifying the "i-box" as "parent of the j-box" above, and we have

$$\tilde{\psi}_{l+1,j}(z) = \psi_{l,i}(z), \quad \text{for } j = 1, 2, 3, 4. \quad (7.74)$$

This relation is the crucial step in building the recursion among the local expansions from $l = 0$ to L, which is done as follows.

Firstly, it should be mentioned that the difference between $\psi_{l,i}(z)$ and $\tilde{\psi}_{l,j}(z)$ is that the former includes more charges from additional 27 boxes which are defined as **"Interaction List"**, marked by "x" in the boxes in Fig. 7.2. If all charges are positive, we have

$$\psi_{l,i}(z) > \tilde{\psi}_{l,i}(z), \quad (7.75)$$

and in general we can interpret the > sign as "more charges are involved" counting both positive and negative charges.

By a recursive manner, we start at level 0, where $\tilde{\psi}_{0,i} = 0$, to form a recursion $\tilde{\psi}_{l,j} \to \tilde{\psi}_{l+1,j}$, and also a recursion for $\psi_{l,i} \to \psi_{l+1,i}$, until the level index $l = L$ is reached. Then we will have obtained the potential of all charges outside the i-box and its neighbors, which will be calculated by direct calculation.

- Step 1: Start at the $l = 0$ level, $\psi_{0,1}(z) = 0$ and $\tilde{\psi}_{0,1} = 0$.
- Step 2: For $l = 0, 1, \ldots, L$,
  we define for all 4^l i-boxes at level l

$$\begin{aligned}\psi_{l,i}(z) &= \tilde{\psi}_{l,i}(z) + \text{potential from the charges } q_i \text{ inside the 27 boxes from the interaction list}\\ &= \tilde{\psi}_{l,i}(z) + \sum_{k=1}^{27} \Phi_{l,k}(z),\end{aligned} \quad (7.76)$$

  where $\Phi_{l,k}(z)$ is the ME of the charges in the kth box from the interaction list, which is assumed as having been converted into a local expansion centered in the i-box.

  Meanwhile, for $l < L$ we update to

$$\tilde{\psi}_{l+1,j}(z) = \psi_{l,i}(z), \qquad j = 1, 2, 3, 4, \quad (7.77)$$

  where the i-box is the parent of the four children j-boxes.

In the updating step (7.77), the local expansion for the i-box will have its center shifted to the center of the four children j-boxes, which is made possible by the following translation operation.

- **Local to local translation**

A local expansion centered at z_0 can be translated to a local expansion centered at 0 by the following identity:

$$\sum_{k=1}^{\infty} a_k (z - z_0)^k = \sum_{l=0}^{n} b_l z^l, \tag{7.78}$$

where

$$b_l = \sum_{k=l}^{n} a_k \binom{k}{l} (-z_0)^{k-l}. \tag{7.79}$$

As seen from (7.76), the key ingredient in completing the recursion for the local expansions is the multipole expansions for boxes in the interaction list on all levels, and those multipole expansions can be obtained with a separate (upward) recursion to be discussed in the next section.

7.3.3 *A Recursion for the Multipole Expansion ($L \to 0$-Level)*

- Step 1: Start at the Lth level, and compute ME $\Phi_{L,j}(z)$, $j = 1, 2, \ldots, 4^L$.
- Step 2: For $l = L - 1, \ldots, 0$, for each i-box on the lth level, translate the center of the ME for each of its four children j-boxes, $j = 1, 2, 3, 4$, to the center of the i-box, i.e.,

$$\Phi_{l,i}(z) = \Phi_{l+1,1}(z) + \Phi_{l+1,2}(z) + \Phi_{l+1,3}(z) + \Phi_{l+1,4}(z), \tag{7.80}$$

where the center of the ME $\Phi_{l+1,j}(z)$, $j = 1, 2, 3, 4$, has been shifted to the center of the i-box, made possible by the following operation.

- **Multipole to multipole translation**

Given a ME centered at z_0:

$$\Phi(z) = a_0 \log(z - z_0) + \sum_{k=1}^{P} \frac{a_k}{(z - z_0)^k}, \tag{7.81}$$

it can be translated into another ME centered at zero:

$$\Phi(z) = a_0 \log(z) + \sum_{l=1}^{\infty} \frac{b_l}{z^l}, \tag{7.82}$$

where

$$b_l = \sum_{k=1}^{l} a_k z_0^{l-k} \binom{l-1}{k-1} - \frac{a_0 z_0^l}{l} \tag{7.83}$$

and

$$\left| \Phi(z) - a_0 \log(z) - \sum_{l=1}^{P} \frac{b_l}{z^l} \right| \le O\left(\frac{|z_0| + R}{z} \right)^{P+1}, \quad |z| > z_0 + R, \tag{7.84}$$

for any $R > 0$.

7.3.4 A Pseudo-Code for FMM

Finally, we summarize the $O(N)$ FMM for electrostatic interactions [3, 22] with the following pseudo-code for the two recursions involved:

- **Upward recursion for multipole expansions (ME)** $\Phi_{l,i}(z)$

Initial: L-level, compute all multipole expansions, $\Phi_{L,i}(z)$, $i = 1, 2, \ldots, 4^L$.

Loop: do

If $L - 1 \ge l \ge 0, i = 1, 2, \ldots, 4^l$,

$$\Phi_{l,i}(z) = \sum_{j=1}^{4} \Phi_{l+1,j}(z) \quad \boxed{\begin{array}{l}\Phi_{l+1,j}(z) \text{ – ME for four children of the i-box} \\ \text{with centers shifted to the center of the i-box}\end{array}} \tag{7.85}$$

- **Downward recursion for local expansions (LE)** $\psi_{l,i}(z)$

Initial: $l = 0$, $\psi_{0,1} = 0$ and $\tilde{\psi}_{0,1} = 0$.

Loop: do

If $0 < l \le L$, for all 4^l i-boxes at level l (7.86)

$$\psi_{l,i}(z) = \tilde{\psi}_{l,i}(z) + \sum_{k=1}^{27} \Phi_{l,k}(z) \quad \boxed{\begin{array}{c}\text{ME } \Phi_{l,k}(z) \text{ for 27 boxes in the} \\ \text{interaction list of the i-box} \\ \text{converted to LE centered at the i-box}\end{array}}$$

$$\tilde{\psi}_{l+1,j} := \psi_{l,i} \quad \boxed{\begin{array}{c}\text{j-box is one of the four children of the i-box,} \\ \text{loop over } 4^{l+1} \text{ j-boxes at level } l+1.\end{array}}$$

7.3.5 Conversion Operators for Electrostatic FMM in $\mathbb{R}^3$

The potential of N charges q_i in $\mathbb{R}^3$ is given by the following form:

$$\Phi(\mathbf{r}) = \sum_{i=1}^{N} \Phi_i(\mathbf{r}) = \frac{1}{4\pi\epsilon_0} \sum_{i=1}^{N} \frac{q_i}{|\mathbf{r} - \mathbf{r}_i|}. \tag{7.87}$$

An L-level oct-tree of nested cubes can be constructed for a cubic box containing all charges. The same upward pass recursion for the ME and downward recursion for the LE can be constructed as in the 2-D case. On the lth level, there will be 8^l cubic boxes, and for any i-box at a given lth level, its interaction list now will contain 189 boxes, which are one box away from the i-box. We will need the following results on the construction of the ME from a group of charges, a multipole to multipole translation, a multipole to local conversion, and a local to local translation [22].

- **Multipole expansion**

Given l charges of strengths $\{q_i\}_{i=1}^{l}$ located at $\{\mathbf{Q}_i = (\rho_i, \alpha_i, \beta_i)\}_{i=1}^{l}$ within a sphere of radius a centered at the origin, the potential at $\mathbf{P} = (r, \theta, \phi)$ with $r > a$ from all l charges can be approximated by the following p-term multipole expansion using spherical harmonics with a truncation error:

$$\left| \Phi(\mathbf{P}) - \sum_{n=0}^{p} \sum_{m=-n}^{n} \frac{M_n^m}{r^{n+1}} Y_n^m(\theta, \phi) \right| \le \frac{A}{r - a} \left(\frac{a}{r} \right)^{p+1}, \tag{7.88}$$

where

$$A = \sum_{l=1}^{l} |q_i| \tag{7.89}$$

and

$$M_n^m = \sum_{i=1}^{l} q_i \rho_i^n Y_n^{-m}(\alpha_i, \beta_i), \tag{7.90}$$

and the spherical harmonics are defined as

$$Y_n^m(\theta, \phi) = \sqrt{\frac{(n - |m|)!}{(n + |m|)!}} P_n^{|m|}(\cos\theta) \mathrm{e}^{\mathrm{i}m\phi}, \tag{7.91}$$

where $P_n^k(x)$ is the associated Legendre function.

- **Translation of multipole expansions**

For l charges of strengths $\{q_i\}_{i=1}^l$ located at $\{\mathbf{Q}_i = (\rho_i, \alpha_i, \beta_i)\}_{i=1}^l$ within a sphere of radius a centered at $\mathbf{Q} = (\rho, \alpha, \beta)$, the potential at $\mathbf{P} = (r, \theta, \phi)$ outside this sphere is represented by a multipole expansion

$$\Phi(\mathbf{P}) = \sum_{n=0}^{\infty}\sum_{m=-n}^{n} \frac{O_n^m}{r'^{n+1}} Y_n^m(\theta', \phi'), \quad r' > a, \tag{7.92}$$

where $\mathbf{P} - \mathbf{Q} = (r', \theta', \phi')$. Then, the ME can be rewritten as another ME centered at the origin

$$\Phi(\mathbf{P}) = \sum_{j=0}^{\infty}\sum_{k=-j}^{j} \frac{M_j^k}{r^{j+1}} Y_j^k(\theta, \phi), \text{ if } r > a + \rho, \tag{7.93}$$

where

$$M_j^k = \sum_{n=0}^{j}\sum_{m=-n}^{n} \frac{O_{j-n}^{k-m} J_m^{k-m} A_n^m A_{j-n}^{k-m} \rho^n Y_n^{-m}(\alpha, \beta)}{A_j^k}, \tag{7.94}$$

and

$$A_n^m = \frac{(-1)^n}{\sqrt{(n-m)!(n+m)!}} \quad \text{and} \quad J_n^m = \begin{cases} (-1)^{\min\{|n|,|m|\}}, & \text{if } mn < 0, \\ 1, & \text{otherwise.} \end{cases} \tag{7.95}$$

Moreover, the series (7.93) can be truncated with the following error estimate:

$$\left| \Phi(\mathbf{P}) - \sum_{j=0}^{p}\sum_{k=-j}^{j} \frac{M_j^k}{r^{j+1}} Y_j^k(\theta, \phi) \right| \le \frac{1}{r - (a+\rho)} \sum_{i=1}^{l} |q_i| \left(\frac{a+\rho}{r} \right)^{p+1}. \tag{7.96}$$

- **Multipole to local expansion conversion**

For l charges of strengths $\{q_i\}_{i=1}^l$ located at $\{\mathbf{Q}_i = (\rho_i, \alpha_i, \beta_i)\}_{i=1}^l$ within a sphere of radius a centered at $\mathbf{Q} = (\rho, \alpha, \beta)$ with $\rho > (c+1)a$ and $c > 1$, the multipole expansion (7.92) can be rewritten as the following local expansion inside the sphere of radius a centered at the origin:

$$\Phi(\mathbf{P}) = \sum_{j=0}^{\infty}\sum_{k=-j}^{j} L_j^k Y_j^k(\theta, \phi) r^j, \quad r < a, \tag{7.97}$$

where $\mathbf{P} = (r, \theta, \phi)$ and

$$L_j^k = \sum_{n=0}^{\infty}\sum_{m=-n}^{n} \frac{O_n^m J_k^m A_n^m A_j^k Y_{j+n}^{m-k}(\alpha, \beta)}{A_{j+n}^{m-k} \rho^{j+n+1}}, \tag{7.98}$$

where

$$J_m^{m'} = \begin{cases} (-1)^{m'}(-1)^{\min\{|m'|,|m|\}}, & \text{if } mm' > 0, \\ (-1)^{m'}, & \text{otherwise,} \end{cases} \tag{7.99}$$

and A_n^m is defined as in (7.95).

Moreover, the series (7.97) can be truncated with the following error estimate:

$$\left| \Phi(\mathbf{P}) - \sum_{j=0}^{p} \sum_{k=-j}^{j} L_j^k Y_j^k(\theta, \phi) r^j \right| \le \frac{1}{(c-1)a} \sum_{i=1}^{l} |q_i| \left(\frac{1}{c}\right)^{p+1}. \tag{7.100}$$

- **Translation of local expansions**

Let $\mathbf{Q} = (\rho, \alpha, \beta)$ be the origin of a local expansion:

$$\Phi(\mathbf{P}) = \sum_{n=0}^{p} \sum_{m=-n}^{n} O_n^m Y_n^m(\theta', \phi') r'^n, \tag{7.101}$$

where $\mathbf{P} = (r, \theta, \phi)$ and $\mathbf{P} - \mathbf{Q} = (r', \theta', \phi')$. Then, the local expansion can be rewritten as another local expansion centered at the origin:

$$\Phi(\mathbf{P}) = \sum_{j=0}^{p} \sum_{k=-j}^{j} L_j^k Y_j^k(\theta, \phi) r^j, \tag{7.102}$$

where

$$L_j^k = \sum_{n=j}^{p} \sum_{m=-n}^{n} \frac{O_n^m J_{n-j,m-k}^m A_{n-j}^{m-k} A_j^k \rho^{n-j} Y_{n-j}^{m-k}(\alpha, \beta)}{A_n^m}, \tag{7.103}$$

and

$$J_{n,m}^{m'} = \begin{cases} (-1)^{n+m}, & \text{if } mm' < 0, \\ (-1)^n(-1)^{m'-m}, & \text{if } mm' > 0, |m'| < |m|, \\ (-1)^n, & \text{otherwise.} \end{cases} \tag{7.104}$$

7.4 Helmholtz FMM of Wideband of Frequencies for N-Current Source Interactions

In this section, we will consider the interaction from N sources distributed at $\mathbf{x}_i$ with strength I_i, $i = 1, 2, \ldots, N$. Each of the (current) sources I_i will generate an oscillating field at a location $\mathbf{x}$ away from $\mathbf{x}_i$ governed by a Helmholtz equation:

$$\text{(2-D)} \quad -(\Delta + k^2)\Phi = \delta(\mathbf{x}), \qquad \Phi(\mathbf{x}) = \frac{\mathrm{i}}{4} H_0^{(2)}(k|\mathbf{x}|), \tag{7.105}$$

and

$$\text{(3-D)} \qquad -(\Delta + k^2)\Phi = \delta(\mathbf{x}), \qquad \Phi(\mathbf{x}) = \frac{e^{-ik|\mathbf{x}|}}{4\pi|\mathbf{x}|}. \tag{7.106}$$

Therefore, in the 2-D case, the interactions between all the sources can be written as (omitting the constant $\frac{i}{4}$)

$$\Phi(\mathbf{x}_i) = \sum_{j=1, j\neq i}^{N} I_j H_0^{(2)}(k|\mathbf{x}_i - \mathbf{x}_j|), \quad i = 1, 2, \ldots, N, \tag{7.107}$$

where $H_0^{(2)}$ denotes the zeroth-order Hankel function of the second kind and k represents the wave number. The FMM will calculate (7.107) with a complexity of $O(N \log N)$ [23] with a hierarchical quad-tree data structure by utilizing two kinds of partial wave expansions corresponding, respectively, to the multipole and local expansions for the electrostatic FMM. For $\mathbf{x} = (x, y)$, denote by ρ the distance between $\mathbf{x}$ and the center of the expansion, by θ the angle between $\mathbf{x}$ and the x-axis, and let H_m and J_m denote the mth-order Hankel and Bessel functions. Then we have

- H- or multipole expansion:

$$\Phi(\mathbf{x}) = \psi(\mathbf{x}) \equiv \sum_{m=-\infty}^{\infty} \beta_m H_m(k\rho) e^{im\theta}, \quad \text{for } |\mathbf{x}| > \max_i |\mathbf{x}_i|, \tag{7.108}$$

where

$$\beta_m = \sum_{i=1}^{N} I_i J_m(k|\mathbf{x}_i|) e^{-im\theta_i}. \tag{7.109}$$

- J- or local expansion:

$$\Phi(\mathbf{x}) = \phi(\mathbf{x}) \equiv \sum_{m=-\infty}^{\infty} \alpha_m J_m(k\rho) e^{im\theta}, \quad \text{for } |\mathbf{x}| < \min_i |\mathbf{x}_i|, \tag{7.110}$$

where

$$\alpha_m = \sum_{i=1}^{N} I_i H_m(k|\mathbf{x}_i|) e^{-im\theta_i}. \tag{7.111}$$

Equations (7.108) and (7.110) will be referred to as the H- and J-expansions, respectively, and they are equivalent, respectively, to multipole and local expansions for the case of the Laplace equation. The H- and J-expansions can be derived from the addition theorem for the Bessel functions [24]. In the high-frequency FMM (HF-FMM), the diagonalized form of translation and conversion operators through far-field forms of wave expansions is used instead; this, however, will fail to converge when k is very small or the level of refinement is high. This is due to a divergent property of the Hankel function in the conversion operator. Specifically, the far-field forms of the H- and J-expansions are defined by

$$F(\theta) = \sum_{m=-\infty}^{\infty} \beta_m \mathrm{e}^{-\mathrm{i}(m\pi/2)} \mathrm{e}^{\mathrm{i}m\theta}, \tag{7.112}$$

$$G(\theta) = \sum_{m=-\infty}^{\infty} \alpha_m \mathrm{e}^{-\mathrm{i}(m\pi/2)} \mathrm{e}^{\mathrm{i}m\theta}, \tag{7.113}$$

respectively, where $\{\beta_m\}_{m=-\infty}^{+\infty}$ and $\{\alpha_m\}_{m=-\infty}^{+\infty}$ are coefficients of the H- and J-expansions from (7.108) and (7.110). Then, $F(\theta)$ is converted to $G(\theta)$ via a diagonalized operator [25] as follows:

$$G_{c_3}(\theta) = \nu_n(\theta) \cdot F_{c_1}(\theta), \tag{7.114}$$

where

$$\nu_n(\theta) = \sum_{m=-n}^{n} \mathrm{e}^{\mathrm{i}m(\theta+\theta_{13}-\pi)} H_m(k\rho_{13}), \tag{7.115}$$

and a subscript in $F(\theta)$ and $G(\theta)$ denotes the center of the expansions, ρ_{13} is the distance between two centers defined by $|c_3 - c_1|$, and θ_{13} is the angle between the x-axis and ρ_{13}. The diagonalized conversion operator $\nu_n(\theta)$ diverges quickly when the order of the Hankel function is larger than its argument ($m > k\rho_{13}$) because of the asymptotic behavior of the Hankel function [24], namely

$$\lim_{m\to\infty} Y_m(z) \left(\frac{\mathrm{e}z}{2m}\right)^m \frac{\sqrt{\pi m}}{\sqrt{2}} = -1, \tag{7.116}$$

where $H_m(z) = J_m(z) + \mathrm{i}Y_m(z)$.

The quad-tree structure given in [26] for the 2-D Helmholtz FMM is divided into a low-frequency part and a high-frequency part based on the value of kR, where R is the size of the box at a level of the tree. In order to overcome the divergence problem for small k, the tree is divided into two parts, with a cut-off level based on the quantity of kR. In the tree level with $kR < 4/\mathrm{e} = 1.471518$ (boxes below the cut-off level), a low-frequency FMM (LF-FMM), which uses the H- and J-expansions and non-diagonalized conversion operator based on the addition theorem [24],

$$\alpha_m = \sum_{j=-n}^{n} \mathrm{e}^{-\mathrm{i}j(\theta_{13}-\pi)} \beta_{m-j} H_j(k\rho_{13}), \tag{7.117}$$

is used directly to avoid divergence of the conversion operator. Then, at and above the cut-off level, coefficients of the H- and J-expansions are converted into far-field forms using (7.112) and (7.113), and the regular HF-FMM is used for the boxes above the cut-off level. In the actual implementation, $kR = 1.5$ is used for a safe determination of the cut-off level. The resulting FMM is applicable for a wide band of frequencies associated with 2-D Helmholtz wave propagations [26, 27], and the 3-D version for the Helmholtz can be found in [23].

7.5 Reaction Field Hybrid Model for Electrostatics

In Sect. 2.3, we constructed efficient image methods to represent the reaction field of dielectric and electrolyte solutions to charges inside a spherical cavity. As a third method that we can use to handle the electrostatic interaction, we consider a hybrid multi-scale approach which combines the explicit atomic Coulomb interaction and the implicit dielectric model of electrostatic interactions. For example, in a hybrid model for biomolecular solvation a central part of the simulated system contains the solute and some solvent, which is considered in atomic detail, while the remaining part is treated as a dielectric continuum. Figure 7.3 (left) describes such a model [28, 29]: the larger sphere contains the particles to be simulated by the molecular dynamics based on Newtonian mechanics with electrostatic and non-electrostatic forces acting on each particle; outside the larger sphere, the solvent medium is represented by a dielectric continuum with dielectric constant ϵ_o, and possibly also with an ionic concentration. Within the larger sphere, a smaller sphere with radius R_c is embedded with a separation τ. The region between the two concentric spheres is a buffer region, which is an important component of the hybrid model to reduce the surface effect and produce a homogeneous environment within the central region of the model. Finally, a simulation region in the shape of a cubic box is embedded inside the smaller sphere, away from the buffer region. Only the particles inside the cubic box will be simulated dynamically; the particles outside the cubic box but inside the larger sphere are in fact periodic image copies of those inside the cubic box, and therefore they are not independent particles and will not be used for calculating structural (radial distribution function, RDF) and dynamical (dielectric constant) quantities for the system under simulation.

In the model shown in Fig. 7.3 (left), the larger sphere surface, where the explicit and implicit model of electrostatic interaction meet, is used for the calculation of the reaction field for all particles inside. The cubic box will also be used for the periodic treatment of non-electrostatic forces such as van der Waals forces, etc. The region inside the cubic box, which is not imaged outside the cubic box, will be considered as the effective production region where statistics on the particle trajectories

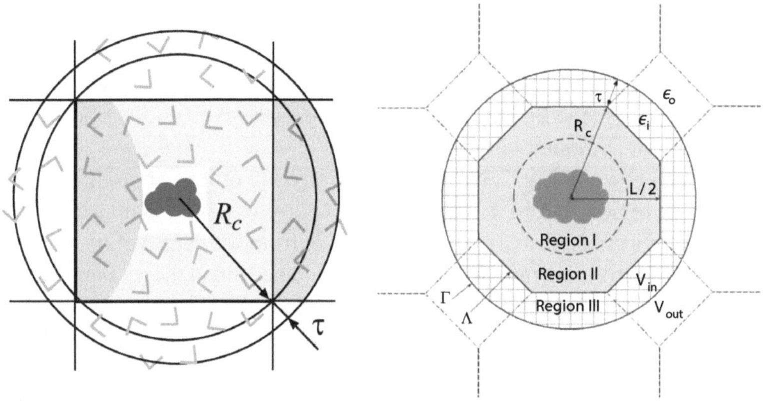

Fig. 7.3 A hybrid model with (left) a cubic simulation box and (right) a truncated octahedron (TO) simulation box. Used with permission of AIP Publishing, from [28]; permission conveyed through Copyright Clearance Center, Inc.

during molecular dynamics simulation will be used for analysis of the biomolecules embedded within the production region.

Two types of potentials are acting on the charges in the explicit solvent part of the larger sphere. Firstly, it is the direct Coulomb potential, through which the charges interact with one another, Φ_S. Secondly, it is an indirect potential that results from the polarization of the continuum solvent region by the explicit charges inside the larger sphere, referred to as the reaction field, Φ_{rf}. The total potential inside the explicit region is expressed as $\Phi = \Phi_S + \Phi_{\text{rf}}$.

Based on the result in Sect. 2.3.2, the reaction field potential can be approximated by $N_i = M_i + 1$ image charges as follows:

$$\Phi_{\text{rf}}(\mathbf{r}) \approx \frac{q_{\text{k}}}{4\pi\epsilon_{\text{i}}\,|\mathbf{r}-\mathbf{r}_{\text{k}}|} + \frac{1}{4\pi\epsilon_{\text{i}}}\sum_{m=0}^{M_i}\frac{q_m}{|\mathbf{r}-\mathbf{x}_m|}, \tag{7.118}$$

where the Kelvin image charge and location q_{k}, $\mathbf{r}_{\text{k}}$ are given in (2.124) and (2.125) and the rest image charges q_m and $\mathbf{x}_m$, $0 \le m \le M_i$, are given in (2.137) and (2.138), respectively.

In summary, let $(q_i, \mathbf{r}_i)$, $1 \le i \le N$, be all the charges and their locations inside the bigger sphere where the electrostatic interaction will be computed. Then, the electrostatic potential acting on any given charge $\mathbf{r}_j$ will be calculated by the following two sums:

$$\Phi(\mathbf{r}_j) = \sum_{i=1, i\neq j}^{N}\frac{1}{4\pi\epsilon_{\text{i}}}\frac{q_i}{|\mathbf{r}_i-\mathbf{r}_j|} + \sum_{i=1}^{N}\Phi_{\text{rf}}^{i}(\mathbf{r}_j), \tag{7.119}$$

where the first sum corresponds to the direct Coulomb interaction from all other charges except $\mathbf{r}_j$ itself, and the second sum is due to the reaction field of all charges

$\mathbf{r}_i$ inside the sphere, and the N_i image charge approximation (7.118) is used to compute all reaction fields.

A periodic boundary condition (PBC) will be used, in reference to the cubic box, or better a truncated octahedron (TO) box, for non-electrostatic interactions to suppress surface effects in computer simulations.

A TO box (shown in Fig. 7.3 (right)) is preferred in order to maximize the size of the production region. In Fig. 7.3 (right), the TO box Λ is built from a cube of length L by cutting eight corners at a distance $L/(4\sqrt{2})$ from the center of its sides, and the figure shows the cross-section of the truncated octahedron in the xz plane if the cubic box from which it is created is centered around the origin. The simulation TO box has 14 faces, including 8 hexagonal faces and 6 square faces, and 24 corners, all of which are equivalent. The distance from the origin to a square face is $L/2$, to a hexagonal face $\sqrt{3}L/4$, and to a corner $R_c = \sqrt{5}L/4$. There are 14 nearest neighbors of the central simulation box, each resulting from a translation through an appropriate side of the TO. Particles in these neighboring boxes that are at a distance $R_c + \tau$ or less from the origin form a buffer layer around the central simulation box, which we refer to as Region III. Together with the simulation box Λ, the buffer layer forms the local volume of a spherical cavity Γ. All charges in Γ will produce the reaction field, to be computed using the image charge method. The particles inside the simulation box that give rise to Region III through periodic imaging are denoted as Region II. The remaining particles in Λ, which are not periodically imaged, are labeled as Region I. This region is the productive part of the simulation box in which a solute may be solvated presumably without experiencing many artificial electrostatic solute–solute interactions. The size of this region, d, can be determined by comparing the points of furthest and nearest separation of particles in Λ from the center of the box. For a TO shown in Fig. 7.3, $d = L(\sqrt{3} - \sqrt{5}/2) - 2\tau = 0.61L - 2\tau$. The same quantity for a cubic box is $d = L(2 - \sqrt{3}) - 2\tau = 0.27L - 2\tau$. Comparing these two quantities, the advantage of using the TO box becomes obvious. For $L = 45$ Å and $\tau = 5$ Å, for instance, the TO box allows simulations of solute molecules with diameter < 17 Å. These same parameters for a cubic box result in $d = 2$ Å, which is not meaningful from the molecular size point of view.

To evaluate the electrostatic forces within the simulation box Λ, the FMM is used for the interaction of all charges in the box Λ, plus their periodic images in Region III and all image charges outside the cavity Γ, with all charges being taken as acting in a homogeneous medium of dielectric permittivity ϵ_{i}. Considering that only the force within the simulation box Λ needs to be evaluated, a simple but more efficient way would be to calculate the interaction between the charges in Λ and the periodic/image charges far away from Λ directly by a local expansion. Specifically, we introduce another reference sphere S_r of radius κR_c centered at the origin with $\kappa > 1$. The evaluation of the field within the TO box Λ due to the charges inside this reference sphere is carried out by the FMM. For all periodic/image charges outside this reference sphere, the potential field at a point $\mathbf{r} = (r, \theta, \phi)$ inside the box Λ generated by these periodic/image charges can be described by a local expansion

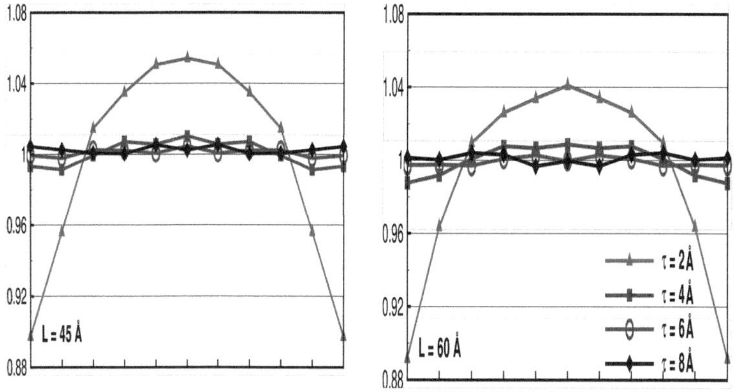

Fig. 7.4 Density across the diagonal of the truncated octahedron (TO) box: (left) 45 Å simulation TO box; (right) 60 Å simulation TO box. Used with permission of AIP Publishing, from [28]; permission conveyed through Copyright Clearance Center, Inc.

$$\Phi(\mathbf{r}) \approx \sum_{j=0}^{p} \sum_{k=-j}^{j} L_j^k \cdot Y_j^k(\theta, \phi) \cdot r^j, \tag{7.120}$$

where p is the local expansion order, $Y_j^k(\theta, \phi)$ are the spherical harmonics, and L_j^k are the local expansion coefficients given by

$$L_j^k = \sum_{l=1}^{L} \hat{q}_l \cdot \frac{Y_j^{-k}(\alpha_l, \beta_l)}{\rho_l^{j+1}}. \tag{7.121}$$

Here, $\hat{q}_l, l = 1, 2, \ldots, L$, are the periodic/image charges outside S_r with $(\rho_l, \alpha_l, \beta_l)$ as their locations. Consequently, the force $\mathbf{f}(\mathbf{r}) = (f_x(\mathbf{r}), f_y(\mathbf{r}), f_z(\mathbf{r}))$ exerted on a particle q at $\mathbf{r} = (r, \theta, \phi)$ inside Λ by these periodic/image charges can be calculated using

$$f_x(\mathbf{r}) = -q\frac{\partial}{\partial x}\Phi(\mathbf{r}) = -q \cdot \mathrm{Re}\,(H_2 - H_3)\,, \tag{7.122}$$

$$f_y(\mathbf{r}) = -q\frac{\partial}{\partial y}\Phi(\mathbf{r}) = -q \cdot \mathrm{Im}\,(H_2 + H_3)\,, \tag{7.123}$$

$$f_z(\mathbf{r}) = -q\frac{\partial}{\partial z}\Phi(\mathbf{r}) = -q \cdot (H_0 + 2\mathrm{Re}(H_1))\,, \tag{7.124}$$

where $\mathrm{Re}(\cdots)$ and $\mathrm{Im}(\cdots)$ represent the real part and the imaginary part of a complex number, respectively, and

$$H_0 = \sum_{j=1}^{p} j L_j^0 P_{j-1}(\cos\theta) r^{j-1},$$

$$H_1 = \sum_{j=1}^{p} \sum_{k=1}^{j-1} (j+k) C_j^k \cdot L_j^k e^{ik\phi} P_{j-1}^k (\cos\theta) r^{j-1},$$

$$H_2 = \sum_{j=1}^{p} L_j^0 e^{i\phi} P_{j-1}^1 (\cos\theta) r^{j-1} + \sum_{j=1}^{p} \sum_{k=1}^{j} C_j^k L_j^k e^{i(k+1)\phi} P_{j-1}^{k+1} (\cos\theta) r^{j-1},$$

$$H_3 = \sum_{j=1}^{p} C_j^{-1} L_j^1 P_{j-1}(\cos\theta) r^{j-1} + \sum_{j=1}^{p} \sum_{k=2}^{j} B_j^k C_j^k L_j^k e^{i(k-1)\phi} P_{j-1}^{k-1} (\cos\theta) r^{j-1},$$

and

$$B_j^k = (j+k)(j+k-1), \qquad C_j^k = \sqrt{\frac{(j-k)!}{(j+k)!}}.$$

The model has been validated by several biological systems [28, 29] including homogeneous pure water and ions solvated inside water. In those studies, the structural and dynamical properties have been shown to agree well with results from the PME, which is known to be accurate for homogeneous systems. Figure 7.4 shows the relative particle density ρ_r as a function of the position in the simulation box with $L = 45$ and 60 Å and $\tau = 2$, 4, 6, and 8 Å (sampled at 11 equi-spaced positions along a line that connects two opposite vertices of the TO box). It is seen that $\tau = 2$ Å produces a noticeably non-uniform density where the density at the edges of the simulation box is about 10% lower than the average density. Buffer size $\tau > 2$ Åleads to a uniform density distribution. Figure 7.4 shows Density across the diagonal of the truncated octahedron (TO) box.

In addition, dielectric properties were evaluated by computing the dielectric permittivity constant ϵ using the formula given in (1.131) in terms of the total dipole moment $\mathbf{M}(R)$ of a spherical sample with radius R. The radius of the sample also has an upper bound, $R_c + \tau$, defined by the geometry of the model. The maximum allowed R cannot be used in the calculations, however, because the sphere then contains periodic images of water molecules. As a linear response approximation, the dielectric constant relies on the quadratic fluctuations of the total dipole moment $\mathbf{M}(R)$, which, as (1.131) shows, scale linearly with the sample volume. If periodic images are present in a sample, the linear scaling of $\langle \mathbf{M}^2(R) \rangle$ is violated, thus invalidating the fluctuation formula. Therefore the radii in the calculations are limited to $R \le R_{\max} = \sqrt{3}L/4$, ensuring that only one copy of each water molecule is considered. The remaining layer of water $R_{\max} < R < R_c + \tau$ acts as part of the dielectric continuum. Its dielectric permittivity $\epsilon' = \epsilon$ is set self-consistently in the calculations using (1.131). It was found that only a few iterations are needed between ϵ and ϵ' to achieve convergence.

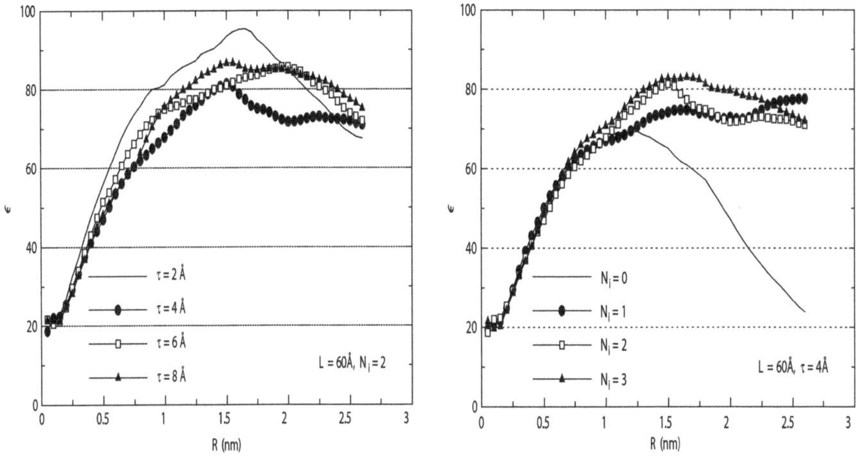

Fig. 7.5 Dielectric constant calculated within a spherical region of radius indicated by the horizontal axis. Left: Dielectric constant versus the thickness of the buffer region τ; right: dielectric constant versus the number of image charges in approximating the reaction field. $N_i = 0$ indicates no reaction field. Used with permission of AIP Publishing, from [28]; permission conveyed through Copyright Clearance Center, Inc.

Figure 7.5 shows the dielectric constant $\epsilon(R)$ with the number of image charges N_i varying from 0 to 3 for $L = 60$ Å and $\tau = 4$ Å, as a function of the radius of the sampled sphere in the total dipole moment fluctuation $\mathbf{M}(R)$ in (1.131). Note that $N_i = 0$ means that no reaction field contribution is included in the model, and the data shown in Fig. 7.5 (right) therefore reveal that the reaction field is essential for maintaining a uniform dielectric response throughout the simulation box [30–32].

7.6 Summary

Many-body interactions form the key computational step in computing the effects of long-range forces. The PME provides an $O(N \log N)$ solution for periodic systems in lattice systems, while the FMM gives an $O(N)$ for electrostatic interactions and $O(N \log N)$ for waves in a finite system with or without periodic boundary conditions. The reaction field approach, together with its image charge representation, provides a multi-scale model for treating electrostatic interactions, and reduces the size of the many-body systems, which can be readily handled by the FMM or other related fast algorithms [21, 33].

References

1. M.P. Allen, D.J. Tildesley, *Computer Simulation of Liquids* (Oxford University Press, Oxford, UK, 1989)
2. P. Ewald, Die Berechnung optischer und elektrostatischer Gitterpotentiale. Ann. Phys. **64**, 253–287 (1921)
3. L. Greengard, V. Rokhlin, A fast algorithm for particle simulations. J. Comput. phys. **73**(2), 325–348 (1987)
4. C. Kittel, *Introduction to Solid State Physics*, 8th edn. (Wiley, New York, 2004)
5. D.M. Heyes, Electrostatic potential and fields in infinite point charge lattice. J. Chem. Phys. **74**, 1924–1929 (1981)
6. L.C. Evans. *Partial Differential Equations*. Graduate Studies in Mathematics, vol. 19 (Proceedings of the American Mathematical Society, Providence, RI, 1998)
7. J.E. Roberts, J. Schnitker, How the unit cell surface charge distribution affects the energetics of ion-solvent interactions in simulations. J. Chem. Phys. **101**, 5024–5031 (1994)
8. T.M. Nymand, P. Linse, Ewald summation and reaction field methods for potentials with atomic charges, dipoles, and polarizabilities. J. Chem. Phys. **112**, 6152–6160 (2000)
9. S.W. de Leeuw, J.W. Perram, E.R. Smith, Simulation of electrostatic systems in periodic boundary conditions. I. Lattice sums and dielectric constants. Proc. R. Soc. Lond. A **373**, 27–56 (1980)
10. A. Redlack, J. Grindlay, The electrostatic potential in a finite ionic crystal. Can. J. Phys. **50**, 2815–2825 (1972)
11. A. Redlack, J. Grindlay, Coulombic potential lattice sums. J. Phys. Chem. Solids **36**, 73–82 (1975)
12. A. Aguado, P.A. Madden, Ewald summation of electrostatic multipole interactions up to the quadrupolar level. J. Chem. Phys. **119**, 7471–7483 (2003)
13. T. Laino, J. Hutter, Notes on 'Ewald summation of electrostatic multipole interactions up to quadrupolar level. J. Chem. Phys. **119**, 7471 (2003). J. Chem. Phys. **129**, 074102 (2008)
14. J. Sala, E. Guàrdia, M. Masia, The polarizable point dipoles method with electrostatic damping: implementation on a model system. J. Chem. Phys. **133**, 234101 (2010)
15. R.W. Hockney, J.W. Eastwood, *Computer Simulation Using Particles* (Taylor & Francis, New York, 1981)
16. T. Darden, D. York, L. Pedersen, Particle mesh Ewald: an $N\log(N)$ method for Ewald sums in large systems. J. Chem. Phys. **98**, 10089–10092 (1993)
17. U. Essmann, L. Perera, M.L. Berkowitz et al., A smooth particle mesh Ewald method. J. Chem. Phys. **103**, 8577–8593 (1995)
18. A.Y. Toukmaji, J.A. Board Jr., Ewald summation techniques in perspective: a survey. Comput. Phys. Commun. **95**, 73–92 (1996)
19. S. Jin, L. Li, Z. Xu, Y. Zhao, A random batch Ewald method for particle systems with Coulomb interactions. SIAM J. Sci. Comput. **43**(4), B937-60 (2021)
20. J. Liang, P. Tan, Y. Zhao, L. Li, S. Jin, L. Hong, Z. Xu. Superscalability of the random batch Ewald method. J. Chem. Phys. **156**(1) (2022)
21. J. Barnes, P. Hut, A hirerachical $o(n \log n)$ force-calculation algorithm. Nature **324**, 446–449 (1986)
22. L. Greengard, *The Rapid Evaluation of Potential Fields in Particle Systems* (MIT Press, Cambridge, MA, 1988)
23. H. Cheng, W. Crutchfield, Z. Gimbutas et al., A wideband fast multipole method for the Helmholtz equation in three dimensions. J. Comput. Phys. **216**, 300–325 (2006)
24. M. Abramowitz, I.A. Stegun, *Handbook of Mathematical Functions with Formulas, Graphs, and Mathematical Tables*, 9th edn. (Dover Publications, New York, 1972)
25. V. Rokhlin, Rapid solution of integral equations of scattering theory in two dimensions. J. Comput. Phys. **86**, 414–439 (1990)
26. M.H. Cho, W. Cai, A wideband fast multipole method for the two-dimensional complex Helmholtz equation. Comput. Phys. Commun. **181**, 2086–2090 (2010)

27. H. Cheng, W. Crutchfield, Z. Gimbutas et al., Remarks on the implementation of the wideband FMM for the Helmholtz equation in two dimensions. Contemp. Math. **408**, 99–110 (2006)
28. Y. Lin, A. Baumketner, S. Deng et al., An image-based reaction field method for electrostatic interactions in molecular dynamics simulations of aqueous solutions. J. Chem. Phys. **131**, 154103 (2009)
29. Y. Lin, A. Baumketner, S. Song et al., Ionic solvation studied by image-charge reaction field method. J. Chem. Phys. **134**, 044105 (2011)
30. L. Wang, J. Hermans, Reaction field molecular dynamics simulation with Friedman's image charge method. J. Phys. Chem. **99**, 12001–12007 (1995)
31. J.M. Rodgers, J.D. Weeks, Interplay of local hydrogen-bonding and long-ranged dipolar forces in simulations of confined water. Proc. Natl. Acad. Sci. USA **105**, 19136–19141 (2008)
32. W. Song, Y. Lin, A. Baumketner et al., Effect of the reaction field on molecular forces and torques revealed by an image-charge solvation model. Commun. Comput. Phys. **13**, 129–149 (2013)
33. L. Ying, G. Biros, D. Zorin, A kernel-independent adaptive fast multipole algorithm in two and three dimensions. J. Comput. Phys. **196**, 591–626 (2004)

Part II
Electromagnetic Scattering

Chapter 8
Maxwell's Equations, Potentials, and Physical/Artificial Boundary Conditions

In this chapter, we will first introduce the time-dependent Maxwell's equations and solution representations using scalar and vector potentials. Then, physical boundary conditions involving interfaces between dielectrics and conductors will be discussed. For computing scattering fields in infinite domains, several types of local artificial boundary conditions for computational domains will be presented, including the Engquist–Majda one-way boundary conditions, the high-order Bayliss–Turkel boundary conditions in auxiliary variable forms, and the uniaxial perfectly matched layered (UPML) boundary conditions.

8.1 Time-Dependent Maxwell's Equations

The electric field **E** and the magnetic flux density (or induction) **B** in media are related by Faraday's law of magnetic induction:

$$\nabla \times \mathbf{E} = -\frac{\partial \mathbf{B}}{\partial t} - \mathbf{J}_m, \tag{8.1}$$

where a magnetic current $\mathbf{J}_m$ is introduced for mathematical symmetry of Maxwell's equations only as there are no naturally found magnetic charges.

The electric field **E** gives the electric force experienced by a test charge Q in (1.2), and the magnetic flux density **B** together with the electric field **E** will produce a Lorentz force on a moving test charge Q with a velocity **v** in the Lorentz force law,

$$\mathbf{F} = Q(\mathbf{E} + \mathbf{v} \times \mathbf{B}), \tag{8.2}$$

and a torque **N** on an atomic magnetic dipole **m**,

W. Cai, *Deterministic, Stochastic, and Deep Learning Methods for Computational Electromagnetics*, https://doi.org/10.1007/978-981-96-0100-4_8

$$\mathbf{N} = \mathbf{m} \times \mathbf{B}, \tag{8.3}$$

where the atomic magnetic dipole moment $\mathbf{m}$ is produced by an electron circulating around the nucleus, the direction of which is normal to the plane of the moving electron. As there are no magnetic charges, the term "magnetic dipole" has no direct analogy with the definition of an electric dipole moment (1.8). In (8.1), $\mathbf{E}$ and $\mathbf{B}$ are considered to be the fundamental physical quantities in electrodynamics, whereas the displacement flux $\mathbf{D}$ and the magnetic field $\mathbf{H}$ are derived quantities.

8.1.1 *Magnetization* M *and Magnetic Field* H

As in the polarization of electric dipole moments under external electric fields in dielectrics discussed in Chap. 1, a similar physical process called magnetization acts on the atomic magnetic dipole moment $\mathbf{m}_i$ inside the material under external magnetic fields, and this is represented by a magnetization density $\mathbf{M}$. In analogy to the polarization density $\mathbf{P}$ discussed in Sect. 1.2, we define the magnetization density $\mathbf{M}$ as

$$\mathbf{M} = \sum_i N_i \langle \mathbf{m}_i \rangle , \tag{8.4}$$

where N_i is the number density of i-type atoms with atomistic magnetic moments $\mathbf{m}_i$. A magnetization current $\mathbf{J}_m = \nabla \times \mathbf{M}$ will be generated to produce magnetic fields similar to those produced by a charge current $\mathbf{J}$; both currents together will form the total magnetic flux density $\mathbf{B}$ inside a material, in a similar manner as for the electric field $\mathbf{E}$ inside a dielectric material, (1.47).

The magnetic flux density $\mathbf{B}$ under a steady free charge current density $\mathbf{J}_e$ along any loop C is given by the Biot–Savart law [1]:

$$\mathbf{B}(\mathbf{r}) = \frac{\mu_0}{4\pi} \int_C \frac{\mathbf{J}_e \times (\mathbf{r} - \mathbf{r}')}{|\mathbf{r} - \mathbf{r}'|^3} \mathrm{d}\mathbf{r}', \tag{8.5}$$

where μ_0 is the vacuum permeability, $\mu_0 = 1.257 \times 10^{-6}\ \mathrm{H} \cdot \mathrm{m}^{-1}$. By using the identity $(\mathbf{r} - \mathbf{r}') / |\mathbf{r} - \mathbf{r}'|^3 = -\nabla(1/|\mathbf{r} - \mathbf{r}'|)$, $\mathbf{B}(\mathbf{r})$ can be rewritten in terms of a vector potential $\mathbf{A}$ as

$$\mathbf{B} = \nabla \times \mathbf{A}, \tag{8.6}$$

where

$$\mathbf{A}(\mathbf{r}) = \frac{\mu_0}{4\pi} \int_C \frac{\mathbf{J}_e}{|\mathbf{r} - \mathbf{r}'|} \mathrm{d}\mathbf{r}'. \tag{8.7}$$

By a simple calculation, Ampère's law,

$$\nabla \times \mathbf{B} = \mu_0 \mathbf{J}_e, \tag{8.8}$$

is obtained, and Gauss's law for the magnetic induction holds (by the virtue of the curl form in (8.6)):

$$\nabla \cdot \mathbf{B} = 0. \tag{8.9}$$

With magnetization $\mathbf{M}$ in a material, the magnetization current $\mathbf{J}_m = \nabla \times \mathbf{M}$ will contribute to the magnetic flux density, which can be included by modifying the vector potential $\mathbf{A}$ in (8.7) to

$$\mathbf{A}(\mathbf{r}) = \frac{\mu_0}{4\pi} \int_C \frac{(\mathbf{J}_e + \nabla \times \mathbf{M})}{|\mathbf{r} - \mathbf{r}'|} \mathbf{dr}'. \tag{8.10}$$

Then, Ampère's law (8.8) becomes

$$\nabla \times \mathbf{B} = \mu_0(\mathbf{J}_e + \nabla \times \mathbf{M}). \tag{8.11}$$

Next, we define a constitutive relation between the magnetic flux density $\mathbf{B}$ and the derived magnetic field $\mathbf{H}$:

$$\mathbf{H} = \frac{1}{\mu_0}\mathbf{B} - \mathbf{M}. \tag{8.12}$$

For isotropic linear media, (8.12) will give a linear relation between $\mathbf{H}$ and $\mathbf{B}$,

$$\mathbf{H} = \frac{1}{\mu}\mathbf{B}, \tag{8.13}$$

which transforms (8.11) into Ampère's law in materials:

$$\nabla \times \mathbf{H} = \mathbf{J}_e. \tag{8.14}$$

For a time-dependent electric field, the displacement current $\partial \mathbf{D}/\partial t$ was added by Maxwell in 1865 [2] to the right-hand side of (8.14), resulting in the Ampère–Maxwell law:

$$\nabla \times \mathbf{H} = \frac{\partial \mathbf{D}}{\partial t} + \mathbf{J}_e, \tag{8.15}$$

where the electric current $\mathbf{J}_e$ measures the flow of free electron charge ρ_e in amperes/square meter (A/m^2) and the unit for the magnetic current density $\mathbf{J}_m$ is the weber/square meter (web/m^2).

We can examine the various current contributions on the right-hand side of (8.15). By using the definition of the electric displacement flux $\mathbf{D}$ in terms of the polarization density $\mathbf{P}$ in (1.25), the Ampère–Maxwell law (8.15) can be rewritten as

$$\nabla \times \mathbf{H} = \epsilon_0 \frac{\partial \mathbf{E}}{\partial t} + \mathbf{J}_p + \mathbf{J}_e, \tag{8.16}$$

where a polarization current $\mathbf{J}_p$ is defined by

$$\mathbf{J}_p = \frac{\partial \mathbf{P}}{\partial t}. \tag{8.17}$$

In addition, we have two Gauss's laws for **D** and **B**, respectively:

$$\nabla \cdot \mathbf{D} = \rho_e, \tag{8.18}$$

$$\nabla \cdot \mathbf{B} = \rho_m (\equiv 0). \tag{8.19}$$

Equations (8.1), (8.15), (8.18), *and* (8.19) *form the complete time-dependent Maxwell's equations in materials.*

As a result of Maxwell's equations, the electric charge ρ_e, the magnetic charge ρ_m, and the corresponding current densities obey the following continuity equations:

$$\nabla \cdot \mathbf{J}_e + \frac{\partial \rho_e}{\partial t} = 0, \tag{8.20}$$

$$\nabla \cdot \mathbf{J}_m + \frac{\partial \rho_m}{\partial t} = 0. \tag{8.21}$$

Inside a conducting material the electric current $\mathbf{J}_e$ is related to the electric field through Ohm's law:

$$\mathbf{J}_e = \sigma \mathbf{E}, \tag{8.22}$$

where σ is the conductivity of the material.

For time-harmonic fields, where the time dependence is assumed to be harmonic, i.e., $\exp(\mathrm{i}\omega t)$, Maxwell's equations (8.1), (8.15), (8.18), and (8.19), with the time-harmonic dependence factored out, and the constitutive relations $\mathbf{D} = \epsilon \mathbf{E}$ and $\mathbf{B} = \mu \mathbf{H}$, can be rewritten for the Fourier transform of the fields (keeping the same notation):

$$\nabla \times \mathbf{E} = -\mathrm{i}\omega\mu \mathbf{H} - \mathbf{J}_m, \tag{8.23}$$

$$\nabla \times \mathbf{H} = \mathrm{i}\omega\epsilon \mathbf{E} + \mathbf{J}_e, \tag{8.24}$$

$$\nabla \cdot \mathbf{D} = \rho_e, \tag{8.25}$$

$$\nabla \cdot \mathbf{B} = \rho_m. \tag{8.26}$$

Then the continuity equations will read

$$\nabla \cdot \mathbf{J}_e + \mathrm{i}\omega \rho_e = 0, \tag{8.27}$$

$$\nabla \cdot \mathbf{J}_m + \mathrm{i}\omega \rho_m = 0. \tag{8.28}$$

- **Vector wave equations**

The electromagnetic fields **E** and **H** can be shown individually to satisfy the following vector wave equations. By dividing (8.23) by μ and applying the curl operator $\nabla\times$, we have

$$\nabla \times \frac{1}{\mu}\nabla \times \mathbf{E} = -\mathrm{i}\omega\nabla \times \mathbf{H} - \nabla \times \frac{1}{\mu}\mathbf{J}_m. \tag{8.29}$$

By using (8.24), the following vector equation for the electric field **E** is obtained:

$$\nabla \times \frac{1}{\mu}\nabla \times \mathbf{E} - \omega^2\epsilon\mathbf{E} = -\mathrm{i}\omega\mathbf{J}_e - \nabla \times \frac{1}{\mu}\mathbf{J}_m. \tag{8.30}$$

Similarly, we can derive a vector equation for the magnetic field **H**:

$$\nabla \times \frac{1}{\epsilon}\nabla \times \mathbf{H} - \omega^2\mu\mathbf{H} = -\mathrm{i}\omega\mathbf{J}_m + \nabla \times \frac{1}{\epsilon}\mathbf{J}_e. \tag{8.31}$$

8.2 Vector and Scalar Potentials

Vector and scalar potentials are quantities useful in constructing solutions to Maxwell's equations, especially when we discuss the integral equation representations of electromagnetic fields in Chap. 11. These potentials are based on the following Helmholtz vector decompositions (the Hodge decompositions).

Theorem 8.1 (Helmholtz vector decomposition) *For a differentiable vector field* $\mathbf{W}(\mathbf{r}) = (W_1(\mathbf{r}), W_2(\mathbf{r}), W_3(\mathbf{r}))^{\mathrm{T}}$, $\mathbf{r} \in \Omega \subset \mathbb{R}^3$, *where* Ω *is simply connected with a Lipschitz-continuous boundary, there exist a unique vector potential* $\Phi(\mathbf{r})$ *and a scalar potential* $q(\mathbf{r})$ *(unique up to a constant) such that the following orthogonal decomposition holds:*

$$\mathbf{W} = \mathbf{W}_1 + \mathbf{W}_2 \equiv \nabla \times \Phi + \nabla q, \tag{8.32}$$

where $\mathbf{W}_1 = \nabla \times \Phi$ *is the solenoid field with zero divergence and* $\mathbf{W}_2 = \nabla q$ *is the irrotational field with zero vorticity, and* $\Phi(\mathbf{r})$ *and* $q(\mathbf{r})$ *satisfy the following elliptic systems with corresponding boundary conditions:*

$$-\nabla^2\Phi = \nabla \times \mathbf{W}, \quad \nabla \cdot \Phi = 0, \tag{8.33}$$

$$\mathbf{n} \times \Phi = \mathbf{0} \quad on \quad \partial\Omega, \tag{8.34}$$

and

$$\nabla^2 q = \nabla \cdot \mathbf{W}, \tag{8.35}$$

$$\nabla q \cdot \mathbf{n} = \mathbf{W} \cdot \mathbf{n} \quad on\ \partial\Omega. \tag{8.36}$$

The proof for the above Helmholtz decomposition can be found in [3, Corollary 3.4] for more general vector fields with weaker smoothness. If $\Omega = \mathbb{R}^3$, then a boundary condition at infinity will be given instead.

8.2.1 Electric and Magnetic Potentials for Time-Harmonic Fields

Based on the Helmholtz vector decomposition (8.32) in $\mathbb{R}^3$, we can express the electromagnetic fields **E** and **H** in $\mathbb{R}^3$ in terms of potential [4]. As the Maxwell's equations (8.23) and (8.24) are linear, by the principle of linear superposition, we will consider the fields caused by the electric currents and charges and those by the magnetic currents and charges separately.

Firstly, we will assume zero magnetic current and charge, i.e., $\mathbf{J}_m = 0$ and $\rho_m = 0$. Then we have from (8.26)

$$\nabla \cdot \mathbf{B} = 0. \tag{8.37}$$

Using (8.35) and (8.36) in $\mathbb{R}^3$ (with the vanishing boundary condition at infinity) with zero right-hand side, we can see that the Helmholtz decomposition (8.32) for the vector **B** implies that there exists a vector potential, say **A**, such that

$$\mathbf{B} = \nabla \times \mathbf{A}. \tag{8.38}$$

Substituting (8.38) into (8.23), we have

$$\nabla \times (\mathbf{E} + \mathrm{i}\omega\mathbf{A}) = \mathbf{0}. \tag{8.39}$$

As a result, using (8.33) and (8.34) with zero right-hand side, the Helmholtz decomposition (8.32) for the vector field $\mathbf{E} + \mathrm{i}\omega\mathbf{A}$ implies that there exists a scalar function, say V_e, such that

$$\mathbf{E} + \mathrm{i}\omega\mathbf{A} = -\nabla V_e. \tag{8.40}$$

Thus, we have

$$\mathbf{E} = -\mathrm{i}\omega\mathbf{A} - \nabla V_e. \tag{8.41}$$

Substituting (8.41) into the vector wave equation (8.30) for **E**, we obtain

$$\nabla \times \frac{1}{\mu}\nabla \times \mathbf{A} - \omega^2\epsilon\mathbf{A} + \mathrm{i}\omega\epsilon\nabla V_e = \mathbf{J}_e, \tag{8.42}$$

and, after using the vector identities

$$\nabla \times c\mathbf{f} = \nabla c \times \mathbf{f} + c\nabla \times \mathbf{f} \tag{8.43}$$

and

$$\nabla \times \nabla \times \mathbf{A} = -\nabla^2\mathbf{A} + \nabla(\nabla \cdot \mathbf{A}), \tag{8.44}$$

Equation (8.42) becomes

$$-\nabla^2\mathbf{A} - k^2\mathbf{A} + \nabla(\nabla \cdot \mathbf{A}) + \mathrm{i}\omega\epsilon\mu\nabla V_e + \mu\left(\nabla\frac{1}{\mu}\right) \times \nabla \times \mathbf{A} = \mu\mathbf{J}_e. \tag{8.45}$$

As (8.38) only specifies the vorticity of the vector field **A**, there is one more degree of freedom on **A**, namely the divergence of **A**, which is set in the Lorenz gauge [5] as

$$\nabla \cdot \mathbf{A} = -\mathrm{i}\omega\epsilon\mu V_e. \tag{8.46}$$

Then, (8.45) for **A** is simplified to

$$\nabla^2\mathbf{A} + k^2\mathbf{A} - \mu\left(\nabla\frac{1}{\mu}\right) \times \nabla \times \mathbf{A} = -\mu\mathbf{J}_e, \tag{8.47}$$

where the wave number k is defined as

$$k^2 = \omega^2\epsilon\mu, \tag{8.48}$$

and, when μ is a constant, (8.47) becomes a Helmholtz equation for the components of **A**:

$$\nabla^2\mathbf{A} + k^2\mathbf{A} = -\mu\mathbf{J}_e. \tag{8.49}$$

By taking the divergence ∇ of (8.49) and using the continuity equation (8.27) and the Lorenz gauge condition (8.46), the scalar potential V_e is shown to satisfy

$$\nabla^2 V_e + k^2 V_e = -\frac{1}{\epsilon}\rho_e. \tag{8.50}$$

Finally, the electric field **E** has the following representation in terms of the magnetic potential **A**:

$$\mathbf{E} = -\mathrm{i}\omega\mathbf{A} + \frac{1}{\mathrm{i}\omega\epsilon\mu}\nabla(\nabla \cdot \mathbf{A}), \tag{8.51}$$

and the magnetic field **H** is given by

$$\mathbf{H} = \frac{1}{\mu}\nabla \times \mathbf{A}. \tag{8.52}$$

Similarly, we can construct the fields in terms of potentials when $\mathbf{J}_e = \mathbf{0}$ and $\rho_e = 0$, where an electric vector potential $\mathbf{F}$ satisfies

$$\nabla^2 \mathbf{F} + k^2 \mathbf{F} - \epsilon \nabla \frac{1}{\epsilon} \times \nabla \times \mathbf{F} = \epsilon \mathbf{J}_m, \tag{8.53}$$

which simplifies to the following equation when ϵ is a constant:

$$\nabla^2 \mathbf{F} + k^2 \mathbf{F} = \epsilon \mathbf{J}_m, \tag{8.54}$$

and the scalar potential given by a similar gauge condition to (8.46) satisfies

$$\nabla^2 V_m + k^2 V_m = \frac{1}{\mu} \rho_m. \tag{8.55}$$

As a result, we have

$$\mathbf{E} = \frac{1}{\epsilon} \nabla \times \mathbf{F}, \tag{8.56}$$

$$\mathbf{H} = \mathrm{i}\omega \mathbf{F} - \frac{1}{\mathrm{i}\omega\epsilon\mu} \nabla(\nabla \cdot \mathbf{F}). \tag{8.57}$$

Finally, for general non-zero electric and magnetic currents, with the principle of superposition, combining (8.51) with (8.56) and (8.52) with (8.57) we have the potential representation of the electromagnetic fields [4] as follows:

$$\mathbf{E} = \frac{1}{\epsilon} \nabla \times \mathbf{F} - \mathrm{i}\omega \mathbf{A} + \frac{1}{\mathrm{i}\omega\epsilon\mu} \nabla(\nabla \cdot \mathbf{A}), \tag{8.58}$$

$$\mathbf{H} = \frac{1}{\mu} \nabla \times \mathbf{A} + \mathrm{i}\omega \mathbf{F} - \frac{1}{\mathrm{i}\omega\epsilon\mu} \nabla(\nabla \cdot \mathbf{F}). \tag{8.59}$$

8.3 Physical Boundary Conditions for E and H

8.3.1 Interface Conditions Between Dielectric Media

Skin depth of conductors and surface currents $\mathbf{J}_{e(s)}$

For a conductor with finite conductivity, the electromagnetic fields can only exist within a thin layer (with a skin depth δ) in the conductor. The skin depth δ depends on the frequency ω of the electromagnetic fields and the permeability μ and conductivity σ of the medium [6]:

$$\delta = \sqrt{\frac{2}{\omega\mu\sigma}} = \frac{1}{\sqrt{\pi f \mu\sigma}}, \tag{8.60}$$

where the frequency $f = \omega/(2\pi)$.

Within the skin depth δ, **E** and **H** decay exponentially into the conductor. For most applications, we could model the effect of the skin depth by means of an effective surface current $\mathbf{J}_{e(s)}$ by ignoring the skin depth, defined as

$$\mathbf{J}_{e(s)}(p) = \int_0^\delta \mathbf{J}_e(p, \xi)\mathrm{d}\xi \quad \text{(A/m)} \tag{8.61}$$

for any point p on the conductor surface S, with the integration along the normal of S. Note that the surface current $\mathbf{J}_{e(s)}$ only has components tangential to S.

As the conductivity increases, the electromagnetic fields of non-zero time frequency ω will tend to zero inside the conductor and will eventually become zero for a perfect electric conductor (PEC), for which $\sigma = \infty$. The static electric and magnetic fields for a PEC are defined by a limiting process of $\omega \to 0$ [1, Sect. 5.13, p. 203] while still maintaining $\omega\mu\sigma \to \infty$; therefore, they are also zero inside the PEC.

Interface conditions between media

When two media with different electric and magnetic properties, (ϵ_1, μ_1) and (ϵ_2, μ_2), meet at an interface S with a normal **n** pointing toward medium 2, Maxwell's equations should be supplemented with the following interface conditions:

$$[\mathbf{n} \times \mathbf{H}] = \mathbf{n} \times (\mathbf{H}_2 - \mathbf{H}_1) = \mathbf{J}_{e(s)}, \tag{8.62}$$

$$[\mathbf{n} \times \mathbf{E}] = \mathbf{n} \times (\mathbf{E}_2 - \mathbf{E}_1) = -\mathbf{J}_{m(s)}, \tag{8.63}$$

$$[\mathbf{n} \cdot \mathbf{B}] = \mathbf{n} \cdot (\mu_2\mathbf{H}_2 - \mu_1\mathbf{H}_1) = \rho_{m(s)}, \tag{8.64}$$

$$[\mathbf{n} \cdot \mathbf{D}] = \mathbf{n} \cdot (\epsilon_2\mathbf{E}_2 - \epsilon_1\mathbf{E}_1) = \rho_{e(s)}. \tag{8.65}$$

In (8.62)–(8.65), possible surface currents and charges are assumed to reside on S; this occurs as an idealization of a metallic thin layer sandwiched between the two dielectric media, as discussed in (8.61), or artificial surface currents could be introduced as equivalent sources on S to represent the field effects from the medium to one side of the interface [4]. This equivalence principle will be discussed in the formulation of integral equations for the electromagnetic fields in Chap. 11.

In particular, if one medium (medium 1) is a PEC in which both the electric and the magnetic fields vanish, we have the following boundary conditions for the field tangential components on the surface of the PEC:

$$\mathbf{n} \times \mathbf{E}_2 = \mathbf{0}, \tag{8.66}$$

$$\mathbf{n} \times \mathbf{H}_2 = \mathbf{J}_{e(s)}, \tag{8.67}$$

where the surface current $\mathbf{J}_{e(s)}$ is supported on the surface of the conductor, and, for the normal components,

$$\mathbf{n} \cdot \mathbf{E}_2 = \rho_{e(s)}, \tag{8.68}$$

$$\mathbf{n} \cdot \mathbf{H}_2 = 0. \tag{8.69}$$

8.3.2 Leontovich Impedance Boundary Conditions for Conductors

Surface impedance $Z_s$and impedance boundary conditions

In many scattering applications, we are interested only in the electromagnetic fields external to the scatterer, without much need for the explicit fields inside. The effect of the interior fields could be taken into account by introducing appropriate boundary conditions on the surface S of the scatterer through a surface impedance quantity z_s [6]. The surface impedance represents the energy dissipation of the electromagnetic fields by the scatterer's surface, and is related to the skin depth δ and the conductivity σ of the scatterer medium by

$$z_s = \frac{(1+\mathrm{i})}{\sigma\delta} = R_s + \mathrm{i}\omega L_i, \tag{8.70}$$

where $R_s = 1/(\sigma\delta)$ and $L_i = R_s/\omega = 1/(\omega\sigma\delta)$ are the surface resistivity and the inner inductance, respectively. For typical metals, at 300 K we have for silver $\sigma = 6.17 \times 10^7$ S/m, $\delta = 0.0642 f^{-1/2}$ m, $R_s = 2.52 \times 10^{-7} f^{1/2}\ \Omega$, and for aluminum $\sigma = 3.72 \times 10^7$ S/m, $\delta = 0.0826 f^{-1/2}$ m, and $R_s = 3.26 \times 10^{-7} f^{1/2}\ \Omega$ [6].

To derive the impedance boundary conditions on a conductor, we consider the interface conditions between two dielectric materials:

$$\begin{aligned}
[\mathbf{n} \times \mathbf{E}] &= \mathbf{0}, \\
[\mathbf{n} \times \mathbf{H}] &= \mathbf{0}, \\
[\epsilon \mathbf{n} \cdot \mathbf{E}] &= 0, \\
[\mu \mathbf{n} \cdot \mathbf{H}] &= 0.
\end{aligned} \tag{8.71}$$

Inside the conductor, from Ohm's law (8.22), the conduction current $\mathbf{J}_e = \sigma\mathbf{E}$ and the time-harmonic Maxwell's equations (8.23) and (8.24) become

$$\begin{aligned}
\nabla \times \mathbf{E} &= -\,\mathrm{i}\omega\mu\mathbf{H}, \\
\nabla \times \mathbf{H} &= \mathrm{i}\omega\tilde{\epsilon}\mathbf{E},
\end{aligned} \tag{8.72}$$

where

$$\tilde{\epsilon} = \epsilon - \frac{\sigma}{\omega}\mathrm{i}. \tag{8.73}$$

Using the vector identity

$$\nabla \times \left(\sqrt{\tilde{\epsilon}}\mathbf{E}\right) = \nabla\left(\sqrt{\tilde{\epsilon}}\right) \times \mathbf{E} + \sqrt{\tilde{\epsilon}}\nabla \times \mathbf{E}, \tag{8.74}$$

we obtain an equation for the scaled electric field [7]:

$$\nabla \times \left(\sqrt{\tilde{\epsilon}}\mathbf{E}\right) + \sqrt{\tilde{\epsilon}}\mathbf{E} \times \nabla \ln \sqrt{\tilde{\epsilon}} = -ik_0 n \sqrt{\mu}\mathbf{H}, \tag{8.75}$$

where the wave number and the refractive index are given by

$$k_0^2 = \epsilon_0\mu_0\omega^2, \qquad n = \sqrt{\frac{\widetilde{\epsilon}\mu}{\epsilon_0\mu_0}}. \tag{8.76}$$

Similarly, we can derive an equation for the scaled magnetic field:

$$\nabla \times \left(\sqrt{\mu}\mathbf{H}\right) + \sqrt{\mu}\mathbf{H}\times\nabla\left(\ln\sqrt{\mu}\right) = \mathrm{i}k_0 n\sqrt{\widetilde{\epsilon}}\mathbf{E}. \tag{8.77}$$

For a good conductor, we can assume that the refractive index n is large, expressed in terms of a small parameter q:

$$n = \frac{\omega}{q}, \quad q \ll 1, \tag{8.78}$$

and the field quantities $\mathbf{E}'$ and $\mathbf{H}'$ inside the conductor are assumed to have the following representations in terms of a phase function ψ:

$$\sqrt{\widetilde{\epsilon}}\mathbf{E}' = \mathbf{A}\mathrm{e}^{-\mathrm{i}k_0\psi/q}, \tag{8.79}$$

$$\sqrt{\mu}\mathbf{H}' = \mathbf{B}\mathrm{e}^{-\mathrm{i}k_0\psi/q}. \tag{8.80}$$

Using the identities

$$\nabla \times \left(\sqrt{\widetilde{\epsilon}}\mathbf{E}'\right) = \left(\nabla \times \mathbf{A} - \frac{\mathrm{i}k_0}{q}\nabla\psi \times \mathbf{A}\right)\mathrm{e}^{-\mathrm{i}k_0\psi/q}, \tag{8.81}$$

$$\nabla \times \left(\sqrt{\mu}\mathbf{H}'\right) = \left(\nabla \times \mathbf{B} - \frac{\mathrm{i}k_0}{q}\nabla\psi \times \mathbf{B}\right)\mathrm{e}^{-\mathrm{i}k_0\psi/q}, \tag{8.82}$$

Equations (8.75) and (8.77) become

$$\omega\mathbf{A} + \nabla\psi \times \mathbf{B} = \frac{q}{\mathrm{i}k_0}\sqrt{\mu}\nabla \times \frac{\mathbf{B}}{\sqrt{\mu}}, \tag{8.83}$$

$$\omega\mathbf{B} - \nabla\psi \times \mathbf{A} = -\frac{q}{\mathrm{i}k_0}\sqrt{\widetilde{\epsilon}}\nabla \times \frac{\mathbf{A}}{\sqrt{\widetilde{\epsilon}}}. \tag{8.84}$$

By considering an asymptotic expansion for both $\mathbf{A}$ and $\mathbf{B}$ in terms of the small parameter q,

$$\mathbf{A} = \mathbf{A}_0 + q\mathbf{A}_1 + q^2\mathbf{A}_2 + \cdots, \tag{8.85}$$

$$\mathbf{B} = \mathbf{B}_0 + q\mathbf{B}_1 + q^2\mathbf{B}_2 + \cdots, \tag{8.86}$$

and substituting the series into (8.83) and (8.84) and equating the coefficients according to the powers of q, the zeroth power of q yields

$$\omega \mathbf{A}_0 + \nabla\psi \times \mathbf{B}_0 = 0, \tag{8.87}$$

$$\omega \mathbf{B}_0 - \nabla\psi \times \mathbf{A}_0 = 0. \tag{8.88}$$

After eliminating $\mathbf{B}_0$, we have

$$\left(\omega^2 - |\nabla\psi|^2\right) \mathbf{A}_0 = \mathbf{0}, \tag{8.89}$$

and the following Eikonal equation is obtained for the phase function:

$$|\nabla\psi| = \omega. \tag{8.90}$$

To proceed, we assume that the interface S is smooth, and that $\mathbf{E}$ and $\mathbf{H}$ vary slowly along the surface S. For a zeroth-order approximation, we can assume that ψ is constant on S, namely that S is a level set curve of the function ψ. Thus,

$$\nabla\psi \parallel \mathbf{n}. \tag{8.91}$$

Therefore, (8.90) implies

$$\nabla\psi = -\omega\mathbf{n}. \tag{8.92}$$

From (8.87) and (8.88), we have

$$\mathbf{A}_0 = \mathbf{n} \times \mathbf{B}_0, \quad \mathbf{B}_0 = -\mathbf{n} \times \mathbf{A}_0, \tag{8.93}$$

which imply

$$\mathbf{A}_0 \perp \mathbf{B}_0 \perp \mathbf{n}. \tag{8.94}$$

Let us introduce local coordinates $(\boldsymbol{\alpha}, \boldsymbol{\beta}, \gamma)$ on the surface S. The continuity condition of the tangential components of $\mathbf{E}$ and $\mathbf{H}$ becomes

$$\begin{aligned} A_{0,(\alpha,\beta)} &= \sqrt{\tilde{\epsilon}} E'_{(\alpha,\beta)} = \sqrt{\tilde{\epsilon}} E_{(\alpha,\beta)}, \\ B_{0,(\alpha,\beta)} &= \sqrt{\mu} H'_{(\alpha,\beta)} = \sqrt{\mu} H_{(\alpha,\beta)}, \end{aligned} \tag{8.95}$$

resulting in

$$E_\alpha = -Z_s H_\beta, \quad E_\beta = Z_s H_\alpha, \tag{8.96}$$

where $Z_s = \sqrt{\mu/\tilde{\epsilon}}$ is the intrinsic surface impedance.

Finally, the zeroth-order impedance boundary condition reads [8]

$$\mathbf{n} \times \mathbf{E} = Z_s \mathbf{n} \times (\mathbf{n} \times \mathbf{H}). \tag{8.97}$$

Higher-order impedance boundary conditions taking into account the surface curvature effects have also been derived by Senior and Volakis [7] in addition to the zeroth-order boundary condition given above.

8.3.3 Sommerfeld and Silver–Müller Radiation Conditions

The behavior of electromagnetic fields at infinity is characterized by a radiation or outgoing wave condition, which embodies the physical reality of scattering waves from finite objects radiating away to infinity without reflection from infinity. Such a condition also ensures mathematically the uniqueness of the solution of the exterior Helmholtz wave equations. In the case of a scalar wave equation, the radiation is given by the Sommerfeld radiation condition [9] while for electromagnetic wave scattering it is given by the Silver–Müller radiation condition [10].

- **Sommerfeld radiation condition** for an exterior time-harmonic scalar wave: consider the Helmholtz equation in the exterior Ω^c of a finite domain Ω,

$$\begin{cases} \Delta u + k^2 u = 0, \mathbf{r} \in \Omega^c, \\ u|_{\partial\Omega} = u_d \quad \text{or} \quad \left.\dfrac{\partial u}{\partial \mathbf{n}}\right|_{\partial\Omega} = u_n, \end{cases} \tag{8.98}$$

  together with the Sommerfeld radiation condition,

$$\left| \frac{\partial u}{\partial r} + \mathrm{i}ku \right| = O(r^{-2}) \quad \text{as } r \to \infty. \tag{8.99}$$

- **Silver–Müller radiation condition** for an exterior time-harmonic electromagnetic scattering: denote by **E** and **H** the time-harmonic electromagnetic fields generated by sources in a finite region. Then, as $r \to \infty$, we have

$$|\sqrt{\mu}\mathbf{H}\times\hat{\mathbf{r}} - \sqrt{\epsilon}\mathbf{E}| = O(r^{-2}), \tag{8.100}$$

$$|\sqrt{\epsilon}\mathbf{E}\times\hat{\mathbf{r}} + \sqrt{\mu}\mathbf{H}| = O(r^{-2}), \tag{8.101}$$

  where $\hat{\mathbf{r}}$ is the unit normal for **r**. By Faraday's and Ampère's laws, (8.100) and (8.101) can be rewritten in terms of **E** and **H** only [11, p. 137] as

$$|\hat{\mathbf{r}} \times \nabla \times \mathbf{E} - \mathrm{i}k\mathbf{E}| = O(r^{-2}), \tag{8.102}$$

$$|\hat{\mathbf{r}} \times \nabla \times \mathbf{H} - \mathrm{i}k\mathbf{H}| = O(r^{-2}). \tag{8.103}$$

The radiation conditions ensure the uniqueness of the solution of the exterior wave problems, as illustrated below for the scalar Helmholtz equations for the scalar wave problem [12].

Theorem 8.2 *The exterior Helmholtz problem has at most one solution.*

Proof To prove the uniqueness, we can let $u_d = 0$ or $u_n = 0$ and integrate (8.98) multiplied by the function $\overline{u}$ and apply Green's formula to arrive at

$$-\int_V \overline{\nabla u}\nabla u \mathrm{d}\mathbf{r} + \int_{|\mathbf{r}|=R} \overline{u}\frac{\partial u}{\partial r}\mathrm{d}s + k^2 \int_V |u|^2 \mathrm{d}\mathbf{r} = 0, \tag{8.104}$$

where $V = \Omega^c \cap \{\mathbf{r} : |\mathbf{r}| \le R\}$.

For $\mathbf{r} \in B_R^c = \{\mathbf{r} : |\mathbf{r}| > R\}$, we have the Fourier expansion for the solution u:

$$u(r, \theta, \phi) = \sum_{l=0}^{\infty}\sum_{m=-l}^{l}\left(\alpha_l^m \frac{h_l^{(1)}(kr)}{h_l^{(1)}(kR)} + \beta_l^m \frac{h_l^{(2)}(kr)}{h_l^{(2)}(kR)}\right) Y_l^m(\theta, \phi),$$

where $h_l^{(1)}(r)$ and $h_l^{(2)}(r)$ are the Hankel functions of the first and second kind of order l, respectively.

We need to show that $\alpha_l^m = \beta_l^m = 0$. Using the asymptotics as $r \to +\infty$:

$$h_l^{(1)}(kr) \approx (-\mathrm{i})^l \frac{\mathrm{e}^{ikr}}{r}, \qquad h_l^{(2)}(kr) \approx (-\mathrm{i})^l \frac{\mathrm{e}^{-ikr}}{r},$$

and

$$\begin{aligned}\frac{\partial u}{\partial r} + \mathrm{i}ku = \sum_{l=0}^{\infty}\sum_{m=-l}^{l}\Bigg[&\frac{k\alpha_l^m}{h_l^{(1)}(kR)}\left(\frac{\partial}{\partial r}h_l^{(1)}(kr) + \mathrm{i}h_l^{(1)}(kr)\right)\\ &+ \frac{k\beta_l^m}{h_l^{(2)}(kR)}\left(\frac{\partial}{\partial r}h_l^{(2)}(kr) + \mathrm{i}h_l^{(2)}(kr)\right)\Bigg] Y_l^m(\theta, \phi),\end{aligned}$$

the Sommerfeld radiation condition implies that

$$\alpha_l^m \equiv 0, \forall l, m.$$

Next, taking the imaginary part of (8.104), we get

$$\mathrm{Im}\left(\int_{|\mathbf{r}|=R} (T_R(u), \overline{u})\mathrm{d}s\right) = 0, \tag{8.105}$$

where $T_R(w)$ is a capacitance operator defined for any $w = \sum\sum \gamma_l^m Y_l^m(\theta, \phi)$:

$$T_R(w) = \sum_{l=0}^{\infty} \sum_{m=-l}^{l} \frac{1}{R} Z_l(kR)\gamma_l^m Y_l^m(\theta, \phi),$$

with $Z_l(r) = r\frac{\partial}{\partial r}\left(h_l^{(2)}(r)\right) / h_l^{(2)}(r)$.

Using the orthogonality property of the basis functions $Y_l^m(\theta, \phi)$ over the sphere $|\mathbf{r}| = R$, for any two functions $u = \sum_l \sum_m \alpha_l^m Y_l^m(\theta, \phi)$ and $v = \sum_l \sum_m \beta_l^m Y_l^m(\theta, \phi)$, we have the following identity [12]:

$$\int_{|\mathbf{r}|=R} (u, \overline{v}) \mathrm{d}s = \sum_l \sum_m \alpha_l^m \beta_l^m.$$

Then we have

$$\mathrm{Im}\left(\sum_l \sum_m \beta_l^m \frac{1}{R} \overline{Z_l(kR)\beta_l^m}\right) = 0.$$

As $\mathrm{Im}(Z_l(kR)) > 0$, we have $|\beta_l^m| \equiv 0$; thus, $u \equiv 0$ for $r \geq R$. As a result of the analyticity of the solution u in Ω^c (assuming that the boundary of Ω is smooth enough), we conclude that $u \equiv 0$ in Ω^c. □

Remark 8.1 The Sommerfeld radiation condition can also be given in a weaker integral form:

$$\int_{|\mathbf{r}|=R} \left|\frac{\partial u}{\partial r} + \mathrm{i}ku\right|^2 \mathrm{d}s \to 0 \quad \text{as } R \to +\infty. \tag{8.106}$$

8.4 Absorbing Boundary Conditions for E and H

In this section, we will discuss artificial absorbing boundary conditions, which are imposed on the boundary of a truncated infinite physical domain within which numerical solutions for wave equations are sought. These boundary conditions ensure minimum non-physical reflection at the boundary and overall stability and accuracy of the numerical schemes, while being easy to implement. There are two types of artificial boundary conditions, local and global. In the former, only local values of the solutions or the derivatives of the solutions are used for the boundary conditions at a given boundary location, whereas the latter involves global information from the solution on the boundary. Here, we will mainly discuss the local boundary conditions due to their easy and cost-efficient way of implementation.

Global boundary conditions can be obtained by the Dirichlet-to-Neumann (DtN) mapping for the time-harmonic wave equations [13–15] and its inverse Fourier or inverse Laplace transform for the time-domain wave equations [16, 17]. This approach is global on the boundary and also involves a time-convolution integral of global nature for transient problems. In order to reduce the cost and memory associated with the time convolutions, efficient calculations have been proposed in [18, 19]. Meanwhile, local in time/global in space boundary conditions have been proposed by Grote and Keller [20] (for spherical boundaries) or with Kirchhoff representation [21, 22] (for general boundaries) or equivalent sources on the boundary [23]. Global boundary conditions based on the DtN mapping for a scalar Helmholtz equation will be discussed in Sect. 18.2.1.

8.4.1 One-Way Wave Engquist–Majda Boundary Conditions

Engquist and Majda [24] introduced the one-way wave differential equation on the solution domain boundary where only waves propagating in one direction (the outgoing direction) will pass the boundary without reflection. The original derivation of the Engquist–Majda boundary condition is based on the factorization of the second-order wave operators with pseudo-differential operators, and the latter is then approximated by the Padé approximations, resulting in a one-way differential operator on the boundary. Here, a derivation based on the wave dispersion relation given by Trefethen and Halpern [25] will be shown instead.

Consider the 2-D scalar wave equation with wave speed $c = 1$ for $u(x, y, t)$:

$$u_{tt} = u_{xx} + u_{yy}. \tag{8.107}$$

The plane wave solution for (8.107) with wave numbers (ξ, η) and frequency ω is

$$u(x, y, t) = \mathrm{e}^{\mathrm{i}(\omega t + \xi x + \eta y)}. \tag{8.108}$$

For (8.108) to be a solution of (8.107), the following dispersion relation should hold between the frequency and the wave numbers:

$$\omega^2 = \xi^2 + \eta^2. \tag{8.109}$$

To examine the waves propagating along the $\pm x$ direction, we solve for the wave number ξ in terms of the frequency ω and wave number η:

$$\xi = \pm\omega\sqrt{1 - s^2}, \tag{8.110}$$

where

$$s = \frac{\eta}{\omega}. \tag{8.111}$$

For a given wave frequency and a fixed wave number along the y-direction, (8.110) clearly shows two wave speeds along the $+x$ and $-x$ directions, respectively. If $x = 0$ is the right-most boundary and we only want to have right-going waves as our physical solution, then we should select the $-$ sign solution of (8.110), namely

$$\xi = -\omega\sqrt{1 - s^2}, \tag{8.112}$$

which can be considered as the dispersion for an ideal one-way wave problem to be imposed on the boundary $x = 0$.

However, the square root in the wave number–frequency space does not correspond to a differential operator in the physical space. To derive an approximation using a differential equation out of (8.112), we approximate the square root by a rational function [24, 25] as

$$\sqrt{1 - s^2} \approx \frac{p_m(s)}{q_n(s)}, \tag{8.113}$$

where $m = n + 2$ or $m = n$ is shown to be able to produce a well-posed boundary value problem for (8.107). In particular, if we have

$$\sqrt{1 - s^2} \approx 1 - \frac{1}{2}s^2, \tag{8.114}$$

then the ideal one-way wave dispersion (8.112) reduces to

$$\xi = \omega\left(1 - \frac{1}{2}\frac{\eta^2}{\omega^2}\right), \tag{8.115}$$

or, equivalently,

$$\xi\omega = \omega^2 - \frac{1}{2}\eta^2, \tag{8.116}$$

which is exactly the dispersion relation for the following differential equation:

$$\frac{1}{c}u_{xt} = \frac{1}{c^2}u_{tt} - \frac{1}{2}u_{yy}, \tag{8.117}$$

where the wave speed c is added back.

Equation (8.117) is the second-order Engquist–Majda equation and is also known as Mur's boundary condition [26]. The first-order boundary condition can be obtained if the approximation in (8.114) is replaced instead by $\sqrt{1 - s^2} \approx 1$, resulting in a linear dispersion

$$\xi = \omega, \tag{8.118}$$

or, equivalently, a first-order one-way wave equation:

$$u_t - cu_x = 0. \tag{8.119}$$

By using different Padé approximations, various one-way wave equation boundary conditions can be derived, including the following Higdon boundary condition, where non-reflection can be achieved for waves incident to the boundary at multiple angles [27]:

$$\prod_{j=1}^{p}\left(\cos\alpha_j\frac{\partial}{\partial t}-c\frac{\partial}{\partial x}\right)u=0, \tag{8.120}$$

which can be implemented with the method of auxiliary variables on the boundary as in (8.135) [28].

8.4.2 *High-Order Local Non-reflecting Bayliss–Turkel Conditions*

Scalar waves

The Bayliss–Turkel local boundary conditions are based on the asymptotic behavior of the wave solutions at infinity (far-field pattern). It can be shown [29] that, consistent with the Sommerfeld radiation condition (8.99), for $r\to\infty$, the following asymptotic expansion holds:

$$u(r,\theta,\phi)=\frac{\mathrm{e}^{-ikr}}{r}\sum_{j=0}^{\infty}\frac{F_j(\theta,\phi)}{(kr)^j}=\frac{\mathrm{e}^{-ikr}}{r}F_0(\theta,\phi)+O\left(\frac{1}{r^2}\right), \tag{8.121}$$

where $0\le\theta\le\pi$ and $0\le\phi\le 2\pi$, and $F_0(\theta,\phi)$ is considered as the far-field pattern of the wave function u.

For $j\ge 1$,

$$F_j(\theta,\phi)=\frac{1}{(2\mathrm{i})^j}\prod_{l=0}^{j}[l(l-1)+\Delta_S]F_0(\theta,\phi), \tag{8.122}$$

where Δ_S is the Laplace–Beltrami operator on the unit sphere:

$$\Delta_S=\frac{1}{\sin\theta}\frac{\partial}{\partial\theta}\left(\sin\theta\frac{\partial}{\partial\theta}\right)+\frac{1}{\sin^2\theta}\frac{\partial^2}{\partial\phi^2}.$$

The Bayliss–Turkel boundary conditions [30] create boundary operators B_m on $\Gamma_R=\{\mathbf{r}:\ |\mathbf{r}|=R\}$, which will eliminate as many terms as possible from the asymptotic expansion (8.121).

Let

$$B_1=\frac{\partial}{\partial r}+\mathrm{i}k+\frac{1}{r}.$$

Then it can be shown that

$$B_1\left(\frac{e^{-ikr}}{r}F_0\right) = 0. \tag{8.123}$$

Thus, we have

$$B_1(u) = B_1\left[\frac{e^{-ikr}}{r}\sum_{j=1}^{\infty}\frac{F_j(\theta,\phi)}{(kr)^j}\right] = O\left(\frac{1}{r^3}\right). \tag{8.124}$$

- First-order Bayliss–Turkel artificial boundary condition:

$$B_1(u) = \left(\frac{\partial}{\partial r} + \mathrm{i}k + \frac{1}{r}\right)u = 0, \quad \mathbf{r} \in \Gamma_R. \tag{8.125}$$

In general, we define the mth-order boundary operator B_m as follows:

$$\begin{aligned} B_m &= \prod_{j=m}^{1}\left(\frac{\partial}{\partial r} + \mathrm{i}k + \frac{2j-1}{r}\right) \\ &= \left(\frac{\partial}{\partial r} + \mathrm{i}k + \frac{2m-1}{r}\right)B_{m-1}. \end{aligned} \tag{8.126}$$

- The mth-order Bayliss–Turkel artificial boundary condition

$$B_m(u) = 0, \quad \mathbf{r} \in \Gamma_R, \tag{8.127}$$

and

$$B_m(u) = O\left(\frac{1}{R^{2m+1}}\right). \tag{8.128}$$

The time-domain version of the Bayliss–Turkel boundary condition can be obtained by replacing ik by $(1/c)\partial/\partial t$ in the following:

$$B_m u = \prod_{j=m}^{1}\left(\frac{\partial}{\partial r} + \frac{1}{c}\frac{\partial}{\partial t} + \frac{2j-1}{r}\right)u(r,t) = 0, \quad \mathbf{r} \in \Gamma_R. \tag{8.129}$$

Bayliss–Turkel boundary condition using auxiliary variables

There are two difficulties associated with the boundary condition in (8.129): the normal derivative and the high-order differential operators. Both difficulties can be addressed by the introduction of auxiliary variables w_k defined on the surface of the sphere Γ_R [31]:

$$w_{k+1} = \left(\frac{\partial}{\partial r} + \frac{1}{c}\frac{\partial}{\partial t} + \frac{2k+1}{R}\right) \cdots \left(\frac{\partial}{\partial r} + \frac{1}{c}\frac{\partial}{\partial t} + \frac{1}{R}\right) u. \tag{8.130}$$

It is clear that we have the following recursion:

$$\left(\frac{\partial}{\partial r} + \frac{1}{c}\frac{\partial}{\partial t} + \frac{2k+1}{R}\right) w_k = w_{k+1}, \quad \text{for} \quad k \geq 1. \tag{8.131}$$

The boundary condition (8.129) is simply given by

$$w_{m+1} = 0, \tag{8.132}$$

together with the recursion (8.131) starting with

$$w_0 = u,\, w_1 = \left(\frac{\partial}{\partial r} + \frac{1}{c}\frac{\partial}{\partial t} + \frac{1}{R}\right) u. \tag{8.133}$$

In addition, the radial derivative $\partial/\partial r$ in (8.131) can be eliminated by using the following identity [32] on Γ_R,

$$\left(\frac{\partial}{\partial r} - \frac{1}{c}\frac{\partial}{\partial t} + \frac{1}{R}\right) w_k = -\frac{1}{R^2}(\Delta_S + k(k-1))w_{k-1}. \tag{8.134}$$

Subtracting (8.134) from (8.131), we have [33]

$$\left(\frac{1}{c}\frac{\partial}{\partial t} + \frac{k}{R}\right) w_k = \frac{1}{2R^2}(\Delta_S + k(k-1))w_{k-1} + \frac{1}{2}w_{k+1}, \tag{8.135}$$

which implies that the auxiliary variables have to be defined only over the spherical surface Γ_R.

Finally, (8.132), (8.133), and (8.135) reformulate the Bayliss–Turkel boundary condition with the auxiliary variables $\{w_k\}_{k=0}^{m+1}$ on Γ_R, which can be implemented with finite element methods [32].

Maxwell's equations Even though the scalar Bayliss–Turkel boundary condition can be applied to each component of the electromagnetic field with auxiliary variables for each component, there is an analog of the Bayliss–Turkel boundary condition (8.126) specifically for electromagnetic waves [34]:

$$\prod_{j=m}^{2}\left[\hat{\mathbf{r}} \times (\nabla \times) - ik - \frac{2(j-1)}{r}\right]\left[\hat{\mathbf{r}} \times (\nabla \times \mathbf{E}) - ik\mathbf{E}_{\tan}\right] = \mathbf{0},\, \mathbf{r} \in \Gamma_R, \tag{8.136}$$

where $\mathbf{E}_{\tan} = -\hat{\mathbf{r}} \times \left(\hat{\mathbf{r}} \times \mathbf{E}\right)$ is the tangential component of $\mathbf{E}$ on the spherical surface. For example, the first-order condition is

$$\hat{\mathbf{r}} \times (\nabla \times \mathbf{E}) - \mathrm{i}k\mathbf{E}_{\tan} = 0, \mathbf{r} \in \Gamma_R, \tag{8.137}$$

which is the tangential component of the Silver–Müller condition (8.102), while the second-order condition is

$$\begin{aligned}\hat{\mathbf{r}} \times (\nabla \times \mathbf{E}) =&\alpha(r)\mathbf{E}_{\tan} + \beta(r)\nabla \times [\hat{\mathbf{r}}(\hat{\mathbf{r}} \cdot \nabla \times \mathbf{E})] \\ &+\frac{\gamma(r)}{r}\nabla_s(\hat{\mathbf{r}} \cdot \mathbf{E}), \mathbf{r} \in \Gamma_R,\end{aligned} \tag{8.138}$$

where ∇_s is defined below in (8.145) and

$$\alpha(r) = \mathrm{i}k,\ \beta(r) = \frac{1}{2\mathrm{i}k + \frac{2}{r}}, \gamma(r) = \frac{\mathrm{i}k}{2\mathrm{i}k + \frac{2}{r}}.$$

Again, the above boundary conditions involve higher-order derivatives on the boundary. To address this problem, a local boundary condition with auxiliary variables is given in [35] for the electromagnetic fields as follows:

$$\hat{\mathbf{r}} \times (\nabla \times \mathbf{E}) = \frac{1}{c}\frac{\partial \mathbf{E}_{\tan}}{\partial t} + \mathbf{w}_1, \tag{8.139}$$

$$\left(\frac{1}{c}\frac{\partial}{\partial t} + \frac{1}{r}\right)\mathbf{w}_1 = \frac{1}{2r^2}\left[\overrightarrow{\mathrm{curl}}_s \times \mathrm{curl}_s\mathbf{E} + \sqrt{\frac{\mu}{\epsilon}}\hat{\mathbf{r}} \times \overrightarrow{\mathrm{curl}}_s\mathrm{curl}_s\mathbf{H}\right] + \mathbf{w}_2, \tag{8.140}$$

$$\left(\frac{1}{c}\frac{\partial}{\partial t} + \frac{j}{r}\right)\mathbf{w}_j = \frac{1}{4r^2}\left(\overrightarrow{\Delta}_S + j(j-1)\right)\mathbf{w}_{j-1} + \mathbf{w}_{j+1}, 2 \le j \le m, \tag{8.141}$$

and the recursion is terminated at $j = m + 1$,

$$\mathbf{w}_{m+1} = \mathbf{0}, \tag{8.142}$$

where the surface vorticity operators are defined as follows:

$$\overrightarrow{\mathrm{curl}}_s u = \frac{1}{\sin\theta}\frac{\partial u}{\partial \phi}\hat{\theta} - \frac{\partial u}{\partial \theta}\hat{\phi}, \tag{8.143}$$

$$\mathrm{curl}_s \mathbf{u} = \frac{1}{\sin\theta}\left(\frac{\partial\left(u_\phi \sin\theta\right)}{\partial\theta} - \frac{\partial u_\theta}{\partial\phi}\right), \tag{8.144}$$

$$\nabla_s u = \frac{\partial u}{\partial \theta}\hat{\theta} + \frac{1}{\sin\theta}\frac{\partial u}{\partial \phi}\hat{\phi}, \tag{8.145}$$

$$\mathrm{div}_s \mathbf{u} = \frac{1}{\sin\theta}\left(\frac{\partial\left(u_\theta \sin\theta\right)}{\partial\theta} + \frac{\partial u_\phi}{\partial\phi}\right), \tag{8.146}$$

and the surface Laplace–Beltrami operator $\overrightarrow{\Delta}_S$ is defined as

$$\overrightarrow{\Delta}_S \mathbf{u} = \nabla_s \mathrm{div}_s \mathbf{u} - \overrightarrow{\mathrm{curl}}_s \mathrm{curl}_s \mathbf{u}. \tag{8.147}$$

The implementation of the boundary condition on the field (8.139) is straightforward, as the weak form for the vector Helmholtz equation will have a boundary term which involves $\hat{\mathbf{r}} \times (\nabla \times \mathbf{E})$, which can be replaced by the right-hand side of (8.139). The auxiliary variables on the boundary can be discretized on the induced mesh on the boundary from the internal mesh, and their time evolution can be calculated using typical time marching schemes.

8.4.3 Uniaxial Perfectly Matched Layer (UPML)

A perfectly matched layer (PML) boundary condition uses a layer of artificial absorbing material designed in such a way that the scattering wave will pass the boundary adjacent to the inner computational domain without reflection, hence perfectly matched. Meanwhile, the absorbing layer attenuates the scattering wave entering the layer. The PML was originally proposed in the ingenious work of Berenger [36] based on a split formulation of the Maxwell's fields. Later, a uniaxial material [37, 38] was constructed, on which the following presentation is based. The designed uniaxial medium can achieve the same effect of absorbing entering waves as Berenger's PML without using non-physical splitting of the electromagnetic fields. A polarization field is involved in the PML region, which makes the medium in the PML region lossy and dispersive such that fields will decay exponentially through the region [39]. Usually, a PEC boundary condition is given at the terminating outer boundary and any reflection from there will not be able to propagate back significantly into the computational region of physical interest. It is shown [38] that the polarization in the PML region can be viewed as produced from a time-derivative Lorentz dispersive medium, where the electron displacement x away from the nucleus satisfies the following oscillator equation:

$$mx'' + m\gamma x' + m\omega_0^2 x = -e\mathbf{E} + \delta \frac{\partial \mathbf{E}}{\partial t}, \tag{8.148}$$

and the x-component of the polarization density is defined by $P_x = nex$, where n is the electron density and e is the electron charge.

Construction of UPML

An important idea of an absorbing layer boundary condition is to design the absorber material such that there is no reflection when the waves coming from the inner physical region enter the absorbing layer, i.e., a perfectly matched layer. The following analysis from Taflove and Hagness [40] demonstrates how the PML can be designed using uniaxial materials, where the dielectric constant and the permeability are diagonal matrices with uniaxial symmetry.

Consider a time-harmonic plane wave $\mathbf{H}^{\mathrm{inc}} = \mathbf{H}_0 \exp(-\mathrm{i}\beta_{1x}x - \mathrm{i}\beta_{1y}y)$ with arbitrary polarization propagating in the isotropic Region 1 ($x < 0$) and impinging on

Region 2 ($x > 0$); the latter is assumed to have uniaxial (along the x-axis) electric and magnetic permittivity tensors:

$$\bar{\bar{\epsilon}}_2 = \epsilon_2 \begin{bmatrix} a & 0 & 0 \\ 0 & b & 0 \\ 0 & 0 & b \end{bmatrix}, \quad \bar{\bar{\mu}}_2 = \mu_2 \begin{bmatrix} c & 0 & 0 \\ 0 & d & 0 \\ 0 & 0 & d \end{bmatrix}. \tag{8.149}$$

In Region 2, the transmitted fields are plane waves also and satisfy the following Maxwell's equations:

$$\boldsymbol{\beta}_2 \times \mathbf{E} = \omega \bar{\bar{\mu}}_2 \mathbf{H}, \quad \boldsymbol{\beta}_2 \times \mathbf{H} = -\omega \bar{\bar{\epsilon}}_2 \mathbf{E}, \tag{8.150}$$

where $\boldsymbol{\beta}_2 = \hat{x}\beta_{2x} + \hat{y}\beta_{2y}$ is the anisotropic wave vector. By eliminating the **E** field, we arrive at the following wave equation:

$$\boldsymbol{\beta}_2 \times \bar{\bar{\epsilon}}_2^{-1} \left(\boldsymbol{\beta}_2 \times \mathbf{H}\right) + \omega^2 \bar{\bar{\mu}}_2 \mathbf{H} = \mathbf{0}, \tag{8.151}$$

which can be rewritten in the following matrix form:

$$\begin{bmatrix} k_2^2 c - (\beta_{2y})^2 b^{-1} & \beta_{2x}\beta_{2y} b^{-1} & 0 \\ \beta_{2x}\beta_{2y} b^{-1} & k_2^2 d - (\beta_{2x})^2 b^{-1} & 0 \\ 0 & 0 & k_2^2 d - (\beta_{2x})^2 b^{-1} - (\beta_{2y})^2 a^{-1} \end{bmatrix} \begin{bmatrix} H_x \\ H_y \\ H_z \end{bmatrix} = 0,$$

where $k_2^2 = \omega^2 \mu_2 \epsilon_2$. The dispersion relation can be derived from the determinant of the above matrix for TE_z and TM_z modes, respectively:

$$k_2^2 - (\beta_{2x})^2 b^{-1} d^{-1} - (\beta_{2y})^2 a^{-1} d^{-1} = 0 : \ \mathrm{TE}_z(H_x, H_y = 0), \tag{8.152}$$

$$k_2^2 - (\beta_{2x})^2 b^{-1} d^{-1} - (\beta_{2y})^2 b^{-1} c^{-1} = 0 : \ \mathrm{TM}_z(H_z = 0). \tag{8.153}$$

If the incident wave in Region 1 is a TE_z wave, then the total field is expressed as the sum of the incident and a reflected wave:

$$\begin{aligned} \mathbf{H}_1 &= \hat{z} H_0 (1 + \Gamma \mathrm{e}^{2\mathrm{i}\beta_{1x}x}) \mathrm{e}^{-\mathrm{i}\beta_{1x}x - \mathrm{i}\beta_{1y}y}, \\ \mathbf{E}_1 &= \left(-\hat{x}\frac{\beta_{1y}}{\omega\epsilon_1}(1 + \Gamma \mathrm{e}^{2\mathrm{i}\beta_{1x}x}) + \hat{y}\frac{\beta_{1x}}{\omega\epsilon_1}(1 - \Gamma \mathrm{e}^{2\mathrm{i}\beta_{1x}x})\right) H_0 \mathrm{e}^{-\mathrm{i}\beta_{1x}x - \mathrm{i}\beta_{1y}y}. \end{aligned} \tag{8.154}$$

After transmitting into Region 2, the wave maintains a TE_z wave, the propagation characteristics are determined by (8.152), and the fields are given by

$$\begin{aligned} \mathbf{H}_2 &= \hat{z} H_0 \tau \mathrm{e}^{-\mathrm{i}\beta_{2x}x - \mathrm{i}\beta_{2y}y}, \\ \mathbf{E}_2 &= \left(-\hat{x}\frac{\beta_{2y}}{\omega\epsilon_2 a} + \hat{y}\frac{\beta_{2x}}{\omega\epsilon_2 b}\right) H_0 \tau \mathrm{e}^{-\mathrm{i}\beta_{2x}x - \mathrm{i}\beta_{2y}y}, \end{aligned} \tag{8.155}$$

where Γ and τ are the reflection and transmission coefficients, respectively, which are given by the continuity of the tangential components of **E** and **H** at the interface $x = 0$:

$$\Gamma = \frac{\beta_{1x} - \beta_{2x} b^{-1}}{\beta_{1x} + \beta_{2x} b^{-1}}, \tau = 1 + \Gamma = \frac{2\beta_{1x}}{\beta_{1x} + \beta_{2x} b^{-1}}. \tag{8.156}$$

Furthermore, at the interface $x = 0$, the field tangential continuity implies that

$$\beta_{2y} = \beta_{1y}. \tag{8.157}$$

After plugging (8.157) into (8.152), we obtain β_{2x}:

$$\beta_{2x} = \sqrt{k_2^2 bd - \beta_{1y}^2 a^{-1} b}. \tag{8.158}$$

If we set $\epsilon_1 = \epsilon_2$, $\mu_1 = \mu_2$, $d = b$, and $a^{-1} = b$, we have $k_1 = k_2$ and

$$\beta_{2x} = \sqrt{k_1^2 b^2 - (\beta_{1y})^2 b^2} = b\sqrt{k_1^2 - (\beta_{1y})^2} = b\beta_{1x}. \tag{8.159}$$

Now plugging (8.159) into (8.156), we obtain a zero reflection

$$\Gamma = 0 \text{ for any } \beta_{1x},$$

namely Regions 1 and 2 are perfectly matched.

The case of the TM_z wave can be treated similarly. As a matter of fact, the reflection coefficient for **E** is dual to (8.156) if we replace b by d and a by c; then the no-reflection condition is achieved. So if we have $b = d$ and $c^{-1} = a$, there will be no reflection for both TE_z and TM_z waves. Therefore, the conditions for no reflection, for all incident angles, polarization, and frequencies, on ϵ and μ tensors, are the following:

$$\bar{\bar{\epsilon}}_2 = \epsilon_1 \bar{\bar{s}}, \quad \bar{\bar{\mu}}_2 = \mu_1 \bar{\bar{s}}, \quad \bar{\bar{s}} = \begin{bmatrix} s_x^{-1} & 0 & 0 \\ 0 & s_x & 0 \\ 0 & 0 & s_x \end{bmatrix}, \tag{8.160}$$

which define the uniaxial perfectly matched layer [37].

Similar to Berenger's PML [36], the no-reflection property of the UPML in Region 2 holds for any s_x. For example, we can set

$$s_x = 1 + \frac{\sigma_x}{\mathrm{i}\omega\epsilon_1} = 1 - \frac{\sigma_x}{\omega\epsilon_1}\mathrm{i}.$$

Then, from (8.159) we have

$$\beta_{2x} = \left(1 - \mathrm{i}\frac{\sigma_x}{\omega\epsilon_1}\right)\beta_{1x}. \tag{8.161}$$

Noting that the real part of β_{2x} is the same as that of β_{1x}, with (8.157) we conclude that the phase velocity of both the incident and transmitted waves are the same in both regions.

If we put (8.157) into (8.155), then the transmitted wave in Region 2 for an incident TE_z wave is given by

$$\begin{aligned}\mathbf{H}_2 &= \hat{z} H_0 \tau \mathrm{e}^{-\mathrm{i}\beta_{1x}x - \mathrm{i}\beta_{1y}y} \mathrm{e}^{-\sigma_x x \eta_1 \cos\theta}, \\ \mathbf{E}_2 &= (-\hat{x} s_x \eta_1 \sin\theta + \hat{y}\eta_1 \cos\theta) H_0 \mathrm{e}^{-\mathrm{i}\beta_{1x}x - \mathrm{i}\beta_{1y}y} \mathrm{e}^{-\sigma_x x \eta_1 \cos\theta},\end{aligned} \tag{8.162}$$

where $\eta_1 = \sqrt{\mu_1/\epsilon_1}$, and θ is the incident angle with respect to the x-axis, which propagates in the UPML region with the same velocity as that of $\mathbf{H}_1$ and $\mathbf{E}_1$ in Region 1, although with an attenuation factor (independent of frequency, though dependent on θ and the conductivity function σ_x in the UPML).

The UPML losses $\sigma_x(x)$ are usually taken to have a polynomial profile [41]:

$$\sigma_x(x) = \left(\frac{l}{\Delta}\right)^m \sigma_{x,\max}, \tag{8.163}$$

where l is the distance into the UPML region measured from the interface between the UPML and the physical solution domain, and Δ is the thickness of the UPML. The definition of $\sigma_y(y)$ is similar. The reflection factor for the UPML is

$$R(\theta) = \exp\left[\frac{-2\eta\sigma_{x,\max}\Delta\cos\theta}{\epsilon_r(m+1)}\right], \tag{8.164}$$

where θ is the incident angle, ϵ_r is the relative electric permittivity of the medium, and η is the UPML's characteristic wave impedance.

UPML in 3-D

In a rectangular truncation for the solution domain, 2-D coordinate planes and 3-D corner blocks will appear in the absorbing region. The construction of the UPML will be carried out in a multiplicative manner, as shown in a 3-D corner block case where the time-harmonic Maxwell's equations can be written as

$$\nabla \times \mathbf{H} = \mathrm{i}\omega\epsilon\bar{\bar{s}}\mathbf{E}, \tag{8.165}$$

$$\nabla \times \mathbf{E} = -\mathrm{i}\omega\epsilon\bar{\bar{s}}\mathbf{H}. \tag{8.166}$$

Here $\bar{\bar{s}}$ is a diagonal tensor defined as follows:

$$\begin{aligned}\bar{\bar{s}} &= \begin{bmatrix} s_x^{-1} & 0 & 0 \\ 0 & s_x & 0 \\ 0 & 0 & s_x \end{bmatrix} \begin{bmatrix} s_y & 0 & 0 \\ 0 & s_y^{-1} & 0 \\ 0 & 0 & s_y \end{bmatrix} \begin{bmatrix} s_z & 0 & 0 \\ 0 & s_z & 0 \\ 0 & 0 & s_z^{-1} \end{bmatrix} \\ &= \begin{bmatrix} s_y s_z s_x^{-1} & 0 & 0 \\ 0 & s_x s_z s_y^{-1} & 0 \\ 0 & 0 & s_x s_y s_z^{-1} \end{bmatrix}.\end{aligned} \tag{8.167}$$

Allowing for general real parts in s, we have

$$s_t = \kappa_t + \frac{\sigma_t}{\mathrm{i}\omega\epsilon}, t = x, y, z. \tag{8.168}$$

With the general definition, we can give the definition for special cases encountered in a typical numerical calculation.

1. Lossless interior region: $\bar{\bar{s}}$ is the identity tensor, and set $s_x = s_y = s_z = 1$.
2. In the UPML region, $\bar{\bar{s}}$ is given by (8.167), with s_x, s_y, and s_z given as follows:
   in $x_{\min} \le x \le x_{\max}$ of the UPML, set $s_y = s_z = 1$;
   in $y_{\min} \le y \le y_{\max}$ of the UPML, set $s_x = s_z = 1$;
   in $z_{\min} \le z \le z_{\max}$ of the UPML, set $s_x = s_y = 1$;
   in the intersection of $x_{\min} \le x \le x_{\max}$ and $y_{\min} \le y \le y_{\max}$ of the UPML, set $s_z = 1$;
   in the intersection of $x_{\min} \le x \le x_{\max}$ and $z_{\min} \le z \le z_{\max}$ of the UPML, set $s_y = 1$;
   in the intersection of $y_{\min} \le y \le y_{\max}$ and $z_{\min} \le z \le z_{\max}$ of the UPML, set $s_x = 1$; and in the corner regions of the UPML, use (8.167).

Time-domain Maxwell's equations in UPML regions

In the following, we will give the time-domain form for the time-harmonic Maxwell's equations (8.165)–(8.168) in the uniaxial perfectly matched layer (UPML).

For simplicity, we just consider the 2-D TM$_z$ case, namely $s_z = 1$ [41]. The time-harmonic Maxwell's equations for the Fourier transform of the TM$_z$ wave in the corner blocks will be

$$\frac{\partial \hat{H}_y}{\partial x} - \frac{\partial \hat{H}_x}{\partial y} = \mathrm{i}\omega\epsilon_0\, s_x s_y \hat{E}_z, \tag{8.169}$$

$$\frac{\partial \hat{E}_z}{\partial y} = -\mathrm{i}\omega\mu_0 \frac{s_y}{s_x} \hat{H}_x, \tag{8.170}$$

$$-\frac{\partial \hat{E}_z}{\partial x} = -\mathrm{i}\omega\mu_0 \frac{s_x}{s_y} \hat{H}_y. \tag{8.171}$$

Equation (8.169) can be written as

$$\frac{\partial \hat{H}_y}{\partial x} - \frac{\partial \hat{H}_x}{\partial y} = \mathrm{i}\omega\epsilon_0 \hat{E}_z + \hat{J}_z(\omega), \tag{8.172}$$

where

$$\hat{J}_z(\omega) = \mathrm{i}\omega\epsilon_0 (s_x s_y - 1)\hat{E}_z.$$

After substituting s_x and s_y from (8.168) with $\kappa_x = \kappa_y = \kappa_z = 1$, we have

$$\begin{aligned}\hat{J}_z(\omega) &= \mathrm{i}\omega\epsilon_0\left[\left(1+\frac{\sigma_x}{\mathrm{i}\omega\epsilon_0}\right)\left(1+\frac{\sigma_y}{\mathrm{i}\omega\epsilon_0}\right)-1\right]\hat{E}_z \\ &= \mathrm{i}\omega\epsilon_0\left[\frac{\sigma_x+\sigma_y}{\mathrm{i}\omega\epsilon_0}+\frac{\sigma_x\sigma_y}{(\mathrm{i}\omega\epsilon_0)^2}\right]\hat{E}_z \\ &= (\sigma_x+\sigma_y)\hat{E}_z+\frac{1}{\mathrm{i}\omega\epsilon_0}\sigma_x\sigma_y\hat{E}_z. \end{aligned} \tag{8.173}$$

Now considering (8.172), we apply the inverse Fourier transform using the identity $\mathrm{i}\omega f(\omega) \to (\partial/\partial t) f(t)$. This yields a time-domain differential equation for (8.172):

$$\frac{\partial H_y}{\partial x} - \frac{\partial H_x}{\partial y} = \epsilon_0\frac{\partial E_z}{\partial t} + J_z(t). \tag{8.174}$$

Next, we derive dynamic equations for all other quantities. The way to obtain a dynamic equation for J_z from (8.173) is first to multiply both sides of this equation by $\mathrm{i}\omega$, which gives

$$\mathrm{i}\omega\hat{J}_z = \mathrm{i}\omega(\sigma_x+\sigma_y)\hat{E}_z + \frac{1}{\epsilon_0}\sigma_x\sigma_y\hat{E}_z. \tag{8.175}$$

Exploiting the differentiation equivalence for the Fourier transform, we perform an inverse Fourier transform of each term in (8.175):

$$\frac{\partial J_z(t)}{\partial t} = (\sigma_x+\sigma_y)\frac{\partial E_z}{\partial t} + \frac{1}{\epsilon_0}\sigma_x\sigma_y E_z. \tag{8.176}$$

We will further simplify the differential equations for the polarization currents J_z so that they will become just ordinary differential equations.

Considering (8.176), we introduce a new parameter

$$P_z = -J_z + (\sigma_x+\sigma_y)E_z, \tag{8.177}$$

or

$$J_z = -P_z + (\sigma_x+\sigma_y)E_z. \tag{8.178}$$

Then (8.176) can be written as

$$\frac{\partial P_z}{\partial t} = -\frac{1}{\epsilon_0}\sigma_x\sigma_y E_z. \tag{8.179}$$

After substituting J_z from (8.178) into (8.174), we have

$$\frac{\partial H_y}{\partial x} - \frac{\partial H_x}{\partial y} = \epsilon_0\frac{\partial E_z}{\partial t} + (\sigma_x+\sigma_y)E_z - P_z. \tag{8.180}$$

Similar operations can be performed on Faraday's law (8.170) and (8.171), and we end up with the following equations:

$$\frac{\partial E_z}{\partial y} = -\mu_0 \frac{\partial H_x}{\partial t} - \mu_0 \frac{\sigma_y - \sigma_x}{\epsilon_0} H_x + Q_x, \tag{8.181}$$

$$-\frac{\partial E_z}{\partial x} = -\mu_0 \frac{\partial H_y}{\partial t} - \mu_0 \frac{\sigma_x - \sigma_y}{\epsilon_0} H_y + Q_y, \tag{8.182}$$

$$\frac{\partial Q_x}{\partial t} = -\frac{\sigma_x}{\epsilon_0} Q_x + \frac{\mu_0 \sigma_x (\sigma_y - \sigma_x)}{\epsilon_0^2} H_x, \tag{8.183}$$

$$\frac{\partial Q_y}{\partial t} = -\frac{\sigma_y}{\epsilon_0} Q_y + \frac{\mu_0 \sigma_y (\sigma_x - \sigma_y)}{\epsilon_0^2} H_y. \tag{8.184}$$

From (8.180), (8.181), (8.182), (8.179), (8.183), and (8.184), we get a new set of equations for E_z, H_x, H_y, P_z, Q_x, and Q_y as follows:

$$\epsilon_0 \frac{\partial E_z}{\partial t} = \frac{\partial H_y}{\partial x} - \frac{\partial H_x}{\partial y} - (\sigma_x + \sigma_y) E_z + P_z, \tag{8.185}$$

$$\mu_0 \frac{\partial H_x}{\partial t} = -\frac{\partial E_z}{\partial y} - \mu_0 \frac{\sigma_y - \sigma_x}{\epsilon_0} H_x + Q_x, \tag{8.186}$$

$$\mu_0 \frac{\partial H_y}{\partial t} = \frac{\partial E_z}{\partial x} - \mu_0 \frac{\sigma_x - \sigma_y}{\epsilon_0} H_y + Q_y, \tag{8.187}$$

$$\frac{\partial P_z}{\partial t} = -\frac{1}{\epsilon_0} \sigma_x \sigma_y E_z, \tag{8.188}$$

$$\frac{\partial Q_x}{\partial t} = -\frac{\sigma_x}{\epsilon_0} Q_x + \frac{\mu_0 \sigma_x (\sigma_y - \sigma_x)}{\epsilon_0^2} H_x, \tag{8.189}$$

$$\frac{\partial Q_y}{\partial t} = -\frac{\sigma_y}{\epsilon_0} Q_y + \frac{\mu_0 \sigma_y (\sigma_x - \sigma_y)}{\epsilon_0^2} H_y. \tag{8.190}$$

Note that (8.188)–(8.190) contain no spatial derivatives and hence they are simply ordinary differential equations for P_z, Q_x, and Q_y. The system (8.185)–(8.187), after dropping the undifferentiated terms, becomes the original 3×3 Maxwell's system in the form of a hyperbolic system, which can be shown to be well-posed [42].

8.5 Summary

The key to obtaining accurate and stable solutions for electromagnetic wave propagation over long time is (and this cannot be over-emphasized) the correct treatment of boundary conditions for material interfaces and numerical artificial boundaries. Discussed in this chapter were the impedance boundary conditions for the conductor boundary, as well as several local absorbing boundary conditions for the computational boundary, due to their ease of implementation and high accuracy. Analytical boundary conditions, such as the Bayliss–Turkel and Engquist–Majda boundary conditions, can be implemented directly on the boundary of the computational domain. Meanwhile, the UPML boundary treatment allows for the more complicated computational regions encountered in engineering applications and, in general, provides better accuracy for waves of a wide range of incident angles on the boundary, though at the expense of additional mesh points in the UPML regions.

References

1. J.D. Jackson, *Classical Electrodynamics*, 3rd edn. (Wiley, Hoboken, NJ, 2001)
2. J.C. Maxwell, *Treatise on Electricity and Magnetism*, vol. 2, 3rd edn. (Oxford University Press, Oxford, UK). Reprinted (Dover Publications, New York, 1891), p. 1954
3. V. Girault, P.-A. Raviart, *Finite Element Methods for Navier-Stokes Equations, Theory and Algorithms* (Springer, New York, 1986)
4. R.F. Harrington, *Time-Harmonic Electromagnetic Fields*, 2nd edn. (Wiley-IEEE Press, Hoboken, NJ, 2001)
5. J.A. Stratton, *Electromagnetic Theory* (McGraw-Hill, New York, 1941)
6. S. Ramo, J.R. Whinnery, T. van Duzer, *Fields and Waves in Communication Electronics*, 3rd edn. (Wiley, Hoboken, NJ, 1994)
7. T.B.A. Senior, J.L. Volakis, *Approximate Boundary Conditions in Electromagnetics*. IEE Electromagnetic Waves, vol. 41 (Institution of Electrical Engineers, Stevenage, UK, 1995)
8. M.A. Leontovich, *Investigations on Radiowave Propagation Part II* (Academy of Sciences, Moscow, 1948)
9. A. Sommerfeld, *Partial Differential Equations in Physics* (Academic Press, New York, 1949)
10. D. Colton, R. Kress, *Inverse Acoustic and Electromagnetic Scattering Theory* (Springer, Berlin, 1992)
11. C. Müller, *Foundations of the Mathematical Theory of Electromagnetic Waves* (Springer, New York, 1969)
12. J.C. Nédélec, *Acoustic and Electromagnetic Equations: Integral Representations for Harmonic Problems* (Springer, Berlin, 2001)
13. J.B. Keller, D. Givoli, Exact non-reflecting boundary conditions. J. Comput. Phys. **82**, 172–192 (1989)
14. D. Givoli, J.B. Keller, Non-reflecting boundary conditions for elastic waves. Wave Motion **12**, 261–279 (1990)
15. H.D. Han, X. Wu, *Artificial Boundary Method* (Tsinghua University Press (in Chinese), Beijing, 2009)
16. T. Hagstrom, S. Lau, Radiation boundary conditions for Maxwell's equations: a review of accurate time domain formulation. J. Comput. Math. **25**, 305–336 (2007)
17. Z. Chen, J.-C. Nédélec, On Maxwell equations with the transparent boundary condition. J. Comput. Math. **26**, 284–296 (2008)
18. B. Alpert, L. Greengard, T. Hagstrom, Rapid evaluation of nonreflecting boundary kernels for time-domain wave propagation. SIAM J. Numer. Anal. **37**, 1138–1164 (2000)
19. B. Alpert, L. Greengard, T. Hagstrom, Nonreflecting boundary conditions for the time-dependent wave equation. J. Comput. Phys. **180**, 270–296 (2002)
20. M.J. Grote, J.B. Keller, Nonreflecting boundary conditions for time-dependent scattering. J. Comput. Phys. **127**, 52–65 (1996)
21. L. Ting, M.J. Miksis, Exact boundary conditions for scattering problems. J. Acoust. Soc. Am. **80**, 1825–1827 (1986)
22. T.-Z. Teng, Exact boundary condition for time-dependent wave equation based on boundary integral. J. Comput. Phys. **190**, 398–418 (2003)

23. S.V. Tsynkov, On the application of lacunae-based methods to Maxwell's equations. J. Comput. Phys. **199**, 126–149 (2004)
24. B. Engquist, A. Majda, Absorbing boundary conditions for the numerical simulation of waves. Math. Comput. **31**, 629–651 (1977)
25. L.N. Trefethen, L. Halpern, Well-posedness of one-way wave equations and absorbing boundary conditions. Math. Comput. **47**, 421–435 (1986)
26. G. Mur, Absorbing boundary conditions for the finite difference approximation of the time-domain electromagnetic-field equations. IEEE Trans. Electromagn. Compat. **23**, 377–382 (1981)
27. R.L. Higdon, Numerical absorbing boundary conditions for the wave equation. Math. Comput. **49**, 65–90 (1987)
28. D. Givoli, B. Neta, High-order non-reflecting boundary scheme for time-dependent waves. J. Comput. Phys. **186**, 24–46 (2003)
29. C.H. Wilcox, An expansion theorem for electromagnetic fields. Commun. Pure Appl. Math. **9**, 115–134 (1956)
30. A. Bayliss, E. Turkel, Radiation boundary conditions for wave-like equations. Commun. Pure Appl. Math. **33**, 707–725 (1980)
31. T. Hagstrom, S.I. Hariharan, A formulation of asymptotic and exact boundary conditions using local operators. Appl. Numer. Math. **27**, 403–416 (1998)
32. R. Huan, L. Thompson, Accurate radiation boundary conditions for the time-dependent wave equation on unbounded domains. Int. J. Numer. Meth. Eng. **47**, 1569–1603 (1999)
33. T. Hagstrom, New results on absorbing layers and radiation boundary conditions, in *Topics in Computational Wave Propagation: Direct and Inverse Problems*. ed. by M. Ainsworth, P. Davies, D. Duncan, P. Martin, B. Rynne (Springer, Berlin, 2003), pp. 1–42
34. A.F. Peterson, Absorbing boundary conditions for the vector wave equation. Microw. Opt. Technol. Lett. **1**, 62–64 (1988)
35. M.J. Grote, Local non-reflecting boundary condition for Maxwell's equations. Comput. Meth. Appl. Mech. Eng. **195**, 3691–3708 (2006)
36. J.-P. Berenger, A perfectly matched layer for the absorption of electromagnetic waves. J. Comput. Phys. **114**, 185–200 (1994)
37. Z.S. Sacks, D.M. Kingsland, R. Lee, J.-F. Lee, A perfectly matched anisotropic absorber for use as an absorbing boundary condition. IEEE Trans. Antenn. Propag. **43**, 1460–1463 (1995)
38. R.W. Ziolkowski, Time derivative Lorentz material model-based absorbing boundary condition. IEEE Trans. Antenn. Propag. **45**, 1530–1535 (1997)
39. S. Abarbanel, D. Gottlieb, On the construction and analysis of absorbing layers in CEM. Appl. Numer. Math. **27**, 331–340 (1998)
40. A. Taflove, S.C. Hagness, *Computational Electromagnetics: The Finite-Difference Time-Domain Method* (Artech House Publishers, Norwood, MA, 2000)
41. T. Lu, P. Zhang, W. Cai, Discontinuous Galerkin methods for dispersive and lossy Maxwell's equations and PML boundary conditions. J. Comput. Phys. **200**, 549–580 (2004)
42. S. Abarbanel, D. Gottlieb, A mathematical analysis of the PML method. J. Comput. Phys. **134**, 357–363 (1997)

Chapter 9
Dyadic Green's Functions in Layered Media

The dyadic Green's functions are the key component in forming an integral representation of electromagnetic fields in Chap. 11. Here, we will derive the dyadic Green's functions for time-harmonic Maxwell's equations in layered media, and spectral forms of Green's functions and their potentials will be given. Fast algorithms for calculating the Hankel transform for Green's functions in the physical domain will be discussed.

9.1 Singular Charge and Current Sources

Charges and currents in concentrated regions are represented by the Dirac delta distribution function $\delta(\mathbf{r})$, and they are considered as singular sources for electromagnetic fields [7].

9.1.1 Singular Charge Sources

A singular charge density $\rho_e(\mathbf{r}) = \delta(\mathbf{r})$ in terms of the Dirac delta function represents the limiting case of a concentrated electric charge $\delta_\Omega(\mathbf{r})$ in a small volume Ω as the volume size $|\Omega|$ tends to zero while the total amount of charge in the volume remains one unit, namely,

$$\rho_e(\mathbf{r}) = \delta(\mathbf{r}) \equiv \lim_{|\Omega|\to 0} \delta_\Omega(\mathbf{r}), \tag{9.1}$$

where

$$\delta_\Omega(\mathbf{r}) = \begin{cases} \dfrac{1}{|\Omega|}, & \text{if } \mathbf{r} \in \Omega, \\ 0, & \text{if } \mathbf{r} \notin \Omega. \end{cases} \tag{9.2}$$

W. Cai, *Deterministic, Stochastic, and Deep Learning Methods for Computational Electromagnetics*, https://doi.org/10.1007/978-981-96-0100-4_9

Therefore, we have

$$\int_{\mathbb{R}^3} \delta(\mathbf{r})\mathrm{d}\mathbf{r} = 1 \quad \text{and} \quad \delta(\mathbf{r}) = 0, \quad \text{for } \mathbf{r} \neq \mathbf{0}, \tag{9.3}$$

and the sifting property of the Dirac delta function $\delta(\mathbf{r})$ for a continuous function $f(\mathbf{r})$:

$$\int_{\mathbb{R}^3} f(\mathbf{r}')\delta(\mathbf{r}' - \mathbf{r})\mathrm{d}\mathbf{r}' = f(\mathbf{r}). \tag{9.4}$$

Consequently, for any volume charge density $\rho(\mathbf{r})$ in V, we have formally

$$\rho(\mathbf{r}) = \int_{\mathbb{R}^3} \rho(\mathbf{r}')\delta(\mathbf{r}' - \mathbf{r})\mathrm{d}\mathbf{r}' = \lim_{|\Delta\mathbf{r}_i'|\to 0} \sum_i \rho(\mathbf{r}_i')\delta(\mathbf{r}_i' - \mathbf{r})\Delta\mathbf{r}_i', \tag{9.5}$$

which heuristically implies that a general charge density can be viewed as a superposition of singular charge density distributions.

Consider a surface $S \subset \mathbb{R}^3$, parameterized by a vector function $\mathbf{r}(u, v)$. Then, the lines of constant u over S, the lines of constant v over S, and the normal direction $\mathbf{n}$ form a local curvilinear coordinate system (u, v, w), with w being the coordinate along the normal direction.

If there is a volume charge density $\rho(u, v, w)$ concentrated around the surface S with dimension d along the normal direction $\mathbf{n}$ being small, then we can define a surface charge density $\rho_s(u, v)$ by integrating $\rho(u, v, w)$ along the normal direction:

$$\rho_s(u, v) = \int_{-d}^{d} \rho(u, v, w)\mathrm{d}w. \tag{9.6}$$

On the other hand, using the 1-D form of (9.3) we have, for all $d > 0$,

$$\rho_s(u, v) = \int_{-d}^{d} \rho_s(u, v)\delta(w)\mathrm{d}w. \tag{9.7}$$

Therefore, as $d \to 0$, the concentrated surface charge density can be represented as follows:

$$\rho(u, v, w) = \rho_s(u, v)\delta(w). \tag{9.8}$$

Similarly, we can consider the charge density $\rho_c(u)$ along a curve C which, for simplicity, is assumed to be the line of constant $v = v_0 = 0$ over the surface S. Suppose that the line charge is distributed within a small interval $[v_0 - d, v_0 + d]$. Then, we can define the line charge density $\rho_c(u)$ as

$$\rho_c(u) = \int_{-d}^{d} \rho_s(u, v)\mathrm{d}v. \tag{9.9}$$

Again using the 1-D form of (9.3) we have, for all $d > 0$,

$$\rho_c(u) = \int_{-d}^{d} \rho_c(u)\delta(v)\mathrm{d}v. \tag{9.10}$$

Hence, as $d \to 0$, the line charge density $\rho_c(u)$ can be seen over the surface S as

$$\rho_s(u, v) = \rho_c(u)\delta(v), \tag{9.11}$$

or, combining with (9.8), in the volume V as

$$\rho(u, v, w) = \rho_c(u)\delta(v)\delta(w). \tag{9.12}$$

In particular, in the Cartesian coordinate system (x, y, z), the three types of singular charge sources are

(a) $\rho_e = \delta(x, y, z)$, a singular volume (point) charge source located at $(0, 0, 0)$;
(b) $\rho_e = \rho_s(x, y)\delta(z)$, a singular surface charge source over the xy-plane, where $\rho_s(x, y)$ is the surface density per unit area over the xy-plane, etc.; and
(c) $\rho_e = \rho_c(x)\delta(y)\delta(z)$, a singular line charge source along the x-axis, where $\rho_c(x)$ is the line density per unit length along the x-axis.

9.1.2 Singular Hertz Dipole Current Sources

An important concentrated current source is the Hertz dipole current, which models a small linear antenna with a length d composed of two time-variant charges $q_1(t) = -q\mathrm{e}^{\mathrm{i}\omega t}$ and $q_2(t) = q\mathrm{e}^{\mathrm{i}\omega t}$, located at $(-d/2, 0, 0)$ and $(d/2, 0, 0)$, respectively. The dipole moment of the antenna is then $\mathbf{p} = \mathbf{d}q\mathrm{e}^{\mathrm{i}\omega t}$, with $\mathbf{d} = (d, 0, 0)$. A Hertz dipole is obtained if we consider the limit of $d \to 0$ while maintaining the dipole moment as $\mu = qd$, resulting in a point dipole density of the form (1.12), namely,

$$\tilde{\mathbf{p}} = \mu\mathrm{e}^{\mathrm{i}\omega t}\delta(\mathbf{r}), \text{ and } \int \tilde{\mathbf{p}}\mathrm{d}\mathbf{r} = \mu\mathrm{e}^{\mathrm{i}\omega t}. \tag{9.13}$$

The corresponding charge density for the Hertz dipole is

$$\begin{aligned}\rho_e(x, y, z) &= \lim_{d \to 0} \left[-q e^{i\omega t} \delta \left(x + \frac{d}{2} \right) \delta(y)\delta(z) + q e^{i\omega t} \delta \left(x - \frac{d}{2} \right) \delta(y)\delta(z) \right] \\ &= -\mu e^{i\omega t} \lim_{d \to 0} \delta(y)\delta(z) \left[\frac{\delta \left(x + \frac{d}{2} \right) - \delta \left(x - \frac{d}{2} \right)}{d} \right] \\ &= -\mu e^{i\omega t} \lim_{d \to 0} \left[\frac{\delta \left(x + \frac{d}{2}, y, z \right) - \delta \left(x - \frac{d}{2}, y, z \right)}{d} \right],\end{aligned}$$

i.e.,

$$\rho_e(x, y, z) = -\mu \delta'(x)\delta(y)\delta(z)e^{i\omega t}. \tag{9.14}$$

Here, the derivative $\delta'(x)$ of the 1-D Dirac delta function $\delta(x)$ defines a distribution such that for $f(x) \in C_0^\infty(-\infty, \infty)$

$$\int_{-\infty}^{\infty} \delta'(x) f(x) \mathrm{d}x = -f'(0). \tag{9.15}$$

The polarization current density $\mathbf{J}_p$ defined in (8.17) is then

$$\mathbf{J}_p = \frac{\partial \widetilde{\mathbf{p}}}{\partial t} = i\omega e^{i\omega t} \boldsymbol{\mu} \delta(\mathbf{r}), \tag{9.16}$$

which is identified as the *Hertz dipole current source*.

Similarly, Hertz dipole type surface current sources over the coordinate planes can be defined by integrating (9.16) along the normal direction of the surfaces (assuming a unit Hertz dipole current source). For instance, we can define the following.

(a) Surface Hertz dipole current over the xy-plane:

$$\mathbf{J}_{e(s)}(x, y, z) = \delta(x)\delta(y)\hat{t}, \hat{t} = \hat{x} \text{ or } \hat{y}, \tag{9.17}$$

where $\hat{t}$ denotes the unit direction of a coordinate axis.

(b) Surface Hertz dipole current over the zx-plane:

$$\mathbf{J}_{e(s)}(x, y, z) = \delta(x)\delta(z)\hat{t}, \hat{t} = \hat{x} \text{ or } \hat{z}. \tag{9.18}$$

The above surface currents flow tangentially to the xy- and zx-planes, respectively. The corresponding surface charge density can be found in a similar way. For instance, over the xy-plane for an $\hat{x}$-directed surface current source,

$$\rho_{e(s)}(x, y, z) = -\frac{1}{i\omega} \delta'(x)\delta(y). \tag{9.19}$$

9.2 Dyadic Green's Functions $\overline{\mathbf{G}}_E(\mathbf{r}|\mathbf{r}')$ and $\overline{\mathbf{G}}_H(\mathbf{r}|\mathbf{r}')$

From the sifting property (9.4) of the Dirac delta function, a general current source $\mathbf{J}(\mathbf{r}) = (J_x, J_y, J_z)$ in a volume V can be written as a superposition of $\hat{x}$-, $\hat{y}$-, or $\hat{z}$-directed Hertz dipole singular sources, namely,

$$\mathbf{J}(\mathbf{r}) = \int_V \left[J_x(\mathbf{r}')\delta(\mathbf{r}'-\mathbf{r})\hat{x} + J_y(\mathbf{r}')\delta(\mathbf{r}'-\mathbf{r})\hat{y} + J_z(\mathbf{r}')\delta(\mathbf{r}'-\mathbf{r})\hat{z}\right]\mathrm{d}\mathbf{r}'. \quad (9.20)$$

Accordingly, the electromagnetic fields generated by $\mathbf{J}(\mathbf{r})$ can be expressed as a superposition of the fields generated by these Hertz dipoles; the latter defines the dyadic Green's functions.

Specifically, let $\mathbf{G}_E^t(\mathbf{r}|\mathbf{r}') = (G_E^{xt}, G_E^{yt}, G_E^{zt})^{\mathrm{T}}$ and $\mathbf{G}_H^t(\mathbf{r}|\mathbf{r}') = (G_H^{xt}, G_H^{yt}, G_H^{zt})^{\mathrm{T}}$, $t = x, y, z$, be the electric and magnetic fields at location $\mathbf{r}$ generated by a $\hat{t}$-directed $-1/(\mathrm{i}\omega\mu)$-Hertz dipole of current moment located at $\mathbf{r}'$, i.e.,

$$\mathbf{J}_e(\mathbf{r}) = -\frac{1}{\mathrm{i}\omega\mu}\delta(\mathbf{r}-\mathbf{r}')\hat{t}, \qquad t \in \{x, y, z\}. \quad (9.21)$$

Then, the dyadic Green's functions for the electric and the magnetic fields are defined, respectively, as

$$\overline{\mathbf{G}}_E(\mathbf{r}|\mathbf{r}') = \left[\mathbf{G}_E^x, \mathbf{G}_E^y, \mathbf{G}_E^z\right] = \sum_{s,t\in(x,y,z)} \hat{s}G_E^{st}\hat{t} \quad (9.22)$$

and

$$\overline{\mathbf{G}}_H(\mathbf{r}|\mathbf{r}') = \left[\mathbf{G}_H^x, \mathbf{G}_H^y, \mathbf{G}_H^z\right] = \sum_{s,t\in(x,y,z)} \hat{s}G_H^{st}\hat{t}. \quad (9.23)$$

It can be seen from (8.30) and (8.31) that the dyadic Green's functions satisfy the following matrix equations (viewed in a column-wise manner):

$$\nabla\times\frac{1}{\mu}\nabla\times\overline{\mathbf{G}}_E(\mathbf{r}|\mathbf{r}') - \omega^2\epsilon\overline{\mathbf{G}}_E(\mathbf{r}|\mathbf{r}') = \frac{1}{\mu}\overline{\mathbf{I}}\delta(\mathbf{r}-\mathbf{r}'), \quad (9.24)$$

$$\nabla\times\frac{1}{\epsilon}\nabla\times\overline{\mathbf{G}}_H(\mathbf{r}|\mathbf{r}') - \omega^2\mu\overline{\mathbf{G}}_H(\mathbf{r}|\mathbf{r}') = -\nabla\times\frac{1}{\mathrm{i}\omega\epsilon\mu}\overline{\mathbf{I}}\delta(\mathbf{r}-\mathbf{r}'), \quad (9.25)$$

where $\overline{\mathbf{I}}$ is the identity matrix. From Maxwell's equation (8.23), we have

$$\overline{\mathbf{G}}_H(\mathbf{r}|\mathbf{r}') = -\frac{1}{\mathrm{i}\omega\mu(\mathbf{r})}\nabla\times\overline{\mathbf{G}}_E(\mathbf{r}|\mathbf{r}'). \quad (9.26)$$

The electric field generated by $\mathbf{J}_e(\mathbf{r})$ (assuming $\mathbf{J}_m(\mathbf{r}) = \mathbf{0}$) can then be obtained through the principle of linear superposition:

$$E_s(\mathbf{r}) = -\mathrm{i}\omega \int\limits_V \mu \left(G_E^{sx} J_{e,x} + G_E^{sy} J_{e,y} + G_E^{sz} J_{e,z} \right) \mathrm{d}\mathbf{r}', \qquad s = x, y, z, \tag{9.27}$$

i.e.,

$$\mathbf{E}(\mathbf{r}) = -\mathrm{i}\omega \int\limits_V \mu(\mathbf{r}') \overline{\mathbf{G}}_E(\mathbf{r}|\mathbf{r}') \cdot \mathbf{J}_e(\mathbf{r}') \mathrm{d}\mathbf{r}'. \tag{9.28}$$

Similarly, the magnetic field can be obtained through

$$\mathbf{H}(\mathbf{r}) = -\mathrm{i}\omega \int\limits_V \mu(\mathbf{r}') \overline{\mathbf{G}}_H(\mathbf{r}|\mathbf{r}') \cdot \mathbf{J}_e(\mathbf{r}') \mathrm{d}\mathbf{r}' = \int\limits_V \nabla \times \overline{\mathbf{G}}_E(\mathbf{r}|\mathbf{r}') \cdot \mathbf{J}_e(\mathbf{r}') \mathrm{d}\mathbf{r}', \tag{9.29}$$

where a constant μ is assumed to obtain the second equality.

9.2.1 Dyadic Green's Functions for Homogeneous Media

For a homogeneous medium, the electromagnetic fields generated by a Hertz dipole (9.21) can be written in terms of a vector potential $\mathbf{A}$ and a scalar potential V_e as in (8.41), for example

$$\mathbf{G}_E^t(\mathbf{r}|\mathbf{r}') = \overline{\mathbf{G}}_E(\mathbf{r}|\mathbf{r}') \cdot \hat{t} = -\mathrm{i}\omega \mathbf{A} - \nabla V_e, \tag{9.30}$$

where the vector potential $\mathbf{A}$, from (8.49), satisfies

$$\nabla^2 \mathbf{A} + k^2 \mathbf{A} = \frac{1}{\mathrm{i}\omega} \delta(\mathbf{r} - \mathbf{r}') \hat{t}. \tag{9.31}$$

Here, $k = \omega\sqrt{\epsilon\mu}$ is the wave number of the homogeneous medium. Then, from (3.4), the solution to the above Helmholtz equation is

$$\mathbf{A} = -\frac{1}{\mathrm{i}\omega} g(\mathbf{r} - \mathbf{r}') \hat{t}, \tag{9.32}$$

where

$$g(\mathbf{r} - \mathbf{r}') = \frac{\mathrm{e}^{-\mathrm{i}k|\mathbf{r}-\mathbf{r}'|}}{4\pi|\mathbf{r} - \mathbf{r}'|}. \tag{9.33}$$

Meanwhile, the Lorenz gauge condition (8.46) implies that

$$V_e = -\frac{1}{\mathrm{i}\omega\epsilon\mu} \nabla \cdot \mathbf{A} = -\frac{1}{k^2} \nabla g(\mathbf{r} - \mathbf{r}') \hat{t}. \tag{9.34}$$

Therefore, from (9.30) we have

$$\overline{\mathbf{G}}_E(\mathbf{r}|\mathbf{r}') \cdot \hat{t} = g(\mathbf{r}-\mathbf{r}')\hat{t} + \frac{1}{k^2}\nabla\nabla g(\mathbf{r}-\mathbf{r}')\hat{t} = \left(1+\frac{\nabla\nabla}{k^2}\right) g(\mathbf{r}-\mathbf{r}')\hat{t},$$

and thus the dyadic Green's function for the electric field in a homogeneous medium is given by

$$\overline{\mathbf{G}}_E(\mathbf{r}|\mathbf{r}') = \left(\overline{\mathbf{I}}+\frac{\nabla\nabla}{k^2}\right) g(\mathbf{r}-\mathbf{r}'). \tag{9.35}$$

From (9.26), the dyadic Green's function for the magnetic field is given by

$$\overline{\mathbf{G}}_H(\mathbf{r}|\mathbf{r}') = -\frac{1}{\mathrm{i}\omega\mu}\nabla \times \left(\overline{\mathbf{I}}+\frac{\nabla\nabla}{k^2}\right) g(\mathbf{r}-\mathbf{r}'). \tag{9.36}$$

9.2.2 *Dyadic Green's Functions for Layered Media*

The layered medium to be considered here is shown in Fig. 11.1; it is a stratified structure consisting of $N+2$ dielectric layers separated by $N+1$ planar interfaces at $z = d_l, l = 0, 1, \ldots, N$. The lth layer of the medium is characterized by permeability μ_l and permittivity ϵ_l. The permittivity is complex if the medium is lossy:

$$\epsilon_l = \epsilon_0\left(\epsilon_{r_l} + \frac{\sigma_l}{\mathrm{i}\omega\epsilon_0}\right), \tag{9.37}$$

where σ_l is the conductivity of the medium in the lth layer.

To derive the dyadic Green's functions for the layered medium, we consider a Hertz dipole current source at $\mathbf{r}' = (x', y', z')$ embedded in the ith layer, whereas the observation point at $\mathbf{r} = (x, y, z)$ is assumed to be in the jth layer. We will find the electromagnetic fields at $\mathbf{r}$ excited by a unit-strength, arbitrarily oriented current Hertz dipole located at $\mathbf{r}'$, namely, Green's functions $\overline{\mathbf{G}}_E(\mathbf{r}|\mathbf{r}')$ and $\overline{\mathbf{G}}_H(\mathbf{r}|\mathbf{r}')$.

9.2.3 *Hankel Transform for Radially Symmetric Functions*

As the multi-layered medium is radially symmetric in the xy-plane, we can use the 2-D Fourier transform $\mathcal{F}$ to solve Maxwell's equations for the dyadic Green's functions in the Fourier transform (spectral) domain, $\widetilde{\overline{\mathbf{G}}}_E$ and $\widetilde{\overline{\mathbf{G}}}_H$. Once the spectral Green's functions are obtained, Green's functions in the physical domain, $\overline{\mathbf{G}}_E(\mathbf{r}|\mathbf{r}')$ and $\overline{\mathbf{G}}_H(\mathbf{r}|\mathbf{r}')$, can be found by using the inverse Fourier transform $\mathcal{F}^{-1}$, where $\mathcal{F}$ and $\mathcal{F}^{-1}$ are defined by

$$\widetilde{f}(k_x, k_y) = \mathcal{F}\{f(x, y)\} = \frac{1}{2\pi} \int_{-\infty}^{\infty} \int_{-\infty}^{\infty} f(x, y) \mathrm{e}^{-\mathrm{i}(k_x x + k_y y)} \mathrm{d}x \mathrm{d}y,$$

$$f(x, y) = \mathcal{F}^{-1}\left\{\widetilde{f}(k_x, k_y)\right\} = \frac{1}{2\pi} \int_{-\infty}^{\infty} \int_{-\infty}^{\infty} \widetilde{f}\left(k_x, k_y\right) \mathrm{e}^{\mathrm{i}(k_x x + k_y y)} \mathrm{d}k_x \mathrm{d}k_y. \tag{9.38}$$

The inverse Fourier integral in (9.38) can be expressed in terms of a Hankel transform if $f(x, y)$ is a radially symmetric function. Introducing polar coordinates in both the space and the transform domains,

$$\begin{aligned} x &= \rho \cos \alpha, & y &= \rho \sin \alpha, \\ k_x &= k_\rho \cos \beta, & k_y &= k_\rho \sin \beta, \end{aligned} \tag{9.39}$$

where

$$\begin{aligned} \rho &= \sqrt{x^2 + y^2}, & \alpha &= \arctan\left(y/x\right), \\ k_\rho &= \sqrt{k_x^2 + k_y^2}, & \beta &= \arctan\left(k_y/k_x\right), \end{aligned} \tag{9.40}$$

it can be shown that

$$f(\rho) = \mathcal{F}^{-1}\left\{\widetilde{f}(k_\rho)\right\} = S_0\left[\widetilde{f}(k_\rho)\right](\rho), \tag{9.41}$$

where the nth-order Hankel transform $S_n\left[\widetilde{f}(k_\rho)\right]$ for an integer $n \geq 0$ is defined as

$$S_n\left[\widetilde{f}(k_\rho)\right](\rho) = \int_0^{\infty} \widetilde{f}(k_\rho) J_n(k_\rho \rho) k_\rho^{n+1} \mathrm{d}k_\rho. \tag{9.42}$$

Here, $J_n(z)$ is the nth-order Bessel function

$$J_n(z) = \frac{1}{\pi} \int_0^{\pi} \cos(n\theta - z \sin\theta) \mathrm{d}\theta. \tag{9.43}$$

Moreover, we have the following identities:

$$\mathcal{F}^{-1}\left\{-\mathrm{i}k_x \widetilde{f}(k_\rho)\right\} = -\cos\alpha\ S_1\left[\widetilde{f}(k_\rho)\right], \tag{9.44}$$

$$\mathcal{F}^{-1}\left\{-\mathrm{i}k_y \widetilde{f}(k_\rho)\right\} = -\sin\alpha\ S_1\left[\widetilde{f}(k_\rho)\right], \tag{9.45}$$

$$\mathcal{F}^{-1}\left\{k_x^2 \widetilde{f}(k_\rho)\right\} = -\frac{1}{2}\left\{\cos 2\alpha\ S_2\left[\widetilde{f}(k_\rho)\right] - S_0\left[k_\rho^2 \widetilde{f}(k_\rho)\right]\right\}, \tag{9.46}$$

$$\mathcal{F}^{-1}\left\{k_y^2 \widetilde{f}(k_\rho)\right\} = \frac{1}{2}\left\{\cos 2\alpha\ S_2\left[\widetilde{f}(k_\rho)\right] + S_0\left[k_\rho^2 \widetilde{f}(k_\rho)\right]\right\}, \tag{9.47}$$

$$\mathcal{F}^{-1}\left\{k_x k_y \widetilde{f}(k_\rho)\right\} = -\frac{1}{2}\sin 2\alpha\ S_2\left[\widetilde{f}(k_\rho)\right]. \tag{9.48}$$

In order to derive the dyadic Green's functions in the spectral domain, we note that (8.30) in a source-free region simplifies to

$$\nabla^2\mathbf{E} + k^2\mathbf{E} = \mathbf{0}. \tag{9.49}$$

Let $\widetilde{\mathbf{E}}$ be the 2-D Fourier transform of $\mathbf{E}$ in the xy-plane, namely, $\widetilde{\mathbf{E}}(k_x, k_y, z) = \mathcal{F}\{\mathbf{E}(x, y, z)\}$. Then, under a Fourier transform, (9.49) yields the following second-order ordinary differential equation for $\widetilde{E}_s$, $s = x, y, z$:

$$\frac{\mathrm{d}^2\widetilde{E}_s}{\mathrm{d}z^2} + \left(k^2 - k_\rho^2\right)\widetilde{E}_s = 0. \tag{9.50}$$

9.2.4 Transverse Versus Longitudinal Field Components

Before we proceed to find all the components of Green's functions $\overline{\mathbf{G}}_E$ and $\overline{\mathbf{G}}_H$, we will show that in layered media the transverse components of the electric and the magnetic fields, E_x, E_y, H_x, and H_y, can be expressed in terms of the longitudinal components E_z and H_z. In fact, applying the 2-D Fourier transform to Maxwell's equations (8.23) and (8.24) in a source-free region ($\mathbf{J}_e = \mathbf{0}$ and $\mathbf{J}_m = \mathbf{0}$), we have

$$\widetilde{\nabla \times \mathbf{E}} = -\mathrm{i}\omega\mu\widetilde{\mathbf{H}}, \tag{9.51}$$

$$\widetilde{\nabla \times \mathbf{H}} = \mathrm{i}\omega\epsilon\widetilde{\mathbf{E}}, \tag{9.52}$$

or

$$\begin{cases} -\mathrm{i}k_y\widetilde{E}_z + \dot{\widetilde{E}}_y = \mathrm{i}\omega\mu\widetilde{H}_x, \\ \mathrm{i}k_x\widetilde{E}_z - \dot{\widetilde{E}}_x = \mathrm{i}\omega\mu\widetilde{H}_y, \\ -\mathrm{i}k_x\widetilde{E}_y + \mathrm{i}k_y\widetilde{E}_x = \mathrm{i}\omega\mu\widetilde{H}_z \end{cases} \tag{9.53}$$

and

$$\begin{cases} \mathrm{i}k_y\widetilde{H}_z - \dot{\widetilde{H}}_y = \mathrm{i}\omega\epsilon\widetilde{E}_x, \\ -\mathrm{i}k_x\widetilde{H}_z + \dot{\widetilde{H}}_x = \mathrm{i}\omega\epsilon\widetilde{E}_y, \\ \mathrm{i}k_x\widetilde{H}_y - \mathrm{i}k_y\widetilde{H}_x = \mathrm{i}\omega\epsilon\widetilde{E}_z, \end{cases} \tag{9.54}$$

where the dot denotes $\partial/\partial z$.

After some calculations, the transverse components $\widetilde{E}_x$, $\widetilde{E}_y$, $\widetilde{H}_x$, and $\widetilde{H}_y$ can be expressed as follows:

$$\widetilde{E}_x = \frac{1}{k_\rho^2}\left(\mathrm{i}k_x\dot{\widetilde{E}}_z + \omega\mu k_y\widetilde{H}_z\right), \quad \widetilde{E}_y = \frac{1}{k_\rho^2}\left(\mathrm{i}k_y\dot{\widetilde{E}}_z - \omega\mu k_x\widetilde{H}_z\right), \tag{9.55}$$

$$\widetilde{H}_x = \frac{1}{k_\rho^2}\left(\mathrm{i}k_x \widetilde{\dot{H}}_z - \omega\epsilon k_y \widetilde{E}_z\right), \qquad \widetilde{H}_y = \frac{1}{k_\rho^2}\left(\mathrm{i}k_y \widetilde{\dot{H}}_z + \omega\epsilon k_x \widetilde{E}_z\right). \tag{9.56}$$

Correspondingly, in the time domain, we have

$$E_x = \frac{\partial}{\partial x}\mathcal{F}^{-1}\left(\frac{\widetilde{\dot{E}}_z}{k_\rho^2}\right) - \mathrm{i}\omega\mu\frac{\partial}{\partial y}\mathcal{F}^{-1}\left(\frac{\widetilde{H}_z}{k_\rho^2}\right), \tag{9.57}$$

$$E_y = \frac{\partial}{\partial y}\mathcal{F}^{-1}\left(\frac{\widetilde{\dot{E}}_z}{k_\rho^2}\right) + \mathrm{i}\omega\mu\frac{\partial}{\partial x}\mathcal{F}^{-1}\left(\frac{\widetilde{H}_z}{k_\rho^2}\right), \tag{9.58}$$

$$H_x = \frac{\partial}{\partial x}\mathcal{F}^{-1}\left(\frac{\widetilde{\dot{H}}_z}{k_\rho^2}\right) + \mathrm{i}\omega\epsilon\frac{\partial}{\partial y}\mathcal{F}^{-1}\left(\frac{\widetilde{E}_z}{k_\rho^2}\right), \tag{9.59}$$

$$H_y = \frac{\partial}{\partial y}\mathcal{F}^{-1}\left(\frac{\widetilde{\dot{H}}_z}{k_\rho^2}\right) - \mathrm{i}\omega\epsilon\frac{\partial}{\partial x}\mathcal{F}^{-1}\left(\frac{\widetilde{E}_z}{k_\rho^2}\right). \tag{9.60}$$

The above relations will hold for the components of Green's functions G_E^{xt} and G_H^{xt} for $t = x, y, z$.

9.2.5 Longitudinal Components of Green's Functions

To calculate the longitudinal components of Green's functions in the spectral domain, $\widetilde{E}_z = \widetilde{G_E^{zt}}$ and $\widetilde{H}_z = \widetilde{G_H^{zt}}$, we will follow the procedures in [8, 10], but with details for the spectral quantities for the field components in [2]. Both $\widetilde{E}_z$ and $\widetilde{H}_z$ satisfy (9.50) with appropriate boundary conditions at $z = d_l, l = 0, 1, \ldots, N$, and $z = z_i'$ (source location). To derive the interface conditions for $\widetilde{E}_z$ and $\widetilde{H}_z$, we use Maxwell's equations and (8.62)–(8.65).

Interface conditions for $\widetilde{E}_z$ and $\widetilde{H}_z$

Let us first consider $\widetilde{H}_z$. From the interface condition on the normal component of the magnetic field, we have

$$\mu_l H_{z,l} = \mu_{l+1} H_{z,l+1} + \rho_{m(s)}, \tag{9.61}$$

where $\rho_{m(s)}$ denotes the surface magnetic charge if any. Meanwhile, the tangential interface condition (8.62) $[\mathbf{n} \times \mathbf{H}] = \mathbf{J}_{e(s)}$, $\mathbf{n} = (0, 0, 1)$, implies that

$$[H_y] = -J_{e(s),x}, \qquad [H_x] = J_{e(s),y}, \tag{9.62}$$

where $\mathbf{J}_{e(s)} = \left(J_{e(s),x}, J_{e(s),y}, 0\right)^{\mathrm{T}}$ is the surface electric current.

From Gauss's law away from an interface, $\nabla \cdot \mathbf{H} = 0$, we see that the jumps of the magnetic field components across the interface satisfy the following condition:

$$\frac{\partial}{\partial x}[H_x] + \frac{\partial}{\partial y}[H_y] + \frac{\partial}{\partial z}[H_z] = 0, \tag{9.63}$$

which in the Fourier spectral domain becomes

$$\left[\dot{\widetilde{H}}_z\right] = -ik_x\left[\widetilde{H}_x\right] - ik_y\left[\widetilde{H}_y\right] = ik_y\widetilde{J}_{e(s),x} - ik_x\widetilde{J}_{e(s),y}. \tag{9.64}$$

As a result, the interface conditions at $z = z'$ for $\widetilde{H}_z$ are given by

$$\mu_l\widetilde{H}_{z,l} = \mu_{l+1}\widetilde{H}_{z,l+1} + \widetilde{\rho}_{m(s)}, \tag{9.65}$$
$$\dot{\widetilde{H}}_{z,l} = \dot{\widetilde{H}}_{z,l+1} + ik_y\widetilde{J}_{e(s),x} - ik_x\widetilde{J}_{e(s),y}. \tag{9.66}$$

Similarly, the interface conditions at $z = z'$ for $\widetilde{E}_z$ are given by

$$\epsilon_l\widetilde{E}_{z,l} = \epsilon_{l+1}\widetilde{E}_{z,l+1} + \widetilde{\rho}_{e(s)}, \tag{9.67}$$
$$\dot{\widetilde{E}}_{z,l} = \dot{\widetilde{E}}_{z,l+1} - ik_y\widetilde{J}_{m(s),x} + ik_x\widetilde{J}_{m(s),y}. \tag{9.68}$$

As neither electric nor magnetic sources are present at $z = d_l$, $\widetilde{J}_{e(s)} = \mathbf{0}$, $\widetilde{J}_{m(s)} = \mathbf{0}$, $\widetilde{\rho}_{e(s)} = 0$, and $\widetilde{\rho}_{m(s)} = 0$. At $z = z_i'$ (source location), where a $\hat{t}$-directed electric Hertz dipole is located, using (9.17) and (9.18) we then have the following boundary conditions at the source location $z = z_i'$:

(a) horizontal electric dipole (HED), $\hat{x}$-directed unit Hertz dipole

$$\widetilde{J}_{e(s)} = \frac{1}{2\pi}\hat{x}, \quad \widetilde{\rho}_{e(s)} = \frac{-ik_x}{2\pi i\omega}, \quad \widetilde{J}_{m(s)} = \mathbf{0}, \quad \widetilde{\rho}_{m(s)} = 0; \tag{9.69}$$

(b) HED, $\hat{y}$-directed unit Hertz dipole

$$\widetilde{J}_{e(s)} = \frac{1}{2\pi}\hat{y}, \quad \widetilde{\rho}_{e(s)} = \frac{-ik_y}{2\pi i\omega}, \quad \widetilde{J}_{m(s)} = \mathbf{0}, \quad \widetilde{\rho}_{m(s)} = 0. \tag{9.70}$$

Now let ψ_l, $l = 0, 1, \ldots, N$, represent either $(-i\omega\mu)\widetilde{E}_z$ or $(-i\omega\mu)\widetilde{H}_z$, satisfying (9.50) in each layer, subject to the following boundary conditions at the interface between layers l and $l + 1$:

$$\alpha_l\psi_l = \alpha_{l+1}\psi_{l+1}, \tag{9.71}$$
$$\dot{\psi}_l - \dot{\psi}_{l+1}, \tag{9.72}$$

with $\alpha_l = \epsilon_l$ for $\widetilde{E}_z$ or $\alpha_l = \mu_l$ for $\widetilde{H}_z$. From Fig. 11.1, we define

$$h_l = d_{l-1} - d_l. \tag{9.73}$$

Introducing a local coordinate for each layer as

$$z_l = z - d_l, \tag{9.74}$$

we can write the fields in a layer without sources as

$$\begin{aligned}\psi_l &= a_l \cosh u_l z_l + b_l \sinh u_l z_l \\ &= (\cosh u_l z_l, \sinh u_l z_l) \begin{pmatrix} a_l \\ b_l \end{pmatrix} \equiv (c_l, s_l)\, V_l,\end{aligned} \tag{9.75}$$

where

$$u_l = \sqrt{k_\rho^2 - k_l^2}, k_l^2 = \omega^2 \epsilon_l \mu_l, \tag{9.76}$$

and

$$c_l = \cosh u_l z_l, \quad s_l = \sinh u_l z_l.$$

Between any two adjacent layers, say layers l and $l+1$, there exists the relation

$$V_l = T_{l,l+1} V_{l+1}, \tag{9.77}$$

where

$$T_{l,l+1} = \begin{bmatrix} \dfrac{\alpha_{l+1}}{\alpha_l} \bar{c}_{l+1}, & \dfrac{\alpha_{l+1}}{\alpha_l} \bar{s}_{l+1} \\ \dfrac{u_{l+1}}{u_l} \bar{s}_{l+1}, & \dfrac{u_{l+1}}{u_l} \bar{c}_{l+1} \end{bmatrix} \tag{9.78}$$

and

$$(\bar{c}_l, \bar{s}_l) = (\cosh u_l h_l, \sinh u_l h_l)\,. \tag{9.79}$$

On the other hand, in the ith layer with the source, the fields can be written as

$$\psi_i = \psi_i^\infty + (c_i, s_i)\, V_i = \begin{cases} \left(c_i^U, s_i^U\right) V_i^U, & \text{if } D \le z_i \le h_i, \\ \left(c_i^L, s_i^L\right) V_i^L, & \text{if } 0 \le z_i \le D, \end{cases} \tag{9.80}$$

where $D = z_i' - d_i$ and

$$\psi_i^\infty = \begin{cases} U_i \mathrm{e}^{-u_i(z_i - D)}, & \text{if } D \le z_i \le h_i, \\ L_i \mathrm{e}^{+u_i(z_i - D)}, & \text{if } 0 \le z_i \le D, \end{cases} \tag{9.81}$$

$$\left(c_i^U, s_i^U\right) = (\cosh\, u_i\, (z_i - D)\,, \sinh\, u_i\, (z_i - D))\,, \tag{9.82}$$

$$\left(c_i^L, s_i^L\right) = (\cosh\, u_i z_i, \sinh\, u_i z_i)\,. \tag{9.83}$$

Table 9.1 U_i, L_i, and S_i

	$\widetilde{G}_H^{zx}$	$\widetilde{G}_H^{zy}$	$\widetilde{G}_E^{zx}$	$\widetilde{G}_E^{zy}$	$\widetilde{G}_E^{zz}$
U_i	$-\frac{\mathrm{i}k_y}{4\pi u_i}$	$\frac{\mathrm{i}k_x}{4\pi u_i}$	$\frac{-\mathrm{i}k_x}{4\pi\mathrm{i}\omega\epsilon_i}$	$\frac{-\mathrm{i}k_y}{4\pi\mathrm{i}\omega\epsilon_i}$	$\frac{-1}{4\pi\mathrm{i}\omega\mu_i u_i}$
L_i	$-\frac{\mathrm{i}k_y}{4\pi u_i}$	$\frac{\mathrm{i}k_x}{4\pi u_i}$	$\frac{\mathrm{i}k_x}{4\pi\mathrm{i}\omega\epsilon_i}$	$\frac{\mathrm{i}k_y}{4\pi\mathrm{i}\omega\epsilon_i}$	$\frac{-1}{4\pi\mathrm{i}\omega\mu_i u_i}$
S_i	$\begin{pmatrix} 0 \\ \frac{\mathrm{i}k_y}{2\pi u_i} \end{pmatrix}$	$\begin{pmatrix} 0 \\ \frac{-\mathrm{i}k_x}{2\pi u_i} \end{pmatrix}$	$\begin{pmatrix} \frac{-\mathrm{i}k_x}{2\pi\mathrm{i}\omega\epsilon_i} \\ 0 \end{pmatrix}$	$\begin{pmatrix} \frac{-\mathrm{i}k_y}{2\pi\mathrm{i}\omega\epsilon_i} \\ 0 \end{pmatrix}$	$\begin{pmatrix} 0 \\ \frac{1}{2\pi\mathrm{i}\omega\mu_i u_i} \end{pmatrix}$

At the location $z = z'$ (or $z_i = D$) inside the ith layer with (ϵ_i, μ_i), similar to (9.65), (9.66) and (9.67), (9.68), we have the following interface conditions for a horizontally directed unit-Hertz dipole source:

$$\widetilde{H}_z^+ = \widetilde{H}_z^-, \tag{9.84}$$

$$\dot{\widetilde{H}}_z^+ = \dot{\widetilde{H}}_z^- + \widetilde{J}_s, \tag{9.85}$$

$$\epsilon_i \widetilde{E}_z^+ = \epsilon_i \widetilde{E}_z^- + \widetilde{\rho}_{e(s)}, \tag{9.86}$$

$$\dot{\widetilde{E}}_z^+ = \dot{\widetilde{E}}_z^-, \tag{9.87}$$

where + indicates the value at z'^+, and

$$\widetilde{J}_s = \mathrm{i}k_y \widetilde{J}_{e(s),x} - \mathrm{i}k_x \widetilde{J}_{e(s),y}, \tag{9.88}$$

and $\widetilde{\rho}_{e(s)}$ is given in (9.69) or (9.70). For a z-directed Hertz dipole, we can solve for $\widetilde{G}_E^{zz}$ from the z-component of the vector wave equation (9.24) with the help of (9.55) and (9.56) and the fact that $\widetilde{G}_H^{zz} = 0$ as discussed at the end of Sect. 9.3.1.

Applying the interface conditions (9.84)–(9.87), we arrive at the following relation for the coefficients of ψ_i in (9.80) at $z_i = D$:

$$V_i^U = C_i V_i^L + S_i, \tag{9.89}$$

$$C_i = \begin{bmatrix} \cosh u_i D, \ \sinh u_i D \\ \sinh u_i D, \ \cosh u_i D \end{bmatrix}, \ S_i = \begin{pmatrix} -L_i + U_i \\ -L_i - U_i \end{pmatrix}. \tag{9.90}$$

Table 9.1 lists the corresponding U_i and L_i in (9.81) and S_i in (9.90) when $\left(-\frac{1}{\mathrm{i}\omega\mu}\right)\cdot$ ψ can represent $\widetilde{G}_H^{zx}$, $\widetilde{G}_H^{zy}$, $\widetilde{G}_E^{zx}$, $\widetilde{G}_E^{zy}$, and $\widetilde{G}_E^{zz}$.

Using the recursive relation (9.77), we arrive at (note that $\widetilde{G}_H^{zz} = 0$):

$$\psi_j = (-\mathrm{i}\omega\mu)\,\{\widetilde{G}_H^{zx}, \widetilde{G}_H^{zy}, \widetilde{G}_E^{zx}, \widetilde{G}_E^{zy}, \widetilde{G}_E^{zz}\} = \left(c_j, s_j\right) V_j, \tag{9.91}$$

where

$$\left(c_j, s_j\right) = \left(\cosh u_j z_j, \sinh u_j z_j\right) \tag{9.92}$$

and

$$V_j = \begin{pmatrix} a_j \\ b_j \end{pmatrix} = \begin{cases} \left(\prod\limits_{l=j+1}^{N} T_{l-1,l} \right) V_N, & j > i, \\ \left(\prod\limits_{l=j+1}^{i} T_{l-1,l} \right) \left[S_i + \left(C_i \prod\limits_{l=i+1}^{N} T_{l-1,l} \right) V_N \right], & j \le i, \end{cases} \tag{9.93}$$

$$V_N = \begin{pmatrix} a_N \\ b_N \end{pmatrix} = \begin{cases} B_N \begin{pmatrix} \eta_H \\ 1 \end{pmatrix}, & \text{for TE waves,} \\ A_N \begin{pmatrix} 1 \\ \eta_E \end{pmatrix}, & \text{for TM waves,} \end{cases} \tag{9.94}$$

with A_N, B_N, η_E, and η_H being defined by the following formulae:

$$A_N = \frac{-(e_1 + e_2)}{\tau_{11} + \tau_{21} + \eta_E (\tau_{12} + \tau_{22})}, \quad B_N = \frac{-(e_1 + e_2)}{\tau_{12} + \tau_{22} + \eta_H (\tau_{11} + \tau_{21})},$$

$$\eta_H = \mathrm{i} Z_s Y_N^{\mathrm{TE}}, \qquad \eta_E = \mathrm{i} Z_s Y_N^{\mathrm{TM}}, \tag{9.95}$$

where

$$Y_N^{\mathrm{TM}} = \frac{\omega \epsilon_N}{u_N}, \qquad Y_N^{\mathrm{TE}} = \frac{u_N}{\omega \mu_N},$$

$$\begin{pmatrix} e_1 \\ e_2 \end{pmatrix} = T^+ S_i, \; T = \begin{bmatrix} \tau_{11} & \tau_{12} \\ \tau_{21} & \tau_{22} \end{bmatrix} = T^+ C_i T^-,$$

$$T^+ = \begin{bmatrix} t_{11}^+ & t_{12}^+ \\ t_{21}^+ & t_{22}^+ \end{bmatrix} = \prod_{l=0}^{i-1} T_{l,l+1}, \; T^- = \begin{bmatrix} t_{11}^- & t_{12}^- \\ t_{21}^- & t_{22}^- \end{bmatrix} = \prod_{l=i}^{N-1} T_{l,l+1}.$$

In (9.95), Z_s represents the boundary intrinsic impedance at $z = d_N$. In particular, if the boundary is a perfect electric conductor (PEC) plane, then $Z_s = 0$.

In addition, some attention should be paid to the definition of $T_{i-1,i}$, since it is different from the definition of other $T_{l,l+1}$ in (9.78):

$$T_{i-1,i} = \begin{bmatrix} \frac{\alpha_i}{\alpha_{i-1}} \cosh u_i (h_i - D), & \frac{\alpha_i}{\alpha_{i-1}} \sinh u_i (h_i - D) \\ \frac{u_i}{u_{i-1}} \sinh u_i (h_i - D), & \frac{u_i}{u_{i-1}} \cosh u_i (h_i - D) \end{bmatrix}. \tag{9.96}$$

The coefficients in the top and bottom layers (if not terminated by a PEC ground plane at $z = 0$ or $z = d_N$) can be determined by the two algebraic equations resulting from (9.93) for $j = 0$ and the fact that $a_0 + b_0 = 0$, $a_N - b_N = 0$ from the boundedness of the fields as $z \to \infty$, $z \to -\infty$, respectively. For a PEC boundary termination at the bottom (or top) layer, we should have $\widetilde{G}_E^{xt} = 0$, $\widetilde{G}_E^{yt} = 0$ and $\widetilde{G}_H^{zt} = 0$, $t = x, y, z$, which also implies from Gauss's law for the electric field that $\frac{\partial}{\partial z}\widetilde{G}_E^{zt} = 0$. Then, boundary conditions in (9.94) can be then used at $z = d_N$.

Finally, we have Green's functions in the spectral domain,

$$\widetilde{\overline{\mathbf{G}}}_E = \begin{bmatrix} \widetilde{G}_E^{xx} & \widetilde{G}_E^{xy} & \widetilde{G}_E^{xz} \\ \widetilde{G}_E^{yx} & \widetilde{G}_E^{yy} & \widetilde{G}_E^{yz} \\ \widetilde{G}_E^{zx} & \widetilde{G}_E^{zy} & \widetilde{G}_E^{zz} \end{bmatrix}, \quad \widetilde{\overline{\mathbf{G}}}_H = \begin{bmatrix} \widetilde{G}_H^{xx} & \widetilde{G}_H^{xy} & \widetilde{G}_H^{xz} \\ \widetilde{G}_H^{yx} & \widetilde{G}_H^{yy} & \widetilde{G}_H^{yz} \\ \widetilde{G}_H^{zx} & \widetilde{G}_H^{zy} & \widetilde{G}_H^{zz} \end{bmatrix}, \tag{9.97}$$

where the components of Green's functions are given as follows.

(a) HED, J_x source:

$$\begin{aligned} &\widetilde{G}_E^{xx} = (-\mathrm{i}k_x)^2 \widetilde{G}_{exx1} - \left(-\mathrm{i}k_y\right)^2 \widetilde{G}_{exx2}, \qquad \widetilde{G}_H^{xx} = (-\mathrm{i}k_x)\left(-\mathrm{i}k_y\right) \widetilde{G}_{azx}, \\ &\widetilde{G}_E^{yx} = (-\mathrm{i}k_x)\left(-\mathrm{i}k_y\right) \widetilde{G}_{eyx}, \qquad \widetilde{G}_H^{yx} = \left(-\mathrm{i}k_y\right)^2 \widetilde{G}_{hyx_1} - \left(-\mathrm{i}k_y\right)^2 \widetilde{G}_{hyx_2}, \\ &\widetilde{G}_E^{zx} = -\mathrm{i}k_x \widetilde{G}_2, \widetilde{G}_H^{zx} = -\mathrm{i}k_y \widetilde{G}_1. \end{aligned} \tag{9.98}$$

(b) HED, J_y source:

$$\begin{aligned} &\widetilde{G}_E^{xy} = \widetilde{G}_E^{yx}, \qquad \widetilde{G}_H^{xy} = -\left(-\mathrm{i}k_y\right)^2 \widetilde{G}_{hyx_1} + \left(-\mathrm{i}k_y\right)^2 \widetilde{G}_{hyx_2}, \\ &\widetilde{G}_E^{yy} = \left(-\mathrm{i}k_y\right)^2 \widetilde{G}_{exx_1} - (-\mathrm{i}k_x)^2 \widetilde{G}_{exx_2}, \qquad \widetilde{G}_H^{yy} = -\left(-\mathrm{i}k_x\right)\left(-\mathrm{i}k_y\right) \widetilde{G}_{azx}, \\ &\widetilde{G}_H^{zy} = +\mathrm{i}k_x \widetilde{G}_1, \qquad \widetilde{G}_E^{zy} = -\mathrm{i}k_y \widetilde{G}_2. \end{aligned} \tag{9.99}$$

(c) VED (vertical electric dipole), J_z source:

$$\begin{aligned} &\widetilde{G}_E^{xz} = -\mathrm{i}k_x \widetilde{G}_{exz}, \qquad \widetilde{G}_H^{xz} = -\mathrm{i}\omega\epsilon\left(-\mathrm{i}k_y\right) \widetilde{G}_{hxz}, \\ &\widetilde{G}_E^{yz} = -\mathrm{i}k_y \widetilde{G}_{exz}, \qquad \widetilde{G}_H^{yz} = -\mathrm{i}\omega\epsilon\left(-\mathrm{i}k_x\right) \widetilde{G}_{hxz}, \\ &\widetilde{G}_H^{zz} = 0, \qquad \widetilde{G}_E^{zz} = \widetilde{G}_3. \end{aligned} \tag{9.100}$$

In the above formulations, $\widetilde{G}_1$, $\widetilde{G}_2$, $\widetilde{G}_3$, $\widetilde{G}_{exx_1}$, $\widetilde{G}_{exx_2}$, $\widetilde{G}_{eyx}$, $\widetilde{G}_{hyx_1}$, $\widetilde{G}_{hyx_2}$, $\widetilde{G}_{azx}$, $\widetilde{G}_{exz}$, and $\widetilde{G}_{hxz}$ are radially symmetric, and their inverse Fourier transforms can be calculated by Hankel transforms. Their explicit formulae [2] are given in the appendix to this chapter, Sect. 9.6.

9.3 Dyadic Green's Functions for Vector Potentials $\overline{\mathbf{G}}_A(\mathbf{r}|\mathbf{r}')$

In this section, we derive the dyadic Green's functions for the vector and the scalar potentials, which will be used for the mixed potential integral equation (MPIE) of scattering problems in Chap. 11. The approach here is to express Green's functions for potentials in terms of the field components, specifically only the z-components of Green's functions for the **E** and **H** fields. It is noted that the vector potentials used to represent the magnetic field **H** in (8.52) are mathematically not unique if no boundary condition on **A** is specified, as $\mathbf{A} + \nabla\phi$, with ϕ being any scalar function, also satisfies (8.52) if **A** does. Therefore, there are many ways of defining the vector potentials. Two popular approaches are the Sommerfeld potentials [11], used to study the Hertz dipole antenna, and transverse potentials [14]. Other potentials include Hertz–Debye potentials [18]. Based on the fact that only two components of the magnetic field are independent, then only two components of the magnetic field are independent in Maxwell's equations; thus only two components of these vector potentials are sufficient. Dyadic Green's functions can also be derived using vector wave functions as in [19].

9.3.1 Sommerfeld Potentials

In the Sommerfeld potential formulation [11], it is stipulated that the electromagnetic fields from a HED can be represented by a horizontal component of **A** in the same direction of the HED and a z-component of **A**, while the fields from a VED can be represented by just the z-component of **A**. So the dyadic Green's function $\overline{\mathbf{G}}_A$ for the Sommerfeld vector potential **A** has the following form:

$$\overline{\mathbf{G}}_A = \left(\hat{x}G_A^{xx} + \hat{z}G_A^{zx}\right)\hat{x} + \left(\hat{y}G_A^{yy} + \hat{z}G_A^{zy}\right)\hat{y} + \hat{z}G_A^{zz}\hat{z}, \tag{9.101}$$

or the matrix form

$$\overline{\mathbf{G}}_A = \begin{bmatrix} G_A^{xx} & 0 & 0 \\ 0 & G_A^{yy} & 0 \\ G_A^{zx} & G_A^{zy} & G_A^{zz} \end{bmatrix}. \tag{9.102}$$

To derive the components of $\overline{\mathbf{G}}_A$, we consider (8.41) in the Fourier spectral domain, namely,

$$\widetilde{\mathbf{E}} = -\mathrm{i}\omega\widetilde{\mathbf{A}} - \left(\mathrm{i}k_x, \mathrm{i}k_y, \frac{\partial}{\partial z}\right)\widetilde{V}_e. \tag{9.103}$$

(a) **HED, J_x source**

In this case, we have

$$\widetilde{\mathbf{A}} = \left(\widetilde{G}_A^{xx}, 0, \widetilde{G}_A^{zx}\right). \tag{9.104}$$

From the Lorenz gauge condition (8.46), we get

$$\widetilde{V}_e = -\frac{1}{\mathrm{i}\omega\epsilon\mu}\left(\mathrm{i}k_x\widetilde{G}_A^{xx} + \frac{\partial}{\partial z}\widetilde{G}_A^{zx}\right). \tag{9.105}$$

Taking the z-component of (9.103) and using (9.105), we have

$$\widetilde{G}_E^{zx} = -\mathrm{i}\omega\widetilde{G}_A^{zx} - \frac{\partial}{\partial z}\widetilde{V}_e = -\mathrm{i}\omega\widetilde{G}_A^{zx} + \frac{k_x}{\omega\epsilon\mu}\frac{\partial}{\partial z}\widetilde{G}_A^{xx} + \frac{1}{\mathrm{i}\omega\epsilon\mu}\frac{\partial^2}{\partial z^2}\widetilde{G}_A^{zx}. \tag{9.106}$$

Meanwhile, (8.49) in the spectral domain for the z-component $\widetilde{G}_A^{zx}$ yields

$$\frac{\partial^2}{\partial z^2}\widetilde{G}_A^{zx} = \left(k_\rho^2 - k^2\right)\widetilde{G}_A^{zx}. \tag{9.107}$$

Substituting (9.107) into (9.106), we obtain

$$k_\rho^2\widetilde{G}_A^{zx} = \mathrm{i}\omega\epsilon\mu\widetilde{G}_E^{zx} - \mathrm{i}k_x\frac{\partial}{\partial z}\widetilde{G}_A^{xx}. \tag{9.108}$$

In the spectral domain, (8.52) becomes

$$\widetilde{\mathbf{H}} = \frac{1}{\mu}\widetilde{\nabla}\times\widetilde{\mathbf{A}}, \tag{9.109}$$

where $\widetilde{\nabla} = (\mathrm{i}k_x, \mathrm{i}k_y, \partial/\partial z)$. From the z-component of this equation, we have

$$\widetilde{G}_A^{xx} = -\frac{\mu}{\mathrm{i}k_y}\widetilde{G}_H^{zx}. \tag{9.110}$$

Substituting (9.110) into (9.108), we obtain

$$k_\rho^2\widetilde{G}_A^{zx} = \mathrm{i}\omega\epsilon\mu\widetilde{G}_E^{zx} + \frac{k_x}{k_y}\mu\frac{\partial}{\partial z}\widetilde{G}_H^{zx}. \tag{9.111}$$

(b) **HED, J_y source**

In this case, we have

$$\widetilde{\mathbf{A}} = \left(0, \widetilde{G}_A^{yy}, \widetilde{G}_A^{zy}\right). \tag{9.112}$$

Similarly, we can express $\widetilde{G}_A^{yy}$ and $\widetilde{G}_A^{zy}$ in terms of the z-components of the **E** and **H** fields as

$$\widetilde{G}_A^{yy} = \frac{\mu}{\mathrm{i}k_x}\widetilde{G}_H^{zy} \tag{9.113}$$

and

$$k_\rho^2\widetilde{G}_A^{zy} = \mathrm{i}\omega\epsilon\mu\widetilde{G}_E^{zy} - \frac{k_y}{k_x}\mu\frac{\partial}{\partial z}\widetilde{G}_H^{zy}. \tag{9.114}$$

(c) **VED, J_z source**

As for a $\hat{z}$-directed Hertz dipole current source, from the vector wave equation for the **H** field (8.31), H_z satisfies a homogeneous scalar Helmholtz equation and continuous interface conditions at layer interfaces; thus we have $H_z = 0$, and, as a result, $\widetilde{G}_H^{zz} = 0$. Therefore, we have

$$\widetilde{\mathbf{A}} = \left(0, 0, \widetilde{G}_A^{zz}\right),$$

which is consistent with the scalar Helmholtz equation (8.49) for components of **A**. The z-component $\widetilde{G}_A^{zz}$ can be similarly derived as in (a) and (b), leading to

$$k_\rho^2 \widetilde{G}_A^{zz} = \mathrm{i}\omega\epsilon\mu \widetilde{G}_E^{zz}. \tag{9.115}$$

9.3.2 Transverse Potentials

In the transverse potential formulation [14], HED-generated electromagnetic fields are represented by two transverse components of **A**, while VED-generated fields are still represented by only the z-component of **A**. So the dyadic Green's function $\overline{\mathbf{G}}_A$ for the transverse vector potential **A** has the following form:

$$\overline{\mathbf{G}}_A = \left(\hat{x}G_A^{xx} + \hat{y}G_A^{yx}\right)\hat{x} + \left(\hat{x}G_A^{xy} + \hat{y}G_A^{yy}\right)\hat{y} + \hat{z}G_A^{zz}\hat{z},$$

or the matrix form

$$\overline{\mathbf{G}}_A = \begin{bmatrix} G_A^{xx} & G_A^{xy} & 0 \\ G_A^{yx} & G_A^{yy} & 0 \\ 0 & 0 & G_A^{zz} \end{bmatrix}. \tag{9.116}$$

By a discussion similar to that for the Sommerfeld potential, we can obtain the following identities.

(a) **HED, J_t source, $t = x, y$**

$$k_\rho^2 \widetilde{G}_A^{xt} = \frac{\omega\epsilon\mu k_x}{u^2}\frac{\partial}{\partial z}\widetilde{G}_E^{zt} + \mathrm{i}\mu k_y \widetilde{G}_H^{zt}, \tag{9.117}$$

$$k_\rho^2 \widetilde{G}_A^{yt} = \frac{\omega\epsilon\mu k_y}{u^2}\frac{\partial}{\partial z}\widetilde{G}_E^{zt} - \mathrm{i}\mu k_x \widetilde{G}_H^{zt}, \tag{9.118}$$

where $u^2 = k_\rho^2 - k^2$.

(b) **VED, J_z source**

$$k_\rho^2 \widetilde{G}_A^{zz} = \mathrm{i}\omega\epsilon\mu \widetilde{G}_E^{zz}. \tag{9.119}$$

9.4 Fast Computation of Dyadic Green's Functions

In this section, we describe fast algorithms for computing the Hankel transform needed for the physical form of the dyadic Green's functions in layered media. The kernel of the Hankel transform contains a Bessel function which oscillates quickly (especially for large ρ) and decays slowly (especially when $z \sim z'$). Moreover, the spectral Green's functions are not well-behaved for high-frequency scattering. General fast algorithms for numerical evaluation of the Hankel transform can be found in [16] and in the references in [17]. Other semi-analytical methods have also been proposed, including the complex image method (CIM) [5], an approach based on the Prony method and the well-known Sommerfeld identity [11], and integration along the steepest-descent path (SDP) for a half-space problem [6]. In this section, we will describe a window-function-based filtering approach [1] to speed up the slow decay of the Hankel kernel, thus reducing the length of the contour of the contour integral in the Hankel transform. With a shortened contour length, the fast Hankel transform techniques [16, 17] can be used to speed up the computations further.

Firstly, a radially symmetric bell-shaped window function $\psi_a(x, y)$ with a support size a is defined as

$$\psi_a(x, y) = \psi_a(\rho) = \begin{cases} \left[1 - \left(\dfrac{\rho}{a}\right)^2\right]^m, & \text{if } \rho \leq a, \\ 0, & \text{otherwise,} \end{cases} \tag{9.120}$$

where $\rho = \sqrt{x^2 + y^2}$, and $m > 0$ is the order of the window function.

The 2-D Fourier transform of the window function $\psi_a(x, y)$ is

$$\widetilde{\psi}_a\left(k_x, k_y\right) = \mathcal{F}\{\psi_a(x, y)\} = \frac{1}{2\pi} \int_{\infty}^{\infty} \int_{\infty}^{\infty} \psi_a(x, y) e^{-i\left(k_x x + k_y y\right)} dx dy, \tag{9.121}$$

which can also be conveniently expressed in terms of the zeroth-order Hankel transform:

$$\widetilde{\psi}_a(k_\rho) = S_0\left[\psi_a(\rho)\right](k_\rho) = \int_0^a \psi_a(\rho) J_0(k_\rho \rho) \rho d\rho. \tag{9.122}$$

It should be noted that the window function $\psi_a(x, y)$ defined above has some unique properties in both the spatial and the spectral domains. In addition to the compact support in the physical space, the window function $\psi_a(x, y)$ is smooth, especially when the order m is large, and its spectral form decays quickly at high frequency and thus gives a low pass filter. Specifically, $\widetilde{\psi}_a(k_\rho)$ has the following property.

Lemma 9.1 *The Hankel transform $\widetilde{\psi}_a(k_\rho)$ for $\psi_a(\rho)$ has the following decaying property:*

$$\widetilde{\psi}_a(k_\rho) = o\left(\frac{1}{k_\rho^m}\right) \quad as\ k_\rho \to +\infty. \tag{9.123}$$

Proof As the window function $\psi_a(x, y)$ is radially symmetric, so is its Fourier transform $\widetilde{\psi}_a\left(k_x, k_y\right)$. Therefore, we have

$$\widetilde{\psi}_a(k_x, k_y) = \widetilde{\psi}_a(k_\rho, 0),$$

namely,

$$\begin{aligned}\widetilde{\psi}_a(k_\rho) = \widetilde{\psi}_a(k_\rho, 0) &= \frac{1}{2\pi}\int_{-\infty}^{\infty}\int_{-\infty}^{\infty}\psi_a(x, y)\mathrm{e}^{-\mathrm{i}k_\rho x}\mathrm{d}x\mathrm{d}y \\ &= \frac{1}{2\pi}\int_{-\infty}^{\infty}\mathrm{d}y\int_{-\infty}^{\infty}\psi_a(x, y)\mathrm{e}^{-\mathrm{i}k_\rho x}\mathrm{d}x.\end{aligned}$$

Due to the compact support of $\psi_a(x, y)$, we then have

$$\begin{aligned}\widetilde{\psi}_a(k_\rho) &= \frac{1}{2\pi}\int_{-\infty}^{\infty}\mathrm{d}y\int_{-\infty}^{\infty}\psi_a(x, y)\mathrm{e}^{-\mathrm{i}k_\rho x}\mathrm{d}x \\ &= \frac{1}{2\pi}\int_{-\infty}^{\infty}\mathrm{d}y\left[\psi_a(x, y)\left(-\frac{\mathrm{e}^{-\mathrm{i}k_\rho x}}{\mathrm{i}k_\rho}\right)\Bigg|_{-\infty}^{\infty} + \frac{1}{\mathrm{i}k_\rho}\int_{-\infty}^{\infty}\frac{\partial\psi_a}{\partial x}\mathrm{e}^{-\mathrm{i}k_\rho x}\mathrm{d}x\right] \\ &= \frac{1}{2\pi}\int_{-\infty}^{\infty}\mathrm{d}y\left(\frac{1}{\mathrm{i}k_\rho}\int_{-\infty}^{\infty}\frac{\partial\psi_a}{\partial x}\mathrm{e}^{-\mathrm{i}k_\rho x}\mathrm{d}x\right).\end{aligned}$$

Continuing to integrate by parts, we get

$$\widetilde{\psi}_a(k_\rho) = \frac{1}{2\pi}\int_{-\infty}^{\infty}\mathrm{d}y\left[\left(\frac{1}{\mathrm{i}k_\rho}\right)^m\int_{-\infty}^{\infty}\frac{\partial^m\psi_a}{\partial x^m}\mathrm{e}^{-\mathrm{i}k_\rho x}\mathrm{d}x\right].$$

We stop at the mth step since the mth derivative of the window function is discontinuous. Now, since

$$\int_{-\infty}^{\infty}\frac{\partial^m\psi_a}{\partial x^m}\mathrm{e}^{-\mathrm{i}k_\rho x}\mathrm{d}x \to 0 \text{ as } k_\rho \to \infty,$$

we have

$$\widetilde{\psi}_a(k_\rho) = o\left(\frac{1}{k_\rho^m}\right).$$

□

In addition, by the convolution theorem of the Fourier transform and (9.41), we immediately have the following lemma.

Lemma 9.2 *For any cylindrically symmetric function $f(x, y)$, the convolution $f * \psi_a$ can be written as*

$$f(x, y) * \psi_a(x, y) = S_0\left[\tilde{f}(k_\rho)\widetilde{\psi}_a(k_\rho)\right](\rho), \tag{9.124}$$

where

$$\tilde{f}(k_\rho) = S_0\left[f(\rho)\right](k_\rho).$$

To recover the value of $f(x, y)$ from its Hankel transform, we also need the following result.

Lemma 9.3 *Let $f(x, y)$ be a C^2 function. Then it can be shown that*

$$f(x, y) * \psi_a(x, y) = M_0 f(x, y) + \frac{M_2}{2}\left[\frac{\partial^2 f}{\partial x^2}(x, y) + \frac{\partial^2 f}{\partial y^2}(x, y)\right] + O(a^6), \tag{9.125}$$

where

$$M_0 = \frac{1}{2\pi}\iint_{\rho\leq a} \psi_a(x, y)\mathrm{d}x\,\mathrm{d}y = \frac{a^2}{2(m+1)},$$

$$M_2 = \frac{1}{2\pi}\iint_{\rho\leq a} \psi_a(x, y)x^2\,\mathrm{d}x\mathrm{d}y = O(a^4).$$

The proof is elementary and thus omitted here.

From (9.125), we have

$$f(x, y) = \frac{1}{M_0} f(x, y) * \psi_a(x, y) + O(a^2). \tag{9.126}$$

So, as a result of (9.124) and (9.126), we can approximate $f(x, y)$ as

$$f(x, y) = \frac{1}{M_0} S_0\left[\tilde{f}(k_\rho)\widetilde{\psi}_a(k_\rho)\right](\rho) + O(a^2), \quad \text{as } a \to 0. \tag{9.127}$$

Window-function-based acceleration algorithms are described in the following.

Algorithm I Fast algorithm for $G(\rho, z; z')$ with $\rho \geq a$.

If $\rho > a$, then, as $a \to 0$, we have

$$G(\rho, z; z') = \frac{1}{M_0} W_0(\rho) + O(a^2), \tag{9.128}$$

where

$$W_0(\rho) = S_0 \left[\widetilde{G} \left(k_\rho, z; z'\right) \widetilde{\psi} \left(k_\rho\right)\right] (\rho). \tag{9.129}$$

Proof Algorithm I is the direct result of (9.127) applied to $G(\rho, z; z')$. □

Algorithm I requires that $\rho \geq a$, as otherwise Green's functions will not be smooth and the estimate in (9.125) will be invalid. Therefore, to apply the approximation (9.127) to the function $G(\rho, z; z')$ for $\rho < a$, we rewrite $G(\rho, z; z')$ as

$$G(\rho, z; z') = \frac{r^2 G(\rho, z; z')}{r^2} \equiv \frac{G_2(\rho, z; z')}{r^2}, \tag{9.130}$$

where $r = \sqrt{\rho^2 + (z - z')^2}$. From the singularity property of Green's functions of the vector and the scalar potentials, we can assume that $G_2(\rho, z; z') = r^2 G(\rho, z; z')$ is a smooth function, and the approximation (9.127) can thus be used.

Algorithm II Fast algorithm for $G(\rho, z; z')$ with $\rho > 0$.

If $\rho > 0$, then, as $a \to 0$, we have

$$G(\rho, z; z') = \frac{1}{M_0 r^2} \left[r^2 W_0(\rho) - 2\rho W_1(\rho) + W_2(\rho)\right] + O(a^2), \tag{9.131}$$

where $r = \sqrt{\rho^2 + (z - z')^2}$, $W_0(\rho)$ is defined in (9.129), and

$$W_1(\rho) = S_1 \left[\widetilde{G}(k_\rho, z; z')\widetilde{\psi}_1(k_\rho)/k_\rho\right] (\rho), \tag{9.132}$$

$$W_2(\rho) = S_0 \left[\widetilde{G}(k_\rho, z; z')\widetilde{\psi}_2(k_\rho)\right] (\rho), \tag{9.133}$$

where

$$\widetilde{\psi}_1(k_\rho) = S_1 \left[\psi(\rho)\right] = \int_0^a \psi(\rho) J_1(k_\rho \rho)\rho^2 \mathrm{d}\rho, \tag{9.134}$$

$$\widetilde{\psi}_2(k_\rho) = S_0 \left[\psi(\rho)\rho^2\right] = \int_0^a \psi(\rho) J_0(k_\rho \rho)\rho^3 \mathrm{d}\rho. \tag{9.135}$$

Proof As $G_2(\rho, z; z') = r^2 G(\rho, z; z')$ and $\Delta z = z - z'$, from the definition of the convolution and using

$$x'^2 = x^2 - 2x(x - x') + (x - x')^2,$$

$$y'^2 = y^2 - 2y(y-y') + (y-y')^2,$$

we have

$$\begin{aligned}
G_2 * \psi_a =& \frac{1}{2\pi}\iint_{\mathbb{R}^2}\left[x'^2 + y'^2 + \Delta z^2\right] G(x', y')\psi_a(x-x', y-y')\mathrm{d}x'\,\mathrm{d}y' \\
=& \left[x^2 + y^2 + \Delta z^2\right]\frac{1}{2\pi}\iint_{\mathbb{R}^2} G(x', y')\psi_a(x-x', y-y')\mathrm{d}x'\mathrm{d}y' \\
& - 2x\frac{1}{2\pi}\iint_{\mathbb{R}^2} G(x', y')(x-x')\psi_a(x-x', y-y')\mathrm{d}x'\,\mathrm{d}y' \\
& - 2y\frac{1}{2\pi}\iint_{\mathbb{R}^2} G(x', y')(y-y')\psi_a(x-x', y-y')\mathrm{d}x'\,\mathrm{d}y' \\
& + \frac{1}{2\pi}\iint_{\mathbb{R}^2} G(x', y')\left[(x-x')^2 + (y-y')^2\right]\psi_a(x-x', y-y')\mathrm{d}x'\mathrm{d}y' \\
=& r^2\left[G * \psi_a\right] - 2x\left[G * (x\psi_a)\right] - 2y\left[G * (y\psi_a)\right] + G * \left[(x^2+y^2)\psi_a\right].
\end{aligned} \tag{9.136}$$

On the one hand, from (9.124), we have

$$G * \psi_a = S_0\left[\widetilde{G}\left(k_\rho, z; z'\right)\widetilde{\psi}\left(k_\rho\right)\right](\rho) = W_0(\rho). \tag{9.137}$$

On the other hand, using the integral definition of the Bessel function

$$J_n(z) = \frac{\mathrm{i}^{-n}}{\pi}\int_0^\pi \mathrm{e}^{\mathrm{i}z\cos\theta}\cos n\theta \mathrm{d}\theta,$$

we can show that

$$V_1 \equiv G * (x\psi_a) = \cos\alpha\int_0^\infty \widetilde{G}(k_\rho, z; z')J_1(\rho k_\rho)\widetilde{\psi}_1(k_\rho)k_\rho \mathrm{d}k_\rho,$$

$$V_2 \equiv G * (y\psi_a) = \sin\alpha\int_0^\infty \widetilde{G}(k_\rho, z; z')J_1(\rho k_\rho)\widetilde{\psi}_1(k_\rho)k_\rho \mathrm{d}k_\rho,$$

and consequently

$$\begin{aligned}
xV_1 + yV_2 &= (x\cos\alpha + y\sin\alpha)\int_0^\infty \widetilde{G}(k_\rho, z; z')J_1(\rho k_\rho)\frac{\widetilde{\psi}_1(k_\rho)}{k_\rho}k_\rho^2 \mathrm{d}k_\rho \\
&= \rho S_1\left[\widetilde{G}(k_\rho, z; z')\widetilde{\psi}_1(k_\rho)/k_\rho\right](\rho) = \rho W_1(\rho).
\end{aligned} \tag{9.138}$$

Similarly, we can show that

$$G * \left[(x^2 + y^2)\psi_a\right] = S_0\left[\widetilde{G}(k_\rho, z; z')\widetilde{\psi}_2(k_\rho)\right] = W_2(\rho). \tag{9.139}$$

Substituting (9.137)–(9.139) into (9.136), we have

$$G_2 * \psi_a = r^2 W_0(\rho) - 2\rho W_1(\rho) + W_2(\rho). \tag{9.140}$$

Finally, combining (9.126) applied to $G_2(\rho, z; z')$ and (9.140), we have the proof of (9.131). □

Lemma 9.4 *The functions $\widetilde{\psi}_1(k_\rho)$ and $\widetilde{\psi}_2(k_\rho)$ have the following decaying property:*

$$\widetilde{\psi}_1(k_\rho) = o\left(\frac{1}{k_\rho^{m-1}}\right), \quad \widetilde{\psi}_2(k_\rho) = o\left(\frac{1}{k_\rho^{m-1}}\right), \text{ as } k_\rho \to +\infty. \tag{9.141}$$

As a result of the fast decay of the functions $\widetilde{\psi}(k_\rho)$, $\widetilde{\psi}_1(k_\rho)$, and $\widetilde{\psi}_2(k_\rho)$, the contour integration in the definition of $W_0(\rho)$, $W_1(\rho)$, and $W_2(\rho)$ can be carried out efficiently with a much shorter contour in the k_ρ space.

Finally, higher-order fast algorithms can be developed using the Richardson extrapolation with two different window sizes in Algorithm I. For instance, we can arrive at the following fourth-order algorithm.

Algorithm III Fourth-order accurate fast algorithm for $G(\rho, z; z')$.

If $\rho > \gamma a$, $1 < \gamma < 2$, then, as $a \to 0$, we have

$$G(\rho, z; z') = \frac{1}{\gamma^2 - 1}\left[\frac{\gamma^2}{M_0(a)} W_0(\rho, a) - \frac{1}{M_0(\gamma a)} W_0(\rho, \gamma a)\right] + O(a^4). \tag{9.142}$$

For $\rho \le \gamma a$, a similar $O(a^4)$ version of Algorithm II can also be obtained.

9.5 Summary

Green's functions and their computation are the main components of integral equation methods for wave scattering, and they also influence the solution procedure for the resulting linear systems. For the dyadic Green's functions of Maxwell's equations in layered media, using the spectral domain is the usual way to derive their analytical forms in terms of Hankel transforms involving Sommerfeld integrals. The slow convergence of these integrals due to the oscillatory Hankel kernel is improved by a window function-based filtering technique in the spectral domain.

9.6 Appendix: Formula for $\widetilde{G}_1$, $\widetilde{G}_2$, and $\widetilde{G}_3$, etc.

We introduce the following shorthand:

$$T_j = \begin{bmatrix} t_{11} & t_{12} \\ t_{21} & t_{22} \end{bmatrix} = \prod_{l=j+1}^{N} T_{l-1,l}, \tag{9.143}$$

$$T_{ij}^{\vee} = \begin{bmatrix} t_{11}^{\vee} & t_{12}^{\vee} \\ t_{21}^{\vee} & t_{22}^{\vee} \end{bmatrix} = \prod_{l=j}^{i-1} T_{l,l+1}, \quad T_{ij}^{\wedge} = \begin{bmatrix} t_{11}^{\wedge} & t_{12}^{\wedge} \\ t_{21}^{\wedge} & t_{22}^{\wedge} \end{bmatrix} = T_{ij}^{\vee} C_i T^{-}. \tag{9.144}$$

In fact, among the 11 quantities for Green's functions in the spectral domain as discussed at the end of Sect. 9.2.5, only three are independent, and they are $\widetilde{G}_1$, $\widetilde{G}_2$, and $\widetilde{G}_3$, which are explicitly listed below [2]:

$$\widetilde{G}_k = (c_j, s_j) \begin{pmatrix} a_{jk} \\ b_{jk} \end{pmatrix}, k = 1, 2, 3,$$

where (c_j, s_j) are given in (9.92), and

$$a_{j1} = \begin{cases} (t_{11}\eta_H + t_{12}) B_N, & j > i, \\ \frac{1}{2\pi u_i} t_{12}^{\vee} + (t_{11}^{\wedge}\eta_H + t_{12}^{\wedge}) B_N, & j \le i, \end{cases} \tag{9.145}$$

$$b_{j1} = \begin{cases} (t_{21}\eta_H + t_{22}) B_N, & j > i, \\ \frac{1}{2\pi u_i} t_{22}^{\vee} + (t_{21}^{\wedge}\eta_H + t_{22}^{\wedge}) B_N, & j \le i, \end{cases} \tag{9.146}$$

$$a_{j2} = \begin{cases} (t_{11} + \eta_E t_{12}) A_N, & j > i, \\ \left(-\frac{1}{2\pi \mathrm{i}\omega\epsilon_i}\right) t_{11}^{\vee} + (t_{11}^{\wedge} + t_{12}^{\wedge}\eta_E) A_N, & j \le i, \end{cases} \tag{9.147}$$

$$b_{j2} = \begin{cases} (t_{21} + \eta_E t_{22}) A_N, & j > i, \\ \left(-\frac{1}{2\pi \mathrm{i}\omega\epsilon_i}\right) t_{21}^{\vee} + (t_{21}^{\wedge} + t_{22}^{\wedge}\eta_E) A_N, & j \le i, \end{cases} \tag{9.148}$$

$$a_{j3} = \begin{cases} (t_{11} + \eta_E t_{12}) A_N, & j > i, \\ \left(-\frac{k_\rho^2}{2\pi \mathrm{i}\omega\epsilon_i u_i}\right) t_{12}^{\vee} + (t_{11}^{\wedge} + t_{12}^{\wedge}\eta_E) A_N, & j \le i, \end{cases} \tag{9.149}$$

$$b_{j3} = \begin{cases} (t_{21} + \eta_E t_{22}) A_N, & j > i, \\ \left(-\frac{k_\rho^2}{2\pi \mathrm{i}\omega\epsilon_i u_i}\right) t_{22}^{\vee} + (t_{21}^{\wedge} + t_{22}^{\wedge}\eta_E) A_N, & j \le i. \end{cases} \tag{9.150}$$

Note that T_j, $T_{ij}^{\vee}$, and $T_{ij}^{\wedge}$ are given in (9.143) and (9.144), A_N, B_N, η_E, and η_H are given in (9.95).

The other eight quantities can be derived from $\widetilde{G}_1$, $\widetilde{G}_2$, and $\widetilde{G}_3$ as follows:

$$\widetilde{G}_{exx1} = \frac{1}{k_\rho^2}\tilde{\tilde{G}}_2 = \frac{u_j}{k_\rho^2}\left(s_j, c_j\right)\begin{pmatrix} a_{j2} \\ b_{j2} \end{pmatrix}, \tag{9.151}$$

$$\widetilde{G}_{exx2} = \frac{\mathrm{i}\omega\mu_j}{k_\rho^2}\widetilde{G}_1 = \frac{\mathrm{i}\omega\mu_j}{k_\rho^2}\left(c_j, s_j\right)\begin{pmatrix} a_{j1} \\ b_{j1} \end{pmatrix}, \tag{9.152}$$

$$\widetilde{G}_{eyx} = \widetilde{G}_{exx_1} + \widetilde{G}_{exx_2}, \tag{9.153}$$

$$\widetilde{G}_{hyx_1} = \frac{1}{k_\rho^2}\tilde{\tilde{G}}_1 = \frac{u_j}{k_\rho^2}\left(s_j, c_j\right)\begin{pmatrix} a_{j1} \\ b_{j1} \end{pmatrix}, \tag{9.154}$$

$$\widetilde{G}_{hyx_2} = \frac{\mathrm{i}\omega\varepsilon_j}{k_\rho^2}\widetilde{G}_2 = \frac{\mathrm{i}\omega\varepsilon_j}{k_\rho^2}\left(c_j, s_j\right)\begin{pmatrix} a_{j2} \\ b_{j2} \end{pmatrix}, \tag{9.155}$$

$$\widetilde{G}_{azx} = \widetilde{G}_{hyx_1} + \widetilde{G}_{hyx_2}, \tag{9.156}$$

$$\widetilde{G}_{exz} = \frac{1}{k_\rho^2}\tilde{\tilde{G}}_3 = \frac{u_j}{k_\rho^2}\left(s_j, c_j\right)\begin{pmatrix} a_{j3} \\ b_{j3} \end{pmatrix}, \tag{9.157}$$

$$\widetilde{G}_{hxz} = \frac{1}{k_\rho^2}\widetilde{G}_3 = \frac{1}{k_\rho^2}\left(c_j, s_j\right)\begin{pmatrix} a_{j3} \\ b_{j3} \end{pmatrix}. \tag{9.158}$$

References

1. W. Cai, T.J. Yu, Fast calculation of dyadic Green's functions for electromagnetic scattering in a multi-layered medium. J. Comput. Phys. **165**, 1–21 (2000)
2. T.J. Yu, W. Cai, FIFA-fast interpolation and Filtering Algorithm for calculating dyadic Green's function in the electromagnetic scattering of multi-layered structures. Commun. Comput. Phys. **1**(2), 228–258 (2006)
3. T.J. Yu, W. Cai, High order window functions and fast algorithms for calculating dyadic electromagnetic Green's functions in multilayered media. Radio Sci. (Jul. 2001)
4. K. Michalski, J.R. Mosig, Multilayered media Green's functions in the integral equation formulations. IEEE Trans. Antennas Propag. **35**(11), 1299 (1987)
5. Y.L. Chow, J.J. Yang, D.G. Fang, G.E. Howard, A closed form spatial Green's function for the thick microstrip substrate. IEEE Trans. Microwave Theory Tech. **39**, 588 (1991)
6. T.J. Cui, W.C. Chew, Fast Evaluation of Sommerfeld Integrals for EM Scattering and Radiation by Three Dimensional Buried Objects, Research Report No. CCEM-34-97 (University of Illinois, Urbana, IL, 1997)
7. J. Van Bladel, *Singular Electromagnetic Fields and Sources* (IEEE Press, New York, 1991)
8. K.A. Michalski, D. Zheng, Electromagnetic scattering and radiation by surfaces of arbitrary shape in layered media, Part I: theory. IEEE Trans. Antennas Propag. **45**, 508 (1997)
9. R.F. Harrington, *Field Computation by Moment Methods* (Macmillian, New York, 1968)
10. J.R. Mosig, Integral equation technique, in *Numerical Techniques for Microwave and Millimeter-Wave Passive Structure*, ed. by T. Itoh (Wiley, New York, 1989), p. 133
11. A. Sommerfeld, *Partial Differential Equations in Physics* (Academic Press, New York, 1964)
12. J.A. Stratton, *Electromagnetic Theory* (McGraw-Hill, New York, 1941)

13. K. Nabors, J. White, Fastcap: a multipole accelerated 3-D capacitance extraction program. IEEE Trans. Comput.-Aided Des. Integr. Circuits Syst. **10**, 1447 (1991)
14. A. Erteza, B.K. Park, Nonuniqueness of resolution of Hertz vector in presence of a boundary, and the horizontal problem. IEEE Trans. Antennas Propag. **17**, 376 (1969)
15. K.A. Michalski, On the scalar potential of a point charge associated with a time-harmonic dipole in a layered medium. IEEE Trans. Antennas Propag. **35**(11), 1299 (1987)
16. A.E. Siegman, Quasi fast Hankel transform. Opt. Lett. **1**(1), 13–15 (1977)
17. J. Markham, J.A. Conchello, Numerical evaluation of Hankel transforms for oscillating functions. J. Opt. Soc. Am. A **20**(4), 621–630 (2003)
18. P.J.W. Debye, Der lichtdruck auf Kugeln von beliegigem material. Ann. Phys. (Leipz) **30**, 57 (1909)
19. W.C. Chew, J.L. Xiong, M.A. Saville, A matrix-friendly formulation of layered medium Green's function. IEEE Antennas Wirel. Propag. Lett. **5**, 490–494 (2006)

Chapter 10
Fast Multipole Methods for Wave Interactions in Layered Media

This chapter, as a continuation of Chap. 7, will present the fast multipole methods for wave source interaction in layered media. The key mathematical foundations include multipole expansion (ME) and local expansion (LE) for wave sources embedded in layered media, extending the ME/LE for free space based on the addition theory of Bessel functions discussed in Sect. 7.4. For Green's functions in layered media corresponding to wave interactions in a layered setting, a plane wave representation of the Green's function is used to derive the ME/LE and their translation operators. The exponential convergence of the ME/LE for the far field depends on a special distance between the field location and the location of an equivalent polarization source for a physical source. The hierarchical framework of the conventional FMM described in Sect. 7.3 can be readily used for the interaction of sources in layered media once the physical sources are given the locations of their equivalent polarization sources. The numerical results show that the FMM for Helmholtz Green's functions in layered media retains both the efficiency and accuracy as the version in free space. Some technical materials in deriving and analyzing the FMM for layered media can be found in Zhang et al. [1, 2] and Wang et al. [3].

10.1 Rederiving ME and LE for the Free Space Helmholtz Green's Function

In this section, we will present an alternative method to the traditional addition theorem approach in deriving the multipole and local expansions for wave sources in free space. This approach employs a plane wave representation of Green's functions, involving so-called Sommerfeld integrals in the Fourier spaces along the horizontal direction(s). It is evident that a similar approach can be applied for Green's functions in layered media.

W. Cai, *Deterministic, Stochastic, and Deep Learning Methods for Computational Electromagnetics*, https://doi.org/10.1007/978-981-96-0100-4_10

10.1.1 ME for the 2-D Free Space Helmholtz Green's Function

For simplicity, we will show an alternative approach for deriving the multipole expansion for wave sources in 2-D free space.

Consider N wave sources with strength q_j placed at locations $\mathbf{x}_j = (x_j, y_j) \in \mathbb{R}^2$, $j = 1, 2, \ldots, N$, within a circle with a radius a centered at $\mathbf{x}_c = (x_c, y_c)$ in free space $\mathbb{R}^2$, and the field located at $\mathbf{x}$ due to all sources is given by

$$u^{\mathrm{f}}(\mathbf{x}) = \sum_{j=1}^{N} q_j G^{\mathrm{f}}(\mathbf{x}, \mathbf{x}_j), \quad \mathbf{x} \in \mathbb{R}^2, \tag{10.1}$$

where G^{f} is the free space Green's function

$$G^{\mathrm{f}}(\mathbf{x}, \mathbf{x}') = \frac{\mathrm{i}}{4} H_0^{(1)}\left(k|\mathbf{x} - \mathbf{x}'|\right), \tag{10.2}$$

and k is the wavenumber, and $H_0^{(1)}$ is the Hankel function of the first kind. A target $\mathbf{x}$ is well-separated from the sources if the distance between $\mathbf{x}$ and the source center $\mathbf{x}_c$ is at least $2a$.

According to Graf's addition theorem [4], the free space Green's function for the well-separated sources $\mathbf{x}_j$ and the target $\mathbf{x}$ can be compressed with a P-term multipole expansion, i.e.,

$$u^{\mathrm{f}}(\mathbf{x}) = \frac{\mathrm{i}}{4} \sum_{p=-\infty}^{\infty} \alpha_p H_p^{(1)}\left(k|\mathbf{x} - \mathbf{x}_c|\right) e^{\mathrm{i}p\theta_c} \approx \frac{\mathrm{i}}{4} \sum_{|p|\le P} \alpha_p H_p^{(1)}(k\rho_c) e^{\mathrm{i}p\theta_c}, \tag{10.3}$$

where $\alpha_p = \sum_{j=1}^{N} q_j J_p(k\rho_j) e^{-\mathrm{i}p\theta_j}$, (ρ_c, θ_c) are the polar coordinates of $\mathbf{x} - \mathbf{x}_c$, (ρ_j, θ_j) are the polar coordinates of $\mathbf{x}_j - \mathbf{x}_c$, and the truncation index P is a constant independent of the number of the source number N [5].

To present the alternative derivation of the ME, we first recall the Bessel generating function g [4, (9.1.41)]. For any $z, v \in \mathbb{C}$ with $v \neq 0$,

$$g(z, v) \equiv \exp\left(\frac{z}{2}(v - v^{-1})\right) = \sum_{p=-\infty}^{\infty} J_p(z) v^p. \tag{10.4}$$

The identity (10.4) expresses a plane wave function in terms of cylindrical functions, in contrast to the Sommerfeld integral representation of the Green's function in (10.5) below, which expresses cylindrical functions in terms of plane waves.

The multipole expansion can be derived in the frequency domain using (10.4) as follows. Consider one source $\mathbf{x}_j$ and suppose that $y - y_j > 0$ and $y - y_c > 0$ for simplicity. The interaction between $\mathbf{x} = (x, y)$ and $\mathbf{x}_j = (x_i, y_j)$ can be represented by a Sommerfeld integral of plane waves [6],

$$G^{\mathrm{f}}(\mathbf{x}, \mathbf{x}_j) = \frac{\mathrm{i}}{4} H_0^{(1)}\left(k|\mathbf{x} - \mathbf{x}_j|\right) = \frac{\mathrm{i}}{4}\frac{1}{\mathrm{i}\pi}\int_{-\infty}^{\infty} \frac{e^{-\sqrt{\lambda^2-k^2}(y-y_j)}}{\sqrt{\lambda^2 - k^2}} e^{\mathrm{i}\lambda(x-x_j)} d\lambda, \tag{10.5}$$

while each term $H_p^{(1)}(k\rho_c)e^{\mathrm{i}p\theta_c}$ in (10.3) has a similar representation [6]

$$H_p^{(1)}(k\rho_c)e^{\mathrm{i}p\theta_c} = \frac{1}{\mathrm{i}\pi}\int_{-\infty}^{\infty} \frac{e^{-\sqrt{\lambda^2-k^2}(y-y_c)}}{\sqrt{\lambda^2 - k^2}} e^{\mathrm{i}\lambda(x-x_c)}(-\mathrm{i})^p\,(w(\lambda))^p\,d\lambda, \tag{10.6}$$

where

$$w(\lambda) = \frac{\lambda - \sqrt{\lambda^2 - k^2}}{k}. \tag{10.7}$$

These integral expressions allow the derivation of the multipole expansion of $\frac{\mathrm{i}}{4}H_0^{(1)}(k|\mathbf{x} - \mathbf{x}_j|) = \frac{\mathrm{i}}{4}H_0^{(1)}(k|\,(\mathbf{x} - \mathbf{x}_c) + \left(\mathbf{x}_c - \mathbf{x}_j\right)|)$ by using separable plane wave factors in the integrands involving $(\mathbf{x} - \mathbf{x}_c)$ and $\left(\mathbf{x}_c - \mathbf{x}_j\right)$ as follows.

$$\begin{aligned}
&\frac{\mathrm{i}}{4} H_0^{(1)}\left(k|\mathbf{x} - \mathbf{x}_j|\right) = \frac{\mathrm{i}}{4}\frac{1}{\mathrm{i}\pi}\int_{-\infty}^{\infty} \frac{e^{-\sqrt{\lambda^2-k^2}(y-y_j)}}{\sqrt{\lambda^2 - k^2}} e^{\mathrm{i}\lambda(x-x_j)} d\lambda \\
&= \frac{\mathrm{i}}{4}\frac{1}{\mathrm{i}\pi}\int_{-\infty}^{\infty} \frac{e^{-\sqrt{\lambda^2-k^2}(y-y_c)}}{\sqrt{\lambda^2 - k^2}} e^{\mathrm{i}\lambda(x-x_c)} \cdot e^{-\sqrt{\lambda^2-k^2}(y_c-y_j)+\mathrm{i}\lambda(x_c-x_j)} d\lambda \\
&= \frac{\mathrm{i}}{4}\frac{1}{\mathrm{i}\pi}\int_{-\infty}^{\infty} \frac{e^{-\sqrt{\lambda^2-k^2}(y-y_c)}}{\sqrt{\lambda^2 - k^2}} e^{\mathrm{i}\lambda(x-x_c)} \cdot g\left(k\rho_j, -\mathrm{i}e^{-\mathrm{i}\theta_j} w(\lambda)\right) d\lambda \\
&= \frac{\mathrm{i}}{4}\frac{1}{\mathrm{i}\pi}\int_{-\infty}^{\infty} \frac{e^{-\sqrt{\lambda^2-k^2}(y-y_c)}}{\sqrt{\lambda^2 - k^2}} e^{\mathrm{i}\lambda(x-x_c)} \cdot \sum_{p=-\infty}^{\infty} J_p(k\rho_j)e^{-\mathrm{i}p\theta_j}\,(-\mathrm{i}w(\lambda))^p\,d\lambda \\
&= \frac{\mathrm{i}}{4}\sum_{p=-\infty}^{\infty} J_p(k\rho_j)e^{-\mathrm{i}p\theta_j} \cdot \frac{1}{\mathrm{i}\pi}\int_{-\infty}^{\infty} \frac{e^{-\sqrt{\lambda^2-k^2}(y-y_c)}}{\sqrt{\lambda^2 - k^2}} e^{\mathrm{i}\lambda(x-x_c)}\,(-\mathrm{i}w(\lambda))^p\,d\lambda \\
&= \frac{\mathrm{i}}{4}\sum_{p=-\infty}^{\infty} J_p(k\rho_j)e^{-\mathrm{i}p\theta_j} \cdot H_p^{(1)}(k\rho_c)e^{\mathrm{i}p\theta_c}.
\end{aligned} \tag{10.8}$$

The interchangeability of the sum and the integration is justified by the validity of the identity itself, i.e., the Graf's addition theorem.

10.1.2 Rederiving h (Multipole) and j (Local)-Expansions of the 3-D Free Space Helmholtz Green's Function, and Translation Operators

In this section, we first review the h- and j-expansions, namely, the multipole and local expansions, for the free-space Helmholtz Green's function in three-dimensional space, and the translation from an h-expansion to an j-expansion. They are the key formulae in the free-space FMM and can be derived by using the well-known addition theorems for these special functions. Nonetheless, an alternative derivation for the h- and j-expansions can be given by using an integral representation of Hankel functions in the Fourier space (i.e., a plane wave representation). Furthermore, this approach will be applied to derive multipole and local expansions for reaction components of layered Green's functions in Sect. 10.2.

10.1.2.1 The h and j-Expansions of Free Space Green's Function

First, we present the addition theorems for regular and singular wave functions. Let $\mathbf{r}_j = (r_j, \theta_j, \varphi_j) \in \mathbb{R}^3$ be the spherical coordinate vector of a point P with respect to given centers O_j for $j = 1, 2$, and $\mathbf{b} = (b, \alpha, \beta)$ is the position vector of O_2 with respect to O_1, such that $\mathbf{r}_2 = \mathbf{r}_1 + \mathbf{b}$. Meanwhile, the spherical harmonics are defined as

$$Y_n^m(\theta, \varphi) = (-1)^m \sqrt{\frac{2n+1}{4\pi}\frac{(n-m)!}{(n+m)!}} P_n^m(\cos\theta)e^{im\varphi} = \widehat{P}_n^m(\cos\theta)e^{im\varphi}, \quad (10.9)$$

where $0 \le |m| \le n$ and $P_n^m(\cos\theta)$ are the associated Legendre functions and $\widehat{P}_n^m(\cos\theta)$ is its normalized version. We use the following addition theorems [7].

Theorem 10.1 *Let $z_\nu(\omega)$ be any spherical Bessel function of order ν, that is*

$$z_\nu(\omega) = j_\nu(\omega),\, y_\nu(\omega),\, h_\nu^{(1)}(\omega),\, or\ h_\nu^{(2)}(\omega).$$

Let $\mathbf{r}_2 = \mathbf{r}_1 + \mathbf{b}$. Then

$$z_0(kr_2) = \begin{cases} 4\pi \sum\limits_{n=0}^{\infty}\sum\limits_{m=-n}^{n} (-1)^n z_n(kb)\overline{Y_n^m(\alpha,\beta)}j_n(kr_1)Y_n^m(\theta_1,\varphi_1), & \text{if } r_1 < b, \\ 4\pi \sum\limits_{n=0}^{\infty}\sum\limits_{m=-n}^{n} (-1)^n j_n(kb)\overline{Y_n^m(\alpha,\beta)}z_n(kr_1)Y_n^m(\theta_1,\varphi_1), & \text{if } r_1 > b. \end{cases} \quad (10.10)$$

The special case of the above theorem when $z_0 = j_0$ was proven by Clebsch in 1863, see [8, p. 363]. According to Watson, another special case with $z_0 = y_0$ was due to Gegenbauer.

Theorem 10.2 *Let* $\mathbf{r}_2 = \mathbf{r}_1 + \mathbf{b}$. *Then*

$$j_n(kr_2)Y_n^m(\theta_2, \varphi_2) = \sum_{\nu=0}^{\infty} \sum_{\mu=-\nu}^{\nu} \widehat{S}_{n\nu}^{m\mu}(\mathbf{b}) j_\nu(kr_1)Y_\nu^\mu(\theta_1, \varphi_1), \tag{10.11}$$

where

$$\widehat{S}_{n\nu}^{m\mu}(\mathbf{b}) = 4\pi(-1)^m \mathrm{i}^{\nu-n} \sum_{q=0}^{\infty} \mathrm{i}^q j_q(kb)\overline{Y_q^{\mu-m}(\alpha, \beta)}\mathcal{G}(n, m; \nu, -\mu; q), \tag{10.12}$$

with $\mathcal{G}(n, m; \nu, -\mu; q)$ *being the Gaunt coefficient given below in* (10.16).

Theorem 10.3 *Let* $\mathbf{r}_2 = \mathbf{r}_1 + \mathbf{b}$. *Then*

$$h_n^{(1)}(kr_2)Y_n^m(\theta_2, \varphi_2) = \sum_{\nu=0}^{\infty} \sum_{\mu=-\nu}^{\nu} S_{n\nu}^{m\mu}(\mathbf{b}) j_\nu(kr_1)Y_\nu^\mu(\theta_1, \varphi_1), \tag{10.13}$$

for $r_1 < b$, *and*

$$h_n^{(1)}(kr_2)Y_n^m(\theta_2, \varphi_2) = \sum_{\nu=0}^{\infty} \sum_{\mu=-\nu}^{\nu} \widehat{S}_{n\nu}^{m\mu}(\mathbf{b}) h_\nu^{(1)}(kr_1)Y_\nu^\mu(\theta_1, \varphi_1), \tag{10.14}$$

for $r_1 > b$, *where* $\widehat{S}_{n\nu}^{m\mu}(\mathbf{b})$ *is given by* (10.12) *and*

$$S_{n\nu}^{m\mu}(\mathbf{b}) = 4\pi(-1)^m \mathrm{i}^{\nu-n} \sum_{q=0}^{\infty} \mathrm{i}^q h_q^{(1)}(kb)\overline{Y_q^{\mu-m}(\alpha, \beta)}\mathcal{G}(n, m; \nu, -\mu; q), \tag{10.15}$$

and $\mathcal{G}(n, m; \nu, -\mu; q)$ *is again the Gaunt coefficient.*

The Gaunt coefficient $\mathcal{G}(n, m; \nu, -\mu; q)$ is defined using the Wigner $3-j$ symbol as follows

$$\mathcal{G}(n, m; \nu, \mu; q) = (-1)^{m+\mu}\mathcal{S} \begin{pmatrix} n & \nu & q \\ 0 & 0 & 0 \end{pmatrix} \begin{pmatrix} n & \nu & q \\ m & \mu & -m-\mu \end{pmatrix}, \tag{10.16}$$

where $\mathcal{S} = \sqrt{(2n+1)(2\nu+1)(2q+1)/(4\pi)}$. Although we have explicit formulae (10.12) and (10.15) for the matrices $\widehat{S}_{n\nu}^{m\mu}(\mathbf{b})$ and $S_{n\nu}^{m\mu}(\mathbf{b})$, recurrence formulae are preferred in practical computations (cf. [9, 10]).

With the addition theorems above, we can present the h- and j-expansions, namely, multipole and local expansions, respectively.

Consider the free space Green's function of the Helmholtz equation with a source and a target at $\mathbf{r}' = (x', y', z')$ and $\mathbf{r} = (x, y, z)$, respectively. By using the addition Theorem 10.1, we have an h-expansion (multipole expansion) with respect to a (source) center $\mathbf{r}_c^s$:

$$h_0^{(1)}(k|\mathbf{r} - \mathbf{r}'|) = \sum_{|n|=0}^{\infty} \sum_{m=-n}^{n} M_{nm} h_n^{(1)}(kr_s) Y_n^m(\theta_s, \varphi_s), \tag{10.17}$$

and a j-expansion (local expansion) with respect to a (target) center $\mathbf{r}_c^t$:

$$h_0^{(1)}(k|\mathbf{r} - \mathbf{r}'|) = \sum_{|n|=0}^{\infty} \sum_{m=-n}^{n} L_{nm} j_n(kr_t) Y_n^m(\theta_t, \varphi_t), \tag{10.18}$$

where

$$M_{nm} = 4\pi j_n(kr_s') \overline{Y_n^m(\theta_s', \varphi_s')}, \quad L_{nm} = 4\pi h_n^{(1)}(kr_t') \overline{Y_n^m(\theta_t', \varphi_t')}, \tag{10.19}$$

$\mathbf{r}_c^s = (x_c^s, y_c^s, z_c^s)$ is the source center close to $\mathbf{r}'$, $\mathbf{r}_c^t = (x_c^t, y_c^t, z_c^t)$ is the target center close to $\mathbf{r}$, $(r_s, \theta_s, \varphi_s)$, $(r_t, \theta_t, \varphi_t)$ are the spherical coordinates of $\mathbf{r} - \mathbf{r}_c^s$ and $\mathbf{r} - \mathbf{r}_c^t$, respectively, and $(r_s', \theta_s', \varphi_s')$, $(r_t', \theta_t', \varphi_t')$ are the spherical coordinates of $\mathbf{r}' - \mathbf{r}_c^s$ and $\mathbf{r}' - \mathbf{r}_c^t$, respectively (Fig. 10.1).

Applying addition Theorem 10.3 to $h_n^{(1)}(kr_s) Y_n^m(\theta_s, \varphi_s)$ in (10.17), the translation from the h-expansion (10.17) to the j-expansion (10.18) is given by

$$L_{nm} = \sum_{|\nu|=0}^{\infty} \sum_{\mu=-\nu}^{\nu} S_{n\nu}^{m\mu}(\mathbf{r}_c^t - \mathbf{r}_c^s) M_{\nu\mu}. \tag{10.20}$$

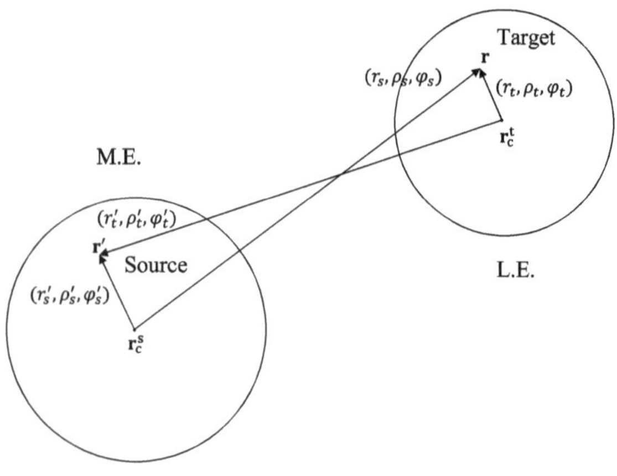

Fig. 10.1 A sketch of the spherical coordinates of the source and target in $\mathbb{R}^3$ with respect to the source center $\mathbf{r}_c^s$ and target center $\mathbf{r}_c^t$ in multipole and local expansions

Similarly, we can shift the centers of the h- and j-expansions via the following translations:

$$\tilde{M}_{nm} = \sum_{\nu=0}^{\infty} \sum_{\mu=-\nu}^{\nu} \overline{\widehat{S}_{n\nu}^{m\mu}(\mathbf{r}_c^s - \tilde{\mathbf{r}}_c^s)} M_{\nu\mu}, \quad \tilde{L}_{nm} = \sum_{n=0}^{\infty} \sum_{\mu=-\nu}^{\nu} \widehat{S}_{\nu n}^{\mu m}(\tilde{\mathbf{r}}_c^t - \mathbf{r}_c^t) L_{\nu\mu}, \tag{10.21}$$

where

$$\tilde{M}_{nm} = 4\pi j_n(k\tilde{r}_s') \overline{Y_n^m(\tilde{\theta}_s', \tilde{\varphi}_s')}, \quad \tilde{L}_{nm} = 4\pi h_n^{(1)}(k\tilde{r}_t') \overline{Y_n^m(\tilde{\theta}_t', \tilde{\varphi}_t')} \tag{10.22}$$

are the h- and j-expansion coefficients at new centers $\tilde{\mathbf{r}}_c^s$ and $\tilde{\mathbf{r}}_c^t$, respectively.

The key feature in (10.17) and (10.18) is that the source and target coordinates are separated, which are used to derive low rank approximations for far fields in the FMM (cf. [11, 12]) due to the fast convergence of the relevant series. Again, as in the 2-D case, this target-source separation can be easily achieved in the Fourier spectral domain through plane waves. An alternative derivation via the Fourier domain for (10.17) and (10.18) is given by using an integral representation of $h_0^{(1)}(k|\mathbf{r}|)$ in the Fourier domain. More importantly, this approach can also be used to derive multipole and local expansions for the reaction components of Helmholtz Green's functions in layered media.

10.1.2.2 A New Derivation of h and j-Expansions

For a spherical wave, we have the well-known Sommerfeld identity

$$h_0^{(1)}(k|\mathbf{r}|) = \frac{1}{2k\pi} \int_0^{\infty} \int_0^{2\pi} k_\rho e^{\mathrm{i}k_\rho(x\cos\alpha + y\sin\alpha)} \frac{e^{\mathrm{i}k_z|z|}}{k_z} d\alpha dk_\rho, \tag{10.23}$$

where $k_z = \sqrt{k^2 - k_\rho^2}$ is a complex wavenumber for $k_\rho > k$. It is necessary to point out that the identity (10.23) is for a lossless medium (i.e., k is real) and needs to be considered as a limiting case of a lossy medium. The contour for the k_ρ-integral is usually set by deforming the real axis to the fourth quadrant. To satisfy the outgoing wave radiation condition in the integrand, we have to ensure that $\mathfrak{Re} k_z > 0$ and $\mathfrak{Im} k_z > 0$. With the Sommerfeld identity (10.23), we perform a source-target separation in the spectral domain as follows

$$\begin{aligned} h_0^{(1)}(k|\mathbf{r} - \mathbf{r}'|) &= \frac{1}{2k\pi} \int_0^{\infty} \int_0^{2\pi} k_\rho \frac{e^{\mathrm{i}\mathbf{k}\cdot(\mathbf{r}-\mathbf{r}_c^s)} e^{-\mathrm{i}\mathbf{k}\cdot(\mathbf{r}'-\mathbf{r}_c^s)}}{k_z} d\alpha dk_\rho, \\ h_0^{(1)}(k|\mathbf{r} - \mathbf{r}'|) &= \frac{1}{2k\pi} \int_0^{\infty} \int_0^{2\pi} k_\rho \frac{e^{\mathrm{i}\mathbf{k}\cdot(\mathbf{r}-\mathbf{r}_c^t)} e^{-\mathrm{i}\mathbf{k}\cdot(\mathbf{r}'-\mathbf{r}_c^t)}}{k_z} d\alpha dk_\rho \end{aligned} \tag{10.24}$$

for $z \geq z'$ where $\mathbf{k} = (k_\rho \cos\alpha, k_\rho \sin\alpha, k_z)$. Without loss of generality, here, we only consider the case $z \geq z'$ for an illustration.

Next, we recall an extended version of the Funk-Hecke formula [7, 8] for complex wave vector $\mathbf{k}$ [3].

Proposition 1 *Given* $\mathbf{r} = (x, y, z) = (r, \theta, \varphi) \in \mathbb{R}^3,\ k > 0,$ *let* $\mathbf{k} = (\sqrt{k^2 - k_z^2}\cos\alpha, \sqrt{k^2 - k_z^2}\sin\alpha, k_z)$ *be a vector of complex entries. Choosing the branch* (10.27) *for* $\sqrt{k^2 - k_z^2}$ *in* $e^{\mathrm{i}\mathbf{k}\cdot\mathbf{r}}$ *and* $\widehat{P}_n^m(\frac{k_z}{k})$*, then*

$$e^{\mathrm{i}\mathbf{k}\cdot\mathbf{r}} = \sum_{n=0}^{\infty}\sum_{m=-n}^{n} \overline{A_n^m(\mathbf{r})}\mathrm{i}^n \widehat{P}_n^m\left(\frac{k_z}{k}\right)e^{\mathrm{i}m\alpha} = \sum_{n=0}^{\infty}\sum_{m=-n}^{n} A_n^m(\mathbf{r})\mathrm{i}^n \widehat{P}_n^m\left(\frac{k_z}{k}\right)e^{-\mathrm{i}m\alpha}, \tag{10.25}$$

holds for all $k_z \in \mathbb{C}$*, where*

$$A_n^m(\mathbf{r}) = 4\pi j_n(kr)Y_n^m(\theta, \varphi). \tag{10.26}$$

Here, $\widehat{P}_n^m\left(\mathbf{z}\right)$ is an analytic extension to the complex plan of the associated Legendre function $\widehat{P}_n^m\left(x\right)$, and the following special branch for $\sqrt{k^2 - k_z^2} = k\sqrt{1 - \left(\frac{k_z}{k}\right)^2}$ is selected. For the polar forms of complex numbers,

$$z + 1 = r_1 e^{\mathrm{i}\theta_1}, \quad -\pi < \theta_1 \leq \pi, \quad z - 1 = r_2 e^{\mathrm{i}\theta_2}, \quad -\pi < \theta_2 \leq \pi,$$

where θ_1 and θ_2 are the principal values of the arguments of complex numbers $z + 1$ and $z - 1$. Then, we cut the complex plane from -1 to $+1$ along the real axis and choose the branch

$$\sqrt{1 - z^2} = -\mathrm{i}\sqrt{r_1 r_2}e^{\mathrm{i}\frac{\theta_1 + \theta_2}{2}}. \tag{10.27}$$

Next, applying the spherical harmonic expansion (10.25) to exponential functions $e^{-\mathrm{i}\mathbf{k}\cdot(\mathbf{r}' - \mathbf{r}_c^s)}$ and $e^{\mathrm{i}\mathbf{k}\cdot(\mathbf{r} - \mathbf{r}_c^t)}$ in (10.24), we have

$$h_0^{(1)}(k|\mathbf{r} - \mathbf{r}'|) = \sum_{n=0}^{\infty}\sum_{m=-n}^{n} \frac{M_{nm}}{2k\pi}\int_0^{\infty}\int_0^{2\pi} k_\rho \frac{e^{\mathrm{i}\mathbf{k}\cdot(\mathbf{r} - \mathbf{r}_c^s)}}{k_z}(-\mathrm{i})^n \widehat{P}_n^m\left(\frac{k_z}{k}\right)e^{\mathrm{i}m\alpha}d\alpha dk_\rho, \tag{10.28}$$

and

$$h_0^{(1)}(k|\mathbf{r} - \mathbf{r}'|) = \sum_{n=0}^{\infty}\sum_{m=-n}^{n} \hat{L}_{nm} j_n(kr_t)Y_n^m(\theta_t, \varphi_t), \tag{10.29}$$

for $z \geq z'$, where M_{nm} is defined in (10.19) and

$$\hat{L}_{nm} = \frac{2\mathrm{i}^n}{k}\int_0^{\infty}\int_0^{2\pi} k_\rho \frac{e^{\mathrm{i}\mathbf{k}\cdot(\mathbf{r}_c^t - \mathbf{r}')}}{k_z}\widehat{P}_n^m\left(\frac{k_z}{k}\right)e^{-\mathrm{i}m\alpha}d\alpha dk_\rho. \tag{10.30}$$

To ensure the convergence of the Sommerfeld-type integrals in the above expansions, we only consider centers such that their z-coordinates satisfy $z_c^s < z$ and $z_c^t > z'$. Using the identity

$$h_n^{(1)}(k|\mathbf{r}|)Y_n^m(\theta,\varphi) = \frac{1}{2k\pi}\int_0^\infty\int_0^{2\pi} k_\rho \frac{e^{\mathrm{i}\mathbf{k}\cdot\mathbf{r}}}{k_z}(-\mathrm{i})^n \widehat{P}_n^m\left(\frac{k_z}{k}\right)e^{\mathrm{i}m\alpha}d\alpha dk_\rho \tag{10.31}$$

for $z \geq 0$, we see that (10.28) and (10.29) are exactly the h-expansions (10.17) and j-expansions (10.18) for the case of $z \geq z'$.

To derive the translation from the h-expansion (10.28) to the j-expansion (10.29), we perform a further splitting in (10.28)

$$e^{\mathrm{i}\mathbf{k}\cdot(\mathbf{r}-\mathbf{r}_c^s)} = e^{\mathrm{i}\mathbf{k}\cdot(\mathbf{r}-\mathbf{r}_c^t)}e^{\mathrm{i}\mathbf{k}\cdot(\mathbf{r}_c^t-\mathbf{r}_c^s)} \tag{10.32}$$

and apply expansion (10.25) to obtain the following translation

$$L_{nm} = \frac{2}{k}\sum_{\nu=0}^{\infty}\sum_{\mu=-\nu}^{\nu} M_{\nu\mu}\cdot \tag{10.33}$$

$$\int_0^\infty\int_0^{2\pi} k_\rho \frac{e^{\mathrm{i}\mathbf{k}(\mathbf{r}_c^t-\mathbf{r}_c^s)}}{k_z}(-1)^\nu \mathrm{i}^{n+\nu}\widehat{P}_\nu^\mu\left(\frac{k_z}{k}\right)\widehat{P}_n^m\left(\frac{k_z}{k}\right)e^{\mathrm{i}(\mu-m)\alpha}d\alpha dk_\rho, \tag{10.34}$$

which implies an integral representation of $S_{n\nu}^{m\mu}(\mathbf{r}_c^t - \mathbf{r}_c^s)$ in (10.20). To ensure the convergence of the Sommerfeld-type integral in the translation operator, the z-coordinates of the centers are also assumed to satisfy $z_c^t > z_c^s$.

10.2 FMM for the 3-D Helmholtz Equation in Layered Media

In this section, the multipole and local expansions for the reaction components of layered media Green's function of the 3-D Helmholtz equation are derived by using the same approach as in the previous section. Based on these expansions and relevant translation operators, FMM for 3-D Helmholtz Green's function in layered media can be developed [3].

10.2.1 Helmholtz Green's Function in 3-D Layered Media

Let us first give the integral representation of the layered Green's function derived in [13]. Consider a $(L+1)$-layer medium consisting of L interfaces located at $z = d_\ell, \ell = 0, 1, \ldots, L-1$ as shown in Fig. 10.2.

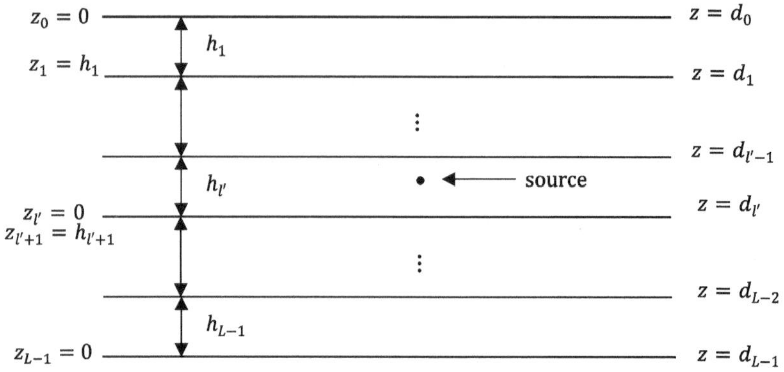

Fig. 10.2 Sketch of a general $(L+1)$-layer medium and a ℓ-th interface coordinate $z_\ell = z - d_\ell$

A point source at $\mathbf{r}' = (x', y', z')$ is located in the ℓ'-th layer ($d_{\ell'} < z' < d_{\ell'-1}$). The layered Green's function for the Helmholtz equation satisfies

$$\boldsymbol{\Delta} u(\mathbf{r}, \mathbf{r}') + k_\ell^2 u(\mathbf{r}, \mathbf{r}') = -\delta(\mathbf{r}, \mathbf{r}'), \quad \mathbf{r}' \in \mathbb{R}, \tag{10.35}$$

at a field point $\mathbf{r} = (x, y, z) \in \mathbb{R}^3$ in the ℓ-th layer ($d_\ell < z < d_\ell - 1$) where $\delta(\mathbf{r}, \mathbf{r}')$ is the Dirac delta function, and k_ℓ is the wavenumber of the ℓ-th layer. The system can be solved analytically in the Fourier (k_x, k_y)-domain for each layer in z by imposing transmission conditions at the interface between the ℓ-th and $(\ell-1)$-th layer ($z = d_{\ell-1}$) as well as the decay conditions in the top and bottom-most layers as $z \to \pm\infty$ [14]. Inside a given layer, e.g., the ℓ-th layer, the solution has the form

$$u(\mathbf{r}, \mathbf{r}') = \begin{cases} u^{\mathrm{r}}_{\ell'\ell'}(\mathbf{r}, \mathbf{r}') + \dfrac{e^{\mathrm{i}k_{\ell'}|\mathbf{r}-\mathbf{r}'|}}{4\pi|\mathbf{r}-\mathbf{r}'|}, & d_{\ell'} < z < d_{\ell'-1}, \\ u^{\mathrm{r}}_{\ell\ell'}(\mathbf{r}, \mathbf{r}'), & d_\ell < z < d_{\ell-1} \text{ and } \ell \neq \ell', \end{cases} \tag{10.36}$$

where the reaction field in the ℓ-th layer

$$u^{\mathrm{r}}_{\ell\ell'}(\mathbf{r}, \mathbf{r}') = \begin{cases} u^{\uparrow}_{0\ell'}(\mathbf{r}, \mathbf{r}'), & \ell = 0, \\ u^{\uparrow}_{\ell\ell'}(\mathbf{r}, \mathbf{r}') + u^{\downarrow}_{\ell\ell'}(\mathbf{r}, \mathbf{r}'), & 0 < \ell < L, \\ u^{\downarrow}_{L\ell'}(\mathbf{r}, \mathbf{r}'), & \ell = L, \end{cases} \tag{10.37}$$

reflects the wave reflections and transmissions by layered media with both upgoing and downgoing components inside intermediate layers ($0 < \ell < L$). However, only the upgoing or downgoing component is required in the top- and bottom-most layers, respectively. The up- and downgoing components have Sommerfeld integral representations

$$
\begin{aligned}
u_{\ell\ell'}^{\uparrow}(\mathbf{r},\mathbf{r}') &= \frac{\mathrm{i}}{8\pi^2}\int_0^{\infty}\int_0^{2\pi} k_\rho e^{\mathrm{i}\mathbf{k}_\alpha\cdot(\rho-\rho')}\frac{e^{\mathrm{i}k_{\ell z}(z-d_\ell)}}{k_{\ell z}}\psi_{\ell\ell'}^{\uparrow}(k_\rho,z')d\alpha dk_\rho, \quad \ell < L,\\
u_{\ell\ell'}^{\downarrow}(\mathbf{r},\mathbf{r}') &= \frac{\mathrm{i}}{8\pi^2}\int_0^{\infty}\int_0^{2\pi} k_\rho e^{\mathrm{i}\mathbf{k}_\alpha\cdot(\rho-\rho')}\frac{e^{\mathrm{i}k_{\ell z}(d_{\ell-1}-z)}}{k_{\ell z}}\psi_{\ell\ell'}^{\downarrow}(k_\rho,z')d\alpha dk_\rho, \quad \ell > 0,
\end{aligned}
\tag{10.38}
$$

where $\mathbf{k}_\alpha = (k_\rho\cos\alpha, k_\rho\sin\alpha) = (k_x, k_y)$, $\rho = (x, y)$, $\rho' = (x', y')$, $k_{\ell z} = \sqrt{k_\ell^2 - k_\rho^2}$,

$$
\begin{cases}
\psi_{\ell 0}^{\uparrow}(k_\rho,z') = e^{\mathrm{i}k_{0z}(z'-d_0)}\sigma_{\ell\ell'}^{\uparrow\uparrow}(k_\rho),\\
\psi_{\ell\ell'}^{\uparrow}(k_\rho,z') = e^{\mathrm{i}k_{\ell' z}(z'-d_{\ell'})}\sigma_{\ell\ell'}^{\uparrow\uparrow}(k_\rho) + e^{\mathrm{i}k_{\ell' z}(d_{\ell'-1}-z')}\sigma_{\ell\ell'}^{\uparrow\downarrow}(k_\rho), \quad 0 < \ell' < L,\\
\psi_{\ell\ell'}^{\downarrow}(k_\rho,z') = e^{\mathrm{i}k_{\ell' z}(z'-d_{\ell'})}\sigma_{\ell\ell'}^{\downarrow\uparrow}(k_\rho) + e^{\mathrm{i}k_{\ell' z}(d_{\ell'-1}-z')}\sigma_{\ell\ell'}^{\downarrow\downarrow}(k_\rho), \quad 0 < \ell' < L,\\
\psi_{\ell L}^{\downarrow}(k_\rho,z') = e^{\mathrm{i}k_{\ell' z}(d_{L-1}-z')}\sigma_{\ell L}^{\downarrow\downarrow}(k_\rho).
\end{cases}
\tag{10.39}
$$

The reaction densities $\sigma_{\ell\ell'}^{\uparrow\uparrow}(k_\rho), \sigma_{\ell\ell'}^{\uparrow\downarrow}(k_\rho), \sigma_{\ell\ell'}^{\downarrow\uparrow}(k_\rho), \sigma_{\ell\ell'}^{\downarrow\downarrow}(k_\rho)$ depend only on the layer structure and wavenumbers. Equations (10.38) and (10.39) are general formulae that are applicable to multilayered media. The reaction densities of a three-layer medium with a source in the middle layer are given here,

$$
\begin{aligned}
\sigma_{01}^{\uparrow\uparrow}(k_\rho) &= \frac{k_1k_{0z}(k_1k_{1z}-k_2k_{2z})e^{-\mathrm{i}d_1k_{1z}}}{k_0k_{0z}\kappa_{11}-\mathrm{i}k_1k_{1z}\kappa_{12}}, \quad \sigma_{01}^{\uparrow\downarrow}(k_\rho) = \frac{k_1k_{0z}k_1k_{1z}+k_2k_{2z}}{k_0k_{0z}\kappa_{11}-\mathrm{i}k_1k_{1z}\kappa_{12}},\\
\sigma_{11}^{\uparrow\uparrow}(k_\rho) &= \frac{(k_1k_{1z}-k_2k_{2z})(k_1k_{1z}+k_0k_{0z})}{2(k_0k_{0z}\kappa_{11}-\mathrm{i}k_1k_{1z}\kappa_{12})},\\
\sigma_{11}^{\downarrow\downarrow}(k_\rho) &= \frac{(k_1k_{1z}-k_0k_{0z})(k_1k_{1z}+k_2k_{2z})}{2(k_0k_{0z}\kappa_{11}-\mathrm{i}k_1k_{1z}\kappa_{12})},\\
\sigma_{11}^{\uparrow\downarrow}(k_\rho) &= \frac{k_1k_{1z}-k_2k_{2z}}{k_0k_{0z}\kappa_{11}-\mathrm{i}k_1k_{1z}\kappa_{12}}\frac{k_1k_{1z}-k_0k_{0z}}{2}e^{-\mathrm{i}d_1k_{1z}},\\
\sigma_{11}^{\downarrow\uparrow}(k_\rho) &= \frac{k_1k_{1z}-k_0k_{0z}}{k_0k_{0z}\kappa_{11}-\mathrm{i}k_1k_{1z}\kappa_{12}}\frac{k_1k_{1z}-k_2k_{2z}}{2}e^{-\mathrm{i}d_1k_{1z}}\\
\sigma_{21}^{\downarrow\uparrow}(k_\rho) &= \frac{k_1k_{2z}(k_1k_{1z}+k_0k_{0z})}{k_0k_{0z}\kappa_{11}-\mathrm{i}k_1k_{1z}\kappa_{12}}, \quad \sigma_{21}^{\downarrow\downarrow}(k_\rho) = \frac{k_1k_{2z}(k_1k_{1z}-k_0k_{0z})e^{-\mathrm{i}d_1k_{1z}}}{k_0k_{0z}\kappa_{11}-\mathrm{i}k_1k_{1z}\kappa_{12}}.
\end{aligned}
$$

where

$$
\begin{aligned}
\kappa_{11} &= \frac{k_1k_{1z}-k_2k_{2z}}{2}e^{-2\mathrm{i}d_1k_{1z}} + \frac{k_1k_{1z}+k_2k_{2z}}{2},\\
\kappa_{12} &= \mathrm{i}\Big(\frac{k_2k_{2z}-k_1k_{1z}}{2}e^{2\mathrm{i}d_1k_{1z}} + \frac{k_1k_{1z}+k_2k_{2z}}{2}\Big).
\end{aligned}
$$

Detailed derivations of the general formulae (10.37)–(10.39) and corresponding reaction densities for Green's function in layered media can be found in [13].

10.2.2 ME and LE for Reaction Components

Consider the reaction field in the middle layers, i.e.,

$$u^{\mathrm{r}}_{\ell\ell'}(\mathbf{r},\mathbf{r}') = u^{\uparrow}_{\ell\ell'}(\mathbf{r},\mathbf{r}') + u^{\downarrow}_{\ell\ell'}(\mathbf{r},\mathbf{r}'). \tag{10.40}$$

By the expressions in (10.38) and (10.39), we can further decompose

$$u^{\uparrow}_{\ell\ell'}(\mathbf{r},\mathbf{r}') = u^{\uparrow\uparrow}_{\ell\ell'}(\mathbf{r},\mathbf{r}') + u^{\uparrow\downarrow}_{\ell\ell'}(\mathbf{r},\mathbf{r}'), \quad u^{\downarrow}_{\ell\ell'}(\mathbf{r},\mathbf{r}') = u^{\downarrow\uparrow}_{\ell\ell'}(\mathbf{r},\mathbf{r}') + u^{\downarrow\downarrow}_{\ell\ell'}(\mathbf{r},\mathbf{r}'), \tag{10.41}$$

and, each component has a Sommerfeld-type integral representation:

$$u^{**}_{\ell\ell'}(\mathbf{r},\mathbf{r}') = \frac{\mathrm{i}}{8\pi^2}\int_0^{\infty}\int_0^{2\pi} k_\rho \frac{\mathcal{E}^{**}_{\ell\ell'}(\mathbf{r},\mathbf{r}')\sigma^{**}_{\ell\ell'}(k_\rho)}{k_{\ell z}} d\alpha dk_\rho, \tag{10.42}$$

where $*$ stands for the arrows $\uparrow$, $\downarrow$ and the plane wave factors $\mathcal{E}^{**}_{\ell\ell'}(\mathbf{r},\mathbf{r}')$ are

$$\begin{aligned}
\mathcal{E}^{\uparrow\uparrow}_{\ell\ell'}(\mathbf{r},\mathbf{r}') &:= e^{\mathrm{i}\mathbf{k}_\alpha\cdot(\boldsymbol{\rho}-\boldsymbol{\rho}')+\mathrm{i}k_{\ell z}(z-d_\ell)+\mathrm{i}k_{\ell' z}(z'-d_{\ell'})},\\
\mathcal{E}^{\uparrow\downarrow}_{\ell\ell'}(\mathbf{r},\mathbf{r}') &:= e^{\mathrm{i}\mathbf{k}_\alpha\cdot(\boldsymbol{\rho}-\boldsymbol{\rho}')+\mathrm{i}k_{\ell z}(z-d_\ell)+\mathrm{i}k_{\ell' z}(d_{\ell'-1}-z')},\\
\mathcal{E}^{\downarrow\uparrow}_{\ell\ell'}(\mathbf{r},\mathbf{r}') &:= e^{\mathrm{i}\mathbf{k}_\alpha\cdot(\boldsymbol{\rho}-\boldsymbol{\rho}')+\mathrm{i}k_{\ell z}(d_{\ell-1}-z)+\mathrm{i}k_{\ell' z}(z'-d_{\ell'})},\\
\mathcal{E}^{\downarrow\downarrow}_{\ell\ell'}(\mathbf{r},\mathbf{r}') &:= e^{\mathrm{i}\mathbf{k}_\alpha\cdot(\boldsymbol{\rho}-\boldsymbol{\rho}')+\mathrm{i}k_{\ell z}(d_{\ell-1}-z)+\mathrm{i}k_{\ell' z}(d_{\ell'-1}-z')}.
\end{aligned} \tag{10.43}$$

Following the same approach for separating the source and target in the exponential $\mathcal{E}^{**}_{\ell\ell'}(\mathbf{r},\mathbf{r}')$ as in (10.24) and using the Funk-Hecke formula (10.25), similarly, we can obtain the following ME for the reaction field components

$$u^{**}_{\ell\ell'}(\mathbf{r},\mathbf{r}') = \sum_{n=0}^{\infty}\sum_{m=-n}^{n} M_{nm}\mathcal{F}^{**}_{nm}(\mathbf{r},\mathbf{r}^s_c), \quad M_{nm} = 4\pi j_n(k_{\ell'}r'_s)\overline{Y^m_n(\theta'_s,\varphi'_s)}, \tag{10.44}$$

with the expansion functions given by Sommerfeld-type integrals

$$\begin{aligned}
\mathcal{F}^{*\uparrow}_{nm}(\mathbf{r},\mathbf{r}^s_c) &= \frac{\mathrm{i}}{8\pi^2}\int_0^{\infty}\int_0^{2\pi} k_\rho \frac{(-1)^m\mathcal{E}^{*\uparrow}_{\ell\ell'}(\mathbf{r},\mathbf{r}^s_c)\sigma^{*\uparrow}_{\ell\ell'}(k_\rho)}{k_{\ell z}} \mathrm{i}^n \widehat{P}^m_n\left(\frac{k_{\ell' z}}{k_{\ell'}}\right)e^{\mathrm{i}m\alpha} d\alpha dk_\rho,\\
\mathcal{F}^{*\downarrow}_{nm}(\mathbf{r},\mathbf{r}^s_c) &= \frac{\mathrm{i}}{8\pi^2}\int_0^{\infty}\int_0^{2\pi} k_\rho \frac{(-1)^n\mathcal{E}^{*\downarrow}_{\ell\ell'}(\mathbf{r},\mathbf{r}^s_c)\sigma^{*\downarrow}_{\ell\ell'}(k_\rho)}{k_{\ell z}} \mathrm{i}^n \widehat{P}^m_n\left(\frac{k_{\ell' z}}{k_{\ell'}}\right)e^{\mathrm{i}m\alpha} d\alpha dk_\rho.
\end{aligned} \tag{10.45}$$

The function $\mathcal{F}^{**}_{nm}(\mathbf{r},\mathbf{r}^s_c)$ is defined only for $\mathbf{r}$ in the ℓ-th layer.

Similarly, we can obtain local expansion

$$u_{\ell\ell'}^{**}(\mathbf{r},\mathbf{r}') = \sum_{n=0}^{\infty}\sum_{m=-n}^{n} L_{nm}^{**} j_n(k_\ell r_t) Y_n^m(\theta_t,\varphi_t), \tag{10.46}$$

with coefficients given by

$$\begin{aligned} L_{nm}^{\uparrow *} &= \frac{\mathrm{i}}{2\pi}\int_0^\infty\int_0^{2\pi} k_\rho \frac{\mathcal{E}_{\ell\ell'}^{\uparrow *}(\mathbf{r}_c^t,\mathbf{r}')\sigma_{\ell\ell'}^{\uparrow *}(k_\rho)}{k_{\ell z}} \mathrm{i}^n \widehat{P}_n^m\Big(\frac{k_{\ell z}}{k_\ell}\Big) e^{-\mathrm{i}m\alpha} d\alpha dk_\rho, \\ L_{nm}^{\downarrow *} &= \frac{(-1)^{n+m}\mathrm{i}}{2\pi}\int_0^\infty\int_0^{2\pi} k_\rho \frac{\mathcal{E}_{\ell\ell'}^{\downarrow *}(\mathbf{r}_c^t,\mathbf{r}')\sigma_{\ell\ell'}^{\downarrow *}(k_\rho)}{k_{\ell z}} \mathrm{i}^n \widehat{P}_n^m\Big(\frac{k_{\ell z}}{k_\ell}\Big) e^{-\mathrm{i}m\alpha} d\alpha dk_\rho. \end{aligned} \tag{10.47}$$

According to the definitions of $\mathcal{E}_{\ell\ell'}^{**}(\mathbf{r},\mathbf{r}_c^s)$ and $\mathcal{E}_{\ell\ell'}^{**}(\mathbf{r}_c^t,\mathbf{r}')$ in (10.43), the centers $\mathbf{r}_c^s$ and $\mathbf{r}_c^t$ must satisfy

$$\begin{aligned} &z_c^s > d_{\ell'},\ z_c^t > d_\ell, \quad \text{for } u_{\ell\ell'}^{\uparrow\uparrow}(\mathbf{r},\mathbf{r}'); \quad z_c^s < d_{\ell'-1},\ z_c^t < d_{\ell-1}, \quad \text{for } u_{\ell\ell'}^{\downarrow\downarrow}(\mathbf{r},\mathbf{r}'); \\ &z_c^s < d_{\ell'-1},\ z_c^t > d_\ell, \quad \text{for } u_{\ell\ell'}^{\uparrow\downarrow}(\mathbf{r},\mathbf{r}'); \quad z_c^s > d_{\ell'},\ z_c^t < d_{\ell-1}, \quad \text{for } u_{\ell\ell'}^{\downarrow\uparrow}(\mathbf{r},\mathbf{r}') \end{aligned} \tag{10.48}$$

to ensure the exponential decay in $\mathcal{E}_{\ell\ell'}^{**}(\mathbf{r},\mathbf{r}_c^s)$ and $\mathcal{E}_{\ell\ell'}^{**}(\mathbf{r}_c^t,\mathbf{r}')$ as $k_\rho \to \infty$ and hence the convergence of the corresponding Sommerfeld-type integrals in (10.45) and (10.47). These restrictions can be met in practice, since we are considering targets in the ℓ-th layer and sources in the ℓ'-th layer, $\ell' \neq \ell$.

10.2.3 Polarization Sources and MEs, LEs, and Their Translations

In this section, we will introduce equivalent polarization sources for sources in layered media, whose distance to targets, instead of the distance between the original source and the targets, will be used in the convergence estimates of the ME and LE. This is due to the following convergence results of the ME at $\mathbf{r}_c^s$ and the LE at $\mathbf{r}_c^t$ for the reaction components [1, 2],

$$\begin{aligned} \Big|u_{\ell\ell'}^{**}(\mathbf{r},\mathbf{r}') - \sum_{n=0}^{p}\sum_{m=-n}^{n} M_{nm}\mathcal{F}_{nm}^{**}(\mathbf{r},\mathbf{r}_c^s)\Big| &= \mathcal{O}\left(\left(\frac{|\mathbf{r}'-\mathbf{r}_c^s|}{d^{**}(\mathbf{r},\mathbf{r}_c^s)}\right)^p\right), \\ \Big|u_{\ell\ell'}^{**}(\mathbf{r},\mathbf{r}') - \sum_{n=0}^{p}\sum_{m=-n}^{n} L_{nm}^{**} j_n(k_\ell r_t) Y_n^m(\theta_t,\varphi_t)\Big| &= \mathcal{O}\left(\left(\frac{|\mathbf{r}-\mathbf{r}_c^t|}{d^{**}(\mathbf{r}_c^t,\mathbf{r}')}\right)^p\right), \end{aligned} \tag{10.49}$$

respectively, where the polarization distances are

$$
\begin{aligned}
d^{\uparrow\uparrow}(\mathbf{r}, \mathbf{r}_c^s) =& \sqrt{(x - x_c^s)^2 + (y - y_c^s)^2 + (z - d_\ell + z_c^s - d_{\ell'})^2}, \\
d^{\uparrow\downarrow}(\mathbf{r}, \mathbf{r}_c^s) =& \sqrt{(x - x_c^s)^2 + (y - y_c^s)^2 + (z - d_\ell + d_{\ell'-1} - z_c^s)^2}, \\
d^{\downarrow\uparrow}(\mathbf{r}, \mathbf{r}_c^s) =& \sqrt{(x - x_c^s)^2 + (y - y_c^s)^2 + (d_{\ell-1} - z + z_c^s - d_{\ell'})^2}, \\
d^{\downarrow\downarrow}(\mathbf{r}, \mathbf{r}_c^s) =& \sqrt{(x - x_c^s)^2 + (y - y_c^s)^2 + (d_{\ell-1} - z + d_{\ell'-1} - z_c^s)^2}.
\end{aligned}
\tag{10.50}
$$

If $\mathbf{r}_c^s$ is a source center close to source $\mathbf{r}'$ with a distance $|\mathbf{r}' - \mathbf{r}_c^s|$, (10.49) states that the error of the truncated multipole expansion (10.44) is not determined by the Euclidean distance between source center $\mathbf{r}_c^s$ and target $\mathbf{r}$, as in the free space case, but rather by the polarization distance defined in (10.50). A similar conclusion can also be drawn for the local expansion (10.46).

There are two special cases in which $d^{\uparrow\downarrow}(\mathbf{r}, \mathbf{r}_c^s) = |\mathbf{r} - \mathbf{r}_c^s|$ if $\mathbf{r}$ and $\mathbf{r}_c^s$ are in the ℓ-th and $(\ell + 1)$-th layers and $d^{\downarrow\uparrow}(\mathbf{r}, \mathbf{r}_c^s) = |\mathbf{r} - \mathbf{r}_c^s|$ if $\mathbf{r}$ and $\mathbf{r}_c^s$ are in the ℓ-th and $(\ell - 1)$-th layers. Therefore, the multipole and local expansions of $u_{\ell\ell+1}^{\uparrow\downarrow}(\mathbf{r}, \mathbf{r}_c^s)$ and $u_{\ell\ell-1}^{\downarrow\uparrow}(\mathbf{r}, \mathbf{r}_c^s)$ have the same convergence behavior as that of the free space Green's function.

Polarization sources Due to (10.49), (10.50), we introduce a polarization source for each original source, and the interaction between the target and source can then be viewed as the interaction between the target and the polarization source. To extend the FMM for free space to the reaction field component of the layered Green's function, the hierarchical tree structure design in free space FMM will be applied to the polarization sources, instead.

Specifically, the polarization sources for the four types of reaction fields (see Fig. 10.3) are defined as follows,

$$
\begin{aligned}
\mathbf{r}'_{\uparrow\uparrow} &= (x', y', d_\ell - (z' - d_{\ell'})), & \mathbf{r}'_{\uparrow\downarrow} &= (x', y', d_\ell - (d_{\ell'-1} - z')), \\
\mathbf{r}'_{\downarrow\uparrow} &= (x', y', d_{\ell-1} + (z' - d_{\ell'})), & \mathbf{r}'_{\downarrow\downarrow} &= (x', y', d_{\ell-1} + (d_{\ell'-1} - z')),
\end{aligned}
\tag{10.51}
$$

which satisfy

$$
d^{\uparrow *}(\mathbf{r}, \mathbf{r}'_{\uparrow *}) = |\mathbf{r} - \mathbf{r}'_{\uparrow *}|, \quad \text{if } z > d_\ell; \quad d^{\downarrow *}(\mathbf{r}, \mathbf{r}'_{\downarrow *}) = |\mathbf{r} - \mathbf{r}'_{\downarrow *}|, \quad \text{if } z < d_{\ell-1}.
\tag{10.52}
$$

The ME and LE in (10.44) and (10.47) for the reaction field can now be re-expressed using the polarization sources. First, we define notations

$$
\widetilde{\mathcal{Z}}_{\ell\ell'}^{\uparrow}(z, z_s) := e^{\mathrm{i}k_{\ell z}(z - d_\ell) + \mathrm{i}k_{\ell' z}(d_\ell - z_s)}, \quad \widetilde{\mathcal{Z}}_{\ell\ell'}^{\downarrow}(z, z_s) := e^{\mathrm{i}k_{\ell z}(d_{\ell-1} - z) + \mathrm{i}k_{\ell' z}(z_s - d_{\ell-1})},
\tag{10.53}
$$

and, then

$$
\widetilde{\mathcal{E}}_{\ell\ell'}^{*}(\mathbf{r}, \mathbf{r}_s) := e^{\mathrm{i}\mathbf{k}_\alpha \cdot (\rho - \rho_s)} \widetilde{\mathcal{Z}}_{\ell\ell'}^{*}(z, z_s).
\tag{10.54}
$$

By recalling the expressions (10.43), we can verify that

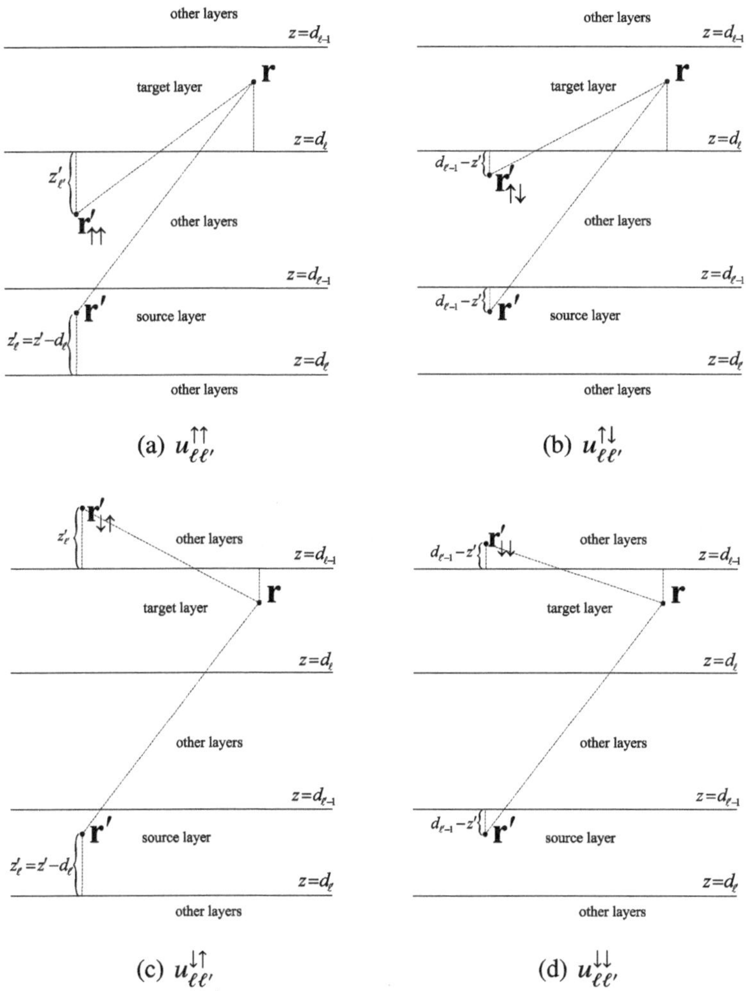

Fig. 10.3 Location of equivalent polarization sources for the computation of $u^{**}_{\ell\ell'}$

$$\mathcal{E}^{\uparrow *}_{\ell\ell'}(\mathbf{r},\mathbf{r}') = \widetilde{\mathcal{E}}^{\uparrow}_{\ell\ell'}(\mathbf{r},\mathbf{r}'_{\uparrow *}), \quad \mathcal{E}^{\downarrow *}_{\ell\ell'}(\mathbf{r},\mathbf{r}') = \widetilde{\mathcal{E}}^{\downarrow}_{\ell\ell'}(\mathbf{r},\mathbf{r}'_{\downarrow *}). \tag{10.55}$$

Therefore, the reaction components can be rewritten using the polarization sources as follows

$$\begin{aligned} u^{\uparrow *}_{\ell\ell'}(\mathbf{r},\mathbf{r}') = \tilde{u}^{\uparrow *}_{\ell\ell'}(\mathbf{r},\mathbf{r}'_{\uparrow *}) &:= \frac{\mathrm{i}}{8\pi^2}\int_0^{\infty}\int_0^{2\pi} k_\rho \frac{\widetilde{\mathcal{E}}^{\uparrow}_{\ell\ell'}(\mathbf{r},\mathbf{r}'_{\uparrow *})\sigma^{\uparrow *}_{\ell\ell'}(k_\rho)}{k_{\ell z}} d\alpha dk_\rho, \\ u^{\downarrow *}_{\ell\ell'}(\mathbf{r},\mathbf{r}') = \tilde{u}^{\downarrow *}_{\ell\ell'}(\mathbf{r},\mathbf{r}'_{\downarrow *}) &:= \frac{\mathrm{i}}{8\pi^2}\int_0^{\infty}\int_0^{2\pi} k_\rho \frac{\widetilde{\mathcal{E}}^{\downarrow}_{\ell\ell'}(\mathbf{r},\mathbf{r}'_{\downarrow *})\sigma^{\downarrow *}_{\ell\ell'}(k_\rho)}{k_{\ell z}} d\alpha dk_\rho. \end{aligned} \tag{10.56}$$

Note that $\tilde{u}_{\ell\ell'}^{\uparrow *}$ and $\tilde{u}_{\ell\ell'}^{\downarrow *}$ have the same form as the two special cases $u_{\ell\ell+1}^{\uparrow\downarrow}$ and $u_{\ell\ell-1}^{\downarrow\uparrow}$, respectively, except for wavenumber $k_{\ell' z}$ and density $\sigma_{\ell\ell'}^{**}$ in the ℓ'-th layer. Following the same previous derivation, we obtain the following multipole expansion

$$\tilde{u}_{\ell\ell'}^{**}(\mathbf{r},\mathbf{r}'_{**})=\sum_{n=0}^{\infty}\sum_{m=-n}^{n}M_{nm}^{**}\widetilde{\mathcal{F}}_{nm}^{**}(\mathbf{r},\mathbf{r}_c^{**}),\quad M_{nm}^{**}=4\pi j_n(k_{\ell'}r_s^{**})\overline{Y_n^m(\theta_s^{**},\varphi_s^{**})},\tag{10.57}$$

at equivalent polarization source centers $\mathbf{r}_c^{**}$ and a local expansion

$$\tilde{u}_{\ell\ell'}^{**}(\mathbf{r},\mathbf{r}'_{**})=\sum_{n=0}^{\infty}\sum_{m=-n}^{n}L_{nm}^{**}j_n(k_\ell r_t)Y_n^m(\theta_t,\varphi_t)\tag{10.58}$$

at the target center $\mathbf{r}_c^t$. Here, $(r_s^{**},\theta_s^{**},\varphi_s^{**})$ are the spherical coordinates of $\mathbf{r}'_{**}-\mathbf{r}_c^{**}$, and $\widetilde{\mathcal{F}}_{nm}^{**}(\mathbf{r},\mathbf{r}_c^{**})$ are given by Sommerfeld-type integrals

$$\begin{aligned}\widetilde{\mathcal{F}}_{nm}^{\uparrow *}(\mathbf{r},\mathbf{r}_c^{\uparrow *})=&\frac{\mathrm{i}}{8\pi^2}\int_0^\infty\int_0^{2\pi}k_\rho\frac{\widetilde{\mathcal{E}}_{\ell\ell'}^{\uparrow}(\mathbf{r},\mathbf{r}_c^{\uparrow *})\sigma_{\ell\ell'}^{\uparrow *}(k_\rho)}{k_{\ell z}}(-\mathrm{i})^n\widehat{P}_n^m\left(\frac{k_{\ell' z}}{k_{\ell'}}\right)e^{\mathrm{i}m\alpha}d\alpha dk_\rho,\\ \widetilde{\mathcal{F}}_{nm}^{\downarrow *}(\mathbf{r},\mathbf{r}_c^{\downarrow *})=&\frac{\mathrm{i}}{8\pi^2}\int_0^\infty\int_0^{2\pi}k_\rho\frac{\widetilde{\mathcal{E}}_{\ell\ell'}^{\downarrow}(\mathbf{r},\mathbf{r}_c^{\downarrow *})\sigma_{\ell\ell'}^{\downarrow *}(k_\rho)}{k_{\ell z}}(-\mathrm{i})^n\widehat{P}_n^m\left(\frac{k_{\ell' z}}{k_{\ell'}}\right)e^{\mathrm{i}m\alpha}d\alpha dk_\rho,\end{aligned}\tag{10.59}$$

and the local expansion coefficients are given by

$$\begin{aligned}L_{nm}^{\uparrow *}=&\frac{\mathrm{i}}{2\pi}\int_0^\infty\int_0^{2\pi}k_\rho\frac{\widetilde{\mathcal{E}}_{\ell\ell'}^{\uparrow}(\mathbf{r}_c^t,\mathbf{r}'_{\uparrow *})\sigma_{\ell\ell'}^{\uparrow *}(k_\rho)}{k_{\ell z}}\mathrm{i}^n\widehat{P}_n^m\left(\frac{k_{\ell z}}{k_\ell}\right)e^{-\mathrm{i}m\alpha}d\alpha dk_\rho,\\ L_{nm}^{\downarrow *}=&\frac{\mathrm{i}}{2\pi}\int_0^\infty\int_0^{2\pi}k_\rho\frac{(-1)^{n+m}\widetilde{\mathcal{E}}_{\ell\ell'}^{\downarrow}(\mathbf{r}_c^t,\mathbf{r}'_{\downarrow *})\sigma_{\ell\ell'}^{\uparrow *}(k_\rho)}{k_{\ell z}}\mathrm{i}^n\widehat{P}_n^m\left(\frac{k_{\ell z}}{k_\ell}\right)e^{-\mathrm{i}m\alpha}d\alpha dk_\rho.\end{aligned}\tag{10.60}$$

Similar to the restrictions in (10.48), the centers r_c^{**} and $\mathbf{r}_c^t$ are required to satisfy

$$z_c^{\uparrow *}<d_\ell,\quad z_c^{\downarrow *}>d_{\ell-1},\quad z_c^t>d_\ell\ \text{ for }\ \tilde{u}_{\ell\ell'}^{\uparrow *}(\mathbf{r},\mathbf{r}'_{\uparrow *});\quad z_c^t<d_{\ell-1}\ \text{ for }\ \tilde{u}_{\ell\ell'}^{\downarrow *}(\mathbf{r},\mathbf{r}'_{\downarrow *}).\tag{10.61}$$

With the convergence results (10.49) and the fact (10.52), we can see that ME (10.57) now becomes

$$\left|\tilde{u}_{\ell\ell'}^{**}(\mathbf{r},\mathbf{r}'_{**})-\sum_{n=0}^{p}\sum_{m=-n}^{n}M_{nm}^{**}\widetilde{\mathcal{F}}_{nm}^{**}(\mathbf{r},\mathbf{r}_c^{**})\right|=\mathcal{O}\left(\left(\frac{|\mathbf{r}'_{**}-\mathbf{r}_c^{**}|}{|\mathbf{r}-\mathbf{r}_c^{**}|}\right)^p\right),\tag{10.62}$$

where the Euclidean distance $|\mathbf{r}-\mathbf{r}'_{**}|$ is used to determine whether multipole expansions (10.57) are good approximations of the far reaction field. Similarly, the local expansion (10.58) for $\tilde{u}^{**}_{\ell\ell'}(\mathbf{r}, \mathbf{r}'_{**})$ has a convergence of order $\mathcal{O}\left(\left(\frac{|\mathbf{r}-\mathbf{r}^t_c|}{|\mathbf{r}^t_c-\mathbf{r}'_{**}|}\right)^p\right)$. These convergence results ensure that the hierarchical tree structure can be applied in the FMM for $\tilde{u}^{**}_{\ell\ell'}(\mathbf{r}, \mathbf{r}'_{**})$ with ME (10.57) and LE (10.58).

Next, we discuss the center shifting and translation for ME (10.57) and LE (10.58). A desirable feature of the expansions of reaction components discussed above is that the formula (10.57) for the ME coefficients and the formula (10.58) for the LE have exactly the same form as the formulae of h-expansion coefficients and j-expansion for the free space Green's function. Therefore, we can see that the center shifts for multipole and local expansions are exactly the same as those in the free space case given in (10.21).

We only need to derive the translation operator from ME (10.57) to LE (10.58). With the definition of exponential functions in (10.54), $\widetilde{\mathcal{E}}^{\uparrow}_{\ell\ell'}(\mathbf{r}, \mathbf{r}^{\uparrow *}_c)$ and $\widetilde{\mathcal{E}}^{\downarrow}_{\ell\ell'}(\mathbf{r}, \mathbf{r}^{\downarrow *}_c)$, we have the following splitting:

$$
\begin{aligned}
\widetilde{\mathcal{E}}^{\uparrow}_{\ell\ell'}(\mathbf{r}, \mathbf{r}^{\uparrow *}_c) &= \widetilde{\mathcal{E}}^{\uparrow}_{\ell\ell'}(\mathbf{r}^t_c, \mathbf{r}^{\uparrow *}_c) e^{\mathrm{i}\mathbf{k}_\alpha\cdot(\rho-\rho^t_c)} e^{\mathrm{i}k_{\ell z}(z-z^t_c)},\\
\widetilde{\mathcal{E}}^{\downarrow}_{\ell\ell'}(\mathbf{r}, \mathbf{r}^{\downarrow *}_c) &= \widetilde{\mathcal{E}}^{\downarrow}_{\ell\ell'}(\mathbf{r}^t_c, \mathbf{r}^{\downarrow *}_c) e^{\mathrm{i}\mathbf{k}_\alpha\cdot(\rho-\rho^t_c)} e^{-\mathrm{i}k_{\ell z}(z-z^t_c)}.
\end{aligned}
\tag{10.63}
$$

Applying the Funk-Hecke formula (10.25), we obtain

$$
e^{\mathrm{i}\mathbf{k}_\alpha\cdot(\rho-\rho^t_c)} e^{\pm\mathrm{i}k_{\ell z}(z-z^t_c)} = 4\pi \sum_{n'=0}^{\infty} \sum_{|m'|=0}^{n'} \mathrm{i}^{n'} (\pm 1)^{\mu} j_{n'}(k_\ell r_t) Y^{m'}_{n'}(\theta_t, \varphi_t) \widehat{P}^{m'}_{n'}\left(\frac{k_{\ell z}}{k_\ell}\right) e^{-\mathrm{i}m'\alpha},
\tag{10.64}
$$

where $\mu = (n'+m')(\text{mod } 2)$. Substituting into (10.57), the multipole expansion is translated to a local expansion (10.58) via

$$
L^{\uparrow *}_{nm} = \sum_{n'=0}^{\infty} \sum_{|m'|=0}^{n'} T^{\uparrow *}_{nm,n'm'} M^{\uparrow *}_{n'm'}, \quad L^{\downarrow *}_{nm} = (-1)^{n+m} \sum_{n'=0}^{\infty} \sum_{|m'|=0}^{n'} T^{\downarrow *}_{nm,n'm'} M^{\downarrow *}_{n'm'}, \tag{10.65}
$$

and the multipole-to-local translation operators are given in integral forms as follows

$$
\begin{aligned}
T^{\uparrow *}_{nm,n'm'} &= \frac{(-1)^{n'}}{2\pi} \int_0^{\infty} \int_0^{2\pi} k_\rho \frac{\widetilde{\mathcal{E}}^{\uparrow}_{\ell\ell'}(\mathbf{r}^t_c, \mathbf{r}^{\uparrow *}_c) \sigma^{\uparrow *}_{\ell\ell'}(k_\rho)}{k_{\ell z}} Q^{n'm'}_{nm}(k_\rho) e^{\mathrm{i}(m'-m)\alpha} d\alpha dk_\rho,\\
T^{\downarrow *}_{nm,n'm'} &= \frac{(-1)^{n'}}{2\pi} \int_0^{\infty} \int_0^{2\pi} k_\rho \frac{\widetilde{\mathcal{E}}^{\downarrow}_{\ell\ell'}(\mathbf{r}^t_c, \mathbf{r}^{\downarrow *}_c) \sigma^{\downarrow *}_{\ell\ell'}(k_\rho)}{k_{\ell z}} Q^{n'm'}_{nm}(k_\rho) e^{\mathrm{i}(m'-m)\alpha} d\alpha dk_\rho,
\end{aligned}
\tag{10.66}
$$

where

$$
Q^{n'm'}_{nm}(k_\rho) = \mathrm{i}^{n+n'+1} \widehat{P}^m_n\left(\frac{k_{\ell z}}{k_\ell}\right) \widehat{P}^{m'}_{n'}\left(\frac{k_{\ell' z}}{k_{\ell'}}\right).
$$

We note that the convergence of the Sommerfeld-type integrals in (10.66) is ensured by the conditions in (10.61).

10.2.4 A FMM Algorithm for Wave Sources in Layered Media

Let $\mathscr{P}_\ell = \{(Q_{\ell j}, \boldsymbol{r}_{\ell j}),\ j = 1, 2, \dots, N_\ell\}$ be a group of source particles distributed in the ℓ-th layer of a multilayered medium with $L+1$ layers (see Fig. 10.2). The interactions between all $N := N_0 + N_1 + \cdots + N_L$ particles are given by the summations

$$\Phi_\ell(\mathbf{r}_{\ell i}) = \Phi_\ell^{\text{free}}(\mathbf{r}_{\ell i}) + \sum_{\ell'=0}^{L-1}[\Phi_{\ell\ell'}^{\uparrow\uparrow}(\mathbf{r}_{\ell i}) + \Phi_{\ell\ell'}^{\downarrow\uparrow}(\mathbf{r}_{\ell i})] + \sum_{\ell'=1}^{L}[\Phi_{\ell\ell'}^{\uparrow\downarrow}(\mathbf{r}_{\ell i}) + \Phi_{\ell\ell'}^{\downarrow\downarrow}(\mathbf{r}_{\ell i})], \tag{10.67}$$

for $i = 1, 2, \dots, N_\ell$, $\ell = 0, 1, \dots, L$, where

$$\Phi_\ell^{\text{free}}(\mathbf{r}_{\ell i}) := \sum_{j=1, j\neq i}^{N_\ell} Q_{\ell j} h_0^{(1)}(k_\ell|\mathbf{r}_{\ell i} - \mathbf{r}_{\ell j}|), \quad \Phi_{\ell\ell'}^{**}(\mathbf{r}_{\ell i}) := \sum_{j=1}^{N_{\ell'}} Q_{\ell' j} u_{\ell\ell'}^{**}(\mathbf{r}_{\ell i}, \mathbf{r}_{\ell' j}). \tag{10.68}$$

Here, $u_{\ell\ell'}^{**}(\mathbf{r}, \mathbf{r}')$ of (10.38)–(10.39) are the reaction components of the layered Green's function in the ℓ-th layer due to a point source in the ℓ'-th layer. We omit the factor $\frac{ik_\ell}{4\pi}$ in $u_{\ell\ell'}^{**}(\mathbf{r}, \mathbf{r}')$ to maintain consistency with the spherical Hankel function used for the free space component.

Since the reaction components of Green's function in multilayer media have different expressions (10.38) for source and target particles in different layers, it is necessary to perform calculations individually for interactions between any two groups of particles among the $L+1$ groups $\{\mathscr{P}_\ell\}_{\ell=0}^{L}$. Here, we only consider the contribution from the reaction components, as the free space components can be calculated using free space FMM. Without loss of generality, let us focus on the computation of the $\uparrow\uparrow$ component in ℓ-th layer due to sources in the ℓ'-th layer, i.e., $\Phi_{\ell\ell'}^{\uparrow\uparrow}(\mathbf{r}_{\ell i})$ for all $i = 1, 2, \dots, N_\ell$ where we have source particles $\mathscr{P}_{\ell'}$ and target particles $\mathscr{P}_\ell$. According to the discussion in the last section, we use equivalent polarization sources (Fig. 10.4)

$$\mathbf{r}_{\ell' j}^{\uparrow\uparrow} := (x_{\ell' j}, y_{\ell' j}, d_\ell - (z_{\ell' j} - d_{\ell'})),$$

and re-express $\Phi_{\ell\ell'}^{\uparrow\uparrow}(\mathbf{r}_{\ell i})$ as

$$\Phi_{\ell\ell'}^{\uparrow\uparrow}(\mathbf{r}_{\ell i}) = \sum_{j=1}^{N_{\ell'}} Q_{\ell' j} \tilde{u}_{\ell\ell'}^{\uparrow\uparrow}(\mathbf{r}_{\ell i}, \mathbf{r}_{\ell' j}^{\uparrow\uparrow}), \quad i = 1, 2, \dots, N_\ell. \tag{10.69}$$

A feature of the equivalent polarization sources $\{\boldsymbol{r}_{\ell' j}^{\uparrow\uparrow}\}_{j=1}^{N_{\ell'}}$ is that they are all separated from the corresponding targets $\{\mathbf{r}_{\ell i}\}_{i=1}^{N_\ell}$ by the interface $z = d_\ell$. In fact, they are located on different sides of the interface $z = d_\ell$, see Fig. 10.4 (right). Equivalent

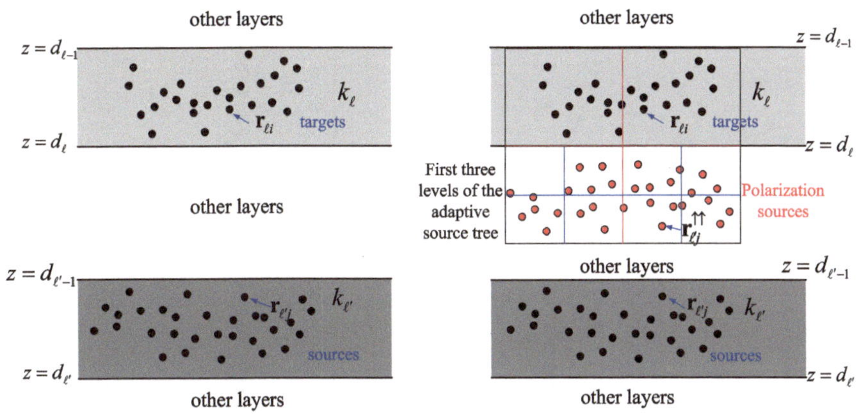

Fig. 10.4 Equivalent polarization sources $\mathbf{r}^{\uparrow\uparrow}_{\ell' j}$ and boxes in the source tree

polarization sources (10.51) defined for other reaction components also have this property.

The framework of the free space FMM with ME (10.57), LE (10.58), M2L translation (10.65) and (10.66) and free-space ME and LE center shifting allow us to use FMM for the computation of the reaction component $\Phi^{\uparrow\uparrow}_{\ell\ell'}(\mathbf{r}_{\ell i})$. The FMM for other reaction components can be treated in the same manner. In the FMM for reaction components, for each reaction component, a corresponding large box is defined to include all equivalent polarization sources and target particles where the adaptive tree structure will be built by a bisection procedure (Fig. 10.4 (right)). Note that the validity of the ME (10.57), LE (10.58) and M2L translation (10.65) used in the algorithm imposes restrictions (10.61) on the centers. This can be ensured by setting the largest box for the specific reaction component to be equally divided by the interface between equivalent polarization sources and targets (Fig. 10.4 (right)). Thus, the largest box for the FMM implementation will be different for different reaction components. With this setting, all source and target boxes of levels higher than the zeroth level in the adaptive tree structure will have centers below or above the interface $z = d_\ell$, respectively.

The fast multipole algorithm for the computation of the reaction component $\Phi^{\uparrow\uparrow}_{\ell\ell'}(\mathbf{r}_{\ell i})$ is summarized as follows.

FMM of H-level for the general component $\Phi^{\uparrow\uparrow}_{\ell\ell'}(\mathbf{r}_{\ell i}), i = 1, 2, \ldots, N_\ell$

- Determine the z-coordinates of the equivalent polarization sources for all the source particles. Generate an adaptive hierarchical tree structure for polarization sources $\{Q_{\ell' j}, \mathbf{r}^{\uparrow\uparrow}_{\ell' j}\}_{j=1}^{N_{\ell'}}$, targets $\{\mathbf{r}_{\ell i}\}_{i=1}^{N_\ell}$.
- **Upward pass**: For $l = H \to 0$: For all boxes j on source tree level l If j is a leaf node form the free-space ME using Eq. (10.57). Else form the free-space ME by merging children's expansions using the free-space center shift translation operator (10.21).

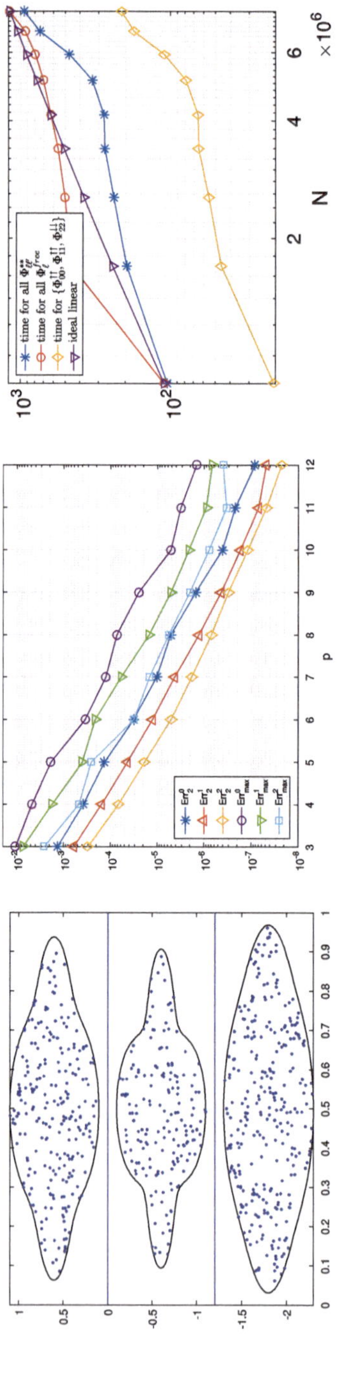

Fig. 10.5 Performance of the FMM for a three layer problem. (Left) Distribution of wave sources; (middle) convergence rates versus order p; (right) CPU time versus N (up to 8 million sources) for the free space part and reaction field parts of the Green's function

- **Downward pass** For $l = 1 \to H$: For all boxes j on target tree level l shift the LE of j's parent to j itself using the free-space shifting (10.21). Collect interaction list contribution using the source box to target box translation operator in Eq. (10.65).
- **Evaluate local expansions**: For each leaf node (childless box) evaluate the local expansion at each particle location.
- **Local Direct interactions**: For$i = 1 \to N$ compute Eq. (10.69) of the target particle i in the neighboring boxes.

All the interactions in (10.67) will be obtained then by calculating all components and summing them up.

Figure 10.5 shows the linear scaling performance of the FMM, for both free space part and reaction field parts, for computing the interactions of up to 8 million wave sources embedded in a 3-layer media [3].

10.3 Summary

The original FMMs for the free-space Greens' functions developed by Greengard and Rokhlin in the 1980s are considered as one of the top 10 algorithms of the 20th century [15]. With multipole and local expansions for the far field of sources in layered media presented here, the FMM is extended for wave interactions in layered media, which has various applications in parameter extractions for microchip designs, geophysical, and medical applications. An extension of the FMM for the Helmholtz equation presented here can be carried out for Maxwell's equations using mixed potential formulations [16]. Moreover, the FMMs for the Laplace equation [17] and Poisson Boltzmann equation [18] in layered media can be obtained via the same approach. Finally, the FMM for the Helmholtz equation based on compression techniques other than multipole expansion, such as Taylor expansions [13] or SVD [19] as well as hierarchical matrix methods [20], are alternative methods for fast computations of wave source interactions in layered media.

References

1. W.Z. Zhang, B. Wang, W. Cai, Exponential convergence for multipole expansion and translation to local expansions for sources in layered media: 2-D acoustic wave. SIAM Numer. Anal. **58**, 1440–1468 (2020)
2. W. Zhang, B. Wang, W. Cai, Exponential convergence theory of the multipole and local expansions for the 3-D Laplace equation in layered media. Ann. Appl. Math. **39**, 1–51 (2023)
3. B. Wang, W.Z. Zhang, W. Cai, Fast multipole method for 3-d Helmholtz equation in layered media. SIAM J. Sci. Comput **41**, A3954–A3981 (2019)
4. M. Abramowitz, I.A. Stegun. *Handbook of Mathematical Functions with Formulas, Graphs, and Mathematical Tables*, 9th edn. (Dover Publications, New York, 1972)
5. V. Rokhlin, Rapid solution of integral equations of scattering theory in two dimensions. J. Comput. Phys. **86**, 414–439 (1990)

6. M.H. Cho, J.F. Huang, D.X. Chen, W. Cai, A heterogeneous FMM for layered media Helmholtz equation I: two layers in R^2. J. Comput. Phys. **369**, 237–251 (2018)
7. P.A. Martin, *Multiple Scattering: Interaction of Time-Harmonic Waves with N Obstacles*, number 107 (Cambridge University Press, 2006)
8. G.N. Watson, *A Treatise of the Theory of Bessel Functions*, 2nd edn. (Cambridge University Press, Cambridge, UK, 1966)
9. W.C. Chew, Recurrence relations for three-dimensional scalar addition theorem. J. Electromagnet. Waves Appl. **6**(1–4), 133–142 (1992)
10. N.A. Gumerov, R. Duraiswami, Recursions for the computation of multipole translation and rotation coefficients for the 3-D Helmholtz equation. SIAM J. Sci. Comput. **25**(4), 1344–1381 (2004)
11. L. Greengard, V. Rokhlin, A fast algorithm for particle simulations. J. Comput. phys. **73**(2), 325–348 (1987)
12. L. Greengard, V. Rokhlin, A new version of the fast multipole method for the Laplace equation in three dimensions. Acta Numer. **6**, 229–269 (1997)
13. B. Wang, D. Chen, B. Zhang, W. Zhang, M.H. Cho, W. Cai, Taylor expansion based fast multipole method for 3-D Helmholtz equations in layered media. J. Comput. Phys. **401**, 109008 (2020)
14. M.H. Cho, W. Cai, A parallel fast algorithm for computing the Helmholtz integral operator in 3-d layered media. J. Comput. Phys. **231**, 5910–5925 (2012)
15. B.A. Cipra, The best of the 20th century: editors name top 10 algorithms. SIAM News **33**(4), 1–2 (2000)
16. W.Z. Zhang, B. Wang, W. Cai, A matrix basis formulation for the dyadic Green's functions of Maxwell's equations in layered media. SIAM J. Appl. Math. **82**(5), 1710–32 (2022)
17. B. Wang, W. Zhang, W. Cai, Fast multipole method for 3-D Laplace equation in layered media. Comput. Phys. Commun. **259**, 107645 (2021)
18. B. Wang, W. Zhang, W. Cai, Fast multipole method for 3-D Poisson-Boltzmann equation in layered electrolyte-dielectric media. J. Comput. Phys. **439**, 110379 (2021)
19. L. Ying, G. Biros, D. Zorin, A kernel-independent adaptive fast multipole algorithm in two and three dimensions. J. Comput. Phys. **196**, 591–626 (2004)
20. W. Hackbusch, *Hierarchical Matrices: Algorithms and Analysis* (Springer, Heidelberg, 2015), p. 14

Chapter 11
High-Order Methods for Surface Electromagnetic Integral Equations

In this chapter, we will present surface integral equations for electromagnetic scattering. Galerkin methods using mixed vector–scalar potentials and Nyström collocation methods using hyper-singular formulation will be discussed. Also, we will discuss the issue of removing resonance from cavity modes with a combined integral equation approach. Finally, the high-order Rao–Wilton–Glisson (RWG) current basis for the Galerkin methods will be given. The volume integral equation method for Maxwell's equations will be discussed in Sect. 14.4.

11.1 Electric and Magnetic Field Surface Integral Equations in Layered Media

11.1.1 Integral Representations

We consider the $N + 2$-layered medium discussed in Sect. 9.2.2 with planar interfaces at $z = d_i, 0 \leq i \leq N$, as shown in Fig. 11.1. The electric and magnetic fields will satisfy boundary conditions at the planar interfaces and also on the bottom terminal layer, or the Sommerfeld radiation conditions at $z \to \pm\infty$ if either the lower or the upper space is not terminated at a finite location. Each of the layers is assumed to be an isotropic material with permittivity $\epsilon = \epsilon(\mathbf{r})$ and permeability $\mu = \mu(\mathbf{r})$. Embedded in this multi-layered medium is a 3-D conducting object with a surface S which has an outwrmal $\mathbf{n}$. Let V_1 be the multi-layered medium outside the scatterer S and V_2 the volume inside the scatterer S. Denote the ith layer in V_1 by $V_1^i = \{(x, y, z) \in V_1 : d_i < z < d_{i-1}\}, 0 \leq i \leq N + 1, d_0 = 0, d_{-1} = \infty$, and $d_{N+1} = -\infty$. Thus

$$V_1 = \bigcup_{0 \leq i \leq N+1} V_1^i, \tag{11.1}$$

W. Cai, *Deterministic, Stochastic, and Deep Learning Methods for Computational Electromagnetics*, https://doi.org/10.1007/978-981-96-0100-4_11

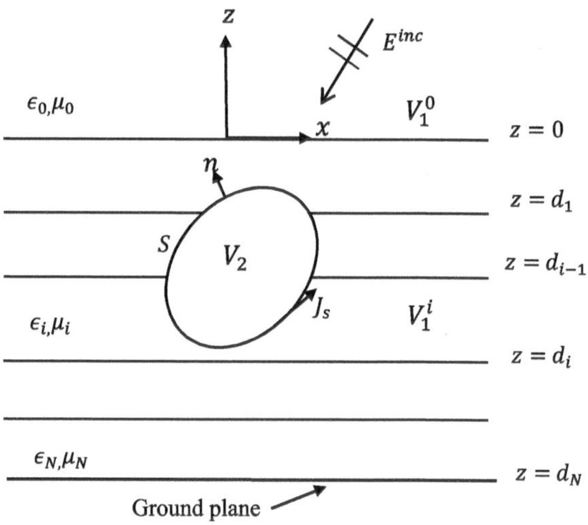

Fig. 11.1 Multi-layered medium with an embedded scatterer

and the interfaces in V_1 are denoted by $\Sigma_i = \{(x, y, z) \in V_1 : z = d_i\}$ and

$$\Sigma = \bigcup_{0 \le i \le N} \Sigma_i. \tag{11.2}$$

Let $\mathbf{J}_e(\mathbf{r})$ be some electric source outside S and in V_1^0, and let $\mathbf{E}_1, \mathbf{H}_1$ and $\mathbf{E}_2, \mathbf{H}_2$ be the electric field and the magnetic field in V_1 and V_2, respectively. From (8.30) and (8.31), they satisfy the following vector equations:

$$\nabla \times \frac{1}{\mu(\mathbf{r})} \nabla \times \mathbf{E}_1(\mathbf{r}) - \omega^2 \epsilon(\mathbf{r}) \mathbf{E}_1(\mathbf{r}) = -\mathrm{i}\omega \mathbf{J}_e(\mathbf{r}), \quad \mathbf{r} \in V_1 \backslash \Sigma, \tag{11.3}$$

$$\nabla \times \frac{1}{\epsilon(\mathbf{r})} \nabla \times \mathbf{H}_1(\mathbf{r}) - \omega^2 \mu(\mathbf{r}) \mathbf{H}_1(\mathbf{r}) = \nabla \times \frac{1}{\epsilon(\mathbf{r})} \mathbf{J}_e(\mathbf{r}), \quad \mathbf{r} \in V_1 \backslash \Sigma, \tag{11.4}$$

with the boundary conditions (8.62)–(8.65) along the interfaces $\Sigma_i, 0 \le i \le N$.

To derive an integral representation for the fields in V_1, we consider the dyadic Green's functions $\overline{\mathbf{G}}_E^1(\mathbf{r}, \mathbf{r}')$ and $\overline{\mathbf{G}}_H^1(\mathbf{r}, \mathbf{r}')$. The former satisfies

$$\nabla \times \frac{1}{\mu(\mathbf{r})} \nabla \times \overline{\mathbf{G}}_E^1(\mathbf{r}, \mathbf{r}') - \omega^2 \epsilon(\mathbf{r}) \overline{\mathbf{G}}_E^1(\mathbf{r}, \mathbf{r}') = \frac{1}{\mu(\mathbf{r})} \overline{\mathbf{I}} \delta(\mathbf{r} - \mathbf{r}'), \quad \mathbf{r} \in V_1 \backslash \Sigma, \tag{11.5}$$

and from Faraday's law in Maxwell's equations the magnetic dyadic Green's function is given by

$$\overline{\mathbf{G}}_H^1(\mathbf{r}, \mathbf{r}') = -\frac{1}{\mathrm{i}\omega\mu(\mathbf{r})} \nabla \times \overline{\mathbf{G}}_E^1(\mathbf{r}, \mathbf{r}'), \tag{11.6}$$

which satisfies

$$\nabla \times \frac{1}{\epsilon(\mathbf{r})} \nabla \times \overline{\mathbf{G}}_H^1(\mathbf{r}, \mathbf{r}') - \omega^2 \mu(\mathbf{r}) \overline{\mathbf{G}}_H^1(\mathbf{r}, \mathbf{r}')$$
$$= - \overline{\mathbf{I}} \nabla \times \frac{1}{\mathrm{i}\omega\mu(\mathbf{r})\epsilon(\mathbf{r})} \delta(\mathbf{r} - \mathbf{r}'), \quad \mathbf{r} \in V_1 \backslash \Sigma. \tag{11.7}$$

Along the interfaces between adjacent layers, both tangential components of each column vector in the dyadic Green's functions are continuous for $\mathbf{r} \in \Sigma_i$ and $\mathbf{r}' \in V_1 \backslash \Sigma$:

$$\mathbf{n} \times \left[\overline{\mathbf{G}}_E^1(\mathbf{r}, \mathbf{r}') \right] = \mathbf{0}, \tag{11.8}$$

$$\mathbf{n} \times \left[\overline{\mathbf{G}}_H^1(\mathbf{r}, \mathbf{r}') \right] = -\mathbf{n} \times \left[\frac{1}{\mu(\mathbf{r})} \nabla \times \overline{\mathbf{G}}_E^1(\mathbf{r}, \mathbf{r}') \right] = \mathbf{0}, \tag{11.9}$$

where [ ] denotes the difference from both sides of Σ_i.

Following Chew [1], we multiply (11.3) by $\overline{\mathbf{G}}_E^1(\mathbf{r}, \mathbf{r}') \cdot \mathbf{e}$, with $\mathbf{e}$ being any of the three coordinate axis unit vectors, and we have

$$\left[\nabla \times \frac{1}{\mu} \nabla \times \mathbf{E}_1(\mathbf{r}) \right] \cdot \overline{\mathbf{G}}_E^1(\mathbf{r}, \mathbf{r}') \cdot \mathbf{e} - \omega^2 \epsilon \mathbf{E}_1(\mathbf{r}) \cdot \overline{\mathbf{G}}_E^1(\mathbf{r}, \mathbf{r}') \cdot \mathbf{e} = -\mathrm{i}\omega \mathbf{J}_e(\mathbf{r}) \cdot \overline{\mathbf{G}}_E^1(\mathbf{r}, \mathbf{r}') \cdot \mathbf{e}. \tag{11.10}$$

Next, we pre-multiply (11.5) by $\mathbf{E}_1(\mathbf{r})$ and then post-multiply it by $\mathbf{e}$ to arrive at the following equation:

$$\mathbf{E}_1(\mathbf{r}) \cdot \nabla \times \frac{1}{\mu} \nabla \times \overline{\mathbf{G}}_E^1(\mathbf{r}, \mathbf{r}') \cdot \mathbf{e} - \omega^2 \epsilon \mathbf{E}_1(\mathbf{r}) \cdot \overline{\mathbf{G}}_E^1(\mathbf{r}, \mathbf{r}') \cdot \mathbf{e} = \frac{1}{\mu(\mathbf{r}')} \mathbf{E}_1(\mathbf{r}) \cdot \mathbf{e} \delta(\mathbf{r} - \mathbf{r}'). \tag{11.11}$$

Subtracting (11.11) from (11.10) gives

$$\left[\nabla \times \frac{1}{\mu} \nabla \times \mathbf{E}_1(\mathbf{r}) \right] \cdot \overline{\mathbf{G}}_E^1(\mathbf{r}, \mathbf{r}') \cdot \mathbf{e} - \mathbf{E}_1(\mathbf{r}) \cdot \nabla \times \frac{1}{\mu} \nabla \times \overline{\mathbf{G}}_E^1(\mathbf{r}, \mathbf{r}') \cdot \mathbf{e}$$
$$= -\mathrm{i}\omega \mathbf{J}_e(\mathbf{r}) \cdot \overline{\mathbf{G}}_E^1(\mathbf{r}, \mathbf{r}') \cdot \mathbf{e} - \frac{1}{\mu} \mathbf{E}_1(\mathbf{r}) \cdot \mathbf{e} \delta(\mathbf{r} - \mathbf{r}'). \tag{11.12}$$

Integrating (11.12) over the region V_1^i yields

$$\int_{V_1^i} d\mathbf{r} \left[\left(\nabla \times \frac{1}{\mu} \nabla \times \right) \mathbf{E}_1(\mathbf{r}) \cdot \overline{\mathbf{G}}_E^1(\mathbf{r}, \mathbf{r}') \cdot \mathbf{e} \right.$$
$$\left. - \mathbf{E}_1(\mathbf{r}) \cdot \left(\nabla \times \frac{1}{\mu} \nabla \times \right) \overline{\mathbf{G}}_E^1(\mathbf{r}, \mathbf{r}') \cdot \mathbf{e} \right]$$
$$= -\mathrm{i}\omega \int_{V_1^i} d\mathbf{r} \left[\mathbf{J}_e(\mathbf{r}) \cdot \overline{\mathbf{G}}_E^1(\mathbf{r}, \mathbf{r}') \cdot \mathbf{e} \right] - c_i \frac{1}{\mu} \mathbf{E}_1(\mathbf{r}') \cdot \mathbf{e}, \tag{11.13}$$

where $c_i = 1$ if $\mathbf{r}' \in V_1^i$, and $c_i = 0$ if $\mathbf{r}' \notin V_1^i$.

The integrand on the left-hand side of (11.13) can be simplified using the following vector identity for the vector operator $\mathcal{L} = \nabla \times \frac{1}{\mu}\nabla\times$:

$$(\mathcal{L}\mathbf{f}) \cdot \mathbf{g} - \mathbf{f} \cdot (\mathcal{L}\mathbf{g}) = \nabla \cdot \frac{1}{\mu}[\,(\nabla \times \mathbf{f}) \times \mathbf{g} + \mathbf{f} \times (\nabla \times \mathbf{g})], \tag{11.14}$$

and thus (11.13) can be rewritten as

$$\int_{V_1^i} \mathrm{d}\mathbf{r}\, \nabla \cdot \frac{1}{\mu}\left[(\nabla \times \mathbf{E}_1(\mathbf{r})) \times \overline{\mathbf{G}}_E^1(\mathbf{r}, \mathbf{r}') \cdot \mathbf{e} + \mathbf{E}_1(\mathbf{r}) \times \left(\nabla \times \overline{\mathbf{G}}_E^1(\mathbf{r}, \mathbf{r}') \cdot \mathbf{e}\right)\right]$$
$$= -\mathrm{i}\omega \int_{V_1^i} \mathrm{d}\mathbf{r}\, \mathbf{J}_e(\mathbf{r}) \cdot \overline{\mathbf{G}}_E^1(\mathbf{r}, \mathbf{r}') \cdot \mathbf{e} - c_i \frac{1}{\mu}\mathbf{E}_1(\mathbf{r}') \cdot \mathbf{e}, \tag{11.15}$$

and the volume integral on the left-hand side of (11.15) could be reduced to a surface integral by Gauss's theorem. So, we have

$$-\int_{S_i \cup \Sigma_{i-1} \cup \Sigma_i} \mathrm{d}s\, \mathbf{n} \cdot \frac{1}{\mu(\mathbf{r})}\left[(\nabla \times \mathbf{E}_1(\mathbf{r})) \times \overline{\mathbf{G}}_E^1(\mathbf{r}, \mathbf{r}') \cdot \mathbf{e}\right.$$
$$\left. + \mathbf{E}_1(\mathbf{r}) \times \left(\nabla \times \overline{\mathbf{G}}_E^1(\mathbf{r}, \mathbf{r}') \cdot \mathbf{e}\right)\right]$$
$$= -\mathrm{i}\omega \int_{V_1^i} \mathrm{d}\mathbf{r}\left[\mathbf{J}_e(\mathbf{r}) \cdot \overline{\mathbf{G}}_E^1(\mathbf{r}, \mathbf{r}') \cdot \mathbf{e}\right] - c_i \frac{1}{\mu(\mathbf{r}')}\mathbf{E}_1(\mathbf{r}') \cdot \mathbf{e}, \tag{11.16}$$

where $S_i = S \cap V_1^i$, and $\mathbf{n}$ is the normal pointing to the interior of V_1^i.

Next, by using the vector identity $\mathbf{u} \cdot (\mathbf{v} \times \mathbf{w}) = (\mathbf{u} \times \mathbf{v}) \cdot \mathbf{w}$ and Faraday's law in Maxwell's equations $\nabla \times \mathbf{E}_1(\mathbf{r}) = -\mathrm{i}\omega\mu\mathbf{H}_1(\mathbf{r})$, we obtain

$$c_i\mathbf{E}_1(\mathbf{r}') \cdot \mathbf{e} = -\mathrm{i}\omega\mu(\mathbf{r}') \int_{V_1^i} \mathrm{d}\mathbf{r}\left[\mathbf{J}_e(\mathbf{r}) \cdot \overline{\mathbf{G}}_E^1(\mathbf{r}, \mathbf{r}') \cdot \mathbf{e}\right] - \mu(\mathbf{r}') \int_{S_i \cup \Sigma_{i-1} \cup \Sigma_i} \mathrm{d}s$$
$$\left[\mathrm{i}\omega\mathbf{n} \times \mathbf{H}_1(\mathbf{r}) \cdot \overline{\mathbf{G}}_E^1(\mathbf{r}, \mathbf{r}') \cdot \mathbf{e} - \mathbf{n} \times \mathbf{E}_1(\mathbf{r}) \cdot \frac{1}{\mu(\mathbf{r})}\nabla \times \overline{\mathbf{G}}_E^1(\mathbf{r}, \mathbf{r}') \cdot \mathbf{e}\right]. \tag{11.17}$$

It can be shown that the surface integrals over each Σ_i will be canceled from adjacent domains V_1^{i+1} and V_1^i. To see this, let us write $\mathbf{g} = \overline{\mathbf{G}}_E^1(\mathbf{r}, \mathbf{r}') \cdot \mathbf{e}$ and $\mathbf{h} = \frac{1}{\mu(\mathbf{r})}\nabla \times \overline{\mathbf{G}}_E^1(\mathbf{r}, \mathbf{r}') \cdot \mathbf{e}$. With the identity

$$\mathbf{f} = \mathbf{n} \times (\mathbf{f} \times \mathbf{n}) + (\mathbf{f} \cdot \mathbf{n})\mathbf{n},$$

the integrands in the surface integral can be written as

$$\mathbf{n}\times\mathbf{H}_1(\mathbf{r})\cdot\mathbf{g}=[\mathbf{n}\times\mathbf{H}_1(\mathbf{r})]\cdot\left[\mathbf{n}\times(\mathbf{g}\times\mathbf{n})\right],$$
$$\mathbf{n}\times\mathbf{E}_1(\mathbf{r})\cdot\mathbf{h}=[\mathbf{n}\times\mathbf{E}_1(\mathbf{r})]\cdot[\mathbf{n}\times(\mathbf{h}\times\mathbf{n})], \quad (11.18)$$

after using the vector identity $(\mathbf{n}\times\mathbf{f})\cdot\mathbf{n}=0$. As the tangential components of Green's functions (11.8) and (11.9) as well as those of $\mathbf{E}_1(\mathbf{r})$ and $\mathbf{H}_1(\mathbf{r})$ are continuous across the interfaces Σ_i, i.e.,

$$[\mathbf{n}\times(\mathbf{g}\times\mathbf{n})]_{\Sigma_i}=\mathbf{0},\ [\mathbf{n}\times(\mathbf{h}\times\mathbf{n})]_{\Sigma_i}=\mathbf{0},$$
$$\mathbf{n}\times[\mathbf{E}_1(\mathbf{r})]_{\Sigma_i}=\mathbf{0},\ \mathbf{n}\times[\mathbf{H}_1(\mathbf{r})]_{\Sigma_i}=\mathbf{0}, \quad (11.19)$$

the surface integrals from both sides of Σ_i will cancel each other when (11.17) is summed up over all V_1^i. As $\mathbf{e}$ is any of the three coordinate axis unit vectors, after all surface integrals over Σ_i cancel out, the following integral representation for the electric field $\mathbf{E}_1(\mathbf{r})$ is obtained (with $\mathbf{r}$ and $\mathbf{r}'$ switched):

$$\mathbf{E}_1(\mathbf{r})=\mathbf{E}^{\mathrm{inc}}(\mathbf{r})-\mu(\mathbf{r})\int_S \mathrm{d}s'\Big[\mathrm{i}\omega\mathbf{n}\times\mathbf{H}_1(\mathbf{r}')\cdot\overline{\mathbf{G}}_E^1(\mathbf{r}',\mathbf{r})$$
$$-\mathbf{n}\times\mathbf{E}_1(\mathbf{r}')\cdot\frac{1}{\mu(\mathbf{r}')}\nabla'\times\overline{\mathbf{G}}_E^1(\mathbf{r}',\mathbf{r})\Big],\quad \mathbf{r}\in V_1, \quad (11.20)$$

where $\mathbf{E}^{\mathrm{inc}}(\mathbf{r})$ would be the incident electric field generated by the source $\mathbf{J}_e(\mathbf{r}')$ in V_1, i.e.,

$$\mathbf{E}^{\mathrm{inc}}(\mathbf{r})=-\mathrm{i}\omega\mu(\mathbf{r})\int_{V_1}\mathrm{d}\mathbf{r}'\,\mathbf{J}_e(\mathbf{r}')\cdot\overline{\mathbf{G}}_E^1(\mathbf{r}',\mathbf{r}). \quad (11.21)$$

From the proceeding derivation, if $\mathbf{r}'\in V_2$, then c_i in (11.17) is replaced with zero, and we have (after switching $\mathbf{r}$ and $\mathbf{r}'$)

$$\mathbf{0}=\mathbf{E}^{\mathrm{inc}}(\mathbf{r})-\mu(\mathbf{r})\int_S \mathrm{d}s'\Big[\mathrm{i}\omega\mathbf{n}\times\mathbf{H}_2(\mathbf{r}')\cdot\overline{\mathbf{G}}_E^1(\mathbf{r}',\mathbf{r})$$
$$-\mathbf{n}\times\mathbf{E}_2(\mathbf{r}')\cdot\frac{1}{\mu(\mathbf{r}')}\nabla'\times\overline{\mathbf{G}}_E^1(\mathbf{r}',\mathbf{r})\Big],\quad \mathbf{r}\in V_2. \quad (11.22)$$

Let us define the equivalent surface magnetic current $\mathbf{J}_{m(s)}$ and the electric current $\mathbf{J}_{e(s)}$ over S, i.e., for $\mathbf{r}\in S$,

$$\mathbf{J}_{e(s)}(\mathbf{r})\equiv\mathbf{n}\times\mathbf{H}_1(\mathbf{r})=\mathbf{n}\times\mathbf{H}_2(\mathbf{r}), \quad (11.23)$$
$$\mathbf{J}_{m(s)}(\mathbf{r})\equiv-\mathbf{n}\times\mathbf{E}_1(\mathbf{r})=-\mathbf{n}\times\mathbf{E}_2(\mathbf{r}). \quad (11.24)$$

Equations (11.20) and (11.22) imply that those currents, together with the volume current source $\mathbf{J}_e(\mathbf{r})$ in V_1^0, will produce the same field as $\mathbf{E}_1(\mathbf{r})$ outside S, but zero field inside S. After using the following reciprocal property for the dyadic Green's function in a layered medium [2, Sect. 4.4]:

$$\frac{1}{k^2(\mathbf{r}')}\left[\nabla' \times \overline{\mathbf{G}}_E^1(\mathbf{r}', \mathbf{r})\right]^{\mathrm{T}} = \frac{1}{k^2(\mathbf{r})}\nabla \times \overline{\mathbf{G}}_E^1(\mathbf{r}, \mathbf{r}'), \tag{11.25}$$

$$\frac{1}{\mu(\mathbf{r}')}\overline{\mathbf{G}}_E^1(\mathbf{r}', \mathbf{r})^{\mathrm{T}} = \frac{1}{\mu(\mathbf{r})}\overline{\mathbf{G}}_E^1(\mathbf{r}, \mathbf{r}'), \tag{11.26}$$

we have finally the following electric field integral representation (assuming $\mu(\mathbf{r}) = \mu(\mathbf{r}')$):

$$\begin{aligned}\mathbf{E}_1(\mathbf{r}) = &\mathbf{E}^{\mathrm{inc}}(\mathbf{r}) - \mu(\mathbf{r})\int_S \mathrm{d}s'\left[\mathrm{i}\omega\overline{\mathbf{G}}_E^1(\mathbf{r}, \mathbf{r}')\cdot\mathbf{J}_{e(s)}(\mathbf{r}')\right.\\ &\left.+\frac{1}{\mu(\mathbf{r})}\frac{\epsilon(\mathbf{r}')}{\epsilon(\mathbf{r})}\nabla\times\overline{\mathbf{G}}_E^1(\mathbf{r}, \mathbf{r}')\cdot\mathbf{J}_{m(s)}(\mathbf{r}')\right], \quad \mathbf{r}\in V_1,\end{aligned} \tag{11.27}$$

where $k^2(\mathbf{r}) = \omega^2\epsilon(\mathbf{r})\mu(\mathbf{r})$, and

$$\begin{aligned}\mathbf{0} = &\mathbf{E}^{\mathrm{inc}}(\mathbf{r}) - \mu(\mathbf{r})\int_S \mathrm{d}s'\left[\mathrm{i}\omega\overline{\mathbf{G}}_E^1(\mathbf{r}, \mathbf{r}')\cdot\mathbf{J}_{e(s)}(\mathbf{r}')\right.\\ &\left.+\frac{1}{\mu(\mathbf{r}')}\nabla\times\overline{\mathbf{G}}_E^1(\mathbf{r}, \mathbf{r}')\cdot\mathbf{J}_{m(s)}(\mathbf{r}')\right], \quad \mathbf{r}\in V_2.\end{aligned} \tag{11.28}$$

Remark 11.1 Equivalent surface currents $\mathbf{J}_{e(s)}$ and $\mathbf{J}_{m(s)}$ produce the total field $\mathbf{E}_1$ and $\mathbf{H}_1$ outside S and zero field inside; the former fact is Huygens' principle [3] and the latter (11.28) is the extinction principle [4].

Similarly, we can obtain the following magnetic field integral representation:

$$\begin{aligned}\mathbf{H}_1(\mathbf{r}) = &\mathbf{H}^{\mathrm{inc}}(\mathbf{r}) + \epsilon(\mathbf{r})\int_S \mathrm{d}s'\left[\frac{1}{\epsilon(\mathbf{r}')}\nabla\times\overline{\mathbf{G}}_E^1(\mathbf{r}, \mathbf{r}')\cdot\mathbf{J}_{e(s)}(\mathbf{r}')\right.\\ &\left.-\mathrm{i}\omega\overline{\mathbf{G}}_E^1(\mathbf{r}, \mathbf{r}')\cdot\mathbf{J}_{m(s)}(\mathbf{r}')\right], \quad \mathbf{r}\in V_1,\end{aligned} \tag{11.29}$$

$$\begin{aligned}\mathbf{0} = &\mathbf{H}^{\mathrm{inc}}(\mathbf{r}) + \epsilon(\mathbf{r})\int_S \mathrm{d}s'\left[\frac{1}{\epsilon(\mathbf{r}')}\nabla\times\overline{\mathbf{G}}_E^1(\mathbf{r}, \mathbf{r}')\cdot\mathbf{J}_{e(s)}(\mathbf{r}')\right.\\ &\left.-\mathrm{i}\omega\overline{\mathbf{G}}_E^1(\mathbf{r}, \mathbf{r}')\cdot\mathbf{J}_{m(s)}(\mathbf{r}')\right], \quad \mathbf{r}\in V_2,\end{aligned} \tag{11.30}$$

where $\mathbf{H}^{\mathrm{inc}}(\mathbf{r})$ would be the incident magnetic field generated by the source $\mathbf{J}_e(\mathbf{r}')$ located within V_1, i.e.,

$$\mathbf{H}^{\mathrm{inc}}(\mathbf{r}) = \epsilon(\mathbf{r})\int_{V_1}\mathrm{d}\mathbf{r}'\frac{1}{\epsilon(\mathbf{r}')}\nabla\times\overline{\mathbf{G}}_E^1(\mathbf{r}, \mathbf{r}')\cdot\mathbf{J}_e(\mathbf{r}'). \tag{11.31}$$

11.1.2 Singular and Hyper-Singular Surface Integral Equations

To derive surface integral equations for electric and magnetic fields $\mathbf{E}_1$ and $\mathbf{H}_1$ over S, we would have to let $\mathbf{r}$ reside on S. As the dyadic Green's functions become singular when $\mathbf{r}$ approaches S, (11.27)–(11.30) for $\mathbf{r}$ on S would be obtained by considering the limits of $\mathbf{r}$ approaching S. As a result, the Cauchy principal value or the Hadamard finite part discussed previously in Sects. 3.1.1 and 3.1.3 will appear for the electromagnetic fields.

Let $\mathbf{r}_0 = (x_0, y_0, z_0) \in S$ not on any of the interfaces Σ_i, which, for simplicity, is assumed to be $(0, 0, 0)$, and let S_a be a circular patch of radius a over S centered at $\mathbf{r}_0$. As the dyadic Green's function $\overline{\mathbf{G}}_E^1(\mathbf{r}', \mathbf{r})$ shares the same type of singularity as the free-space Green's function, we can assume that, near the singularity, $\overline{\mathbf{G}}_E^1(\mathbf{r}, \mathbf{r}')$ is given by (9.35).

Denote the first and second integrals in (11.27) as the contribution by the electric and the magnetic currents, respectively, by

$$\mathbf{E}^{\text{elec}}(\mathbf{r}) = \mathrm{i}\omega\mu(\mathbf{r}) \int_S \mathrm{d}s'\, \overline{\mathbf{G}}_E^1(\mathbf{r}, \mathbf{r}') \cdot \mathbf{J}_{e(s)}(\mathbf{r}'), \tag{11.32}$$

$$\mathbf{E}^{\text{mag}}(\mathbf{r}) = \mu(\mathbf{r}) \int_S \mathrm{d}s' \frac{1}{\mu(\mathbf{r}')} \nabla \times \overline{\mathbf{G}}_E^1(\mathbf{r}, \mathbf{r}') \cdot \mathbf{J}_{m(s)}(\mathbf{r}'). \tag{11.33}$$

Therefore, (11.27) can be rewritten as

$$\mathbf{E}_1(\mathbf{r}) = \mathbf{E}^{\text{inc}}(\mathbf{r}) - \mathbf{E}^{\text{elec}}(\mathbf{r}) - \mathbf{E}^{\text{mag}}(\mathbf{r}), \quad \mathbf{r} \in V_1. \tag{11.34}$$

Principal value of $\mathbf{E}^{\text{mag}}(\mathbf{r})$

By using the identity

$$\begin{aligned} \nabla \times \overline{\mathbf{G}}_E^1(\mathbf{r}, \mathbf{r}') &= \nabla \times \left(\overline{\mathbf{I}} + \frac{1}{k^2} \nabla\nabla \right) g(\mathbf{r}, \mathbf{r}') \\ &= \nabla \times \left[g(\mathbf{r}, \mathbf{r}') \overline{\mathbf{I}} \right], \end{aligned} \tag{11.35}$$

the integral over S_a can be written as

$$\begin{aligned} &\int_{S_a} \mathrm{d}s'\, \nabla \times \overline{\mathbf{G}}_E^1(\mathbf{r}, \mathbf{r}') \cdot \mathbf{J}_{m(s)}(\mathbf{r}') \\ &= \int_{S_a} \mathrm{d}s'\, \nabla \times \left[g(\mathbf{r}, \mathbf{r}') \overline{\mathbf{I}} \right] \cdot \mathbf{J}_{m(s)}(\mathbf{r}') = \int_{S_a} \mathrm{d}s'\, \nabla g(\mathbf{r}, \mathbf{r}') \times \mathbf{J}_{m(s)}(\mathbf{r}') \\ &\approx \left[\int_{S_a} \mathrm{d}s'\, \nabla g(\mathbf{r}, \mathbf{r}') \right] \times \mathbf{J}_{m(s)}(\mathbf{r}_0). \end{aligned} \tag{11.36}$$

If we assume that S_a is flat with local coordinates $(x', y', z' = 0)$, $\mathbf{J}_{m(s)}(\mathbf{r}_0) = (J_{sx}, J_{sy}, 0)^{\mathrm{T}}$, then the contribution of the tangential component above is

$$\int_{S_a} \nabla_s g(\mathbf{r}, \mathbf{r}')\mathrm{d}s' = \int_0^a \int_0^{2\pi} \left(\frac{\partial}{\partial x'}, \frac{\partial}{\partial y'}\right) g(R)\rho\, \mathrm{d}\rho\, \mathrm{d}\phi$$
$$= \int_0^a \int_0^{2\pi} g'(R)\left(\frac{x'}{R}, \frac{y'}{R}\right)\rho\, \mathrm{d}\rho\, \mathrm{d}\phi = \mathbf{0}, \tag{11.37}$$

where $R = \sqrt{\rho^2 + (z - z')^2}$.

On the other hand, along the normal z-direction, we have

$$\int_{S_a} \frac{\partial}{\partial \mathbf{n}} g(\mathbf{r}, \mathbf{r}')\mathrm{d}s' = \int_0^a \int_0^{2\pi} \frac{\partial}{\partial z} g(R)\,|_{z'=0}\, \rho\, \mathrm{d}\rho\, \mathrm{d}\phi$$
$$= \int_0^a \int_0^{2\pi} g'(R)\frac{z}{R}\rho\, \mathrm{d}\rho\, \mathrm{d}\phi = -\frac{z}{2}\int_0^a \frac{(\mathrm{i}kR + 1)\mathrm{e}^{-\mathrm{i}kR}}{R^3}\rho\, \mathrm{d}\rho, \tag{11.38}$$

and if a is small and $z \to z' = 0$, then $R = \sqrt{\rho^2 + z^2} \to 0$, and we have

$$\int_{S_a} \frac{\partial}{\partial \mathbf{n}} g(\mathbf{r}, \mathbf{r}')\mathrm{d}s' \approx \frac{1}{2}\left(\frac{z}{\sqrt{a^2 + z^2}} - \frac{z}{|z|}\right) \to -\frac{1}{2}. \tag{11.39}$$

Therefore, substituting (11.37) and (11.39) into (11.36), we have

$$\mu(\mathbf{r}) \int_{S_a} \mathrm{d}s' \left[\frac{1}{\mu(\mathbf{r}')}\nabla \times \overline{\mathbf{G}}_E^1(\mathbf{r}, \mathbf{r}') \cdot \mathbf{J}_{m(s)}(\mathbf{r}')\right] \approx \begin{pmatrix} 0 \\ 0 \\ \int_{S_a} \frac{\partial g}{\partial \mathbf{n}}\, \mathrm{d}s' \end{pmatrix} \times \mathbf{J}_{m(s)}(\mathbf{r}_0)$$
$$= (-J_{sy}, J_{sx}, 0)^{\mathrm{T}} \int_{S_a} \frac{\partial}{\partial \mathbf{n}} g(\mathbf{r}, \mathbf{r}')\, \mathrm{d}s' \to -\frac{1}{2}(-J_{sy}, J_{sx}, 0)^{\mathrm{T}} = \frac{1}{2}\mathbf{J}_{m(s)}(\mathbf{r}_0) \times \mathbf{n}. \tag{11.40}$$

Let $\mathbf{r} \to \mathbf{r}_0$ (along the normal direction $\mathbf{n}$) in (11.33), so we have (after replacing $\mathbf{r}_0$ by $\mathbf{r}$)

$$
\begin{aligned}
\mathbf{E}^{\text{mag}}(\mathbf{r}) &= \mu(\mathbf{r})\text{p.v.} \int_S \mathrm{d}s' \frac{1}{\mu(\mathbf{r}')} \nabla \times \overline{\mathbf{G}}_E^1(\mathbf{r}, \mathbf{r}') \cdot \mathbf{J}_{m(s)}(\mathbf{r}') + \frac{1}{2}\mathbf{J}_{m(s)}(\mathbf{r}) \times \mathbf{n} \\
&= \mu(\mathbf{r}) \text{ p.v.} \int_S \mathrm{d}s' \frac{1}{\mu(\mathbf{r}')} \nabla g(\mathbf{r}, \mathbf{r}') \times \mathbf{J}_{m(s)}(\mathbf{r}') + \frac{1}{2}\mathbf{J}_{m(s)}(\mathbf{r}) \times \mathbf{n}.
\end{aligned} \tag{11.41}
$$

Hadamard finite part of $\mathbf{E}^{\text{elec}}(\mathbf{r})$ and its CPV representation

Here, we have

$$
\begin{aligned}
\mathbf{E}^{\text{elec}}(\mathbf{r}) &= \mathrm{i}\omega\mu(\mathbf{r}) \int_S \mathrm{d}s' \left(\bar{\mathbf{I}} + \frac{1}{k^2}\nabla\nabla\right) g(\mathbf{r}, \mathbf{r}') \cdot \mathbf{J}_{e(s)}(\mathbf{r}') \\
&= \mathrm{i}\omega\mu(\mathbf{r}) \int_S \mathrm{d}s' g(\mathbf{r}, \mathbf{r}')\mathbf{J}_{e(s)}(\mathbf{r}') + \frac{\mathrm{i}\omega\mu(\mathbf{r})}{k^2} \int_S \mathrm{d}s' \, \nabla'\nabla' g(\mathbf{r}, \mathbf{r}') \cdot \mathbf{J}_{e(s)}(\mathbf{r}') \\
&\equiv \mathbf{S}_1(\mathbf{r}) + \mathbf{S}_2(\mathbf{r}).
\end{aligned} \tag{11.42}
$$

Firstly, let us examine the limit of $\mathbf{S}_1(\mathbf{r})$ as $\mathbf{r} \to \mathbf{r}_0$:

$$
\mathbf{S}_1(\mathbf{r}) = \mathrm{i}\omega\mu \left(\int_{S\setminus S_a} + \int_{S_a} \right) \mathrm{d}s' g(\mathbf{r}, \mathbf{r}')\mathbf{J}_{e(s)}(\mathbf{r}'). \tag{11.43}
$$

The term $\mathbf{J}_{e(s)}(\mathbf{r}')$ in the second integral in (11.43) can be approximated as $\mathbf{J}_{e(s)}(\mathbf{r}_0)$; thus

$$
\begin{aligned}
\mathrm{i}\omega\mu \int_{S_a} \mathrm{d}s' g(\mathbf{r}, \mathbf{r}')\mathbf{J}_{e(s)}(\mathbf{r}') &\approx \mathrm{i}\omega\mu\mathbf{J}_{e(s)}(\mathbf{r}_0) \int_{S_a} g(\mathbf{r}, \mathbf{r}')\mathrm{d}s' \\
&= \frac{\mathrm{i}\omega\mu\mathbf{J}_{e(s)}(\mathbf{r}_0)}{4\pi} \int_0^a \int_0^{2\pi} \frac{\mathrm{e}^{-\mathrm{i}kR}}{R} \rho \, \mathrm{d}\rho \, \mathrm{d}\phi \\
&\approx \frac{\mathrm{i}\omega\mu\mathbf{J}_{e(s)}(\mathbf{r}_0)}{4\pi} \int_0^a \int_0^{2\pi} \frac{\rho}{\sqrt{\rho^2 + z^2}} \mathrm{d}\rho \, \mathrm{d}\phi \\
&= \frac{\mathrm{i}\omega\mu\mathbf{J}_{e(s)}(\mathbf{r}_0)}{2} \left(\sqrt{a^2 + z^2} - |z|\right) = O(a), \quad \text{as } z \to z' = 0.
\end{aligned} \tag{11.44}
$$

Therefore, as $\mathbf{r} \to \mathbf{r}_0$ and $a \to 0$, we have

$$
\lim_{\mathbf{r} \to \mathbf{r}_0} \mathbf{S}_1(\mathbf{r}) = \mathrm{i}\omega\mu \text{ p.v.} \int_S \mathrm{d}s' g(\mathbf{r}, \mathbf{r}')\mathbf{J}_{e(s)}(\mathbf{r}'). \tag{11.45}
$$

Next, we consider the limiting value of $\mathbf{S}_2(\mathbf{r})$ above as $\mathbf{r} \to \mathbf{r}_0 \in S$. As the kernel is hyper-singular, the limit will be of a Hadamard finite part for the hyper-singular integral [5] as discussed in Sect. 3.1.1. Now we have

$$\mathbf{S}_2(\mathbf{r}) = \frac{\mathrm{i}\omega\mu(\mathbf{r})}{k^2} \int_S \mathrm{d}s' \, \nabla'\nabla' g(\mathbf{r}, \mathbf{r}') \cdot \mathbf{J}_{e(s)}(\mathbf{r}')$$
$$= \frac{\mathrm{i}}{\omega\epsilon} \left(\int_{S \setminus S_a} + \int_{S_a} \right) \mathrm{d}s' \, \nabla'\nabla' g(\mathbf{r}, \mathbf{r}') \cdot \mathbf{J}_{e(s)}(\mathbf{r}'). \tag{11.46}$$

We will show that the Hadamard finite part of $\mathbf{S}_2(\mathbf{r})$ can be computed by the Cauchy principal values of a weaker singular kernel. Let us first examine the integral over the patch S_a and denote the integrand as

$$\mathbf{f} = \nabla'\nabla' g(\mathbf{r}, \mathbf{r}') \cdot \mathbf{J}_{e(s)}(\mathbf{r}'), \tag{11.47}$$

which in component form reads

$$f_i = \frac{\partial^2}{\partial x_i' \partial x_j'} g(\mathbf{r}, \mathbf{r}') J_j, \tag{11.48}$$

where $J_j = J_{e(s),j}$. We can rewrite f_i as

$$f_i = \frac{\partial}{\partial x_j'} \left(\frac{\partial g}{\partial x_i'} J_j \right) - \frac{\partial g}{\partial x_i'} \frac{\partial J_j}{\partial x_j'} = \nabla' \cdot \left(\frac{\partial g}{\partial x_i'} \mathbf{J}_{e(s)} \right) - \frac{\partial g}{\partial x_i'} \left(\nabla' \cdot \mathbf{J}_{e(s)} \right)$$
$$= \nabla' \cdot \left(\frac{\partial g}{\partial x_i'} \mathbf{J}_{e(s)} \right) + \mathrm{i}\omega \rho_{e(s)} \frac{\partial g}{\partial x_i'},$$

where the continuity equation $\mathrm{i}\omega\rho_{e(s)} + \nabla_s' \cdot \mathbf{J}_{e(s)} = 0$ has been used for the final equality.

Using the fact that $\nabla' = \nabla_s' + \mathbf{n}\frac{\partial}{\partial \mathbf{n}}$ and $\mathbf{n} \perp \mathbf{J}_{e(s)}$, we have

$$f_i = \nabla_s' \cdot \left(\frac{\partial g}{\partial x_i'} \mathbf{J}_{e(s)} \right) + \mathrm{i}\omega \rho_{e(s)} \frac{\partial g}{\partial x_i'}. \tag{11.49}$$

Using Gauss's theorem, we can compute the integral over S_a in (11.46) as follows:

$$\int_{S_a} f_i \, \mathrm{d}s' = \int_{S_a} \left[\nabla_s' \cdot \left(\frac{\partial g}{\partial x_i'} \mathbf{J}_{e(s)} \right) + \mathrm{i}\omega \rho_{e(s)} \frac{\partial g}{\partial x_i'} \right] \mathrm{d}s'$$
$$= \int_{\partial S_a} \frac{\partial g}{\partial x_i'} \mathbf{J}_{e(s)}(\mathbf{r}') \cdot \boldsymbol{\tau} \, \mathrm{d}l' + \mathrm{i}\omega \int_{S_a} \rho_{e(s)} \frac{\partial g}{\partial x_i'} \, \mathrm{d}s', \tag{11.50}$$

where $\boldsymbol{\tau}$ is the outer normal of ∂S_a:

$$\boldsymbol{\tau} = (\cos\phi, \sin\phi, 0). \tag{11.51}$$

Using (11.37) and (11.39), we obtain

$$\begin{aligned}\int_{S_a} \mathbf{f}\, \mathrm{d}s' &\approx \int_{\partial S_a} \nabla' g \left[\mathbf{J}_{e(s)}(\mathbf{r}') \cdot \boldsymbol{\tau}\right] \mathrm{d}l' + \frac{\mathrm{i}\omega}{2}\rho_{e(s)}(\mathbf{r}_0)\mathbf{n} \\ &\approx \int_{\partial S_a} g'(R)\mathbf{a}_R \left[\mathbf{J}_{e(s)}(\mathbf{r}_0) \cdot \boldsymbol{\tau}\right] \mathrm{d}l' + \frac{\mathrm{i}\omega}{2}\rho_{e(s)}(\mathbf{r}_0)\mathbf{n},\end{aligned} \tag{11.52}$$

where

$$\mathbf{a}_R = \nabla' R = \frac{\mathbf{r}' - \mathbf{r}}{|\mathbf{r}' - \mathbf{r}|} = -\mathbf{n}\cos\theta + \boldsymbol{\tau}\sin\theta, \tag{11.53}$$

and θ is the angle between vector $\mathbf{r}' - \mathbf{r}$ and $-\mathbf{n}$ at $\mathbf{r}_0$.

Assuming that

$$\mathbf{J}_{e(s)}(\mathbf{r}_0) = |\mathbf{J}_{e(s)}(\mathbf{r}_0)|(\cos\phi_0, \sin\phi_0, 0), \tag{11.54}$$

then the line integral in (11.52) can be simplified as follows:

$$\begin{aligned}&\int_{\partial S_a} g'(R)\mathbf{a}_R \left[\mathbf{J}_{e(s)}(\mathbf{r}_0) \cdot \boldsymbol{\tau}\right] \mathrm{d}l' \\ &= a|\mathbf{J}_{e(s)}(\mathbf{r}_0)| g'(R) \int_0^{2\pi} (-\mathbf{n}\cos\theta + \boldsymbol{\tau}\sin\theta)\cos(\phi - \phi_0)\mathrm{d}\phi \\ &= a|\mathbf{J}_{e(s)}(\mathbf{r}_0)| g'(R)\sin\theta \int_0^{2\pi} \boldsymbol{\tau}\cos(\phi - \phi_0)\mathrm{d}\phi \\ &= \pi a|\mathbf{J}_{e(s)}(\mathbf{r}_0)| g'(R)\sin\theta(\cos\phi_0, \sin\phi_0, 0) \\ &= \pi a\mathbf{J}_{e(s)}(\mathbf{r}_0) g'(R)\sin\theta \to -\frac{\mathbf{J}_{e(s)}(\mathbf{r}_0)}{4a} \quad \text{as } z \to z' = 0, \quad \theta \to \frac{\pi}{2}.\end{aligned}$$

Finally, we have the integral over S_a:

$$\int_{S_a} \mathrm{d}s'\, \nabla'\nabla' g(\mathbf{r}, \mathbf{r}') \cdot \mathbf{J}_{e(s)}(\mathbf{r}') = \int_{S_a} \mathbf{f}\, \mathrm{d}s' = -\frac{\mathbf{J}_{e(s)}(\mathbf{r}_0)}{4a} + \frac{\mathrm{i}\omega}{2}\rho_{e(s)}(\mathbf{r}_0)\mathbf{n}. \tag{11.55}$$

The term $\mathbf{J}_{e(s)}(\mathbf{r}_0)/4a$ becomes unbounded as a goes to zero, but we can use the idea of the Hadamard finite part since a similar unbounded term with an opposite sign, as for all finite parts of Hadamard integrals, will come from the following integral:

$$\int_{S\setminus S_a} ds'\, \nabla'\nabla' g(\mathbf{r},\mathbf{r}')\cdot \mathbf{J}_{e(s)}(\mathbf{r}') = \mathrm{i}\omega \int_{S\setminus S_a} ds'\, \nabla' g(\mathbf{r},\mathbf{r}')\rho_{e(s)}(\mathbf{r}') + \frac{\mathbf{J}_{e(s)}(\mathbf{r}_0)}{4a}. \tag{11.56}$$

Therefore, $\lim_{\mathbf{r}\to\mathbf{r}_0} \mathbf{S}_2(\mathbf{r})$ as the finite part (p.f.) of the Hadamard integral

$$\lim_{\mathbf{r}\to\mathbf{r}_0} \mathbf{S}_2(\mathbf{r}) = \text{p.f.}\frac{\mathrm{i}\omega\mu(\mathbf{r})}{k^2}\int_S ds'\, \nabla'\nabla' g(\mathbf{r},\mathbf{r}')\cdot \mathbf{J}_{e(s)}(\mathbf{r}') \tag{11.57}$$

can be computed using the following Cauchy principal integral after combining (11.55) and (11.56) and canceling the divergent term $\mathbf{J}_{e(s)}(\mathbf{r}_0)/4a$:

$$\mathbf{S}_2(\mathbf{r}) = -\text{p.v.}\frac{1}{\epsilon}\int_S ds'\, \nabla' g(\mathbf{r},\mathbf{r}')\rho_{e(s)}(\mathbf{r}') - \frac{1}{2\epsilon}\rho_{e(s)}(\mathbf{r}_0)\mathbf{n}. \tag{11.58}$$

As a result, we have two equivalent representations for $\mathbf{E}^{\text{elec}}(\mathbf{r})$ with $\mathbf{r} \in S$: one by a Hadamard finite part integral

$$\begin{aligned}\mathbf{E}^{\text{elec}}(\mathbf{r}) &= \mathrm{i}\omega\mu(\mathbf{r})\ \text{p.f.}\int_S ds' \left(\bar{\mathbf{I}} + \frac{1}{k^2}\nabla\nabla\right) g(\mathbf{r},\mathbf{r}')\cdot \mathbf{J}_{e(s)}(\mathbf{r}') \\ &= \mathrm{i}\omega\mu\text{p.f.}\int_S ds'\, \overline{\mathbf{G}}^1_E(\mathbf{r},\mathbf{r}')\cdot \mathbf{J}_{e(s)}(\mathbf{r}'),\end{aligned} \tag{11.59}$$

and the other one by a Cauchy principal value integral:

$$\begin{aligned}\mathbf{E}^{\text{elec}}(\mathbf{r}) =& \mathrm{i}\omega\mu\ \text{p.v.}\int_S ds'\, g(\mathbf{r},\mathbf{r}')\mathbf{J}_{e(s)}(\mathbf{r}') \\ &- \text{p.v.}\frac{1}{\epsilon}\int_S ds'\, \nabla' g(\mathbf{r},\mathbf{r}')\rho_{e(s)}(\mathbf{r}') - \frac{1}{2\epsilon}\rho_{e(s)}(\mathbf{r}_0)\mathbf{n}.\end{aligned} \tag{11.60}$$

Electric field integral equation (EFIE)

Substituting (11.41) and (11.59) or (11.41) and (11.60) into (11.34) and taking the tangential component (using the fact that $\mathbf{n}\times\left(\mathbf{J}_{m(s)}(\mathbf{r})\times\mathbf{n}\right) = \mathbf{J}_{m(s)}(\mathbf{r})$), we arrive at the EFIE:

$$\begin{aligned}-\frac{1}{2}\mathbf{J}_{m(s)}(\mathbf{r}) =& \mathbf{n}\times\mathbf{E}^{\text{inc}}(\mathbf{r}) - \mathrm{i}\omega\mu\mathbf{n}\times\text{p.f.}\int_S ds'\, \overline{\mathbf{G}}^1_E(\mathbf{r},\mathbf{r}')\cdot\mathbf{J}_{e(s)}(\mathbf{r}') \\ &- \mu\mathbf{n}\times\text{p.v.}\int_S ds' \frac{1}{\mu(\mathbf{r}')}\nabla\times\overline{\mathbf{G}}^1_E(\mathbf{r},\mathbf{r}')\cdot\mathbf{J}_{m(s)}(\mathbf{r}'),\quad \mathbf{r}\in S.\end{aligned} \tag{11.61}$$

The first singular integral should be understood in terms of the finite part of the Hadamard integral and, as shown in (11.60), for homogeneous media it can be converted into a regular Cauchy principal value for the charge density (noting that the term $(\rho_{e(s)}(\mathbf{r}_0)/2\epsilon)\mathbf{n}$ in (11.60) is orthogonal to the surface tangents) as follows:

$$
\begin{aligned}
-\frac{1}{2}\mathbf{J}_{m(s)}(\mathbf{r}) = \mathbf{n}\times\mathbf{E}^{\mathrm{inc}}(\mathbf{r}) - \mathbf{n}\times\Bigg[& \mathrm{i}\omega\mu \mathrm{p.v.}\int_S \mathrm{d}s'\,\mathbf{J}_{e(s)}(\mathbf{r}')g(\mathbf{r},\mathbf{r}') \\
& - \mathrm{p.v.}\frac{1}{\epsilon(\mathbf{r})}\int_S \mathrm{d}s'\nabla' g(\mathbf{r},\mathbf{r}')\rho_{e(s)}(\mathbf{r}') \\
& + \mu(\mathbf{r})\,\mathrm{p.v.}\int_S \mathrm{d}s'\frac{1}{\mu(\mathbf{r}')}\nabla g(\mathbf{r},\mathbf{r}')\times\mathbf{J}_{m(s)}(\mathbf{r}')\Bigg], \quad \mathbf{r}\in S. \qquad (11.62)
\end{aligned}
$$

This integral equation is attributed to [6] and is also known as the Stratton–Chu representation (without 1/2 if $\mathbf{r}\notin S$) [7].

Magnetic field integral equation (MFIE)

Similarly, a surface MFIE can be obtained:

$$
\begin{aligned}
\frac{1}{2}\mathbf{J}_{e(s)}(\mathbf{r}) = & \,\mathbf{n}\times\mathbf{H}^{\mathrm{inc}}(\mathbf{r}) \\
& + \epsilon(\mathbf{r})\mathbf{n}\times\mathrm{p.v.}\int_S \mathrm{d}s'\left[\frac{1}{\epsilon(\mathbf{r}')}\nabla\times\overline{\mathbf{G}}_E^1(\mathbf{r},\mathbf{r}')\cdot\mathbf{J}_{e(s)}(\mathbf{r}')\right. \\
& \left. - \mathrm{i}\omega\overline{\mathbf{G}}_E^1(\mathbf{r},\mathbf{r}')\cdot\mathbf{J}_{m(s)}(\mathbf{r}')\right], \quad \mathbf{r}\in S. \qquad (11.63)
\end{aligned}
$$

From (11.41) and (11.59), we can define integral operators $\mathcal{T}(\mathbf{J})$ and $\mathcal{K}(\mathbf{J})$ [8] as follows:

$$
\mathcal{T}(\mathbf{J}) = -ik\mathbf{n}\times\mathrm{p.f.}\int_S \mathrm{d}s'\,\overline{\mathbf{G}}_E^1(\mathbf{r},\mathbf{r}')\cdot\mathbf{J}(\mathbf{r}'), \qquad (11.64)
$$

$$
\begin{aligned}
\mathcal{K}(\mathbf{J}) &= \mathbf{n}\times\mathrm{p.v.}\int_S \mathrm{d}s'\nabla\times\overline{\mathbf{G}}_E^1(\mathbf{r},\mathbf{r}')\cdot\mathbf{J}(\mathbf{r}') \\
&= \mathbf{n}\times\mathrm{p.v.}\int_S \mathrm{d}s'\nabla g(\mathbf{r},\mathbf{r}')\times\mathbf{J}(\mathbf{r}'). \qquad (11.65)
\end{aligned}
$$

As shown in (11.62), the electric integral operator $\mathcal{T}(\mathbf{J})$ using the Hadamard finite part can also be defined using the Cauchy principal value of the integral with weaker singularity:

$$\mathcal{T}(\mathbf{J}) = - \, ik\mathbf{n} \times \text{p.v.} \int_S ds' \, \mathbf{J}(\mathbf{r}')g(\mathbf{r}, \mathbf{r}') - \frac{i}{k}\mathbf{n} \times \text{p.v.} \int_S ds' \nabla' g(\mathbf{r}, \mathbf{r}') \nabla'_s \cdot \mathbf{J}(\mathbf{r}'). \tag{11.66}$$

The surface integral equations (11.61) and (11.63) become

$$\left(\frac{1}{2}I - \mathcal{K}\right) \mathbf{J}_{m(s)} + \mathcal{T}\left(Z\mathbf{J}_{e(s)}\right) = - \, \mathbf{n} \times \mathbf{E}^{\text{inc}}(\mathbf{r}) \text{ (EFIE for exterior scattering)}, \tag{11.67}$$

$$\left(\frac{1}{2}I - \mathcal{K}\right) \left(Z\mathbf{J}_{e(s)}\right) - \mathcal{T}\mathbf{J}_{m(s)} = Z\mathbf{n} \times \mathbf{H}^{\text{inc}}(\mathbf{r}) \text{ (MFIE for exterior scattering)}, \tag{11.68}$$

where the impedance of the exterior medium $Z = \sqrt{\mu/\epsilon}$.

According to the impedance boundary condition in (8.97) for conductors, we have a relation between these two currents:

$$\mathbf{J}_{m(s)} = -Z_s\mathbf{n} \times \mathbf{J}_{e(s)}, \tag{11.69}$$

which can be used with either (11.67) or (11.68) to form a closed system of equations.

For a PEC (perfect electric conductor), the intrinsic surface impedance $Z_s = 0$, so we have $\mathbf{J}_{m(s)} = \mathbf{0}$; the surface integral equations for the electric current $\mathbf{J} = \mathbf{J}_{e(s)}$ are given by

$$\mathcal{T}\left(Z\mathbf{J}\right) = - \, \mathbf{n} \times \mathbf{E}^{\text{inc}}(\mathbf{r}), \tag{11.70}$$

$$\left(\frac{1}{2}I - \mathcal{K}\right) \mathbf{J} = \mathbf{n} \times \mathbf{H}^{\text{inc}}(\mathbf{r}). \tag{11.71}$$

Meanwhile, for the electromagnetic fields inside the cavity, similar surface integral equations can be obtained with Green's function $\overline{\mathbf{G}}_E^2(\mathbf{r}', \mathbf{r})$ for the definition of the surface integral operators in (11.64) and (11.65):

$$\left(\frac{1}{2}I + \mathcal{K}\right) \mathbf{J}_{m(s)} - \mathcal{T}\left(Z\mathbf{J}_{e(s)}\right) = -\mathbf{n} \times \mathbf{E}^{\text{inc}}(\mathbf{r}) \text{ (EFIE for interior scattering)}, \tag{11.72}$$

$$\left(\frac{1}{2}I + \mathcal{K}\right) \left(Z\mathbf{J}_{e(s)}\right) + \mathcal{T}\mathbf{J}_{m(s)} = Z\mathbf{n} \times \mathbf{H}^{\text{inc}}(\mathbf{r}) \text{ (MFIE for interior scattering)}, \tag{11.73}$$

where

$$\mathbf{J}_{e(s)}(\mathbf{r}) = \mathbf{n} \times \mathbf{H}_2(\mathbf{r}), \tag{11.74}$$
$$\mathbf{J}_{m(s)}(\mathbf{r}) = -\mathbf{n} \times \mathbf{E}_2(\mathbf{r}), \tag{11.75}$$

and the incident waves $\mathbf{E}^{\text{inc}}(\mathbf{r})$ and $\mathbf{H}^{\text{inc}}(\mathbf{r})$ come from sources inside V_2:

$$\mathbf{E}^{\text{inc}}(\mathbf{r}) = -\mathrm{i}\omega\mu(\mathbf{r}) \int_{V_2} \mathrm{d}\mathbf{r}' \,\overline{\mathbf{G}}_E^2(\mathbf{r}, \mathbf{r}') \cdot \mathbf{J}_e(\mathbf{r}'), \tag{11.76}$$

$$\mathbf{H}^{\text{inc}}(\mathbf{r}) = \epsilon(\mathbf{r}) \int_{V_2} \mathrm{d}\mathbf{r}' \frac{1}{\epsilon(\mathbf{r}')} \nabla' \times \overline{\mathbf{G}}_E^2(\mathbf{r}, \mathbf{r}') \cdot \mathbf{J}_e(\mathbf{r}'). \tag{11.77}$$

If the cavity has a PEC boundary, then the surface electric current $\mathbf{J} = \mathbf{J}_{e(s)}$ will satisfy the following integral equations:

$$\mathcal{T}(Z\mathbf{J}) = -\mathbf{n} \times \mathbf{E}^{\text{inc}}(\mathbf{r}), \tag{11.78}$$
$$\left(\frac{1}{2}I + \mathcal{K}\right)\mathbf{J} = \mathbf{n} \times \mathbf{H}^{\text{inc}}(\mathbf{r}). \tag{11.79}$$

Remark 11.2 (*Surface integral equations on dielectric scatterers*) As an electromagnetic field can penetrate into dielectric media, both interior and exterior field scattering will occur. By using the continuity of the tangential components of the electromagnetic fields, we can see that $\mathbf{J}_{m(s)}$ and $\mathbf{J}_{e(s)}$ in (11.67) and (11.72) are the same. Therefore, combining these two equations will give the surface EFIE for a dielectric scatterer. Similarly, (11.68) and (11.73) will form the surface MFIE for a dielectric scatterer.

11.2 Resonance and Combined Integral Equations

The surface EFIE and MFIE have unique solutions except at resonant frequencies k when V_2 (as a cavity) has non-zero solutions with the PEC boundary condition on S, i.e., $\mathbf{n} \times \mathbf{E}|_S = \mathbf{0}$. These solutions are called resonant eigen-modes of the cavity and k is the eigen-frequency of the cavity. We will discuss these modes for a spherical cavity V_2 with radius a.

According to the general solution for Maxwell's equations using the magnetic and electric vector potentials (8.58) and (8.59), together with the identity

$$\nabla(\nabla \cdot \mathbf{A}) = \nabla \times \nabla \times \mathbf{A} + \nabla^2 \mathbf{A}$$

and the fact that

$$\nabla^2 \mathbf{A} + k^2 \mathbf{A} = 0$$

(the same for $\mathbf{F}$ in (8.58)) in source-free cases, we have

$$\mathbf{E} = -\frac{1}{\epsilon}\nabla \times \mathbf{F} + \frac{1}{\mathrm{i}\omega\epsilon\mu}\nabla \times \nabla \times \mathbf{A}, \tag{11.80}$$

$$\mathbf{H} = \frac{1}{\mu}\nabla \times \mathbf{A} + \frac{1}{\mathrm{i}\omega\epsilon\mu}\nabla \times \nabla \times \mathbf{F}. \tag{11.81}$$

To find the cavity resonant modes inside the sphere, we consider the following vector potential [4, p. 267]

$$\mathbf{A} = \mathbf{r}A_r, \quad \mathbf{F} = \mathbf{r}F_r.$$

Then, the electric and the magnetic fields are given as

$$\begin{aligned} E_r &= \frac{1}{\epsilon}\frac{1}{\alpha}\left(\frac{\partial^2}{\partial r^2} + k^2\right)A_r, \\ E_\theta &= \frac{1}{\epsilon}\left(\frac{-1}{r\sin\theta}\frac{\partial F_r}{\partial\phi} + \frac{1}{\alpha r}\frac{\partial^2 A_r}{\partial r\partial\theta}\right), \\ E_\phi &= \frac{1}{\epsilon}\left(\frac{1}{r}\frac{\partial F_r}{\partial\theta} + \frac{1}{\alpha r\sin\theta}\frac{\partial^2 A_r}{\partial r\partial\phi}\right), \end{aligned} \tag{11.82}$$

$$\begin{aligned} H_r &= \frac{1}{\mu}\frac{1}{\beta}\left(\frac{\partial^2}{\partial r^2} + k^2\right)F_r, \\ H_\theta &= \frac{1}{\mu}\left(\frac{1}{r\sin\theta}\frac{\partial A_r}{\partial\phi} + \frac{1}{\beta r}\frac{\partial^2 F_r}{\partial r\partial\theta}\right), \\ H_\phi &= \frac{1}{\mu}\left(-\frac{1}{r}\frac{\partial A_r}{\partial\theta} + \frac{1}{\beta r\sin\theta}\frac{\partial^2 F_r}{\partial r\partial\phi}\right), \end{aligned} \tag{11.83}$$

where $\alpha = \mathrm{i}\omega\mu$ and $\beta = \mathrm{i}\omega\epsilon$.

- TE eigen-modes. By setting $A_r = 0$, then $E_r = 0$, so $\mathbf{E}$ has only transverse (θ, ϕ) components inside the sphere and takes

$$F_r = \widehat{j}_n(kr)P_n^m(\cos\theta)\begin{Bmatrix}\cos m\phi \\ \sin m\phi\end{Bmatrix}, \tag{11.84}$$

where

$$\widehat{j}_n(kr) = \sqrt{\frac{2\pi kr}{2}}\, j_{n+\frac{1}{2}}(kr), \tag{11.85}$$

and $j_{n+\frac{1}{2}}(x)$ is the Bessel function of fractional order and $\widehat{j}_n(kr)$ satisfies the following differential equation [4]:

$$\left(\frac{\mathrm{d}^2}{\mathrm{d}r^2} + k^2 - \frac{n(n+1)}{r^2}\right)\widehat{j}_n(kr) = 0. \tag{11.86}$$

We can see that the PEC boundary condition is satisfied, namely, that the tangential components E_θ and E_ϕ of $\mathbf{E}$ will vanish if ka are the roots of the spherical Bessel functions

$$\widehat{j}_n(ka) = 0. \tag{11.87}$$

- TM eigen-modes. By setting $F_r = 0$, then $H_r = 0$, so $\mathbf{H}$ has only transverse (θ, ϕ) components inside the sphere, and takes A_r in the same form as (11.84). Again, in this case, we can see that the PEC boundary condition is satisfied, namely that the tangential components E_θ and E_ϕ of $\mathbf{E}$ will vanish if ka are the roots of the derivatives of the spherical Bessel functions

$$\widehat{j}_n'(ka) = 0. \tag{11.88}$$

As $\widehat{j}_n(ka)$ and $\widehat{j}_n'(ka)$ cannot be zero simultaneously, the transverse (θ, ϕ) components of the magnetic field $\widehat{\mathbf{r}} \times \mathbf{H}$ are non-zero for both TE and TM modes above. Because there is no source inside the cavity, the non-zero electric current on the surface of the sphere,

$$\mathbf{J}_{\text{mode}}(\mathbf{r}) = \widehat{\mathbf{r}} \times \mathbf{H}_{\text{eigen-mode}}, \tag{11.89}$$

will satisfy a homogeneous version of (11.78) and (11.79):

$$\mathcal{T}(\mathbf{J}_{\text{mode}}) = \mathbf{0}, \tag{11.90}$$

$$\left(\frac{1}{2}I + \mathcal{K}\right)\mathbf{J}_{\text{mode}} = \mathbf{0}. \tag{11.91}$$

Thus, from (11.90), we conclude that the surface integral equation for the exterior scattering surface EFIE is not uniquely solvable, as the same operator $\mathcal{T}$ is used in (11.70) and (11.90). Also, (11.91) implies that $\lambda = -1/2$ is an eigenvalue for the operator $\mathcal{K}$, and it can be proven [8] that $1/2$ will also be an eigenvalue for $\mathcal{K}$. Therefore, the surface MFIE (11.71) for the scattering problem does not have a unique solution either for a perfectly conducting scatterer when k is an eigen-frequency.

The mathematical non-uniqueness of the surface integral equations for the exterior scattering problem for a PEC is found to be related to the fact that the solution of the EFIE fails to give the zero normal magnetic field on the boundary: $\mathbf{n} \cdot \mathbf{H} = 0$. There are various remedies to resolve the non-unique solution problem for the EFIE and MFIE. One is to supplement the EFIE with an explicit condition $\mathbf{n} \cdot \mathbf{H} = 0$ on S, the so-called augmented EFIE (A-EFIE) [9]. The most popular solution is to use the combined field integral equation approach by Mautz and Harrington [10], which combines the operators $\mathcal{T}$ and $\mathcal{K}$ such that the spectra of the new operator stays away from zero. The combined field operator $\mathcal{C}$ is defined as

$$\mathcal{C} = \left(\frac{1}{2}I - \mathcal{K}\right) + \eta \mathbf{n} \times \mathcal{T}. \tag{11.92}$$

The eigenvalues and the eigenfunctions for $\mathcal{C}$ for a unit PEC sphere ($a = 1$) can be explicitly computed as follows [8]:

$$\mathcal{C}(\nabla_s Y_n^m) = \left(\frac{1}{2} + \lambda_n - \eta\alpha_n\right) \nabla_s Y_n^m, \tag{11.93}$$

$$\mathcal{C}(\nabla_s \times Y_n^m) = \left(\frac{1}{2} - \lambda_n + \eta\beta_n\right) \nabla_s \times Y_n^m, \tag{11.94}$$

where

$$\lambda_n = \frac{ik}{2}\left\{j_n(k)[kh_n^{(1)}(k)]' + h_n^{(1)}(k)[kj_n(k)]'\right\}, \tag{11.95}$$

$$\alpha_n = [kj_n(k)]'[kh_n^{(1)}(k)]', \tag{11.96}$$

$$\beta_n = -k^2 j_n(k)h_n^{(1)}(k). \tag{11.97}$$

Therefore, the eigenvalues for the combined field operator $\mathcal{C}$ cluster around $1/2$ away from zero for appropriate choices of η.

The eigen-equations (11.93) and (11.94) can be easily verified using the following identities for the operators $\mathcal{T}$ and $\mathcal{K}$ on the unit spherical surface [8]:

$$\mathcal{T}(\nabla_s Y_n^m) = \alpha_n \nabla_s \times Y_n^m, \tag{11.98}$$

$$\mathcal{T}(\nabla_s \times Y_n^m) = \beta_n \nabla_s Y_n^m, \tag{11.99}$$

and

$$\mathcal{K}(\nabla_s Y_n^m) = -\lambda_n \nabla_s Y_n^m, \tag{11.100}$$

$$\mathcal{K}(\nabla_s \times Y_n^m) = \lambda_n \nabla_s \times Y_n^m, \tag{11.101}$$

as well as the identities $\nabla_s \times f = \nabla_s f \times \mathbf{n}$ and $\mathbf{n}\times(\nabla_s f \times \mathbf{n}) = -\nabla_s f$.

With the combined integral operator $\mathcal{C}$, Mautz and Harrington [10] proposed the following CFIE for general scattering surfaces:

$$\mathcal{C}(\mathbf{J}) = \mathbf{n} \times \mathbf{H}^{\text{inc}} - \eta\mathbf{n} \times \mathbf{n} \times \mathbf{E}^{\text{inc}}, \quad \text{on } S \qquad \text{(CFIE)}. \tag{11.102}$$

11.3 Nyström Collocation Methods for Maxwell's Equations

11.3.1 Surface Differential Operators

Assume that the surface S in $\mathbb{R}^3$ is smooth and parameterized by $\mathbf{r} = \mathbf{r}(u_1, u_2)$. Firstly, we define the metric tensors and various surface differential operators.

Tangential vectors: $\partial_i \mathbf{r}$, $i = 1, 2$, are defined as

$$\partial_i \mathbf{r} = \frac{\partial \mathbf{r}}{\partial u_i}, \quad i = 1, 2. \tag{11.103}$$

Metric tensors: the distance between two points on S parameterized by (u_1, u_2) and $(u_1 + \mathrm{d}u_1, u_2 + \mathrm{d}u_2)$ is given by

$$(\mathrm{d}s)^2 = g_{\mu\nu}(u)\ \mathrm{d}u_\mu \mathrm{d}u_\nu, \tag{11.104}$$

where the repeated indices imply a summation and

$$g_{\mu\nu} = \frac{\partial \mathbf{r}}{\partial u_\mu} \cdot \frac{\partial \mathbf{r}}{\partial u_\nu}, \quad 1 \le \mu, \nu \le 2, \tag{11.105}$$

and $\{g_{\mu\nu}\}$ is defined as the covariant tensor [11]. The contra-variant tensor $\{g^{\alpha\beta}\}$ is defined by

$$g_{\alpha\nu} g^{\nu\beta} = \delta_{\alpha\beta}. \tag{11.106}$$

The determinant of $\{g_{\mu\nu}\}$ is denoted by

$$g = \det\{g_{\mu\nu}\} = g_{11}g_{22} - g_{12}^2 = |\partial_1 \mathbf{r} \times \partial_2 \mathbf{r}|^2. \tag{11.107}$$

Normal to the surface:

$$\mathbf{n} = \frac{\partial_1 \mathbf{r} \times \partial_2 \mathbf{r}}{|\partial_1 \mathbf{r} \times \partial_2 \mathbf{r}|}. \tag{11.108}$$

Differential operators on S: ∇_s.

- Scalar $\nabla_s \phi$ – the surface gradient of a scalar function $\phi(u_1, u_2)$:

$$\nabla_s \phi = g^{\mu\nu} \frac{\partial \phi}{\partial u_\mu} \frac{\partial \mathbf{r}}{\partial u_\nu}, \tag{11.109}$$

and $g_{12} = 0$ if we have orthogonal parametric coordinates, i.e., the u_1 and u_2 coordinate axes are orthogonal, and

$$\nabla_s \phi = g^{11} \frac{\partial \phi}{\partial u_1} \frac{\partial \mathbf{r}}{\partial u_1} + g^{22} \frac{\partial \phi}{\partial u_2} \frac{\partial \mathbf{r}}{\partial u_2} = g^{11} \partial_1 \phi \partial_1 \mathbf{r} + g^{22} \partial_2 \phi \partial_2 \mathbf{r}. \tag{11.110}$$

- Divergence $\nabla_s \cdot \mathbf{f}$ – the surface divergence of a vector field $\mathbf{f} = \mathbf{f}(u_1, u_2)$ on S:

$$\nabla_s \cdot \mathbf{f} = g^{\mu\nu} \partial_\mu \mathbf{f} \cdot \partial_\nu \mathbf{r} \tag{11.111}$$

11.3.2 Locally Corrected Nyström Method for Hyper-Singular EFIE

Consider an integral equation for an unknown function $\psi(\mathbf{r})$,

$$\int_S G(\mathbf{r}-\mathbf{r}')\psi(\mathbf{r}')\mathrm{d}s' = \phi(\mathbf{r}), \tag{11.112}$$

where $\phi(\mathbf{r})$ is assumed given. The conventional Nyström method for (11.112) is based on a quadrature formula to discretize the integral operator, i.e.,

$$\int_S f(\mathbf{r}')\mathrm{d}s' \approx \sum_{n=1}^{N} \omega_n f(\mathbf{r}_n). \tag{11.113}$$

The weights and the abscissas are obtained by first transforming the integral to a regular reference (square) parameter domain where the usual Gauss quadratures $\{\varpi_n, \mathbf{u}_n\}$ are used. Then, we have

$$\omega_n = \sqrt{g(\mathbf{u}_n)}\,\varpi_n, \tag{11.114}$$

$$\mathbf{r}_n = \mathbf{r}(\mathbf{u}_n), \tag{11.115}$$

where $\mathbf{r}(\mathbf{u})$ is the parameterization mapping between the reference domain and S, and $g(\mathbf{u})$ is the determinant of the mapping function. The discretized system for the integral equation is simply

$$\sum_{n=1}^{N} \omega_n G(\mathbf{r}_m - \mathbf{r}_n)\psi(\mathbf{r}_n) = \phi(\mathbf{r}_m),\ \ m = 1, 2, \ldots, N. \tag{11.116}$$

The accuracy of the solution of (11.116) is determined by that of the quadrature formula and the smoothness of the surface, namely the mapping function $\mathbf{r}(\mathbf{u})$. However, the accuracy of (11.116) will be lost once the kernel function G becomes singular or even hyper-singular. To retain the high-order accuracy, new types of specially designed quadrature formulae will be needed to account for the specific nature of the kernel function when $\mathbf{r}_m$ and $\mathbf{r}_n$ are close. This is achieved by the local correction strategy in [12] where the local corrections are introduced:

$$\widetilde{G}_{mn} \equiv \begin{cases} L_{mn}, & \text{if } \mathbf{r}_n \in D_m, \\ G(\mathbf{r}_m - \mathbf{r}_n), & \text{otherwise}, \end{cases} \tag{11.117}$$

where D_m is a neighborhood of $\mathbf{r}_m$.

The correction L_{mn} is obtained by constructing a quadrature formula for a G-weighted integral, which is required to be exact for some classes of test functions.

Therefore, L_{mn} depends on $G(\mathbf{r})$ for each m. For each $\mathbf{r}_m$ and its neighborhood D_m, the new quadrature for the G-weighted integral is found by satisfying K constraints:

$$\int\limits_{D_m} G(\mathbf{r}_m - \mathbf{r}') f^{(k)}(\mathbf{r}') \mathrm{d}s' = \sum_{n=1}^{J} L_{mn} f^{(k)}(\mathbf{r}_m - \mathbf{r}_n), \quad k = 1, 2, \ldots, K, \tag{11.118}$$

where the test functions $f^{(k)}(\mathbf{r})$ are pre-selected functions, usually polynomials. As K and J are small, the correction coefficients can be pre-calculated for each $\mathbf{r}_m$ with small cost.

Now, the high-order Nyström method [13] for the singular kernel G becomes

$$\sum_{n=1}^{N} \omega_n \widetilde{G}_{mn} \psi(\mathbf{r}_n) = \phi(\mathbf{r}_m), \quad m = 1, 2, \ldots, N. \tag{11.119}$$

The locally corrected quadrature L_{mn} in (11.117) requires accurate evaluation of the left-hand side of (11.118), which is singular or even hyper-singular. In treating the hyper-singular integrals in the EFIE (11.61), there are methods of direct calculations as discussed in Sect. 3.1.3 for general curved surfaces or see [14] for flat patches. There is also a reformulation of the singular integral in terms of weaker singular integrals as in (11.66). A similar approach was used in [13] where the hyper-singular terms are handled by using Stokes' theorem (see the review in [15]) to avoid integrating the hyper-singular terms directly, as reviewed below. Three main types of singularities for the EFIE are listed as follows.

- Type 1:
$$G(R)\mathbf{t}(\mathbf{r}) \cdot \mathbf{t}'(\mathbf{r}'); \tag{11.120}$$
- Type 2:
$$\mathbf{t}(\mathbf{r}) \cdot \left(\nabla G(R) \times \mathbf{t}'(\mathbf{r}')\right); \tag{11.121}$$
- Type 3:
$$\left(\mathbf{t}(\mathbf{r}) \cdot \nabla\right) \left(\nabla' G(R) \cdot \mathbf{t}'(\mathbf{r}')\right), \tag{11.122}$$

where
$$G(R) = \frac{\mathrm{e}^{-\mathrm{i}kR}}{R} = G^{\mathrm{r}}(R) - \mathrm{i}\frac{\sin kR}{R}$$

and
$$G^{\mathrm{r}}(R) = G^{\mathrm{r}}(\mathbf{r}, \mathbf{r}') = \frac{\cos kR}{R}, \quad R = |\mathbf{r} - \mathbf{r}'|. \tag{11.123}$$

We will show how to handle Type 3 hyper-singularities for a given field point $\mathbf{r} = \mathbf{r}_0 \in S$, and the other two types can be treated similarly [13]. The key is to convert the integral into smooth or less singular integrals as field point $\mathbf{r}$ approaches $\mathbf{r}_0$ on S so it can be computed accurately for selected test functions:

$$
\begin{aligned}
&(\mathbf{t}(\mathbf{r})\cdot\nabla)\left(\nabla' G(R)\cdot\mathbf{t}'(\mathbf{r}')\right)\\
&= -\mathrm{i}k^3\left[\frac{\frac{\sin kR}{kR}-\cos kR}{(kR)^2}\mathbf{t}\cdot\mathbf{t}'+k^2\frac{\frac{\sin kR}{kR}-3\left(\frac{\sin kR}{kR}-\cos kR\right)/(kR)^2}{(kR)^2}(\mathbf{t}\cdot\mathbf{r})(\mathbf{t}'\cdot\mathbf{r})\right]\\
&+(\mathbf{t}\cdot\nabla)\left(\nabla' G^{\mathrm{r}}(R)\cdot\mathbf{t}'\right).
\end{aligned}
\tag{11.124}
$$

whereas the first term in the right-hand side of (11.124) is regular, the second term is hyper-singular to be simplified using Gauss's theorem on the surface patch:

$$
\begin{aligned}
&\int_S ds'\ (\mathbf{t}(\mathbf{r})\cdot\nabla)\left(\nabla' G^{\mathrm{r}}(\mathbf{r},\mathbf{r}')\cdot\mathbf{t}'(\mathbf{r}')\right)\\
&=\int_S ds'\,\mathbf{t}'(\mathbf{r}')\cdot\nabla_s'\left(\mathbf{t}(\mathbf{r})\cdot\nabla G^{\mathrm{r}}(\mathbf{r},\mathbf{r}')\right)\\
&=\int_S ds'\,\nabla_s'\cdot\left[\mathbf{t}'(\mathbf{r}')\left(\mathbf{t}(\mathbf{r})\cdot\nabla G^{\mathrm{r}}(\mathbf{r},\mathbf{r}')\right)\right]-\int_S ds'\left[\mathbf{t}(\mathbf{r})\cdot\nabla G^{\mathrm{r}}(\mathbf{r},\mathbf{r}')\right]\left(\nabla_s'\cdot\mathbf{t}'(\mathbf{r}')\right)\\
&=\int_{\partial S} dl\left(\boldsymbol{\tau}\cdot\mathbf{t}'(\mathbf{r}')\right)\left(\mathbf{t}(\mathbf{r})\cdot\nabla G^{\mathrm{r}}(\mathbf{r},\mathbf{r}')\right)+\int_S ds'\,\nabla' G^{\mathrm{r}}(\mathbf{r},\mathbf{r}')\cdot\left[\mathbf{t}(\mathbf{r})\left(\nabla_s'\cdot\mathbf{t}'(\mathbf{r}')\right)\right],
\end{aligned}
\tag{11.125}
$$

where $\boldsymbol{\tau}$ still denotes the outer normal of the curve ∂S tangential to the surface S.

The first term will be void of singularities as long as the field point $\mathbf{r}$ is not on the boundary ∂S, whereas the second term can be regularized at $\mathbf{r}-\mathbf{r}'$ by the following splitting

$$
\mathbf{t}(\mathbf{r})\nabla_s'\cdot\mathbf{t}(\mathbf{r}')=\left[\mathbf{t}(\mathbf{r})\left(\nabla_s'\cdot\mathbf{t}(\mathbf{r}')\right)-\alpha^{\mu}\frac{\partial_\mu'\mathbf{r}'}{\sqrt{g(\mathbf{u})}}\right]+\alpha^{\mu}\frac{\partial_\mu'\mathbf{r}'}{\sqrt{g(\mathbf{u})}},
\tag{11.126}
$$

where the coefficient vector $=(\alpha^{\mu})$ is selected to make the first term vanish at $\mathbf{r}'=\mathbf{r}$, namely

$$
\begin{aligned}
\alpha^{\mu}=\alpha^{\mu}(\mathbf{r})&\equiv\sqrt{g(\mathbf{u})}g^{\mu\nu}(\mathbf{t}(\mathbf{r})\cdot\partial_\mu'\mathbf{r}')\left(\nabla_s'\cdot\mathbf{t}'(\mathbf{r}')\right)\\
&=\sqrt{g(\mathbf{u})}g^{\mu\nu}(\mathbf{t}(\mathbf{r})\cdot\partial_\mu'\mathbf{r}')\left(g^{\kappa\sigma}\partial_\kappa'\mathbf{t}'\cdot\partial_\sigma'\mathbf{r}'\right)|_{\mathbf{r}'=\mathbf{r}}.
\end{aligned}
\tag{11.127}
$$

Then, the second term in (11.125) can be split as follows:

$$\int_S ds' \, \nabla' G^{\mathrm{r}}(\mathbf{r}, \mathbf{r}') \cdot \left[\mathbf{t}(\mathbf{r}) \nabla'_s \cdot \mathbf{t}'(\mathbf{r}')\right]$$
$$= \int_S ds' \, \nabla' G^{\mathrm{r}}(\mathbf{r}, \mathbf{r}') \cdot \left[\mathbf{t}(\mathbf{r}) \nabla'_s \cdot \mathbf{t}(\mathbf{r}') - \mathbf{c}(\mathbf{r}, \mathbf{r}')\right] + \int_S ds' \, \nabla' G^{\mathrm{r}}(\mathbf{r}, \mathbf{r}') \cdot \mathbf{c}(\mathbf{r}, \mathbf{r}'), \tag{11.128}$$

where

$$\mathbf{c}(\mathbf{r}, \mathbf{r}') \equiv \alpha^{\mu}(\mathbf{r}) \frac{\partial'_{\mu} \mathbf{r}'}{\sqrt{g(\mathbf{u})}}.$$

Due to the choice in (11.127), the first term has a singularity at most of $1/|\mathbf{r} - \mathbf{r}'|$, whereas the second term can be converted into a boundary integral as follows:

$$\int_S ds' \, \nabla' G^{\mathrm{r}}(\mathbf{r}, \mathbf{r}') \cdot \mathbf{c}(\mathbf{r}, \mathbf{r}') = \int_S ds' \, \nabla'_s G^{\mathrm{r}}(\mathbf{r}, \mathbf{r}') \cdot \mathbf{c}(\mathbf{r}, \mathbf{r}')$$
$$= \int_{\partial S} dl' \, \boldsymbol{\tau} \cdot G^{\mathrm{r}}(\mathbf{r}, \mathbf{r}') \mathbf{c}(\mathbf{r}, \mathbf{r}'). \tag{11.129}$$

Finally, the integral in (11.125) can be computed as

$$\int_S ds' \, (\mathbf{t}(\mathbf{r}) \cdot \nabla) \left(\nabla' G^{\mathrm{r}}(\mathbf{r}, \mathbf{r}') \cdot \mathbf{t}'(\mathbf{r}')\right)$$
$$= \int_S ds' \, \nabla' G^{\mathrm{r}}(\mathbf{r}, \mathbf{r}') \cdot \left[\mathbf{t}(\mathbf{r}) \nabla'_s \cdot \mathbf{t}'(\mathbf{r}') - \mathbf{c}(\mathbf{r}, \mathbf{r}')\right]$$
$$+ \int_{\partial S} dl' \, \boldsymbol{\tau} \cdot \left[\mathbf{t}'(\mathbf{r}') \left(\mathbf{t}(\mathbf{r}) \cdot \nabla G^{\mathrm{r}}(\mathbf{r}, \mathbf{r}')\right) + G^{\mathrm{r}}(\mathbf{r}, \mathbf{r}') \mathbf{c}(\mathbf{r}, \mathbf{r}')\right]. \tag{11.130}$$

The test function used in defining L_{mn} in (11.118) is taken as

$$\mathbf{t}'(\mathbf{r}') = \mathbf{t}'^{(k)}(\mathbf{r}') = \frac{\partial_{\mu} \mathbf{r}'}{\sqrt{g(\mathbf{u})}} f^{(k)}(\mathbf{u}), \tag{11.131}$$

where $\mathbf{u}$ is the parameterization of $\mathbf{r}' \in S$, i.e., $\mathbf{r}' = \mathbf{r}'(\mathbf{u})$ while, for a fixed field point $\mathbf{r} = \mathbf{r}(\mathbf{u_0})$, the basis function is given by

$$\mathbf{t}_{\mu}(\mathbf{r}) = \partial_{\mu} \mathbf{r}(\mathbf{u})|_{\mathbf{u}=\mathbf{u}_0}. \tag{11.132}$$

Thus, (11.125) becomes [13]

$$\int_S ds' \, (\mathbf{t}(\mathbf{r}) \cdot \nabla) \left(\nabla' G^{\mathrm{r}}(\mathbf{r}, \mathbf{r}') \cdot \mathbf{t}'(\mathbf{r}')\right)$$
$$= \int_S ds' \, \nabla' G^{\mathrm{r}}(\mathbf{r}, \mathbf{r}') \left[\partial_\mu \boldsymbol{r}(\boldsymbol{u}) \partial'_\nu f^{(\mathbf{k})}(\mathbf{u}) - \partial'_\mu \mathbf{r}' \partial'_\nu f^{(\mathbf{k})}(\mathbf{u}_0)\right] / \sqrt{g(\mathbf{u})}$$
$$+ \int_{\partial S} dl' \boldsymbol{\tau} \cdot \left[\partial_\mu \mathbf{r}(\mathbf{u}) \cdot \nabla G^{\mathrm{r}}(\mathbf{r}, \mathbf{r}') f^{(\mathbf{k})}(\mathbf{u}) \partial'_\nu \mathbf{r}' + G^{\mathrm{r}}(\mathbf{r}, \mathbf{r}') \partial'_\nu f^{(\mathbf{k})}(\mathbf{u}_0) \partial'_\mu \mathbf{r}'\right]. \tag{11.133}$$

Here the field point $\mathbf{r} = \mathbf{r}(\mathbf{u_0})$ as $\mathbf{r}$ tends to the surface S.

The first term in (11.133) has at most a singularity of $1/|\mathbf{r} - \mathbf{r}'|$ due to the selection of (11.127) at $\mathbf{r} - \mathbf{r}'$, which can be handled by a Duffy mapping (see Sect. 11.3.3), and the boundary integral will be finite as long as the field point $\mathbf{r}$ is not on the boundary ∂S in the limiting process, and thus can be evaluated by a regular quadrature on ∂S.

Remark 11.3 The Nyström collocation method [16] discussed here requires the surface EFIE to hold at some specific points on the surface S. The resulting algebraic equation can be conditioned by the Calderon identity [8], which is addressed in various literatures [17–20].

11.3.3 Nyström Method for Mixed Potential EFIE

A high-order Nyström method by Tong and Chew [21] is based on the Stratton–Chu representation (11.62), where only singular integrals are involved, in contrast to the hyper-singular integrals in (11.61) with the dyadic Green's functions. Here we will only consider the free-space Green's function and the media inside or outside the scatterer will be homogeneous. We consider the following singular term in the EFIE (11.62) over a patch S:

$$\mathbf{n} \times \int_S ds' \, \nabla' g(\mathbf{r}, \mathbf{r}') \rho_s(\mathbf{r}')$$
$$= \int_S ds' \, \mathbf{n} \times \nabla' \mathbf{g}(\mathbf{r}, \mathbf{r}') \rho_{\mathbf{s}}(\mathbf{r}') = \int_S ds' \, \mathbf{n} \times \nabla'_s \mathbf{g}(\mathbf{r}, \mathbf{r}') \rho_s(\mathbf{r}')$$
$$= \int_S ds' \mathbf{n} \times \nabla'_s \mathbf{g}(\mathbf{r}, \mathbf{r}') \left[\rho_s(\mathbf{r}') - \rho_{\mathbf{s}}(\mathbf{r})\right] + \rho_{\mathbf{s}}(\mathbf{r}) \int_S ds' \mathbf{n} \times \nabla'_s \mathbf{g}(\mathbf{r}, \mathbf{r}')$$
$$= \int_S ds' \, \mathbf{n} \times \nabla'_s \mathbf{g}(\mathbf{r}, \mathbf{r}') \left[\rho_s(\mathbf{r}') - \rho_{\mathbf{s}}(\mathbf{r})\right] + \rho_{\mathbf{s}}(\mathbf{r}) \int_{\partial S} dl' \, g(\mathbf{r}, \mathbf{r}'). \tag{11.134}$$

Now the first term has only a $1/|\mathbf{r}-\mathbf{r}'|$ singularity which can be treated by using the Duffy transform [22]. This transform maps a triangular domain $T = \{(u_1, u_2) : 0 \le u_1 \le 1, 0 \le u_2 \le u_1\}$ in the parametric variables of the surface S into a square $Q = \{(\zeta, \eta) : 0 \le \zeta, \eta \le 1\}$ through a collapsing transform:

$$u_1 = \zeta, \qquad u_2 = u_1\eta = \zeta\eta, \tag{11.135}$$

and the Jacobian of the mapping (11.135), $J = \partial(u_1, u_2)/\partial(\zeta, \eta) = \zeta$, will then be able to cancel the singularity $1/|\mathbf{r}-\mathbf{r}'| = 1/|\mathbf{r}(u_1, u_2) - \mathbf{r}(u_1', u_2')|$ once the integration of the first term in (11.134) is carried out in the (ζ, η) variables. Alternatively, a local polar coordinate system on the surface where $(\zeta, \eta) \equiv (\rho, \theta)$ can also achieve the same effect of singularity cancelation [23]. Meanwhile, the second term in (11.134) remains finite as $\mathbf{r}$ is away from ∂S as it approaches S.

Similarly, in the MFIE, we need to compute the following integral accurately:

$$\begin{aligned}&\int_S \mathrm{d}s'\, \mathbf{n}\times \left(\nabla' g(\mathbf{r}, \mathbf{r}') \times \mathbf{J}(\mathbf{r}')\right)\\ &= \int_S \mathrm{d}s' \left[\left(\mathbf{n}\cdot \mathbf{J}(\mathbf{r}')\right) \mathbf{R} - (\mathbf{n}\cdot \mathbf{R})\, \mathbf{J}(\mathbf{r}')\right] \times \frac{(-ikR-1)}{R^2} \mathrm{e}^{-ikR},\end{aligned} \tag{11.136}$$

where the integrand is of $1/R$ singularity because $\mathbf{R} = \mathbf{r} - \mathbf{r}'$ becomes orthogonal to $\mathbf{n}$ as $\mathbf{r}$ tends to the surface S.

11.4 Galerkin Methods and High-Order RWG Current Basis

11.4.1 Galerkin Method Using Vector–Scalar Potentials

The Galerkin method for the electric field integral equation is usually based on a vector and scalar potential representation of the electric field [24], which involves a weaker singular kernel in (11.62), and the unknowns will be the surface current. Such a procedure was first proposed in [25].

For a PEC (perfect electric conductor), $Z_s = 0$, so (11.62) becomes

$$\begin{aligned}&\mathrm{i}\omega\mu\mathbf{n} \times \text{p.v.} \int_S \mathrm{d}s'\, \mathbf{J}_{e(s)}(\mathbf{r}')g(\mathbf{r}, \mathbf{r}') - \mathbf{n} \times \text{p.v.} \frac{1}{\epsilon(\mathbf{r})} \int_S \mathrm{d}s'\, \nabla' g(\mathbf{r}, \mathbf{r}')\rho_{e(s)}(\mathbf{r}')\\ &= \mathbf{n} \times \mathbf{E}^{\text{inc}}(\mathbf{r}).\end{aligned} \tag{11.137}$$

Let $\mathbf{J}_l(\mathbf{r}), l = 1, 2, \ldots, N$, be the RWG current basis functions given in Sects. 11.4.3 and 11.4.4, and let the current $\mathbf{J}_s$ and the charge ρ_s be expressed in terms of these basis functions, namely

$$\mathbf{J}_s(\mathbf{r}) = \sum_{l=1}^{N} I_l \mathbf{J}_l(\mathbf{r}), \tag{11.138}$$

$$\rho_s(\mathbf{r}) = \sum_{l=1}^{N} I_l \rho_l(\mathbf{r}), \tag{11.139}$$

where

$$\rho_l(\mathbf{r}) = -\frac{1}{\mathrm{i}\omega} \nabla_s \cdot \mathbf{J}_l(\mathbf{r}). \tag{11.140}$$

We apply the Galerkin projection procedure to (11.137) by multiplying it with a test function $\mathbf{J}_k(\mathbf{r}')$ and then integrating over the whole surface S to obtain the following algebraic equation:

$$\sum_{l=1}^{N} Z_{kl} I_l = V_k, \quad k = 1, 2, \ldots, N, \tag{11.141}$$

where, after using integration by parts to transfer the gradient operator ∇ in (11.137) to the test function $\mathbf{J}_l(\mathbf{r})$, and using the continuity equation for the surface charge (11.140), we have

$$Z_{kl} = \mathrm{i}\omega\mu \int_S \int_S g(R) \mathbf{J}_k(\mathbf{r}') \cdot \mathbf{J}_l(\mathbf{r}) \mathrm{d}s' \mathrm{d}s + \frac{\mathrm{i}\omega}{\epsilon} \int_S \int_S g(R) \rho_k(\mathbf{r}') \rho_l(\mathbf{r})\, \mathrm{d}s' \mathrm{d}s, \tag{11.142}$$

and

$$V_k = \int_S \mathbf{n} \times \mathbf{E}^{\mathrm{inc}}(\mathbf{r}) \cdot \mathbf{J}_k(\mathbf{r}) \mathrm{d}s. \tag{11.143}$$

11.4.2 Functional Space for Surface Current $\mathbf{J}(\mathbf{r})$

In order to transfer the ∇ operator in the second term in (11.137) to the current function $\mathbf{J}(\mathbf{r})$, the divergence of the current basis functions should be L^2 integrable, which implies a continuous normal component of the basis functions across the common interface of any two patches when S is decomposed into either triangular or quadrilateral patches. The normal continuity of the current basis function insures no non-physical accumulation of charges on the patch interfaces.

Let us examine more closely the smoothness property of the surface currents. From the tangential continuity of the electric and magnetic fields, we have $\{\mathbf{E}, \mathbf{H}\} \in H_{\mathrm{loc}}(\mathrm{curl}, V)$, where

$$H_{\text{loc}}(\text{curl}, V) = \left\{ \mathbf{u} : \int_{\Omega} |\operatorname{curl} \mathbf{u}|^2 \, d\mathbf{r} < \infty, \int_{\Omega} |\mathbf{u}|^2 \, d\mathbf{r} < \infty, \forall \Omega \subset V \right\}. \quad (11.144)$$

Similarly, we can define the space $H_{\text{loc}}(\text{div}, V)$ as

$$H_{\text{loc}}(\text{div}, V) = \left\{ \mathbf{u} : \int_{\Omega} |\operatorname{div} \mathbf{u}|^2 \, d\mathbf{r} < \infty, \int_{\Omega} |\mathbf{u}|^2 \, d\mathbf{r} < \infty, \forall \Omega \subset V \right\}. \quad (11.145)$$

Meanwhile, the Sobolve space $H_1(V)$ is defined as

$$H_1(V) = \left\{ \mathbf{u} : \int_{V} |\mathbf{u}|^2 \, d\mathbf{r} + \sum_{i=1}^{3} \int_{V} |\nabla u_i|^2 \, d\mathbf{r} < \infty \right\}, \quad (11.146)$$

which can be shown to have the following property:

$$H_1(V) \subset H_{\text{loc}}(\text{curl}, V) \cap H_{\text{loc}}(\text{div}, V). \quad (11.147)$$

The fact that $\{\mathbf{E}, \mathbf{H}\} \in H_{\text{loc}}(\text{curl}, V)$ implies that their traces [26, lemma 4.2, p. 341] satisfy

$$\mathbf{n} \times \mathbf{E}|_s, \mathbf{n} \times \mathbf{H}|_s \in H^{-1/2}(S) = (H^{1/2}(S))^*. \quad (11.148)$$

The definition of the trace of $H_{\text{loc}}(\text{curl}, V)$ for smooth domains can be found in [27] while that for non-smooth domains are in [28, 29]. On applying the surface divergence to $\mathbf{n} \times \mathbf{E}$, we have

$$\nabla_s \cdot (\mathbf{n} \times \mathbf{E}) = -\mathbf{n} \cdot (\nabla \times \mathbf{E}) = \mathbf{n} \cdot (\mathrm{i}\omega\mu\mathbf{H}), \quad (11.149)$$

and by using Faraday's law in the second equation in (11.149) and the following gradient operator identity in the first equation:

$$\nabla_s = \nabla - \mathbf{n}\frac{\partial}{\partial \mathbf{n}}. \quad (11.150)$$

Therefore, assuming that $\mathbf{n} \cdot \mathbf{H}$ has a trace in $H^{-1/2}(S)$, namely that the field has no singularity near the surface S, then we have

$$\nabla_s \cdot (\mathbf{n} \times \mathbf{E}) \in H^{-1/2}(S). \quad (11.151)$$

Together with (11.148), we can see that

$$\mathbf{J}_{m(s)} = -\mathbf{n} \times \mathbf{E} \in H^{-1/2}(\text{div}_s, S) = \left\{ \mathbf{u} : \mathbf{u}, \nabla_s \mathbf{u} \in H^{-1/2}(S) \right\}, \quad (11.152)$$

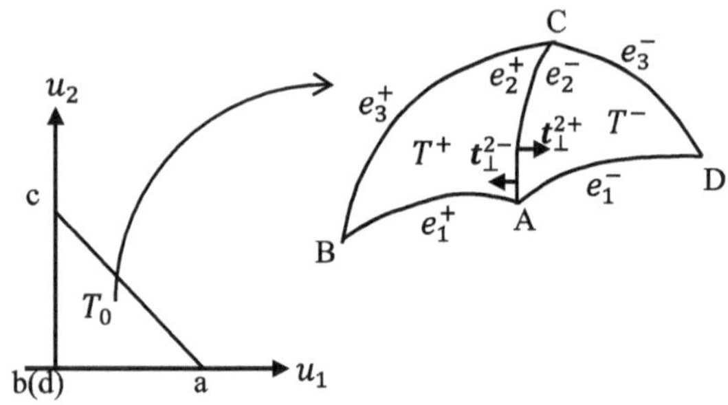

Fig. 11.2 Two triangular patches

to which

$$H(\mathrm{div}_s, S) = \left\{ \mathbf{u} : \int_S |\mathbf{u}|^2 \,\mathrm{d}s < \infty, \int_S |\mathrm{div}_s \mathbf{u}|^2 \,\mathrm{d}s < \infty \right\} \tag{11.153}$$

is a dense subspace.

In the following, we present the construction of high-order current basis functions in $H(\mathrm{div}_s, S)$ [30], which generalizes the zeroth-order RWG basis functions [31].

11.4.3 Basis Functions Over Triangular–Triangular Patches

Consider two curved triangular patches T^+ and T^- with a common interface AC with length l in Fig. 11.2. Let T^+ and T^- be parameterized, respectively, by

$$\begin{aligned} \mathbf{r} &= \mathbf{r}^+(u_1, u_2) : T_0 \to T^+, \\ \mathbf{r} &= \mathbf{r}^-(u_1, u_2) : T_0 \to T^-. \end{aligned} \tag{11.154}$$

We assume that the interface AC in both T^+ and T^- is parameterized by $u_1 + u_2 = 1$ and is labeled as side e_2^+ in T^+ and side e_2^- in T^-. In [30], the following basis functions with continuous normal components are constructed.

High-order basis functions for a triangular–triangular patch in Fig. 11.2 can be written in terms of the tangential vectors $\partial_1 \mathbf{r}$ and $\partial_2 \mathbf{r}$ defined in (11.103) with variable coefficients in the following form [32]:

$$\mathbf{f}(\mathbf{r}) = \begin{cases} \dfrac{l}{\sqrt{g^+}} \left(P_1^+(u_1, u_2)\, \partial_1 \mathbf{r} + P_2^+(u_1, u_2)\, \partial_2 \mathbf{r} \right), \\ \qquad \text{if} \quad \mathbf{r} = \mathbf{r}^+(u_1, u_2) \in T^+, \\ \dfrac{l}{\sqrt{g^-}} \left(P_1^-(u_1, u_2)\, \partial_1 \mathbf{r} + P_2^-(u_1, u_2)\, \partial_2 \mathbf{r} \right), \\ \qquad \text{if}\, \mathbf{r} = \mathbf{r}^-(u_1, u_2) \in T^-. \end{cases} \tag{11.155}$$

On the common interface AC : $u_1 + u_2 = 1$, we have the tangential direction $\mathbf{t}_\parallel$ of $\overrightarrow{AC}$ as

$$\mathbf{t}_\parallel = \frac{\partial_1 \mathbf{r} - \partial_2 \mathbf{r}}{|\partial_1 \mathbf{r} - \partial_2 \mathbf{r}|} = \frac{\partial_1 \mathbf{r} - \partial_2 \mathbf{r}}{\sqrt{g_{11} + g_{22} - 2g_{12}}}, \tag{11.156}$$

and the direction $\mathbf{t}_\perp$ normal to AC and $\mathbf{n}$ is given by

$$\begin{aligned}\mathbf{t}_\perp &= -\mathbf{t}_\parallel \times \mathbf{n} = -\frac{\partial_1 \mathbf{r} - \partial_2 \mathbf{r}}{\sqrt{g_{11} + g_{22} - 2g_{12}}} \times \frac{(\partial_1 \mathbf{r} \times \partial_2 \mathbf{r})}{\sqrt{g}} \\ &= \frac{\partial_2 \mathbf{r} \times (\partial_1 \mathbf{r} \times \partial_2 \mathbf{r}) - \partial_1 \mathbf{r} \times (\partial_1 \mathbf{r} \times \partial_2 \mathbf{r})}{\sqrt{g}\sqrt{g_{11} + g_{22} - 2g_{12}}} \\ &= \frac{\partial_1 \mathbf{r}(g_{22} - g_{12}) + \partial_2 \mathbf{r}(g_{11} - g_{12})}{\sqrt{g}\sqrt{g_{11} + g_{22} - 2g_{12}}}.\end{aligned} \tag{11.157}$$

Therefore, on the interface AC $= T^+ \cap T^-$, the projection of the basis function $\mathbf{f}(\mathbf{r})$ along the normal direction $\mathbf{t}_\perp$ is

$$\begin{aligned}\mathbf{f} \cdot \mathbf{t}_\perp &= \frac{l}{\sqrt{g}} (P_1 \partial_1 \mathbf{r} + P_2 \partial_2 \mathbf{r}) \cdot \mathbf{t}_\perp \\ &= \frac{l}{\sqrt{g_{11} + g_{22} - 2g_{12}}} (P_1(u_1, u_2) + P_2(u_1, u_2)),\end{aligned} \tag{11.158}$$

where the length element $\sqrt{g_{11} + g_{22} - 2g_{12}}$ is the same for both triangles.

Since $\mathbf{t}_\perp^+ = -\mathbf{t}_\perp^-$, the continuity of the normal component of the vector basis function $\mathbf{f}$ over the surface S implies that

$$\mathbf{f} \cdot \mathbf{t}_\perp^+ = \mathbf{f} \cdot \mathbf{t}_\perp^-, \tag{11.159}$$

namely

$$P_1^+ + P_2^+ = -(P_1^- + P_2^-), \tag{11.160}$$

for $u_1 + u_2 = 1$ on AC.

In order to satisfy the constraint in (11.160) for coefficients $P_1(u_1, u_2)$ and $P_2(u_1, u_2)$ taken in the polynomial spaces, we will employ the hierarchical polynomial basis over the reference triangle T_0 whose vertices are a $= (1, 0)$, b $= (0, 0)$, and c $= (0, 1)$. We group (u_1, u_2) polynomials into three modes: vertex modes, edge modes, and internal modes [33].

- Vertex modes:

$$\begin{aligned}g_a(u_1, u_2) &= u_1, \\ g_b(u_1, u_2) &= 1 - u_1 - u_2, \\ g_c(u_1, u_2) &= u_2.\end{aligned} \tag{11.161}$$

Each vertex mode will take value 1 at one vertex and zero at the other two vertices.

- Edge modes: for $2 \le l \le M$,

$$
\begin{aligned}
g_l^{\mathrm{ab}}(u_1, u_2) &= g_{\mathrm{a}}(u_1, u_2) g_{\mathrm{b}}(u_1, u_2) P_{l-2}(g_{\mathrm{b}} - g_{\mathrm{a}}),\\
g_l^{\mathrm{bc}}(u_1, u_2) &= g_{\mathrm{b}}(u_1, u_2) g_{\mathrm{c}}(u_1, u_2) P_{l-2}(g_{\mathrm{c}} - g_{\mathrm{b}}),\\
g_l^{\mathrm{ca}}(u_1, u_2) &= g_{\mathrm{c}}(u_1, u_2) g_{\mathrm{a}}(u_1, u_2) P_{l-2}(g_{\mathrm{a}} - g_{\mathrm{c}}),
\end{aligned} \tag{11.162}
$$

where $P_l(\xi), \xi \in [-1, 1]$, is the lth-order Legendre polynomial.

Each of the edge modes is non-zero only along one edge of the triangle T_0.

- Internal modes: $0 \le k + l \le M - 3$,

$$
g_{l,k}^{\mathrm{int}}(u_1, u_2) = g_{\mathrm{a}}(u_1, u_2) g_{\mathrm{b}}(u_1, u_2) g_{\mathrm{c}}(u_1, u_2) P_k(2g_{\mathrm{c}} - 1) P_l(g_{\mathrm{b}} - g_{\mathrm{a}}). \tag{11.163}
$$

Each of the internal modes will vanish over all edges of T_0.

Now we set the coefficients $P_1^{\pm}$ and $P_2^{\pm}$ in the following form in terms of the hierarchical polynomials in (11.161)–(11.163):

$$
\begin{aligned}
P_1^{+}(u_1, u_2) &= I_n^{\mathrm{a}} g_{\mathrm{a}}(u_1, u_2) + \sum_{m=2}^{M} \frac{I_n^{(m)} - \widetilde{I}_t^{(m)}}{2} g_m^{e_2^+}(u_1, u_2) + \sum_{(l,m)\in L_\Delta} c_{lm}^1 g_{lm}^{\mathrm{int}},\\
P_2^{+}(u_1, u_2) &= I_n^{\mathrm{c}} g_{\mathrm{c}}(u_1, u_2) + \sum_{m=2}^{M} \frac{I_n^{(m)} + \widetilde{I}_t^{(m)}}{2} g_m^{e_2^+}(u_1, u_2) + \sum_{(l,m)\in L_\Delta} c_{lm}^2 g_{lm}^{\mathrm{int}},
\end{aligned} \tag{11.164}
$$

and

$$
\begin{aligned}
P_1^{-}(u_1, u_2) &= -I_n^{\mathrm{a}} g_{\mathrm{a}}(u_1, u_2) + \sum_{m=2}^{M} \frac{-I_n^{(m)} - \widehat{I}_t^{(m)}}{2} g_m^{e_2^-}(u_1, u_2) + \sum_{(l,m)\in L_\Delta} d_{lm}^1 g_{lm}^{\mathrm{int}},\\
P_2^{-}(u_1, u_2) &= -I_n^{\mathrm{c}} g_{\mathrm{c}}(u_1, u_2) + \sum_{m=2}^{M} \frac{-I_n^{(m)} + \widehat{I}_t^{(m)}}{2} g_m^{e_2^-}(u_1, u_2) + \sum_{(l,m)\in L_\Delta} d_{lm}^2 g_{lm}^{\mathrm{int}},
\end{aligned} \tag{11.165}
$$

with

$$
L_\Delta = \{(l, m) : 0 \le l + m \le M - 3\}. \tag{11.166}
$$

Unknowns for each edge are

$$
I_n^{\mathrm{a}}, I_n^{\mathrm{c}}, I_n^{(m)}, \widetilde{I}_t^{(m)}, \widehat{I}_t^{(m)}, \quad 2 \le m \le M, \tag{11.167}
$$

and interior unknowns for each triangular patch are

$$c^1_{lm}, c^2_{lm}, d^1_{lm}, d^2_{lm}, \quad (l, m) \in L_\Delta. \tag{11.168}$$

- **RWG basis**

If we assume that the normal components of the current basis function remain constant, $I^{\mathrm{a}}_n = I^{\mathrm{c}}_n = I_n$, along the common edge AC, we have

$$\mathbf{f}(\mathbf{r}) = I_n \begin{cases} \frac{l}{\sqrt{g^+}}(g_{\mathrm{a}}(u_1, u_2)\partial_1\mathbf{r} + g_{\mathrm{c}}(u_1, u_2)\partial_2\mathbf{r}), & \text{if } \mathbf{r} = \mathbf{r}^+(u_1, u_2) \in T^+, \\ -\frac{l}{\sqrt{g^-}}(g_{\mathrm{a}}(u_1, u_2)\partial_1\mathbf{r} + g_{\mathrm{c}}(u_1, u_2)\partial_2\mathbf{r}), & \text{if } \mathbf{r} = \mathbf{r}^-(u_1, u_2) \in T^-, \end{cases} \tag{11.169}$$

and the unknown for each edge AC is just I_n.

In particular, for flat triangular patches, we have in T^+ the following:

$$\mathbf{r} = \mathbf{r}^+(u_1, u_2) = g_{\mathrm{a}}(u_1, u_2)\mathbf{r}_{\mathrm{A}} + g_{\mathrm{b}}(u_1, u_2)\mathbf{r}_{\mathrm{B}} + g_{\mathrm{c}}(u_1, u_2)\mathbf{r}_{\mathrm{C}}. \tag{11.170}$$

Then the tangential vectors are

$$\begin{aligned} \partial_1\mathbf{r} &= \mathbf{r}_{\mathrm{A}} - \mathbf{r}_{\mathrm{B}}, \\ \partial_2\mathbf{r} &= \mathbf{r}_{\mathrm{C}} - \mathbf{r}_{\mathrm{B}}. \end{aligned} \tag{11.171}$$

Similarly, we have in T^-:

$$\mathbf{r} = \mathbf{r}^-(u_1, u_2) = g_{\mathrm{a}}(u_1, u_2)\mathbf{r}_{\mathrm{A}} + g_{\mathrm{d}}(u_1, u_2)\mathbf{r}_{\mathrm{D}} + g_{\mathrm{c}}(u_1, u_2)\mathbf{r}_{\mathrm{C}}, \tag{11.172}$$

where $g_{\mathrm{d}}(u_1, u_2) = g_{\mathrm{b}}(u_1, u_2)$, and the tangential vectors are

$$\begin{aligned} \partial_1\mathbf{r} &= \mathbf{r}_{\mathrm{A}} \quad \mathbf{r}_{\mathrm{D}}, \\ \partial_2\mathbf{r} &= \mathbf{r}_{\mathrm{C}} - \mathbf{r}_{\mathrm{D}}. \end{aligned} \tag{11.173}$$

Substituting (11.171) and (11.173) into (11.169), we get the original RWG basis function [31]:

$$\mathbf{f}(\mathbf{r}) = I_n \begin{cases} \frac{l}{2A^+}(\mathbf{r} - \mathbf{r}_{\mathrm{B}}), & \text{if } \mathbf{r} = \mathbf{r}^+(u_1, u_2) \in T^+, \\ -\frac{l}{2A^-}(\mathbf{r} - \mathbf{r}_{\mathrm{D}}), & \text{if } \mathbf{r} = \mathbf{r}^-(u_1, u_2) \in T^-, \end{cases} \tag{11.174}$$

where A^+ and A^- are the areas of the triangles T^+ and T^-, respectively.

From (11.174), it can easily be seen that the normal component of $\mathbf{f}(\mathbf{r})$ from each triangle is exactly the height of the triangle from vertex B (or D) to the common edge AC, which cancels the area A^+ (or A^-) of each triangle exactly after being multiplied by the length l of the common edge AC.

- **First-order basis**

From (11.155) and (11.164)–(11.165), we get two first-order basis functions for each edge:

$$vertexmodes, edgemodes\mathbf{f}_1(\mathbf{r}) = I_n^{\mathrm{a}} \begin{cases} \dfrac{1}{\sqrt{g^+}} u_1 \partial_1 \mathbf{r}_s^+, & \text{if } \mathbf{r} = \mathbf{r}_s^+(u_1, u_2) \in T^+, \\ -\dfrac{1}{\sqrt{g^-}} u_1 \partial_1 \mathbf{r}_s^-, & \text{if } \mathbf{r} = \mathbf{r}_s^-(u_1, u_2) \in T^-, \end{cases} \tag{11.175}$$

$$\mathbf{f}_2(\mathbf{r}) = I_n^{\mathrm{c}} \begin{cases} \dfrac{1}{\sqrt{g^+}} u_2 \partial_2 \mathbf{r}_s^+, & \text{if } \mathbf{r} = \mathbf{r}_s^+(u_1, u_2) \in T^+, \\ -\dfrac{1}{\sqrt{g^-}} u_2 \partial_2 \mathbf{r}_s^-, & \text{if } \mathbf{r} = \mathbf{r}_s^-(u_1, u_2) \in T^-, \end{cases} \tag{11.176}$$

where the unknowns for each edge AC are I_n^{a} for $\mathbf{f}_1(\mathbf{r})$ and I_n^{c} for $\mathbf{f}_2(\mathbf{r})$.

- **Second-order basis**

From (11.155) and (11.164)–(11.165), we get five second-order basis functions for each edge. Note that $\mathbf{f}_1(\mathbf{r})$ and $\mathbf{f}_2(\mathbf{r})$ are given in (11.175) and (11.176), and

$$\mathbf{f}_3(\mathbf{r}) = \widetilde{I}_t^{(2)} \begin{cases} \dfrac{1}{\sqrt{g^+}} u_1 u_2 \left(-\partial_1 \mathbf{r}_s^+ + \partial_2 \mathbf{r}_s^+\right), & \text{if } \mathbf{r} = \mathbf{r}_s^+(u_1, u_2) \in T^+, \\ \mathbf{0}, & \text{if } \mathbf{r} = \mathbf{r}_s^-(u_1, u_2) \in T^-, \end{cases} \tag{11.177}$$

$$\mathbf{f}_4(\mathbf{r}) = \widehat{I}_t^{(2)} \begin{cases} \mathbf{0}, & \text{if } \mathbf{r} = \mathbf{r}_s^+(u_1, u_2) \in T^+, \\ \dfrac{1}{\sqrt{g^-}} u_1 u_2 \left(-\partial_1 \mathbf{r}_s^- + \partial_2 \mathbf{r}_s^-\right), & \text{if } \mathbf{r} = \mathbf{r}_s^-(u_1, u_2) \in T^-, \end{cases} \tag{11.178}$$

and

$$\mathbf{f}_5(\mathbf{r}) = I_n^{(2)} \begin{cases} \dfrac{1}{\sqrt{g^+}} u_1 u_2 \partial_1 \mathbf{r}_s^+, & \text{if } \mathbf{r} = \mathbf{r}_s^+(u_1, u_2) \in T^+, \\ -\dfrac{1}{\sqrt{g^-}} u_1 u_2 \partial_1 \mathbf{r}_s^-, & \text{if } \mathbf{r} = \mathbf{r}_s^-(u_1, u_2) \in T^-. \end{cases} \tag{11.179}$$

11.4.4 Basis Functions over Triangular–Quadrilateral Patches

Consider a curved quadrilateral patch Ω and a curved triangular patch T which are parameterized separately by two mappings $\mathbf{r}_i(u_1, u_2)$, $i = 1, 2$, i.e.,

$$\begin{aligned}\mathbf{r}_1(u_1,u_2):\Omega_0\to\Omega \qquad (u_1,u_2)\in\Omega_0,\\ \mathbf{r}_2(u_1,u_2):T_0\to T \qquad (u_1,u_2)\in T_0.\end{aligned}\tag{11.180}$$

The edges of Ω and T are labeled as in Fig. 11.3. The common interface is BC, which is parameterized by $u_1 = 0$.

We consider mixed high-order basis functions for a triangular–quadrilateral patch in the following form:

$$\mathbf{f}(\mathbf{r}) = \begin{cases} \frac{l}{\sqrt{g^\Omega}}(Q_1(u_1,u_2)\,\partial_1\mathbf{r} + Q_2(u_1,u_2)\,\partial_2\mathbf{r}), & \text{if } \mathbf{r}=\mathbf{r}_1(u_1,u_2)\in\Omega,\\ -\frac{l}{\sqrt{g^T}}(P_1(u_1,u_2)\,\partial_1\mathbf{r} + P_2(u_1,u_2)\,\partial_3\mathbf{r}) & \\ = \frac{l}{\sqrt{g^T}}(-(P_1+P_2)\partial_1\mathbf{r} + P_2(u_1,u_2)\,\partial_2\mathbf{r}), & \text{if } \mathbf{r}=\mathbf{r}_2(u_1,u_2)\in T,\end{cases}\tag{11.181}$$

where for $\mathbf{r} = \mathbf{r}_2(u_1, u_2) \in T$ we have

$$\partial_3\mathbf{r} = \partial_1\mathbf{r} - \partial_2\mathbf{r}.\tag{11.182}$$

Along the common edge BC, the vector $\mathbf{t}_{\parallel}^{\Omega}$ tangential to the common edge is

$$\mathbf{t}_{\parallel}^{\Omega} = \frac{\partial_2\mathbf{r}}{|\partial_2\mathbf{r}|} = \frac{\partial_2\mathbf{r}}{\sqrt{g_{22}^{\Omega}}},\tag{11.183}$$

while the vector $\mathbf{t}_{\perp}^{\Omega}$ orthogonal to both the edge and the normal $\mathbf{n}^{\Omega}$ of the surface S is

$$\begin{aligned}\mathbf{t}_{\perp}^{\Omega} = \mathbf{t}_{\parallel}^{\Omega}\times\mathbf{n}^{\Omega} &= \frac{\partial_2\mathbf{r}}{\sqrt{g_{22}^{\Omega}}}\times\frac{\partial_1\mathbf{r}\times\partial_2\mathbf{r}}{\sqrt{g^{\Omega}}}\\ &= \frac{g_{22}^{\Omega}\partial_1\mathbf{r}-g_{12}^{\Omega}\partial_2\mathbf{r}}{\sqrt{g_{22}^{\Omega}}\sqrt{g_{11}g_{22}-g_{12}^2}}.\end{aligned}\tag{11.184}$$

Therefore, from the side of Ω, the normal component of $\mathbf{f}(\mathbf{r})$ satisfies

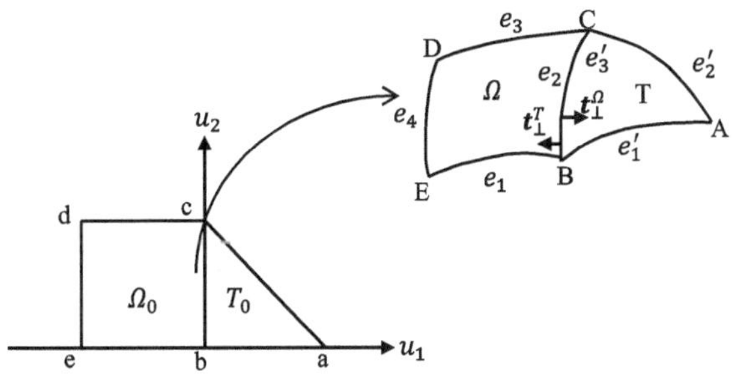

Fig. 11.3 A triangular patch and a quadrilateral patch

$$\mathbf{f}(\mathbf{r}) \cdot \mathbf{t}_{\perp}^{\Omega} = \frac{l}{\sqrt{g_{22}^{\Omega}}} Q_1(u_1, u_2). \tag{11.185}$$

Meanwhile, the vector $\mathbf{t}_{\parallel}^{T}$ tangential to the common edge is

$$\mathbf{t}_{\parallel}^{\Omega} = \frac{\partial_2 \mathbf{r}}{|\partial_2 \mathbf{r}|} = \frac{\partial_2 \mathbf{r}}{\sqrt{g_{22}^{T}}}, \tag{11.186}$$

while the vector $\mathbf{t}_{\perp}^{T}$ orthogonal to both the edge and the normal $\mathbf{n}^{T}$ of the surface S is

$$\begin{aligned} \mathbf{t}_{\perp}^{T} &= -\mathbf{t}_{\parallel}^{T} \times \mathbf{n}^{T} = -\frac{\partial_2 \mathbf{r}}{\sqrt{g_{22}^{T}}} \times \frac{\partial_1 \mathbf{r} \times \partial_2 \mathbf{r}}{\sqrt{g^{T}}} \\ &= \frac{g_{12}^{T} \partial_2 \mathbf{r} - g_{22}^{T} \partial_1 \mathbf{r}}{\sqrt{g_{22}^{T}} \sqrt{g_{11}^{T} g_{22}^{T} - \left(g_{12}^{T}\right)^2}}. \end{aligned} \tag{11.187}$$

Thus, from the side of T, the normal component of $\mathbf{f}(\mathbf{r})$ satisfies

$$\mathbf{f}(\mathbf{r}) \cdot \mathbf{t}_{\perp}^{T} = \frac{l}{\sqrt{g_{22}^{T}}} (P_1(u_1, u_2) + P_2(u_1, u_2)). \tag{11.188}$$

Along the common edge BC, we have the identity of the two parameterizations of the triangular and the quadrilateral patches. Therefore,

$$g_{22}^{T} = \sqrt{\partial_2 \mathbf{r} \cdot \partial_2 \mathbf{r}} = g_{22}^{\Omega}. \tag{11.189}$$

As a result, in order to have continuity of the normal component of the current $\mathbf{f}(\mathbf{r})$, namely

$$\mathbf{f}(\mathbf{r}) \cdot \mathbf{t}_{\perp}^{\Omega} = \mathbf{f}(\mathbf{r}) \cdot \mathbf{t}_{\perp}^{T}, \tag{11.190}$$

we require that along BC ($u_1 = 0$)

$$P_1(u_1, u_2) + P_2(u_1, u_2) = -Q_1(u_1, u_2). \tag{11.191}$$

In order to satisfy the constraint in (11.191) for coefficients $Q_1(u_1, u_2)$ taken in the polynomial spaces, we employ the hierarchical polynomial basis over the reference square Ω_0 with vertices b, c, d, and e, as in Fig. 11.3. We group (u_1, u_2) hierarchical polynomials of order M over Ω_0 into three modes: vertex modes, edge modes, and internal modes [33].

- Vertex modes:

$$\begin{aligned} N_{\rm b}(u_1,u_2) &= (1+u_1)(1-u_2),\\ N_{\rm c}(u_1,u_2) &= (1+u_1)u_2,\\ N_{\rm d}(u_1,u_2) &= -u_1u_2,\\ N_{\rm e}(u_1,u_2) &= -u_1(1-u_2). \end{aligned} \tag{11.192}$$

Each vertex mode will take value 1 at one vertex and zero at the other three vertices.

- Edge modes: $2 \le l \le M$

$$\begin{aligned} N_l^{\rm eb}(u_1,u_2) &= (1-u_2)\phi_l(2u_1+1),\\ N_l^{\rm bc}(u_1,u_2) &= (1+u_1)\phi_l(2u_2-1),\\ N_l^{\rm cd}(u_1,u_2) &= u_2\phi_l(2u_1+1),\\ N_l^{\rm de}(u_1,u_2) &= -u_1\phi_l(2u_2-1), \end{aligned} \tag{11.193}$$

where

$$\phi_l(\xi) = \frac{1}{4}\left(1-\xi^2\right)P_{l-2}(\xi). \tag{11.194}$$

Each of the edge modes is non-zero only along one edge of the rectangle Ω_0.

- Internal modes: $2 \le k, l \le M$

$$N_{l,k}^{\rm int}(u_1,u_2) = \phi_l(2u_1+1)\phi_k(2u_2-1). \tag{11.195}$$

Each of the internal modes will vanish over all edges of Ω_0.

Now we set the coefficients in (11.181) in the following form [34]:

$$\begin{aligned} Q_1(u_1,u_2) =& I_n^{\rm b}N_{\rm b}(u_1,u_2) + I_n^{\rm c}N_{\rm c}(u_1,u_2)\\ &+\sum_{l=2}^{M} I_n^{(l)}\,N_l^{e_2}(u_1,u_2) + \sum_{2\le l,m\le M}\gamma_{lm}^1 N_{lm}^{\rm int},\\ Q_2(u_1,u_2) =& \sum_{l=2}^{M}\widehat{I}_t^{(l)}\,N_l^{e_2}(u_1,u_2) + \sum_{2\le l,m\le M}\gamma_{lm}^2 N_{lm}^{\rm int}, \end{aligned} \tag{11.196}$$

and

$$
\begin{aligned}
P_1(u_1,u_2) &= -I_n^{\mathrm{b}} g_{\mathrm{b}}(u_1,u_2) - \sum_{l=2}^{M} \frac{I_n^{(l)} - \widetilde{I}_t^{(l)}}{2} g_l^{e_3'}(u_1,u_2) + \sum_{(l,m)\in L_\Delta} c_{lm}^1 g_{lm}^{\mathrm{int}}, \\
P_2(u_1,u_2) &= -I_n^{\mathrm{c}} g_{\mathrm{c}}(u_1,u_2) - \sum_{l=2}^{M} \frac{I_n^{(l)} + \widetilde{I}_t^{(l)}}{2} g_l^{e_3'}(u_1,u_2) + \sum_{(l,m)\in L_\Delta} c_{lm}^2 g_{lm}^{\mathrm{int}},
\end{aligned}
\tag{11.197}
$$

where $e_3' = e_2 = \mathrm{BC}$. The unknowns for each edge BC are

$$
I_n^{\mathrm{b}}, I_n^{\mathrm{c}}, I_n^{(l)}, \widetilde{I}_t^{(l)}, \widehat{I}_t^{(l)}, \quad 2 \le l \le M, \tag{11.198}
$$

and the interior unknowns for each triangular element are

$$
c_{lm}^1, c_{lm}^2, \quad (l,m) \in L_\Delta, \tag{11.199}
$$

while the interior unknowns for each quadrilateral element are

$$
\gamma_{lm}^1, \gamma_{lm}^2, \quad 2 \le l, m \le M. \tag{11.200}
$$

- **Mixed RWG basis**

If we assume that the normal components of the current basis function remain constant, then we have

$$
\mathbf{f}(\mathbf{r}) = I_n \begin{cases} \dfrac{l}{\sqrt{g^\Omega}}[N_{\mathrm{b}}(u_1,u_2) + N_{\mathrm{c}}(u_1,u_2)]\partial_1\mathbf{r}, & \text{if } \mathbf{r} = \mathbf{r}_1(u_1,u_2) \in \Omega, \\ -\dfrac{l}{\sqrt{g^T}}[-(g_{\mathrm{b}}(u_1,u_2) + g_{\mathrm{c}}(u_1,u_2))\partial_1\mathbf{r} \\ \quad + g_{\mathrm{c}}(u_1,u_2)\partial_2\mathbf{r}], & \text{if } \mathbf{r} = \mathbf{r}_2(u_1,u_2) \in T. \end{cases} \tag{11.201}
$$

For flat triangular and quadrilateral patches, $\sqrt{g^T} = 2A^T$, where A^T denotes the area of T. In Ω,

$$
\begin{aligned}
\partial_1\mathbf{r} = \partial_1\mathbf{r}_1 &= (1-u_2)(\mathbf{r}_{\mathrm{B}} - \mathbf{r}_{\mathrm{E}}) + u_2(\mathbf{r}_{\mathrm{C}} - \mathbf{r}_{\mathrm{D}}), \\
\partial_2\mathbf{r} = \partial_2\mathbf{r}_1 &= -u_1(\mathbf{r}_{\mathrm{D}} - \mathbf{r}_{\mathrm{E}}) + (1+u_1)(\mathbf{r}_{\mathrm{C}} - \mathbf{r}_{\mathrm{B}}),
\end{aligned}
\tag{11.202}
$$

and in T

$$
\begin{aligned}
\partial_1\mathbf{r} &= \partial_1\mathbf{r}_2 = \mathbf{r}_{\mathrm{A}} - \mathbf{r}_{\mathrm{B}}, \\
\partial_2\mathbf{r} &= \partial_2\mathbf{r}_2 = \mathbf{r}_{\mathrm{C}} - \mathbf{r}_{\mathrm{B}}, \\
\partial_3\mathbf{r} &= \partial_3\mathbf{r}_2 = \partial_1\mathbf{r}_2 - \partial_2\mathbf{r}_2 = \mathbf{r}_{\mathrm{A}} - \mathbf{r}_{\mathrm{C}}.
\end{aligned}
\tag{11.203}
$$

Thus, we have the mixed RWG basis functions

$$
\mathbf{f}(\mathbf{r}) = I_n \begin{cases} \frac{l(1+u_1)}{\sqrt{g^\Omega}}[(1-u_2)(\mathbf{r}_\mathrm{B}-\mathbf{r}_\mathrm{E}) \\ \quad +u_2(\mathbf{r}_\mathrm{C}-\mathbf{r}_\mathrm{D})], & \text{if } \mathbf{r}=\mathbf{r}_1(u_1,u_2)\in\Omega, \\ -\frac{l}{2A^T}(\mathbf{r}-\mathbf{r}_\mathrm{A}), & \text{if } \mathbf{r}=\mathbf{r}_2(u_1,u_2)\in T. \end{cases} \tag{11.204}
$$

The unknown for each edge BC is I_n.

- **Mixed first-order basis**

From (11.181) and (11.196)–(11.197), we get two first-order basis functions:

$$
\mathbf{f}_1(\mathbf{r}) = I_n^\mathrm{b} \begin{cases} \frac{l}{\sqrt{g^\Omega}} N_\mathrm{b}(u_1,u_2)\partial_1\mathbf{r}, & \text{if } \mathbf{r}=\mathbf{r}_1(u_1,u_2)\in\Omega, \\ \frac{l}{\sqrt{g^T}} g_\mathrm{b}(u_1,u_2)\partial_1\mathbf{r}, & \text{if } \mathbf{r}=\mathbf{r}_2(u_1,u_2)\in T, \end{cases} \tag{11.205}
$$

$$
\mathbf{f}_2(\mathbf{r}) = I_n^\mathrm{c} \begin{cases} \frac{l}{\sqrt{g^\Omega}} N_\mathrm{c}(u_1,u_2)\partial_1\mathbf{r}, & \text{if } \mathbf{r}=\mathbf{r}_1(u_1,u_2)\in\Omega, \\ \frac{l}{\sqrt{g^T}} g_\mathrm{c}(u_1,u_2)\partial_3\mathbf{r}, & \text{if } \mathbf{r}=\mathbf{r}_2(u_1,u_2)\in T. \end{cases} \tag{11.206}
$$

The unknowns for each edge BC are I_n^b and I_n^c.

- **Mixed second-order basis**

In this case, there are seven basis functions (five associated with the common edge and two for the quadrilateral patch). Again, $\mathbf{f}_1(\mathbf{r})$ and $\mathbf{f}_2(\mathbf{r})$ are given in (11.205) and (11.206), and

$$
\mathbf{f}_3(\mathbf{r}) = I_n^{(2)} \begin{cases} \frac{l}{\sqrt{g^\Omega}} N_2^{e_2}(u_1,u_2)\partial_1\mathbf{r}, & \text{if } \mathbf{r}=\mathbf{r}_1(u_1,u_2)\in\Omega, \\ \frac{l}{2\sqrt{g^T}}\left[g_2^{e_3'}(u_1,u_2)\partial_1\mathbf{r} \right. \\ \quad \left. +\, g_2^{e_3'}(u_1,u_2)\partial_3\mathbf{r}\right], & \text{if } \mathbf{r}=\mathbf{r}_2(u_1,u_2)\in T, \end{cases} \tag{11.207}
$$

$$
\mathbf{f}_4(\mathbf{r}) = \widehat{I}_t^{(2)} \begin{cases} \frac{l}{\sqrt{g^\Omega}} N_2^{e_2}(u_1,u_2)\partial_2\mathbf{r}, & \text{if } \mathbf{r}=\mathbf{r}_1(u_1,u_2)\in\Omega, \\ \mathbf{0}, & \text{if } \mathbf{r}=\mathbf{r}_2(u_1,u_2)\in T, \end{cases} \tag{11.208}
$$

$$
\mathbf{f}_5(\mathbf{r}) = \widetilde{I}_t^{(2)} \begin{cases} \mathbf{0}, & \text{if } \mathbf{r}=\mathbf{r}_1(u_1,u_2)\in\Omega, \\ \frac{l}{2\sqrt{g^T}}\left[-g_2^{e_3'}(u_1,u_2)\partial_1\mathbf{r} \right. \\ \quad \left. +\, g_2^{e_3'}(u_1,u_2)\partial_3\mathbf{r}\right], & \text{if } \mathbf{r}=\mathbf{r}_2(u_1,u_2)\in T, \end{cases} \tag{11.209}
$$

$$\mathbf{f}_6(\mathbf{r}) = \gamma_{22}^1 \begin{cases} \frac{l}{\sqrt{g^\Omega}} N_{22}^{\text{int}}(u_1, u_2)\partial_1 \mathbf{r}, & \text{if } \mathbf{r} = \mathbf{r}_1(u_1, u_2) \in \Omega, \\ 0, & \text{if } \mathbf{r} = \mathbf{r}_2(u_1, u_2) \in T, \end{cases} \tag{11.210}$$

$$\mathbf{f}_7(\mathbf{r}) = \gamma_{22}^2 \begin{cases} \frac{l}{\sqrt{g^\Omega}} N_{22}^{\text{int}}(u_1, u_2)\partial_2 \mathbf{r}, & \text{if } \mathbf{r} = \mathbf{r}_1(u_1, u_2) \in \Omega, \\ 0, & \text{if } \mathbf{r} = \mathbf{r}_2(u_1, u_2) \in T, \end{cases} \tag{11.211}$$

where $e_2 = e_3' = \text{BC}$.

The unknowns for each edge BC are I_n^{b}, I_n^{c}, $I_n^{(2)}$, $\widehat{I}_t^{(2)}$, and $\widetilde{I}_t^{(2)}$, and the unknowns for each quadrilateral are γ_{22}^1 and γ_{22}^2.

11.5 Summary

Surface integral representation of electromagnetic fields using equivalent surface currents (i.e., tangential components of electric and magnetic fields) is a useful method for both the theoretical and computational study of scattering off conducting or dielectric scatterers. The Nyström collocation method and the Galerkin method can be used to discretize the boundary integral equations (BIEs) derived from the impedance boundary conditions on conductors or continuities of field tangential components on the boundary of dielectric scatterers. In both cases, the treatment of the hyper-singularity of the electric Green's function is critical for the success of the BIEs. For the Nyström collocation method, the singularity of the integrals is handled by using locally corrected quadratures for specific singular kernels. For the Galerkin method, however, the singularity of the integral kernel can be reduced by using a mixed (vector and scalar) potential representation of the electric field and an integration by parts, provided that the current unknowns have normal continuity. Basis functions with such normal continuity are provided by the traditional RWG basis and its higher-order extensions discussed here. Either the Nyström collocation or the Galerkin method can achieve high-order convergence if the surface is given with a parameterization with sufficient smoothness. For surface triangulations with limited smoothness such as piecewise flat triangular patches or spline patches, Galerkin methods with normal continuous current basis functions of appropriate order should be preferred. As a matter of fact, maintaining good accuracy of either Nyström or Galerkin methods near geometric singularities such as corners and edges remains a challenging issue for integral equation methods and special techniques such as graded meshes are used near the singularities to improve numerical accuracy [35–37].

Many important topics for integral equation methods for Maxwell's equations have not been addressed here, including pre-conditioning of the hyper-singular BIEs or the combined integral equations by Calderon operators [17, 18, 38], and fast solutions of the resulting dense linear system [39–42], and the breakdown of the EFIE at low frequency [43] (ill-conditioning of the matrix system in (11.141) as

$\omega \to 0$). Some solutions to address these problems can be found in the references cited here.

References

1. W.C. Chew, *Waves and Fields in Inhomogeneous Media* (Van Nostrand Reinhold, New York, 1990)
2. C.-T. Tai, *Dyadic Green Functions in Electromagnetic Theory*, 2nd edn. (IEEE Press, Piscataway, NJ, 1994)
3. A. Sommerfeld. *Optics*, Sect. 27 (Academic Press, New York, 1954)
4. R.F. Harrington, *Time-Harmonic Electromagnetic Fields*, 2nd edn. (Wiley-IEEE Press, Hoboken, NJ, 2001)
5. J. Hadamard, *Lectures on Cauchy Problems in Linear Partial Differential Equations* (Dover Publications, New York, 2003)
6. A.W. Maue, Zur Formulierung eines allgemeinen Beugungsproblems durch eine Integralgleichung. Z. Physik. **126**, 601–618 (1949)
7. J.A. Stratton, L.J. Chu, Diffraction theory of electromagnetic waves. Phys. Rev. **56**, 99–107 (1939)
8. G.C. Hsiao, R.E. Kleinman, Mathematical foundations for error estimation in numerical solutions of integral equations in electromagnetics. IEEE Trans. Antenn. Propag. **45**, 316–328 (1997)
9. A.D. Yaghjian, Augmented electric- and magnetic-field integral equations. Radio Sci. **16**, 987–1001 (1981)
10. J.R. Mautz, R.F. Harrington, H-field, E-field, and combined-field solutions for conducting bodies of resolution. Arch. Electron. Ubertragungstech. (Electron. Commun.) **32**, 157–164 (1978)
11. E. Kreyszig, *Differential Geometry* (Dover Publications, New York, 1991)
12. J. Strain, Locally corrected multidimensional quadrature rules for singular functions. SIAM J. Sci. Comput. **16**, 992–1017 (1995)
13. L.F. Canino, J.J. Ottusch, M.A. Stalzer, J.L. Visher, S.M. Wandzura, Numerical solution of the Helmholtz equations in 2D and 3D using a high-order Nyström discretization. J. Comput. Phys. **146**, 627–663 (1998)
14. M.S. Tong, W.C. Chew, Super-hyper singularity treatment for solving 3D electric field integral equations. Microw. Opt. Technol. Lett. **49**, 1383–1388 (2007)
15. L.J. Gray, J.M. Glaeser, T. Kaplan, Direct evaluation of hypersingular Galerkin surface integrals. SIAM J. Sci. Comput. **25**, 1534–1556 (2004)
16. E.J. Nyström, Über die praktische Auflösung von Integralgleichungen mit Anwendungen auf Randwertaufgaben. Acta Math. **54**, 185–204 (1930)
17. H. Contopanagos, B. Dembart, M. Epton et al., Well-conditioned boundary integral equations for three-dimensional electromagnetic scattering. IEEE Trans. Antenn. Propag. **50**, 1824–1830 (2002)
18. S.H. Christiansen, J.-C. Nédélec, A preconditioner for the electric field integral equation based on Calderon formulas. SIAM J. Numer. Anal. **40**, 1100–1135 (2002)
19. S. Borel, D.P. Levadoux, F. Alouges, A new well-conditioned integral formulation for Maxwell equations in three dimensions. IEEE Trans. Antenn. Propag. **53**, 2995–3004 (2005)
20. F.P. Andriulli, K. Cools, H. Bagci et al., A multiplicative Calderon preconditioner for the electric field integral equation. IEEE Trans. Antenn. Propag. **56**, 2398–2412 (2008)
21. M.S. Tong, W.C. Chew, A higher order Nyström scheme for electromagnetic scattering by arbitrarily shaped surfaces. IEEE Antenn. Wire. Propag. Lett. **4**, 277–280 (2005)
22. M.G. Duffy, Quadrature over a pyramid or cube of integrands with a singularity at a vertex. SIAM J. Numer. Anal. **19**, 1260–1262 (1982)

23. W. Cai, Y. Yu, X.C. Yuan, Singularity treatment and high-order RWG basis functions for integral equations of electromagnetic scattering. Int. J. Numer. Meth. Eng. **53**, 31–47 (2002)
24. J.R. Mosig, Integral equation technique, in *Numerical Techniques for Microwave and Millimeter-Wave Passive Structure*. ed. by T. Itoh (Wiley, New York, 1989), pp.133–214
25. R.F. Harrington, *Field Computation by Moment Methods* (Wiley-IEEE Press, Hoboken, NJ, 1993)
26. G. Duvaut, J.L. Lions, *Inequalities in Mechanics and Physics*, Lemma 4.2, p. 341 (Springer, New York, 1976)
27. J.C. Nédélec, *Acoustic and Electromagnetic Equations: Integral Representations for Harmonic Problems* (Springer, Berlin, 2001)
28. A. Buffa, M. Costabel, D. Sheen, On traces for H(Curl, ω) in Lipschitz domains. J. Math. Anal. Appl. **276**, 845–867 (2002)
29. A. Buffa, M. Costabel, C. Schwab, Boundary element methods for Maxwell's equations on non-smooth domains. Numer. Math. **92**, 679–710 (2002)
30. W. Cai, T. Yu, H. Wang, Y. Yu, High-order mixed RWG basis functions for electromagnetic applications. IEEE Trans. Microw. Theory Tech. **49**, 1295–1303 (2001)
31. S.M. Rao, D.R. Wilton, A.W. Glisson, Electromagnetic scattering by surfaces of arbitrary shape. IEEE Trans. Antenn. Propag. **30**, 409–418 (1982)
32. S.M. Wandzura, Electric current basis functions for curved surfaces. Electromagnetics **12**, 77–91 (1992)
33. B. Szabó, I. Babuska, *Finite Element Analysis* (Wiley-Interscience, New York, 1991)
34. W. Cai, High-order mixed current basis functions for electromagnetic scattering of curved surfaces. J. Sci. Comput. **14**, 73–105 (1999)
35. G. Chandler, Galerkin's method for boundary integral equations on polygonal domains. J. Austral. Math. Soc., Ser. B **26**, 1–13 (1984)
36. K.E. Atkinson, I. Graham, Iterative variants of the nyström method for second kind boundary integral operators. SIAM J. Sci. Stat. Comput. **13**, 694–722 (1990)
37. R.A. Kress, A nyström method for boundary integral equations in domains with corners. Numer. Math. **58**, 145–161 (1990)
38. O. Bruno, T. Elling, R. Paffenroth, C. Turc, Electromagnetic integral equations requiring small numbers of Krylov-subspace iterations. J. Comput. Phys. **228**, 6169–6183 (2009)
39. A. Brandt, A. Lubrecht, Multilevel matrix multiplication and fast solution of integral equations. J. Comput. Phys. **90**, 348–370 (1990)
40. E. Bleszynski, M. Bleszynski, T. Jaroszewicz, Aim: adaptive integral method for solving large scale electromagnetic scattering and radiation problems. Radio Sci. **31**, 1225–1251 (1996)
41. W.C. Chew, J.-M. Jin, E. Michielssen, J. Song, *Fast and Efficient Algorithms in Computational Electromagnetics* (Artech House Publishers, Norwood, MA, 2001)
42. M.H. Cho, W. Cai, A parallel fast algorithm for computing the helmholtz integral operator in 3-d layered media. J. Comput. Phys. **231**, 5910–5925 (2012)
43. W.C. Chew, M.S. Tong, B. Hu, *Integral Equation Methods for Electromagnetic and Elastic Waves* (Morgan & Claypool, San Francisco, CA, 2008)

Chapter 12
High-Order Hierarchical Nédélec Edge Elements

The Nédélec edge elements [1, 2] form the natural choices of basis functions to approximate electromagnetic fields, as the field tangential continuity is built into the basis functions. The edge elements have been applied extensively in microwave and optical devices [3]. In this chapter, we will first present the original construction of the Nédélec edge element basis including some mathematical detail. Then, explicit high-order hierarchical versions of the Nédélec basis with good conditioning properties will be presented.

12.1 Nédélec Edge Elements in H(Curl)

In the two seminal papers [1, 2], Nédélec proposed to use quantities (moments of tangential components of vector fields) on edges and faces to define the finite dimensional space in $H(\mathrm{curl})$, thus the name edge element. This work lays the foundation of vector finite elements for solving Maxwell's equations.

It can be shown that the tangential continuities of the electric and magnetic fields in a domain Ω imply that their appropriate solution spaces should be $H(\mathrm{curl}, \Omega)$. This fact can be illustrated with a simple example in a 2-D problem. For a more precise mathematical argument, refer to [4].

In a finite element method, the solution domain Ω will be tessellated with a partition $\mathcal{T}_h$ as follows:

$$\mathcal{T}_h = \{K_i : K_i \cap K_j = \emptyset,\ i \neq j,\ \bigcup_i K_i = \Omega\}, \tag{12.1}$$

where each K_i may be a triangle or quadrilateral in 2-D or a tetrahedron or hexahedron in 3-D. The finite element space consists of a triplet (K, P, A), where P defines the solution space over each K (polynomials or other specially selected functions) and

W. Cai, *Deterministic, Stochastic, and Deep Learning Methods for Computational Electromagnetics*, https://doi.org/10.1007/978-981-96-0100-4_12

A specifies the degree of freedom (d.o.f.), which uniquely defines the function in the finite element space. For illustration purposes, consider the 2-D problem of a TE wave $\mathbf{E} = (E_1(x, y), E_2(x, y), 0)$ and $\mathbf{H} = (0, 0, H_z(x, y))$, where $(x, y) \in \mathbb{R}^2$. Maxwell's equations are simplified to

$$\epsilon \frac{\partial \mathbf{E}}{\partial t} - \overrightarrow{\text{curl}}_s H_z = -\mathbf{J}, \tag{12.2}$$

$$\mu \frac{\partial H_z}{\partial t} + \text{curl}_s \mathbf{E} = 0, \tag{12.3}$$

where two types of the 2-D curl operators in the Cartesian coordinates are used:

- $\overrightarrow{\text{curl}}_s$:

$$\overrightarrow{\text{curl}}_s H_z = \left(\frac{\partial H_z}{\partial y}, -\frac{\partial H_z}{\partial x}, 0 \right), \tag{12.4}$$

- curl_s :

$$\text{curl}_s \mathbf{E} = \frac{\partial E_2}{\partial x} - \frac{\partial E_1}{\partial y}. \tag{12.5}$$

The finite element solution $\mathbf{E}$ is assumed to be a polynomial function on each element K. Let the interface between two elements K_i and K_j, $S = K_i \cap K_j$, be on the x-axis. Then, E_1 will be the tangential component of $\mathbf{E}$. If E_1 is discontinuous across S, then $\text{curl}_s \mathbf{E} \notin L^2(K_i \cup K_j)$ as $\partial E_1/\partial y = c\delta(y)$ for some constant c. In other words, if E_1 is continuous, we will have $\text{curl}_s \mathbf{E} \in L^2(K_i \cup K_j)$, namely, $\mathbf{E} \in H(\text{curl}_s, K_i \cup K_j)$, which shows that the correct function space for $\mathbf{E}$ is indeed $H(\text{curl}_s, \Omega)$.

12.1.1 Finite Element Method for E *or* H *Wave Equations*

From (8.30), by replacing $i\omega$ by $\partial/\partial t$, we have the following time-dependent vector wave equation for $\mathbf{E}(\mathbf{x}, t)$:

$$\epsilon \frac{\partial^2 \mathbf{E}}{\partial t^2} + \nabla \times \frac{1}{\mu} \nabla \times \mathbf{E} = -\frac{\partial \mathbf{J}}{\partial t}, \ \mathbf{x} \in \Omega, \tag{12.6}$$

with the initial conditions

$$\begin{aligned} \mathbf{E}(\mathbf{x}, 0) &= \mathbf{E}_0(\mathbf{x}), \\ \mathbf{E}_t(\mathbf{x}, 0) &= \frac{1}{\epsilon} \left(-\mathbf{J}(\mathbf{x},0) + \nabla \times \mathbf{H}(\mathbf{x},0) \right), \end{aligned} \tag{12.7}$$

where (12.7) results from the Ampère–Maxwell equation. Meanwhile, a PEC boundary condition is set on $\Gamma = \partial\Omega$:

$$(\mathbf{n} \times \mathbf{E})|_\Gamma = \mathbf{0}. \tag{12.8}$$

The weak formulation of Maxwell's equations can now be defined with the following Sobolev spaces:

$$H(\text{curl}, \Omega) = \left\{\mathbf{u} \in L_2(\Omega)^3 : \nabla \times \mathbf{u} \in L_2(\Omega)^3\right\} \tag{12.9}$$

and

$$H_0(\text{curl}, \Omega) = \left\{\mathbf{u} \in L_2(\Omega)^3 : \nabla \times \mathbf{u} \in L_2(\Omega)^3, (\mathbf{n} \times \mathbf{u})|_\Gamma = \mathbf{0}\right\}. \tag{12.10}$$

The weak formulation of Maxwell's equations in $\mathbf{E}$ is given as follows. Find $\mathbf{E} \in H_0(\text{curl}, \Omega)$ such that $\forall \mathbf{v} \in H_0(\text{curl}, \Omega)$,

$$\epsilon \frac{\partial^2}{\partial t^2}(\mathbf{E}, \mathbf{v}) + \left(\frac{1}{\mu} \nabla \times \mathbf{E}, \nabla \times \mathbf{v}\right) = -\left(\frac{\partial \mathbf{J}}{\partial t}, \mathbf{v}\right). \tag{12.11}$$

The finite element method will be a finite dimensional analog of (12.11) in a subspace $U_{h0} \subset H_0(\text{curl}, \Omega)$, resulting in a semi-discretization method.

Semi-discretization finite element method

Find $\mathbf{E}_h \in U_{h0}$ such that $\forall \mathbf{v}_h \in U_{h0}$,

$$\epsilon \frac{\partial^2}{\partial t^2}(\mathbf{E}_h, \mathbf{v}) + \left(\frac{1}{\mu} \nabla \times \mathbf{E}_h, \nabla \times \mathbf{v}_h\right) = -\left(\frac{\partial \mathbf{J}}{\partial t}, \mathbf{v}_h\right). \tag{12.12}$$

Let $\{\mathbf{v}_h^i\}_{i=1}^N$ be a basis for U_{h0}, where $\{\mathbf{v}_h^i\}_{i=1}^{N_0}$ are the internal basis functions associated with interior edges and faces of the tessellation of the domain, and $\{\mathbf{v}_h^i\}_{i=N_0+1}^N$ are the basis functions associated with the domain boundary Γ.

After using the zero boundary condition $\mathbf{E}_h \times \mathbf{n} = \mathbf{0}$, we have

$$\mathbf{E}_h(\mathbf{x}, t) = \sum_{i=1}^{N_0} \mathbf{e}_i(t) \mathbf{v}_h^i(\mathbf{x}). \tag{12.13}$$

Setting $\mathbf{e} = \left(\mathbf{e}_1(t), \ldots, \mathbf{e}_{N_0}(t)\right)^{\mathrm{T}}$, the finite element method produces the following matrix equation:

$$M \frac{\mathrm{d}^2 \mathbf{e}}{\mathrm{d}t^2} + A\mathbf{e} = \mathbf{j}, \tag{12.14}$$

where the mass matrix M is defined as

$$M_{ij} = \left(\epsilon \mathbf{v}_h^i, \mathbf{v}_h^j\right), \tag{12.15}$$

and the stiffness matrix A is given by

$$A_{ij} = \left(\frac{1}{\mu} \nabla \times \mathbf{v}_h^i, \nabla \times \mathbf{v}_h^j \right). \tag{12.16}$$

The right-hand side $\mathbf{j}$ is defined by

$$j_i = -\left(\frac{\partial \mathbf{J}}{\partial t}, \mathbf{v}_h^i \right). \tag{12.17}$$

The system of ordinary differential equations (12.14) for the unknown vector $\mathbf{e}^n$, which approximates $\mathbf{e}(t^n)$ at time step $t^n = n\Delta t$, can be solved by various time marching algorithms, for example a second-order central scheme:

$$M \frac{\mathbf{e}^{n+1} - 2\mathbf{e}^n + \mathbf{e}^{n-1}}{\Delta t^2} + A\mathbf{e}^n = \mathbf{j}^n. \tag{12.18}$$

The explicit time discretization requires a CFL type condition,

$$\Delta t = O\left(\frac{1}{\sqrt{\lambda_{\max}}} \right), \tag{12.19}$$

where $\lambda_{\max}$ is the largest eigenvalue of the matrix $M^{-1}A$. Implicit time discretization can also be used for better stability and larger time step but with the additional cost of solving a linear matrix system at each time step.

12.1.2 Reference Elements and Piola Transformations

In defining the Nédélec elements, moments of the solution's tangential components along the edges and the faces of the elements are used as degrees of freedom (d.o.f.). To ensure the $H(\text{curl})$ conformity in the physical space, an appropriate mapping between the reference element and the physical element is required such that the moments of the tangential components of solutions in the physical space will be identified with the correct moments' d.o.f. on the reference element, and basis functions thus constructed will have matching tangential moments along the shared faces/edges in the physical space. Also, the tangential vectors for the physical elements and the reference element, required in the definition of the d.o.f., along the faces and the edges, will be mapped correspondingly. The Piola transformations are designed to meet such a requirement [5].

Consider a trilinear mapping $F_K : \hat{K} \to \mathbb{R}^3$, such that $K = F_K(\hat{K})$, whose Jacobian is $\mathbf{DF} = DF_K(\hat{\mathbf{x}})/D\hat{\mathbf{x}}$ with $JF = \det(\mathbf{DF})$. For a vector valued function $\hat{\mathbf{u}} : \hat{K} \to \mathbb{R}^3$, we define $\mathbf{u} = \mathbf{P}_F\hat{\mathbf{u}} : K \to \mathbb{R}^3$ by

$$\mathbf{u}(\mathbf{x}) = \mathbf{P}_F\hat{\mathbf{u}} \equiv (\mathbf{DF})^{-\mathrm{T}}(\hat{\mathbf{x}})\hat{\mathbf{u}}(\hat{\mathbf{x}}), \quad \text{for} \quad \mathbf{x} = \mathbf{F}(\hat{\mathbf{x}}). \tag{12.20}$$

The above mapping relates the moments of the tangential components of functions $\mathbf{u}(\mathbf{x})$ and $\hat{\mathbf{u}}(\hat{\mathbf{x}})$ in the following manner. Moreover, if $\mathbf{t}$ and $\hat{\mathbf{t}}$ are the tangential vectors on a corresponding edge of the elements K and $\hat{K}$, respectively, and $\mathbf{n}$ and $\hat{\mathbf{n}}$ are the normal vectors of a corresponding face of K and $\hat{K}$, respectively, we have

$$\mathbf{n}(\mathbf{x}) = \frac{\left[DF(\hat{\mathbf{x}})\right]^{-\mathrm{T}} \hat{\mathbf{n}}(\hat{\mathbf{x}})}{|\left[DF(\hat{\mathbf{x}})\right]^{-\mathrm{T}} \hat{\mathbf{n}}(\hat{\mathbf{x}})|}, \qquad \mathbf{t}(\mathbf{x}) = \frac{\left[DF(\hat{\mathbf{x}})\right]^{\mathrm{T}} \hat{\mathbf{t}}(\hat{\mathbf{x}})}{|\left[DF(\hat{\mathbf{x}})\right]^{\mathrm{T}} \hat{\mathbf{t}}(\hat{\mathbf{x}})|}. \tag{12.21}$$

For test functions $\hat{p}$, $\hat{\mathbf{q}}$, and $\hat{\mathbf{w}}$ defined on $\hat{K}$, we then have

$$\int_e \mathbf{u} \cdot \mathbf{t} p \, \mathrm{d}l = \operatorname{sign}(\det(\mathbf{DF})) \int_{\hat{e}} \hat{\mathbf{u}} \cdot \hat{\mathbf{t}} \hat{p} \, \mathrm{d}\hat{l}, \quad \text{for } p = \hat{p} \circ F^{-1}, \tag{12.22}$$

$$\int_f \mathbf{u} \times \mathbf{n} \cdot \mathbf{w} \, \mathrm{d}s = \operatorname{sign}(\det(\mathbf{DF})) \int_{\hat{f}} \hat{\mathbf{u}} \times \hat{\mathbf{n}} \cdot \hat{\mathbf{w}} \, \mathrm{d}\hat{s}, \quad \text{for } \mathbf{w} = P_F \hat{\mathbf{w}}, \tag{12.23}$$

$$\int_K \mathbf{u} \cdot \mathbf{q} \, \mathrm{d}\Omega = \operatorname{sign}(\det(\mathbf{DF})) \int_{\hat{K}} \hat{\mathbf{u}} \cdot \hat{\mathbf{q}} \, \mathrm{d}\hat{\Omega}, \quad \text{for } \mathbf{q} = R_F \hat{\mathbf{q}}, \tag{12.24}$$

where

$$\mathbf{q} = \mathbf{R}_F \hat{\mathbf{q}} \equiv (JF)^{-1} (\hat{\mathbf{x}}) \mathbf{DF}(\hat{\mathbf{x}}) \hat{\mathbf{q}}(\hat{\mathbf{x}}), \quad \text{for } \mathbf{x} = \mathbf{F}(\hat{\mathbf{x}}). \tag{12.25}$$

Relations (12.22), (12.23), and (12.24) map the edge and face moments of tangential components of a solution and the element moments of the solution between K and $\hat{K}$, respectively.

12.1.3 Nédélec Finite Element Basis in $H(\text{Curl})$

In this section, we present the original construction of the Nédélec elements following closely the presentations in [1, 2]. The reference element is denoted by K.

2-D edge elements in a rectangle

Denote $Q_{p,q}$ as the polynomial space in (x, y) of degree p in x and q in y, respectively, i.e.,

$$Q_{p,q} = \operatorname{span}\{x^i y^j : 0 \le i \le p, \; 0 \le j \le q\}, \tag{12.26}$$

and P_k as the polynomial space of degree k in one variable,

$$P_k - \operatorname{span}\{x^i : 0 \le i \le k\}. \tag{12.27}$$

Consider the finite element triplet (K, P, A), where

$$
\begin{aligned}
K &= \text{a unit square in 2-D} = \{(x, y) : 0 \le x, y \le 1\},\\
P &= \{\mathbf{u} = (u_1, u_2) : u_1 \in Q_{k,k+1}, u_2 \in Q_{k+1,k}\},\\
A &= \{\alpha_i(\mathbf{u}),\ \text{degree of freedom (d.o.f.)}\},
\end{aligned}
\tag{12.28}
$$

and $\dim(P) = 2(k+1)(k+2)$. If e denotes an edge of K and $\mathbf{t}$ denotes the unit tangential vector of the edge, the following moments of a vector solution will be used as degrees of freedom.

- Edge d.o.f. with edge moments:

$$\alpha_e(\mathbf{u}) = \int_e \mathbf{u} \cdot \mathbf{t} q \, \mathrm{d}l, \quad q \in P_k(e), \tag{12.29}$$

  with $k+1$ d.o.f. for each edge and a total of $4(k+1)$ d.o.f. for four edges.
- Quadrilateral d.o.f. with element moments:

$$\alpha_K(\mathbf{u}) = \int_K \mathbf{u} \cdot \mathbf{q} \, \mathrm{d}\Omega, \quad \mathbf{q} \in Q_{k,k-1} \times Q_{k-1,k}, \tag{12.30}$$

  with a total of $2(k+1)k$ d.o.f. for the element.

Altogether the total number of all degrees of freedom is

$$4(k+1) + 2(k+1)k = 2(k+1)(k+2) = \dim\left(Q_{k,k+1} \times Q_{k+1,k}\right) = \dim(P).$$

Next, we will show the uni-solvent property of the above d.o.f., namely, functions in the finite element space over each element K can be uniquely defined by using those $2(k+1)(k+2)$ d.o.f. Equivalently, we want to show the following.

Lemma 12.1 (Uni-solvence) *If $\alpha_e(\mathbf{u}) = 0$ and $\alpha_K(\mathbf{u}) = 0$, then $\mathbf{u} \equiv 0$ in K.*

Proof Let

$$\mathbf{u} \in Q_{k,k+1} \times Q_{k+1,k}$$

for the edge $e_1 = \{(x, 0) : 0 \le x \le 1\}$ whose tangential is $\mathbf{t} = (1, 0)$. Then

$$\mathbf{u} \cdot \mathbf{t} = u_1 \in Q_{k,k+1},$$

and thus

$$\mathbf{u} \cdot \mathbf{t}|_{e_1} \in P_k(e_1).$$

With $\alpha_{e_1}(\mathbf{u}) = 0$, we have

$$\int_{e_1} \mathbf{u} \cdot \mathbf{t} q \, \mathrm{d}l = \int_{e_1} u_1 q \, \mathrm{d}l = 0, \quad \forall q \in P_k(e_1). \tag{12.31}$$

Set $q = u_1$ in (12.31). Then $\int_{e_1} u_1^2 \,\mathrm{d}l = 0$, which implies $u_1 = 0$ on e_1. Similarly, on $e_3 = \{(x, 1) : 0 \le x \le 1\}$, $u_1 = 0$. Together, we conclude that $u_1 = y(1 - y)v_1(x, y)$ for some $v_1 \in Q_{k,k-1}$.

Similarly, we can show that $u_2 = x(1 - x)v_2(x, y)$ for some $v_2 \in Q_{k-1,k}$.

Next, consider the condition that the element d.o.f. $\alpha_K(\mathbf{u}) = 0$. As $\alpha_K(\mathbf{u}) = \int_K \mathbf{u} \cdot \mathbf{q} \,\mathrm{d}\Omega = \int_K (u_1 q_1 + u_2 q_2)\mathrm{d}\Omega$ with $(q_1, q_2) \in Q_{k,k-1} \times Q_{k-1,k}$, we have

$$0 = \alpha_K(\mathbf{u}) = \int_K (u_1 q_1 + u_2 q_2)\mathrm{d}\Omega = \int_K [y(1 - y)v_1 q_1 + x(1 - x)v_2 q_2]\,\mathrm{d}\Omega.$$

Let $q_i = v_i$ for $i = 1, 2$, and we get

$$\int_K \left[y(1 - y)v_1^2 + x(1 - x)v_2^2\right]\mathrm{d}\Omega = 0,$$

which means $v_1 = v_2 = 0$. Therefore, we have $\mathbf{u} \equiv 0$. □

Lemma 12.2 *If $\alpha_e(\mathbf{u}^+) = \alpha_e(\mathbf{u}^-)$, then $\mathbf{u}^+ \cdot \mathbf{t} = \mathbf{u}^- \cdot \mathbf{t}$ (the tangential component is continuous), i.e., $\mathbf{u} \in H(\text{curl}, K^+ \cup K^-)$ with $\mathbf{u}|_{K^+} = \mathbf{u}^+$ and $\mathbf{u}|_{K^-} = \mathbf{u}^-$.*

Proof Since $\alpha_e(\mathbf{u}^+) = \int_e \mathbf{u}^+ \cdot \mathbf{t}q \,\mathrm{d}l$ and $\alpha_e(\mathbf{u}^-) = \int_e \mathbf{u}^- \cdot \mathbf{t}q \,\mathrm{d}l$, we have

$$\alpha_e(\mathbf{u}^+) - \alpha_e(\mathbf{u}^-) = \int_e (\mathbf{u}^+ - \mathbf{u}^-) \cdot \mathbf{t}q \,\mathrm{d}l.$$

Consider the case of $\mathbf{t} = (0, 1)$. We have $(\mathbf{u}^+ - \mathbf{u}^-) \cdot \mathbf{t} \in P_k(e)$ on e. So we can set $\mathbf{q} = (\mathbf{u}^+ - \mathbf{u}^-) \cdot \mathbf{t}$:

$$\int_e (\mathbf{u}^+ - \mathbf{u}^-) \cdot \mathbf{t}q \,\mathrm{d}l = \int_e |(\mathbf{u}^+ - \mathbf{u}^-) \cdot \mathbf{t}|^2 \,\mathrm{d}l = 0,$$

which means $\mathbf{u}^+ \cdot \mathbf{t} = \mathbf{u}^- \cdot \mathbf{t}$, i.e., the tangential component is continuous, which in turn implies [4] that $\mathbf{u} \in H(\text{curl}, K^+ \cup K^-)$, as illustrated at the beginning of this section. □

2-D edge elements in a triangle

Let $K = \{(x, y) : 0 \le x, y, x + y \le 1\}$ be the reference triangle in Fig. 11.2 and let the finite element be (K, P, A), where

$$P = (P_k(K))^2, \tag{12.32}$$

where

$$P_k(K) = \text{span}\{x^i y^j : 0 \le i + j \le k\} \tag{12.33}$$

and

$$\dim(P) = (k+2)(k+1). \tag{12.34}$$

We also define the following spaces to be used in the definition of moments [2]:

$$\mathcal{D}_k = (P_{k-1}(K))^2 \oplus \widetilde{P}_{k-1} \cdot \mathbf{r}, \tag{12.35}$$

where $\widetilde{P}_k$ is the kth homogeneous polynomial space in $\mathbf{r} = (x_1, x_2)$,

$$\widetilde{P}_k = \text{span}\{x^i y^j \, : \, i + j = k\}, \tag{12.36}$$

$\dim(\widetilde{P}_k) = k + 1$ in 2-D, and the dimension of the space $\mathcal{D}_k$ is

$$\dim(\mathcal{D}_k) = k(k+2). \tag{12.37}$$

Due to the special design of $\mathcal{D}_k$, the image of $\mathcal{D}_k$ by the divergence operator is exactly P_{k-1}, namely

$$\nabla(\mathcal{D}_k) = P_{k-1}(K). \tag{12.38}$$

Now the degrees of freedom for the element $A = \{\alpha_i(\mathbf{u})\}$ are grouped with respect to the geometric identities of the triangle.

- Edge d.o.f. with edge moments:

$$\int_e \mathbf{u} \cdot \mathbf{t} q \, dl, \quad q \in P_k(e), \tag{12.39}$$

  with a total of $3(k+1)$ edge d.o.f.
- Element d.o.f. with element moments:

$$\int_K \mathbf{u} \cdot \mathbf{q} \, d\Omega, \quad \mathbf{q} \in \mathcal{D}_{k-1}, \tag{12.40}$$

  with a total of $(k-1)(k+1)$ element d.o.f.

The total number of all degrees of freedom is then

$$\mathcal{N}_k = 3(k+1) + (k-1)(k+1) = (k+2)(k+1) = \dim(P). \tag{12.41}$$

The proof of the uni-solvence for this set of d.o.f., A, is a 2-D version of the 3-D tetrahedron edge element given later, and thus is omitted here.

3-D edge elements in a cube

Consider the finite element triplet (K, P, A), where

$$\begin{aligned}
K &= \text{a cube in 3-D} = \{(x, y, z) : 0 \le x, y, z \le 1\},\\
P &= \{\mathbf{u} = (u_1, u_2, u_3) : u_1 \in Q_{k-1,k,k}, u_2 \in Q_{k,k-1,k}, u_3 \in Q_{k,k,k-1}\},\\
A &= \{\text{d.o.f.}\},
\end{aligned} \tag{12.42}$$

and $\dim(P) = 3k(k+1)^2$. The following moments of the solution will be used as degrees of freedom.

- Edge d.o.f. with edge moments:

$$\alpha(\mathbf{u}) = \int_e \mathbf{u} \cdot \mathbf{t} q \, dl, \quad q \in P_{k-1}(e), \tag{12.43}$$

  with a total of $12k$ edge d.o.f. for 12 edges.
- Face d.o.f. with face moments:

$$\alpha(\mathbf{u}) = \int_f \mathbf{u} \times \mathbf{n} \cdot \mathbf{q} \, ds, \quad q = (q_1, q_2) \in Q_{k-2,k-1} \times Q_{k-1,k-2}, \tag{12.44}$$

  with $2k(k-1)$ d.o.f. for each face and a total of $12k(k-1)$ face d.o.f.

- Cube d.o.f. with element moments:

$$\alpha(\mathbf{u}) = \int_K \mathbf{u} \cdot \mathbf{q} \, d\Omega,$$

$$\text{for } \mathbf{q} = (q_1, q_2, q_3) \in Q_{k-1,k-2,k-2} \times Q_{k-2,k-1,k-2} \times Q_{k-2,k-2,k-1}, \tag{12.45}$$

  with a total of $3k(k-1)^2$ element d.o.f. as each of the three Q subspaces has a dimension $k(k-1)^2$.

In summary, the total degrees of freedom for the cubic element is

$$12k + 12k(k-1) + 3k(k-1)^2 = 3k(k+1)^2 = \dim(P). \tag{12.46}$$

3-D edge elements in a tetrahedron

Let K be the reference tetrahedron shown in Fig. 12.1, and let the finite element be (K, P, A), where

$$P = (P_k(K))^3, \tag{12.47}$$

where

$$P_k(K) = \text{span}\{x^l y^m z^n : 0 \le l + m + n \le k\} \tag{12.48}$$

and

$$\dim(P) = \frac{(k+3)(k+2)(k+1)}{2}. \tag{12.49}$$

As in (12.35), we define the following spaces for the definition of moments:

$$\mathcal{D}_k = (P_{k-1})^3 \oplus \widetilde{P}_{k-1} \cdot \mathbf{r}, \tag{12.50}$$

where $\widetilde{P}_k$ is the kth homogeneous polynomial space, $\dim(\widetilde{P}_k) = (k+1)(k+2)/2$ in 3-D, $\mathbf{r} = (x_1, x_2, x_3)$, and the dimension of the space $\mathcal{D}_k$ [2] is given by

$$\dim(\mathcal{D}_k) = \frac{(k+3)(k+1)k}{2}. \tag{12.51}$$

Again, it is easy to verify that the image of $\mathcal{D}_k$ by the divergence operator is exactly P_{k-1}, namely

$$\nabla(\mathcal{D}_k) = P_{k-1}(K). \tag{12.52}$$

Now, the degrees of freedom for the element $A = \{\alpha_i(\mathbf{u})\}_{i=1}^{n}$ are grouped with respect to the geometric identities of the tetrahedron.

- Edge d.o.f. with edge moments:

$$\int_e \mathbf{u} \cdot \mathbf{t} q \, \mathrm{d}l, \quad q \in P_k(e), \tag{12.53}$$

  with a total of $6(k+1)$ edge d.o.f.
- Face d.o.f. with face moments:

$$\int_f \mathbf{u} \times \mathbf{n} \cdot \mathbf{q} \, \mathrm{d}s, \quad \mathbf{q} \in \mathcal{D}_{k-1}(f), \tag{12.54}$$

  with a total of $4(k-1)(k+1)$ face d.o.f.
- Volume d.o.f. with element moments:

$$\int_K \mathbf{u} \cdot \mathbf{q} \, \mathrm{d}\Omega, \quad \mathbf{q} \in \mathcal{D}_{k-2}(K), \tag{12.55}$$

  with a total of $(k+1)(k-1)(k-2)/2$ volume d.o.f.

The total number of all degrees of freedom in A is then

$$\begin{aligned} \mathcal{N}_k &= 6(k+1) + 4(k-1)(k+1) + \frac{(k+1)(k-1)(k-2)}{2} \\ &= \frac{(k+3)(k+2)(k+1)}{2} = \dim(P). \end{aligned} \tag{12.56}$$

In the following, we will prove the uni-solvence for the finite element and $H(\text{curl})$ conformity, following the work of [2]. To prove the $H(\text{curl})$ conformity, we only need to show the continuity of the tangential components $\mathbf{u}_f = \mathbf{n} \times (\mathbf{u} \times \mathbf{n})$ of the finite element solutions on any common interface $f = K^+ \cap K^-$ with a normal $\mathbf{n}$ if

$$\alpha_i(\mathbf{u}^+) = \alpha_i(\mathbf{u}^-) \quad \text{for d.o.f. } \alpha_i.$$

Equivalently, due to the linearity of the functional $\alpha_i(\mathbf{u})$, we need only show that

$$\mathbf{u}_f = 0 \text{ if } \alpha_i(\mathbf{u}) = 0 \text{ for all edge and face d.o.f. on face f.} \tag{12.57}$$

Theorem 12.1 *The finite element (K, P, A) defined in (12.47) and (12.53)–(12.55) is conforming in the space H(curl).*

Proof Given a common face $f = K^+ \cap K^-$ between two neighboring elements, we assume all edge and face d.o.f. associated with f to be zero. As

$$\mathbf{u} \cdot \mathbf{t} \in P_k(e), \tag{12.58}$$

by setting $q = \mathbf{u} \cdot \mathbf{t} \in P_k(e)$ in the edge moment d.o.f. in (12.53) on each face f, we have

$$\alpha(\mathbf{u}) = \int_e (\mathbf{u} \cdot \mathbf{t})^2 \, \mathrm{d}l = 0. \tag{12.59}$$

Therefore, $\mathbf{u} \cdot \mathbf{t} = 0$ on $e \in \partial f$. Applying Stokes' formula on the tangential component $\mathbf{u}_f = \mathbf{n} \times (\mathbf{u} \times \mathbf{n}) \equiv (u_1, u_2)$ over a face f, we have

$$\int_f \left(\overrightarrow{\text{curl}}_s q \cdot \mathbf{u}_f - q \, \text{curl}_s \, \mathbf{u}_f \right) \mathrm{d}s = \int_{\partial f} (\mathbf{u}_f \cdot \mathbf{t}) q \, \mathrm{d}l = \int_{\partial f} (\mathbf{u} \cdot \mathbf{t}) q \, \mathrm{d}l = 0, \tag{12.60}$$

where in the second to last equality we have used the facts that $\mathbf{t} \perp \mathbf{n}$ and

$$\mathbf{u} = (\mathbf{u} \cdot \mathbf{n})\mathbf{n} + \mathbf{u}_f = (\mathbf{u} \cdot \mathbf{n})\mathbf{n} + \mathbf{n} \times (\mathbf{u} \times \mathbf{n}),$$

$$\mathbf{u} \cdot \mathbf{t} = \mathbf{u}_f \cdot \mathbf{t}.$$

Next, for $q \in P_{k-1}(f)$, $\overrightarrow{\text{curl}}_s q \in (P_{k-2})^2 \subset \mathcal{D}_{k-1}(f)$, and, along with the fact that the following face moments vanish:

$$\int_f \overrightarrow{\text{curl}}_s q \cdot \mathbf{u}_f \, \mathrm{d}s = 0, \tag{12.61}$$

and (12.60), we have

$$\int_f q \operatorname{curl}_s \mathbf{u}_f \, ds = 0 \quad \text{for } q \in P_{k-1}(f). \tag{12.62}$$

As $\mathbf{u}_f \in P_k$, we can set in (12.62)

$$q = \operatorname{curl}_s \mathbf{u}_f \in P_{k-1},$$

yielding

$$\int_f \left(\operatorname{curl}_s \mathbf{u}_f\right)^2 ds = 0 \quad \text{on } f. \tag{12.63}$$

Thus, we get

$$\operatorname{curl}_s \mathbf{u}_f = 0 \quad \text{on } f. \tag{12.64}$$

Therefore, on each face f, we have

$$\mathbf{u}_f = \nabla_f \phi \text{ for some } \phi \in P_{k+1}(f), \tag{12.65}$$

with $\nabla_f = (\partial/\partial\xi_1, \partial/\partial\xi_2)$ being the surface gradient operator on the face f.

However, from vanishing tangential components of $\mathbf{u}$ along each edge of f,

$$\mathbf{u}_f \cdot \mathbf{t} = 0 \text{ on } \partial f, \quad \text{namely } \left(\nabla_f \phi\right) \cdot \mathbf{t} = 0, \tag{12.66}$$

we have

$$\frac{d\phi}{d\mathbf{t}} = 0 \quad \text{on } \partial f. \tag{12.67}$$

Then $\phi|_{\partial f} = \text{constant}$. As the constant here can be set to be zero, we obtain the following form for ϕ:

$$\phi = \lambda_1 \lambda_2 \lambda_3 \psi, \qquad \psi \in P_{k-2}, \quad \phi \in P_{k+1}, \tag{12.68}$$

where λ_1, λ_2, and λ_3 are the area coordinates on the face f.

Now consider the vanishing face moment d.o.f.

$$\int_f \mathbf{u}_f \cdot \mathbf{q} \, ds = \int_f \nabla_f \phi \cdot \mathbf{q} \, ds = 0, \qquad \forall \mathbf{q} \in \mathcal{D}_{k-1}(f), \tag{12.69}$$

which, using integration by parts, can be rewritten as

$$0 = \int_f \phi(\nabla_f \cdot \mathbf{q}) ds - \int_{\partial f} \phi \mathbf{q} \cdot \mathbf{n} \, dl = \int_f \lambda_1 \lambda_2 \lambda_3 \psi (\nabla_f \cdot \mathbf{q}) ds, \tag{12.70}$$

where $\phi|_{\partial f} = 0$ is used in the second equality.

As $\psi \in P_{k-2}$ and $\nabla_f \mathcal{D}_{k-1}(f) = P_{k-2}$, we can always find a $\mathbf{q} \in \mathcal{D}_{k-1}(f)$ such that $\nabla_f \cdot \mathbf{q} = \psi$. Then (12.70) becomes

$$\int_f \lambda_1\lambda_2\lambda_3\psi^2\,\mathrm{d}s = 0.$$

Since $\lambda_i \geq 0$ for $i = 1, 2, 3$, $\psi = 0$. Therefore, $\phi = 0$, which implies that $\mathbf{u}_f = 0$, or equivalently $\mathbf{u} \in H(\mathrm{curl})$. □

Theorem 12.2 *(K, P, A) defined in (12.47) and (12.53)–(12.55) is uni-solvent, i.e.,*

$$\alpha_i(\mathbf{u}) = 0,\ \forall \alpha_i(\mathbf{u}) \ \textit{if and only if}\ \mathbf{u} = \mathbf{0} \textit{ on } K.$$

Proof We need only to prove the uni-solvence on a reference tetrahedron K, due to the invariant transforms (12.22) and (12.23).

Step 1: First we prove $\mathbf{v} = \mathrm{curl}(\mathbf{u}) = \mathbf{0}$ on K.

Applying Stokes' theorem on f, we have

$$\int_f \left(\overrightarrow{\mathrm{curl}}_s q \cdot \mathbf{u}_f - q\,\mathrm{curl}_s\,\mathbf{u}_f\right)\mathrm{d}s = \int_{\partial f} (\mathbf{u}_f \cdot \mathbf{t}) q\,\mathrm{d}l = 0,\ q \in P_k(f). \tag{12.71}$$

Let us orientate f to be on the xy-plane. Then, for $u = (u_1, u_2, u_3)^{\mathrm{T}}$,

$$\mathbf{u}_f = \begin{bmatrix} u_1 \\ u_2 \end{bmatrix},\ \mathrm{curl}\,\mathbf{u} = \mathbf{k}\left(\frac{\partial u_2}{\partial x} - \frac{\partial u_1}{\partial y}\right) + \mathbf{i}(\star) + \mathbf{j}(\star),$$

where the last two terms are not relevant for our discussion. Note that $\mathrm{curl}_s\,\mathbf{u}_f = \partial u_2/\partial x - \partial u_1/\partial y$, which is exactly the normal component of $\mathrm{curl}\,\mathbf{u}$, i.e., $\mathrm{curl}_s\,\mathbf{u}_f = \mathrm{curl}\,\mathbf{u}\cdot\mathbf{k} = \mathbf{v}\cdot\mathbf{n}$. So (12.71) implies that

$$\int_f \left(\overrightarrow{\mathrm{curl}}_s q \cdot \mathbf{u}_f - q\mathbf{v}\cdot\mathbf{n}\right)\mathrm{d}s = 0, \qquad q \in P_k(f). \tag{12.72}$$

As the first term vanishes for $q \in P_{k-1}(f)$ because $\overrightarrow{\mathrm{curl}}_s q \in (P_{k-2})^2 \subset \mathcal{D}_{k-1}$, we have

$$0 = \int_f q\mathbf{v}\cdot\mathbf{n}\,\mathrm{d}s,\ \forall q \in P_{k-1}(f). \tag{12.73}$$

Setting $q = \mathbf{v}\cdot\mathbf{n} \in P_{k-1}(f)$ in (12.73) gives

$$\int_f (\mathbf{v}\cdot\mathbf{n})^2\,\mathrm{d}s = 0,$$

i.e., $\mathbf{v} \cdot \mathbf{n} = 0$ on f for $\mathbf{v} \in (P_{k-1})^3$, implying that

$$\begin{aligned} v_3 &= z\psi_3 = x_3\psi_3, \\ v_2 &= y\psi_2 = x_2\psi_2, \\ v_1 &= x\psi_1 = x_1\psi_1, \end{aligned}$$

where $\psi_i \in P_{k-2}$ for $i = 1, 2, 3$ as $\mathbf{v} \in (P_{k-1})^3$.

Finally, we apply Stokes' theorem on K:

$$\int_K (\text{curl}\mathbf{u} \cdot \mathbf{q} - \text{curl}\, \mathbf{q} \cdot \mathbf{u}) \mathrm{d}\Omega = \int_{\partial K} \mathbf{q} \cdot \mathbf{n} \times \mathbf{u}\, \mathrm{d}s = 0, \ \ \forall \mathbf{q} \in (P_{k-2})^3,$$

where the last equality is due to the fact that $\mathbf{q}|_f \in (P_{k-2}(f))^2 \subset \mathcal{D}_{k-1}(f)$ and the vanishing of the face d.o.f. in (12.54). As curl $\mathbf{q} \in (P_{k-3})^3 \subset \mathcal{D}_{k-2}$, the second integral on the left-hand side vanishes from the vanishing of the element d.o.f. in (12.55); thus

$$\int_K \mathbf{v} \cdot \mathbf{q}\, \mathrm{d}\Omega = \int_K (x_1\psi_1 q_1 + x_2\psi_2 q_2 + x_3\psi_3 q_3) \mathrm{d}\Omega = 0, \forall \mathbf{q} \in (P_{k-2})^3 \subset \mathcal{D}_{k-1}(f). \tag{12.74}$$

On setting $q_i = \psi_i \in P_{k-2}, i = 1, 2, 3$, in (12.74), we have

$$\int_K \left(x_1\psi_1^2 + x_2\psi_2^2 + x_3\psi_3^2 \right) \mathrm{d}\Omega = 0$$

and

$$\psi_i = 0, \quad i = 1, 2, 3,$$

which proves that $\mathbf{v} = \text{curl}(\mathbf{u}) = \mathbf{0}$ on K.

Step 2: We now show that $\mathbf{u} = \mathbf{0}$.

As $\text{curl}(\mathbf{u}) = \mathbf{0}$ on K, we get

$$\mathbf{u} = \nabla\phi, \quad \text{for some} \ \ \phi \in P_{k+1}(K).$$

Because $\alpha_i(\mathbf{u}) = 0$, $\forall \alpha_i$ (face and edge d.o.f.), from the proof of Theorem 12.1 we have $\mathbf{u}_f = 0$ on all faces. On the xy-plane,

$$\mathbf{u}_f = (u_1, u_2)|_f = 0;$$

thus

$$u_1 = \frac{\partial \phi}{\partial x} = 0, \ u_2 = \frac{\partial \phi}{\partial y} = 0,$$

on f, which implies that ϕ = constant, possibly set to be zero on all faces from the continuity of the function ϕ on K. In terms of the barycentric coordinates $\lambda_1, \ldots, \lambda_4$:

$$\lambda_1 = \frac{|V_1|}{|V|}, \quad \lambda_2 = \frac{|V_2|}{|V|}, \quad \lambda_3 = \frac{|V_3|}{|V|}, \quad \lambda_4 = \frac{|V_4|}{|V|},$$

through sub-volume V_i, which is formed by the point $\mathbf{x} \in K$ with three vertices of K other than the ith vertex. We have

$$\phi = \left(\prod_{i=1}^{4} \lambda_i\right) \psi, \qquad \psi \in P_{k-3}. \tag{12.75}$$

For $\mathbf{q} \in \mathcal{D}_{k-2}$, the vanishing of the element d.o.f. in (12.55) with an integration by parts says that

$$0 = \int_K \mathbf{u} \cdot \mathbf{q}\, d\Omega = \int_K \nabla\phi \cdot \mathbf{q}\, d\Omega = \int_K \phi\nabla \cdot \mathbf{q}\, d\Omega - \int_{\partial K} \phi\mathbf{q} \cdot \mathbf{n}\, ds,$$

together with $\phi|_{\partial K} = 0$ and (12.75), resulting in

$$0 = \int_K \phi\nabla \cdot \mathbf{q}\, d\Omega = \int_K (\lambda_1\lambda_2\lambda_3\lambda_4\psi\nabla \cdot \mathbf{q}) d\Omega, \qquad \mathbf{q} \in \mathcal{D}_{k-2}. \tag{12.76}$$

Finally, from (12.52), we can find a $\mathbf{q}$ such that $\nabla \cdot \mathbf{q} = \psi$, and then

$$\int_K \lambda_1\lambda_2\lambda_3\lambda_4(\psi)^2\, d\Omega = 0, \qquad \lambda_i \geq 0 \quad (i = 1, \ldots, 4),$$

which implies that $\psi = 0$, so $\mathbf{u} = \mathbf{0}$ in K. □

12.2 Hierarchical Nédélec Basis Functions

In this section, we present the explicit formula for the Nédélec basis for different elements in 2-D and 3-D spaces. For the convenience of the adaptive p-refinement [6] (i.e., adaptively adjusting the degree of the polynomials on individual elements on a fixed mesh), the hierarchical Nédélec basis will be presented in this section. Since the proposal of the Nédélec curl-conforming basis, many types of hierarchical bases have been developed, including the work of [7–15]. The constructed basis functions can span or contain either of the polynomial spaces proposed in [1] or [2], and a detailed classification was given in [15] based on the type of spaces generated by the basis functions. Meanwhile, using the perspective of differential forms, [16] laid a general framework for canonical construction of $H(\text{curl})$- and $H(\text{div})$-conforming finite elements. More details can be found in [17–19], and the monograph [20].

The basis functions described in this section try to achieve the maximal partial orthogonality among the basis functions to produce a better conditioned mass matrix and stiffness matrices when applied to Maxwell's systems [21, 22]. Their constructions are based upon the studies by [10, 11] and the orthogonal polynomials of several variables in [23].

12.2.1 Construction on 2-D Quadrilaterals

Using the notation in [10], we consider a reference square element K,

$$K := \left\{(\xi, \eta) \in \mathbb{R}^2 : -1 \le \xi,\ \eta \le 1\right\}, \tag{12.77}$$

and the edges are denoted by $\gamma_i, i = 1, \ldots, 4$, with γ_1 as the bottom edge ($\eta = -1$), γ_2 as the top edge ($\eta = 1$), γ_3 as the left edge ($\xi = -1$), and γ_4 as the right edge ($\xi = 1$), respectively.

Edge basis

- **Lowest-order basis**

The lowest-order $H(\text{curl})$-conforming basis consists of four shape functions with one on each edge:

$$\begin{aligned}\Phi_0^{\gamma_1} &= \frac{1}{2}(1-\eta)\,\mathbf{t}_{\gamma_1},\ \Phi_0^{\gamma_2} = \frac{1}{2}(1+\eta)\,\mathbf{t}_{\gamma_2},\\ \Phi_0^{\gamma_3} &= \frac{1}{2}(1-\xi)\,\mathbf{t}_{\gamma_3},\ \Phi_0^{\gamma_4} = \frac{1}{2}(1+\xi)\,\mathbf{t}_{\gamma_4},\end{aligned} \tag{12.78}$$

where $\mathbf{t}_{\gamma_i}$ is the unit tangential vector along the edge γ_i.

- **(pth)-order basis** $p \ge 1$:

$$\begin{aligned}\Phi_p^{\gamma_1} &= \frac{1}{2}(1-\eta)P_p(\xi)\mathbf{t}_{\gamma_1},\ \Phi_p^{\gamma_2} = \frac{1}{2}(1+\eta)P_p(-\xi)\mathbf{t}_{\gamma_2},\\ \Phi_p^{\gamma_3} &= \frac{1}{2}(1-\xi)P_p(-\eta)\mathbf{t}_{\gamma_3},\ \Phi_p^{\gamma_4} = \frac{1}{2}(1+\xi)P_p(\eta)\mathbf{t}_{\gamma_4}.\end{aligned} \tag{12.79}$$

Interior basis For $j = 0, \ldots, p$ and $k = 2, \ldots, p+1$, the interior basis is given by

$$\begin{aligned}\Phi_{j,k}^{I\xi} &= P_j(\xi)l_k(\eta)\hat{\boldsymbol{e}}_\xi,\\ \Phi_{j,k}^{I\eta} &= P_j(\eta)l_k(\xi)\hat{\boldsymbol{e}}_\eta,\end{aligned} \tag{12.80}$$

where $\hat{\boldsymbol{e}}_\xi$ and $\hat{\boldsymbol{e}}_\eta$ are the unit vectors along ξ and η axes, respectively.

Here $P_p(\eta)$ is the pth-order Legendre polynomial, and the polynomials $l_k(x)$ are defined as the integrated Legendre polynomials [24] as follows:

$$l_0(x) = \frac{1-x}{2},\ l_0(-1) = 1,\ l_0(1) = 0,$$
$$l_1(x) = \frac{1+x}{2},\ l_1(-1) = 0,\ l_1(1) = 1,$$
$$l_k(x) = \sqrt{\frac{2k-1}{2}} \int_{-1}^{x} P_{k-1}(\sigma)\mathrm{d}\sigma, \quad k \geq 2, \tag{12.81}$$

where $P_{k-1}(\sigma)$ is the Legendre polynomial of degree $k-1$.

The expression in (12.81), except for a scaling factor, was used by Jørgensen et al. [25] to construct hierarchical Legendre basis functions. It is clear that the newly defined polynomials have the following properties [24]. By using the symmetry and differentiation properties of the Legendre polynomials, the above integral can be readily shown to take the form

$$l_k(x) = \frac{P_k(x) - P_{k-2}(x)}{\sqrt{2(2k-1)}}, \quad k \geq 2, \tag{12.82}$$

and has the property

$$l_k(-1) = l_k(1) = 0,\ k \geq 2,$$

$$\int_{-1}^{1} \frac{\mathrm{d}l_{i+1}(\tau)}{\mathrm{d}\tau} \frac{\mathrm{d}l_{j+1}(\tau)}{\mathrm{d}\tau} \mathrm{d}\tau = \delta_{ij}, \quad i, j \geq 1, \tag{12.83}$$

where δ_{ij} is the Kronecker delta.

12.2.2 Construction on 2-D Triangles

In this case, the following reference triangle K is used to construct the basis functions

$$K := \left\{(\xi, \eta) \in \mathbb{R}^2 : 0 \leq \xi,\ \eta,\ \xi + \eta \leq 1\right\}. \tag{12.84}$$

The coordinates for the vertices are $V_1(1, 0)$, $V_2(0, 1)$, and $V_3(0, 0)$. Each edge is the directed line segment which is labeled in terms of the opposite vertex, i.e., $\mathbf{e}_1 := V_2 \rightarrow V_3$, $\mathbf{e}_2 := V_3 \rightarrow V_1$, and $\mathbf{e}_3 := V_1 \rightarrow V_2$, and their corresponding unit direction vectors are denoted as $\hat{\Gamma}_i$, $i = 1, 2, 3$.

The barycentric coordinates of the reference element are simply

$$\lambda_1 = \xi, \quad \lambda_2 = \eta, \quad \lambda_3 = 1 - \xi - \eta. \tag{12.85}$$

In terms of the barycentric coordinates, each edge can be parameterized as

$$\tau_1|_{\mathbf{e}_1} = \lambda_3 - \lambda_2, \quad \tau_2|_{\mathbf{e}_2} = \lambda_1 - \lambda_3, \quad \tau_3|_{\mathbf{e}_3} = \lambda_2 - \lambda_1. \tag{12.86}$$

The parameter varies in the range $\tau_i = [-1, 1], i = 1, 2, 3$. The normal vector on each edge is

$$\mathbf{n}_1 = \nabla\lambda_1 = \begin{bmatrix} 1 \\ 0 \end{bmatrix}, \quad \mathbf{n}_2 = \nabla\lambda_2 = \begin{bmatrix} 0 \\ 1 \end{bmatrix}, \quad \mathbf{n}_3 = \nabla\lambda_3 = \begin{bmatrix} -1 \\ -1 \end{bmatrix}. \tag{12.87}$$

We now construct basis functions for the $H(\text{curl})$-conforming elements on the reference element as in [10, 21].

First-order basis

The lowest-order elements first constructed by Whitney, now called the Whitney elements [26], consist of three shape functions with one on each edge, namely,

$$\Phi_0^{\mathbf{e}_j} = |\mathbf{e}_j|(\lambda_{j_1}\mathbf{n}_{j_2} - \lambda_{j_2}\mathbf{n}_{j_1}), \quad j = 1, 2, 3, \tag{12.88}$$

where

$$\begin{aligned} j_1 &= \begin{cases} \operatorname{mod}(j+1, 3), & \text{if } j+1 \neq 3, \\ 3, & \text{otherwise,} \end{cases} \\ j_2 &= \begin{cases} \operatorname{mod}(j+2, 3), & \text{if } j+2 \neq 3, \\ 3, & \text{otherwise.} \end{cases} \end{aligned} \tag{12.89}$$

The following functions will complete the first-order basis:

$$\Phi_1^{\mathbf{e}_j} = |\mathbf{e}_j|(\lambda_{j_1}\mathbf{n}_{j_2} + \lambda_{j_2}\mathbf{n}_{j_1}), \quad j = 1, 2, 3. \tag{12.90}$$

It is easy to verify that the above basis functions have the following property:

$$\mathbf{e}_k \cdot \Phi_0^{\mathbf{e}_j} = \mathbf{e}_k \cdot \Phi_1^{\mathbf{e}_j} = \delta_{jk}, \quad j, k = 1, 2, 3. \tag{12.91}$$

So each basis function has a constant *unit* tangential component on its associated edge and a zero tangential component on the other two edges.

Higher-order basis

The functions for a higher-order basis can be grouped into two classes according to their associated geometric identities on the reference element.

- **Edge-based functions**

The higher-order edge-based functions are given in [10] as

$$\Phi_{i+1}^{\mathbf{e}_j} = \frac{2i+1}{i+1} P_i(\tau_j)\Phi_1^{\mathbf{e}_j} - \frac{i}{i+1} P_{i-1}(\tau_j)\Phi_0^{\mathbf{e}_j}, \quad i = 1, 2, \ldots, p-1, \; j = 1, 2, 3, \tag{12.92}$$

where again each basis function has a non-zero tangential component on its associated edge and has a zero tangential component on the other two edges.

- **Interior functions**

For a complete polynomial approximation of the Nédélec space with degree $p \geq 2$, interior functions are needed. The interior functions are separated into two groups: *normal* functions and *bubble* functions [9, 10, 21]. All interior functions have no tangential contribution along any edge. However, the normal functions will have normal components on their associated edges, whereas the bubble functions are free of normal and tangential on all edges.

Normal functions
Using the Jacobi polynomials , the following interior normal functions are orthonormal on the reference element:

$$\Phi_{k+2}^{\mathbf{e}_j,n} = 8\sqrt{k+3}\left(1-\lambda_j\right)^k J_k^{(2,2)}\left(\frac{\tau_j}{1-\lambda_j}\right)\lambda_{j_1}\lambda_{j_2}\frac{\mathbf{n}_j}{|\mathbf{n}_j|}, \quad k = 0, 1, 2, \ldots, p-2, \tag{12.93}$$

where $J_k^{(\alpha,\beta)}(\cdot)$ is the normalized Jacobi polynomial of degree k on interval $[-1, 1]$ defined in (5.4), same normalization is assumed below for the 2-D triangle case. The subscripts j_1 and j_2 are defined in (12.89), and the scaling constant $|\mathbf{n}_j|$ for each edge is given by

$$|\mathbf{n}_1| = 1, \quad |\mathbf{n}_2| = 1, \quad |\mathbf{n}_3| = \sqrt{2}. \tag{12.94}$$

The interior normal functions (12.93) have two important properties:

$$\mathbf{e}_j \cdot \Phi_{k+2}^{\mathbf{e}_i,n} = 0, \quad i, j = 1, 2, 3; \quad k = 0, 1, 2, \ldots, p-2, \tag{12.95}$$

and

$$\left\langle \Phi_{i+2}^{\mathbf{e}_j,n}, \Phi_{k+2}^{\mathbf{e}_j,n} \right\rangle\Big|_K = \delta_{ik}, \quad j = 1, 2, 3; \quad i, k = 0, 1, 2, \ldots, p-2. \tag{12.96}$$

The property (12.95), i.e., free of tangential component, can be readily seen as the normal functions (12.93) vanish on two edges and are perpendicular to the third one. The orthonormal property (12.96) can be proved directly by using [23], [23, Prop. 2.3.8], which is cited in Theorem 13.1.

Bubble functions
Using the orthogonal polynomials of several variables of Theorem 13.1, the following bubble functions are orthonormal on the reference triangle:

$$\Phi^b_{i,j} = \{\mathbf{e}_\xi, \mathbf{e}_\eta\} \otimes h^K_{i,j}\, \lambda_1\lambda_2\lambda_3\,(1-\lambda_1)^i\, J_i^{(2,2)}\left(\frac{\lambda_2-\lambda_3}{1-\lambda_1}\right) J_j^{(2i+5,2)}\,(2\lambda_1-1)\,, \tag{12.97}$$

where

$$h^K_{i,j} = 2^{i+13/2}, \quad 0 \le i, j, i+j \le p-3. \tag{12.98}$$

The interior bubble functions have the following two properties:

$$\Phi^b_{i,j}|_{\partial K} = 0, \quad 0 \le i, j, i+j \le p-3, \tag{12.99}$$

and

$$\left\langle \Phi^b_{i,j}, \Phi^b_{k,\ell}\right\rangle\Big|_K = \delta_{ik}\,\delta_{j\ell}, \quad 0 \le i, j, k, \ell, i+j, k+\ell \le p-3. \tag{12.100}$$

The first property, (12.99), which states that the bubble functions have vanishing tangential and normal components on the boundary (three edges) of the reference element, can be seen by noting that the factor $\lambda_1\lambda_2\lambda_3$ is included with each shape function in (12.97). The orthonormal property (12.100) can be proved by using the result in Theorem 13.1 by identifying $\lambda_1 = x_1, \lambda_2 = x_2$, and $\lambda_3 = 1 - x_1 - x_2$. Also, by following the argument in [10], it can be shown that the basis above is a hierarchical basis for triangular $H(\text{curl})$-conforming elements.

12.2.3 Construction on 3-D Cubes

The reference cube K is given by [27]

$$K := \left\{(\xi, \eta, \zeta) \in \mathbb{R}^3 : -1 \le \xi, \eta, \zeta \le 1\right\}. \tag{12.101}$$

The basis functions in this case can be simply grouped in three coordinate directions, where each group has components along only one coordinate direction.

- $\hat{\xi}$-directed basis:

$$\Phi^\xi_{ijk}(\xi, \eta, \zeta) = P_i(\xi) l_j(\eta) l_k(\zeta)\hat{\xi},\ 0 \le i \le p-1,\ 0 \le j \le p,\ 0 \le k \le p. \tag{12.102}$$

- $\hat{\eta}$-directed basis:

$$\Phi^\eta_{ijk}(\xi, \eta, \zeta) = l_i(\xi) P_j(\eta) l_k(\zeta)\hat{\eta},\ 0 \le i \le p,\ 0 \le j \le p-1,\ 0 \le k \le p. \tag{12.103}$$

- $\hat{\zeta}$-directed basis:

$$\Phi^{\zeta}_{ijk}(\xi, \eta, \zeta) = l_i(\xi)l_j(\eta)P_k(\zeta)\hat{\zeta},\ 0 \le i \le p,\ 0 \le j \le p,\ 0 \le k \le p-1. \tag{12.104}$$

12.2.4 Construction on 3-D Tetrahedra

The shape functions are again grouped into several categories based upon their topological entities on the reference tetrahedron in Fig. 12.1. If possible, the basis functions in each category are constructed so that they are orthonormal on the reference element. The vertices are numbered as $\mathbf{v}_0(0, 0, 0)$, $\mathbf{v}_1(1, 0, 0)$, $\mathbf{v}_2(0, 1, 0)$, and $\mathbf{v}_3(0, 0, 1)$. The barycentric coordinates are given as

$$\lambda_0 = 1 - \xi - \eta - \zeta, \quad \lambda_1 = \xi, \quad \lambda_2 = \eta, \quad \lambda_3 = \zeta. \tag{12.105}$$

A generic edge can be uniquely identified with

$$\mathbf{e}_j = [j_1, j_2], \quad j_1 = 0, 1, 2, \quad j_1 < j_2 \le 3, \quad j = j_1 + j_2 + \mathrm{sign}(j_1), \tag{12.106}$$

where $\mathrm{sign}(0) = 0$. Specifically,

$$\mathbf{e}_1 = \mathbf{v}_1 - \mathbf{v}_0,\ \mathbf{e}_2 = \mathbf{v}_2 - \mathbf{v}_0,\ \mathbf{e}_3 = \mathbf{v}_3 - \mathbf{v}_0, \tag{12.107}$$

$$\mathbf{e}_4 = \mathbf{v}_2 - \mathbf{v}_1,\ \mathbf{e}_5 = \mathbf{v}_3 - \mathbf{v}_1,\ \mathbf{e}_6 = \mathbf{v}_3 - \mathbf{v}_2. \tag{12.108}$$

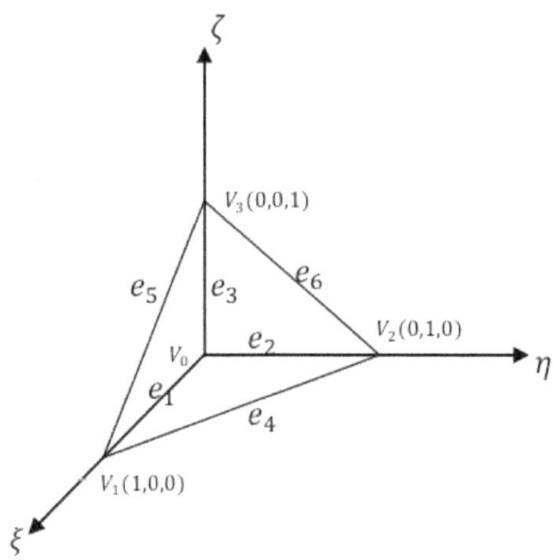

Fig. 12.1 Reference tetrahedron

The directed tangent on a generic edge $\mathbf{e}_j = [j_1, j_2]$ is defined as

$$\tau^{\mathbf{e}_j} := \tau^{[j_1, j_2]} = \mathbf{v}_{j_2} - \mathbf{v}_{j_1}, \quad j_1 < j_2. \tag{12.109}$$

The edge is parameterized by

$$\gamma_{\mathbf{e}_j} := \lambda_{j_2} - \lambda_{j_1}, \quad j_1 < j_2. \tag{12.110}$$

Edge functions

- **First-order basis**

The shape functions for the lowest order, also called the Whitney element [20, 26], are given as

$$\Phi_0^{\mathbf{e}_j} = |\tau^{\mathbf{e}_j}| \left(\lambda_{j_2} \nabla \lambda_{j_1} - \lambda_{j_1} \nabla \lambda_{j_2} \right), \tag{12.111}$$

and, together with the following additional function, they will complete the first-order basis,

$$\Phi_1^{\mathbf{e}_j} = |\tau^{\mathbf{e}_j}| \left(\lambda_{j_2} \nabla \lambda_{j_1} + \lambda_{j_1} \nabla \lambda_{j_2} \right). \tag{12.112}$$

The tangential component of the function $\Phi_0^{\mathbf{e}_j}$ and $\Phi_1^{\mathbf{e}_j}$ on its associated edge is of unit size and vanishes on other edges, namely, it has the property

$$\mathbf{e}_k \cdot \left(\Phi_0^{\mathbf{e}_j} \right) = \mathbf{e}_k \cdot \left(\Phi_1^{\mathbf{e}_j} \right) = \delta_{jk}, \quad j, k = 1, 2, \ldots, 6, \tag{12.113}$$

where δ_{jk} is the Kronecker delta.

- **Higher-order basis**

The shape functions for higher-order approximation are given in [11] for $i = 1, 2, \ldots, p-1$, $j = 1, 2, \ldots, 6$, as

$$\Phi_{i+1}^{\mathbf{e}_j} = \frac{2i+1}{i+1} P_i(\gamma_{\mathbf{e}_j}) \Phi_1^{\mathbf{e}_j} - \frac{i}{i+1} P_{i-1}(\gamma_{\mathbf{e}_j}) \Phi_0^{\mathbf{e}_j}, \tag{12.114}$$

where the property (12.113) also holds for $\Phi_{i+1}^{\mathbf{e}_j}$.

For each edge, there are $p+1$ basis functions; therefore, altogether there are $6(p+1)$ edge basis functions per element.

Face functions

Each face on the tetrahedron is uniquely defined as

$$\mathbf{f}_{j_1} = [j_2, j_3, j_4], \quad 0 \le j_1, j_2, j_3, j_4 \le 3, \quad j_2 < j_3 < j_4. \tag{12.115}$$

The face functions are further grouped into two categories: edge-based face functions and face bubble functions [11, 22].

• **Edge-based face functions**

These functions are associated with the three edges of a certain face $\mathbf{f}_{j_1}$, and have non-zero tangential components only on the associated face $\mathbf{f}_{j_1}$. By using the results in Theorem 13.1, the orthonormal shape functions are given by

$$\Phi_{\mathbf{e}[k_1,k_2]}^{\mathbf{f}_{j_1},i} = C_i \lambda_{k_1} \lambda_{k_2} (1-\lambda_{k_1})^i J_i^{(1,2)} \left(\frac{2\lambda_{k_2}}{1-\lambda_{k_1}} - 1 \right) \frac{\nabla \lambda_{k_3}}{|\nabla \lambda_{k_3}|}, \tag{12.116}$$

where $C_i = (i+3)\sqrt{(2i+4)(2i+5)(2i+7)/(i+1)}$, $0 \le i \le p-2$, and $k_1 = \{j_2, j_3\}$, $k_2 = \{j_3, j_4\}$, $k_1 < k_2$, $k_3 = \{j_2, j_3, j_4\} \setminus \{k_1, k_2\}$.

In (12.116), the function $J_i^{(1,2)}(\cdot)$ is the un-normalized Jacobi polynomial of degree i with a single variable defined in (5.4), same un-normalization is assumed for occurrences in other basis functions below in the tetrahedron case. Again, by using the results in Theorem 13.1, one can prove the orthonormal property of the edge-based face functions,

$$\left\langle \Phi_{\mathbf{e}=[k_1,k_2]}^{\mathbf{f}_{j_1},m}, \Phi_{\mathbf{e}=[k_1,k_2]}^{\mathbf{f}_{j_1},n} \right\rangle \Big|_{K^3} = \delta_{mn}, \quad m, n = 0, 1, \ldots, p-2. \tag{12.117}$$

For each face, there are $3(p-1)$ edge-based face functions; therefore, altogether there are $12(p-1)$ edge-based face functions for the four faces of an element.

• **Face bubble functions**

The face bubble functions, which belong to each specific set and are associated with a particular face $\mathbf{f}_{j_1}$, vanish on all other three faces. In view of the results in Theorem 13.1, the explicit formulae are given by

$$\begin{aligned} \Phi_{m,n}^{\mathbf{f}_{j_1},j_3} &= \Lambda (1-\lambda_{j_2})^m (1-\lambda_{j_2}-\lambda_{j_3})^n J_m^{(2n+3,2)} \left(\frac{2\lambda_{j_3}}{1-\lambda_{j_2}} - 1 \right) \\ &\quad \cdot J_n^{(0,2)} \left(\frac{2\lambda_{j_4}}{1-\lambda_{j_2}-\lambda_{j_3}} - 1 \right) \frac{\tau^{[j_2,j_3]}}{|\tau^{[j_2,j_3]}|}, \end{aligned} \tag{12.118}$$

$$\begin{aligned} \Phi_{m,n}^{\mathbf{f}_{j_1},j_4} &= \Lambda (1-\lambda_{j_2})^m (1-\lambda_{j_2}-\lambda_{j_3})^n J_m^{(2n+3,2)} \left(\frac{2\lambda_{j_3}}{1-\lambda_{j_2}} - 1 \right) \\ &\quad \cdot J_n^{(0,2)} \left(\frac{2\lambda_{j_4}}{1-\lambda_{j_2}-\lambda_{j_3}} - 1 \right) \frac{\tau^{[j_2,j_4]}}{|\tau^{[j_2,j_4]}|}, \end{aligned} \tag{12.119}$$

where $0 \le m, n, m+n \le p-3$, and

$$\Lambda = C_m^{n,1} C_m^{n,2} \lambda_{j_2} \lambda_{j_3} \lambda_{j_4}, \tag{12.120}$$

where

$$C_m^{n,1} = \sqrt{(2n+3)(m+n+3)(m+2n+4)(m+2n+5)} \tag{12.121}$$

and

$$C_m^{n,2} = \frac{\sqrt{(2m+2n+7)(2m+2n+8)(2m+2n+9)}}{\sqrt{(m+1)(m+2)}}. \tag{12.122}$$

The face bubble functions again share the orthonormal property on the reference tetrahedron for $0 \le m_1, m_2, n_1, n_2, m_1+n_1, m_2+n_2 \le p-3$:

$$\left\langle \Phi_{m_1,n_1}^{\mathbf{f}_{j_1},j_3}, \Phi_{m_2,n_2}^{\mathbf{f}_{j_1},j_3} \right\rangle\Big|_{K^3} = \delta_{m_1m_2}\delta_{n_1n_2}, \tag{12.123}$$

$$\left\langle \Phi_{m_1,n_1}^{\mathbf{f}_{j_1},j_4}, \Phi_{m_2,n_2}^{\mathbf{f}_{j_1},j_4} \right\rangle\Big|_{K^3} = \delta_{m_1m_2}\delta_{n_1n_2}. \tag{12.124}$$

For each face, there are $(p-1)(p-2)$ face bubble functions; therefore, altogether there are $4(p-1)(p-2)$ face bubble functions for the four faces of an element.

Interior functions

The interior functions are also classified into two categories: face-based interior functions and interior bubble functions [11, 22].

- **Face-based interior functions**

The face-based interior functions that are associated with a particular face $\mathbf{f}_{j_1}$ have non-zero normal components on the associated face, and have zero tangential components on all four faces. The formulae of these functions are given by

$$\begin{aligned}\Phi_{m,n}^{\mathbf{t},\mathbf{f}_{j_1}} &= \Lambda(1-\lambda_{j_2})^m(1-\lambda_{j_2}-\lambda_{j_3})^n J_m^{(2n+3,2)}\left(\frac{2\lambda_{j_3}}{1-\lambda_{j_2}}-1\right)\\ &\cdot J_n^{(0,2)}\left(\frac{2\lambda_{j_4}}{1-\lambda_{j_2}-\lambda_{j_3}}-1\right)\frac{\nabla\lambda_{j_1}}{|\nabla\lambda_{j_1}|},\end{aligned} \tag{12.125}$$

where $0 \le m, n, m+n \le p-3$.

The face-based interior functions enjoy the orthonormal property on the reference 3-simplex for $0 \le m_1, m_2, n_1, n_2, m_1+n_1, m_2+n_2 \le p-3$:

$$\left\langle \Phi_{m_1,n_1}^{\mathbf{t},\mathbf{f}_{j_1}}, \Phi_{m_2,n_2}^{\mathbf{t},\mathbf{f}_{j_1}} \right\rangle\Big|_{K^3} = \delta_{m_1m_2}\delta_{n_1n_2}. \tag{12.126}$$

For each element, there are $2(p-1)(p-2)$ face-based interior functions.

- **Interior bubble functions**

The interior bubble functions have both vanishing tangential and normal components on all four faces of the reference 3-simplex. Similarly, by using the results in Theorem 13.1, the formulae of these functions are given by

$$\Phi_{\ell,m,n}^{\mathbf{t},\mathbf{e}_d} = \Gamma J_\ell^{(2m+2n+8,2)}(2\lambda_1-1)\, J_m^{(2n+5,2)}\left(\frac{2\lambda_2}{1-\lambda_1}-1\right)$$

$$\cdot J_n^{(2,2)}\left(\frac{2\lambda_3}{1-\lambda_1-\lambda_2}-1\right)\mathbf{e}_d,\ d=1,2,3, \tag{12.127}$$

where

$$\Gamma = C_{\ell,m,n}\lambda_0\lambda_1\lambda_2\lambda_3(1-\lambda_1)^m(1-\lambda_1-\lambda_2)^n, \tag{12.128}$$

where $0 \le \ell, m, n, \ell+m+n \le p-4,\ d=1,2,3,\ C_{\ell,m,n} = C^1_{\ell,m,n}C^2_{\ell,m,n}$, and

$$C^1_{\ell,m,n} = \sqrt{\frac{(\ell+2m+2n+9)(\ell+2m+2n+10)(2\ell+2m+2n+11)(m+2n+6)}{(\ell+1)(m+1)(n+1)}}, \tag{12.129}$$

$$C^2_{\ell,m,n} = \sqrt{\frac{(m+2n+7)(2m+2n+8)(n+3)(n+4)(2n+5)}{(\ell+2)(m+2)(n+2)}}. \tag{12.130}$$

Again, one can show the orthonormal property of the interior bubble functions:

$$\left\langle \Phi^{\mathbf{t},\mathbf{e}_{d_1}}_{\ell_1,m_1,n_1}, \Phi^{\mathbf{t},\mathbf{e}_{d_2}}_{\ell_2,m_2,n_2}\right\rangle\Big|_{K^3} = \delta_{\ell_1\ell_2}\delta_{m_1m_2}\delta_{n_1n_2}, \tag{12.131}$$

where

$$0 \le \ell_1, \ell_2, m_1, m_2, n_1, n_2, \ell_1+m_1+n_1, \ell_2+m_2+n_2 \le p-4,\ d_1, d_2 = 1,2,3.$$

For each element, there are $(p-1)(p-2)(p-3)/2$ interior bubble functions.

Finally, by a simple calculation, the number of all the basis functions for each element adds to $(p+1)(p+2)(p+3)/2 = \dim(P_p(K))^3$.

12.3 Summary

The Nédélec edge element forms the conforming basis for the electric and magnetic fields where tangential continuities of the fields are required on material interfaces. The explicit form of 2-D and 3-D hierarchical high-order Nédélec edge elements presented here facilitates their use in arbitrary orders when high-frequency wave phenomena demand higher resolution of the wave structures.

The fast solution of the linear system by iterative solvers such as multigrid methods is not addressed in this book, and work in this important area can be found in [28]. Also, the mathematical properties such as the discrete de Rham commutativity and the convergence analysis of the Nédélec elements for Maxwell's equations including cavity problems can be found in [4, 29–32].

References

1. J.C. Nédélec, Mixed finite elements in $\mathbb{R}^3$. Numer. Math. **35**, 315–341 (1980)
2. J.C. Nédélec, A new family of mixed finite elements in $\mathbb{R}^3$. Numer. Math. **50**, 57–81 (1986)
3. J.-M. Jin, *The Finite Element Method in Electromagnetics*, 2nd edn. (Wiley, New York, 2002)
4. P. Monk, *Finite Element Methods for Maxwell's Equations* (Oxford University Press, Oxford, UK, 2003)
5. V. Girault, P.-A. Raviart, *Finite Element Methods for Navier-Stokes Equations, Theory and Algorithms* (Springer, New York, 1986)
6. W. Rachowicz, L. Demkowicz, An hp-adaptive finite element method for electromagnetics—part II. A 3D implementation. Int. J. Numer. Meth. Eng. **53**, 147–180 (2002)
7. R.D. Graglia, D.R. Wilton, A.F. Peterson, Higher order interpolatory vector bases for computational electromagnetics. IEEE Trans. Antenn. Propag. **45**, 329–342 (1997)
8. L.S. Andersen, J.L. Volakis, Development and application of a novel class of hierarchical tangential vector finite elements for electromagnetics. IEEE Trans. Antenn. Propag. **47**, 112–120 (1999)
9. J.P. Webb, Hierarchical vector basis functions of arbitrary order for triangular and tetrahedral finite elements. IEEE Trans. Antenn. Propag. **47**, 1244–1253 (1999)
10. M. Ainsworth, J. Coyle, Hierarchic hp-edge element families for Maxwell's equations on hybrid quadrilateral/triangular meshes. Comput. Meth. Appl. Mech. Eng. **190**, 6709–6733 (2001)
11. M. Ainsworth, J. Coyle, Hierarchic finite element bases on unstructured tetrahedral meshes. Int. J. Numer. Meth. Eng. **58**, 2103–2130 (2003)
12. J. Schöberl, S. Zaglmayr, High order Nédélec elements with local complete sequence properties. COMPEL **24**, 374–384 (2005)
13. P. Ingelström, A new set of H(curl)-conforming hierarchical basis functions for tetrahedral meshes. IEEE Trans. Microw. Theory Tech. **54**, 106–114 (2006)
14. R. Abdul-Rahman, M. Kasper, Orthogonal hierarchical Nédélec elements. IEEE Trans. Mag. **44**, 1210–1213 (2008)
15. R.D. Graglia, A.F. Peterson, F.P. Andriulli, Curl-conforming hierarchical vector bases for triangles and tetrahedra. IEEE Trans. Antenn. Propag. **59**, 950–959 (2011)
16. R. Hiptmair, Canonical construction of finite elements. Math. Comput. **68**, 1325–1346 (1999)
17. R. Hiptmair, Higher order Whitney forms. PIER **32**, 271–299 (2001)
18. F. Rapetti, High order edge elements on simplicial meshes. *ESAIM: M2AN*, **41**, 1001–1020 (2007)
19. F. Rapetti, A. Bossavit, Whitney forms of higher degree. SIAM J. Numer. Anal. **47**, 2369–2386 (2009)
20. A. Bossavit, I.D. Mayergoyz, *Computational Electromagnetism: Variational Formulations, Complementarity, Edge Elements* (Academic Press, New York, 1997)
21. J. Xin, W. Cai, A well-conditioned hierarchical basis for triangular H(curl)-conforming elements. Commun. Comput. Phys. **9**, 780–806 (2011)
22. J. Xin, N. Guo, W. Cai, On the construction of well-conditioned hierarchical bases for tetrahedral H(curl)-conforming Nédélec elements. J. Comput. Math. **29**, 526–542 (2011)
23. C.F. Dunkl, Y. Xu, *Orthogonal Polynomials of Several Variables* (Cambridge University Press, Cambridge, UK, 2001)
24. B. Szabó, I. Babuska, *Finite Element Analysis* (Wiley-Interscience, New York, 1991)
25. E. Jørgensen, J.L. Volakis, P. Meincke, O. Breinbjerg, Higher order hierarchical Legendre basis functions for electromagnetic modeling. IEEE Trans. Antenn. Propag. **52**, 2985–2995 (2004)
26. H. Whitney, *Geometric Integration Theory* (Princeton University Press, Princeton, NJ, 1957)
27. M.M. Ilić, B.M. Notaroš, Higher order hierarchical curved hexahedral vector finite elements for electromagnetic modeling. IEEE Trans. Microw. Theory Tech. **51**, 1026–1033 (2003)
28. R. Hiptmair, J. Xu, Nodal auxiliary space preconditioning in H(curl) and H(div) spaces. SIAM J. Numer. Anal. **45**, 2483–2509 (2007)
29. F. Kikuchi, On a discrete compactness property for the Nédélec finite elements. J. Fac. Sci. Univ. Tokyo Sect. I A Math. **36**, 479–490 (1989)

30. D. Boffi, P. Fernandes, L. Gastaldi, I. Perugia, Computational models of electromagnetic resonators: analysis of edge element approximation. SIAM J. Numer. Anal. **36**, 1264–1290 (1999)
31. S. Caorsi, P. Fernandes, M. Raffetto, On the convergence of Galerkin finite element approximations of electromagnetic eigenproblems. SIAM J. Numer. Anal. **38**, 580–607 (2000)
32. S. Caorsi, P. Fernandes, M. Raffetto, Spurious-free approximations of electromagnetic eigenproblems by means of Nédélec-type elements. Math. Model Numer. Anal. **35**, 331–354 (2001)

Chapter 13
Time-Domain Methods: Discontinuous Galerkin Method and Yee Scheme

Time-domain solutions of Maxwell's equations provide information on wave interactions involving multiple frequencies and can describe nonlinear phenomena such as second harmonic generation and parametric amplification in nonlinear optical materials. In this chapter, we will discuss two time-domain numerical methods for solving transient Maxwell's equations. The first one is a high-order discontinuous Galerkin (DG) method on unstructured finite element meshes. High-order hierarchical basis functions for the DG discretization will be presented. The second is the popular finite difference Yee scheme on staggered Cartesian grids.

13.1 Weak Formulation of Maxwell's Equations

Maxwell's equations for non-dispersive materials can be written in the following conservative form:

$$\mathbf{u}_t + \nabla \cdot \overline{\mathbf{F}} = \mathbf{S}, \quad \mathbf{r} \in \Omega, \tag{13.1}$$

where $\mathbf{u} = \begin{pmatrix} \mathbf{B} \\ \mathbf{D} \end{pmatrix}$, $\overline{\mathbf{F}}(\mathbf{u}) = (\mathbf{f}, \mathbf{g}, \mathbf{h})$, and

$$\nabla \cdot \overline{\mathbf{F}} \equiv \frac{\partial}{\partial x}\mathbf{f} + \frac{\partial}{\partial y}\mathbf{g} + \frac{\partial}{\partial z}\mathbf{h}, \tag{13.2}$$

$$\mathbf{f} = \begin{pmatrix} \hat{x} \times \mathbf{E} \\ -\hat{x} \times \mathbf{H} \end{pmatrix}, \mathbf{g} = \begin{pmatrix} \hat{y} \times \mathbf{E} \\ -\hat{y} \times \mathbf{H} \end{pmatrix}, \mathbf{h} = \begin{pmatrix} \hat{z} \times \mathbf{E} \\ -\hat{z} \times \mathbf{H} \end{pmatrix}, \tag{13.3}$$

and $\hat{x}$, $\hat{y}$, and $\hat{z}$ are the unit vectors along the x-, y-, and z-axes, respectively. The source term $\mathbf{S} = (-\mathbf{J}_e, -\mathbf{J}_m)^{\mathrm{T}}$ contains currents in a conducting material.

W. Cai, *Deterministic, Stochastic, and Deep Learning Methods for Computational Electromagnetics*, https://doi.org/10.1007/978-981-96-0100-4_13

Vector functions $(\mathbf{B}, \mathbf{D})$ are weak solutions to (13.1), without a source $(\mathbf{S} = \mathbf{0})$, if $\forall \boldsymbol{\phi}(\mathbf{r}) \in (C_0^1(\Omega))^3$:

$$\int_{\Omega} \frac{\partial \mathbf{u}}{\partial t} \cdot \boldsymbol{\phi} \mathrm{d}\mathbf{r} - \int_{\Omega} \overline{\mathbf{F}}(\mathbf{u}) \nabla \cdot \boldsymbol{\phi} \mathrm{d}\mathbf{r} + \int_{\partial\Omega} \left(\overline{\mathbf{F}}(\mathbf{u}) \cdot \mathbf{n}\right) \cdot \boldsymbol{\phi} \mathrm{d}s = 0, \tag{13.4}$$

where $\overline{\mathbf{F}}(\mathbf{u})\nabla = \mathbf{f}\frac{\partial}{\partial x} + \mathbf{g}\frac{\partial}{\partial y} + \mathbf{h}\frac{\partial}{\partial z}$. Equation (13.4) can be rewritten for $\mathbf{B}$ and $\mathbf{D}$ separately as follows:

$$\int_{\Omega} \frac{\partial \mathbf{B}}{\partial t} \cdot \boldsymbol{\phi} \mathrm{d}\mathbf{r} - \int_{\Omega} \mathbf{E} \cdot (\nabla \times \boldsymbol{\phi}) \mathrm{d}\mathbf{r} + \int_{\partial\Omega} \mathbf{n} \times \mathbf{E} \cdot \boldsymbol{\phi} \, \mathrm{d}s = 0, \tag{13.5}$$

$$\int_{\Omega} \frac{\partial \mathbf{D}}{\partial t} \cdot \boldsymbol{\phi} \mathrm{d}\mathbf{r} + \int_{\Omega} \mathbf{H} \cdot (\nabla \times \boldsymbol{\phi}) \mathrm{d}\mathbf{r} - \int_{\partial\Omega} \mathbf{n} \times \mathbf{H} \cdot \boldsymbol{\phi} \, \mathrm{d}s = 0, \tag{13.6}$$

with appropriate boundary conditions for $\mathbf{B}$ and $\mathbf{D}$.

Usually, (13.1) and (13.4) are formulated in a computational domain whose coordinates are, say (ξ, η, ζ). By introducing a new conservative quantity

$$\widetilde{\mathbf{u}} = J\mathbf{u}, \tag{13.7}$$

Maxwell's equations (13.1) are recast as

$$\widetilde{\mathbf{u}}_t + \nabla \cdot \widetilde{\mathbf{F}} = \widetilde{\mathbf{S}}, \tag{13.8}$$

where $J = \partial(x, y, z)/\partial(\xi, \eta, \varsigma)$ is the Jacobian for the transformation between the physical domain and the computational domain $\mathbf{r} = \mathbf{r}(\xi, \eta, \zeta)$, the flux function $\widetilde{\mathbf{F}} = (\widetilde{\mathbf{f}}, \widetilde{\mathbf{g}}, \widetilde{\mathbf{h}})$ is given by

$$\begin{aligned}
\widetilde{\mathbf{f}} &= J(\xi_x \mathbf{f} + \xi_y \mathbf{g} + \xi_z \mathbf{h}), \\
\widetilde{\mathbf{g}} &= J(\eta_x \mathbf{f} + \eta_y \mathbf{g} + \eta_z \mathbf{h}), \\
\widetilde{\mathbf{h}} &= J(\zeta_x \mathbf{f} + \zeta_y \mathbf{g} + \zeta_z \mathbf{h}),
\end{aligned} \tag{13.9}$$

and the source term $\widetilde{\mathbf{S}}$ is

$$\widetilde{\mathbf{S}} = J\mathbf{S}.$$

13.2 Discontinuous Galerkin (DG) Discretization

Let $\mathcal{T}_h$ be a discretization of the solution domain Ω. For each element $K \in \mathcal{T}_h$, ϵ_r and μ_r are assumed constant. We denote a finite dimensional space of smooth functions defined on the element K by $\mathcal{P}(K)$. This space will be used to approximate the variable $\mathbf{u}$. Set

$$V_h := \{v \in L^2(\Omega): \quad v|_K \in \mathcal{P}(K) \quad \forall K \in \mathcal{T}_h\}. \tag{13.10}$$

Let $\mathbf{u}_h = \begin{pmatrix} \mathbf{B}_h \\ \mathbf{D}_h \end{pmatrix}$ be the approximate solution to $\mathbf{u}$ belonging to the finite element space

$$V_h^6 := \underbrace{V_h \times V_h \times \cdots \times V_h}_{6}. \tag{13.11}$$

For each $K_j \in \mathcal{T}_h$, $\forall \boldsymbol{\phi}_h \in (\mathcal{P}(K_j))^3$,

$$\int_{K_j} \frac{\partial \mathbf{B}_h}{\partial t} \cdot \boldsymbol{\phi}_h \mathrm{d}\mathbf{r} - \int_{K_j} \mathbf{E}_h \cdot (\nabla \times \boldsymbol{\phi}_h) \mathrm{d}\mathbf{r} + \int_{\partial K_j} \mathbf{n} \times \mathbf{E}_h \cdot \boldsymbol{\phi}_h \mathrm{d}s = 0, \tag{13.12}$$

$$\int_{K_j} \frac{\partial \mathbf{D}_h}{\partial t} \cdot \boldsymbol{\phi}_h \mathrm{d}\mathbf{r} + \int_{K_j} \mathbf{H}_h \cdot (\nabla \times \boldsymbol{\phi}_h) \mathrm{d}\mathbf{r} - \int_{\partial K_j} \mathbf{n} \times \mathbf{H}_h \cdot \boldsymbol{\phi}_h \mathrm{d}s = 0. \tag{13.13}$$

Thus, the weak form for the DG discretization is formed on each individual element for the element-supported test function $\boldsymbol{\phi}_h$, and the numerical solution $\mathbf{u}_h$ will comprise discontinuous functions made of piecewise polynomials, hence the name discontinuous Galerkin (DG) method [1, 2]. The solution in a DG approximation will have two different values on a common interface between two elements, and the DG method will use a common numerical flux $\mathbf{h}(\mathbf{u}^-, \mathbf{u}^+)$ to couple the solutions from different elements while guaranteeing the conservative property of the numerical solution. For consistency, the numerical flux $\mathbf{h}(\mathbf{u}^-, \mathbf{u}^+)$ should approximate the exact flux

$$\overline{\mathbf{F}}(\mathbf{u}) \cdot \mathbf{n} = \begin{pmatrix} \mathbf{n} \times \mathbf{E} \\ -\mathbf{n} \times \mathbf{H} \end{pmatrix}; \tag{13.14}$$

namely if $\mathbf{u}^- = \mathbf{u}^+ = \mathbf{u}$, then

$$\mathbf{h}(\mathbf{u}, \mathbf{u}) = \overline{\mathbf{F}}(\mathbf{u}) \cdot \mathbf{n}. \tag{13.15}$$

The finite element solution $\mathbf{u}_h = \begin{pmatrix} \mathbf{B}_h \\ \mathbf{D}_h \end{pmatrix}$ is then required to satisfy the following weak form. For $j = 1, 2, \ldots, N$, $\forall \boldsymbol{\phi}_h \in (\mathcal{P}(K_j))^3$,

$$\int_{K_j} \frac{\partial \mathbf{B}_h}{\partial t} \cdot \boldsymbol{\phi}_h \mathrm{d}\mathbf{r} - \int_{K_j} \mathbf{E}_h \cdot (\nabla \times \boldsymbol{\phi}_h) \mathrm{d}\mathbf{r} + \int_{\partial K_j} \mathbf{h}_B(\mathbf{u}^-, \mathbf{u}^+) \cdot \boldsymbol{\phi}_h \mathrm{d}s = 0, \tag{13.16}$$

$$\int_{K_j} \frac{\partial \mathbf{D}_h}{\partial t} \cdot \boldsymbol{\phi}_h \mathrm{d}\mathbf{r} + \int_{K_j} \mathbf{H}_h \cdot (\nabla \times \boldsymbol{\phi}_h) \mathrm{d}\mathbf{r} \quad \int_{\partial K_j} \mathbf{h}_D(\mathbf{u}^-, \mathbf{u}^+) \cdot \boldsymbol{\phi}_h \mathrm{d}s = 0, \tag{13.17}$$

where $\mathbf{h}_B(\mathbf{u}^-, \mathbf{u}^+)$ and $\mathbf{h}_D(\mathbf{u}^-, \mathbf{u}^+)$ are the components of the numerical flux:

$$\mathbf{h}(\mathbf{u}^-, \mathbf{u}^+) = \begin{pmatrix} \mathbf{h}_B(\mathbf{u}^-, \mathbf{u}^+) \\ \mathbf{h}_D(\mathbf{u}^-, \mathbf{u}^+) \end{pmatrix}.$$

13.3 Numerical Flux $\mathbf{h}(\mathbf{u}^-, \mathbf{u}^+)$

Let us assume that $x = 0$ is a point on the interface with an external normal $\mathbf{n} = \hat{x}$, and the material constants are

$$\epsilon = \begin{cases} \epsilon^-, \text{ if } x < 0, \\ \epsilon^+, \text{ if } x > 0, \end{cases} \qquad \mu = \begin{cases} \mu^-, \text{ if } x < 0, \\ \mu^+, \text{ if } x > 0. \end{cases} \tag{13.18}$$

The projection of the analytical flux function $\overline{\mathbf{F}}(\mathbf{u}) \cdot \mathbf{n}$ along the normal of an interface $\mathbf{n}$ can be shown to be

$$\mathbf{f}_n(\mathbf{u}) = \overline{\mathbf{F}}(\mathbf{u}) \cdot \mathbf{n} = \begin{cases} A^-\mathbf{u}, \text{ if } x < 0, \\ A^+\mathbf{u}, \text{ if } x > 0, \end{cases} \tag{13.19}$$

where

$$A^\pm = \begin{pmatrix} 0 & 0 & 0 & 0 & 0 & 0 \\ 0 & 0 & 0 & 0 & 0 & -\frac{1}{\epsilon^\pm} \\ 0 & 0 & 0 & 0 & \frac{1}{\epsilon^\pm} & 0 \\ 0 & 0 & 0 & 0 & 0 & 0 \\ 0 & 0 & \frac{1}{\mu^\pm} & 0 & 0 & 0 \\ 0 & -\frac{1}{\mu^\pm} & 0 & 0 & 0 & 0 \end{pmatrix}. \tag{13.20}$$

In order to derive the numerical flux along the normal direction $\mathbf{n}$, the following 1-D Riemann problem for the linear system will be considered as, for any conservation law with discontinuous solutions,

$$\mathbf{u}_t + (\mathbf{f}_n(\mathbf{u}))_x = 0, \qquad (x, t) \in (-\infty, +\infty) \times (0, T), \tag{13.21}$$

with the initial condition

$$\mathbf{u}(x, 0) = \begin{cases} \mathbf{u}^-, \text{ if } x < 0, \\ \mathbf{u}^+, \text{ if } x > 0, \end{cases}$$

where $\mathbf{f}_n(\mathbf{u})$ is defined in (13.19).

For simplicity of derivation, we introduce the following scaling of the fields:

$$\tilde{\mathbf{B}} = \sqrt{\mu}\mathbf{B}, \quad \tilde{\mathbf{D}} = \sqrt{\epsilon}\mathbf{D}, \tag{13.22}$$

and denote $\tilde{\mathbf{u}} = \begin{pmatrix} \tilde{\mathbf{B}} \\ \tilde{\mathbf{D}} \end{pmatrix}$. Then (13.21) can be rewritten as

$$\tilde{\mathbf{u}}_t + \left(\tilde{\mathbf{f}}_n(\tilde{\mathbf{u}})\right)_x = 0, \qquad (x, t) \in (-\infty, +\infty) \times (0, T), \tag{13.23}$$

with the initial condition

$$\tilde{\mathbf{u}}(x, 0) = \begin{cases} \tilde{\mathbf{u}}^-, \text{ if } x < 0, \\ \tilde{\mathbf{u}}^+, \text{ if } x > 0, \end{cases}$$

where $\tilde{\mathbf{f}}_n(\tilde{\mathbf{u}})$ is defined as

$$\tilde{\mathbf{f}}_n(\tilde{\mathbf{u}}) = \begin{cases} \tilde{A}^-\tilde{\mathbf{u}}, \text{ if } x < 0, \\ \tilde{A}^+\tilde{\mathbf{u}}, \text{ if } x > 0. \end{cases} \tag{13.24}$$

Here $\tilde{A}^-$ and $\tilde{A}^+$ are given by

$$\tilde{A}^\pm = \begin{pmatrix} 0 & 0 & 0 & 0 & 0 & 0 \\ 0 & 0 & 0 & 0 & 0 & -c^\pm \\ 0 & 0 & 0 & 0 & c^\pm & 0 \\ 0 & 0 & 0 & 0 & 0 & 0 \\ 0 & 0 & c^\pm & 0 & 0 & 0 \\ 0 & -c^\pm & 0 & 0 & 0 & 0 \end{pmatrix}, \tag{13.25}$$

respectively, where $c^\pm = 1/\sqrt{\epsilon^\pm \mu^\pm}$. At $x = 0$, $\tilde{\mathbf{u}}$ is discontinuous and a classical solution to (13.21) does not exist. Therefore, a weak solution will have to be defined in the following weak sense [3, 4]: $\forall \boldsymbol{\Phi}(x, t) \in (C_0^1(\mathbb{R} \times \mathbb{R}^+))^6$, where $\mathbb{R}^+ = (0, \infty)$ and $\boldsymbol{\Phi}(x, t = 0)$ has a non-empty compact support in $\mathbb{R}$:

$$\int_0^\infty \int_{-\infty}^\infty (\boldsymbol{\Phi} \cdot \tilde{\mathbf{u}} - \boldsymbol{\Phi}_x \cdot \mathbf{f}_n(\tilde{\mathbf{u}}) \mathrm{d}x\mathrm{d}t = -\int_{-\infty}^\infty \boldsymbol{\Phi} \cdot \tilde{\mathbf{u}}(x, 0)\mathrm{d}x. \tag{13.26}$$

Noting that

$$\tilde{\mathbf{f}}_n(\tilde{\mathbf{u}}) = \tilde{A}(x)\tilde{\mathbf{u}} = A(x)\mathbf{u} = \mathbf{f}_n(\mathbf{u}), \tag{13.27}$$

we see that (13.21) and (13.23) are equivalent. Therefore, we will only have to consider (13.23). Now A^+ and A^- both have six eigenvalues and eigenvectors, where three of the eigenvalues are distinct. For instance, A^-'s eigenvalues are

$$\lambda_1 = -c^-, \;\; \lambda_2 = 0, \; \lambda_3 = c^-. \tag{13.28}$$

Each of the eigenvalues defines one of the three characteristics (see Fig. 13.1) across which the solution will, in general, be discontinuous in the (x, t)-space. If we assume the solutions are constant in the regions bounded by the characteristics [5], the weak formulation (13.26) will result in the following Rankine–Hugoniot condition:

$$s[\tilde{\mathbf{u}}] = [\mathbf{f}_n(\tilde{\mathbf{u}})], \tag{13.29}$$

Fig. 13.1 Characteristics for Rankine–Hugoniot conditions

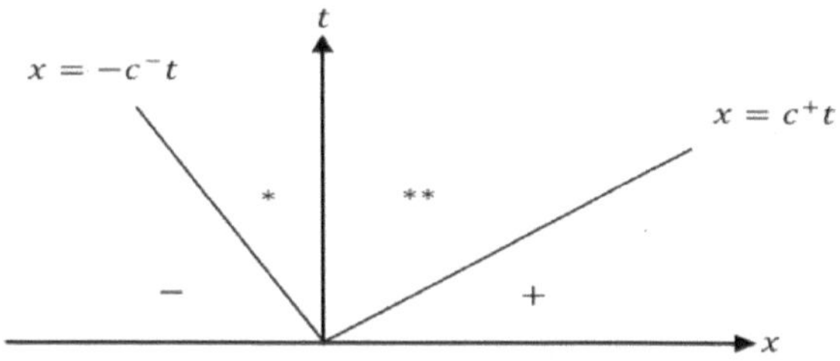

where $s = \pm c$ and 0 are the inverse reciprocals of the slopes of the characteristics in the (x, t)-plane.

In the region of $x < 0$ and $x + c^- t < 0$, $\tilde{\mathbf{u}}(x, t) = \tilde{\mathbf{u}}^-$ assumes the value given by the initial condition at $t = 0$. For the region $0 < x + c^- t$ and $x < 0$, we let $\tilde{\mathbf{u}}(x, t) = \tilde{\mathbf{u}}^*$, $t > 0$; then $\tilde{\mathbf{u}}^-$ and $\tilde{\mathbf{u}}^*$ will satisfy the Rankine–Hugoniot condition, namely

$$-c^-(\tilde{\mathbf{u}}^- - \tilde{\mathbf{u}}^*) = A^-(\tilde{\mathbf{u}}^- - \tilde{\mathbf{u}}^*). \tag{13.30}$$

Similarly, for the region $x > 0$ and $x - c^+ t > 0$, $\mathbf{u}(x, t) = \mathbf{u}^+$ assumes the value given by the initial condition at $t = 0$, $\tilde{\mathbf{u}}(x, t) = \tilde{\mathbf{u}}^+$. In the region $x > 0$ and $x - c^+ t < 0$, we let $\tilde{\mathbf{u}}(x, t) = \tilde{\mathbf{u}}^{**}$, $t > 0$; therefore, $\tilde{\mathbf{u}}^+$ and $\tilde{\mathbf{u}}^{**}$ at the interface $x - c^+ t = 0$ will also satisfy the Rankine–Hugoniot condition:

$$c^+(\tilde{\mathbf{u}}^+ - \tilde{\mathbf{u}}^{**}) = A^+(\tilde{\mathbf{u}}^+ - \tilde{\mathbf{u}}^{**}). \tag{13.31}$$

Next, we decide the conditions to be satisfied by $\tilde{\mathbf{u}}^*$ and $\tilde{\mathbf{u}}^{**}$ at $x = 0$ by the Rankine–Hugoniot condition:

$$\mathbf{0} = A^- \tilde{\mathbf{u}}^* - A^+ \tilde{\mathbf{u}}^{**}. \tag{13.32}$$

From (13.24), we can see that the numerical flux should be defined by

$$\mathbf{h}(\tilde{\mathbf{u}}^-, \tilde{\mathbf{u}}^+) \equiv A^- \tilde{\mathbf{u}}^* = A^+ \tilde{\mathbf{u}}^{**}. \tag{13.33}$$

To find the intermediate states $\tilde{\mathbf{u}}^*$ and $\tilde{\mathbf{u}}^{**}$, we solve (13.30), (13.31), and (13.32), which can be rewritten as follows:

$$\begin{aligned} -Z^-(\mathbf{H}^- - \mathbf{H}^*) &= \hat{x} \times (\mathbf{E}^- - \mathbf{E}^*), \\ -Y^-(\mathbf{E}^- - \mathbf{E}^*) &= \hat{x} \times (\mathbf{H}^- - \mathbf{H}^*), \end{aligned} \tag{13.34}$$

$$\begin{aligned} Z^+(\mathbf{H}^+ - \mathbf{H}^{**}) &= \hat{x} \times (\mathbf{E}^+ - \mathbf{E}^{**}), \\ Y^+(\mathbf{E}^+ - \mathbf{E}^{**}) &= \hat{x} \times (\mathbf{H}^+ - \mathbf{H}^{**}), \end{aligned} \tag{13.35}$$

$$\begin{aligned} \mathbf{0} &= \hat{x} \times (\mathbf{E}^* - \mathbf{E}^{**}), \\ \mathbf{0} &= \hat{x} \times (\mathbf{H}^* - \mathbf{H}^{**}), \end{aligned} \tag{13.36}$$

where the local impedance Z and the admittance Y are defined as $Z = 1/Y = \sqrt{\mu/\epsilon}$. By applying the vector cross product $\hat{x}\ \times$ on both sides of (13.34) and (13.35), we transform (13.34)–(13.36) involving only quantities $\hat{x} \times \mathbf{E}^*$, $\hat{x} \times \mathbf{H}^*$, $\hat{x} \times \mathbf{E}^{**}$, and $\hat{x} \times \mathbf{H}^{**}$. From these new equations, we can obtain

$$\begin{aligned} \hat{x} \times \mathbf{E}^* &= \hat{x} \times \frac{(Y\mathbf{E} - \hat{x} \times \mathbf{H})^- + (Y\mathbf{E} + \hat{x} \times \mathbf{H})^+}{Y^- + Y^+}, \\ \hat{x} \times \mathbf{H}^* &= \hat{x} \times \frac{(Z\mathbf{H} + \hat{x} \times \mathbf{E})^- + (Z\mathbf{H} - \hat{x} \times \mathbf{E})^+}{Z^- + Z^+}. \end{aligned} \tag{13.37}$$

Using (13.36) and (13.37), we obtain the two states $\tilde{\mathbf{u}}^*$ and $\tilde{\mathbf{u}}^{**}$ in (13.34) and (13.35). Moreover, from (13.33), (13.27), and (13.14), we can define the numerical flux as

$$\mathbf{h}(\mathbf{u}^-, \mathbf{u}^+) = \mathbf{h}(\tilde{\mathbf{u}}^-, \tilde{\mathbf{u}}^+) = \begin{pmatrix} \hat{x} \times \mathbf{E}^* \\ -\hat{x} \times \mathbf{H}^* \end{pmatrix}, \tag{13.38}$$

which can be shown to be consistent with $\mathbf{f}(\mathbf{u}) \cdot \hat{x}$. For a general normal direction $\mathbf{n}$ we can show that a consistent numerical flux is defined as

$$\mathbf{h}(\mathbf{u}^-, \mathbf{u}^+) = \begin{pmatrix} \mathbf{n} \times \mathbf{E}^* \\ -\mathbf{n} \times \mathbf{H}^* \end{pmatrix}, \tag{13.39}$$

where $\mathbf{n} \times \mathbf{E}^*$ and $\mathbf{n} \times \mathbf{H}^*$ are defined as in (13.37) with $\hat{x}$ replaced by $\mathbf{n}$, respectively [6].

13.4 Orthonormal Hierarchical Basis for DG Methods

13.4.1 Orthonormal Hierarchical Basis on Quadrilaterals or Hexahedra

Let $Q^n = [-1, 1]^n$ be the reference cube in $\mathbb{R}^n$, $n = 2, 3$. A set of orthogonal hierarchical basis functions in 2-D will be

$$\mathbb{Q}_{p,p} = \text{span}\{P_i(\xi)P_j(\eta) \colon 0 \le i, j \le p\}, \tag{13.40}$$

where $P_i(\cdot)$ is the Legendre polynomial of order i.

Similarly, the orthogonal hierarchical basis functions in 3-D will be

$$\mathbb{Q}_{p,p,p} = \text{span}\{P_i(\xi)P_j(\eta)P_k(\zeta) :\ 0 \le i, j, k \le p\}. \tag{13.41}$$

13.4.2 Orthonormal Hierarchical Basis on Triangles or Tetrahedra

We construct orthonormal hierarchical basis functions on the reference simplicial elements. Let K^n be the simplex in $\mathbb{R}^n$, i.e.,

$$K^n := \left\{ \mathbf{x} \in \mathbb{R}^n : 0 \le x_i;\ \sum_{i=1}^{n} x_i \le 1 \right\}. \tag{13.42}$$

The notation $|\mathbf{x}|$ means the discrete ℓ^1 norm for a generic point $\mathbf{x} \in K^n$, i.e.,

$$|\mathbf{x}| = \sum_{i=1}^{n} |x_i|. \tag{13.43}$$

Denote by $\mathbf{x}_i$ the truncation or projection of the point $\mathbf{x}$ in the first i dimensions, i.e.,

$$\mathbf{x}_0 \equiv 0, \quad \mathbf{x}_i \equiv (x_1, x_2, \ldots, x_i), \quad 1 \le i \le n. \tag{13.44}$$

For a point $\boldsymbol{\alpha} \in \mathbb{N}_0^n$, $\mathbb{N}_0 = \{0, 1, 2, \ldots, \}$, denote by $\boldsymbol{\alpha}^i$ the truncation or projection of the point $\boldsymbol{\alpha}$ from the ith dimension, i.e.,

$$\boldsymbol{\alpha}^i \equiv (\alpha_i, \alpha_{i+1}, \ldots, \alpha_n), \quad 1 \le i \le n. \tag{13.45}$$

For a point $\boldsymbol{\tau} \in \mathbb{R}^{n+1}$, the notation $\boldsymbol{\tau}^i$ is similarly defined as in $\boldsymbol{\alpha}^i$, namely

$$\boldsymbol{\tau}^i \equiv (\tau_i, \tau_{i+1}, \ldots, \tau_{n+1}), \quad 1 \le i \le n+1. \tag{13.46}$$

Orthogonal basis functions for DG methods will be constructed using orthogonal polynomials over an n-simplex K^n given in the following theorem from [7] under the following weight function:

$$W_{\boldsymbol{\tau}}^{(K^n)}(\mathbf{x}) = (1 - |\mathbf{x}|)^{\tau_{n+1} - \frac{1}{2}} \prod_{i=1}^{n} x_i^{\tau_i - \frac{1}{2}}, \quad \mathbf{x} \in K^n, \ \tau_i \ge -\frac{1}{2}, \ i = 1, 2, \ldots, n+1. \tag{13.47}$$

Theorem 13.1 *The polynomials*

$$P_{\boldsymbol{\alpha}}\left(W_\tau^{(K^n)}; \mathbf{x}\right) = [h_{\boldsymbol{\alpha}}]^{-1} \prod_{i=1}^{n} \left(\frac{1 - |\mathbf{x}_i|}{1 - |\mathbf{x}_{i-1}|} \right)^{|\boldsymbol{\alpha}^{i+1}|} J_{\alpha_i}^{(\rho_i^1, \rho_i^2)} \left(\frac{2x_i}{1 - |\mathbf{x}_{i-1}|} - 1 \right), \tag{13.48}$$

where $J_{\alpha_i}^{(\rho_i^1, \rho_i^2)}(x)$ are the normalized version of the Jacobi polynomials, keeping the same notation, defined in (5.4), $\rho_i^1 = 2|\boldsymbol{\alpha}^{i+1}| + |\boldsymbol{\tau}^{i+1}| + (n - i - 1)/2$ and $\rho_i^2 = \tau_i - 1/2$, are orthonormal over K^n, the normalization constant $h_\alpha^{(K^n)}$ is given by

$$[h_\alpha]^{-2} = \prod_{i=1}^{n} 2^{\rho_i^1+\rho_i^2+1}, \tag{13.49}$$

and the weight function takes the form in (13.47).

Using the result in Theorem 13.1, the orthonormal hierarchical shape functions on the reference element in two and three dimensions are given as follows [8].

- Orthonormal hierarchical basis in a 2-D triangle: for $0 \le i, j, i+j \le p$,

$$\Phi_{i,j} = \kappa(1-x_1)^j J_i^{(2j+1,0)}(2x_1-1)\, J_j^{(0,0)}\left(\frac{2x_2}{1-x_1}-1\right), \tag{13.50}$$

where $\kappa = \sqrt{2(i+j+1)(2j+1)}$.

- Orthonormal hierarchical basis in a 3-D tetrahedron: for $0 \le i, j, k,\ i+j+k \le p$,

$$\Phi_{i,j,k} = \lambda(1-x_1)^j(1-x_1-x_2)^k \times J_i^{(2j+2k+2,0)}(2x_1-1)\, J_j^{(2k+1,0)}\left(\frac{2x_2}{1-x_1}-1\right) J_k^{(0,0)}\left(\frac{2x_3}{1-x_1-x_2}-1\right), \tag{13.51}$$

where $\lambda = \sqrt{(2k+1)(2j+2k+2)(2i+2j+2k+3)}$.

From Theorem 13.1, we have the following orthonormal conditions of the basis functions:

$$M_{\ell_1,\ell_2} := \left\langle \Phi_{\ell_1}, \Phi_{\ell_2} \right\rangle|_{K^d} = \delta_{\ell_1,\ell_2} \tag{13.52}$$

where $\ell = (i, j)$ in 2-D and $\ell = (i, j, k)$ in 3-D.

In deriving (13.50) and (13.51), the parameter τ_i in Theorem 13.1 takes the value of 1/2. The function $J_n^{(\alpha,\beta)}(x)$ is the normalized Jacobi polynomials of a single variable defined in (5.4).

Remark 13.1 A special case of the orthonormal basis functions above can be found in [9], where the parameter of the weight function in (13.47) takes the particular value $\tau_i = 0$ [7] for the Dubiner basis [9, 10]. In the construction presented here, this parameter has the value $\tau_i = 1/2$.

Theorem 13.2 *Let $k \in \mathbb{N}_0$. The bases for the spaces $P_k(K^n), n = 2, 3,$ of polynomials of total degree at most k are given in (13.50) and (13.51), respectively, namely*

$$P_k(K^2) = \text{span}\{\Phi_{i,j} : 0 \leq i,\, j,\, i + j \leq k\}, \tag{13.53}$$

$$P_k(K^3) = \text{span}\{\Phi_{l,m,n} : 0 \leq l, m, n, l + m + n \leq k\}. \tag{13.54}$$

Proof Firstly, for each basis function, we have $\Phi_{i,j} \in P_k(K^2)$ and $\Phi_{i,j,k} \in P_k(K^3)$. Further, it is noticed that with different indexes the shape functions given in (13.50) and (13.51) are linearly independent. Secondly, the numbers of independent shape functions are $(k+1)(k+2)/2$ and $(k+1)(k+2)(k+3)/6$ for two and three dimensions, respectively, which coincide with the respective dimensions of $P_k(K^2)$ and $P_k(K^3)$. □

13.5 Explicit Formulae of Basis Functions

Polynomial basis functions up to third order and in terms of the coordinates for the reference element are given in the following.

Basis functions in 2-D triangles

- Zeroth-order:

$$\Phi_{0,0} = \sqrt{2}.$$

- First-order:

$$\Phi_{1,0} = (6x_1 - 2), \quad \Phi_{0,1} = 2\sqrt{3}(x_1 + 2x_2 - 1).$$

- Second-order:

$$\begin{aligned}
\Phi_{2,0} &= \sqrt{6}(1 - 8x_1 + 10x_1^2),\\
\Phi_{0,2} &= \sqrt{30}(1 - 2x_1 + x_1^2 - 6x_2 + 6x_1x_2 + 6x_2^2),\\
\Phi_{1,1} &= 3\sqrt{2}(x_1 + 2x_2 - 1)(5x_1 - 1).
\end{aligned}$$

- Third-order:

$$\begin{aligned}
\Phi_{3,0} &= 2\sqrt{2}(15x_1 - 45x_1^2 + 35x_1^3 - 1),\\
\Phi_{0,3} &= 2\sqrt{14}(3x_1 - 3x_1^2 + 12x_2 + x_1^3 - 24x_1x_2 - 30x_2^2 + 20x_2^3\\
&\quad + 12x_2x_1^2 + 30x_1x_2^2 - 1),\\
\Phi_{2,1} &= 2\sqrt{6}(x_1 + 2x_2 - 1)(1 - 12x_1 + 21x_1^2),\\
\Phi_{1,2} &= 2\sqrt{10}(7x_1 - 1)(1 - 2x_1 + x_1^2 - 6x_2 + 6x_1x_2 + 6x_2^2).
\end{aligned}$$

Basis functions in 3-D tetrahedra

- Zeroth-order:

$$\Phi_{0,0,0} = \sqrt{6}.$$

- First-order:

$$\Phi_{1,0,0} = \sqrt{10}(4x_1 - 1), \quad \Phi_{0,1,0} = 2\sqrt{5}(x_1 + 3x_2 - 1),$$
$$\Phi_{0,0,1} = 2\sqrt{15}(x_1 + x_2 + 2x_3 - 1).$$

- Second-order:

$$\begin{aligned}\Phi_{2,0,0} &= \sqrt{14}(1 - 10x_1 + 15x_1^2),\\ \Phi_{0,2,0} &= \sqrt{42}(1 - 2x_1 + x_1^2 - 8x_2 + 8x_1x_2 + 10x_2^2),\\ \Phi_{0,0,2} &= \sqrt{210}(1 - 2x_1 - 2x_2 + x_1^2 + 2x_1x_2\\ &\quad + x_2^2 - 6x_3 + 6x_1x_3 + 6x_2x_3 + 6x_3^2),\end{aligned}$$

$$\begin{aligned}\Phi_{1,1,0} &= 2\sqrt{7}(x_1 + 3x_2 - 1)(6x_1 - 1),\\ \Phi_{1,0,1} &= 2\sqrt{21}(x_1 + x_2 + 2x_3 - 1)(6x_1 - 1),\\ \Phi_{0,1,1} &= 3\sqrt{14}(x_1 + 5x_2 - 1)(x_1 + x_2 + 2x_3 - 1).\end{aligned}$$

- Third-order:

$$\begin{aligned}\Phi_{3,0,0} &= 3\sqrt{2}(18x_1 - 63x_1^2 + 56x_1^3 - 1),\\ \Phi_{0,3,0} &= 6\sqrt{2}(3x_1 + 15x_2 + x_1^3 + 35x_2^3 - 30x_1x_2 - 45x_2^2 - 3x_1^2\\ &\quad + 15x_2x_1^2 + 45x_1x_2^2 - 1),\\ \Phi_{0,0,3} &= 6\sqrt{14}\left(12x_3(1 - x_1 - x_2)^2 + 30g(\mathbf{x})^2(1 - x_1 - x_2) + 20g(\mathbf{x})^3\right.\\ &\quad \left. -11(1 - x_1 - x_2)^3\right),\end{aligned}$$

$$\begin{aligned}\Phi_{2,1,0} &= 6(x_1 + 3x_2 - 1)(1 - 14x_1 + 28x_1^2),\\ \Phi_{2,0,1} &= 6\sqrt{3}(2x_3 + x_1 + x_2 - 1)(1 - 14x_1 + 28x_1^2),\\ \Phi_{1,2,0} &= 3\sqrt{6}(8x_1 - 1)(1 - 2x_1 + x_1^2 - 8x_2 + 8x_1x_2 + 10x_2^2),\\ \Phi_{0,2,1} &= 6\sqrt{6}(2x_3 + x_1 + x_2 - 1)\\ &\quad \cdot (1 - 2x_1 + x_1^2 - 12x_2 + 12x_1x_2 + 21x_2^2),\end{aligned}$$

$$
\begin{aligned}
\Phi_{1,0,2} &= 3\sqrt{30}(8x_1 - 1)(1 - 2x_1 - 2x_2 + x_1^2 + 2x_1x_2 + x_2^2 - 6x_3 \\
&\quad + 6x_1x_3 + 6x_2x_3 + 6x_3^2), \\
\Phi_{0,1,2} &= 6\sqrt{10}(1 - 2x_1 - 2x_2 + x_1^2 + 2x_1x_2 + x_2^2 - 6x_3 + 6x_1x_3 \\
&\quad + 6x_2x_3 + 6x_3^2)(x_1 + 7x_2 - 1), \\
\Phi_{1,1,1} &= 9\sqrt{2}(8x_1 - 1)(x_1 + 5x_2 - 1)(2x_3 + x_1 + x_2 - 1),
\end{aligned}
$$

where $g(\mathbf{x}) = x_1 + x_2 + x_3 - 1$.

13.6 Computation of Whispering Gallery Modes (WGMs) with DG Methods

In this section, we demonstrate the high accuracy of the DG methods by simulating whispering gallery modes (WGMs) in coupled resonator optical waveguides (CROWs) made of 2-D cylinders [11]. WGMs are traveling electromagnetic resonances confined within dielectric media of circular symmetric structure such as circular rods, micro-disks, and micro-spheres. In the case of a dielectric rod, the WGMs were first studied by Lord Rayleigh [12] who tried to understand the acoustic waves clinging to the dome of St. Paul's Cathedral; the waves were shown to be trapped between the cylindrical boundary and a caustic inside the rod [13].

13.6.1 WGMs in Dielectric Cylinders

Electromagnetic WGMs exist in a circular dielectric cylinder, which is assumed to be of radius a and infinite length with dielectric constant ϵ_1 and magnetic permeability μ_1, embedded in an infinite homogeneous medium of material parameters ϵ_2 and μ_2. In the cylindrical coordinate system (r, θ, z), the components of the magnetic field $\mathbf{H} = (H_r, H_\theta, H_z)$ and the electric field $\mathbf{E} = (E_r, E_\theta, E_z)$ of time-harmonic WGMs (with time dependence as $\exp(-i\omega t)$) are given by the following equations [14]:

$$
\begin{aligned}
H_r &= \left[a_n \frac{nk^2}{\mu\omega\lambda^2 r} G_n(\lambda r) + b_n \frac{ih}{\lambda} G_n'(\lambda r) \right] F_n, \\
H_\theta &= \left[a_n \frac{ik^2}{\mu\omega\lambda} G_n'(\lambda r) - b_n \frac{nh}{\lambda^2 r} G_n(\lambda r) \right] F_n, \\
H_z &= b_n G_n(\lambda r) F_n,
\end{aligned}
\tag{13.55}
$$

$$
\begin{aligned}
E_r &= \left[a_n \frac{ih}{\lambda} G_n'(\lambda r) - b_n \frac{\mu \omega n}{\lambda^2 r} G_n(\lambda r) \right] F_n, \\
E_\theta &= - \left[a_n \frac{nh}{\lambda^2 r} G_n(\lambda r) + b_n \frac{\mathrm{i}\mu\omega}{\lambda} G_n'(\lambda r) \right] F_n, \\
E_z &= a_n G_n(\lambda r) F_n,
\end{aligned} \tag{13.56}
$$

where $F_n = \exp(\mathrm{i}n\theta + \mathrm{i}hz - \mathrm{i}\omega t)$, with h being the axial propagation constant. The function $G_n \equiv J_n$ for $r < a$ and $H_n^{(1)}$ for $r > a$, where J_n is the Bessel function of the first kind and $H_n^{(1)}$ is the Hankel function of the first kind. Also, for $r < a$, $k = k_1 = \omega\sqrt{\epsilon_1 \mu_1}, \lambda = \lambda_1$, where $\lambda_1^2 = k_1^2 - h^2$, and, for $r > a, k = k_2 = \omega\sqrt{\epsilon_2 \mu_2}$, $\lambda = \lambda_2$, where $\lambda_2^2 = k_2^2 - h^2$. The coefficients a_n and b_n are determined by the tangential continuity boundary condition at the cylindrical boundary $r = a$. For a nontrivial solution, the axial propagation constant h satisfies the following characteristic equation [13]:

$$
\begin{aligned}
&\left[\frac{\mu_1}{u} \frac{J_n'(u)}{J_n(u)} - \frac{\mu_2}{v} \frac{H_n^{(1)'}(v)}{H_n^{(1)}(v)} \right] \left[\frac{k_1^2}{\mu_1 u} \frac{J_n'(u)}{J_n(u)} - \frac{k_2^2}{\mu_2 v} \frac{H_n^{(1)'}(v)}{H_n^{(1)}(v)} \right] \\
&= n^2 h^2 \left(\frac{1}{v^2} - \frac{1}{u^2} \right)^2,
\end{aligned} \tag{13.57}
$$

where $u = \lambda_1 a$ and $v = \lambda_2 a$. For a given mode number n, (13.57) does not have a unique solution, and the electromagnetic WGMs are represented by solutions of (13.57) when n is of the order of $\lambda_1 a$. Note that the mode number n is also the number of maxima in the field intensity in the azimuthal direction and is thus called the azimuthal number of the WGMs. We will confine ourselves to WGMs with an axial propagation constant h between k_1 and k_2, i.e., $k_1 > h > k_2$. In this case, $\lambda_2 = -\mathrm{i}|\lambda_2|$ and $\lambda_1 = |\lambda_1|$, which prevents any ohmic losses, and the WGMs would be un-attenuated along a perfectly straight cylinder [13].

13.6.2 *Optical Energy Transfer in Coupled Micro-Cylinders*

We now study the optical energy coupling by WGMs between micro-cylinders [15]. For a WGM with axial propagation constant h, the magnetic field $\mathbf{H} = (H_x, H_y, H_z)$ and the electric field $\mathbf{E} = (E_x, E_y, E_z)$ in a rectangular coordinate system (x, y, z) may be expressed as

$$
\mathbf{H}(x, y, z, t) = \mathbf{H}(x, y, t) \exp(\mathrm{i}hz), \quad \mathbf{E}(x, y, z, t) = \mathbf{E}(x, y, t) \exp(\mathrm{i}hz). \tag{13.58}
$$

Substituting (13.58) into the sourceless Maxwell's equations (13.1), we obtain the following system of equations in matrix form:

$$\frac{\partial \mathbf{u}}{\partial t} + A(\epsilon,\mu)\frac{\partial \mathbf{u}}{\partial x} + B(\epsilon,\mu)\frac{\partial \mathbf{u}}{\partial y} = \mathbf{S}, \tag{13.59}$$

where

$$\mathbf{u} = \begin{bmatrix} \mu\mathbf{H} \\ \epsilon\mathbf{E} \end{bmatrix},$$

$$A(\epsilon,\mu) = \begin{bmatrix} 0 & 0 & 0 & 0 & 0 & 0 \\ 0 & 0 & 0 & 0 & 0 & -\frac{1}{\epsilon} \\ 0 & 0 & 0 & 0 & \frac{1}{\epsilon} & 0 \\ 0 & 0 & 0 & 0 & 0 & 0 \\ 0 & 0 & \frac{1}{\mu} & 0 & 0 & 0 \\ 0 & -\frac{1}{\mu} & 0 & 0 & 0 & 0 \end{bmatrix}, \quad B(\epsilon,\mu) = \begin{bmatrix} 0 & 0 & 0 & 0 & 0 & \frac{1}{\epsilon} \\ 0 & 0 & 0 & 0 & 0 & 0 \\ 0 & 0 & 0 & -\frac{1}{\epsilon} & 0 & 0 \\ 0 & 0 & -\frac{1}{\mu} & 0 & 0 & 0 \\ 0 & 0 & 0 & 0 & 0 & 0 \\ \frac{1}{\mu} & 0 & 0 & 0 & 0 & 0 \end{bmatrix}, \tag{13.60}$$

and $\mathbf{S} = (\mathrm{i}hE_y, -\mathrm{i}hE_x, 0, -\mathrm{i}hH_y, \mathrm{i}hH_x, 0)^{\mathrm{T}}$.

Here, for the purpose of non-dimensionalization, the reference length is chosen as the free space wave length of an electromagnetic field with frequency 100 THz; that is 3 μm. Therefore, one unit of length corresponds to 3 μm and one unit of time corresponds to 10 fs. We consider a model CROW system of two identical circular dielectric cylinders of infinite length in contact with each other. The radii of the cylinders are $r_1 = r_2 = 0.5775$ and the cylinders have material refractive index 3.2, i.e., $\epsilon_1 = 10.24$ and $\mu_1 = 1$, and the external medium is a vacuum.

By setting the angular frequency $\omega = 2\pi$ and the azimuthal number $n = 8$, we find that the characteristic equation (13.57) has a solution $h = 6.80842739$ between $k_1 = 6.4\pi$ and $k_2 = 2\pi$; the resulting WGM is denoted by $\mathrm{WGM}_{8,1,0}$.

To investigate the optical energy transport by WGMs from one cylinder to the other, we assume that initially there exists a WGM in the left cylinder and that no fields exist inside the right cylinder. As initial conditions, the exact values of $\mathrm{WGM}_{8,1,0}$ in the left cylinder are taken in the entire computational domain, except for the inside of the right cylinder, where a zero field is initialized. To ensure that the initial field satisfies the interface condition on the surface of the right cylinder, we impose artificial surface magnetic and electric currents over the surface of the right cylinder for a short duration:

$$\mathbf{J}_{\mathbf{m}}(\mathbf{r},t) = \mathbf{J}^{\mathbf{0}}_{\mathbf{m}}(\mathbf{r})\exp(-\alpha t), \qquad \mathbf{J}_{\mathbf{e}}(\mathbf{r},t) = \mathbf{J}^{\mathbf{0}}_{\mathbf{e}}(\mathbf{x})\exp(-\alpha t), \tag{13.61}$$

where the constant $\alpha > 0$ is chosen so the surface currents become negligible in a short time, and $\mathbf{J}^{\mathbf{0}}_{\mathbf{m}}$ and $\mathbf{J}^{\mathbf{0}}_{\mathbf{e}}$ are calculated from the initial fields $\mathbf{E}(\mathbf{r},0)$ and $\mathbf{H}(\mathbf{r},0)$ as follows:

$$\mathbf{J}^{\mathbf{0}}_{\mathbf{m}}(\mathbf{r}) = -\mathbf{n}\times[\mathbf{E}^{+}(\mathbf{r},0) - \mathbf{E}^{-}(\mathbf{r},0)], \quad \mathbf{J}^{\mathbf{0}}_{\mathbf{e}}(\mathbf{r}) = \mathbf{n}\times[\mathbf{H}^{+}(\mathbf{r},0) - \mathbf{H}^{-}(\mathbf{r},0)].$$

For such boundary currents, the numerical normal flux will have to be modified on both sides of the surface. Given two states $\mathbf{u}^-$ and $\mathbf{u}^+$, the numerical normal flux can be written as

$$(\mathbf{F} \cdot \mathbf{n})^- = \begin{bmatrix} \mathbf{n} \times \frac{(Y\mathbf{E}-\mathbf{n}\times\mathbf{H})^- + (Y\mathbf{E}+\mathbf{n}\times\mathbf{H})^+ - \mathbf{J_e}}{Y^- + Y^+} + \frac{Y^+}{Y^-+Y^+}\mathbf{J_m} \\ -\mathbf{n} \times \frac{(Z\mathbf{H}+\mathbf{n}\times\mathbf{E})^- + (Z\mathbf{H}-\mathbf{n}\times\mathbf{E})^+ - \mathbf{J_m}}{Z^- + Z^+} + \frac{Z^+}{Z^-+Z^+}\mathbf{J_e} \end{bmatrix}, \tag{13.62}$$

for the − side, and

$$(\mathbf{F} \cdot \mathbf{n})^+ = \begin{bmatrix} \mathbf{n} \times \frac{(Y\mathbf{E}-\mathbf{n}\times\mathbf{H})^- + (Y\mathbf{E}+\mathbf{n}\times\mathbf{H})^+ - \mathbf{J_e}}{Y^- + Y^+} - \frac{Y^-}{Y^-+Y^+}\mathbf{J_m} \\ -\mathbf{n} \times \frac{(Z\mathbf{H}+\mathbf{n}\times\mathbf{E})^- + (Z\mathbf{H}-\mathbf{n}\times\mathbf{E})^+ - \mathbf{J_m}}{Z^- + Z^+} - \frac{Z^-}{Z^-+Z^+}\mathbf{J_e} \end{bmatrix}, \tag{13.63}$$

for the + side, respectively.

The computational domain will be decomposed into quadrilateral elements where the cylindrical boundary will conform to the triangulation. On each quadrilateral element, the order of the polynomial basis in (13.40) is $p = 10$, while the constant $\alpha = 10$ in (13.61). To demonstrate the dynamics of the optical energy transport by WGMs from the left cylinder to the right cylinder, the snapshots of the E_z component at four different times are given in Fig. 13.2. The initial state of the system is represented by a counterclockwise circulating wave, i.e., the fundamental mode $\mathrm{WGM}_{8,1,0}$ in the left cylinder. The four sequential snapshots in Fig. 13.2a–d then illustrate the generation of a clockwise WGM in the right cylinder due to the optical coupling, which indicates an optical energy transport from the left cylinder to the right cylinder.

13.7 Finite Difference Yee Scheme

The differential form of Maxwell's equations has an equivalent and general integral form. For any given surface S, on integrating Faraday's law (8.1) over S, we have

$$\frac{\mathrm{d}}{\mathrm{d}t}\int_S \mathbf{B} \cdot \mathrm{d}\mathbf{s} + \int_S \nabla \times \mathbf{E} \cdot \mathrm{d}\mathbf{s} = 0, \tag{13.64}$$

and using Stokes' theorem we arrive at

$$\frac{\mathrm{d}}{\mathrm{d}t}\int_S \mathbf{B} \cdot \mathrm{d}\mathbf{s} + \int_{\partial S} \mathbf{E} \cdot \mathrm{d}\mathbf{l} = 0. \tag{13.65}$$

Similarly, the integral form of the Ampère–Maxwell law (8.15) is

$$\frac{\mathrm{d}}{\mathrm{d}t}\int_S \mathbf{D} \cdot \mathrm{d}\mathbf{s} - \int_{\partial S} \mathbf{H} \cdot \mathrm{d}\mathbf{l} = -\int_S \mathbf{J} \cdot \mathrm{d}\mathbf{s}. \tag{13.66}$$

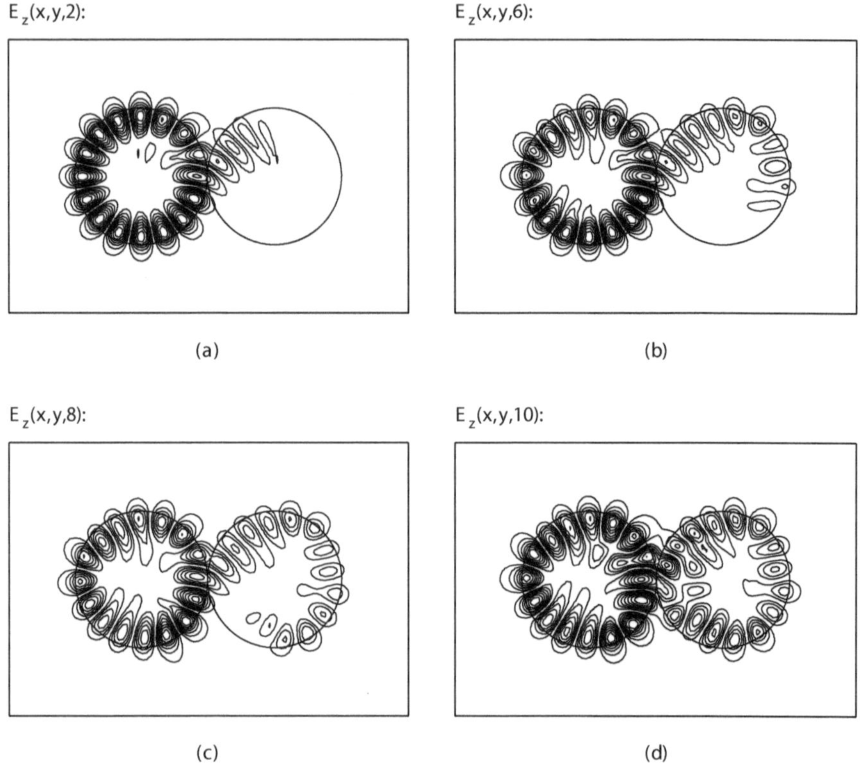

Fig. 13.2 Optical energy transport by WGMs between two identical micro-cylinders in contact. The four sequential snapshots at $t = 2, 6, 8$, and 10 (fs) illustrate the generation of a clockwise WGM in the right cylinder due to resonant optical coupling. Reprinted with permission from [15], © (2005) by Optical Society of America

The Yee scheme is based on a dual grid on a rectangular lattice consisting of $(x_i, y_j, z_k) = (i\Delta x, j\Delta y, k\Delta z)$, $1 \le i \le N_x$, $1 \le j \le N_y$, $1 \le k \le N_z$; the unknowns are located on the faces and the edges of the primary cell centered at (x_i, y_j, z_k). Namely, the **E**-field unknowns are on the boundaries of the faces of the cell, and the **H**-field unknowns are at the centers of the faces of the primary cell (see Fig. 13.3). A dual cell can be obtained by translating the primary cell one-half a cell unit in each direction, which creates the staggered grid for the Yee scheme.

- Cell edge unknowns for the electric-field component tangential along the edges:

$$E_{x,i,j+\frac{1}{2},k+\frac{1}{2}},\ E_{y,i+\frac{1}{2},j,k+\frac{1}{2}},\ E_{z,i+\frac{1}{2},j+\frac{1}{2},k}. \tag{13.67}$$

- Cell face unknowns for the magnetic-field components in the face normal:

$$H_{x,i+\frac{1}{2},j,k},\ H_{y,i,j+\frac{1}{2},k},\ H_{z,i,j,k+\frac{1}{2}}. \tag{13.68}$$

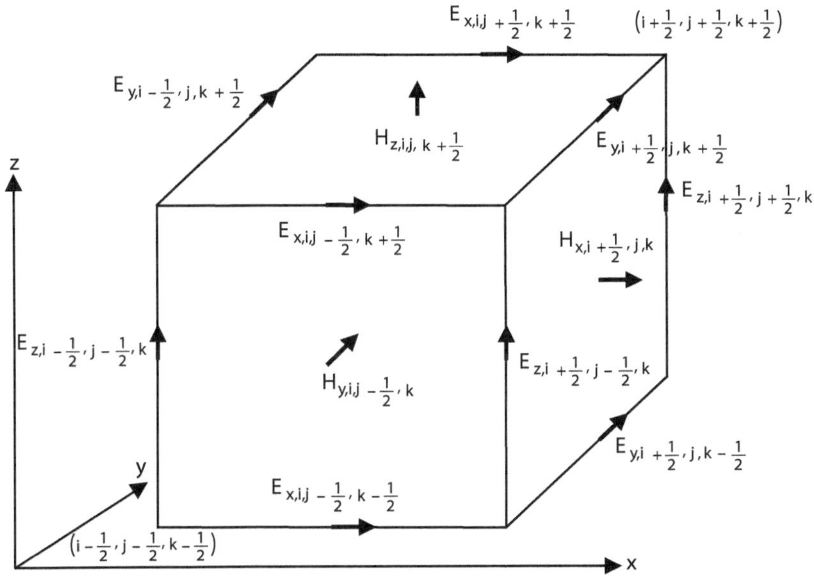

Fig. 13.3 **E**- and **H**-field unknowns in a Yee cell

It should be noted that the magnetic normal components on the faces of the primary cells are also tangential components along the edges of the dual cell. To obtain a finite difference equation for Faraday's law, we apply Eq. (13.69) on S (the right-most face with $+\hat{x}$ as normal):

$$\frac{\mathrm{d}}{\mathrm{d}t}\int_S \mathbf{B}\cdot \mathrm{d}\mathbf{s} + \int_\Gamma \mathbf{E}\cdot \mathrm{d}\mathbf{l} = 0. \tag{13.69}$$

Firstly, we approximate the surface integral with a middle point rule,

$$\int_S \mathbf{B}\cdot \mathrm{d}\mathbf{s} = \mathbf{B}_{i+\frac{1}{2},j,k}\cdot(1,0,0)\Delta y\Delta z + O((\Delta y\Delta_z)^4) \approx B_{x,i+\frac{1}{2},j,k}\Delta y\Delta z. \tag{13.70}$$

Next, we approximate the line integral along the boundary $\Gamma = \partial S$ of S, using again the middle point rule for each of the four edges of the surface S:

$$\begin{aligned}\int_\Gamma \mathbf{E}\cdot \mathrm{d}\mathbf{l} =& \int_{\Gamma_1} \mathbf{E}\cdot(0,1,0)\mathrm{d}y + \int_{\Gamma_2} \mathbf{E}\cdot(0,0,1)\mathrm{d}z\\ &+ \int_{\Gamma_3} \mathbf{E}\cdot(0,-1,0)\mathrm{d}y + \int_{\Gamma_4} \mathbf{E}\cdot(0,0,-1)\mathrm{d}z\\ &\approx E_{y,i+\frac{1}{2},j,k-\frac{1}{2}}\Delta y + E_{z,i+\frac{1}{2},j+\frac{1}{2},k}\Delta z\end{aligned}$$

$$- E_{y,i+\frac{1}{2},j,k+\frac{1}{2}}\Delta y - E_{z,i+\frac{1}{2},j-\frac{1}{2},k}\Delta z. \tag{13.71}$$

Combining (13.70) and (13.71), we obtain the semi-discretized second-order accurate Yee scheme:

$$\begin{aligned}\frac{\mathrm{d}}{\mathrm{d}t}B_{x,i+\frac{1}{2},j,k} = &+\frac{1}{\Delta z}\left(E_{y,i+\frac{1}{2},j,k+\frac{1}{2}} - E_{y,i+\frac{1}{2},j,k-\frac{1}{2}}\right)\\ &-\frac{1}{\Delta y}\left(E_{z,i+\frac{1}{2},j+\frac{1}{2},k} - E_{z,i+\frac{1}{2},j-\frac{1}{2},k}\right).\end{aligned} \tag{13.72}$$

Repeating the same procedure on the other faces with $\hat{y}$ and $\hat{z}$ as normals, we obtain similar equations for $\hat{y}$- and $\hat{z}$-components for the magnetic induction:

$$\begin{aligned}\frac{\mathrm{d}}{\mathrm{d}t}B_{y,i,j+\frac{1}{2},k} = &+\frac{1}{\Delta x}\left(E_{z,i+\frac{1}{2},j+\frac{1}{2},k} - E_{z,i-\frac{1}{2},j+\frac{1}{2},k}\right)\\ &-\frac{1}{\Delta z}\left(E_{x,i,j+\frac{1}{2},k+\frac{1}{2}} - E_{x,i,j+\frac{1}{2},k-\frac{1}{2}}\right),\end{aligned} \tag{13.73}$$

$$\begin{aligned}\frac{\mathrm{d}}{\mathrm{d}t}B_{z,i,j,k+\frac{1}{2}} = &-\frac{1}{\Delta x}\left(E_{y,i+\frac{1}{2},j,k+\frac{1}{2}} - E_{y,i-\frac{1}{2},j,k+\frac{1}{2}}\right)\\ &+\frac{1}{\Delta y}\left(E_{x,i,j+\frac{1}{2},k+\frac{1}{2}} - E_{x,i,j-\frac{1}{2},k+\frac{1}{2}}\right).\end{aligned} \tag{13.74}$$

To obtain the finite difference equations for Ampère's law, we select the surface S on the dual cells and repeat the same numerical quadrature of the related surface and edge integrals, which will give the following:

$$\begin{aligned}\frac{\mathrm{d}}{\mathrm{d}t}D_{x,i,j+\frac{1}{2},k+\frac{1}{2}} = &-\frac{1}{\Delta z}\left(H_{y,i,j+\frac{1}{2},k+1} - H_{y,i,j+\frac{1}{2},k}\right)\\ &+\frac{1}{\Delta y}\left(H_{z,i,j+1,k+\frac{1}{2}} - H_{z,i,j,k+\frac{1}{2}}\right) - J_{x,i,j+\frac{1}{2},k+\frac{1}{2}},\end{aligned} \tag{13.75}$$

$$\begin{aligned}\frac{\mathrm{d}}{\mathrm{d}t}D_{y,i+\frac{1}{2},j,k+\frac{1}{2}} = &-\frac{1}{\Delta x}\left(H_{z,i+1,j,k+\frac{1}{2}} - H_{z,i,j,k+\frac{1}{2}}\right)\\ &+\frac{1}{\Delta z}\left(H_{x,i+\frac{1}{2},j,k+1} - H_{x,i+\frac{1}{2},j,k}\right) - J_{y,i+\frac{1}{2},j,k+\frac{1}{2}},\end{aligned} \tag{13.76}$$

$$\begin{aligned}\frac{\mathrm{d}}{\mathrm{d}t}D_{z,i+\frac{1}{2},j+\frac{1}{2},k} = &+\frac{1}{\Delta x}\left(H_{y,i+1,j+\frac{1}{2},k} - H_{y,i,j+\frac{1}{2},k}\right)\\ &-\frac{1}{\Delta y}\left(H_{x,i+\frac{1}{2},j+1,k} - H_{x,i+\frac{1}{2},j,k}\right) - J_{z,i+\frac{1}{2},j+\frac{1}{2},k}.\end{aligned} \tag{13.77}$$

To obtain the fully discretized Maxwell's equations, we will use a leap-frog type discretization in time for the **B** and **D** variables with a staggered grid in time, i.e.,

$$\left.\frac{\partial \mathbf{B}}{\partial t}\right|_{t^n} = \frac{\mathbf{B}^{n+1/2} - \mathbf{B}^{n-1/2}}{\Delta t}, \tag{13.78}$$

$$\left.\frac{\partial \mathbf{D}}{\partial t}\right|_{t^{n+1/2}} = \frac{\mathbf{D}^{n+1} - \mathbf{D}^{n}}{\Delta t}. \tag{13.79}$$

Finally, we have the fully discretized Yee scheme as follows.

Yee scheme

Given $\mathbf{B}^{n-1/2}$ at $t^{n-1/2} = (n-1/2)\Delta t$, and $\mathbf{D}^n$ at $t^n = n\Delta t$, the following time marching scheme calculates $\mathbf{B}^{n+1/2}$ and $\mathbf{D}^{n+1}$:

$$\begin{aligned} B^{n+1/2}_{x,i+\frac{1}{2},j,k} &= B^{n-1/2}_{x,i+\frac{1}{2},j,k} + \frac{\Delta t}{\Delta z}\left(E^n_{y,i+\frac{1}{2},j,k+\frac{1}{2}} - E^n_{y,i+\frac{1}{2},j,k-\frac{1}{2}}\right) \\ &\quad - \frac{\Delta t}{\Delta y}\left(E^n_{z,i+\frac{1}{2},j+\frac{1}{2},k} - E^n_{z,i+\frac{1}{2},j-\frac{1}{2},k}\right), \end{aligned} \tag{13.80}$$

$$\begin{aligned} B^{n+1/2}_{y,i,j+\frac{1}{2},k} &= B^{n-1/2}_{y,i,j+\frac{1}{2},k} + \frac{\Delta t}{\Delta x}\left(E^n_{z,i+\frac{1}{2},j+\frac{1}{2},k} - E^n_{z,i-\frac{1}{2},j+\frac{1}{2},k}\right) \\ &\quad - \frac{\Delta t}{\Delta z}\left(E^n_{x,i,j+\frac{1}{2},k+\frac{1}{2}} - E^n_{x,i,j+\frac{1}{2},k-\frac{1}{2}}\right), \end{aligned} \tag{13.81}$$

$$\begin{aligned} B^{n+1/2}_{z,i,j,k+\frac{1}{2}} &= B^{n-1/2}_{z,i,j,k+\frac{1}{2}} - \frac{\Delta t}{\Delta x}\left(E_{y,i+\frac{1}{2},j,k+\frac{1}{2}} - E_{y,i-\frac{1}{2},j,k+\frac{1}{2}}\right) \\ &\quad + \frac{\Delta t}{\Delta y}\left(E_{x,i,j+\frac{1}{2},k+\frac{1}{2}} - E_{x,i,j-\frac{1}{2},k+\frac{1}{2}}\right), \end{aligned} \tag{13.82}$$

$$\begin{aligned} D^{n+1}_{x,i,j+\frac{1}{2},k+\frac{1}{2}} &= D^{n}_{x,i,j+\frac{1}{2},k+\frac{1}{2}} - \frac{\Delta t}{\Delta z}\left(H^{n+\frac{1}{2}}_{y,i,j+\frac{1}{2},k+1} - H^{n+\frac{1}{2}}_{y,i,j+\frac{1}{2},k}\right) \\ &\quad + \frac{\Delta t}{\Delta y}\left(H^{n+\frac{1}{2}}_{z,i,j+1,k+\frac{1}{2}} - H^{n+\frac{1}{2}}_{z,i,j,k+\frac{1}{2}}\right) - \Delta t J^{n+\frac{1}{2}}_{x,i,j+\frac{1}{2},k+\frac{1}{2}}, \end{aligned} \tag{13.83}$$

$$\begin{aligned} D^{n+1}_{y,i+\frac{1}{2},j,k+\frac{1}{2}} &= D^{n}_{y,i+\frac{1}{2},j,k+\frac{1}{2}} - \frac{\Delta t}{\Delta x}\left(H^{n+\frac{1}{2}}_{z,i+1,j,k+\frac{1}{2}} - H^{n+\frac{1}{2}}_{z,i,j,k+\frac{1}{2}}\right) \\ &\quad + \frac{\Delta t}{\Delta z}\left(H^{n+\frac{1}{2}}_{x,i+\frac{1}{2},j,k+1} - H^{n+\frac{1}{2}}_{x,i+\frac{1}{2},j,k}\right) - \Delta t J^{n+\frac{1}{2}}_{y,i+\frac{1}{2},j,k+\frac{1}{2}}, \end{aligned} \tag{13.84}$$

$$
\begin{aligned}
D^{n+1}_{z,i+\frac{1}{2},j+\frac{1}{2},k} = D^{n}_{z,i+\frac{1}{2},j+\frac{1}{2},k} &+ \frac{\Delta t}{\Delta x}\left(H^{n+\frac{1}{2}}_{y,i+1,j+\frac{1}{2},k} - H^{n+\frac{1}{2}}_{y,i,j+\frac{1}{2},k}\right) \\
&- \frac{\Delta t}{\Delta y}\left(H^{n+\frac{1}{2}}_{x,i+\frac{1}{2},j+1,k} - H^{n+\frac{1}{2}}_{x,i+\frac{1}{2},j,k}\right) - \Delta t J^{n+\frac{1}{2}}_{z,i+\frac{1}{2},j+\frac{1}{2},k}.
\end{aligned} \tag{13.85}
$$

One of the most important features of the Yee scheme lies in the fact that the divergence of the magnetic field of the numerical solution will remain zero if it is zero initially. The proof is given in Sect. 21.4 when such a condition is critical for the magneto-hydrodynamic (MHD) equations of plasmas. The proof of the second-order convergence of the Yee scheme has been obtained in [16].

Remark 13.2 (*Treatment of curved boundaries*) The Yee scheme suffers a loss of accuracy near curved boundaries or material interfaces due to the phenomena of "staircases" from the lattice representation of boundaries not aligned with coordinate lines. Various numerical techniques have been proposed to handle the difficulties associated with the curved boundaries, including, to list a few, the contour path finite difference method derived as above but with a contour conforming to the curved boundary [17], the local conformal finite difference method [18], the embedded boundary upwinding finite difference method [19, 20], and the locally modified finite difference method [21].

13.8 Summary

The discontinuous Galerkin method for Maxwell's equations offers a highly parallel algorithm for computing electromagnetic wave propagation in complex media, including dispersive materials, with the help of auxiliary differential equations to handle the frequency-dependent dielectric constants. A well-conditioned hierarchical basis for the DG method is also introduced. Meanwhile, the simple Yee scheme offers a second-order approximation that satisfies the divergence-free constraint for the magnetic field (a condition not addressed explicitly in the DG approximation), but, however, suffers a degeneracy in accuracy near boundaries not aligned with mesh coordinate lines.

References

1. B. Cockburn, C.-W. Shu, The local discontinuous Galerkin method for time-dependent convection-diffusion systems. SIAM J. Numer. Anal. **35**, 2440–2463 (1998)
2. J.S. Hesthaven, T. Warburton, *Nodal Discontinuous Galerkin Methods: Algorithms, Analysis, and Applications* (Springer, Berlin, 2008)
3. P.D. Lax, Hyperbolic systems of conservation laws and the mathematical theory of shock waves, in *SIAM Regional Conference Series in Applied Mathematics*, vol. 11 (SIAM, Philadelphia, PA, 1972)
4. J. Smoller, *Shock Waves and Reaction-Diffusion Equations* (Springer, New York, 1983)

5. E. Godlewski, P.-A. Raviart, *Numerical Approximation of Hyperbolic Systems of Conservation Laws* (Springer, Berlin, 1996)
6. A.H. Mahammadian, V. Shankar, W.F. Hall, Computation of electromagnetic scattering and radiation using a time-domain finite-volume discretization procedure. Comput. Phys. Commun. **68**, 175–196 (1991)
7. C.F. Dunkl, Y. Xu, *Orthogonal Polynomials of Several Variables* (Cambridge University Press, Cambridge, UK, 2001)
8. J. Xin, W. Cai, Well-conditioned orthonormal hierarchical $\mathcal{L}_2$ bases on $\mathbb{R}^n$ simplicial elements. J. Sci. Comput. **50**, 446–461 (2011)
9. M. Dubiner, Spectral methods on triangles and other domains. J. Sci. Comput. **6**, 345–390 (1991)
10. G.E. Karniadakis, S.J. Sherwin, *Spectral/hp Element Methods for Computational Fluid Dynamics* (Oxford University Press, Oxford, UK, 2005)
11. A. Yariv, Y. Xu, R.K. Lee, A. Scherer, Coupled resonator optical wave-guide: a proposal and analysis. Opt. Lett. **24**, 711–713 (1999)
12. Lord Rayleigh, Further applications of Bessel's functions of high order to the whispering gallery and allied problems. Phil. Mag. **27**, 100–109 (1914)
13. J.R. Wait, Electromagnetic whispering gallery modes in a dielectric rod. Radio Sci. **2**, 1005–1017 (1967)
14. J.A. Stratton, *Electromagnetic Theory* (McGraw-Hill, New York, 1941)
15. S. Deng, W. Cai, Discontinuous spectral element method modeling of optical coupling by whispering gallery modes between microcylinders. J. Opt. Soc. Am. A **22**, 952–960 (2005)
16. P. Monk, E. Süli, A convergence analysis of Yee's scheme on nonuniform grids. SIAM J. Numer. Anal. **31**, 393–412 (1994)
17. T.G. Jurgens, A. Taflove, K. Umashankar, T.G. Moore, Finite difference time domain modeling of curved surfaces. IEEE Trans. Antenn. Propag. **40**, 357–366 (1992)
18. S. Dey, R. Mittra, A locally conformal finite difference time-domain (FDTD) algorithm for modeling three-dimensional perfectly conducting objects. IEEE Microw. Guided Wave Lett. **7**, 273–275 (1997)
19. W. Cai, S. Deng, An upwinding embedded boundary method for Maxwell's equations in media with material interfaces: 2D case. J. Comput. Phys. **190**, 159–183 (2003)
20. T. Xiao, Q.H. Liu, A staggered upwind embedded boundary (SUEB) method to eliminate the FDTD staircasing error. IEEE Trans. Antenn. Propag. **52**, 730–741 (2004)
21. A. Ditkowski, K. Dridi, J.S. Hesthaven, Convergent Cartesian grid methods for Maxwell's equations in complex geometries. J. Comput. Phys. **170**, 39–80 (2001)

Chapter 14
Scattering in Periodic Structures and Surface Plasmons

In this chapter, we will discuss numerical methods for computing wave scattering in periodic structures in photonics, and surface plasmons of electron density waves at interfaces between metallic materials and dielectrics. Both types of electromagnetic phenomena have many applications in nano-photonics and near-field optics. Firstly, we present the general Bloch theory and Bloch wave expansions for electromagnetic waves in periodic structures. Several numerical methods are discussed that can be used for the calculation of the photonic band structure, including a plane wave method and a calculation of transmission spectra by Fourier methods. Then, a modified Nédélec edge element for periodic structure eigenvalue problems in the frequency domain, and a time-domain finite element with Bloch wave transparent boundary conditions, are presented. Finally, we present a volume integral equation method for surface plasmons through nano-holes in thin films, and a time-domain discontinuous Galerkin method, using auxiliary differential equations (ADEs) to treat dispersive materials, for a resonant coupling of surface plasmons in nano-silver wires.

14.1 Bloch Theory and Band Gap for Periodic Structures

The electric and magnetic properties in a 3-D periodic structure are described by the following eigenvalue problems for the vector Helmholtz operator $\mathcal{L}_E(\mathbf{E})$ or $\mathcal{L}_H(\mathbf{H})$.

- Electric field $\mathbf{E}(\mathbf{r})$:

$$\mathcal{L}_E(\mathbf{E}) \equiv -\nabla \times \left(\frac{1}{\mu}\nabla \times \mathbf{E}\right) + \omega^2 \epsilon\, \mathbf{E} = 0. \tag{14.1}$$

W. Cai, *Deterministic, Stochastic, and Deep Learning Methods for Computational Electromagnetics*, https://doi.org/10.1007/978-981-96-0100-4_14

- Magnetic field $\mathbf{H}(\mathbf{r})$:

$$\mathcal{L}_H(\mathbf{H}) \equiv -\nabla \times \left(\frac{1}{\epsilon}\nabla \times \mathbf{H}\right) + \omega^2 \mu \, \mathbf{H} = 0. \tag{14.2}$$

Here, ω is the frequency of the electromagnetic wave and ϵ is the dielectric constant of the structure, respectively.

The 3-D periodic structure of the medium is defined by its Bravais vectors $\{\mathbf{a}_i, i = 1, 2, 3\}$ and their corresponding elementary reciprocal lattice vectors $\{\mathbf{b}_j, j = 1, 2, 3\}$ defined in (7.2). The dielectric constant ϵ is assumed to be a periodic function over the periodic structure, namely

$$\epsilon(\mathbf{r} + \mathbf{a}_i) = \epsilon(\mathbf{r}), \quad i = 1, 2, 3. \tag{14.3}$$

The Bloch theory of eigenvalue problems (14.1) and (14.2) states that the eigenfunctions, for (14.1) for instance, should take the following form:

$$\mathbf{E}_{\mathbf{k}}(\mathbf{r}) = \mathrm{e}^{\mathrm{i}\mathbf{k}\cdot\mathbf{r}} \mathbf{u}_{\mathbf{k}}(\mathbf{r}), \tag{14.4}$$

where the Bloch wave vector (continuous) $\mathbf{k}$ is given by

$$\mathbf{k} = m_1 \mathbf{b}_1 + m_2 \mathbf{b}_2 + m_3 \mathbf{b}_3, \quad m_j \in (-\infty, \infty), \; j = 1, 2, 3, \tag{14.5}$$

and the function $\mathbf{u}_{\mathbf{k}}$ is a periodic function over the lattice, which can be represented in the following form of a Fourier series:

$$\mathbf{u}_{\mathbf{k}}(\mathbf{r}) = \sum_{\mathbf{G}} \hat{\mathbf{E}}(\mathbf{k} - \mathbf{G}) \mathrm{e}^{-\mathrm{i}\mathbf{G}\cdot\mathbf{r}}, \tag{14.6}$$

where the reciprocal vector $\mathbf{G}$ is defined in (7.3), and

$$\mathbf{u}_{\mathbf{k}}(\mathbf{r} + \mathbf{a}_i) = \mathbf{u}_{\mathbf{k}}(\mathbf{r}), \quad i = 1, 2, 3. \tag{14.7}$$

The proof of the Bloch theory for the 3-D case follows the same procedure as for the 1-D case given below by utilizing the Fourier series of periodic functions over lattices.

14.1.1 Bloch Theory for 1-D Periodic Helmholtz Equations

The Helmholtz equation for a 1-D periodic structure of a lattice period a is

$$\mathcal{L}(E) = \frac{\mathrm{d}^2 E}{\mathrm{d}x^2} + \frac{\omega^2}{c^2} \epsilon(x) E = 0, \qquad x \in (-\infty, \infty), \tag{14.8}$$

where $c = 1/\sqrt{\epsilon_0 \mu_0}$ is the speed of light in the vacuum, and the relative dielectric constant $\epsilon(x)$ is an a-periodic function, i.e.,

$$\epsilon(x + a) = \epsilon(x). \tag{14.9}$$

The Bloch theory in this case implies that the solution of (14.8) takes the following form:

$$E_k(x) = \mathrm{e}^{\mathrm{i}kx} u_k(x), \qquad k \in (-\infty, \infty), \tag{14.10}$$

where $u_k(x)$ is an a-periodic function, namely

$$u_k(x + a) = u_k(x). \tag{14.11}$$

The proof of the above statement is based on Fourier expansions of periodic functions. As the electric field is defined for the whole interval, we consider the solution of (14.8) on a finite domain $[-L/2, L/2]$ with $L = Na$, where N is a positive integer, satisfying a periodic boundary condition, i.e.,

$$E\left(-\frac{L}{2}\right) = E\left(\frac{L}{2}\right). \tag{14.12}$$

So, the electric field $E(x)$ can be expressed by the following Fourier series:

$$E(x) = \sum_{k' \in \mathbb{Z}} \hat{E}(k) \mathrm{e}^{\mathrm{i}kx}, \quad k = \frac{2\pi}{L} k', \qquad x \in \left[-\frac{L}{2}, \frac{L}{2}\right]. \tag{14.13}$$

Meanwhile, the dielectric constant $\epsilon(x)$ also admits the following Fourier series with reciprocal index g:

$$\epsilon(x) = \sum_{g \in G} \hat{\epsilon}(g) \mathrm{e}^{\mathrm{i}gx}, \qquad x \in \left[-\frac{L}{2}, \frac{L}{2}\right], \tag{14.14}$$

where $G = \{g = (2\pi/a)n : n \in \mathbb{Z}\}$. Substituting (14.13) and (14.14) into (14.8), we have

$$\sum_{k' \in \mathbb{Z}} \left[-k^2 \hat{E}(k) \mathrm{e}^{\mathrm{i}kx} + \frac{\omega^2}{c^2} \sum_{g \in G} \hat{E}(k) \hat{\epsilon}(g) \mathrm{e}^{\mathrm{i}(k+g)x} \right] = 0, \tag{14.15}$$

which can be rearranged as follows by changing the summation index $k + g \to k$ (and $k \to k - g$):

$$\sum_{k' \in \mathbb{Z}} \left[-k^2 \hat{E}(k) + \frac{\omega^2}{c^2} \sum_{g \in G} \hat{E}(k - g) \hat{\epsilon}(g) \right] \mathrm{e}^{\mathrm{i}kx} = 0, \qquad x \in \left[-\frac{L}{2}, \frac{L}{2}\right]. \tag{14.16}$$

Due to the orthogonality of the trigonometric functions, we have

$$-k^2\hat{E}(k)+\frac{\omega^2}{c^2}\sum_{g\in G}\hat{E}(k-g)\hat{\epsilon}(g)=0,\quad k=\frac{2\pi}{L}k',\quad k'\in\mathbb{Z}. \tag{14.17}$$

Equation (14.17) imposes coupling conditions on the Fourier coefficients $\hat{E}(k)$ in (14.13), implying that an eigenfunction can be completely defined with only those coefficients related through the reciprocal index g, whereas other coefficients could be set independently to zero. Namely, for each given k, the following $E(x)$ makes a solution for (14.8) for $x \in [-L/2, L/2]$ with "periodic boundary conditions" (14.12):

$$E(x)=\sum_{g\in G}\hat{E}(k-g)\mathrm{e}^{\mathrm{i}(k-g)x}, \tag{14.18}$$

which in turn can be rewritten as

$$E(x)=\mathrm{e}^{\mathrm{i}kx}\sum_{g\in G}\hat{E}(k-g)\mathrm{e}^{-\mathrm{i}gx}\equiv\mathrm{e}^{\mathrm{i}kx}u_k(x),\quad x\in\left[-\frac{L}{2},\frac{L}{2}\right], \tag{14.19}$$

where $u_k(x)$ satisfies the a-periodicity condition, i.e.,

$$u_k(x+a)=\sum_{g\in G}\hat{E}(k-g)\mathrm{e}^{-\mathrm{i}g(x+a)}=\sum_{g\in G}\hat{E}(k-g)\mathrm{e}^{-\mathrm{i}gx}=u_k(x), \tag{14.20}$$

using the fact that $\mathrm{e}^{-\mathrm{i}ga}=1$. Next, let the interval length $L\to\infty$. Then $u_k(x)$ will continue to be an a-periodic function, while $\{k=(2\pi/L)k' : k'\in\mathbb{Z}\}$ becomes a dense subset of the continuous wave number $k\in(-\infty,\infty)$. This concludes the proof of (14.10) and (14.11).

If two wave numbers k and k' are related as $k'=k\pm n(\pi/a)$, they apparently correspond to the same plane wave function $\mathrm{e}^{\mathrm{i}kx}$ for $x\in[0,a]$ (see (14.26) below). Therefore we only have to consider $k\in B=[-\pi/a,\pi/a]$, which is defined as the *first Brillouin zone* [1].

14.1.2 Bloch Wave Expansions

The eigenfunctions $\psi_n(x,k)$ for the operator $\mathcal{L}$ for almost all k (except for a set of zero measure) in the first Brillouin zone $B=[-\pi/a,\pi/a]$ are defined as

$$\mathcal{L}(\psi_n)=\lambda_n\psi_n, \tag{14.21}$$

where $0<\lambda_1<\lambda_2<\cdots<\lambda_n<\cdots$, and, based on (14.10), the Bloch generalized eigenfunctions ψ_n have the form

$$\psi_n(x,k) = e^{ikx}\phi_n(x), \tag{14.22}$$

with an a-periodic function $\phi_n(x)$.

The eigenfunctions $\psi_n(x,k)$ form an orthonormal sequence [2], i.e.,

$$\int_0^a \psi_m^*(x,k)\psi_n(x,k)\mathrm{d}x = \int_0^a \phi_m^*(x)\phi_n(x)\mathrm{d}x = \delta_{mn}. \tag{14.23}$$

Also, it can be shown that the union of all eigenfunctions $\{\psi_n(x,k)\}_{k\in B,n=1,2,\ldots}$ expands $L^2(-\infty,\infty)$. Namely, for any function $f(x) \in L^2(-\infty,\infty)$, the following Fourier series with the Bloch generalized eigenfunctions $\psi_n(x,k)$ holds in the L^2 limit:

$$f(x) = \sum_{n=1}^{\infty}\int_B \hat{f}_n(k)\psi_n(x,k)\mathrm{d}k, \tag{14.24}$$

where the "Fourier" coefficients are defined as

$$\hat{f}_n(k) = \int_{-\infty}^{\infty} f(x)\psi_n^*(x,k)\mathrm{d}x. \tag{14.25}$$

This Fourier series can be used to build a spectral method for problems with periodic potentials [3].

14.1.3 Band Gaps of Photonic Structures

For each n, $\lambda_n(k)$ defines one of many dispersion curves for the periodic structure, while the set $\{\lambda_n(k), k \in B\}_{n=1}^{\infty}$ defines the band structure for the periodic potential operator $\mathcal{L}$ (refer to Fig. 14.1). One of the most important properties of a periodic dielectric medium is the existence of a band gap in the dispersion curve for the eigenfunctions of (14.1) and (14.2). If the frequency of the electromagnetic wave ω falls within one of the band gaps, then propagation through the photonic structure will be forbidden. Therefore, the calculation of the band gap of a given periodic structure is one of the most important tasks in the study of periodic systems.

As shown above, the eigenfunctions are enumerated with a $\mathbf{k}$-vector in (14.4) for the photonic structure of infinite extent. For all practical purposes, considering a photonic structure of a finite dimension L with $L = Na$, $N \gg 1$, will be sufficient to determine the band gap. In this case, the eigenfunctions for (14.1) with a periodic boundary condition $\mathbf{E}(0) = \mathbf{E}(L)$ are given by (14.4), (14.6), and (14.7). The selection of $\mathbf{k}$ is then given by (14.5); we can show that two $\mathbf{k}$ vectors differing by a reciprocal vector will yield the same eigenfunctions. Let $\mathbf{k}' = \mathbf{k} + \mathbf{G}_0$ for some reciprocal vector $\mathbf{G}_0$. By definition, the eigenfunction for $\mathbf{k}'$ is given as

$$\begin{aligned}\mathbf{E}_{\mathbf{k}'}(\mathbf{r}) &= \mathrm{e}^{\mathrm{i}\mathbf{k}'\cdot\mathbf{r}}\sum_{\mathbf{G}}\hat{\mathbf{E}}(\mathbf{k}'-\mathbf{G})\mathrm{e}^{-\mathrm{i}\mathbf{G}\cdot\mathbf{r}} = \mathrm{e}^{\mathrm{i}\mathbf{k}'\cdot\mathbf{r}}\sum_{\mathbf{G}}\hat{\mathbf{E}}(\mathbf{k}+\mathbf{G}_0-\mathbf{G})\mathrm{e}^{-\mathrm{i}\mathbf{G}\cdot\mathbf{r}} \\ &= \mathrm{e}^{\mathrm{i}\mathbf{k}\cdot\mathbf{r}}\sum_{\mathbf{G}'}\hat{\mathbf{E}}(\mathbf{k}-\mathbf{G}')\mathrm{e}^{-\mathrm{i}\mathbf{G}'\cdot\mathbf{r}} = \mathbf{E}_{\mathbf{k}}(\mathbf{r}).\end{aligned} \tag{14.26}$$

As a result, only those $\mathbf{k}$ that belong to the range

$$\mathbf{B} = \left\{\mathbf{k} = \alpha_1\mathbf{b}_1 + \alpha_2\mathbf{b}_2 + \alpha_3\mathbf{b}_3 : \alpha_i \in \left[-\frac{1}{2}, \frac{1}{2}\right], i = 1, 2, 3\right\} \tag{14.27}$$

will yield distinctive eigenfunctions, and this $\mathbf{k}$-range $\mathbf{B}$ is the *first Brillouin zone* for the 3-D periodic structure.

14.1.4 Plane Wave Method for Band Gap Calculations

The problem of finding the band gap of a photonic crystal can be reduced to an eigenvalue problem based on (14.1) or (14.2) and the Bloch theory. Let us consider the Hermitian eigenvalue problem (14.2) and let $\mathbf{H}_{\mathbf{k}}(\mathbf{r}) = \mathrm{e}^{\mathrm{i}\mathbf{k}\cdot\mathbf{r}}\mathbf{u}_{\mathbf{k}}(\mathbf{r})$. Then, the function $\mathbf{u}_{\mathbf{k}}(\mathbf{r})$ satisfies the following eigenvalue problem with periodic boundary conditions:

$$-(\nabla + \mathrm{i}\mathbf{k}) \times \left(\frac{1}{\epsilon}(\nabla + \mathrm{i}\mathbf{k}) \times \mathbf{u}_{\mathbf{k}}(\mathbf{r})\right) + \frac{\omega^2}{c^2}\mathbf{u}_{\mathbf{k}}(\mathbf{r}) = \mathbf{0}, \tag{14.28}$$

with $\mathbf{k} = k_1\mathbf{b}_1 + k_2\mathbf{b}_2 + k_3\mathbf{b}_3$, $-1/2 \le k_i \le 1/2$, $i = 1, 2, 3$.

The periodic eigenfunctions $\mathbf{u}_{\mathbf{k}}(\mathbf{r})$ at $\mathbf{r} = \sum_{k=1}^{3} n_k\mathbf{a}_k/N_k$ will be expanded in terms of plane waves due to the fact that they are eigenfunctions of the differential operators [4]:

$$\begin{aligned}\mathbf{u}_{\mathbf{k}}(\mathbf{r}) &= \sum_{\mathbf{m}=\{m_j\}} \mathbf{h}_{\mathbf{m}} \exp\left(\mathrm{i}\sum_{j,k} m_j\mathbf{b}_j \cdot n_k\mathbf{a}_k/N_k\right) \\ &= \sum_{\mathbf{m}=\{m_j\}} \mathbf{h}_{\mathbf{m}} \exp\left(2\pi\mathrm{i}\sum_{j} m_j n_j/N_j\right).\end{aligned} \tag{14.29}$$

Here, the reciprocal indices $m_j = -N_j/2 + 1, \ldots, N_j/2$, $j = 1, 2, 3$, and $n_k = 0, 1, \ldots, N_k - 1$, $k = 1, 2, 3$ are the spatial coordinates on an $N_1 \times N_2 \times N_3$ affine grid defined by the Bravais lattice vectors. Summation in (14.29) can be implemented by the discrete fast Fourier transform (FFT). Plugging (14.29) into (14.28) results in a linear algebraic system for the expansion coefficients $\mathbf{h}_{\mathbf{m}}$:

$$A\mathbf{h} = \frac{\omega^2}{c^2}B\mathbf{h}, \tag{14.30}$$

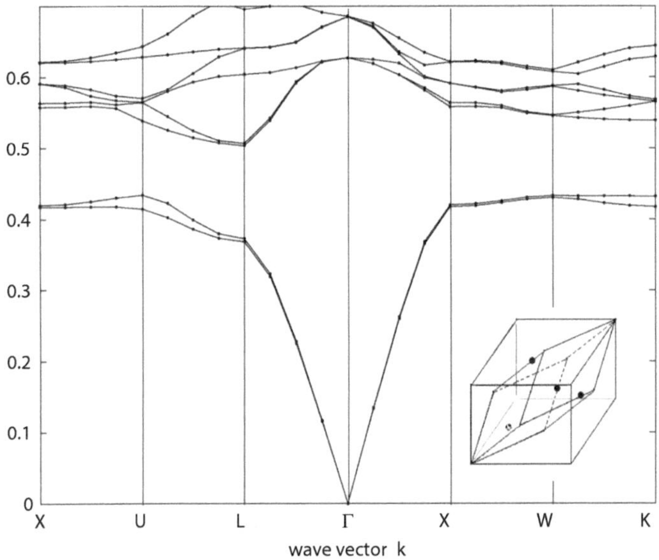

Fig. 14.1 The band gap structure of a 3-D diamond lattice using 434 plane waves. The insert shows the unit cell of the diamond lattice. Reprinted with permission from [5], © (2003) by Optical Society of America

where A and B are $N \times N$ matrices, with N being the number of basis functions. This system will, in general, be solved by an iterative method such as the GMRES method [6]. The main operation in the iteration is the matrix–vector product, which in this case can be achieved straightforwardly using the FFT in $O(N \log N)$ operations in the following manner. As the operation of the curl operator is equivalent to a cross-product in the **k**-vector space, i.e., $(\nabla + i\mathbf{k})\times \leftrightarrow (\mathbf{k} + \mathbf{b}_m)\times$, we can compute $A\mathbf{h}$ in the following manner:

$$(\mathbf{k} + \mathbf{b}_l) \times \cdots \mathrm{FFT}^{-1} \cdots \widetilde{\epsilon^{-1}} \cdots \mathrm{FFT} \cdots (\mathbf{k} + \mathbf{b}_m) \times \mathbf{h}. \tag{14.31}$$

Figure 14.1 shows the band gap structure of a 3-D diamond lattice using 434 plane waves, and the insert shows the unit cell of the diamond lattice [5]. The letters, X, U, L, etc., on the **k**-vector axis indicate points of symmetry in the first Brillouin zone resulting from the symmetry of the unit cell in the photonic crystal structure [1, Sect. 3.7].

14.1.5 Rayleigh–Bloch Waves and Band Gaps by Transmission Spectra

The transmittance spectrum of a photonic crystal of periodic dielectric structure is closely related to its band gap as the latter contains the range of wave frequencies with no transmission through the structure. In this section, we present a plane-wave-based calculation of the transmittance spectrum by [7, 8], which is similar in nature to the rigorous coupled wave analysis (RCWA) method [9]. Consider a photonic crystal slab with an incident wave $\mathbf{E}^{\rm inc}$ impinging from the top. Region 1 is above the top of the photonic crystal ($y \geq 0$), Region 2 is the crystal itself ($-L \leq y \leq 0$), and Region 3 is below the photonic crystal ($y \leq -L$) where the transmitted wave $\mathbf{E}^{\rm t}$ is measured for the transmittance spectrum $T(\omega)$, which is defined as

$$T(\omega) = \left|\mathbf{E}^{\rm t}\right|^2 / \left|\mathbf{E}^{\rm inc}\right|^2 . \tag{14.32}$$

Each region is described by a periodic relative dielectric function $\epsilon_i(\mathbf{r})$; however, $\epsilon_1(\mathbf{r}) \equiv \epsilon_1$ and $\epsilon_3(\mathbf{r}) \equiv \epsilon_3$ are assumed to be constant. Let us consider a transverse electric (TE) wave where $\mathbf{H} = (0, 0, H_z)$ and $\mathbf{E} = (E_x, E_y, 0)$. Note that H_z satisfies the following scalar Helmholtz equation (with no z-dependence, $\nabla = (\partial/\partial x, \partial/\partial y)$, $\mathbf{x} = (x, y)$):

$$\mathcal{L}(H_z) = \nabla \cdot \left(\frac{1}{\epsilon(\mathbf{x})}\nabla H_z\right) + \frac{\omega^2}{c^2} H_z = 0. \tag{14.33}$$

- **Region 1**: $y \geq 0$

Here, for a given incident wave with a directional wave number $\mathbf{k}_{\rm i} = (k_x, k_{1y})$, there will be multiple reflection directions $\mathbf{k}_{\rm r}^{(n)}$ with individual reflection coefficients as shown in Fig. 14.2, a situation different from the single specular reflection from a half-plane space in the classical Snell's law. Consequently, the magnetic field in this region can be expressed as

$$H_{1z}(x, y) = H_0 e^{i\mathbf{k}_{\rm i}\cdot\mathbf{x}} + \sum_n R_n e^{i\mathbf{k}_{\rm r}^{(n)}\cdot\mathbf{x}}, \tag{14.34}$$

where R_n is called the nth-order Bragg reflection coefficient.

- **Region 3**: $y \leq -L$

Similarly, the transmitted wave in Region 3 has multiple transmitted directions $\mathbf{k}_{\rm t}^{(n)}$ with individual transmission coefficients, as shown in Fig. 14.2:

$$H_{3z}(x, y) = \sum_n T_n e^{i\mathbf{k}_{\rm t}^{(n)}\cdot(\mathbf{x}-\mathbf{L})}, \quad \mathbf{L} = (0, -L), \tag{14.35}$$

where T_n is called the nth-order Bragg transmission coefficient.

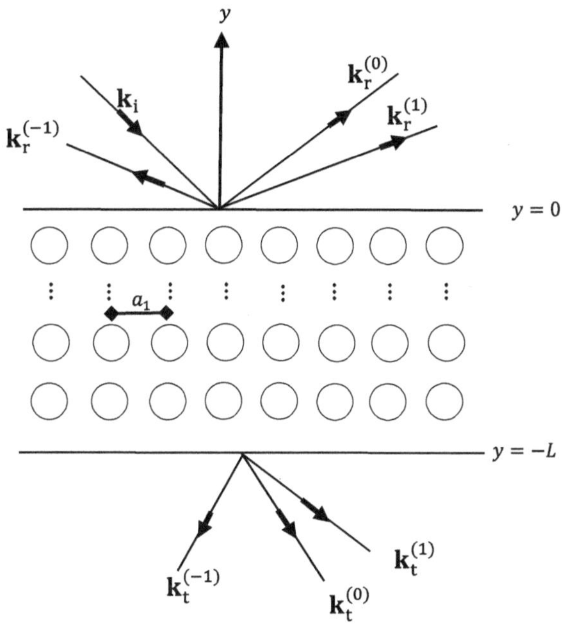

Fig. 14.2 Rayleigh–Bloch reflections on the top and transmissions at the bottom

In order to satisfy the Helmholtz equation (14.33), the y-component of the **k**-vectors will be related to their x-component as follows:

$$k_{\mathrm{r},y}^{(n)} = \begin{cases} +\sqrt{k_1^2 - \left(k_{\mathrm{r},x}^{(n)}\right)^2}, & \text{if } k_1 > \left|k_{\mathrm{r},x}^{(n)}\right|, \\ +\mathrm{i}\sqrt{\left(k_{\mathrm{r},x}^{(n)}\right)^2 - k_1^2}, & \text{otherwise,} \end{cases} \tag{14.36}$$

and

$$k_{\mathrm{t},y}^{(n)} = \begin{cases} -\sqrt{k_3^2 - \left(k_{\mathrm{t},x}^{(n)}\right)^2}, & \text{if } k_3 > \left|k_{\mathrm{t},x}^{(n)}\right|, \\ -\mathrm{i}\sqrt{\left(k_{\mathrm{t},x}^{(n)}\right)^2 - k_3^2}, & \text{otherwise,} \end{cases} \tag{14.37}$$

where $k_1 = \sqrt{\epsilon_1}\omega/c$ and $k_3 = \sqrt{\epsilon_3}\omega/c$, respectively, and ω is the frequency of the incident field.

- **Region 2**: $-L < y < 0$

In this region, the field H_{2z} satisfies the following scalar Helmholtz equation:

$$\mathcal{L}(H_{2z}) = \frac{\partial}{\partial x}\left(\frac{1}{\epsilon(x,y)}\frac{\partial H_{2z}}{\partial x}\right) + \frac{\partial}{\partial y}\left(\frac{1}{\epsilon(x,y)}\frac{\partial H_{2z}}{\partial y}\right) + \frac{\omega^2}{c^2}H_{2z} = 0, \tag{14.38}$$

where H_{2z} is periodic in x and continuous at $y = 0$ and $y = -L$ with H_{1z} and H_{3z}, respectively. To find the solution of (14.38) in Region 2, let us handle the inhomogeneous interface boundary conditions at $y = 0$ and $y = -L$ by introducing a function

$$f(x, y) = \frac{1}{L} \sum_{n=-\infty}^{\infty} [-yT_n + (y + L)(\delta_{n0} H_0 + R_n)] \, \mathrm{e}^{\mathrm{i}k_x^{(n)} x}. \tag{14.39}$$

Then, the difference between $H_{2z}(x, y)$ and $f(x, y)$ is given by

$$\psi(x, y) = H_{2z}(x, y) - f(x, y), \tag{14.40}$$

which vanishes at $y = 0$ and $y = -L$ and satisfies the following equation:

$$\mathcal{L}(\psi) = -\mathcal{L}(f). \tag{14.41}$$

As a result of its vanishing boundary values, $\psi(x, y)$ has the following sine series expansion in the interval $[-L, 0]$:

$$\psi(x, y) = \sum_{n=-\infty}^{\infty} \sum_{m=1}^{\infty} A_{nm} \mathrm{e}^{\mathrm{i}k_x^{(n)} x} \sin\left(\frac{m\pi}{L} y\right). \tag{14.42}$$

The continuity of the tangential component of the **H** field (H_z) imposes the following conditions on the x-component of the **k**-vectors in all three regions:

$$k_{\mathrm{r},x}^{(n)} = k_{\mathrm{t},x}^{(n)} = k_x^{(n)} = k_x + G_n, \tag{14.43}$$

where $G_n = (2\pi/a_1)n$, $n \in \mathbb{Z}$, is the x-component of the reciprocal vectors in Region 2. Each of the nth **k**-vectors defines a *Rayleigh–Bloch* (R–B) wave for the reflection and transmission waves in the form of (14.19). For instance, the reflection wave is expressed in terms of the R–B reflection mode for each $k_{r,y}^{(n)}$:

$$H_{1z}(x, y) = \mathrm{e}^{\mathrm{i}(k_x,0)\cdot\mathbf{x}} \left[H_0 \mathrm{e}^{\mathrm{i}k_{1y} y} + \sum_n R_n \mathrm{e}^{\mathrm{i}(G_n, k_{r,y}^{(n)})\cdot\mathbf{x}} \right] = \mathrm{e}^{\mathrm{i}(k_x,0)\cdot\mathbf{x}} u(x, y),$$

where $u(x, y)$ is an a-periodic function in x.

Meanwhile, the periodic inverse dielectric function $1/\epsilon(x, y)$ can also be expressed as

$$\frac{1}{\epsilon(x, y)} = \sum_{m,n=-\infty}^{\infty} \kappa_{nm} \mathrm{e}^{\mathrm{i}(G_n x + \frac{m\pi}{L} y)}, \quad -L \le y \le 0, \tag{14.44}$$

where the expansion coefficients κ_{nm} are defined in [7]. Plugging (14.42) and (14.44) into (14.41), and using the following Fourier series expansions of functions (y and 1) over the interval $[-L, 0]$ in $f(x, y)$:

$$\frac{2L}{\pi}\sum_{m=1}^{\infty}\frac{(-1)^{m-1}}{m}\sin\left(\frac{m\pi}{L}y\right) = y, \quad -L < y < L,$$

$$\frac{2}{\pi}\sum_{m=1}^{\infty}\frac{1-(-1)^{m}}{m}\sin\left(\frac{m\pi}{L}y\right) = \begin{cases} 1, & 0 < y < L, \\ -1, & -L < y < 0, \end{cases} \tag{14.45}$$

we obtain the following algebraic equations for the coefficients A_{nm} [10]:

$$\frac{\omega^2}{c^2}A_{nm} - \sum_{n'=-\infty}^{\infty}\sum_{m'=1}^{\infty}\left[\begin{array}{l}\left(\frac{mm'\pi^2}{L^2} + k_x^{(n)}k_x^{(n')}\right)\kappa_{n-n',|m-m'|} \\ +\left(\frac{mm'\pi^2}{L^2} - k_x^{(n)}k_x^{(n')}\right)\kappa_{n-n',m+m'}\end{array}\right]A_{n'm'} = B_{mn}, \tag{14.46}$$

where

$$B_{mn} = -\frac{2\omega^2}{\pi c^2}\frac{(-1)^{m-1}T_n + R_n + \delta_{n0}H_0}{m} + \frac{2m\pi}{L^2}\sum_{n'=-\infty}^{\infty}(T_{n'} - R_{n'} - \delta_{n'0}H_0)\kappa_{n-n',m}$$
$$+\frac{2k_x^{(n)}}{\pi}\sum_{n'=-\infty}^{\infty}\sum_{m'=1}^{\infty}k_x^{(n')}(\kappa_{n-n',|m-m'|} - \kappa_{n-n',m+m'})\cdot\frac{(-1)^{m'-1}T_{n'} + R_{n'} + \delta_{n'0}H_0}{m'}. \tag{14.47}$$

The infinite summations in (14.46) and (14.47) will be truncated to a finite sum for $n \in [-N, N]$ and $m \in [1, M]$, to yield a total of $(2N + 1)M$ equations for the $(2N + 1)(M + 2)$ unknowns $R_n, T_n, A_{nm}, n = -N, \ldots, N, m = 1, \ldots, M$. To close the system of equations, $2(2N + 1)$ more equations will be needed, which can be derived from the boundary conditions at $y = 0$ and $y = -L$ for the electric field **E**.

From the Ampère–Maxwell equations

$$\nabla \times \mathbf{H} = \mathrm{i}\omega\epsilon\mathbf{E}, \tag{14.48}$$

we have that the tangential component of the electric field, E_x, satisfies

$$\mathrm{i}\omega E_x = \frac{1}{\epsilon}\frac{\partial H_z}{\partial y}.$$

The continuity of E_x at the interfaces $y = 0$ and $y = -L$ implies that

$$\frac{1}{\epsilon_1}\frac{\partial H_{1z}}{\partial y} = \frac{1}{\epsilon_2}\frac{\partial H_{2z}}{\partial y}, \quad y = 0, \tag{14.49}$$

$$\frac{1}{\epsilon_2}\frac{\partial H_{2z}}{\partial y} = \frac{1}{\epsilon_3}\frac{\partial H_{3z}}{\partial y}, \quad y = -L. \tag{14.50}$$

Equation (14.49) implies that, for $-N \le n \le N$,

$$\pi\epsilon_1 \sum_{m=1}^{M} m A_{nm} = \left(\mathrm{i}\epsilon_2 L k_{\mathrm{r},y}^{(n)} - \epsilon_1\right) R_n + \epsilon_1 T_n + H_0 \left(\mathrm{i}\epsilon_2 L k_{1,y} - \delta_{n0}\epsilon_1\right), \tag{14.51}$$

while (14.50) gives

$$\pi\epsilon_3 \sum_{m=1}^{M} m(-1)^m A_{nm} = -\epsilon_3 R_n + \left(\mathrm{i}\epsilon_2 L k_{\mathrm{t},y}^{(n)} + \epsilon_3\right) T_n - \delta_{n0} H_0 \epsilon_3. \tag{14.52}$$

Thus, (14.51) and (14.52) will provide the needed additional $2(2N + 1)$ equations, which allow the solution of A_{nm} in (14.42).

14.2 Finite Element Methods for Periodic Structures

The Fourier plane wave approximations employed in Sect. 14.1 suffer accuracy degeneracy for a large contrast of dielectric discontinuities; an alternative method is the finite element method for solving Maxwell's equations in the frequency- or time-domain.

14.2.1 Nédélec Edge Element for Eigen-Mode Problems

The calculation of eigen-modes for Maxwell's systems by finite element methods has been an active research topic of both engineering and mathematical interest [11, 12]. Initial applications of nodal finite element methods, where the components of the electromagnetic fields are approximated by continuous finite element bases whose degrees of freedom are nodal values of the solutions at vertices or selected points on edges or interior of elements, have generated non-physical "spurious modes" in the calculation of resonant modes of cavities with perfectly conducting boundaries [13–15]. This phenomenon is mostly related to the treatment of the zero frequency $\omega = 0$, though the spurious modes can also pollute the positive eigen-mode spectrum [11].

Consider the **E**-field eigenvalue problem (14.1) in a cavity Ω with the perfectly conducting boundary condition

$$\mathbf{n} \times \mathbf{E}|_{\partial\Omega} = \mathbf{0}. \tag{14.53}$$

The electric field **E** belongs to the following Sobolev space:

$$H_0(\text{curl}, \Omega) = \left\{ \mathbf{u} \in L^2(\Omega)^3 : \nabla \times \mathbf{u} \in L^2(\Omega)^3, \ \mathbf{n} \times \mathbf{u}|_{\partial\Omega} = \mathbf{0} \right\}, \tag{14.54}$$

and, in addition, in the absence of a source, as in a typical cavity resonant mode problem, we have

$$\nabla \cdot \epsilon \mathbf{E} = 0, \tag{14.55}$$

namely **E** also belongs to the space of $H(\text{div}^0, \Omega; \epsilon)$,

$$H(\text{div}^0, \Omega; \epsilon) = \{ \mathbf{u} \in L^2(\Omega)^3 : \nabla \cdot \mathbf{u} \in L^2(\Omega), \ \nabla \cdot \epsilon \mathbf{u} = 0 \}. \tag{14.56}$$

Thus, the weak form for the eigenvalue problem, on which a standard conforming Galerkin method can be constructed, can be posed as follows:

find $\omega \in \mathbb{R}$, s.t. there exists $\mathbf{0} \neq \mathbf{u} \in H_0(\text{curl}, \ \Omega) \cap H(\text{div}^0, \Omega; \epsilon)$:

$$(\mu^{-1} \nabla \times \mathbf{u}, \nabla \times \mathbf{v}) = \omega^2 (\epsilon \mathbf{u}, \mathbf{v}) \quad \forall \mathbf{v} \in H_0(\text{curl}, \ \Omega) \cap H(\text{div}^0, \Omega; \epsilon). \tag{14.57}$$

It can be shown [16] that the bilinear form on the left-hand side of (14.57) is symmetric, continuous, and coercive on the space $H_0(\text{curl}; \ \Omega) \cap H(\text{div}^0, \Omega; \epsilon)$, and the operator associated with the problem (14.57) is compact and self-adjoint. Therefore the eigenvalue problem (14.57) (or (14.1) and (14.55)) has a countable set of real and positive eigenvalues, and each eigenspace is finite dimensional. It should be emphasized that 0 is in fact not an eigenvalue of the full Maxwell's problem (14.57). If $\omega = 0$ happens in (14.57), then, by setting $\mathbf{v} = \mathbf{u}$, we have $\nabla \times \mathbf{u} = \mathbf{0}$. Then, using the zero-divergence condition $\nabla \cdot \mathbf{E} = 0$ and the boundary condition (14.53), the Helmholtz decomposition theorem implies that $\mathbf{u} = \mathbf{0}$ [17].

The problem of spurious modes appears if we ignore the divergence-free condition on the electric field and consider the weak form of (14.1) in the larger space $H_0(\text{curl}; \Omega)$ instead. Namely, the following eigenvalue problem is considered.

Find $\omega \in \mathbb{R}$, s.t. there exists $\mathbf{0} \neq \mathbf{u} \in H_0(\text{curl}, \Omega)$:

$$(\mu^{-1} \nabla \times \mathbf{u}, \nabla \times \mathbf{v}) = \omega^2 (\epsilon \mathbf{u}, \mathbf{v}) \quad \forall \mathbf{v} \in H_0(\text{curl}, \ \Omega). \tag{14.58}$$

It is evident that an additional zero eigenvalue $\omega = 0$ is created whose eigenspace is denoted as

$$K = \{ \mathbf{u} \in H_0(\text{curl}, \ \Omega) : \ \nabla \times \mathbf{u} = \mathbf{0} \} \equiv H_0(\text{curl}^0, \Omega), \tag{14.59}$$

which consists of the gradient of all scalar functions for a simply connected domain,

$$K \subseteq \{ \mathbf{u} - \nabla p : \ p \in H_0^1(\Omega) \} \text{ (equality holds for simply connected } \Omega). \tag{14.60}$$

As a matter of fact, K is the kernel space of the differential operator $\mathcal{L} = \nabla \times \mu^{-1} \nabla \times$, i.e., $K = \text{Null}(\mathcal{L})$. Therefore, finite element spaces $V_h \subset H_0(\text{curl}; \ \Omega)$, to be free of spurious modes based on the weak form in the space $H_0(\text{curl}, \Omega)$, will be the ones that can in fact reproduce this 0-eigenspace (or kernel) K. The Nédélec edge element is shown to contain the gradient of the first-order nodal element basis [18], which is dense in $H_0^1(\Omega)$ as the mesh is refined, and the lowest order of the Nédélec element on tetrahedra in fact does produce a spurious-mode-free approximation to the non-zero eigen-spectrum of the full Maxwell's eigen-problem [16]. Later, a complete analysis on the Nédélec edge elements of both types and any orders on tetrahedra shows the spurious-mode-free approximations to the non-zero spectrum using the concept of *discrete compactness* of finite element spaces [19–21].

An alternative way to enforce the divergence free condition of the eigen-modes (14.55) is to use a penalty type approximation in a mixed type method proposed by Kikuchi [22–24], thus getting rid of the issue of the kernel space. Here, we consider the discrete form of the mixed formulation.

Find $(\mathbf{u}_h, p_h) \in V_h \times P_h$, such that $\forall\ (\mathbf{v}_h, q_h) \in V_h \times P_h$:

$$\begin{cases} \int\limits_\Omega \left[\frac{1}{\mu}\nabla \times \mathbf{u}_h \cdot \nabla \times \mathbf{v}_h + \nabla p_h \cdot \mathbf{v}_h\right] \mathbf{dr} = \omega_h^2 \int\limits_\Omega \mathbf{u}_h \cdot \mathbf{v}_h \, \mathbf{dr}, \\ \int\limits_\Omega \mathbf{u}_h \cdot \nabla q_h \, \mathbf{dr} = 0. \end{cases} \tag{14.61}$$

The mixed formulation (14.61) can be shown to approximate Maxwell's eigenvalues of (14.57) without spurious modes [25, 26].

For domains with corners and edges, the electromagnetic fields possess singularities [27], whose singular behavior is related to a corresponding Dirichlet or Neumann problem for Laplace operators in irregular domains [28]. The existence of the singularities has also to be taken care of when eigenvalues of Maxwell's equations are calculated [11].

Next, for the computation of band gaps of periodic structures, the Nédélec element can be modified [29, 30] based on the Bloch modes of the electromagnetic fields. For instance, the magnetic field $\mathbf{H}$ will be assumed to be in the form of (14.4) and the periodic part of the Bloch wave will satisfy the following Helmholtz equations:

$$\nabla_{\mathbf{k}} \times \frac{1}{\epsilon} \nabla_{\mathbf{k}} \times \boldsymbol{u} = \omega^2 \mathbf{u} \tag{14.62}$$

and

$$\nabla_{\mathbf{k}} \cdot \mathbf{u} = 0, \tag{14.63}$$

where the shift gradient operator $\nabla_{\mathbf{k}}$ is defined as

$$\nabla_{\mathbf{k}} = \nabla + i\mathbf{k}. \tag{14.64}$$

To set up the finite element approximation of (14.62) and (14.63), we need the following spaces for periodic functions for the unit cell Ω:

$$\begin{aligned}
H_p^1(\Omega) &= \left\{ v \in L^2(\Omega) : \nabla v \in L^2(\Omega)^3 \right\}, \\
\mathbf{H}_p(\text{curl}, \Omega) &= \left\{ \mathbf{v} \in L^2(\Omega)^3 : \nabla \times \mathbf{v} \in L^2(\Omega)^3 \right\}, \\
\mathbf{H}_p(\text{div}, \Omega) &= \left\{ \mathbf{v} \in L^2(\Omega)^3 : \nabla \cdot \mathbf{v} \in L^2(\Omega) \right\}, \\
\mathbf{H}_p(\text{div}_{\mathbf{k}}^0, \Omega) &= \left\{ \mathbf{v} \in \mathbf{H}_p(\text{div}, \Omega) : \nabla_{\mathbf{k}} \cdot \mathbf{v} = 0 \right\}.
\end{aligned} \tag{14.65}$$

The weak formulation for (14.62) will be: find $\omega^2 \in \mathbb{R}$, $(\mathbf{0},0) \neq (\mathbf{u}, p) \in \mathbf{H}_p(\text{curl}, \Omega) \times H_p^1(\Omega)$, such that $\forall(\mathbf{v}, q) \in \mathbf{H}_p(\text{curl}, \Omega) \times H_p^1(\Omega)$,

$$\begin{cases} a(\mathbf{u}, \mathbf{v}) + b(p, \mathbf{v}) = \omega^2(\mathbf{u}, \mathbf{v}), \\ \qquad\quad \overline{b(q, \mathbf{u})} \;= 0, \end{cases} \tag{14.66}$$

where the over-bar denotes the complex conjugate, and

$$a(\mathbf{u}, \mathbf{v}) = \int_{\Omega} \frac{1}{\epsilon} \nabla_{\mathbf{k}} \times \mathbf{u} \cdot \overline{\nabla_{\mathbf{k}} \times \mathbf{v}} \mathrm{d}\mathbf{r}, \tag{14.67}$$

$$b(p, \mathbf{v}) = \int_{\Omega} \nabla_{\mathbf{k}} p \cdot \overline{\mathbf{v}} \, \mathrm{d}\mathbf{r}. \tag{14.68}$$

The unit cell Ω is assumed to be triangulated into tetrahedra, i.e., $\Omega = \cup K$, and the finite element space used to discretize (14.66) is based on a modified Nédélec element [29, 30] of mth-order on the tetrahedra of Sect. 12.1. The space for the electric field will be $\mathbf{U}_h$:

$$\mathbf{U}_h = \left\{ \mathbf{u} \in \mathbf{H}_p(\text{curl}, \Omega) : \; \mathbf{u}|_K = \mathrm{e}^{-\mathrm{i}\mathbf{k}\cdot\mathbf{x}} \widetilde{\mathbf{u}}, \text{ for some } \widetilde{\mathbf{u}} \in (P_m(K))^3 \right\}, \tag{14.69}$$

and the modified space for the scalar Lagrange multiplier q is Q_h:

$$Q_h = \left\{ q \in H_p^1(\Omega) : \; q|_K = \mathrm{e}^{-\mathrm{i}\mathbf{k}\cdot\mathbf{x}} \widetilde{q}, \text{ for some } \widetilde{q} \in P_{m+1}(K) \right\}. \tag{14.70}$$

For a periodic vector function $\mathbf{v} \in H_p^1(\Omega)$, its projection $\Pi_h \mathbf{v} \in \mathbf{U}_h$ is defined by the following degrees of freedom (refer to (12.53)–(12.55)).

- Edge moments:

$$\int_e \left[\mathrm{e}^{\mathrm{i}\mathbf{k}\cdot(\mathbf{x}-\mathbf{x}_e)} (\mathbf{v} - \Pi_h \mathbf{v}) \cdot \mathbf{t} \right] q \, \mathrm{d}l = 0, \quad \forall q \in P_m(e), \tag{14.71}$$

  where $\mathbf{x}_e$ is selected as the center of the edge e and $\mathbf{t}$ is the tangential direction of the edge e.

- Face moments:

$$\int_f \left[e^{i\mathbf{k}\cdot(\mathbf{x}-\mathbf{x}_f)}(\mathbf{v}-\Pi_h\mathbf{v})\times\mathbf{n}\right]\cdot\mathbf{q}\,\mathrm{d}s=0,\quad \forall\mathbf{q}\in\mathcal{D}_{m-1}(f), \tag{14.72}$$

  where $\mathbf{x}_f$ is selected as the barycenter of the face f and $\mathbf{n}$ is the normal direction of face f.
- Element moments:

$$\int_K \left[e^{i\mathbf{k}\cdot(\mathbf{x}-\mathbf{x}_K)}(\mathbf{v}-\Pi_h\mathbf{v})\right]\cdot\mathbf{q}\,\mathrm{d}\Omega=0,\quad \forall\mathbf{q}\in\mathcal{D}_{m-2}(K), \tag{14.73}$$

  where $\mathbf{x}_K$ is selected as the barycenter of the element K.

For edges and faces on the periodic boundaries of Ω, the edge and the face moments will be the same to enforce the periodic condition.

The discrete form for the mixed formulation of the eigenvalue problem is:

find $\omega_h^2\in\mathbb{R}$, $(\mathbf{0},0)\neq(\mathbf{u}_h,p_h)\in\mathbf{U}_h\times Q_h$, such that $\forall(\mathbf{v}_h,q_h)\in\mathbf{U}_h\times Q_h$,

$$\begin{cases} a(\mathbf{u}_h,\mathbf{v}_h)+b(p_h,\mathbf{v}_h) &= \omega_h^2(\mathbf{u}_h,\mathbf{v}_h),\\ \overline{b(q_h,\mathbf{u}_h)} &= 0, \end{cases} \tag{14.74}$$

which will result in the following matrix system:

$$\begin{pmatrix} A & B^* \\ B & \mathbf{0} \end{pmatrix}\begin{pmatrix} \mathbf{U} \\ \mathbf{p} \end{pmatrix}=\omega_h^2\begin{pmatrix} M & \mathbf{0} \\ \mathbf{0} & \mathbf{0} \end{pmatrix}\begin{pmatrix} \mathbf{U} \\ \mathbf{p} \end{pmatrix}, \tag{14.75}$$

where $\mathbf{U}$ contains all the degrees of freedom for the numerical solution $\mathbf{u}_h$ defined in (14.71)–(14.73), $\mathbf{p}$ contains the nodal values of p_h, and the hermitian stiffness matrix A, the matrix B, and the mass matrix M are given as follows:

$$\begin{aligned} A_{ij} &= a(\psi_i,\psi_j),\\ B_{ij} &= b(\psi_i,\phi_j),\\ M_{ij} &= (\psi_i,\psi_j), \end{aligned} \tag{14.76}$$

respectively. Note that $\mathbf{U}_h=\mathrm{span}\{\psi_i\}_{i=1}^{N(h)}$, where the basis function ψ_i is the hierarchical basis defined in Sect. 12.2 and $Q_h=\mathrm{span}\{\phi_j\}$, where ϕ_j is the normal nodal finite element basis.

It is proven [29, 30] that the finite elements used in the mixed formulation (14.74) satisfy the LBB (Ladyzhenskaya–Babuska–Brezzi) condition [24] for stability of the finite element methods, and the linear system will have exact $N(h)=\dim(\mathbf{U}_h)$ real and positive eigenvalues, which will approximate the Maxwell's eigenvalues of (14.62) and (14.63).

14.2.2 Time-Domain Finite Element Methods for Periodic Array Antennas

A phased array antenna consists of a periodic array of patch antennas, where the electromagnetic fields will have a given phase shift on opposite sides of a unit cell, which can be described by the Bloch modes. Due to the large contrast of dielectric constants in this system or some photonic crystals, the Fourier method introduced for periodic structure in Sect. 14.1.5 may suffer degeneracy of accuracy as the series expansion for the inverse dielectric constant (14.44) and the field variables (14.42) may converge too slowly. In this case, the time-domain finite element can be used for the scattering of an electromagnetic wave by a unit cell in a periodic structure with periodicity (L_x, L_y), which will yield the band structure property of the periodic structure. Here, we will describe such a time-domain finite element for the unit cell problem [31, 32], where the computational domain will be the unit cell in the horizontal direction with periodic boundary conditions, while the vertical direction will be truncated and equipped with local transparent boundary conditions for Rayleigh–Bloch waves. Alternatively, Dirichlet-to-Neumann mapping type transparent boundary conditions, which are global on the boundary, can be used for the truncation of the computational domain in the vertical direction for the diffraction of periodic gratings [33].

The time-Fourier-transformed electric field $\mathbf{E}(x, y, z; \omega)$ satisfies the vector Helmholtz equation (8.30) when the electric current $\mathbf{J}_e$ is related to the electric field by Ohm's law (8.22); in addition, source currents $\mathbf{J}_{e,\text{imp}}$ and $\mathbf{J}_{m,\text{imp}}$ may exist from the feeds to the antennas. The vector wave equation (8.30) becomes

$$\nabla \times \left(\frac{1}{\mu_r} \nabla \times \mathbf{E} \right) - k_0^2 \epsilon_r \mathbf{E} + \mathrm{i}\omega\mu_0\sigma\mathbf{E} = -\mathrm{i}\omega\mu_0 \mathbf{J}_{e,\text{imp}} - \nabla \times \left(\frac{1}{\mu_r} \mathbf{J}_{m,\text{imp}} \right), \tag{14.77}$$

where $k_0 = \omega/c = \omega\sqrt{\epsilon_0\mu_0}$ is the wave number, $Z_0 = \sqrt{\mu_0/\epsilon_0}$ is the wave impedance in free space, σ is the conductivity, ϵ is the relative permittivity, and μ is the relative permeability.

The electric field above the structure is in the Bloch wave form given by (14.4). For a given scan angle of the antenna array at (θ_s, ϕ_s) (with θ_s as the angle with the z-axis, which is assumed to be perpendicular to the periodic layer structure), the transverse wave vector in the x- and the y-directions is given by

$$\mathbf{k}_t^s = \left(k_x^s, k_y^s\right) \equiv k_0(\sin\theta_s \cos\phi_s, \sin\theta_s \sin\phi_s, 0) = k_0 \hat{\mathbf{k}}_t^s,$$

and the Bloch wave mode for the electric field is then

$$\mathbf{E}(x, y, z; \omega) = \mathrm{e}^{-\mathrm{i}(k_x^s x + k_y^s y)} \mathbf{P}(x, y, z; \omega), \tag{14.78}$$

where $\mathbf{P}(x, y, z; \omega)$ is a periodic function in (x, y) coordinates, i.e.,

$$\mathbf{P}(x+L_x, y, z; \omega) = \mathbf{P}(x, y, z; \omega), \quad \mathbf{P}(x, y+L_y, z; \omega) = \mathbf{P}(x, y, z; \omega). \tag{14.79}$$

Substituting (14.78) into (14.77), and recasting the resulting equation for $\mathbf{P}$ back in the time domain with the replacement of $i\omega \to \partial/\partial t$, we can show that $\mathbf{P}(x, y, z; t)$ satisfies the following wave equation [31, 32, 34]:

$$\begin{aligned}&\frac{\epsilon_r}{c^2}\frac{\partial^2 \mathbf{P}}{\partial t^2} + \frac{1}{c^2}\hat{\mathbf{k}}_\mathrm{t}^\mathrm{s} \times \frac{1}{\mu_r}\left(\hat{\mathbf{k}}_\mathrm{t}^\mathrm{s} \times \frac{\partial^2 \mathbf{P}}{\partial t^2}\right) + \mu_0\sigma\frac{\partial \mathbf{P}}{\partial t} - \frac{1}{c}\nabla \times \left(\frac{\hat{\mathbf{k}}_\mathrm{t}^\mathrm{s}}{\mu_r} \times \frac{\partial \mathbf{P}}{\partial t}\right)\\ &- \frac{1}{c\mu_r}\hat{\mathbf{k}}_\mathrm{t}^\mathrm{s} \times \left(\nabla \times \frac{\partial \mathbf{P}}{\partial t}\right) + \nabla \times \left(\frac{1}{\mu_r}\nabla \times \mathbf{P}^\mathrm{s}\right) = \mathbf{G}(\mathbf{J}_{e,\mathrm{imp}}, \mathbf{J}_{m,\mathrm{imp}}),\end{aligned} \tag{14.80}$$

where

$$\begin{aligned}\mathbf{G}(\mathbf{J}_{e,\mathrm{imp}}, \mathbf{J}_{m,\mathrm{imp}}) = &-\mu_0\frac{\partial}{\partial t}\mathbf{J}_{e,\mathrm{imp}}\left(t + \frac{\hat{\mathbf{k}}_\mathrm{t}^\mathrm{s}\cdot \mathbf{r}}{c}\right) - \nabla \times \left[\frac{1}{\mu_r}\mathbf{J}_{m,\mathrm{imp}}\left(t + \frac{\hat{\mathbf{k}}_\mathrm{t}^\mathrm{s}\cdot \mathbf{r}}{c}\right)\right]\\ &+ \frac{\hat{\mathbf{k}}_\mathrm{t}^\mathrm{s}}{c} \times \left[\frac{1}{\mu_r}\frac{\partial}{\partial t}\mathbf{J}_{m,\mathrm{imp}}\left(t + \frac{\hat{\mathbf{k}}_\mathrm{t}^\mathrm{s}\cdot \mathbf{r}}{c}\right)\right].\end{aligned}$$

The weak form of (14.80) can be derived by multiplying it by a test vector function $\mathbf{N}$ and performing an integration by parts as follows:

$$\begin{aligned}&\int_\Omega \Bigg\{(\nabla \times \mathbf{N}) \cdot \frac{1}{\mu_r}(\nabla \times \mathbf{P}) + \frac{1}{c^2}\epsilon_r \mathbf{N} \cdot \frac{\partial^2 \mathbf{P}}{\partial t^2} - \frac{1}{c^2}\left(\hat{\mathbf{k}}_\mathrm{t}^\mathrm{s} \times \mathbf{N}\right)\frac{1}{\mu_r}\cdot\left(\hat{\mathbf{k}}_\mathrm{t}^\mathrm{s} \times \frac{\partial^2 \mathbf{P}}{\partial t^2}\right)\\ &+ \mu_0\sigma\mathbf{N}\cdot\frac{\partial \mathbf{P}}{\partial t} + \frac{1}{c}\left[\left(\hat{\mathbf{k}}_\mathrm{t}^\mathrm{s} \times \mathbf{N}\right)\cdot\frac{1}{\mu_r}\left(\nabla \times \frac{\partial \mathbf{P}}{\partial t}\right) - (\nabla \times \mathbf{N})\cdot\frac{1}{\mu_r}\left(\hat{\mathbf{k}}_\mathrm{t}^\mathrm{s} \times \frac{\partial \mathbf{P}}{\partial t}\right)\right]\mathrm{d}\mathbf{r}\\ &+ \int_{\partial\Omega}\mathbf{N}\cdot\left[\mathbf{n} \times \left(\frac{1}{\mu_r}\nabla \times \mathbf{P} - \frac{1}{c}\hat{\mathbf{k}}_\mathrm{t}^\mathrm{s} \times \frac{\partial \mathbf{P}}{\partial t}\right)\right]\mathrm{d}s = \int_\Omega \mathbf{N}\cdot\mathbf{G}(\mathbf{J}_{e,\mathrm{imp}}, \mathbf{J}_{m,\mathrm{imp}})\mathrm{d}\mathbf{r}.\end{aligned} \tag{14.81}$$

Equation (14.81) will be solved in a computational domain Ω made of a unit cell in the (x, y) domain $[0, L_x] \times [0, L_y]$ and a truncated interval $z \in [z_1, z_2]$, namely, $\Omega = [0, L_x] \times [0, L_y] \times [z_1, z_2]$. Naturally, the periodic boundary conditions will be used on the side walls of the domain,

$$\mathbf{P}(0, y, z; t) = \mathbf{P}(L_x, y, z; t), \quad \mathbf{P}(x, 0, z; t) = \mathbf{P}(x, L_y, z; t), \tag{14.82}$$

while transparent boundary conditions at the top and the bottom of the computational domain will be designed for Rayleigh–Bloch waves discussed for the slab photonics in (14.43).

Rayleigh–Bloch waves and transparent boundary conditions

To derive the transparent boundary conditions, we first express the periodic function $\mathbf{P}$ in a Fourier series:

$$\mathbf{P}(x, y, z = z_1; \omega) = \sum_{n,m=-\infty}^{\infty} \hat{\mathbf{P}}_{mn}(\omega, z = z_1) \exp\left[\mathrm{i}\left(\frac{2\pi m}{L_x}x + \frac{2\pi n}{L_y}y\right)\right]. \quad (14.83)$$

The z-dependence in $\hat{\mathbf{P}}_{mn}(\omega, z)$ is assumed to be of the form $\exp(\pm \mathrm{i}k_{z,mn}(z - z_1))$, which can be shown to be related to the wave numbers in the x- and y-directions by

$$k_{z,mn} = \begin{cases} \sqrt{k_0^2 - |\mathbf{k}_{\mathrm{t},mn}|^2}, & \text{if } |\mathbf{k}_{\mathrm{t},mn}| \le k_0, \\ -\mathrm{i}\sqrt{|\mathbf{k}_{\mathrm{t},mn}|^2 - k_0^2}, & \text{otherwise,} \end{cases} \quad (14.84)$$

where $\mathbf{k}_{\mathrm{t},mn} = (k_{xm}, k_{yn}, 0)$.

The coefficients $\hat{\mathbf{P}}_{mn}(\omega)$ can be related to the values of $\mathbf{P}$ at the top and the bottom boundaries of the computational domain $z = z_1$ and $z = z_2$ as follows:

$$\hat{\mathbf{P}}_{mn}(\omega, z_1) = \frac{1}{L_x L_y} \int_0^{L_y}\int_0^{L_x} \mathbf{P}(x, y, z_1; \omega) \exp\left[-\mathrm{i}\left(\frac{2\pi m}{L_x}x + \frac{2\pi n}{L_y}y\right)\right] \mathrm{d}x\, \mathrm{d}y. \quad (14.85)$$

Combining (14.78) and (14.83), the electric field takes the Rayleigh–Bloch wave form,

$$\mathbf{E}(x, y, z; \omega) = \sum_{n,m=-\infty}^{\infty} \hat{\mathbf{E}}_{mn}(\omega) \exp\left\{\mathrm{i}\left[k_{xm}x + k_{yn}y \pm k_{z,mn}(z - z_1)\right]\right\}, \quad (14.86)$$

where

$$k_{xm} = \frac{2\pi m}{L_x} - k_x^{\mathrm{s}}, \quad k_{yn} = \frac{2\pi n}{L_y} - k_y^{\mathrm{s}}. \quad (14.87)$$

Moreover, due to the transverse wave nature of the electric field in free space as a result of Gauss's law

$$\nabla \cdot \mathbf{E} = 0, \quad (14.88)$$

we should have

$$\mathbf{k}_{mn}^{\mathrm{s}} \cdot \hat{\mathbf{E}}_{mn} = 0, \qquad \mathbf{k}_{mn}^{\mathrm{s}} = \mathbf{k}_{\mathrm{t},mn} + (0, 0, k_{z,mn}). \quad (14.89)$$

Based on the expansion (14.86), it was shown that the following approximate boundary condition [34],

$$\hat{\mathbf{z}} \times (\nabla \times \mathbf{E}) = \frac{ik_0}{\cos\theta_s}\hat{\mathbf{k}}_t^s\left(\hat{\mathbf{k}}_t^s \cdot \mathbf{E}\right) - ik_0\cos\theta_s\hat{\mathbf{z}} \times \left(\hat{\mathbf{z}} \times \mathbf{E}\right), \quad z = z_1, z_2, \tag{14.90}$$

absorbs perfectly the fundamental Bloch mode $(m, n) = (0, 0)$ in (14.86). Higher-order absorbing boundary conditions can also be designed to absorb higher-order Rayleigh–Bloch modes, however, which will involve expensive time convolutions.

Again, substituting (14.78) into (14.90), we have

$$\hat{\mathbf{z}} \times \left(\nabla \times \mathbf{P} - ik_0\hat{\mathbf{k}}_t^s \times \mathbf{P}\right) = \frac{ik_0}{\cos\theta_s}\hat{\mathbf{k}}_t^s\left(\hat{\mathbf{k}}_t^s \cdot \mathbf{P}\right) - ik_0\cos\theta_s\hat{\mathbf{z}} \times \left(\hat{\mathbf{z}} \times \mathbf{P}\right), \quad z = z_1, z_2, \tag{14.91}$$

which can be recast in the time domain with the replacement $i\omega \to \partial/\partial t$ as, for $z = z_1, z_2$,

$$\hat{\mathbf{z}} \times \left(\nabla \times \mathbf{P} - \frac{1}{c}\hat{\mathbf{k}}_t^s \times \frac{\partial \mathbf{P}}{\partial t}\right) = \frac{1}{c\cos\theta_s}\hat{\mathbf{k}}_t^s\left(\hat{\mathbf{k}}_t^s \cdot \frac{\partial \mathbf{P}}{\partial t}\right) - \frac{1}{c}\cos\theta_s\hat{\mathbf{z}} \times \left(\hat{\mathbf{z}} \times \frac{\partial \mathbf{P}}{\partial t}\right). \tag{14.92}$$

Equation (14.81) can be solved by a Galerkin finite element method by expanding $\mathbf{P}(x, y, z; t)$ in terms of the Nédélec edge element basis function $\mathbf{N}_i$,

$$\mathbf{P}(x, y, z; t) = \sum_{j=1}^{N} u_j(t)\mathbf{N}_j(x, y, z), \tag{14.93}$$

and setting $\mathbf{N} = \mathbf{N}_i$ in (14.81); we obtain a second-order ordinary differential equation (ODE) in time for the unknown coefficients $\mathbf{U}(t) = (u_1(t), \ldots, u_N(t))^{\mathrm{T}}$:

$$M\frac{\mathrm{d}^2\mathbf{U}}{\mathrm{d}t^2} + (R + R^{\mathrm{ABC}})\frac{\mathrm{d}\mathbf{U}}{\mathrm{d}t} + S\mathbf{U} = \mathbf{f}, \tag{14.94}$$

where

$$M_{ij} = \frac{1}{c^2}\int_\Omega \left[\epsilon_r\mathbf{N}_i \cdot \mathbf{N}_j - \left(\hat{\mathbf{k}}_t^s \times \mathbf{N}_i\right)\frac{1}{\mu_r}\left(\hat{\mathbf{k}}_t^s \times \mathbf{N}_j\right)\right]\mathrm{d}\mathbf{r}, \tag{14.95}$$

$$\begin{aligned} R_{ij} &= \frac{1}{c}\int_\Omega \left[\left(\hat{\mathbf{k}}_t^s \times \mathbf{N}_i\right) \cdot \frac{1}{\mu_r}(\nabla \times \mathbf{N}_j) - (\nabla \times \mathbf{N}_i) \cdot \frac{1}{\mu_r}\left(\hat{\mathbf{k}}_t^s \times \mathbf{N}_j\right)\right]\mathrm{d}\mathbf{r} \\ &\quad + \int_\Omega \mu_0\sigma\mathbf{N}_i \cdot \mathbf{N}_j\,\mathrm{d}\mathbf{r}, \end{aligned} \tag{14.96}$$

$$\begin{aligned} R^{\mathrm{ABC}} &= \int_{\partial\Omega'} \left[\frac{\cos\theta_s}{c}(\mathbf{n} \times \mathbf{N}_i) \cdot (\mathbf{n} \times \mathbf{N}_j)\right. \\ &\quad \left. + \frac{1}{c\cos\theta_s}\left(\hat{\mathbf{k}}_t^s \cdot \mathbf{N}_i\right)\left(\hat{\mathbf{k}}_t^s \cdot \mathbf{N}_j\right)\right]\mathrm{d}s, \end{aligned} \tag{14.97}$$

$$S_{ij} = \int_{\Omega} \frac{1}{\mu} (\nabla \times \mathbf{N}_i) \cdot (\nabla \times \mathbf{N}_j) \, d\mathbf{r}, \tag{14.98}$$

$$f_i = \int_{\Omega} \mathbf{N}_i \cdot \mathbf{G}(\mathbf{J}_{e,\mathrm{imp}}, \mathbf{J}_{m,\mathrm{imp}}) d\mathbf{r}, \tag{14.99}$$

with $\partial\Omega'$ consisting of the top and bottom surfaces of the computational domain (ABC boundaries), and the boundary condition (14.92) has been used to derive (14.97).

14.3 Physics of Surface Plasmon Waves

Surface plasmon waves are electron density waves confined to the interface between a metallic material and a dielectric medium [35], which are transverse electric (TE) wave solutions to Maxwell's equations for a flat interface. There are two types of plasmon waves, propagating and localized [36, 37].

14.3.1 *Propagating Plasmons on Planar Surfaces*

Let x be the normal direction of the interface, let y be along the interface, and let z point into the xy-plane as in Fig. 14.3. A TE$_z$ wave has the following surface evanescent nature:

$$\mathbf{E}^{\alpha} = (E_x^{\alpha}, E_y^{\alpha}, 0) e^{-\delta_\alpha |x|} e^{i(k_\alpha y - \omega t)}, \tag{14.100}$$

$$\mathbf{H}^{\alpha} = (0, 0, H_z^{\alpha}) e^{-\delta_\alpha |x|} e^{i(k_\alpha y - \omega t)}, \tag{14.101}$$

where $\alpha = 1$ is for the field in the metallic medium ($x < 0$) with relative dielectric constant ϵ_1, and $\alpha = 2$ for the field inside the dielectric material ($x > 0$) with relative dielectric constant ϵ_2.

The Ampère–Maxwell law (8.15) implies that

$$-\delta_1 H_z^1 = i\omega\epsilon_1\epsilon_0 E_y^1, \tag{14.102}$$

$$\delta_2 H_z^2 = i\omega\epsilon_2\epsilon_0 E_y^2, \tag{14.103}$$

and the Helmholtz equations for each field component imply that

$$\delta_\alpha = \sqrt{k_\alpha^2 - \epsilon_\alpha k_0^2}, \quad \alpha = 1, 2, \tag{14.104}$$

where the wave number in the vacuum is defined as

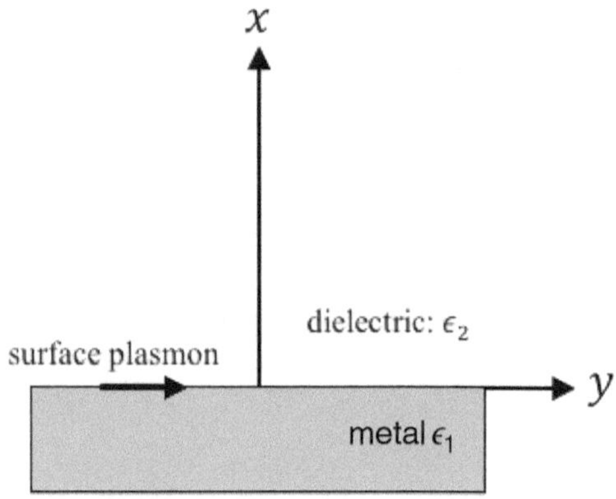

Fig. 14.3 A metal and dielectric interface

$$k_0^2 = \frac{\omega^2}{c^2}. \tag{14.105}$$

Meanwhile, we have the continuity of the tangential fields E_y^α and H_z^α, i.e.,

$$E_y^1 = E_y^2, \tag{14.106}$$

$$H_z^1 = H_z^2. \tag{14.107}$$

Using (14.102) and (14.103) in (14.106) to replace the E fields yields

$$\frac{\delta_1}{\epsilon_1} H_z^1 + \frac{\delta_2}{\epsilon_2} H_z^2 = 0. \tag{14.108}$$

Therefore, (14.107) and (14.108) imply that

$$\frac{\delta_1}{\epsilon_1} + \frac{\delta_2}{\epsilon_2} = 0, \tag{14.109}$$

and the continuity of H_z in (14.107) also means that the wave number k should be the same for both media:

$$k_1 = k_2 \equiv k. \tag{14.110}$$

Solving (14.109) and (14.110) gives the dispersion relation $\omega = \omega(k)$ for the surface plasmon:

$$k^2 = \frac{\omega^2}{c^2} \frac{\epsilon_1 \epsilon_2}{\epsilon_1 + \epsilon_2}. \tag{14.111}$$

For a propagating wave along the metal/dielectric interface, a positive k value is required. Due to the negative real part of the metallic dielectric constant for the frequency range considered, we have the following conditions for the existence of the surface plasmon:

$$\epsilon_1 \epsilon_2 < 0, \tag{14.112}$$

$$\epsilon_1 + \epsilon_2 < 0. \tag{14.113}$$

The exact dispersion curve for the surface plasmon depends on the specific frequency-dependent dielectric constant for the metallic material. For instance, in a Drude–Sommerfeld model [38, Chap. 10] based on a free electron gas (no correlation effects among electrons considered), we have

$$\epsilon_1 = \epsilon_1(\omega) = \epsilon_0 \left(1 - \frac{\omega_p^2}{\omega^2 + \Gamma\omega}\right), \tag{14.114}$$

where ω_p is the plasmon frequency for an electron gas of charge e and density n with electron mass m_e, given by:

$$\omega_p = \sqrt{\frac{ne^2}{m_e}}. \tag{14.115}$$

Thus, for a metal of the Drude-Sommerfeld model of (14.114) with damping parameter $\Gamma = 0$, the dispersion relation for the surface plasmon is given by

$$\omega_{sp}(k) = \sqrt{\omega_s^2 + k^2c^2 - \sqrt{\omega_s^4 + k^4c^4}}, \tag{14.116}$$

as shown in Fig. 14.4 and the surface plasmon frequency ω_s is defined as

$$\omega_s = \frac{\omega_p}{\sqrt{2}}. \tag{14.117}$$

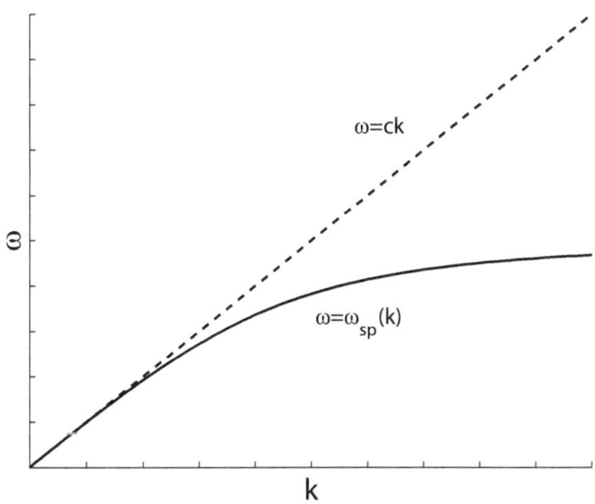

Fig. 14.4 Dispersion curves: (solid) surface plasmon $\omega = \omega_{sp}(k)$ and (dashed) light in the vacuum $\omega = ck$

As seen in Fig. 14.4, the dispersion curve for the surface plasmon is to the right of the light dispersion curve in the dielectric medium (or vacuum), so the surface plasmon on a smooth flat interface cannot be excited by the light radiation from the dielectric medium (or vacuum). In order to cause the excitation of the surface plasmon in the metallic material, two approaches can be used to produce the intersection of the light dispersion line with that of the surface plasmon. The first one is to pass the light through a prism such that a total internal reflection at the base of the prism will generate an evanescent light wave to the outside of the prism base, for which the dispersion curve of the light will be effectively rotated downward to intersect with that of the surface plasmon. The prism settings are achieved by either the Otto [39] or Kretschmann [40] methods, as seen in Fig. 14.5. The other way to excite the surface plasmon is to incorporate a periodic grating structure on the metallic surface to produce Bloch waves whose dispersion curve will be a folding over of (14.116). The resulting surface plasmon wave excited by the light radiation is called the surface plasmon polariton (SPP) [37].

14.3.2 Localized Surface Plasmons

Localized plasmon waves can be found in a metallic object of spherical or cylindrical shape, and they are easily excited by incident light radiation due to the curvature of the boundary of the object.

Consider a sphere of radius a with a complex dielectric constant ϵ_1 and permeability μ_1 embedded in a dielectric medium of dielectric constant ϵ_2 and permeability μ_2. The sphere is impacted by an incident plane wave polarized in the x-direction:

$$\mathbf{E}^{\text{inc}} = E_0 \mathrm{e}^{ikr\cos\theta} \mathbf{e}_x, \tag{14.118}$$

or, in terms of spherical coordinate vectors,

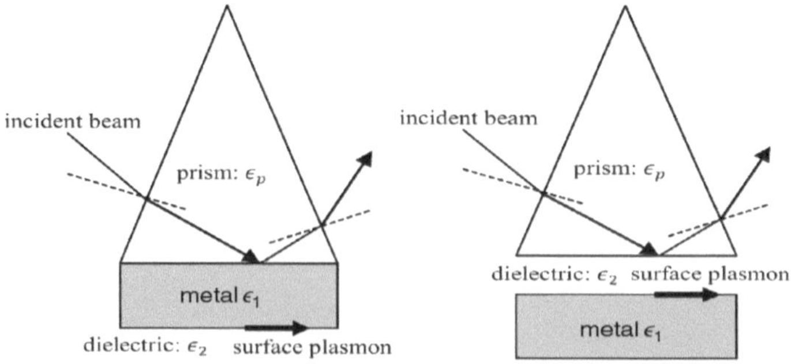

Fig. 14.5 Excitation of surface plasmon via a prism: (left) Kretschmann configuration, and (right) Otto configuration

$$\mathbf{e}_x = \sin\theta\cos\phi\mathbf{e}_r + \cos\theta\cos\phi\mathbf{e}_\theta - \sin\phi\mathbf{e}_\phi. \tag{14.119}$$

The scattering wave off the sphere can be expressed [41, 42] as follows:

$$\mathbf{E}^s = \sum_{n=1}^{\infty} E_n(\mathrm{i}a_n\mathbf{N}_{e1n} - b_n\mathbf{M}_{o1n}), \tag{14.120}$$

$$\mathbf{H}^s = \frac{k_2}{\omega\mu_2}\sum_{n=1}^{\infty} E_n(\mathrm{i}b_n\mathbf{N}_{o1n} + a_n\mathbf{M}_{e1n}), \tag{14.121}$$

where $k_2 = \omega\sqrt{\epsilon_2\mu_2}$, the vector functions $\mathbf{M}$ and $\mathbf{N}$ are defined through scalar potentials as

$$\mathbf{M}_{emn} = \nabla\times(\mathbf{r}\psi_{emn}), \qquad \mathbf{M}_{omn} = \nabla\times(\mathbf{r}\psi_{omn}), \tag{14.122}$$

$$\mathbf{N}_{emn} = \frac{1}{k_2}\nabla\times\mathbf{M}_{emn}, \qquad \mathbf{N}_{omn} = \frac{1}{k_2}\nabla\times\mathbf{M}_{omn}, \tag{14.123}$$

and

$$\psi_{emn} = \cos m\phi P_n^m(\cos\theta)h_n^{(2)}(k_2r), \tag{14.124}$$

$$\psi_{omn} = \sin m\phi P_n^m(\cos\theta)h_n^{(2)}(k_2r), \tag{14.125}$$

where $h_n^{(2)}(r)$ is the spherical Hankel function of the second kind.

The scattering coefficients in (14.120) and (14.121) can be obtained by the interface condition of the electromagnetic fields:

$$a_n = \frac{\mu_2 m^2 j_n(mx)[xj_n(x)]' - \mu_1 j_n(x)[mxj_n(mx)]'}{\mu_2 m^2 j_n(mx)[xh_n^{(2)}(x)]' - \mu_1 h_n^{(2)}(x)[mxj_n(mx)]'}, \tag{14.126}$$

$$b_n = \frac{\mu_1 j_n(mx)[xj_n(x)]' - \mu_2 j_n(x)[mxj_n(mx)]'}{\mu_1 j_n(mx)[xh_n^{(2)}(x)]' - \mu_2 h_n^{(2)}(x)[mxj_n(mx)]'}, \tag{14.127}$$

where

$$x = k_2 a, \quad m = \frac{k_1}{k_2}, \tag{14.128}$$

and the wave numbers are defined as $k_\alpha^2 = \omega^2\epsilon_\alpha\mu_\alpha$, $\alpha = 1, 2$.

A specific nth mode in the scattering field will be dominant if the denominator in either a_n or b_n becomes vanishingly small. Using the asymptotics of the spherical Hankel function for small argument (small sphere $x \sim 0$), the denominator of a_n will vanish approximately once

$$\epsilon_1 n + \epsilon_2(n+1) = 0, \tag{14.129}$$

which yields a Mie plasmon frequency

$$\omega_n = \omega_p \sqrt{\frac{n}{2n+1}}, \tag{14.130}$$

provided that the dielectric constant of the sphere is given by the Drude formula (14.114).

It can also be easily checked [42, Sect. 12.1] that the radial component of the Mie plasmon mode inside the sphere has the following profile along the radial direction:

$$\mathbf{E}^1(r) \propto r^{n-1}, \tag{14.131}$$

which is confined to the spherical surface; therefore, these Mie plasmon modes are identified as the *localized surface plasmons*.

14.4 Desingularized Volume Integral Equation (VIE) for Maxwell's Equations

In this section, we present a volume integral equation method based on the dyadic Green's function from Chap. 9 for plasmon waves in layered metallic structures. A typical structure is given in Fig. 14.6, where the inclusion Ω in the form of holes or impurities of nano-scales is made into background layered metallic materials. The derivation of the volume integral equation is based on the vector identity (11.14) in Sect. 11.1.1.

Firstly, we have

$$\mathcal{L}\mathbf{E}(\mathbf{r}) - \omega^2\epsilon(\mathbf{r})\mathbf{E}(\mathbf{r}) = -\mathrm{i}\omega\mathbf{J}_e(\mathbf{r}), \quad \mathbf{r} \in \mathbb{R}^3 \backslash (\Sigma \cup \partial\Omega), \tag{14.132}$$

where Σ consists of all the interfaces of the layered materials,

$$\mathcal{L} = \nabla \times \frac{1}{\mu} \nabla \times,$$

and $\mathbf{J}_e(\mathbf{r})$ is the far-field source (assumed to be away from the layered structure), which produces the incident waves impinging on the layered structure from the top, i.e.,

$$\mathbf{E}^{\mathrm{inc}}(\mathbf{r}) = -\mathrm{i}\omega\mu(\mathbf{r}) \int_{\mathbb{R}^3} \overline{\mathbf{G}}_E(\mathbf{r}, \mathbf{r}') \cdot \mathbf{J}_e(\mathbf{r}')d\mathbf{r}', \tag{14.133}$$

and $\overline{\mathbf{G}}_E(\mathbf{r}, \mathbf{r}')$ is the dyadic Green's function for the layered media. The inclusion Ω is characterized by a different dielectric constant from the layered background dielectrics, i.e.,

$$\epsilon(\mathbf{r}) = \epsilon_L(\mathbf{r}) + \Delta\epsilon(\mathbf{r}), \tag{14.134}$$

where $\Delta\epsilon(\mathbf{r}) = 0, \mathbf{r} \notin \Omega$. Then, (14.132) can be rewritten as

$$\mathcal{L}\mathbf{E}(\mathbf{r}) - \omega^2\epsilon_L(\mathbf{r})\mathbf{E}(\mathbf{r}) = -\mathrm{i}\omega\mathbf{J}(\mathbf{r}), \tag{14.135}$$

where

$$\mathbf{J}(\mathbf{r}) = \mathbf{J}_e(\mathbf{r}) + \mathbf{J}_{\mathrm{eq}}(\mathbf{r}), \tag{14.136}$$

and the equivalent current source $\mathbf{J}_{\mathrm{eq}}(\mathbf{r})$ is defined to reflect the existence of the inclusion Ω:

$$\mathbf{J}_{\mathrm{eq}}(\mathbf{r}) = \mathrm{i}\omega\Delta\epsilon(\mathbf{r})\mathbf{E}(\mathbf{r}). \tag{14.137}$$

Let us consider any interior point inside the inclusion $\mathbf{r}' \in \Omega$ and a small volume $V_\delta = V_\delta(\mathbf{r}') \subset \Omega$ centered at $\mathbf{r}'$. The dyadic Green's function $\overline{\mathbf{G}}_E(\mathbf{r}, \mathbf{r}')$ satisfies

$$\mathcal{L}\overline{\mathbf{G}}_E(\mathbf{r}, \mathbf{r}') - \omega^2\epsilon_L(\mathbf{r})\overline{\mathbf{G}}_E(\mathbf{r}, \mathbf{r}') = \frac{1}{\mu(\mathbf{r})}\overline{\mathbf{I}}\delta(\mathbf{r} - \mathbf{r}'),\ \mathbf{r} \in \mathbb{R}^3. \tag{14.138}$$

On multiplying (14.135) by $\overline{\mathbf{G}}_E(\mathbf{r}, \mathbf{r}')$ and (14.138) by $\mathbf{E}(\mathbf{r})$ and forming the difference, and then integrating over the domain $\mathbb{R}^3 \backslash V_\delta$, with the help of the identity (11.14) and Gauss's theorem (following the same procedure leading to (11.28) in Sect. 11.1.1), we arrive at the following (after switching $\mathbf{r}$ and $\mathbf{r}'$):

$$\begin{aligned} &-\mathrm{i}\omega\mu(\mathbf{r}) \int_{\mathbb{R}^3\backslash V_\delta} d\mathbf{r}'\, \overline{\mathbf{G}}_E(\mathbf{r}, \mathbf{r}') \cdot \mathbf{J}(\mathbf{r}') - \mu(\mathbf{r}) \int_{S_\delta} ds' \Big[\mathrm{i}\omega\, \overline{\mathbf{G}}_E(\mathbf{r}, \mathbf{r}') \cdot \big(\mathbf{n} \times \mathbf{H}(\mathbf{r}')\big) \\ &- \frac{1}{\mu(\mathbf{r}')}\nabla \times \overline{\mathbf{G}}_E(\mathbf{r}, \mathbf{r}') \cdot \big(\mathbf{n} \times \mathbf{E}(\mathbf{r}')\big)\Big] = \mathbf{0}, \qquad \mathbf{r} \in \Omega, \end{aligned} \tag{14.139}$$

where $S_\delta = \partial V_\delta(\mathbf{r})$, $\mathbf{n}$ is the normal of S_δ pointing out of $V_\delta(\mathbf{r})$, and the right-hand side being zero is due to the fact that the singularity of the delta function is outside the integration domain.

As $\delta \to 0$, the first integral will approach the Cauchy principal value of a singular integral, while the boundary integrals will in fact depend on the geometric shape of the volume V_δ. Without loss of generality, we can assume the dyadic Green's function on S_δ to be the dyadic Green's function for a homogeneous medium:

$$\overline{\mathbf{G}}_{\mathrm{E}}(\mathbf{r}, \mathbf{r}') = \overline{\mathbf{G}}_{\mathrm{E}}(\mathbf{r}', \mathbf{r}) = \left(\overline{\mathbf{I}} + \frac{1}{k^2}\nabla\nabla\right) g(\mathbf{r}, \mathbf{r}'), \tag{14.140}$$

where $k^2 = \omega^2\epsilon_i\mu_i$, $\epsilon_i = \epsilon_L(\mathbf{r})$,

$$g(\mathbf{r}, \mathbf{r}') = \frac{1}{4\pi}\frac{e^{-ikR}}{R}, \qquad R = |\mathbf{r} - \mathbf{r}'|. \tag{14.141}$$

In order to estimate the surface integrals, we have the following asymptotics for small $kR \ll 1$:

$$\overline{\mathbf{G}}_{\mathrm{E}}(\mathbf{r},\mathbf{r}') = \frac{1}{4\pi k^2 R^3}(\mathbf{I} - 3\mathbf{u}_R\mathbf{u}_R) + O\left(\frac{1}{R^2}\right), \tag{14.142}$$

$$\nabla' \times \overline{\mathbf{G}}_{\mathrm{E}}(\mathbf{r},\mathbf{r}') = \frac{1}{4\pi R^2}\mathbf{u}_R \times \mathbf{I} + O\left(\frac{1}{R}\right), \tag{14.143}$$

where $\mathbf{u}_R = (\mathbf{r}' - \mathbf{r})/R$, which implies that [43], [44, p. 87]:

$$\lim_{\delta\to 0}\int_{S_\delta} \mathrm{d}s'\, \mathbf{n}\times\mathbf{E}(\mathbf{r}')\cdot\nabla\times\overline{\mathbf{G}}_E(\mathbf{r}',\mathbf{r}) = -\left[\mathbf{I} - \mathbf{L}_{V_\delta}\right]\cdot\mathbf{E}(\mathbf{r}), \tag{14.144}$$

$$\lim_{\delta\to 0}\int_{S_\delta} \mathrm{d}s'\, \mathbf{n}\times\mathbf{H}(\mathbf{r}')\cdot\overline{\mathbf{G}}_E(\mathbf{r}',\mathbf{r}) = -\frac{1}{k^2}\mathbf{L}_{V_\delta}\cdot\nabla\times\mathbf{H}(\mathbf{r}), \tag{14.145}$$

and the **L**-dyadics for V_δ of various geometric shapes are given as follows [45]:

$$\mathbf{L}_{V_\delta} = \frac{1}{3}\mathbf{I} \tag{14.146}$$

for a sphere;

$$\mathbf{L}_{V_\delta} = \mathrm{diag}(\alpha, \alpha, 1-\alpha) \tag{14.147}$$

for a cylinder with radius a and height $2h$, $\alpha = h/(2\sqrt{a^2+h^2})$; and

$$\mathbf{L}_{V_\delta} = \frac{1}{4\pi}\mathrm{diag}(\Omega_x, \Omega_y, \Omega_z) \tag{14.148}$$

for a rectangular box where Ω_x, Ω_y, and Ω_z are twice the solid angle subtended at point **r** by the side of the rectangular box perpendicular to the x-, y-, and z-directions, respectively.

Substituting (14.144) and (14.145) into (14.139), we have

$$\begin{aligned}\mathbf{0} =& -\mathrm{i}\omega\mu(\mathbf{r})\int_{\mathbb{R}^3\setminus V_\delta} \mathrm{d}\mathbf{r}'\,\mathbf{J}(\mathbf{r}')\cdot\overline{\mathbf{G}}_E(\mathbf{r}',\mathbf{r}) \\ &+ \mathrm{i}\omega\mu(\mathbf{r})\frac{1}{k^2}\mathbf{L}_{\mathbf{V}_\delta}\cdot\nabla\times\mathbf{H}(\mathbf{r}) - \left[\mathbf{I} - \mathbf{L}_{\mathbf{V}_\delta}\right]\cdot\mathbf{E}(\mathbf{r}), \quad \mathbf{r}\in\Omega,\end{aligned} \tag{14.149}$$

or

$$\begin{aligned}\mathbf{0} =& \mathbf{E}^{\mathrm{inc}}(\mathbf{r}) - \mathrm{i}\omega\mu(\mathbf{r})\mathrm{p.v.}\int_{\mathbb{R}^3} \mathrm{d}\mathbf{r}'\mathbf{J}_{eq}(\mathbf{r}')\cdot\overline{\mathbf{G}}_E(\mathbf{r}',\mathbf{r}) \\ &+ \mathbf{L}_{\mathbf{V}_\delta}\cdot\left[\mathrm{i}\omega\mu(\mathbf{r})\frac{1}{k^2}\nabla\times\mathbf{H}(\mathbf{r}') + \mathbf{E}(\mathbf{r})\right] - \mathbf{E}(\mathbf{r}), \quad \mathbf{r}\in\Omega.\end{aligned} \tag{14.150}$$

After using Ampère's law (8.24), we obtain

$$\mathbf{0} = \mathbf{E}^{\text{inc}}(\mathbf{r}) - \mathrm{i}\omega\mu(\mathbf{r})\text{p.v.} \int_{\mathbb{R}^3} \mathrm{d}\mathbf{r}' \, \mathbf{J}_{eq}(\mathbf{r}') \cdot \overline{\mathbf{G}}_E(\mathbf{r}', \mathbf{r})$$
$$+ \mathbf{L}_{\mathbf{V}_\delta} \cdot \left(\frac{\mathrm{i}}{\omega} \mathbf{J} \right) - \mathbf{E}(\mathbf{r}), \quad \mathbf{r} \in \Omega. \tag{14.151}$$

Finally, noting that $\mathbf{J}(\mathbf{r}) = \mathbf{J}_{\text{eq}}(\mathbf{r}) = \mathrm{i}\omega\Delta\epsilon(\mathbf{r})\mathbf{E}(\mathbf{r})$ inside Ω, we have the volume integral equation for $\mathbf{r} \in \Omega$:

$$\mathbf{C} \cdot \mathbf{E}(\mathbf{r}) = \mathbf{E}^{\text{inc}}(\mathbf{r}) + \mu(\mathbf{r})\omega^2 \text{ p.v.} \int_{\Omega} \mathrm{d}\mathbf{r}' \Delta\epsilon(\mathbf{r}')\mathbf{E}(\mathbf{r}') \cdot \overline{\mathbf{G}}_E(\mathbf{r}', \mathbf{r}), \tag{14.152}$$

where the coefficient matrix is given by

$$\mathbf{C} = \mathbf{I} + \mathbf{L}_{V_\delta} \cdot \Delta\epsilon(\mathbf{r}). \tag{14.153}$$

The Cauchy principle integral in (14.152) is defined through a limiting process of taking singularity exclusion volume $V_\delta(\mathbf{r})$ to zero. In practical computation, a finite size V_δ is usually used, and the VIE (14.152) can be reformulated with the singularity removed as follows [46]: for $\mathbf{r} \in \Omega$

$$\begin{aligned}\mathbf{C} \cdot \mathbf{E}(\mathbf{r}) = \mathbf{E}^{\text{inc}}(\mathbf{r}) &+ \mu\omega^2 \int_{\Omega \backslash V_\delta} d\mathbf{r}' \Delta\epsilon(\mathbf{r}')\mathbf{E}(\mathbf{r}') \cdot \overline{\mathbf{G}}^*_{\mathbf{E}}(\mathbf{r}, \mathbf{r}') \\
&+ \mu\omega^2 \int_{\Omega} \mathrm{d}\mathbf{r}' \Delta\epsilon(\mathbf{r}')\mathbf{E}(\mathbf{r}') \cdot \overline{\mathbf{G}}^{\text{reg}}_{\mathbf{E}}(\mathbf{r}, \mathbf{r}') \\
&+ \mu\omega^2 \int_{V_\delta(\mathbf{r})} \mathrm{d}\mathbf{r}' \Delta\epsilon(\mathbf{r}')\mathbf{E}(\mathbf{r}') \cdot \left(\overline{\mathbf{G}}^*_{\mathbf{E}}(\mathbf{r}, \mathbf{r}') - \frac{1}{k^2} \nabla\nabla g_0 \right) \\
&+ \frac{\mu\omega^2}{k^2} \int_{V_\delta(\mathbf{r})} \mathrm{d}\mathbf{r}' \nabla\nabla g_0(\mathbf{r}, \mathbf{r}') \left[\Delta\epsilon(\mathbf{r}')\mathbf{E}(\mathbf{r}') - \Delta\epsilon(\mathbf{r})\mathbf{E}(\mathbf{r}) \right],\end{aligned} \tag{14.154}$$

where

$$g_0(\mathbf{r}, \mathbf{r}') = \frac{1}{4\pi|\mathbf{r} - \mathbf{r}'|}, \tag{14.155}$$

and the dyadic Green's function $\overline{\mathbf{G}}_{\mathbf{E}}(\mathbf{r}, \mathbf{r}')$ is separated into a smooth regularized part and a singular part, i.e.,

$$\overline{\mathbf{G}}_{\mathbf{E}}(\mathbf{r}, \mathbf{r}') = \overline{\mathbf{G}}^*_{\mathbf{E}}(\mathbf{r}, \mathbf{r}') + \overline{\mathbf{G}}^{\text{reg}}_{\mathbf{E}}(\mathbf{r}, \mathbf{r}'), \tag{14.156}$$

with $\overline{\mathbf{G}}^*_{\mathrm{E}}(\mathbf{r}, \mathbf{r}')$ being the free space Green's function

$$\overline{\mathbf{G}}^*_{\mathrm{E}}(\mathbf{r}, \mathbf{r}') = \left(\overline{\mathbf{I}} + \frac{1}{k^2}\nabla\nabla\right) g(\mathbf{r}, \mathbf{r}'), \tag{14.157}$$

and

$$g(\mathbf{r}, \mathbf{r}') = \frac{1}{4\pi}\frac{\mathrm{e}^{-\mathrm{i}kR}}{R}, \qquad R = |\mathbf{r} - \mathbf{r}'|. \tag{14.158}$$

The VIE (14.154) holds for any finite $\delta > 0$ as long as $V_\delta(\mathbf{r}) \subset \Omega$, and all the involved integrals are well-defined provided that $\Delta\epsilon(\mathbf{r})\mathbf{E}(\mathbf{r})$ is Hölder continuous. The last two terms in Eq. (14.154) are the correction terms for computing the Cauchy principal value with a finite-sized exclusion volume $V_\delta(\mathbf{r})$, and both are weakly singular integrals of removable singularities by a spherical coordinate transform and of order $O(\delta^2)$ [47].

14.5 Extraordinary Optical Transmission (EOT) in Thin Metallic Films

The volume integral equation (14.152) can be applied to simulate the transmission of optical waves through corrugated thin metallic films and subwavelength-aperture arrays in metal plates, which have been shown to have extraordinary transmissions due to the surface plasmons [48, 49] . A typical 2-D structure (assuming a long dimension in the in-plane y-direction) is shown in Fig. 14.6, in which (plasmon) pits are etched on both sides or one side of the metal films. Plasmon pits can be described by the perturbation $\Delta\epsilon(\mathbf{r})$ of their dielectric constants in (14.134) from the background layered media (composed of air and silver with a refractive index $n_{\mathrm{Ag}} = 0.05 - 2.87\mathrm{i}$).

The transmission of the power of the incident optical signal (taken to be a Gaussian beam of 500 nm wavelength) is measured by means of the optical transmission coefficient T:

$$T = \frac{\displaystyle\int_{-\infty}^{\infty} S_z \,\mathrm{d}x}{\displaystyle Y_0 \int_{-\infty}^{\infty} |\mathbf{E}^{\mathrm{inc}}(x, z)| \mathrm{d}x},$$

where $Y_0 = \sqrt{\epsilon_0/\mu_0}$ is the impedance of the air and S_z is the z-component of the time-averaged Poynting vector $\mathbf{S}$ measured at the top exit side, where

$$\langle \mathbf{S} \rangle = \frac{1}{2}\,\mathrm{Re}[\mathbf{E} \times \mathbf{H}^*].$$

Figure 14.7 shows the increased transmission for the three- and five-layer thin silver films with plasmon pits (left) compared with the single layer of silver plate

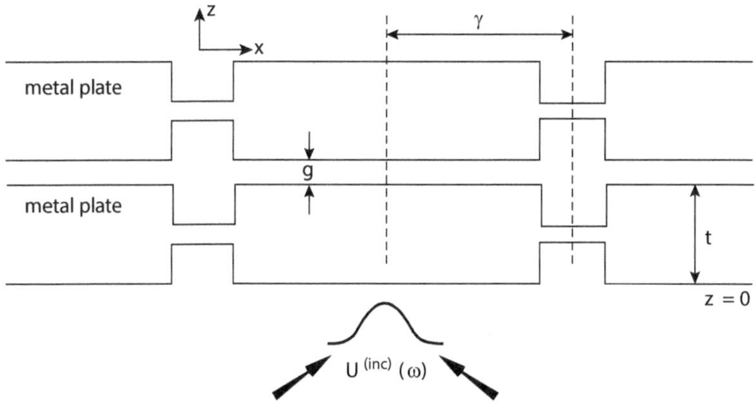

Fig. 14.6 Two thin silver films of thickness $t = 100$ nm (with a separation $g = 20$ nm) with pits to excite surface plasmons: the pits are 40 nm in width and height and separated by a spacing $\gamma = 80$ nm. Reprinted with permission from [50], © (2009) by Optical Society of America

with equivalent metal volume and thickness (right) [50]; the former shows a much larger transmission of optical power.

14.6 Discontinuous Galerkin Method for Resonant Plasmon Couplings

In a linear dispersive material, the constitutive relation between the displacement flux **D** and the electric field **E** and that between the magnetic inductance flux **B** and the magnetic field **H** are given as follows:

$$\mathbf{D} = \epsilon_0 \epsilon_{r,\infty} \mathbf{E} + \mathbf{P}, \tag{14.159}$$

$$\mathbf{B} = \mu_0 \mu_{r,\infty} \mathbf{H} + \mathbf{M}, \tag{14.160}$$

where ϵ_0 and μ_0 are the electric permittivity and the magnetic permeability of free space, $\epsilon_{r,\infty}$ and $\mu_{r,\infty}$ are the relative electric permittivity and the relative magnetic permeability of the medium at infinite frequency, and **P** and **M** are the electric polarization and the magnetization densities, respectively.

As discussed in (1.15), in a linear and isotropic medium,

$$\hat{\mathbf{P}} = \epsilon_0 \chi(\omega) \hat{\mathbf{E}}, \tag{14.161}$$

where $\chi(\omega)$ is the electric susceptibility of the medium in the frequency domain and $\hat{\mathbf{P}}$ denotes the time-Fourier transform of $\mathbf{P}(\mathbf{t})$. Here we assume that the magnetization **M** is zero, i.e., there are no magnetic effects.

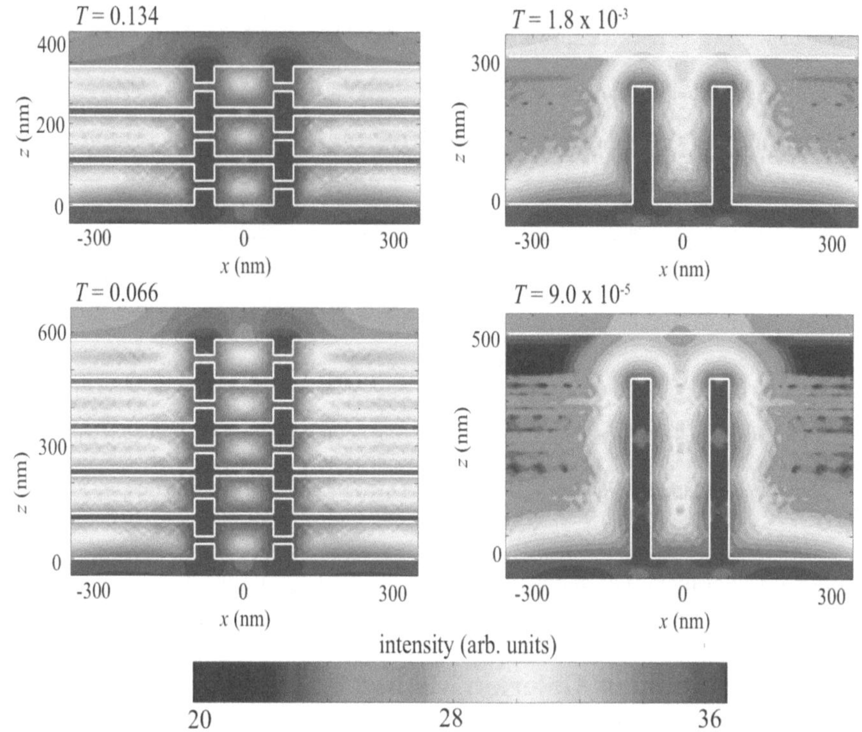

Fig. 14.7 Left: the transmission of three or five thin silver films with the same spacing as in Fig. 14.6; right: the transmission of one silver plate with the same metal volume and the thickness. The darker color above the grating indicates a stronger field strength. Reprinted with permission from [50], © (2009) by Optical Society of America

Maxwell's equations in dispersive media can be recast for the field variables $\mathbf{U}^{(1)}$ with auxiliary polarization current variables $\mathbf{U}^{(2)}$ as

$$\frac{\partial \mathbf{U}}{\partial t} + \nabla \cdot (\bar{\mathbb{A}}\mathbf{U}) = \mathbf{S}, \tag{14.162}$$

where $\mathbf{U} = (\mathbf{U}^{(1)}, \mathbf{U}^{(2)})^{\mathrm{T}}$ is defined explicitly in Sect. 14.7. We can divide the above conservation system into two parts:

$$\frac{\partial \mathbf{U}^{(1)}}{\partial t} + \nabla \cdot (\mathbb{A}\mathbf{U}^{(1)}) = \mathbf{S}^{(1)}, \tag{14.163}$$

$$\frac{\partial \mathbf{U}^{(2)}}{\partial t} = \mathbf{S}^{(2)}, \tag{14.164}$$

and $\bar{\mathbb{A}}$, $\mathbb{A}$, and $\mathbf{S} = (\mathbf{S}^{(1)}, \mathbf{S}^{(2)})^{\mathrm{T}}$ are given in (14.212)–(14.216) or (14.218)–(14.222) of the appendix (Sect. 14.7). The conservation law (14.162) can be solved by the DG method described in Sect. 13.2.

Resonant coupling of local plasmons in nano-silver wires

Firstly, we study the cross-section and the time-domain behavior of coupled plasmon resonant modes for coupled Ag nano-wires [51]. The integration path for the calculation of the cross-section [42] is selected to be a circle of large radius r_∞, which will encircle all the Ag cylindrical nano-wires. The circle does not have to be very large as all Ag nano-wires are placed in the non-absorbing free space.

After choosing the radius r_∞, we compute the time-averaged Poynting vector $\mathbf{S}$ along this circle using

$$\mathbf{S}_{\mathrm{sca}} = \frac{1}{2}\mathrm{Re}\{\mathbf{E}_{\mathrm{sca}} \times \mathbf{H}_{\mathrm{sca}}^{*}\}, \quad \mathbf{S}_{\mathrm{ext}} = \frac{1}{2}\mathrm{Re}\{\mathbf{E}_{\mathrm{i}} \times \mathbf{H}_{\mathrm{sca}}^{*} + \mathbf{E}_{\mathrm{sca}} \times \mathbf{H}_{\mathrm{i}}^{*}\},$$

where the footnote "sca" represents "scattering", "i" represents "incident", and "ext" represents "extinction"; all the variables are functions of frequency, obtained by the Fourier transform from the time domain. Then we compute the following values:

$$W_{\mathrm{sca}} = \int_A \mathbf{S}_{\mathrm{sca}} \cdot \hat{\mathbf{e}}_r \mathrm{d}s, \quad W_{\mathrm{ext}} = -\int_A \mathbf{S}_{\mathrm{ext}} \cdot \hat{\mathbf{e}}_r \mathrm{d}s, \quad W_{\mathrm{abs}} = W_{\mathrm{ext}} - W_{\mathrm{sca}},$$

where "abs" represents "absorption", the integration is performed along the circle, and $\hat{\mathbf{e}}_r$ is the outward unit normal to the circle. Finally, we get the cross-sections:

$$C_{\mathrm{ext}} = \frac{W_{\mathrm{ext}}}{I_{\mathrm{i}}}, \quad C_{\mathrm{abs}} = \frac{W_{\mathrm{abs}}}{I_{\mathrm{i}}}, \quad C_{\mathrm{sca}} = \frac{W_{\mathrm{sca}}}{I_{\mathrm{i}}},$$

where I_{i} is the incident irradiance.

We use curved triangles to describe the circle exactly; these curved triangles can be mapped by an isoparametric transformation [52] onto a reference triangle $\{(x, y) : 0 \le x, y, x + y \le 1\}$. We set the UPML parameters $m = 3$, $R(0) = \exp(-16)$ in (8.163) and (8.164), fourth-order basis functions are used for the space discretization, and a fourth-order Runge–Kutta method is used for the time integration. A Gaussian pulsed source modulating a carrier wave is used to excite the plasmon modes inside the nano-wires, and the pulse can be obtained as follows [53].

1. Assume the time dependence for the pulse is

$$f(t) = \exp[\mathrm{i}\omega_c(t - t_0)] \exp\left[\left(\frac{t - t_0}{t_{\mathrm{decay}}}\right)^2\right],$$

where ω_c is the central frequency.

2. Take the Fourier transform of $f(t)$:

$$\hat{f}(\omega) = \frac{t_{\text{decay}}}{\sqrt{2}} \exp\left(-\frac{(\omega_c - \omega)^2 t_{\text{decay}}^2}{4}\right) \exp(-\mathrm{i}t_0\omega).$$

3. Set

$$f(x, y, z, t) = \frac{1}{\sqrt{2\pi}} \int F(x, y, z, \omega) \hat{f}(\omega) \exp(\mathrm{i}\omega t) \mathrm{d}\omega,$$

   where $F(x, y, z, \omega)$ is the distribution of the desired mode in the case of the sinusoidal excitation mentioned above; $f(x, y, z, t)$ so defined will satisfy Maxwell's equations and contain a range of frequencies around ω_c.

Normal illumination of two nano-wires

We consider TE scattering off two Ag nano-wires. The Drude parameters for the Ag nano-wires are taken from [54]:

$$\epsilon_{r,\infty} = 8.926, \quad \omega_p = 11.585 \text{ eV}, \quad \gamma = 0.203 \text{ eV}.$$

The computed cross-sections [53] for two nano-wires with a radius $r = 15$ nm show that in addition to the one weak maximum close to the resonance of an isolated nano-wire, an additional stronger resonance due to the interaction of the nano-wires is also present, i.e., a distinct secondary resonance.

14.7 Appendix: Auxiliary Differential Equation (ADE) DG Methods for Dispersive Maxwell's Equations

The ADE formulation [55] for the dispersive Maxwell's equations will be derived for the discontinuous Galerkin method.

14.7.1 Debye Material

Consider an inhomogeneous, conductive, and electrically dispersive medium with relative magnetic permeability μ_r and conductivity σ. For a single-pole Debye medium, the electric susceptibility in the frequency domain can be expressed as the relative permittivity with a single pole:

$$\epsilon_r(\omega) = \epsilon_{r,\infty} + \frac{\epsilon_{r,s} - \epsilon_{r,\infty}}{1 + \mathrm{i}\omega\tau}, \tag{14.165}$$

where $\epsilon_{r,s}$ is the static zero-frequency relative electric permittivity and τ is the pole relaxation time. Then, a general time-harmonic form (with $\mathrm{e}^{\mathrm{i}\omega t}$ time dependence) of Maxwell's equations in a Debye medium including the artificial material in a uniaxial perfectly matched layer (UPML) can be written in a unified form as follows [56]:

$$\nabla \times \hat{\mathbf{H}} = \mathrm{i}\omega\epsilon_0 \left(\epsilon_r + \frac{\sigma}{\mathrm{i}\omega\epsilon_0}\right) \bar{\bar{\epsilon}}\, \hat{\mathbf{E}}, \tag{14.166}$$

$$\nabla \times \hat{\mathbf{E}} = -\mathrm{i}\omega\mu_0\mu_r \bar{\bar{\mu}}\, \hat{\mathbf{H}}, \tag{14.167}$$

where

$$\bar{\bar{\epsilon}} = \bar{\bar{\mu}} = \begin{bmatrix} \frac{s_y s_z}{s_x} & 0 & 0 \\ 0 & \frac{s_x s_z}{s_y} & 0 \\ 0 & 0 & \frac{s_x s_y}{s_z} \end{bmatrix} \tag{14.168}$$

and

$$s_i = 1 + \frac{\sigma_i}{\mathrm{i}\omega\epsilon_0}, \quad i = x, y, z. \tag{14.169}$$

Here, ϵ_r is the relative permittivity of the dispersive medium, $\sigma_i = 0$ corresponds to a physical dispersive medium, and $\sigma_i \neq 0$ is used for the UPML region.

For simplicity, we will consider the 2-D transverse magnetic (TM) case ($s_z = 1$); Maxwell's equations for the TM wave for the Debye medium can be written as

$$\frac{\partial \hat{H}_y}{\partial x} - \frac{\partial \hat{H}_x}{\partial y} = \mathrm{i}\omega\epsilon_0 \left(\epsilon_{r,\infty} + \frac{\epsilon_{r,s} - \epsilon_{r,\infty}}{1 + \mathrm{i}\omega\tau} + \frac{\sigma}{\mathrm{i}\omega\epsilon_0}\right) s_x s_y \hat{E}_z, \tag{14.170}$$

$$\frac{\partial \hat{E}_z}{\partial y} = -\mathrm{i}\omega\mu_0\mu_r \frac{s_y}{s_x} \hat{H}_x, \tag{14.171}$$

$$-\frac{\partial \hat{E}_z}{\partial x} = -\mathrm{i}\omega\mu_0\mu_r \frac{s_x}{s_y} \hat{H}_y. \tag{14.172}$$

Equation (14.170) can be written as

$$\frac{\partial \hat{H}_y}{\partial x} - \frac{\partial \hat{H}_x}{\partial y} = \mathrm{i}\omega\epsilon_0\epsilon_{r,\infty}\hat{E}_z + \sigma\hat{E}_z + \hat{J}_{z,1}(\omega) + \hat{J}_{z,2}(\omega) + \hat{J}_{z,3}(\omega) + \hat{J}_{z,4}(\omega), \tag{14.173}$$

where

$$\hat{J}_{z,1}(\omega) = \mathrm{i}\omega\epsilon_0\epsilon_{r,\infty}(s_x s_y - 1)\hat{E}_z, \tag{14.174}$$

$$\hat{J}_{z,2}(\omega) = \mathrm{i}\omega\epsilon_0 \frac{\epsilon_{r,s} - \epsilon_{r,\infty}}{1 + \mathrm{i}\omega\tau} \hat{E}_z, \tag{14.175}$$

$$\begin{aligned} \hat{J}_{z,3}(\omega) &= \mathrm{i}\omega\epsilon_0 \frac{\epsilon_{r,s} - \epsilon_{r,\infty}}{1 + \mathrm{i}\omega\tau}(s_x s_y - 1)\hat{E}_z \\ &= (s_x s_y - 1)\hat{J}_{z,2}(\omega) = \frac{\epsilon_{r,s} - \epsilon_{r,\infty}}{\epsilon_{r,\infty}(1 + \mathrm{i}\omega\tau)} \hat{J}_{z,1}(\omega), \end{aligned} \tag{14.176}$$

$$\hat{J}_{z,4}(\omega) = \sigma(s_x s_y - 1)\hat{E}_z = \frac{\sigma}{\mathrm{i}\omega\epsilon_0\epsilon_{r,\infty}}\hat{J}_{z,1}(\omega). \tag{14.177}$$

After substituting s_x and s_y from (14.169), we have

$$\begin{aligned}
\hat{J}_{z,1}(\omega) &= \mathrm{i}\omega\epsilon_0\epsilon_{r,\infty}\left[\left(1+\frac{\sigma_x}{\mathrm{i}\omega\epsilon_0}\right)\left(1+\frac{\sigma_y}{\mathrm{i}\omega\epsilon_0}\right)-1\right]\hat{E}_z \\
&= \mathrm{i}\omega\epsilon_0\epsilon_{r,\infty}\left(\frac{\sigma_x+\sigma_y}{\mathrm{i}\omega\epsilon_0}+\frac{\sigma_x\sigma_y}{(\mathrm{i}\omega\epsilon_0)^2}\right)\hat{E}_z \\
&= \epsilon_{r,\infty}(\sigma_x+\sigma_y)\hat{E}_z + \frac{\epsilon_{r,\infty}}{\mathrm{i}\omega\epsilon_0}\sigma_x\sigma_y\hat{E}_z, \qquad (14.178)
\end{aligned}$$

$$\hat{J}_{z,2}(\omega) = \mathrm{i}\omega\epsilon_0\frac{\epsilon_{r,s}-\epsilon_{r,\infty}}{1+\mathrm{i}\omega\tau}\hat{E}_z, \tag{14.179}$$

$$\hat{J}_{z,3}(\omega) = \frac{\epsilon_{r,s}-\epsilon_{r,\infty}}{\epsilon_{r,\infty}(1+\mathrm{i}\omega\tau)}\hat{J}_{z,1}, \tag{14.180}$$

$$\hat{J}_{z,4}(\omega) = \frac{\sigma}{\mathrm{i}\omega\epsilon_0\epsilon_{r,\infty}}\hat{J}_{z,1}. \tag{14.181}$$

Now considering (14.173), we apply the inverse Fourier transform using the identity $\mathrm{i}\omega f(\omega) \to (\partial/\partial t)f(t)$. This yields an equivalent equation of the time-domain differential equation for (14.173):

$$\frac{\partial H_y}{\partial x} - \frac{\partial H_x}{\partial y} = \epsilon_0\epsilon_{r,\infty}\frac{\partial E_z}{\partial t} + \sigma E_z + J_{z,1}(t) + J_{z,2}(t) + J_{z,3}(t) + J_{z,4}(t). \tag{14.182}$$

Next we derive dynamic equations for (14.174)–(14.177). The way to obtain a dynamic equation for $J_{z,1}$ is first to multiply both sides of (14.174) by $\mathrm{i}\omega$, which gives

$$\mathrm{i}\omega\hat{J}_{z,1} = \mathrm{i}\omega\epsilon_{r,\infty}(\sigma_x+\sigma_y)\hat{E}_z + \frac{\epsilon_{r,\infty}}{\epsilon_0}\sigma_x\sigma_y\hat{E}_z. \tag{14.183}$$

Exploiting the differentiation equivalence for the Fourier transform, we perform an inverse Fourier transform of each term in (14.183):

$$\frac{\partial J_{z,1}(t)}{\partial t} = \epsilon_{r,\infty}(\sigma_x+\sigma_y)\frac{\partial E_z}{\partial t} + \frac{\epsilon_{r,\infty}}{\epsilon_0}\sigma_x\sigma_y E_z. \tag{14.184}$$

To obtain the dynamic equation for $J_{z,2}$ from (14.175), we again multiply both sides of this equation by $(1+\mathrm{i}\omega\tau)$:

$$\hat{J}_{z,2} + \mathrm{i}\omega\tau\hat{J}_{z,2} = \mathrm{i}\omega\epsilon_0(\epsilon_{r,s}-\epsilon_{r,\infty})\hat{E}_z. \tag{14.185}$$

Similarly, using the inverse Fourier transform, we get

$$J_{z,2} + \tau \frac{\partial J_{z,2}}{\partial t} = \epsilon_0(\epsilon_{r,s} - \epsilon_{r,\infty}) \frac{\partial E_z}{\partial t}, \tag{14.186}$$

and, by the same token, from (14.176) and (14.177) we have

$$J_{z,3} + \tau \frac{\partial J_{z,3}}{\partial t} = \frac{\epsilon_{r,s} - \epsilon_{r,\infty}}{\epsilon_{r,\infty}} J_{z,1}, \tag{14.187}$$

$$\frac{\partial J_{z,4}}{\partial t} = \frac{\sigma}{\epsilon_0 \epsilon_{r,\infty}} J_{z,1}. \tag{14.188}$$

Therefore, (14.184), (14.186), (14.187), and (14.188) are the time-domain differential equations for all J_z.

We further simplify the differential equations for the polarization currents J_z so that they will become simply ordinary differential equations.

Considering (14.184), we introduce a new parameter

$$P_{z,1} = -J_{z,1} + \epsilon_{r,\infty}(\sigma_x + \sigma_y)E_z. \tag{14.189}$$

Then (14.184) can be written as

$$\frac{\partial P_{z,1}}{\partial t} = -\frac{\epsilon_{r,\infty}}{\epsilon_0} \sigma_x \sigma_y E_z. \tag{14.190}$$

Similarly for (14.186), on introducing a new parameter

$$P_{z,2} = -J_{z,2} + \frac{\epsilon_0(\epsilon_{r,s} - \epsilon_{r,\infty})}{\tau} E_z, \tag{14.191}$$

we have

$$\frac{\partial P_{z,2}}{\partial t} = \frac{1}{\tau} J_{z,2}. \tag{14.192}$$

From (14.189) and (14.191), we can obtain

$$J_{z,1} = -P_{z,1} + \epsilon_{r,\infty}(\sigma_x + \sigma_y)E_z, \tag{14.193}$$

$$J_{z,2} = -P_{z,2} + \frac{\epsilon_0(\epsilon_{r,s} - \epsilon_{r,\infty})}{\tau} E_z. \tag{14.194}$$

After substituting for $J_{z,1}$ and $J_{z,2}$ from (14.193) and (14.194) into (14.182), we have

$$\frac{\partial H_y}{\partial x} - \frac{\partial H_x}{\partial y} = \epsilon_0\epsilon_{r,\infty}\frac{\partial E_z}{\partial t} + \sigma E_z - P_{z,1} + \epsilon_{r,\infty}(\sigma_x + \sigma_y)E_z - P_{z,2} + \frac{\epsilon_0(\epsilon_{r,s} - \epsilon_{r,\infty})}{\tau}E_z + J_{z,3}(t) + J_{z,4}(t)$$

$$= \epsilon_0\epsilon_{r,\infty}\frac{\partial E_z}{\partial t} + \left[\sigma + \epsilon_{r,\infty}(\sigma_x + \sigma_y) + \frac{\epsilon_0(\epsilon_{r,s} - \epsilon_{r,\infty})}{\tau}\right]E_z - P_{z,1} - P_{z,2} + J_{z,3} + J_{z,4}. \tag{14.195}$$

Next, considering (14.192), after substituting for $J_{z,2}$ from (14.194), we have

$$\frac{\partial P_{z,2}}{\partial t} = -\frac{1}{\tau}P_{z,2} + \frac{\epsilon_0(\epsilon_{r,s} - \epsilon_{r,\infty})}{\tau^2}E_z. \tag{14.196}$$

Similarly for (14.187), after substituting for $J_{z,1}$ from (14.193), we have

$$\frac{\partial J_{z,3}}{\partial t} = -\frac{1}{\tau}J_{z,3} + \frac{\epsilon_{r,s} - \epsilon_{r,\infty}}{\epsilon_{r,\infty}\tau}\left[-P_{z,1} + \epsilon_{r,\infty}(\sigma_x + \sigma_y)E_z\right]$$
$$= -\frac{1}{\tau}J_{z,3} - \frac{\epsilon_{r,s} - \epsilon_{r,\infty}}{\epsilon_{r,\infty}\tau}P_{z,1} + \frac{(\epsilon_{r,s} - \epsilon_{r,\infty})(\sigma_x + \sigma_y)}{\tau}E_z. \tag{14.197}$$

Finally, for (14.188), after substituting for $J_{z,1}$ from (14.193), we have

$$\frac{\partial J_{z,4}}{\partial t} = -\frac{\sigma}{\epsilon_0\epsilon_{r,\infty}}P_{z,1} + \frac{\sigma(\sigma_x + \sigma_y)}{\epsilon_0}E_z. \tag{14.198}$$

Similar operations can be applied to Faraday's law (14.171) and (14.172), and we have the following equations:

$$\frac{\partial E_z}{\partial y} = -\mu_0\mu_r\frac{\partial H_x}{\partial t} - \mu_0\mu_r\frac{\sigma_y - \sigma_x}{\epsilon_0}H_x + Q_x, \tag{14.199}$$
$$-\frac{\partial E_z}{\partial x} = -\mu_0\mu_r\frac{\partial H_y}{\partial t} - \mu_0\mu_r\frac{\sigma_x - \sigma_y}{\epsilon_0}H_y + Q_y, \tag{14.200}$$
$$\frac{\partial Q_x}{\partial t} = -\frac{\sigma_x}{\epsilon_0}Q_x + \frac{\mu_0\mu_r\sigma_x(\sigma_y - \sigma_x)}{\epsilon_0^2}H_x, \tag{14.201}$$
$$\frac{\partial Q_y}{\partial t} = -\frac{\sigma_y}{\epsilon_0}Q_y + \frac{\mu_0\mu_r\sigma_y(\sigma_x - \sigma_y)}{\epsilon_0^2}H_y. \tag{14.202}$$

For consistency of notation, we let $P_{z,3} = J_{z,3}$ and $P_{z,4} = J_{z,4}$; then, from (14.195), (14.199), (14.200), (14.190), (14.196), (14.197), (14.198), (14.201), and (14.202), we get a new set of equations for E_z, H_x, H_y, $P_{z,1}$, $P_{z,2}$, $P_{z,3}$, $P_{z,4}$, Q_x, and Q_y as follows:

$$\begin{aligned}
\epsilon_0\epsilon_{r,\infty}\frac{\partial E_z}{\partial t} &= \frac{\partial H_y}{\partial x} - \frac{\partial H_x}{\partial y} \\
&\quad - \left[\sigma + \epsilon_{r,\infty}(\sigma_x + \sigma_y) + \frac{\epsilon_0(\epsilon_{r,s} - \epsilon_{r,\infty})}{\tau}\right] E_z \\
&\quad + P_{z,1} + P_{z,2} - P_{z,3} - P_{z,4}, \qquad (14.203)\\
\mu_0\mu_r\frac{\partial H_x}{\partial t} &= -\frac{\partial E_z}{\partial y} - \mu_0\mu_r\frac{\sigma_y - \sigma_x}{\epsilon_0}H_x + Q_x, \qquad (14.204)\\
\mu_0\mu_r\frac{\partial H_y}{\partial t} &= \frac{\partial E_z}{\partial x} - \mu_0\mu_r\frac{\sigma_x - \sigma_y}{\epsilon_0}H_y + Q_y, \qquad (14.205)
\end{aligned}$$

$$\begin{aligned}
\frac{\partial P_{z,1}}{\partial t} &= -\frac{\epsilon_{r,\infty}}{\epsilon_0}\sigma_x\sigma_y E_z, \qquad (14.206)\\
\frac{\partial P_{z,2}}{\partial t} &= -\frac{1}{\tau}P_{z,2} + \frac{(\epsilon_{r,s} - \epsilon_{r,\infty})}{\tau^2}E_z, \qquad (14.207)\\
\frac{\partial P_{z,3}}{\partial t} &= -\frac{1}{\tau}P_{z,3} - \frac{\epsilon_{r,s} - \epsilon_{r,\infty}}{\epsilon_{r,\infty}\tau}P_{z,1} + \frac{(\epsilon_{r,s} - \epsilon_{r,\infty})(\sigma_x + \sigma_y)}{\tau}E_z, \qquad (14.208)\\
\frac{\partial P_{z,4}}{\partial t} &= -\frac{\sigma}{\epsilon_0\epsilon_{r,\infty}}P_{z,1} + \frac{\sigma(\sigma_x + \sigma_y)}{\epsilon_0}E_z, \qquad (14.209)\\
\frac{\partial Q_x}{\partial t} &= -\frac{\sigma_x}{\epsilon_0}Q_x + \frac{\mu_0\mu_r\sigma_x(\sigma_y - \sigma_x)}{\epsilon_0^2}H_x, \qquad (14.210)\\
\frac{\partial Q_y}{\partial t} &= -\frac{\sigma_y}{\epsilon_0}Q_y + \frac{\mu_0\mu_r\sigma_y(\sigma_x - \sigma_y)}{\epsilon_0^2}H_y. \qquad (14.211)
\end{aligned}$$

Note that (14.206)–(14.211) contain no spatial derivatives and hence they are simply ODEs for $P_{z,1}$, $P_{z,2}$, $P_{z,3}$, $P_{z,4}$, Q_x, and Q_y. Abarbanel and Gottlieb [57] pointed out that the system using the split-field PML terminating the computational domain was only *weakly* well-posed for the initial value problem. In contrast, the system (14.203)–(14.211), after dropping the undifferentiated terms, becomes the original 3×3 Maxwell's system, which is symmetric and hyperbolic and therefore *strongly* well-posed.

The new auxiliary polarization currents **P** and **Q** have decoupled the frequency-dependent constitutive relations (14.159) and (14.161). Note that $P_{z,1}$, Q_x, and Q_y are auxiliary variables introduced by the UPML, $P_{z,2}$ is introduced by the medium dispersion, $P_{z,3}$ is introduced by the medium dispersion and the PML, $P_{z,4}$ is introduced by the medium loss and the UPML, σ is the relative electric conductivity, and σ_x and σ_y are the parameters for the UPML. In the case that $\sigma_x = \sigma_y = 0$, the above equations reduce to the original Maxwell's equations in the physical dispersive region.

We can put (14.203)–(14.211) in the general form of (14.162)–(14.164) by defining $\mathbf{U}^{(1)} = (\epsilon_0\epsilon_{r,\infty}E_z, \mu_0\mu_r H_x, \mu_0\mu_r H_y)^{\mathrm{T}}$, $\mathbf{U}^{(2)} = (P_{z,1}, P_{z,2}, P_{z,3}, P_{z,4}, Q_x, Q_y)^{\mathrm{T}}$, $\mathbf{U} = (\mathbf{U}^{(1)}, \mathbf{U}^{(2)})^{\mathrm{T}}$, $\mathbf{S} = (\mathbf{S}^{(1)}, \mathbf{S}^{(2)})^{\mathrm{T}}$, and $\mathbb{A}$, $\bar{\mathbb{A}}$ are given as

$$\mathbb{A} = (A_x, A_y),\ \bar{\mathbb{A}} = (\bar{A}_x, \bar{A}_y), \tag{14.212}$$

where

$$A_x = \begin{pmatrix} 0 & 0 & -1/\mu_r \\ 0 & 0 & 0 \\ -1/\epsilon_{r,\infty} & 0 & 0 \end{pmatrix},\quad A_y = \begin{pmatrix} 0 & 1/\mu_r & 0 \\ 1/\epsilon_{r,\infty} & 0 & 0 \\ 0 & 0 & 0 \end{pmatrix}, \tag{14.213}$$

and

$$\bar{A}_x = \begin{pmatrix} A_x & 0_{3\times 6} \\ 0_{6\times 3} & 0_{6\times 6} \end{pmatrix},\quad \bar{A}_y = \begin{pmatrix} A_y & 0_{3\times 6} \\ 0_{6\times 3} & 0_{6\times 6} \end{pmatrix}. \tag{14.214}$$

Here $0_{n\times m}$ denotes the zero matrix with n rows and m columns. The source terms $\mathbf{S}^{(1)}$ and $\mathbf{S}^{(2)}$ represent body forces, e.g., currents:

$$\mathbf{S}^{(1)} = \begin{pmatrix} -\left[\sigma + \epsilon_{r,\infty}(\sigma_x + \sigma_y) + \frac{\epsilon_{r,s}-\epsilon_{r,\infty}}{\tau}\right] E_z + P_{z,1} + P_{z,2} - P_{z,3} - P_{z,4} \\ -\mu_r\mu_0(\sigma_y - \sigma_x)H_x/\epsilon_0 + Q_x \\ -\mu_r\mu_0(\sigma_x - \sigma_y)H_y/\epsilon_0 + Q_y \end{pmatrix}, \tag{14.215}$$

$$\mathbf{S}^{(2)} = \begin{pmatrix} -\epsilon_{r,\infty}\sigma_x\sigma_y E_z \\ -\frac{1}{\tau}P_{z,2} + \frac{\epsilon_{r,s}-\epsilon_{r,\infty}}{\tau^2}E_z \\ -\frac{1}{\tau}P_{z,3} - \frac{\epsilon_{r,s}-\epsilon_{r,\infty}}{\epsilon_{r,\infty}\tau}P_{z,1} + \frac{(\epsilon_{r,s}-\epsilon_{r,\infty})(\sigma_x+\sigma_y)}{\tau}E_z \\ -\frac{\sigma}{\epsilon_{r,\infty}}P_{z,1} + \sigma(\sigma_x + \sigma_y)E_z/\epsilon_0 \\ -\sigma_x Q_x + \mu_r\mu_0\sigma_x(\sigma_y - \sigma_x)H_x/\epsilon_0^2 \\ -\sigma_y Q_y + \mu_r\mu_0\sigma_y(\sigma_x - \sigma_y)H_y/\epsilon_0^2 \end{pmatrix}. \tag{14.216}$$

14.7.2 Drude Material

We consider a dispersive material described by a single-pole Drude medium, whose relative electric permittivity is rewritten here as

$$\epsilon_r(\omega) = \epsilon_{r,\infty} - \frac{\omega_p^2}{\omega^2 + \mathrm{i}\Gamma\omega}, \tag{14.217}$$

where ω_p is the plasmon frequency, Γ is the damping constant, and $\epsilon_{r,\infty}$ is the relative electric permittivity at infinite frequency.

The augmented Maxwell's equations for a transverse electric (TE) wave with auxiliary polarization currents for the new augmented variables $\mathbf{U}^{(1)} = (\mu_0\mu_r H_z, \epsilon_0\epsilon_{r,\infty}E_x, \epsilon_0\epsilon_{r,\infty}E_y)^{\mathrm{T}}$, $\mathbf{U}^{(2)} = (Q_z, P_{x,2}, P_{x,3}, P_{x,4}, P_{y,2}, P_{y,3}, P_{y,4})^{\mathrm{T}}$, $\mathbf{U} = (\mathbf{U}^{(1)}, \mathbf{U}^{(2)})^{\mathrm{T}}$, and $\mathbf{S} = (\mathbf{S}^{(1)}, \mathbf{S}^{(2)})^{\mathrm{T}}$ can be cast in the form of (14.162)–(14.164), and $\mathbb{A}$, $\bar{\mathbb{A}}$ are given as

$$\mathbb{A} = (A_x, A_y),\ \bar{\mathbb{A}} = (\bar{A}_x, \bar{A}_y), \tag{14.218}$$

where

$$A_x = \begin{pmatrix} 0 & 0 & 1/\epsilon_{r,\infty} \\ 0 & 0 & 0 \\ 1/\mu_r & 0 & 0 \end{pmatrix}, \quad A_y = \begin{pmatrix} 0 & -1/\epsilon_{r,\infty} & 0 \\ -1/\mu_r & 0 & 0 \\ 0 & 0 & 0 \end{pmatrix}, \tag{14.219}$$

and

$$\bar{A}_x = \begin{pmatrix} A_x & 0_{3\times 7} \\ 0_{7\times 3} & 0_{7\times 7} \end{pmatrix}, \quad \bar{A}_y = \begin{pmatrix} A_y & 0_{3\times 7} \\ 0_{7\times 3} & 0_{7\times 7} \end{pmatrix}. \tag{14.220}$$

Meanwhile, the source terms $\mathbf{S}^{(1)}$ and $\mathbf{S}^{(2)}$ represent body forces, e.g., polarization currents [51]:

$$\mathbf{S}^{(1)} = -\begin{pmatrix} \mu_r d_{0,z} H_z + \mu_r d_{1,z} Q_z \\ \epsilon_{r,\infty} c_{1,x} E_x + \omega_p^2 P_{x,4} + \omega_p^2 c_{1,x} P_{x,3} + c_{2,x} P_{x,2} \\ \epsilon_{r,\infty} c_{1,y} E_y + \omega_p^2 P_{y,4} + \omega_p^2 c_{1,y} P_{y,3} + c_{2,y} P_{y,2} \end{pmatrix}, \tag{14.221}$$

$$\mathbf{S}^{(2)} = \begin{pmatrix} H_z \\ \omega_p^2 P_{x,3} + \epsilon_{r,\infty} E_x - \sigma_x P_{x,2} \\ P_{x,4} \\ E_x - \gamma P_{x,4} \\ \omega_p^2 P_{y,3} + \epsilon_{r,\infty} E_y - \sigma_y P_{y,2} \\ P_{y,4} \\ E_y - \gamma P_{y,4} \end{pmatrix}, \tag{14.222}$$

where

$$\begin{aligned} c_{1,x} &= \sigma_y - \sigma_x, \quad c_{2,x} = -\sigma_x(\sigma_y - \sigma_x), \\ c_{1,y} &= \sigma_x - \sigma_y, \quad c_{2,y} = -\sigma_y(\sigma_x - \sigma_y), \\ d_{0,z} &= \sigma_x + \sigma_y, \quad d_{1,z} = \sigma_x \sigma_y. \end{aligned} \tag{14.223}$$

14.8 Summary

Wave scattering and propagation in periodic media offer a wide range of interesting physical and mathematical properties, including the band gap of dielectric photonics, the excitation of plasmon waves, and the field enhancement near metallic surfaces. In the frequency domain, plane wave methods and transmission spectra calculations using Fourier series and Nédélec edge elements are discussed for the study of band-gap properties of periodic structures. While the first two methods give a diagonal representation of the Laplace operator in the eigen-mode problem, they may suffer Gibbs phenomenon oscillations when the material contrast becomes large. The Nédélec edge element and the volume integral methods offer flexibility in handling complex material interfaces and large material contrasts. In the time domain, finite element methods with the conforming Nédélec basis and discontinuous Galerkin implementations can be used to study the transient wave interactions, including nonlinear phenomena.

Nonlinear optical properties of materials which could occur under strong incident light sources such as lasers have not been discussed in this book, and important nonlinear behaviors, such as second harmonic generations and parametric amplifications and stimulated Raman scattering, are discussed in [58, 59].

References

1. E. Kaxiras, *Atomic and Electronic Structure of Solids* (Cambridge University Press, Cambridge, UK, 2003)
2. C.H. Wilcox, Theory of Bloch waves. J. D'anal. Math. **33**, 146–167 (1978)
3. Z. Huang, S. Jin, P.A. Markowich, C. Sparber, On the Bloch decomposition based spectral method for wave propagation in periodic media. Wave Motion **46**, 15–28 (2009)
4. S.G. Johnson, J.D. Joannopoulos, Block-iterative frequency-domain methods for Maxwell's equations in a planewave basis. Opt. Express **8**, 173–190 (2001)
5. S. Guo, S. Albin, Simple plane wave implementation for photonic crystal calculations. Opt. Express **11**, 167–175 (2003)
6. Y. Saad, *Iterative Methods for Sparse Linear Systems*, 2nd edn. (SIAM, Philadelphia, PA, 2003)
7. K. Sakoda, Optical transmittance of a two-dimensional triangular photonic lattice. Phys. Rev. B **51**, 4672–4675 (1995)
8. K. Sakoda, Transmittance and Bragg reflectivity of two-dimensional photonic lattices. Phys. Rev. B **52**, 8992–9002 (1995)
9. M.G. Moharam, T.K. Gaylord, Rigorous coupled-wave analysis of planar-grating diffraction. J. Opt. Soc. Am. **71**, 811–818 (1981)
10. K. Sakoda, *Optical Properties of Photonic Crystals* (Springer, Berlin, 2001)
11. M. Costabel, M. Dauge, Computation of resonance frequencies for Maxwell equations in non smooth domains, in *Topics in Computational Wave Propagation: Direct and Inverse Problems*. ed. by M. Ainsworth, P. Davies, D. Duncan, P. Martin, B. Rynne (Springer, Berlin, 2003), pp.127–164
12. D. Boffi, Finite element approximation of eigenvalue problems. Acta Numer. **19**, 1–120 (2010)
13. Z.J. Csendes, P. Silvester, Numerical solution of dielectric loaded wave-guides: I—finite-element analysis. IEEE Trans. Microw. Theory Tech. **18**, 1124–1131 (1970)
14. J.B. Davies, F.A. Fernandez, G.Y. Philippou, Finite element and analysis of all modes in cavities with circular symmetry. IEEE Trans. Microw. Theory Tech. **30**, 1975–1980 (1982)
15. M. Hara, T. Wada, T. Fukasawa, F. Kikuchi, A three dimensional analysis of RF electromagnetic fields by the finite element method. IEEE Trans. Mag. **19**, 2417–2420 (1983)
16. D. Boffi, P. Fernandes, L. Gastaldi, I. Perugia, Computational models of electromagnetic resonators: analysis of edge element approximation. SIAM J. Numer. Anal. **36**, 1264–1290 (1999)
17. V. Girault, P.-A. Raviart, *Finite Element Methods for Navier-Stokes Equations, Theory and Algorithms* (Springer, New York, 1986)
18. A. Bossavit, Solving Maxwell equations in a closed cavity, and the question of 'spurious modes'. IEEE Trans. Mag. **26**, 702–705 (1990)
19. F. Kikuchi, On a discrete compactness property for the Nédélec finite elements. J. Fac. Sci. Univ. Tokyo Sect. I A Math. **36**, 479–490 (1989)
20. S. Caorsi, P. Fernandes, M. Raffetto, On the convergence of Galerkin finite element approximations of electromagnetic eigenproblems. SIAM J. Numer. Anal. **38**, 580–607 (2000)

21. S. Caorsi, P. Fernandes, M. Raffetto, Spurious-free approximations of electromagnetic eigenproblems by means of Nédélec-type elements. Math. Model Numer. Anal. **35**, 331–354 (2001)
22. F. Brezzi, On the existence, uniqueness and approximation of saddle-point problems arising from Lagrangian multipliers. RAIRO **8**, 129–151 (1974)
23. F. Kikuchi, Mixed and penalty formulations for finite element analysis of an eigenvalue problem in electromagnetism. Comput. Meth. Appl. Mech. Eng. **64**, 509–521 (1987)
24. M. Fortin, F. Brezzi, *Mixed and Hybrid Finite Element Methods* (Springer, New York, 1991)
25. D. Boffi, F. Brezzi, L. Gastaldi, On the convergence of eigenvalues for mixed formulations. Ann. Scuola Norm. Sup. Pisa Cl. Sci. **25**(4), 131–154 (1997)
26. D. Boffi, F. Brezzi, L. Gastaldi, On the problem of spurious eigenvalues in the approximation of linear elliptic problems in mixed form. Math. Comput. **69**, 121–140 (2000)
27. M. Costabel, M. Dauge, S. Nicaise, Singularities of Maxwell interface problems. Math. Model Numer. Anal. **33**, 627–649 (1999)
28. M. Costabel, M. Dauge, Singularities of electromagnetic fields in polyhedral domains. Arch. Ration. Mech. Anal. **151**, 221–276 (1997)
29. D.C. Dobson, J.E. Pasciak, Analysis of an algorithm for computing electromagnetic Bloch modes using Nédélec spaces. Comput. Meth. Appl. Math. **1**, 138–153 (2001)
30. D. Boffi, M. Conforti, L. Gastaldi, Modified edge finite elements for photonic crystals. Numer. Math. **105**, 249–266 (2006)
31. L.E. Rickard Petersson, J.-M. Jin, A three-dimensional time-domain finite-element formulation for periodic structures. IEEE Trans. Antenn. Propag. **54**, 12–19 (2006)
32. L.E. Rickard Petersson, J.-M. Jin, Analysis of periodic structures via a time-domain finite-element formulation with a Floquet ABC. IEEE Trans. Antenn. Propag. **54**, 933–944 (2006)
33. G. Bao, D.C. Dobson, J.A. Cox, Mathematical studies in rigorous grating theory. J. Opt. Soc. Am. A **12**, 1029–1042 (1995)
34. J.-M. Jin, D.J. Riley, *Finite Element Analysis of Antennas and Arrays* (IEEE Press, Piscataway, NJ, 2008)
35. R.H. Ritchie, Plasma losses by fast electrons in thin films. Phys. Rev. **106**, 874–881 (1957)
36. S.A. Maier, *Plasmonics: Fundamentals and Applications* (Springer, New York, 2007)
37. J.M. Pitarke, V.M. Silkin, E.V. Chulkov, P.M. Echenique, Theory of surface plasmons and surface-plasmon polaritons. Rep. Prog. Phys. **70**, 1–87 (2007)
38. C. Kittel, *Introduction to Solid State Physics*, 8th edn. (Wiley, New York, 2004)
39. A. Otto, Excitation of nonradiative surface plasma waves in silver by the method of frustrated total reflection. Zeitschrift Physik **216**, 398–410 (1968)
40. E. Kretschmann, H. Raether, Radiative decay of non-radiative surface plasmons excited by light. Z. Naturforsh. **23**, 2135–2136 (1968)
41. J.A. Stratton, *Electromagnetic Theory* (McGraw-Hill, New York, 1941)
42. C.F. Bohren, D.R. Huffman, *Absorption and Scattering of Light by Small Particles* (Wiley-VCH, Hoboken, NJ, 1998)
43. R.E. Collin, *Field Theory of Guided Waves* (Wiley-IEEE Press, Hoboken, NJ, 1990)
44. J. van Bladel, *Singular Electromagnetic Fields and Sources* (Clarendon Press & Oxford University Press, Oxford, UK, 1991)
45. A.D. Yaghjian, Electric dyadic Green's functions in the source region. Proc. IEEE **68**, 248–263 (1980)
46. D. Chen, M.H. Cho, W. Cai, Accurate and efficient Nystrom volume integral equation method for electromagnetic scattering of 3-D metamaterials in layered media. SIAM J. Sci. Comput. **40**(1), B259–B282 (2018)
47. D. Chen, B. Zinser, W. Cai, M.H. Cho, Accurate and efficient Nystrom volume integral equation method for the Maxwell equations for multiple 3-D scatterers. J. Comput. Phys. **321**, 303–320 (2016)
48. T.W. Ebbesen, H.J. Lezec, H.F. Ghaemi, T. Thio, P.A. Wolff, Extra-ordinary optical transmission through sub-wavelength hole arrays. Nature **391**, 667–669 (1998)

49. H. Liu, P. Lalanne, Microscopic theory of the extraordinary optical transmission. Nature **452**, 728–731 (2008)
50. C.H. Gan, G. Gbur, Extraordinary optical transmission through multi-layered systems of corrugated metallic thin films. Opt. Express **17**, 20553–20566 (2009)
51. X. Ji, W. Cai, P. Zhang, High order DGTD methods for dispersive Maxwell's equations and modelling of silver nanowire coupling. Int. J. Numer. Meth. Eng. **69**, 308–325 (2007)
52. B. Szabó, I. Babuska, *Finite Element Analysis* (Wiley-Interscience, New York, 1991)
53. X. Ji, T. Lu, W. Cai, P. Zhang, Discontinuous Galerkin time domain (DGTD) methods for the study of 2-D waveguide coupled microring resonators. J. Lightwave Technol. **23**, 3864–3874 (2005)
54. D.W. Lynch, W.R. Hunter, Comments on the optical constants of metals and an introduction to the data for several metals, in *Handbook of Optical Constants of Solids*. ed. by E.D. Palik (Academic Press, New York, 1985), pp.275–367
55. A. Taflove, S.C. Hagness, *Computational Electromagnetics: The Finite-Difference Time-Domain Method* (Artech House Publishers, Norwood, MA, 2000)
56. T. Lu, P. Zhang, W. Cai, Discontinuous Galerkin methods for dispersive and lossy Maxwell's equations and PML boundary conditions. J. Comput. Phys. **200**, 549–580 (2004)
57. S. Abarbanel, D. Gottlieb, On the construction and analysis of absorbing layers in CEM. Appl. Numer. Math. **27**, 331–340 (1998)
58. A. Yariv, *Quantum Electronics*, 3rd edn. (Wiley, Hoboken, NJ, 1989)
59. P.N. Butcher, D. Cotter, *The Elements of Nonlinear Optics* (Cambridge University Press, Cambridge, UK, 1991)

Chapter 15
Schrödinger Equations for Waveguides and Quantum Dots

Schrödinger equations are used for paraxial approximations of Maxwell's equations in optical waveguides or for describing electron density wave functions in quantum dots embedded in layered media. In this chapter, we will discuss numerical methods for computing their solutions, which may be discontinuous in values or derivatives due to the existence of discontinuities in material dielectric properties inside the waveguides or electric potentials experienced by the electrons confined in quantum dots. In the case of optical waveguides, the Schrödinger equations are reformulated with generalized functions (distributions) by using Dirac δ functions as source terms to represent the discontinuities in the solutions. Then, a generalized discontinuous Galerkin (DG) beam propagation method is discussed for the guided waves in optical waveguides. For quantum dots, a volume integral equation for the density wave functions is presented with Green's functions for layered media.

15.1 Generalized DG (GDG) Methods for Schrödinger Equations

In this section, we introduce a generalized discontinuous Galerkin method for Schrödinger equations with discontinuities in either the solution or the derivative of the solution, which are modeled as Dirac $\delta(x)$-distribution sources of the PDEs.

W. Cai, *Deterministic, Stochastic, and Deep Learning Methods for Computational Electromagnetics*, https://doi.org/10.1007/978-981-96-0100-4_15

15.1.1 One-Dimensional Schrödinger Equations

Let us consider a model 1-D problem:

$$\mathrm{i}\frac{\partial \varphi(x,t)}{\partial t} = \frac{\partial^2 \varphi(x,t)}{\partial x^2}, \quad \text{for } x \in [a,b]\backslash\{\tau\}, \tag{15.1}$$

where φ is a complex-valued wave function.

With the shorthand notation

$$[u(\tau,t)] = u(\tau^+,t) - u(\tau^-,t), \tag{15.2}$$

we denote the jump conditions at the interface τ by

$$[\varphi(\tau,t)] = f(t), \tag{15.3}$$

$$\left[\frac{\partial \varphi(\tau,t)}{\partial x}\right] = g(t). \tag{15.4}$$

We can incorporate the jump conditions into (15.1) using δ and δ' source terms at τ, namely

$$\mathrm{i}\frac{\partial \varphi(x,t)}{\partial t} = \frac{\partial^2 \varphi(x,t)}{\partial x^2} - g(t)\delta(x-\tau) - f(t)\delta'(x-\tau), \quad \text{for } x \in [a,b]. \tag{15.5}$$

The use of the δ and δ' source terms compensates for the singularity introduced by the jump conditions at the interface. Furthermore, to avoid dealing with δ' in (15.5), we introduce an auxiliary distributional variable p to rewrite (15.5) as

$$\mathrm{i}\frac{\partial \varphi}{\partial t} = \frac{\partial p}{\partial x} - g(t)\delta(x-\tau), \tag{15.6a}$$

$$p = \frac{\partial \varphi}{\partial x} - f(t)\delta(x-\tau). \tag{15.6b}$$

Now, both φ and p are piecewise continuous functions over $[a,b]$, while $\partial\varphi/\partial x$ and $\partial p/\partial x$ are treated as distributions.

To derive a finite element approximation of (15.6a) and (15.6b), we first divide $\Omega = [a,b]$ into N elements,

$$\{a = x_0 < \cdots < x_{k^*} = \tau < \cdots < x_N = b\}, \tag{15.7}$$

and denote a general element by

$$K = [x_k, x_{k+1}], \quad \text{for } k = 0, 1, \ldots, N-1. \tag{15.8}$$

To proceed, we introduce the concept of an evenly split δ function:

$$\int_{\pm a}^{0} v(x)\delta(x)\mathrm{d}x = \mp\frac{1}{2}v(0), \quad \text{for } a > 0,\ v(x) \in C(\pm a, 0], \tag{15.9}$$

and the following integration-by-parts identities for distributional variables $\partial\varphi/\partial x$ and $\partial p/\partial x$ over a closed interval:

$$\int_{\tau}^{\tau+h} \frac{\partial\varphi}{\partial x}v(x)\mathrm{d}x = \varphi(\tau+h)v(\tau+h) - \{\varphi\}v(\tau) - \int_{\tau}^{\tau+h} \frac{\partial v}{\partial x}\varphi(x)\mathrm{d}x, \tag{15.10}$$

$$\int_{\tau}^{\tau+h} \frac{\partial p}{\partial x}v(x)\mathrm{d}x = p(\tau+h)v(\tau+h) - \{p\}v(\tau) - \int_{\tau}^{\tau+h} \frac{\partial v}{\partial x}p(x)\mathrm{d}x, \tag{15.11}$$

where $\{u\} = (u(\tau^+) + u(\tau^-))/2$ denotes the average of the values of the function u at the interface τ. The proof of (15.10) and (15.11) is given in [1].

Let $P_J(K)$ be the space of polynomials in K of degree at most J, and let $v \in L^1[a, b]$ be the test function, where $v|_K \in P_J(K)$. First, we consider the element adjacent to the right-hand side of the interface, say $K = [x_{k^*}, x_{k^*+1}]$, $x_{k^*} = \tau$. By multiplying (15.6a) by v and then using integration-by-parts formula (15.10) on K, we get

$$\begin{aligned} \mathrm{i}\int_K \frac{\partial\varphi}{\partial t}v\,\mathrm{d}x &= pv|_{x^-_{k^*+1}} - \{p\}(\tau)v(x^+_{k^*}) - \int_K p\frac{\mathrm{d}v}{\mathrm{d}x}\,\mathrm{d}x - g(t)\int_K \delta(x-\tau)v\,\mathrm{d}x \\ &= pv|_{x^-_{k^*+1}} - \left(\{p\}(\tau) + \frac{1}{2}g(t)\right)v(x^+_{k^*}) - \int_K p\frac{\mathrm{d}v}{\mathrm{d}x}\,\mathrm{d}x, \end{aligned} \tag{15.12}$$

where the factor $1/2$ in front of $g(t)$ comes from (15.9). Equation (15.12) suggests that we should define the fluxes on the right-hand side of the interface $x^+_{k^*} = \tau^+$ as

$$h_\varphi(x^+_{k^*}) = \{p\}(x_{k^*}) + \frac{1}{2}g(t), \tag{15.13}$$

and those at $x^-_{k^*+1}$ as

$$h_\varphi(x^-_{k^*+1}) = p(x^-_{k^*+1}). \tag{15.14}$$

Repeating the above procedure for the element $K = [x_{k^*-1}, x_{k^*}]$, one obtains the fluxes on the left-hand side of the interface $x^-_{k^*} = \tau^-$ as

$$h_\varphi(x^-_{k^*}) = \{p\}(x_{k^*}) - \frac{1}{2}g(t), \tag{15.15}$$

and those at $x^+_{k^*-1}$ as

$$h_\varphi(x^+_{k^*-1}) = p(x^+_{k^*-1}). \tag{15.16}$$

As the exact solution is continuous at $x_{k^*\pm 1}$, we can replace (15.14) and (15.16) by

$$h_\varphi(x^\pm_{k^*\pm 1}) = \{p\}(x_{k^*\pm 1}). \tag{15.17}$$

Similarly, we can define the fluxes for p as

$$h_p(x^\pm_{k^*}) = \{\varphi\}(x_{k^*}) \pm \frac{1}{2} f(t), \tag{15.18}$$

$$h_p(x^\pm_{k^*\pm 1}) = \{\varphi\}(x_{k^*\pm 1}). \tag{15.19}$$

Finally, for all elements $K = [x_k, x_{k+1}]$, we will have the following method for the 1-D Schrödinger equation (15.6a) and (15.6b).

1-D Generalized Discontinuous Galerkin (GDG) Method

We have

$$\mathrm{i}\int_K \frac{\partial\varphi}{\partial t} v\,\mathrm{d}x = h_\varphi(x^-_{k+1})v(x^-_{k+1}) - h_\varphi(x^+_k)v(x^+_k) - \int_K p\frac{\mathrm{d}v}{\mathrm{d}x}\,\mathrm{d}x, \tag{15.20}$$

$$\int_K pv\,\mathrm{d}x = h_p(x^-_{k+1})v(x^-_{k+1}) - h_p(x^+_k)v(x^+_k) - \int_K \varphi\frac{\mathrm{d}v}{\mathrm{d}x}\,\mathrm{d}x, \tag{15.21}$$

where, if $x_k \neq \tau$,

$$h_\varphi(x^\pm_k) = \{p\}(x_k), \tag{15.22}$$

$$h_p(x^\pm_k) = \{\varphi\}(x_k), \tag{15.23}$$

and at τ

$$h_\varphi(\tau^\pm) = \{p\}(\tau) \pm \frac{1}{2} g(t), \tag{15.24}$$

$$h_p(\tau^\pm) = \{\varphi\}(\tau) \pm \frac{1}{2} f(t). \tag{15.25}$$

Let $\phi_j(x)$, $j = 0, 1, \ldots, J$, be the basis functions, and let φ and p be expanded as

$$\varphi(x,t) = \sum_{j=0}^{J} \alpha_j(t)\phi_j(x), \quad p(x,t) = \sum_{j=0}^{J} p_j(t)\phi_j(x). \tag{15.26}$$

In each K, by choosing the test function $v = \phi_l(x)$ for $l = 0, 1, \ldots, J$, and denoting m_{lj} and m_{lj}^x as

$$m_{lj} = \int_K \phi_l \phi_j \, \mathrm{d}x, \qquad m_{lj}^x = \int_K \frac{\mathrm{d}\phi_l}{\mathrm{d}x} \phi_j \, \mathrm{d}x, \tag{15.27}$$

we have

$$\mathrm{i} \sum_j m_{lj} \frac{\mathrm{d}\alpha_j}{\mathrm{d}t} = h_\varphi(x_{k+1})\phi_l(x_{k+1}^-) - h_\varphi(x_k)\phi_l(x_k^+) - \sum_j m_{lj}^x p_j, \tag{15.28a}$$

$$\sum_j m_{lj} p_j = h_p(x_{k+1})\phi_l(x_{k+1}^-) - h_p(x_k)\phi_l(x_k^+) - \sum_j m_{lj}^x \alpha_j. \tag{15.28b}$$

Equation (15.28b) can be used to eliminate the variables $\{p_j\}$ in terms of $\{\alpha_j\}$ on each element locally.

15.1.2 Two-Dimensional Schrödinger Equations

Let the solution domain Ω be decomposed into regions Ω_i with jumps across their interfaces, i.e., $\Omega = \cup_i \Omega_i$. We consider the following time-dependent scalar 2-D Schrödinger equation: for $(x, y) \notin \Gamma = \cup_i \partial\Omega_i$,

$$\mathrm{i} \frac{\partial \varphi}{\partial t} = \frac{\partial^2 \varphi}{\partial x^2} + \frac{\partial^2 \varphi}{\partial y^2} + S(\varphi), \tag{15.29}$$

where S is a source term. The jumps at $(x^*, y^*) \in \Gamma$ are given by

$$f(x^*, y^*, t) = [\varphi(x^*, y^*, t)] = \varphi(x^{*+}, y^{*+}, t) - \varphi(x^{*-}, y^{*-}, t), \tag{15.30}$$

$$g(x^*, y^*, t) = \left[\frac{\partial \varphi(x^*, y^*, t)}{\partial \mathbf{n}} \right] = \frac{\partial \varphi(x^{*+}, y^{*+}, t)}{\partial \mathbf{n}} - \frac{\partial \varphi(x^{*-}, y^{*-}, t)}{\partial \mathbf{n}}, \tag{15.31}$$

where $\mathbf{n}$ is the normal to the interface Γ pointing to the $+$ side.

On the interface Γ, local coordinates (ξ, η) will be introduced, where ξ is along the normal direction and η is along the tangential direction(s). Following the same procedure as for (15.6a) and (15.6b), we can rewrite (15.29) as follows [1]:

$$\mathrm{i} \frac{\partial \varphi}{\partial t} = \frac{\partial p}{\partial x} + \frac{\partial q}{\partial y} - \delta(\xi - \xi^*)|\nabla \xi|^2 g + S, \tag{15.32a}$$

$$p = \frac{\partial \varphi}{\partial x} - \delta(\xi - \xi^*) f \frac{\partial \xi}{\partial x}, \tag{15.32b}$$

$$q = \frac{\partial \varphi}{\partial y} - \delta(\xi - \xi^*) f \frac{\partial \xi}{\partial y}. \tag{15.32c}$$

As in the 1-D case, for each element K in the discretization of Ω, let $P_J(K)$ denote the space of polynomials in K of degree at most J, and let $v \in L^1(\Omega)$ be the test function, where $v|_K \in P_J(K)$. Multiplying (15.32a)–(15.32c) by v and integrating by parts in K, we obtain the following.

2-D Generalized Discontinuous Galerkin (GDG) Method

We have

$$\begin{aligned} \mathrm{i}\int_K \frac{\partial \varphi}{\partial t}\,\mathrm{d}x\,\mathrm{d}y &= \int_{\partial K} h_\varphi^x v n_x\,\mathrm{d}s - \int_K p\frac{\partial v}{\partial x}\,\mathrm{d}x\mathrm{d}y \\ &\quad + \int_{\partial K} h_\varphi^y v n_y\,\mathrm{d}s - \int_K q\frac{\partial v}{\partial y}\,\mathrm{d}x\,\mathrm{d}y + \int_K Sv\,\mathrm{d}x\,\mathrm{d}y, \end{aligned} \tag{15.33a}$$

$$\int_K pv\,\mathrm{d}x\,\mathrm{d}y = \int_{\partial K} h_p v n_x \mathrm{d}s - \int_K \varphi\frac{\partial v}{\partial x}\mathrm{d}x\,\mathrm{d}y, \tag{15.33b}$$

$$\int_K qv\,\mathrm{d}x\,\mathrm{d}y = \int_{\partial K} h_q v n_y\,\mathrm{d}s - \int_K \varphi\frac{\partial v}{\partial y}\,\mathrm{d}x\,\mathrm{d}y, \tag{15.33c}$$

where (n_x, n_y) is the external normal of ∂K and $(h_\varphi^x, h_\varphi^y, h_p = h_q)$ are numerical fluxes which relate to (p, q, φ) at ∂K and are given by, for $\mathbf{x} = (x, y) \in \partial K$,

$$h_\varphi^x(\mathbf{x}^\pm) = \{p\} \pm a^x,\ h_\varphi^y(\mathbf{x}^\pm) = \{q\} \pm a^y,\ h_p(\mathbf{x}^\pm) = \{\varphi\} \pm b, \tag{15.34}$$

where $+$ indicates the exterior side of ∂K, and a^x, a^y, and b are from the jump conditions and are defined as

$$(a^x,\ a^y,\ b) = \begin{cases} \left(\frac{1}{2}g|\nabla\xi|n_x,\ \frac{1}{2}g|\nabla\xi|n_y,\ \frac{1}{2}f\right), & \text{if } \Gamma \cap K \neq \emptyset, \\ (0,\ 0,\ 0), & \text{if } \Gamma \cap K = \emptyset. \end{cases} \tag{15.35}$$

As in (15.22)–(15.25) for the 1-D case, simple averages are used in (15.34) for all element boundaries except the material interface, where averages plus/minus one half of the jump are used [1].

Let $\phi_j(x, y)$, $j = 0, 1, \ldots, n_J$, be the basis functions, where $n_J + 1$ is the number of basis functions required for the Jth-order approximation. We expand φ, p, and q as follows:

$$\varphi = \sum_{j=0}^{n_J} \alpha_j(t)\phi_j(x, y),\ q = \sum_{j=0}^{n_J} q_j(t)\phi_j(x, y),\ p = \sum_{j=0}^{n_J} p_j(t)\phi_j(x, y), \tag{15.36}$$

and choose the test function $v(x, y) = \phi_l(x, y)$ for $l = 0, 1, \ldots, n_J$, and we get

$$\mathrm{i}\sum_j m_{lj}\frac{\mathrm{d}\alpha_j}{\mathrm{d}t} = \int_{\partial K}(h_\varphi^x n_x + h_\varphi^y n_y)\phi_l \mathrm{d}s - \sum_j (m_{lj}^x p_j + m_{lj}^y q_j) + s_l, \tag{15.37a}$$

$$\sum_j m_{lj} p_j = \int_{\partial K} h_p n_x \phi_l \,\mathrm{d}s - \sum_j m_{lj}^x \alpha_j, \tag{15.37b}$$

$$\sum_j m_{lj} q_j = \int_{\partial K} h_q n_y \phi_l \,\mathrm{d}s - \sum_j m_{lj}^y \alpha_j, \tag{15.37c}$$

where $s_l = \int_K S\phi_l \,\mathrm{d}x \,\mathrm{d}y$ and

$$m_{lj} = \int_K \phi_l \phi_j \,\mathrm{d}x \,\mathrm{d}y, \quad m_{lj}^x = \int_K \frac{\partial \phi_l}{\partial x}\phi_j \mathrm{d}x \,\mathrm{d}y, \quad m_{lj}^y = \int_K \frac{\partial \phi_l}{\partial y}\phi_j \,\mathrm{d}x \,\mathrm{d}y. \tag{15.38}$$

Again, the local variables $\{p_j\}$ and $\{q_j\}$ can be eliminated in terms of $\{\alpha_j\}$ on each element K locally by using (15.37b) and (15.37c).

15.2 GDG Beam Propagation Methods (BPMs) for Optical Waveguides

Beam propagation methods (BPMs) [2, 3] are based on paraxial approximations for wave propagations in optical waveguides, where time-harmonic Maxwell's equations are approximated by Schrödinger equations and the propagation direction is treated as the time axis. By propagating a general wave of complex cross-section mode profile down the waveguide for some distance using the BPMs, we can identify the guided modes for a specific waveguide. Due to the mismatch of refractive indices in the cross-section of the waveguides, the electromagnetic fields are discontinuous solutions to the Schrödinger equations. Different variants of the BPM, using various types of spatial discretization in the cross-section of waveguides, exist, such as the finite element (FE)-BPM [4], the fast Fourier transform (FFT)-BPM [5], and the finite difference (FD)-BPM [6]. In this section, we discuss GDG-BPMs to address specifically the discontinuities in fields across dielectric interfaces [7].

15.2.1 Guided Modes in Optical Waveguides

Light at optical frequencies can be confined within the core of an optical fiber or the inner layer of layered dielectrics through the mechanism of total internal reflection, when the refractive index of the interior dielectric n_1 is larger than that of the sur-

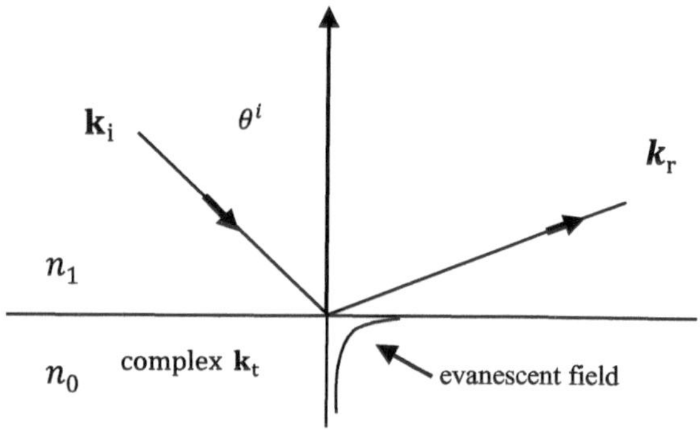

Fig. 15.1 Total internal reflection

rounding medium (called cladding in the case of optical fibers) n_0 [8]. The refractive index of a dielectric medium is defined by

$$n = \sqrt{\epsilon_r}, \tag{15.39}$$

where ϵ_r is the relative dielectric constant. An incident plane wave with a wave vector $\mathbf{k}_i$ from the side with larger refractive index (Fig. 15.1) will have its energy reflected completely if the incident angle

$$\theta^i \geq \theta_c, \tag{15.40}$$

where θ_c is the critical angle defined by

$$\theta_c = \sin^{-1}\left(\frac{n_0}{n_1}\right). \tag{15.41}$$

In this case, the transmitted plane wave on the side with the smaller refractive index will have a complex wave vector $\mathbf{k}_t$, which corresponds to an evanescent wave.

Once condition (15.40) is satisfied, guided wave modes can exist in optical fibers such as that shown in Fig. 15.2 with cylindrical core Ω_1 and cladding Ω_2:

$$\Omega_1 = \{(x, y) : x^2 + y^2 \leq a\}, \quad \Omega_2 = \{(x, y) : a \leq x^2 + y^2 < \infty\}, \tag{15.42}$$

where we have assumed the cladding Ω_2 is of infinite extent for simplicity.

The guided modes are assumed to be in the following forms:

$$\mathbf{E} \text{ or } \mathbf{H} = (\varphi_1(x, y), \varphi_2(x, y), \varphi_3(x, y))e^{-i\beta z}, \tag{15.43}$$

where $e^{-i\beta z}$ indicates the propagation nature of the modes along the z-direction.

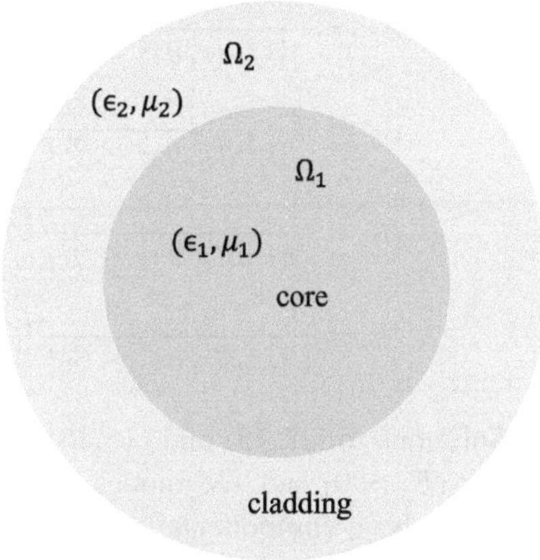

Fig. 15.2 Optical fiber with cylindrical core and cladding of finite thickness

Maxwell's equations in the cylindrical coordinates (r, θ) are

$$\begin{aligned}
\frac{1}{r}\frac{\partial E_z}{\partial \theta} + \mathrm{i}\beta E_\theta &= -\mathrm{i}\omega\mu_0 H_r, \\
-\mathrm{i}\beta E_r - \frac{\partial E_z}{\partial r} &= -\mathrm{i}\omega\mu_0 H_\theta, \\
\frac{1}{r}\frac{\partial (r E_\theta)}{\partial r} - \frac{1}{r}\frac{\partial E_r}{\partial \theta} &= -\mathrm{i}\omega\mu_0 H_z, \\
\frac{1}{r}\frac{\partial H_z}{\partial \theta} + \mathrm{i}\beta H_\theta &= \mathrm{i}\omega\epsilon_0 n^2 E_r, \\
-\mathrm{i}\beta H_r - \frac{\partial H_z}{\partial r} &= \mathrm{i}\omega\epsilon_0 n^2 E_\theta, \\
\frac{1}{r}\frac{\partial (r H_\theta)}{\partial r} - \frac{1}{r}\frac{\partial H_r}{\partial \theta} &= \mathrm{i}\omega\epsilon_0 n^2 E_z,
\end{aligned} \tag{15.44}$$

where the dielectric constant $\epsilon(r) = \epsilon_0 n^2 = \epsilon_0 \epsilon_r$ is assumed to be a function of r only for optical fibers.

Meanwhile, the vector Helmholtz equations (8.30) and (8.31) for the z-components of the fields (with $e^{-\mathrm{i}\beta z}$ factored out) in the cylindrical coordinates are given by

$$\frac{\partial^2 E_z}{\partial z^2} + \frac{1}{r}\frac{\partial E_z}{\partial r} + \frac{1}{r^2}\frac{\partial^2 E_z}{\partial \theta^2} + \left[k^2 n(r,\theta)^2 - \beta^2\right] E_z = 0, \tag{15.45}$$

$$\frac{\partial^2 H_z}{\partial z^2} + \frac{1}{r}\frac{\partial H_z}{\partial r} + \frac{1}{r^2}\frac{\partial^2 H_z}{\partial \theta^2} + \left[k^2 n(r,\theta)^2 - \beta^2\right] H_z = 0, \tag{15.46}$$

and the transverse components can be expressed in terms of the z-components by

$$E_r = -\frac{\mathrm{i}}{[k^2 n(r,\theta)^2 - \beta^2]}\left(\beta\frac{\partial E_z}{\partial r} + \frac{\omega\mu_0}{r}\frac{\partial H_z}{\partial \theta}\right), \tag{15.47}$$

$$E_\theta = -\frac{\mathrm{i}}{[k^2 n(r,\theta)^2 - \beta^2]}\left(\frac{\beta}{r}\frac{\partial E_z}{\partial \theta} - \frac{\omega\mu_0}{r}\frac{\partial H_z}{\partial r}\right), \tag{15.48}$$

$$H_r = -\frac{\mathrm{i}}{[k^2 n(r,\theta)^2 - \beta^2]}\left(\beta\frac{\partial H_z}{\partial r} - \frac{\omega\epsilon_0 n(r)^2}{r}\frac{\partial E_z}{\partial \theta}\right), \tag{15.49}$$

$$H_\theta = -\frac{\mathrm{i}}{[k^2 n(r,\theta)^2 - \beta^2]}\left(\frac{\beta}{r}\frac{\partial H_z}{\partial \theta} + \frac{\omega\epsilon_0 n(r)^2}{r}\frac{\partial E_z}{\partial r}\right). \tag{15.50}$$

Solutions to (15.45) and (15.46) can be found in transverse modes such as TE modes ($E_z = 0$) and TM modes ($H_z = 0$) as well as hybrid modes, or HE modes (both of the z-components are non-zero). To find the hybrid modes, methods of separation of variables are used by looking for the z-components in the following form (with continuity implied at $r = a$) [8]:

$$E_z(r,\theta) = \begin{cases} A J_n\left(\frac{u}{a}r\right)\cos(n\theta + \Psi), & \text{if } 0 \le r \le a, \\ A\frac{J_n(u)}{K_n(w)} K_n\left(\frac{w}{a}r\right)\cos(n\theta + \Psi), & \text{if } r \ge a, \end{cases} \tag{15.51}$$

and

$$H_z(r,\theta) = \begin{cases} B J_n\left(\frac{u}{a}r\right)\sin(n\theta + \Psi), & \text{if } 0 \le r \le a, \\ B\frac{J_n(u)}{K_n(w)} K_n\left(\frac{w}{a}r\right)\sin(n\theta + \Psi), & \text{if } r \ge a, \end{cases} \tag{15.52}$$

where $J_n(r)$ is the nth-order Bessel function of the first kind, $K_n(r)$ is the nth-order modified Bessel function of the second kind, Ψ is a phase parameter defining the polarization of the HE mode, and the following shorthand notation is used:

$$u = a\sqrt{k^2 n_1^2 - \beta^2}, \quad w = a\sqrt{\beta^2 - k^2 n_0^2}. \tag{15.53}$$

The coefficients A and B, as well as the propagation constant β, will be determined by the continuity condition of the tangential components E_θ and H_θ, given by (15.48) and (15.50), at the core/cladding interface $r = a$, yielding the following relations, respectively:

$$A\beta\left(\frac{1}{u^2} + \frac{1}{w^2}\right) = -B\omega\mu_0\left[\frac{J_n'(u)}{uJ_n(u)} + \frac{K_n'(w)}{wK_n(w)}\right] \tag{15.54}$$

and

$$A\omega\epsilon_0\left[n_1^2\frac{J_n'(u)}{uJ_n(u)} + n_0^2\frac{K_n'(w)}{wK_n(w)}\right] = -B\beta\left(\frac{1}{u^2} + \frac{1}{w^2}\right)n. \tag{15.55}$$

Combining (15.54) and (15.55), we obtain the following nonlinear equation (characteristic equation) for the propagation constant β:

$$\left[\frac{J_n'(u)}{uJ_n(u)} + \frac{K_n'(w)}{wK_n(w)}\right]\left[\frac{J_n'(u)}{uJ_n(u)} + \left(\frac{n_0}{n_1}\right)^2 \frac{K_n'(w)}{wK_n(w)}\right]$$
$$= n^2\left(\frac{1}{u^2} + \frac{1}{w^2}\right)\left[\frac{1}{u^2} + \left(\frac{n_0}{n_1}\right)^2 \frac{1}{w^2}\right]. \tag{15.56}$$

With β obtained, the coefficients A and B are found from (15.54) to be related by

$$B = -A\frac{\beta}{\omega\mu_0}s, \tag{15.57}$$

where

$$s = \frac{n(1/u^2 + 1/w^2)}{\left[\frac{J_n'(u)}{uJ_n(u)} + \frac{K_n'(w)}{wK_n(w)}\right]}. \tag{15.58}$$

For $n = 1$, the HE mode is called the fundamental hybrid mode, denoted by HE_{11}, whose field components in (x, y)-coordinates can be computed by

$$E_x = E_r\cos\theta - E_\theta\sin\theta,$$
$$E_y = E_r\sin\theta + E_\theta\cos\theta,$$

along with a similar relation for H_x and H_y, resulting in

$$E_x = \begin{cases} -\mathrm{i}A\beta\frac{a}{u}\begin{bmatrix} \frac{1-s}{2}J_0(\frac{u}{a}r)\cos(\Psi) \\ -\frac{1+s}{2}J_2(\frac{u}{a}r)\cos(2\theta+\Psi)\end{bmatrix}, & \text{in } \Omega_1, \\ -\mathrm{i}A\beta\frac{aJ_1(u)}{wK_1(w)}\begin{bmatrix} \frac{1-s}{2}K_0(\frac{w}{a}r)\cos(\Psi) \\ +\frac{1+s}{2}K_2(\frac{w}{a}r)\cos(2\theta+\Psi)\end{bmatrix}, & \text{in } \Omega_2, \end{cases} \tag{15.59}$$

$$E_y = \begin{cases} \mathrm{i}A\beta\frac{a}{u}\begin{bmatrix} \frac{1-s}{2}J_0(\frac{u}{a}r)\sin(\Psi) \\ +\frac{1+s}{2}J_2(\frac{u}{a}r)\sin(2\theta+\Psi)\end{bmatrix}, & \text{in } \Omega_1, \\ \mathrm{i}A\beta\frac{aJ_1(u)}{wK_1(w)}\begin{bmatrix} \frac{1-s}{2}K_0(\frac{w}{a}r)\sin(\Psi) \\ -\frac{1+s}{2}K_2(\frac{w}{a}r)\sin(2\theta+\Psi)\end{bmatrix}, & \text{in } \Omega_2, \end{cases} \tag{15.60}$$

$$E_z = \begin{cases} AJ_1(\frac{u}{a}r)\cos(\theta+\Psi), & \text{in } \Omega_1, \\ A\frac{J_1(u)}{K_1(w)}K_1(\frac{w}{a}r)\cos(\theta+\Psi), & \text{in } \Omega_2, \end{cases} \tag{15.61}$$

and

$$H_x = \begin{cases} -\mathrm{i}A\omega\epsilon_0 n_1^2 \frac{a}{u} \begin{bmatrix} \frac{1-s_1}{2} J_0(\frac{u}{a}r)\sin(\Psi) \\ +\frac{1+s_1}{2} J_2(\frac{u}{a}r)\sin(2\theta+\Psi) \end{bmatrix}, & \text{in } \Omega_1, \\ \\ -\mathrm{i}A\omega\epsilon_0 n_0^2 \frac{aJ_1(u)}{wK_1(w)} \begin{bmatrix} \frac{1-s_0}{2} K_0(\frac{w}{a}r)\sin(\Psi) \\ -\frac{1+s_0}{2} K_2(\frac{w}{a}r)\sin(2\theta+\Psi) \end{bmatrix}, & \text{in } \Omega_2, \end{cases} \tag{15.62}$$

$$H_y = \begin{cases} -\mathrm{i}A\omega\epsilon_0 n_1^2 \frac{a}{u} \begin{bmatrix} \frac{1-s_1}{2} J_0(\frac{u}{a}r)\cos(\Psi) \\ -\frac{1+s_1}{2} J_2(\frac{u}{a}r)\cos(2\theta+\Psi) \end{bmatrix}, & \text{in } \Omega_1, \\ \\ -\mathrm{i}A\omega\epsilon_0 n_0^2 \frac{aJ_1(u)}{wK_1(w)} \begin{bmatrix} \frac{1-s_0}{2} K_0(\frac{w}{a}r)\cos(\Psi) \\ +\frac{1+s_0}{2} K_2(\frac{w}{a}r)\cos(2\theta+\Psi) \end{bmatrix}, & \text{in } \Omega_2, \end{cases} \tag{15.63}$$

$$H_z = \begin{cases} -A\frac{\beta s}{\omega\mu_0} J_1(\frac{u}{a}r)\sin(\theta+\Psi), & \text{in } \Omega_1, \\ \\ -A\frac{\beta s}{\omega\mu_0}\frac{J_1(u)}{K_1(w)} K_1(\frac{w}{a}r)\sin(\theta+\Psi), & \text{in } \Omega_2, \end{cases} \tag{15.64}$$

where

$$s_0 = \frac{\beta^2}{k^2 n_0^2} s, \; s_1 = \frac{\beta^2}{k^2 n_1^2} s. \tag{15.65}$$

Here $\Psi = 0$ and $\Psi = \pi/2$ correspond to the x-polarized (HE_{11}^x) mode and the y-polarized (HE_{11}^y) mode, respectively.

15.2.2 *Discontinuities in Envelopes of Guided Modes*

In a BPM for optical waveguides, we assume that the electric or magnetic field takes the envelope form (15.43). The envelope functions φ_j, $j = 1, 2, 3$, are assumed to vary slowly along the *propagation direction* z, which will be denoted as the time variable t [2, 3].

Figure 15.2 shows the cross-section Ω of an optical waveguide with a core Ω_1, a cladding Ω_2, and an interface at $\Gamma = \Omega_1 \cap \Omega_2$ with an exterior normal direction $\mathbf{n} = (n_x, n_y)$. Let us denote the jump of a function φ at position τ along $\mathbf{n} = (n_x, n_y)$ on Γ as

$$[\varphi(\tau, t)] = \varphi(\tau^+, t) - \varphi(\tau^-, t). \tag{15.66}$$

Then, for any point (x^*, y^*) on the interface, for $j = 1, 2, 3$, the jump data

$$\begin{aligned} f_j(x^*, y^*, t) &= [\varphi_j(x^*, y^*, t)] = \varphi_j(x^{*+}, y^{*+}, t) - \varphi_j(x^{*-}, y^{*-}, t), \\ g_j(x^*, y^*, t) &= \left[\frac{\partial\varphi_j(x^*, y^*, t)}{\partial\mathbf{n}}\right] = \frac{\partial\varphi_j(x^{*+}, y^{*+}, t)}{\partial\mathbf{n}} - \frac{\partial\varphi_j(x^{*-}, y^{*-}, t)}{\partial\mathbf{n}}, \end{aligned} \tag{15.67}$$

can be shown to satisfy identities based on the interface conditions for the electromagnetic fields and Maxwell's equations as follows.

To derive the jump data f_j of the electric field, we start from the interface conditions of the electric field,

$$\epsilon_1 E_\xi^- = \epsilon_2 E_\xi^+, \quad E_\eta^- = E_\eta^+, \quad E_z^- = E_z^+, \tag{15.68}$$

where ξ and η are the local normal and the tangential coordinates on the interface Γ, and $E_\xi = E_x n_x + E_y n_y$ and $E_\eta = -E_x n_y + E_y n_x$ are the normal and the tangential electric field components, respectively. Using the envelope assumption of (15.43), and denoting

$$\varphi_\xi = \varphi_1 n_x + \varphi_2 n_y, \quad \varphi_\eta = -\varphi_1 n_y + \varphi_2 n_x, \tag{15.69}$$

$$\gamma_f = \frac{\epsilon_1}{\epsilon_2}, \tag{15.70}$$

we can obtain

$$\gamma_f \varphi_\xi^- = \varphi_\xi^+, \quad \varphi_\eta^- = \varphi_\eta^+, \quad \varphi_3^- = \varphi_3^+. \tag{15.71}$$

From (15.71), f_j can be written in terms of $(\varphi_\xi^-, \varphi_\eta^-, \varphi_3^-)$ and $(\varphi_\xi^+, \varphi_\eta^+, \varphi_3^+)$ in a symmetric form:

$$\begin{cases} \begin{pmatrix} f_1 \\ f_2 \end{pmatrix} = \frac{1}{2} \begin{bmatrix} n_x & -n_y \\ n_y & n_x \end{bmatrix} \begin{pmatrix} (\gamma_f - 1)\varphi_\xi^- + (1 - \gamma_f^{-1})\varphi_\xi^+ \\ 0 \end{pmatrix}, \\ f_3 = 0, \end{cases} \tag{15.72}$$

which recasts the interface conditions (15.68) of the electric field in terms of the envelope functions.

To derive the jump data g_j for the normal derivative of the electric field, we consider the interface conditions of the magnetic field,

$$\mu_1 H_\xi^- = \mu_2 H_\xi^+, \quad H_\eta^- = H_\eta^+, \quad H_z^- = H_z^+. \tag{15.73}$$

With the envelope assumption (15.43) and Ampère's law

$$\nabla \times \mathbf{E} = -\mathrm{i}\omega\mu\mathbf{H}, \tag{15.74}$$

defining

$$\gamma_g = \frac{\mu_2}{\mu_1}, \tag{15.75}$$

we get

$$\begin{cases} \left[\frac{\partial\varphi_3}{\partial\eta} - \left(\frac{\partial\varphi_\eta}{\partial t} - \mathrm{i}\beta\varphi_\eta\right)\right]^- = \left[\frac{\partial\varphi_3}{\partial\eta} - \left(\frac{\partial\varphi_\eta}{\partial t} - \mathrm{i}\beta\varphi_\eta\right)\right]^+, \\ \gamma_g\left[\left(\frac{\partial\varphi_\xi}{\partial t} - \mathrm{i}\beta\varphi_\xi\right) - \frac{\partial\varphi_3}{\partial\xi}\right]^- = \left[\left(\frac{\partial\varphi_\xi}{\partial t} - \mathrm{i}\beta\varphi_\xi\right) - \frac{\partial\varphi_3}{\partial\xi}\right]^+, \\ \gamma_g\left(\frac{\partial\varphi_\eta}{\partial\xi} - \frac{\partial\varphi_\xi}{\partial\eta}\right)^- = \left(\frac{\partial\varphi_\eta}{\partial\xi} - \frac{\partial\varphi_\xi}{\partial\eta}\right)^+. \end{cases} \tag{15.76}$$

From the second equation, g_3 can be expressed in terms of $\partial\varphi_3^-/\partial\xi$ and $\partial\varphi_3^+/\partial\xi$ in a symmetric form:

$$\begin{aligned} g_3 = \frac{1}{2}\Bigg[&(1+\gamma_g^{-1})\left(\frac{\partial\varphi_\xi^+}{\partial t} - \mathrm{i}\beta\varphi_\xi^+\right) + (\gamma_g - 1)\frac{\partial\varphi_3^-}{\partial\xi} \\ &-(1+\gamma_g)\left(\frac{\partial\varphi_\xi^-}{\partial t} - \mathrm{i}\beta\varphi_\xi^-\right) + (1-\gamma_g^{-1})\frac{\partial\varphi_3^+}{\partial\xi}\Bigg]. \end{aligned} \tag{15.77}$$

The right-hand side of (15.77) actually involves time derivatives of φ_1 and φ_2 on both sides of the interface, which can be replaced with spatial derivatives by using the time-dependent Schrödinger equations for φ_1 and φ_2.

Next, using the identity

$$n_x g_1 + n_y g_2 = \frac{\partial\varphi_\xi^+}{\partial\xi} - \frac{\partial\varphi_\xi^-}{\partial\xi}, \tag{15.78}$$

and the third equation in (15.76), and after some manipulation, we find g_1 and g_2 in terms of $(\partial\varphi_\xi^-/\partial\xi, \partial\varphi_\eta^-/\partial\xi)$ and $(\partial\varphi_\xi^+/\partial\xi, \partial\varphi_\eta^+/\partial\xi)$ in a symmetric form:

$$\begin{pmatrix} g_1 \\ g_2 \end{pmatrix} = \frac{1}{2}\begin{bmatrix} n_x & -n_y \\ n_y & n_x \end{bmatrix}\begin{pmatrix} 2\frac{\partial\varphi_\xi^+}{\partial\xi} - 2\frac{\partial\varphi_\xi^-}{\partial\xi} \\ (\gamma_g - 1)\frac{\partial\varphi_\eta^-}{\partial\xi} + \left(1 - \frac{1}{\gamma_g}\right)\frac{\partial\varphi_\eta^+}{\partial\xi} + \\ \left(1 + \frac{1}{\gamma_g}\right)\frac{\partial\varphi_\xi^+}{\partial\eta} - (1+\gamma_g)\frac{\partial\varphi_\xi^-}{\partial\eta} \end{pmatrix}. \tag{15.79}$$

Similarly, the jump data f_j for the magnetic field components and g_j for their normal derivatives can be shown to satisfy (15.72), (15.77), and (15.79), with

$$\gamma_f = \frac{\mu_1}{\mu_2}, \qquad \gamma_g = \frac{\epsilon_2}{\epsilon_1}. \tag{15.80}$$

15.2.3 GDG-BPM for Electric Fields

In this section, we use the GDG method to obtain a full vectorial GDG-BPM for optical waveguides where the electromagnetic fields and/or their derivatives can be discontinuous across material interfaces. To illustrate the GDG-BPM, we consider the paraxial approximation of a standard cylindrical optical fiber (Fig. 15.2).

Assuming that the field is time-harmonic with frequency ω and that there are no charge or current sources, the vector wave equation for $\mathbf{E}(x, y, z = t)$ in (8.30) simplifies to

$$\nabla \times \nabla \times \mathbf{E} = \omega^2 \epsilon \mu \mathbf{E}, \tag{15.81}$$

which leads to

$$\nabla^2 \mathbf{E} + \omega^2 \epsilon \mu \mathbf{E} = \nabla(\nabla \cdot \mathbf{E}). \tag{15.82}$$

Since $\nabla \cdot (\epsilon \mathbf{E}) = 0$, we have

$$\nabla \cdot \mathbf{E} = \nabla \cdot \left(\frac{1}{\epsilon} \epsilon \mathbf{E} \right) = -\frac{1}{\epsilon} (\nabla \epsilon \cdot \mathbf{E}) = -\nabla \hat{\epsilon} \cdot \mathbf{E}, \tag{15.83}$$

where

$$\hat{\epsilon} \equiv \ln \epsilon. \tag{15.84}$$

Assuming that $\epsilon = \epsilon(x, y)$ is uniform along the fiber propagation direction, the vector Helmholtz equation (15.82) becomes

$$\nabla^2 \mathbf{E} + \omega^2 \epsilon \mu \mathbf{E} = -\nabla \left(\frac{\partial \hat{\epsilon}}{\partial x} E_x + \frac{\partial \hat{\epsilon}}{\partial y} E_y \right). \tag{15.85}$$

Based on the slow envelope assumption for (15.13) (paraxial approximation, [9]), i.e.,

$$\left| \frac{\partial^2 \varphi_j}{\partial t^2} \right| \ll 2\beta \left| \frac{\partial \varphi_j}{\partial t} \right|, \quad j = 1, 2, 3, \tag{15.86}$$

we can ignore the second-order derivative in t to obtain the following coupled equations at $(x, y) \notin \Gamma$:

$$\begin{aligned} i2\beta \frac{\partial \varphi_1}{\partial t} =& \frac{\partial^2 \varphi_1}{\partial x^2} + \frac{\partial^2 \varphi_1}{\partial y^2} + (\omega^2 \epsilon \mu - \beta^2) \varphi_1 \\ &+ \frac{\partial^2 \hat{\epsilon}}{\partial x^2} \varphi_1 + \frac{\partial^2 \hat{\epsilon}}{\partial x \partial y} \varphi_2 + \frac{\partial \hat{\epsilon}}{\partial x} \frac{\partial \varphi_1}{\partial x} + \frac{\partial \hat{\epsilon}}{\partial y} \frac{\partial \varphi_2}{\partial x}, \end{aligned} \tag{15.87}$$

$$\begin{aligned} i2\beta \frac{\partial \varphi_2}{\partial t} =& \frac{\partial^2 \varphi_2}{\partial x^2} + \frac{\partial^2 \varphi_2}{\partial y^2} + (\omega^2 \epsilon \mu - \beta^2) \varphi_2 \\ &+ \frac{\partial^2 \hat{\epsilon}}{\partial x \partial y} \varphi_1 + \frac{\partial^2 \hat{\epsilon}}{\partial y^2} \varphi_2 + \frac{\partial \hat{\epsilon}}{\partial x} \frac{\partial \varphi_1}{\partial y} + \frac{\partial \hat{\epsilon}}{\partial y} \frac{\partial \varphi_2}{\partial y}, \end{aligned} \tag{15.88}$$

$$\begin{aligned} \mathrm{i}2\beta\frac{\partial\varphi_3}{\partial t} =& \frac{\partial^2\varphi_3}{\partial x^2} + \frac{\partial^2\varphi_3}{\partial y^2} + (\omega^2\epsilon\mu - \beta^2)\varphi_3 \\ &+ \frac{\partial\hat{\epsilon}}{\partial x}\left(\frac{\partial\varphi_1}{\partial t} - \mathrm{i}\beta\varphi_1\right) + \frac{\partial\hat{\epsilon}}{\partial y}\left(\frac{\partial\varphi_2}{\partial t} - \mathrm{i}\beta\varphi_2\right). \end{aligned} \tag{15.89}$$

For convenience, we define the jump data for the dielectric constant ϵ as

$$\begin{aligned} f_{\hat{\epsilon}}(x^*, y^*, t) &= \left[\hat{\epsilon}(x^*, y^*, t)\right] = \hat{\epsilon}(x^{*+}, y^{*+}, t) - \hat{\epsilon}(x^{*-}, y^{*-}, t), \\ g_{\hat{\epsilon}}(x^*, y^*, t) &= \left[\frac{\partial\hat{\epsilon}(x^*, y^*, t)}{\partial\mathbf{n}}\right] = \frac{\partial\hat{\epsilon}(x^{*+}, y^{*+}, t)}{\partial\mathbf{n}} - \frac{\partial\hat{\epsilon}(x^{*-}, y^{*-}, t)}{\partial\mathbf{n}}. \end{aligned}$$

Following the procedure proposed in Sect. 15.1 [1], we can rewrite the system (15.87)–(15.89) using Dirac δ functions as follows.

Formulation A

For $j = 1, 2, 3$,

$$\mathrm{i}2\beta\frac{\partial\varphi_j}{\partial t} = \frac{\partial p_j}{\partial x} + \frac{\partial q_j}{\partial y} - \delta(\xi - \xi^*)|\nabla\xi|^2 g_j + S_j, \tag{15.90}$$

$$p_j = \frac{\partial\varphi_j}{\partial x} - \delta(\xi - \xi^*) f_j \frac{\partial\xi}{\partial x}, \tag{15.91}$$

$$q_j = \frac{\partial\varphi_j}{\partial y} - \delta(\xi - \xi^*) f_j \frac{\partial\xi}{\partial y}, \tag{15.92}$$

where the jump data f_j and g_j are given by (15.72), (15.77), and (15.79) to enforce the physical jump conditions for the electromagnetic field components. The lower-order terms above are given by

$$\begin{aligned} S_1(\varphi_1, \varphi_2, p_1, p_2) &= (\omega^2\epsilon\mu - \beta^2)\varphi_1 + p_{\hat{\epsilon}}^x\varphi_1 + q_{\hat{\epsilon}}^x\varphi_2 + p_{\hat{\epsilon}}p_1 + q_{\hat{\epsilon}}p_2, \\ S_2(\varphi_1, \varphi_2, q_1, q_2) &= (\omega^2\epsilon\mu - \beta^2)\varphi_2 + p_{\hat{\epsilon}}^y\varphi_1 + q_{\hat{\epsilon}}^y\varphi_2 + p_{\hat{\epsilon}}q_1 + q_{\hat{\epsilon}}q_2, \\ S_3(\varphi_1, \varphi_2, \varphi_3) &= (\omega^2\epsilon\mu - \beta^2)\varphi_3 + p_{\hat{\epsilon}}\left(\frac{\partial\varphi_1}{\partial t} - \mathrm{i}\beta\varphi_1\right) + q_{\hat{\epsilon}}\left(\frac{\partial\varphi_2}{\partial t} - \mathrm{i}\beta\varphi_2\right), \end{aligned}$$

where $\partial\varphi_1/\partial t$ and $\partial\varphi_2/\partial t$ in S_3 can be replaced by (15.90), with $j = 1, 2$, and

$$\begin{aligned} p_{\hat{\epsilon}} &= \frac{\partial\hat{\epsilon}}{\partial x} - \delta(\xi - \xi^*) f_{\hat{\epsilon}}\frac{\partial\xi}{\partial x}, \\ q_{\hat{\epsilon}} &= \frac{\partial\hat{\epsilon}}{\partial y} - \delta(\xi - \xi^*) f_{\hat{\epsilon}}\frac{\partial\xi}{\partial y}, \\ p_{\hat{\epsilon}}^x &= \frac{\partial p_{\hat{\epsilon}}}{\partial x} - \delta(\xi - \xi^*)\left(\frac{\partial f_{\hat{\epsilon}}}{\partial x} + g_{\hat{\epsilon}}\frac{\partial\xi}{\partial x}\right)\frac{\partial\xi}{\partial x}, \end{aligned}$$

$$
\begin{aligned}
p_{\hat{\epsilon}}^y &= \frac{\partial p_{\hat{\epsilon}}}{\partial y} - \delta(\xi - \xi^*)\left(\frac{\partial f_{\hat{\epsilon}}}{\partial x} + g_{\hat{\epsilon}}\frac{\partial \xi}{\partial x}\right)\frac{\partial \xi}{\partial y}, \\
q_{\hat{\epsilon}}^x &= \frac{\partial p_{\hat{\epsilon}}}{\partial x} - \delta(\xi - \xi^*)\left(\frac{\partial f_{\hat{\epsilon}}}{\partial y} + g_{\hat{\epsilon}}\frac{\partial \xi}{\partial y}\right)\frac{\partial \xi}{\partial x}, \\
q_{\hat{\epsilon}}^y &= \frac{\partial p_{\hat{\epsilon}}}{\partial y} - \delta(\xi - \xi^*)\left(\frac{\partial f_{\hat{\epsilon}}}{\partial y} + g_{\hat{\epsilon}}\frac{\partial \xi}{\partial y}\right)\frac{\partial \xi}{\partial y}.
\end{aligned}
$$

Here, $p_{\hat{\epsilon}}, q_{\hat{\epsilon}}, p_{\hat{\epsilon}}^x, p_{\hat{\epsilon}}^y, q_{\hat{\epsilon}}^x$, and $q_{\hat{\epsilon}}^y$ will be zero if ϵ is piecewise constant.

In the above derivations, partial derivatives of $f(x, y)$ on Γ are used, while $f(x, y)$ is only given on the interface Γ. Therefore, some types of smooth extension away from the interface will be needed to yield those partial derivatives. The simplest one is to use a constant extension locally along the normal direction to the interface Γ, i.e., assuming $\partial f/\partial \xi = 0$. Then we have

$$
\begin{aligned}
\frac{\partial f}{\partial x} &= \frac{\partial f}{\partial \xi}\frac{\partial \xi}{\partial x} + \frac{\partial f}{\partial \eta}\frac{\partial \eta}{\partial x} = \frac{\partial f}{\partial \eta}\frac{\partial \eta}{\partial x}, \\
\frac{\partial f}{\partial y} &= \frac{\partial f}{\partial \xi}\frac{\partial \xi}{\partial y} + \frac{\partial f}{\partial \eta}\frac{\partial \eta}{\partial y} = \frac{\partial f}{\partial \eta}\frac{\partial \eta}{\partial y}.
\end{aligned}
$$

The extension is by no means unique. However, the accuracy of the resulting numerical methods will not be affected by a specific choice of the extension as long as the extension produces a locally smooth function.

Alternatively, we can use Gauss's law $\nabla \cdot (\epsilon \mathbf{E}) = 0$ to solve for the E_z component, which results in the following equation for $\varphi_3(x, y, z \equiv t)$:

$$
\epsilon\frac{\partial \varphi_3}{\partial t} = -\epsilon\frac{\partial \varphi_1}{\partial x} - \epsilon\frac{\partial \varphi_2}{\partial y} - \epsilon_x\varphi_1 - \epsilon_y\varphi_2 - (\epsilon_t - \mathrm{i}\beta\epsilon)\varphi_3. \tag{15.93}
$$

If the evolution equation (15.93) is used for φ_3, instead of (15.90) with $j = 3$, we have the following alternative version of the GDG-BPM for the electric field.

Formulation B

For $j = 1, 2$,

$$
\begin{aligned}
\mathrm{i}2\beta\frac{\partial \varphi_j}{\partial t} &= \frac{\partial p_j}{\partial x} + \frac{\partial q_j}{\partial y} - \delta(\xi - \xi^*)|\nabla\xi|^2 g_j + S_j, && (15.94) \\
\epsilon\frac{\partial \varphi_3}{\partial t} &= -\epsilon p_1 - \epsilon q_2 - \epsilon_x\varphi_1 - \epsilon_y\varphi_2 - (\epsilon_t - \mathrm{i}\beta\epsilon)\varphi_3, && (15.95) \\
p_j &= \frac{\partial \varphi_j}{\partial x} - \delta(\xi - \xi^*)f_j\frac{\partial \xi}{\partial x}, && (15.96) \\
q_j &= \frac{\partial \varphi_j}{\partial y} - \delta(\xi - \xi^*)f_j\frac{\partial \xi}{\partial y}. && (15.97)
\end{aligned}
$$

The time evolution equation for φ_3 in formulation B is a simple ODE, thus requiring less computational cost compared to the corresponding equation for φ_3 in formulation A.

15.2.4 GDG-BPM for Magnetic Fields

We obtain a wave equation for $\mathbf{H}$ as follows:

$$\nabla \times \left(\frac{1}{\epsilon}\nabla \times \mathbf{H}\right) = \omega^2\mu\mathbf{H}. \tag{15.98}$$

Assuming that μ is a constant (implying $\nabla \cdot \mathbf{H} = 0$), we get

$$\nabla \times \left(\frac{1}{\epsilon}\nabla \times \mathbf{H}\right) = -\frac{1}{\epsilon}\nabla^2\mathbf{H} + \left(\nabla\frac{1}{\epsilon}\right) \times (\nabla \times \mathbf{H}). \tag{15.99}$$

Therefore, we have

$$\frac{1}{\epsilon}\nabla^2\mathbf{H} = \nabla\frac{1}{\epsilon} \times (\nabla \times \mathbf{H}) - \omega^2\mu\mathbf{H}. \tag{15.100}$$

Again, we assume $\mathbf{H}$ has an envelope formulation:

$$\begin{aligned}\mathbf{H} &= (H_x(x, y, z), H_y(x, y, z), H_z(x, y, z))^{\mathrm{T}} \\ &= (\varphi_1(x, y, z), \varphi_2(x, y, z), \varphi_3(x, y, z))^{\mathrm{T}}\mathrm{e}^{-\mathrm{i}\beta z}.\end{aligned} \tag{15.101}$$

On dropping the term $\partial^2\varphi_j/\partial z^2$ based on the paraxial approximation, replacing z by t, and assuming that $\hat{\epsilon} \equiv \ln \epsilon$ is independent of z, we get the following coupled equations at $(x, y) \notin \Gamma$:

$$\mathrm{i}2\beta\frac{\partial\varphi_1}{\partial t} = \frac{\partial^2\varphi_1}{\partial x^2} + \frac{\partial^2\varphi_1}{\partial y^2} + (\omega^2\epsilon\mu - \beta^2)\varphi_1 + \frac{\partial\hat{\epsilon}}{\partial y}\left(\frac{\partial\varphi_2}{\partial x} - \frac{\partial\varphi_1}{\partial y}\right), \tag{15.102}$$

$$\mathrm{i}2\beta\frac{\partial\varphi_2}{\partial t} = \frac{\partial^2\varphi_2}{\partial x^2} + \frac{\partial^2\varphi_2}{\partial y^2} + (\omega^2\epsilon\mu - \beta^2)\varphi_2 - \frac{\partial\hat{\epsilon}}{\partial x}\left(\frac{\partial\varphi_2}{\partial x} - \frac{\partial\varphi_1}{\partial y}\right), \tag{15.103}$$

$$\begin{aligned}\mathrm{i}2\beta\frac{\partial\varphi_3}{\partial t} &= \frac{\partial^2\varphi_3}{\partial x^2} + \frac{\partial^2\varphi_3}{\partial y^2} + (\omega^2\epsilon\mu - \beta^2)\varphi_3 \\ &\quad + \frac{\partial\hat{\epsilon}}{\partial x}\left(\frac{\partial\varphi_1}{\partial t} - \mathrm{i}\beta\varphi_1 - \frac{\partial\varphi_3}{\partial x}\right) + \frac{\partial\hat{\epsilon}}{\partial y}\left(\frac{\partial\varphi_2}{\partial t} - \mathrm{i}\beta\varphi_2 - \frac{\partial\varphi_3}{\partial y}\right).\end{aligned} \tag{15.104}$$

Similar to Sect. 15.2.3, for the interface conditions, we use f_j and g_j to denote the jumps of φ_j and $\partial\varphi_j/\partial\mathbf{n}$ on the interface, respectively.

Now, we use the δ function and the auxiliary variables p and q to rewrite (15.102)–(15.104) as follows:

Formulation C

For $j = 1, 2, 3$,

$$i2\beta \frac{\partial \varphi_j}{\partial t} = \frac{\partial p_j}{\partial x} + \frac{\partial q_j}{\partial y} - \delta(\xi - \xi^*)|\nabla \xi|^2 g_j + S_j, \tag{15.105}$$

$$p_j = \frac{\partial \varphi_j}{\partial x} - \delta(\xi - \xi^*) f_j \frac{\partial \xi}{\partial x}, \tag{15.106}$$

$$q_j = \frac{\partial \varphi_j}{\partial y} - \delta(\xi - \xi^*) f_j \frac{\partial \xi}{\partial y}, \tag{15.107}$$

where the jump data f_j and g_j are again given by (15.72), (15.77), and (15.79) with (15.80) to enforce the physical interface conditions for the electromagnetic field components. The lower-order source terms are given by

$$\begin{aligned}
S_1(\varphi_1, \varphi_2, p_2, q_1) &= (\omega^2 \epsilon\mu - \beta^2)\varphi_1 + q_{\hat{\epsilon}}(p_2 - q_1), \\
S_2(\varphi_1, \varphi_2, p_2, q_1) &= (\omega^2 \epsilon\mu - \beta^2)\varphi_2 - p_{\hat{\epsilon}}(p_2 - q_1), \\
S_3(\varphi_1, \varphi_2, \varphi_3, p_3, q_3) &= (\omega^2 \epsilon\mu - \beta^2)\varphi_3 + p_{\hat{\epsilon}} \left(\frac{\partial \varphi_1}{\partial t} - i\beta\varphi_1 - p_3 \right) \\
&\quad + q_{\hat{\epsilon}} \left(\frac{\partial \varphi_2}{\partial t} - i\beta\varphi_2 - q_3 \right),
\end{aligned}$$

where $\partial\varphi_1/\partial t$ and $\partial\varphi_2/\partial t$ in S_3 can be replaced by (15.105), and

$$p_{\hat{\epsilon}} = \frac{\partial \hat{\epsilon}}{\partial x} - \delta(\xi - \xi^*) f_{\hat{\epsilon}} \frac{\partial \xi}{\partial x}, \quad q_{\hat{\epsilon}} = \frac{\partial \hat{\epsilon}}{\partial y} - \delta(\xi - \xi^*) f_{\hat{\epsilon}} \frac{\partial \xi}{\partial y}.$$

Note that $p_{\hat{\epsilon}}$ and $q_{\hat{\epsilon}}$ will be zero if ϵ is piecewise constant.

Similarly, we can solve the H_z component in terms of H_x and H_y using $\nabla \cdot \mathbf{H} = 0$ to obtain

$$\frac{\partial \varphi_3}{\partial t} = -\frac{\partial \varphi_1}{\partial x} - \frac{\partial \varphi_2}{\partial y} + i\beta\varphi_3. \tag{15.108}$$

As a result, we have the following alternative formulation for the magnetic field.

Formulation D

For $j = 1, 2$,

$$i2\beta \frac{\partial \varphi_j}{\partial t} = \frac{\partial p_j}{\partial x} + \frac{\partial q_j}{\partial y} - \delta(\xi - \xi^*)|\nabla \xi|^2 g_j + S_j, \tag{15.109}$$

$$\frac{\partial \varphi_3}{\partial t} = -p_1 - q_2 + \mathrm{i}\beta\varphi_3, \tag{15.110}$$

$$p_j = \frac{\partial \varphi_j}{\partial x} - \delta(\xi - \xi^*) f_j \frac{\partial \xi}{\partial x}, \tag{15.111}$$

$$q_j = \frac{\partial \varphi_j}{\partial y} - \delta(\xi - \xi^*) f_j \frac{\partial \xi}{\partial y}. \tag{15.112}$$

Again, the time evolution equation for φ_3 in formulation D is a simple ODE, thus requiring less computational cost compared with the corresponding equation for φ_3 in formulation C.

Propagation of HE_{11} modes can be simulated by the proposed GDG-BPM accurately [7].

15.3 Volume Integral Equations for Quantum Dots

15.3.1 One-Particle Schrödinger Equation for Electrons

The motion of a quantum particle such as an electron with charge q and mass m under the influence of an electric potential $V(\mathbf{x})$ is described by the probability wave function $\Psi(\mathbf{x}, t)$, which satisfies the following

- one-electron Schrödinger equation:

$$\mathrm{i}\hbar\frac{\partial \Psi}{\partial t} = \left(-\frac{\hbar^2}{2m}\nabla^2 + U\right)\Psi(\mathbf{x}, t), \tag{15.113}$$

where $2\pi\hbar = 4.135667516 \times 10^{-15}$ eV· s is the Planck constant and the potential energy U is related to the potential V by

$$U = qV. \tag{15.114}$$

The potential in such a one-electron description consists of the effect from the periodic lattice potentials $V_L(\mathbf{x})$ due to the ions of the nucleus, the potential energy due to the other electrons in the system, which is treated by an average potential energy $U_S(\mathbf{x})$ such as the Kohn–Sham potential energy in the density functional theory [10], discussed in Chap. 17, and the external macroscopic potential energy $U_E(\mathbf{x})$. Namely, we should in general have

$$U(\mathbf{x}) = U_E(\mathbf{x}) + U_L(\mathbf{x}) + U_S(\mathbf{x}). \tag{15.115}$$

The details of each potential energy term are given in Sect. 17.1 and $U_L(\mathbf{x}) = E_{\mathrm{ext}}(\rho, \mathbf{R})$, $U_S(\mathbf{x}) = E_{\mathrm{H}}(\rho) + E_{\mathrm{xc}}(\text{æ})$ in (17.23).

The solution for a periodic potential $U_L(\mathbf{x})$ only is given by the Bloch theory discussed in Chap. 14 as

$$\Psi(\mathbf{x}, t) = \exp\left(-\frac{\mathrm{i}E(\mathbf{k})t}{\hbar}\right) \mathrm{e}^{\mathrm{i}\mathbf{k}\cdot\mathbf{x}} u_{\mathbf{k}}(\mathbf{x}), \tag{15.116}$$

where the dispersion relation between the energy E and wavenumber $\mathbf{k}$ of the electrons for the nth band under the given periodic potential is denoted by

$$E = E(\mathbf{k}) = E_n(\mathbf{k}), \tag{15.117}$$

the computation of which by numerical methods has been discussed in Chap. 14. The effect of the periodic potential can be modeled by an effective mass Schrödinger equation.

- Effective mass Schrödinger equation:

$$\mathrm{i}\hbar\frac{\partial\psi}{\partial t} = E(-\mathrm{i}\nabla)\psi(\mathbf{x}, t) + (U_E(\mathbf{x}) + U_S(\mathbf{x}))\,\psi(\mathbf{x}, t). \tag{15.118}$$

The differential operator $E(-\mathrm{i}\nabla)$ is obtained by replacing $\mathbf{k}$ in (15.117) by $(-\mathrm{i}\nabla)$. For a parabolic dispersion at the edge of the lowest conduction band ($n = 1$), we have

$$E(\mathbf{k}) \approx E_C + \frac{\hbar^2}{2m^*}(k_x^2 + k_y^2 + k_z^2), \tag{15.119}$$

where E_C is the conduction band-edge, and m^* is related to the curvature of the dispersion curve, considered to be the effective mass of the electron while moving through the periodic lattice potential:

$$\frac{1}{m^*} = \frac{1}{\hbar^2}\frac{\partial^2 E}{\partial k^2}. \tag{15.120}$$

Therefore, the effective mass equation for the one-electron Schrödinger equation (15.113) becomes

$$\mathrm{i}\hbar\frac{\partial\psi}{\partial t} = \left[-\frac{\hbar^2}{2m^*}\nabla^2 + E_C + U_E(\mathbf{x}) + U_S(\mathbf{x})\right]\psi(\mathbf{x}, t). \tag{15.121}$$

It can been shown [11] that the original wave function $\Psi(x, t)$ is related to the Bloch wave solutions as follows:

$$\Psi(\mathbf{x}, t) = \psi(\mathbf{x}, t)u_{\mathbf{k}}(\mathbf{x}). \tag{15.122}$$

As a result, the probability of finding the electron in the region $[\mathbf{x}, \mathbf{x}+\mathbf{dx}]$ is given by the squared modulus of the wave function, i.e.,

$$|\Psi(\mathbf{x},t)|^2\,\mathbf{dx} = |\psi(\mathbf{x},t)|^2|u_{\mathbf{k}}(\mathbf{x})|^2 \tag{15.123}$$

$$\propto \text{probability of finding the particle inside the region } [\mathbf{x}, \mathbf{x}+\mathbf{dx}],$$

and the electron density is

$$\begin{aligned}\rho(\mathbf{x})\mathbf{dx} &= q|\psi(\mathbf{x},t)|^2|u_{\mathbf{k}}(\mathbf{x})|^2\,\mathbf{dx} \\ &\propto \text{charge density in } [\mathbf{x}, \mathbf{x}+\mathbf{dx}].\end{aligned} \tag{15.124}$$

- Charge continuity equation

The current density $\mathbf{J}$ is defined as [11]

$$\mathbf{J} = \frac{\hbar q}{2im}\left(\psi^*\frac{\partial}{\partial x}\psi - \psi\frac{\partial}{\partial x}\psi^*\right), \tag{15.125}$$

and the conservation of charge can be obtained from the Schrödinger equation (15.121):

$$\frac{\partial\rho}{\partial t} + \nabla\cdot\mathbf{J} = 0. \tag{15.126}$$

15.3.2 VIE for Electrons in Quantum Dots

We consider a quantum dot embedded in a layered medium with different dielectric constant as in Fig. 15.3, which is under the impact of an incident plane wave $\psi^{\mathrm{inc}}(\mathbf{r},t) = \exp(-\mathrm{i}\omega t)\psi^{\mathrm{inc}}(\mathbf{r})$. The scattering wave function and the transmitted wave function will satisfy the following time-harmonic Schrödinger equation:

$$\left(\nabla\frac{1}{m(\mathbf{r})}\nabla + \bar{k}^2\right)\psi(\mathbf{r}) = 0, \tag{15.127}$$

where the electron energy $E = \hbar\omega$ and

$$\bar{k}^2 = \frac{2(E - V(\mathbf{r}))}{\hbar^2}. \tag{15.128}$$

The electron potential $V(\mathbf{r})$ is determined by the potentials given at the top and the bottom contacts and the band structure of the layered medium. In principle, a self-consistent potential via a Poisson equation should be used. Here, we will assume that $V(\mathbf{r})$ is provided a priori and that constant V_i is in the ith layer outside the quantum dot.

As the quantum dot is embedded in the $(N+1)$-homogeneous-layered medium with piecewise $V(\mathbf{r})$, the Schrödinger equation in the ith layer outside the quantum dot can be written as

$$\left(\frac{1}{m_i}\nabla^2 + \bar{k}_i^2\right)\psi_i(\mathbf{r}) = 0, \quad \mathbf{r} \in \Omega_{\mathrm{QD}}^c, \tag{15.129}$$

where $\psi_i(\mathbf{r})$ denotes the wave function in the ith layer, and

$$\bar{k}_i^2 = \frac{2(E - V_i)}{\hbar^2}. \tag{15.130}$$

Meanwhile, inside the quantum dot region Ω_{QD}, the Schrödinger equation can be written as

$$\left[\frac{1}{m_i}\nabla^2 + \left(\bar{k}_i^2 + \Delta k_i^2\right)\right]\psi_i(\mathbf{r}) = 0, \quad \mathbf{r} \in \Omega_{\mathrm{QD}}, \tag{15.131}$$

where

$$\Delta k_i^2 = \frac{2(E - V_{\mathrm{QD},i})}{\hbar^2}\frac{m_{\mathrm{QD}}}{m_i} - \frac{2(E - V_i)}{\hbar^2}. \tag{15.132}$$

Altogether, the Helmholtz equation can be cast into one equation as

$$\left[\frac{1}{m_i}\nabla^2 + \left(\bar{k}_i^2 + \Delta k_i^2\right)\right]\psi_i(\mathbf{r}) = 0, \tag{15.133}$$

where

$$\Delta k_i^2 = \begin{cases} \dfrac{2(E - V_{\mathrm{QD},i})}{\hbar^2}\dfrac{m_{\mathrm{QD}}}{m_i} - \bar{k}_i^2, & \text{if } \mathbf{r} \in \Omega_{\mathrm{QD}}, \\ 0, & \text{if } \mathbf{r} \in \Omega_{\mathrm{QD}}^c. \end{cases} \tag{15.134}$$

Introducing the differential operator

$$\hat{\mathcal{L}}_i = \frac{1}{m_i}\nabla^2 + \bar{k}_i^2, \tag{15.135}$$

Equation (15.133) becomes

$$\hat{\mathcal{L}}_i\psi_i(\mathbf{r}) = -\Delta k_i^2\psi_i(\mathbf{r}). \tag{15.136}$$

To derive a volume integral equation for $\psi(\mathbf{r})$, we use Green's function $G(\mathbf{r}, \mathbf{r}')$ for the layered medium,

$$\hat{\mathcal{L}}_i G(\mathbf{r}, \mathbf{r}') = -\frac{1}{m_i}\delta(\mathbf{r}, \mathbf{r}'), \tag{15.137}$$

and the computation of $G(\mathbf{r}, \mathbf{r}')$ has been detailed in Sect. 9.2.5.

It can be shown in Sect. 15.3.3 that the wave function inside the quantum dot satisfies the following volume integral equation with a special surface contribution:

$$\left(\frac{1}{m_i(\mathbf{r})} - S - V\right)\psi^{\text{int}}(\mathbf{r}) = f(\mathbf{r}), \quad \mathbf{r} \in \Omega_{\text{QD}}, \tag{15.138}$$

where

$$S\left(\psi^{\text{int}}\right)(\mathbf{r}) = \int_{\partial\Omega_{\text{QD}}} \left(\frac{1}{m_i(\mathbf{r}')} - \frac{1}{m_{\text{QD}}}\right) G(\mathbf{r}, \mathbf{r}') \frac{\partial \psi^{\text{int}}(\mathbf{r}')}{\partial \mathbf{n}_{\text{QD}}} \mathrm{d}s', \tag{15.139}$$

$$V\left(\psi^{\text{int}}\right)(\mathbf{r}) = \text{p.v.} \int_{\Omega_{\text{QD}}} G(\mathbf{r}, \mathbf{r}')\Delta k_i^2(\mathbf{r}')\psi^{\text{int}}(\mathbf{r}')\mathrm{d}\mathbf{r}', \tag{15.140}$$

$$f(\mathbf{r}) = \int_{S_0} \frac{1}{m_0}\left(G_0(\mathbf{r}', \mathbf{r})\frac{\partial \psi^{\text{inc}}(\mathbf{r}')}{\partial z'} - \psi^{\text{inc}}(\mathbf{r}')\frac{\partial G_0(\mathbf{r}', \mathbf{r})}{\partial z'}\right) \mathrm{d}s'. \tag{15.141}$$

15.3.3 Derivation of the VIE for Quantum Dots Embedded in Layered Media

By subtracting (15.137) $\times\, \psi(\mathbf{r})$ from (15.136) $\times\, G(\mathbf{r}, \mathbf{r}')$, we obtain

$$G(\mathbf{r}, \mathbf{r}')\hat{\mathcal{L}}_i\psi(\mathbf{r}) - \psi(\mathbf{r})\hat{\mathcal{L}}_i G(\mathbf{r}, \mathbf{r}') = -G(\mathbf{r}, \mathbf{r}')\Delta k_i^2\psi(\mathbf{r}) + \frac{1}{m_i}\delta(\mathbf{r}, \mathbf{r}')\psi(\mathbf{r}). \tag{15.142}$$

Excluding a small sphere $\Omega_\epsilon(\mathbf{r}')$ as depicted in Fig. 15.3, we integrate over $\mathbb{R}^3 \backslash \Omega_\epsilon(\mathbf{r}')$ to obtain

$$\int_{\mathbb{R}^3\backslash\Omega_\epsilon(\mathbf{r}')} \{G(\mathbf{r}, \mathbf{r}')\hat{\mathcal{L}}_i\psi(\mathbf{r}) - \psi(\mathbf{r})\hat{\mathcal{L}}_i G(\mathbf{r}, \mathbf{r}')\}\mathrm{d}\mathbf{r} = -\int_{\Omega_{\text{QD}}\backslash\Omega_\epsilon(\mathbf{r}')} G(\mathbf{r}, \mathbf{r}')\Delta k_i^2\psi(\mathbf{r})\mathrm{d}\mathbf{r}. \tag{15.143}$$

Denote the reflected wave in the contact ($z > 0$) by ψ^{r}, the wave exterior to the quantum dot ($z < 0$) by ψ^{ext}, and the wave interior to the quantum dot ($z < 0$) by ψ^{int}, respectively. We will derive a volume integral equation for ψ^{int}.

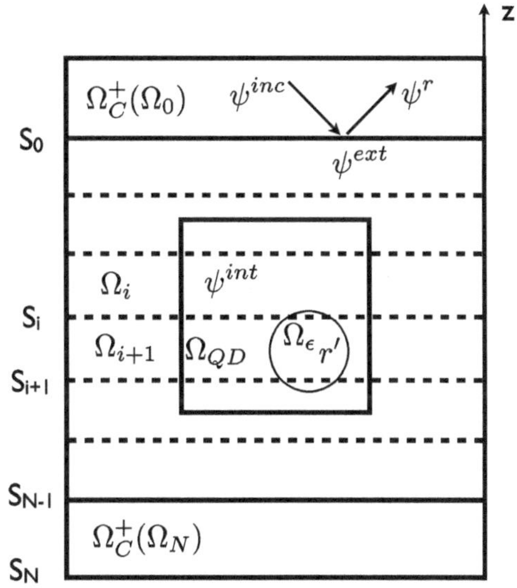

Fig. 15.3 Quantum dot in a layered medium: Ω_C^+ denotes a contact; and Ω_{QD} represents a quantum dot. Note that S_i and S_{i+1} are the boundaries of Ω_{i+1}, and Ω_ϵ is a small circle around r' with radius ϵ used for deriving the volume integral equation

Equation (15.143) can be rewritten as follows:

$$
\begin{aligned}
&- \int_{\Omega_{QD}\setminus\Omega_\epsilon(\mathbf{r}')} G(\mathbf{r},\mathbf{r}')\Delta k_i^2 \psi^{\mathrm{int}}(\mathbf{r}) d\mathbf{r} \\
&= \int_{z>0} \left[G(\mathbf{r},\mathbf{r}')\hat{\mathcal{L}}_i \psi^{\mathrm{r}}(\mathbf{r}) - \psi^{\mathrm{r}}(\mathbf{r})\hat{\mathcal{L}}_i G(\mathbf{r},\mathbf{r}') \right] d\mathbf{r} \\
&\quad + \int_{z<0,\Omega_{QD}^c} \left[G(\mathbf{r},\mathbf{r}')\hat{\mathcal{L}}_i \psi^{\mathrm{ext}}(\mathbf{r}) - \psi^{\mathrm{ext}}(\mathbf{r})\hat{\mathcal{L}}_i G(\mathbf{r},\mathbf{r}') \right] d\mathbf{r} \\
&\quad + \int_{z<0,\Omega_{QD}\setminus\Omega_\epsilon(\mathbf{r}')} \left[G(\mathbf{r},\mathbf{r}')\hat{\mathcal{L}}_i \psi^{\mathrm{int}}(\mathbf{r}) - \psi^{\mathrm{int}}(\mathbf{r})\hat{\mathcal{L}}_i G(\mathbf{r},\mathbf{r}') \right] d\mathbf{r}.
\end{aligned}
$$

Replacing $\hat{\mathcal{L}}_i$ by ∇^2/m_i, introducing a contour Γ_∞ made of two semi-spherical surfaces Γ_∞ over the interface (S_0) $z = 0$, and under the interface (S_N) $z = z_N$, respectively, and applying Green's second identity, we have

$$
\begin{aligned}
&- \int_{\Omega_{QD}\setminus\Omega_\epsilon(\mathbf{r}')} G(\mathbf{r},\mathbf{r}')\Delta k_i^2 \psi^{\mathrm{int}}(\mathbf{r}) d\mathbf{r} \\
&= \int_{S_0} \left[-\frac{1}{m_0} G_0(\mathbf{r},\mathbf{r}') \frac{\partial \psi^{\mathrm{r}}(\mathbf{r})}{\partial z} + \frac{1}{m_0} \psi^{\mathrm{r}}(\mathbf{r}) \frac{\partial G_0(\mathbf{r},\mathbf{r}')}{\partial z} \right] ds
\end{aligned}
$$

$$+\int_{\Gamma_\infty}\left[\frac{1}{m_0}G_0(\mathbf{r},\mathbf{r}')\frac{\partial\psi^{\mathrm{r}}(\mathbf{r})}{\partial\mathbf{n}}-\frac{1}{m_0}\psi^{\mathrm{r}}(\mathbf{r})\frac{\partial G_0(\mathbf{r},\mathbf{r}')}{\partial\mathbf{n}}\right]\mathrm{d}s$$
$$+\int_{S_0}\left[\frac{1}{m_1}G_1(\mathbf{r},\mathbf{r}')\frac{\partial\psi_1^{\mathrm{ext}}(\mathbf{r})}{\partial z}-\frac{1}{m_1}\psi_1^{\mathrm{ext}}(\mathbf{r})\frac{\partial G_1(\mathbf{r},\mathbf{r}')}{\partial z}\right]\mathrm{d}s$$
$$+\int_{\partial\Omega_{\mathrm{QD}}}\left[-\frac{1}{m_i}G_i(\mathbf{r},\mathbf{r}')\frac{\partial\psi_i^{\mathrm{ext}}(\mathbf{r})}{\partial\mathbf{n}_{\mathrm{QD}}}+\frac{1}{m_i}\psi_i^{\mathrm{ext}}(\mathbf{r})\frac{\partial G_i(\mathbf{r},\mathbf{r}')}{\partial\mathbf{n}_{\mathrm{QD}}}\right]\mathrm{d}s$$
$$+\int_{\Gamma_\infty}\left[\frac{1}{m_N}G_N(\mathbf{r},\mathbf{r}')\frac{\partial\psi_N^{\mathrm{ext}}(\mathbf{r})}{\partial\mathbf{n}}-\frac{1}{m_N}\psi_N^{\mathrm{ext}}(\mathbf{r})\frac{\partial G_N(\mathbf{r},\mathbf{r}')}{\partial\mathbf{n}}\right]\mathrm{d}s$$
$$+\sum_{i=1}^{N-1}\int_{S_i}\left[-\frac{1}{m_i}G_i(\mathbf{r},\mathbf{r}')\frac{\partial\psi_i^{\mathrm{ext}}(\mathbf{r})}{\partial z}+\frac{1}{m_i}\psi_i^{\mathrm{ext}}(\mathbf{r})\frac{\partial G_i(\mathbf{r},\mathbf{r}')}{\partial z}\right]\mathrm{d}s$$
$$+\sum_{i=1}^{N-1}\int_{S_i}\left[\frac{1}{m_{i+1}}G_{i+1}(\mathbf{r},\mathbf{r}')\frac{\partial\psi_{i+1}^{\mathrm{ext}}(\mathbf{r})}{\partial z}-\frac{1}{m_{i+1}}\psi_{i+1}^{\mathrm{ext}}(\mathbf{r})\frac{\partial G_{i+1}(\mathbf{r},\mathbf{r}')}{\partial z}\right]\mathrm{d}s$$
$$+\int_{\partial\Omega_{\mathrm{QD}}}\left[\frac{1}{m_i}G_i(\mathbf{r},\mathbf{r}')\frac{\partial\psi_i^{\mathrm{int}}(\mathbf{r})}{\partial\mathbf{n}_{\mathrm{QD}}}-\frac{1}{m_i}\psi_i^{\mathrm{int}}(\mathbf{r})\frac{\partial G_i(\mathbf{r},\mathbf{r}')}{\partial\mathbf{n}_{\mathrm{QD}}}\right]\mathrm{d}s$$
$$+\int_{\partial\Omega_\epsilon}\left[-\frac{1}{m_i}G_i(\mathbf{r},\mathbf{r}')\frac{\partial\psi_i^{\mathrm{int}}(\mathbf{r})}{\partial\mathbf{n}_{\Omega_\epsilon}}+\frac{1}{m_i}\psi_i^{\mathrm{int}}(\mathbf{r})\frac{\partial G_i(\mathbf{r},\mathbf{r}')}{\partial\mathbf{n}_{\Omega_\epsilon}}\right]\mathrm{d}s,$$

where S_i is the planar interface between layers, and $\mathbf{n}_{\Omega_\epsilon}$ and $\mathbf{n}_{\mathrm{QD}}$ are the outer normal of the region Ω_ϵ and Ω_{QD}, respectively. So, we have

$$-\int_{\Omega_{\mathrm{QD}}\setminus\Omega_\epsilon(\mathbf{r}')}G(\mathbf{r},\mathbf{r}')\Delta k_i^2\psi^{\mathrm{int}}(\mathbf{r})\mathrm{d}\mathbf{r}=I+II+III+IV+V, \tag{15.144}$$

where

$$I=\sum_{i=1}^{N-1}\int_{S_i}\left[-\frac{1}{m_i}G_i(\mathbf{r},\mathbf{r}')\frac{\partial\psi_i^{\mathrm{ext}}(\mathbf{r})}{\partial z}+\frac{1}{m_i}\psi_i^{\mathrm{ext}}(\mathbf{r})\frac{\partial G_i(\mathbf{r},\mathbf{r}')}{\partial z}\right]\mathrm{d}s$$
$$+\sum_{i=1}^{N-1}\int_{S_i}\left[\frac{1}{m_{i+1}}G_{i+1}(\mathbf{r},\mathbf{r}')\frac{\partial\psi_{i+1}^{\mathrm{ext}}(\mathbf{r})}{\partial z}-\frac{1}{m_{i+1}}\psi_{i+1}^{\mathrm{ext}}(\mathbf{r})\frac{\partial G_{i+1}(\mathbf{r},\mathbf{r}')}{\partial z}\right]\mathrm{d}s,$$

$$II = \int_{S_0} \left[-\frac{1}{m_0} G_0(\mathbf{r}, \mathbf{r}') \frac{\partial \psi^{\mathrm{r}}(\mathbf{r})}{\partial z} + \frac{1}{m_0} \psi^{\mathrm{r}}(\mathbf{r}) \frac{\partial G_0(\mathbf{r}, \mathbf{r}')}{\partial z} \right] \mathrm{d}s$$
$$+ \int_{S_0} \left[\frac{1}{m_1} G_1(\mathbf{r}, \mathbf{r}') \frac{\partial \psi_1^{\mathrm{ext}}(\mathbf{r})}{\partial z} - \frac{1}{m_1} \psi_1^{\mathrm{ext}}(\mathbf{r}) \frac{\partial G_1(\mathbf{r}, \mathbf{r}')}{\partial z} \right] \mathrm{d}s,$$

$$III = \int_{\Gamma_\infty} \left[\frac{1}{m_0} G_0(\mathbf{r}, \mathbf{r}') \frac{\partial \psi^{\mathrm{r}}(\mathbf{r})}{\partial \mathbf{n}} - \frac{1}{m_0} \psi^{\mathrm{r}}(\mathbf{r}) \frac{\partial G_0(\mathbf{r}, \mathbf{r}')}{\partial \mathbf{n}} \right] \mathrm{d}s$$
$$+ \int_{\Gamma_\infty} \left[\frac{1}{m_N} G_N(\mathbf{r}, \mathbf{r}') \frac{\partial \psi_N^{\mathrm{ext}}(\mathbf{r})}{\partial \mathbf{n}} - \frac{1}{m_N} \psi_N^{\mathrm{ext}}(\mathbf{r}) \frac{\partial G_N(\mathbf{r}, \mathbf{r}')}{\partial \mathbf{n}} \right] \mathrm{d}s,$$

$$IV = \int_{\partial\Omega_{\mathrm{QD}}} \left[-\frac{1}{m_i} G_i(\mathbf{r}, \mathbf{r}') \frac{\partial \psi_i^{\mathrm{ext}}(\mathbf{r})}{\partial \mathbf{n}_{\mathrm{QD}}} + \frac{1}{m_i} \psi_i^{\mathrm{ext}}(\mathbf{r}) \frac{\partial G_i(\mathbf{r}, \mathbf{r}')}{\partial \mathbf{n}_{\mathrm{QD}}} \right] \mathrm{d}s$$
$$+ \int_{\partial\Omega_{\mathrm{QD}}} \left[\frac{1}{m_i} G_i(\mathbf{r}, \mathbf{r}') \frac{\partial \psi_i^{\mathrm{int}}(\mathbf{r})}{\partial \mathbf{n}_{\mathrm{QD}}} - \frac{1}{m_i} \psi_i^{\mathrm{int}}(\mathbf{r}) \frac{\partial G_i(\mathbf{r}, \mathbf{r}')}{\partial \mathbf{n}_{\mathrm{QD}}} \right] \mathrm{d}s,$$

$$V = \int_{\partial\Omega_\epsilon} \left[-\frac{1}{m_i} G_i(\mathbf{r}, \mathbf{r}') \frac{\partial \psi_i^{\mathrm{int}}(\mathbf{r})}{\partial \mathbf{n}_{\Omega_\epsilon}} + \frac{1}{m_i} \psi_i^{\mathrm{int}}(\mathbf{r}) \frac{\partial G_i(\mathbf{r}, \mathbf{r}')}{\partial \mathbf{n}_{\Omega_\epsilon}} \right] \mathrm{d}s.$$

Using the interface condition at the interface S_i, we have

$$\psi_i^{\mathrm{ext}} = \psi_{i+1}^{\mathrm{ext}}, \tag{15.145}$$
$$\frac{1}{m_i} \frac{\partial \psi_i^{\mathrm{ext}}}{\partial z} = \frac{1}{m_{i+1}} \frac{\partial \psi_{i+1}^{\mathrm{ext}}}{\partial z}, \tag{15.146}$$
$$G_i = G_{i+1}, \tag{15.147}$$
$$\frac{1}{m_i} \frac{\partial G_i^{\mathrm{ext}}}{\partial z} = \frac{1}{m_{i+1}} \frac{\partial G_{i+1}^{\mathrm{ext}}}{\partial z}. \tag{15.148}$$

It is clear that $I = 0$. The Sommerfeld radiation condition implies that $III = 0$. To simplify integral II, we use the interface condition at S_0, $z = 0$, between the total wave $\psi^{\mathrm{r}} + \psi^{\mathrm{inc}}$ for $z > 0$ and ψ^{ext} for $z < 0$:

$$\psi^{\mathrm{ext}} = \psi^{\mathrm{r}} + \psi^{\mathrm{inc}}, G_0 = G_1, \tag{15.149}$$
$$\frac{1}{m_1} \frac{\partial \psi^{\mathrm{ext}}}{\partial z} = \frac{1}{m_0} \frac{\partial \psi^{\mathrm{inc}}}{\partial z} + \frac{1}{m_0} \frac{\partial \psi^{\mathrm{r}}}{\partial z}, \tag{15.150}$$

$$\frac{1}{m_0}\frac{\partial G_0}{\partial z} = \frac{1}{m_1}\frac{\partial G_1}{\partial z}. \tag{15.151}$$

We obtain

$$\begin{aligned}
II &= \int_{S_0} \left[\frac{1}{m_1} G_1(\mathbf{r}, \mathbf{r}')\frac{\partial \psi_1^{\text{ext}}(\mathbf{r})}{\partial z} - \frac{1}{m_0} G_0(\mathbf{r}, \mathbf{r}')\frac{\partial \psi^{\text{r}}(\mathbf{r})}{\partial z}\right] \mathrm{d}s \\
&+ \int_{S_0} \left[-\frac{1}{m_1}\psi_1^{\text{ext}}(\mathbf{r})\frac{\partial G_1(\mathbf{r}, \mathbf{r}')}{\partial z} + \frac{1}{m_0}\psi^{\text{r}}(\mathbf{r})\frac{\partial G_0(\mathbf{r}, \mathbf{r}')}{\partial z}\right] \mathrm{d}s \\
&= \int_{S_0} \left[G_0(\mathbf{r}, \mathbf{r}')\left(\frac{1}{m_1}\frac{\partial \psi_1^{\text{ext}}(\mathbf{r})}{\partial z} - \frac{1}{m_0}\frac{\partial \psi^{\text{r}}(\mathbf{r})}{\partial z}\right)\right. \\
&\quad \left. -\frac{1}{m_0}\frac{\partial G_0(\mathbf{r}, \mathbf{r}')}{\partial z}\left(\psi_1^{\text{ext}}(\mathbf{r}) - \psi^{\text{r}}(\mathbf{r})\right)\right] \mathrm{d}s \\
&= \int_{S_0} \left[G_0(\mathbf{r}, \mathbf{r}')\frac{1}{m_0}\frac{\partial \psi^{\text{inc}}(\mathbf{r})}{\partial z} - \frac{1}{m_0}\frac{\partial G_0(\mathbf{r}, \mathbf{r}')}{\partial z}\psi^{\text{inc}}(\mathbf{r})\right] \mathrm{d}s \\
&= \int_{S_0} \frac{1}{m_0}\left(G_0\frac{\partial \psi^{\text{inc}}(\mathbf{r})}{\partial z} - \psi^{\text{inc}}(\mathbf{r})\frac{\partial G_0(\mathbf{r}, \mathbf{r}')}{\partial z}\right) \mathrm{d}s \equiv f(\mathbf{r}).
\end{aligned}$$

For integral V, we let ϵ approach zero. Then

$$V = \frac{1}{m_i}\psi^{\text{int}}(\mathbf{r}'). \tag{15.152}$$

Finally, along the boundary of the quantum dot, we have

$$\psi_i^{\text{ext}} = \psi_i^{\text{int}}, \tag{15.153}$$

$$\frac{1}{m_i}\frac{\partial \psi_i^{\text{ext}}}{\partial n_{\text{QD}}} = \frac{1}{m_{\text{QD}}}\frac{\partial \psi_i^{\text{int}}}{\partial n_{\text{QD}}}, \tag{15.154}$$

so integral IV becomes

$$\begin{aligned}
IV &= \int_{\partial\Omega_{\text{QD}}} \left[-\frac{1}{m_i} G_i(\mathbf{r}, \mathbf{r}')\frac{m_i}{m_{\text{QD}}}\frac{\partial \psi_i^{\text{int}}(\mathbf{r})}{\partial \mathbf{n}_{\text{QD}}} + \frac{1}{m_i}\psi_i^{\text{int}}(\mathbf{r})\frac{\partial G_i(\mathbf{r}, \mathbf{r}')}{\partial \mathbf{n}_{\text{QD}}}\right] \mathrm{d}s \\
&+ \int_{\partial\Omega_{\text{QD}}} \left[\frac{1}{m_i} G_i(\mathbf{r}, \mathbf{r}')\frac{\partial \psi_i^{\text{int}}(\mathbf{r})}{\partial \mathbf{n}_{\text{QD}}} - \frac{1}{m_i}\psi_i^{\text{int}}(\mathbf{r})\frac{\partial G_i(\mathbf{r}, \mathbf{r}')}{\partial \mathbf{n}_{\text{QD}}}\right] \mathrm{d}s \\
&= \int_{\partial\Omega_{\text{QD}}} \frac{1}{m_i}\left(1 - \frac{m_i}{m_{\text{QD}}}\right) G_i(\mathbf{r}, \mathbf{r}')\frac{\partial \psi_i^{\text{int}}(\mathbf{r})}{\partial \mathbf{n}_{\text{QD}}} \mathrm{d}s \equiv S\left(\psi^{\text{int}}\right)(\mathbf{r}).
\end{aligned}$$

Therefore, the volume integral equations (15.138)–(15.141) are obtained for a quantum dot embedded in a layered medium, illuminated by an incident electron wave $\psi^{\rm inc}(\mathbf{r}')$, where the effective mass of the quantum dot is different from that of the surrounding layered medium.

15.4 Summary

Schrödinger equations occur in both optical waveguides and quantum dots, where the solution may be not smooth due to the material inhomogeneity in the former case and potential jumps in the latter. A generalized DG method, based on the distributional representation of the solution to the Schrödinger equation, designs the DG approximation accordingly to provide accurate treatment of material interfaces inside the waveguides. For quantum dots, a volume integral equation allows the treatment of dissimilar effective mass and potential discontinuities, in addition to the natural enforcement of the boundary conditions at infinity.

References

1. K. Fan, W. Cai, X. Ji, A generalized discontinuous Galerkin (GDG) method for Schrödinger equations with nonsmooth solutions. J. Comput. Phys. **227**, 2387–2410 (2008)
2. M.D. Feit, J.A. Fleck Jr., Light propagation in graded-index optical fibers. Appl. Opt. **17**, 3990–3998 (1978)
3. M.D. Feit, J.A. Fleck Jr., Computation of mode properties in optical fiber waveguide by a propagating beam method. Appl. Opt. **19**, 1154–1164 (1980)
4. Y. Tsuji, M. Koshiba, N. Takimoto, Finite element beam propagation method for anisotropic optical waveguides. J. Lightwave Technol. **17**, 723–728 (1999)
5. L. Thylen, D. Yevick, Beam propagation method in anisotropic media. Appl. Opt. **21**, 2751–2754 (1982)
6. C.L. Xu, W.P. Huang, J. Chrostowski, S.K. Chaudhuri, A full-vectorial beam propagation method for anisotropic waveguide. J. Lightwave Technol. **12**, 1926–1931 (1994)
7. K. Fan, W. Cai, X. Ji, A full vectorial generalized discontinuous Galerkin beam propagation method (GDG-BPM) for nonsmooth electromagnetic fields in waveguides. J. Comput. Phys. **227**, 7178–7191 (2008)
8. K. Okamoto, *Fundamentals of Optical Waveguides*, 2nd edn. (Academic Press, New York, 2005)
9. A.W. Snyder, J. Love, *Optical Waveguide Theory* (Springer, Berlin, 1983)
10. R.G. Parr, W.T. Yang, *Density Functional Theory of Atoms and Molecules* (Oxford University Press, Oxford, 1989)
11. S. Datta, *Quantum Phenomena* (Addison-Wesley, Reading, 1989)

Part III
Electron Transport and Electronic Structure

Chapter 16
Quantum Electron Transport in Semiconductors

In this and the three chapters following, electronic structure and quantum transport of electron in nano-electronics, subjects normally not considered in the classical electromagnetics but closely related nonetheless due to the Coulomb interaction potentials in the Schrodinger equations, will be discussed. The transport of carriers (electrons and holes) in semiconductor devices such as hetero-junctions, MOSFETs, and superlattices can be described by classical hydrodynamics, drift diffusion models, or semi-classical models (Boltzmann equations) when the devices are of micron or sub-micron scales, and by quantum models (such as Wigner distributions) for devices at nano-scales. The key factor in selecting a specific proper transport model is the mean free path of the carriers inside a device in comparison to the size of the device.

In this chapter, we present an overview of quantum transport models, firstly by deriving the Fermi–Dirac distribution for electrons in an equilibrium system using quantum ensemble theories. Secondly, we define the density matrix for nano-devices, Wigner distributions, and Wigner–Moyal expansions. The Landauer transmission theory for quantum transport models will be reviewed. Finally, the non-equilibrium Green's function (NEGF) method is introduced for computing the transmission coefficients of quantum devices. The semi-classical Boltzmann model, the classical hydrodynamic models, and their numerical methods will be discussed in Chap. 20. For Part III of this book, E will denote energy (not to be confused with the expectation notation in Chap. 4), and the time dependence of a time-harmonic wave function in the Schrödinger equation is in the form $\mathrm{e}^{-\mathrm{i}Et/\hbar}$ for convention.

W. Cai, *Deterministic, Stochastic, and Deep Learning Methods for Computational Electromagnetics*, https://doi.org/10.1007/978-981-96-0100-4_16

16.1 Ensemble Theory for Quantum Systems

16.1.1 Thermal Equilibrium of a Quantum System

For a quantum subsystem S with given macroscopic state variables such as the particle number N, the overall energy E, and the volume V, $\Omega(N, V, E)$ denotes the number of quantum states (also called microstates) accessible to S. Each of the quantum states can be obtained as the stationary eigenstate solution $\Psi(\mathbf{r}_1, \mathbf{r}_2, \ldots, \mathbf{r}_N, t) = e^{-iEt/\hbar}\psi(\mathbf{r}_1, \mathbf{r}_2, \ldots, \mathbf{r}_N)$ of a many-particle Schrödinger equation

$$H\psi = E\psi, \tag{16.1}$$

where

$$H(\mathbf{r}_1, \mathbf{r}_2, \ldots, \mathbf{r}_N) = -\frac{\hbar^2}{2m}\sum_{i=1}^{N}\nabla_i^2 + U(\mathbf{r}_1, \mathbf{r}_2, \ldots, \mathbf{r}_N), \tag{16.2}$$

and the detailed form of the electrostatic interaction energy $U(\mathbf{r}_1, \mathbf{r}_2, \ldots, \mathbf{r}_N)$ is given in (17.1).

In general, the system S will be found among the $\Omega(N, V, E)$ quantum microstates, i.e., the eigenstates of the Schrödinger equation (16.1) with some given energy E.

If we bring two systems S_1 and S_2 into contact, allowing exchange of energy and particles, the combined system $S^{(0)}$ is assumed to be isolated (i.e., there is no exchange of energy or particles with the outside environment), and eventually, it reaches its own equilibrium [1]. The combined system will have constant total energy and number of particles:

$$E^{(0)} = E_1 + E_2, \tag{16.3}$$

$$N^{(0)} = N_1 + N_2. \tag{16.4}$$

If we ignore the interaction energy between the two subsystems S_1 and S_2, the number of quantum states for the combined system $S^{(0)}$ is given by

$$\Omega^{(0)}(E_1, N_1; E_2, N_2) = \Omega_1(E_1, N_1)\Omega_2(E_2, N_2),$$

or

$$\Omega^{(0)}(E_1, N_1; E^{(0)} - E_1, N^{(0)} - N_1) = \Omega_1(E_1, N_1)\Omega_2(E^{(0)} - E_1, N^{(0)} - N_1). \tag{16.5}$$

As E_1 and N_1 are variables due to the exchanges between the two systems, the number of quantum microstates $\Omega^{(0)}$ will then take on different values. When the macroscopic system $S^{(0)}$ settles down to equilibrium, we expect that $\Omega^{(0)}$ will be maximized, namely, that the entropy $\log \Omega^{(0)}$ will be maximized (i.e., the system $S^{(0)}$ arrives at its maximum entropy state). Thus, when $S^{(0)}$ is at equilibrium we have

$$0 = \left.\frac{\partial \Omega^{(0)}}{\partial E_1}\right|_{E_1=\overline{E}_1} = \left.\frac{\partial \Omega_1}{\partial E_1}\right|_{E_1=\overline{E}_1} \Omega_2(\overline{E}_2) - \Omega_1(\overline{E}_1) \left.\frac{\partial \Omega_2}{\partial E_2}\right|_{E_2=\overline{E}_2}, \tag{16.6}$$

after using the fact that $\partial E_2/\partial E_1 = -1$, resulting in the following equality for the subsystems S_1 and S_2:

$$\left.\frac{\partial \ln \Omega_1}{\partial E_1}\right|_{E_1=\overline{E}_1} = \left.\frac{\partial \ln \Omega_2}{\partial E_2}\right|_{E_2=\overline{E}_2}, \tag{16.7}$$

where $\overline{E}_1$ and $\overline{E}_2$ are the equilibrium values of the variables E_1 and E_2, respectively.

The above equality for the two systems in contact allows us to define a thermodynamic temperature T of a macroscopic system in equilibrium with its environment as follows:

$$\frac{1}{k_B T} = \beta = \left.\frac{\partial \ln \Omega(N, V, E)}{\partial E}\right|_{N,V,E=\overline{E}}. \tag{16.8}$$

A similar calculation can show that the quantity $\left.\frac{\partial \ln \Omega(N,V,E)}{\partial N}\right|_{N=\overline{N},V,E}$ at equilibrium should also become equal for the two subsystems, and correspondingly the chemical potential μ for the system is defined as

$$\mu = -k_B T \left.\frac{\partial \ln \Omega(N, V, E)}{\partial N}\right|_{N=\overline{N},V,E}. \tag{16.9}$$

16.1.2 Microcanonical Ensembles

An isolated Hamiltonian system, i.e., a system without exchange of energy or particles with the outside environment, is supposed to occupy any of its microstates with a prior equal probability (this is the sole assumption of equilibrium classical statistical mechanics):

$$P_0 = \frac{1}{\Omega(N, V, E)}, \tag{16.10}$$

where the energy E is assumed to be fixed (i.e., we invoke the conservation of energy).

Moreover, the system is assumed to be ergodic, namely, starting from any given microstate at the initial time $t = 0$, it will eventually come arbitrarily close to any other microstates. Its dynamics, given by the time-dependent Schrödinger equations for the quantum system or the Hamiltonian equations for the classical system, allows one to define a time average of any dynamical variable operator A (for example, the momentum operator $A = -i\hbar\nabla$):

$$\overline{A} = \lim_{T\to\infty} \frac{1}{T} \int_0^T \langle\psi|A|\psi\rangle \, \mathrm{d}t, \tag{16.11}$$

where $\langle\psi|A|\psi\rangle$ denotes the quantum average with respect to a given state ψ:

$$\langle\psi|A|\psi\rangle \equiv \int_{\mathbb{R}^{3N}} \psi^*(\mathbf{r}_1, \ldots, \mathbf{r}_N, t) A\psi(\mathbf{r}_1, \ldots, \mathbf{r}_N, t) \, \mathrm{d}^{3N}\mathbf{r}. \tag{16.12}$$

The time average of a dynamical variable is an experimentally measurable quantity that can be obtained by solving the dynamic equations with computer solutions of the Schrödinger equations or the Hamiltonian equations, as in molecular dynamics simulations.

Meanwhile, Gibbs [2] proposed an ensemble formulation to describe the isolated system, which samples the high-dimensional phase space with the given probability (16.10), and the resulting collection of $\mathcal{N}$ microstates,

$$\{\psi^{(k)}\}_{k=1}^{\mathcal{N}} \quad \text{(an ensemble of microstates)}, \tag{16.13}$$

is termed "an ensemble" of the isolated system. For an N-particle classical system, the phase space is simply the $6N$-dimensional space

$$\Gamma = \{(\mathbf{q}_1, \ldots, \mathbf{q}_N; \mathbf{p}_1, \ldots, \mathbf{p}_N)\}, \tag{16.14}$$

for position $\mathbf{q}$ and conjugate momentum variables $\mathbf{p}$. For quantum systems, the phase space can be defined as the expansion coefficients of the wave functions under some orthonormal basis $\{\phi_i\}|_{i=1}^{\infty}$ [3], i.e.,

$$\Gamma = \{(a_1, \ldots, a_k, \ldots)\}, \tag{16.15}$$

where

$$\psi = \sum_{k=1}^{\infty} a_k \phi_k. \tag{16.16}$$

The ensemble should sample the phase space with a uniform distribution given by (16.10) for a finite microstate number Ω, and such an ensemble is called a "microcanonical ensemble", particularly for an isolated system where the energy is at a fixed value. In the Gibbs ensemble theory, a microcanonical ensemble average is defined as

$$\langle A\rangle_{\text{me}} = \frac{1}{\mathcal{N}} \sum_{k=1}^{\infty} \left\langle\psi^{(k)}|A|\psi^{(k)}\right\rangle. \tag{16.17}$$

Within the Gibbs ensemble theory, for an ergodic system, we can expect that the time average and the microcanonical ensemble average should agree, namely

$$\overline{A} = \langle A \rangle_{\text{me}} \tag{16.18}$$

for any initial microstate in the definition of the time average.

16.1.3 Canonical Ensembles

Let A be a subsystem of a bigger system $A^{(0)}$ at thermal equilibrium of temperature T, which does not exchange particles with the rest of $A^{(0)}$, which is considered isolated. To construct a Gibbs ensemble to represent such a subsystem at the given temperature T in its environment, identified by the macrostate variables (N, V, T), we examine its microstates from the perspective of the bigger system $A^{(0)}$. For any $\mathcal{N}$ microstates of $A^{(0)}$, we will have a corresponding set of $\mathcal{N}$ microstates for the subsystem A; the latter will be called a "*canonical ensemble*" (C.E.) of the system A. It is also referred to as the NVT ensemble.

Due to the exchange of energy between the system A and its environment $A^{(0)}\backslash A$, the energy of A will fluctuate while the energy of the bigger system $A^{(0)}$ remains constant. It can be shown that, in fact, the energy fluctuation is very small around some average value. Let us denote by n_r the number of microstates in a canonical ensemble of A whose energy is E_r. Then

$$\sum_r n_r = \mathcal{N}. \tag{16.19}$$

Finding the exact energy level E_r for a given N-particle system in a volume V is a many-body eigenvalue problem, as in (16.1), which is usually too complicated to be solved directly. Therefore, approximations such as the density functional theory (DFT) in Sect. 17.1 will be used. However, our discussion on the canonical ensemble can proceed, assuming the energy levels are E_r, $r = 1, 2, \ldots,$ as given.

We denote the total energy of the ensemble by $\mathcal{E}$. We would like to see how this amount of energy $\mathcal{E}$ will be shared by the microstates in this ensemble. Firstly, we should have

$$\sum_r n_r E_r = \mathcal{E}. \tag{16.20}$$

As expected, the average energy within this ensemble $\overline{E} = \mathcal{E}/\mathcal{N}$ will correspond to the real physical subsystem under study, where $\overline{E}$ should approach the energy of the specific system A as the ensemble size $\mathcal{N} \to \infty$. Therefore, we could consider the total energy of the ensemble $\mathcal{E}$ as a constant for a given $\mathcal{N}$.

The canonical ensemble of A characterized by $\{n_r\}$ with the constraints (16.19) and (16.20) is not unique, and the possible number of such canonical ensembles will be given by

$$W\{n_r\} = \frac{\mathcal{N}!}{n_1!(\mathcal{N}-n_1)!}\frac{(\mathcal{N}-n_1)!}{n_2!(\mathcal{N}-n_1-n_2)!}\cdots = \frac{\mathcal{N}!}{n_1!n_2!n_3!\cdots}. \tag{16.21}$$

The first term in (16.21), $\mathcal{N}!/(n_1!(\mathcal{N}-n_1)!)$, gives the probability of n_1 out of the $\mathcal{N}$ microstates taking the energy E_1, and the second term, $(\mathcal{N}-n_1)!/(n_2!\,(\mathcal{N}-n_1-n_2)!)$, gives the probability of n_2 out of the remaining $(\mathcal{N}-n_1)$ microstates taking the energy E_2, etc. The occurrence of a specific $\{n_r\}$ will be determined by how the initial $\mathcal{N}$ microscopic states of $A^{(0)}$ are selected. As each selection of the $\mathcal{N}$ microstates of $A^{(0)}$ will have the same probability according to the equal probability postulate for the isolated system $A^{(0)}$, the most likely choice for $\{n_r\}$, denoted by $\{n_r^*\}$, will be the one for which $W\{n_r\}$ is maximized. For the canonical ensemble with an energy distribution given by this $\{n_r^*\}$, we will have the most likely energy for the system A, i.e.,

$$E^* = \sum_r \frac{n_r^*}{\mathcal{N}} E_r. \tag{16.22}$$

To find $\{n_r^*\}$, we consider $\ln W$ instead, where

$$\ln W = \ln(\mathcal{N}!) - \sum_r \ln(n_r!). \tag{16.23}$$

Using the Stirling formula, $\ln n! \approx n \ln n - n$, (16.23) can be rewritten as

$$\ln W = \mathcal{N} \ln \mathcal{N} - \sum_r n_r \ln n_r. \tag{16.24}$$

Thus, if the set $\{n_r\}$ is shifted to a slightly different set $\{n_r + \delta n_r\}$, then the variation of $\ln W$ is given by

$$\delta(\ln W) = -\sum_r (\ln n_r + 1)\delta n_r. \tag{16.25}$$

At an extreme value of $\ln W$, we should have

$$\delta(\ln W) = -\sum_r (\ln n_r^* + 1)\delta n_r = 0. \tag{16.26}$$

In addition, from the constraints (16.19) and (16.20), we have

$$\sum_r \delta n_r = 0, \tag{16.27}$$

$$\sum_r E_r \delta n_r = 0. \tag{16.28}$$

Equations (16.26)–(16.28) form a constrained maximization problem, which can be reformulated in terms of two Lagrange multipliers,

$$\sum_r [-(\ln n_r^* + 1) - \alpha - \beta E_r]\delta n_r = 0, \tag{16.29}$$

where δn_r is now unrestricted. Therefore, we have

$$\ln n_r^* = -(\alpha + 1) - \beta E_r, \tag{16.30}$$

which implies

$$n_r^* = C \exp(-\beta E_r), \tag{16.31}$$

where the constant C can be determined from (16.19):

$$C = \frac{\mathcal{N}}{\sum_r \exp(-\beta E_r)}. \tag{16.32}$$

Thus, the most likely energy for the system A will be

$$E^* = \sum_r \frac{n_r^*}{\mathcal{N}} E_r = \frac{\sum_r E_r \exp(-\beta E_r)}{\sum_r \exp(-\beta E_r)}. \tag{16.33}$$

The Lagrange multiplier β can be shown [1] to be related to the thermodynamic temperature T as follows:

$$\beta = \frac{1}{k_B T}. \tag{16.34}$$

On the other hand, we should consider the mean value, the expectation value $\langle n_r \rangle$, among all possible choices of $\{n_r\}$ subject to the constraints (16.19) and (16.20) defined as

$$\langle n_s \rangle = \frac{\sum'_{\{n_r\}} n_s W\{n_r\}}{\sum'_{\{n_r\}} W\{n_r\}}, \tag{16.35}$$

where $\sum'_{\{n_r\}}$ indicates the summation only over $\{n_r\}$ satisfying the constraints (16.19) and (16.20).

Considering $\langle n_r \rangle$ as a fraction of the total ensemble number $\mathcal{N}$, the probability P_r that a macro-system of (N, V, T) is found in a microstate with energy E_r should be just

$$P_r \equiv \frac{\langle n_r \rangle}{\mathcal{N}}. \tag{16.36}$$

When $\mathcal{N}$ is large, the only dominant term in (16.35) is $W\{n_r^*\}$, and all the other terms in the summations can be ignored. It may be shown with mathematical rigor [1] that, as $\mathcal{N} \to \infty$, we indeed have

$$\langle n_r \rangle = n_r^*. \tag{16.37}$$

Therefore, we have

$$P_r = \frac{n_r^*}{\mathcal{N}} = \frac{\exp(-\beta E_r)}{\sum_r \exp(-\beta E_r)} = \frac{\exp(-\beta E_r)}{Z_N(T)}, \tag{16.38}$$

where a partition function $Z_N(T)$ for the canonical ensemble is defined as

$$Z_N(T) = \sum_r \exp(-\beta E_r). \tag{16.39}$$

Now, the canonical ensemble (C.E.) average energy $\langle E \rangle$ for a sufficiently large C.E. of the system A is given by

$$\langle E \rangle = \langle E \rangle_{\mathrm{ce}} \equiv \sum_r P_r E_r = \frac{\sum_r E_r \exp(-\beta E_r)}{Z_N(T)}, \tag{16.40}$$

which is the same as E^*; namely, the most likely energy E^* of the system A is in fact the same as the canonical ensemble average $\langle E \rangle$ of the system—the latter is what experimental measurement yields for the system A. Finally, the constant β can be found from (16.40) for a measured energy $\langle E \rangle$ of the system A.

16.1.4 Grand Canonical Ensembles

In device transport, the particle number in a device subsystem A is not necessarily a fixed quantity as it is in the canonical ensemble theory. To accommodate this situation, a grand canonical ensemble theory will be used. In the grand ensemble, the number of particles N_r and the energy E_s in the subsystem A are both variables, and $\Omega(N_r, E_s)$ denotes the number of microstates (quantum states) available to the subsystem A. Meanwhile, N_r' and E_s' are the number of particles and the energy in $A' = A^{(0)} \backslash A$, respectively – the rest of an isolated system $A^{(0)}$. And we have

$$N_r + N_r' = N^{(0)}, \tag{16.41}$$

$$E_s + E_s' = E^{(0)}. \tag{16.42}$$

We consider an ensemble of $\mathcal{N}$ copies of the macroscopic system A with N_r particles and energy E_s. For each possible choice of the particle number N_r, we denote the number of copies at energy E_s by $n_{r,s}$; then we have

$$\sum_{r,s} n_{r,s} = \mathcal{N}, \tag{16.43}$$

$$\sum_{r,s} n_{r,s} N_r = \mathcal{N}\overline{N}, \tag{16.44}$$

$$\sum_{r,s} n_{r,s} E_s = \mathcal{N}\,\overline{E}, \tag{16.45}$$

where $\overline{E}$ and $\overline{N}$ denote the average energy and the average particle number of the ensemble, respectively, which are supposed to approach those of the system A during the experimental measurement of the system under study as $\mathcal{N} \to \infty$.

Following the same Lagrange multiplier technique as in the canonical ensemble, we can show that the probability $P_{r,s}$ that the system is described by (N_r, E_s) is given by the following Boltzmann factor:

$$P_{r,s} = \frac{\exp(-\alpha N_r - \beta E_s)}{Z_G(\mu, V, T)}, \tag{16.46}$$

where the grand canonical partition function is defined as

$$Z_G(\mu, V, T) = \sum_{r,s} \exp(-\alpha N_r - \beta E_s), \tag{16.47}$$

and

$$\alpha = -\frac{\mu}{k_B T}. \tag{16.48}$$

16.1.5 Bose–Einstein and Fermi–Dirac Distributions

The N identical particles in a quantum system differ from those in a classical system due to the formers' indistinguishability, and also because of the Pauli exclusion principle for fermion particles (such as electrons). As a result, the ensemble distributions for the canonical and the grand canonical will be modified to reflect these unique features of many-particle quantum systems.

Due to the indistinguishability of the identical quantum particles, no individuality such as an enumeration and distinct labeling can be associated with each particle in the quantum system. Therefore, an occupation number representation $\mathbf{n}_s = \{n_s, s = 1, 2, \ldots\}$ will be used to describe the N-particle system, where n_s denotes the number of particles with energy ε_s:

$$\sum_{s=1}^{\infty} n_s = N, \tag{16.49}$$

and the total energy for the system is then given by

$$E_{\mathbf{n}_s} = \sum_{s=1}^{\infty} n_s \varepsilon_s. \tag{16.50}$$

The N-particle wave function $\phi_N(\mathbf{r}_1, \mathbf{r}_2, \ldots, \mathbf{r}_N)$ for a quantum system of identical and indistinguishable particles satisfies an anti-symmetric or symmetric property in terms of the permutation of the particle labels in the wave function [4].

- Fermion particles (anti-symmetric): $\phi_N(\mathbf{r}_1, \mathbf{r}_2, \ldots, \mathbf{r}_N) = -\phi_N(\mathbf{r}_2, \mathbf{r}_1, \ldots, \mathbf{r}_N)$, e.g., electrons, protons, neutrons.
- Boson particles (symmetric): $\phi_N(\mathbf{r}_1, \mathbf{r}_2, \ldots, \mathbf{r}_N) = \phi_N(\mathbf{r}_2, \mathbf{r}_1, \ldots, \mathbf{r}_N)$, e.g., photons, ^{4}He.

As an illustration [4], let us consider a box of size L with N free non-interacting particles of momentum $\mathbf{p}$ represented by the wave function

$$\phi_{\mathbf{p}}(\mathbf{r}) = \frac{1}{\sqrt{L^3}} e^{\mathrm{i}\mathbf{p}\cdot\mathbf{r}/\hbar}, \tag{16.51}$$

with periodic boundary conditions and $\mathbf{p} = (p_x, p_y, p_z)^{\mathrm{T}}$:

$$p_i = \frac{2\pi\hbar}{L} n_i, \quad n_i = 0, \pm 1, \pm 2, \ldots, \quad i = x, y, z. \tag{16.52}$$

The N-particle total wave function labeled by its momentum vector for the Hamiltonian (16.2) with $U = 0$ takes the following form in coordinate representation:

$$\phi_{\mathbf{p}_1,\ldots,\mathbf{p}_N}(\mathbf{r}_1, \ldots, \mathbf{r}_N) = \prod_{s=1}^{N} \phi_{\mathbf{p}_s}(\mathbf{r}_s), \tag{16.53}$$

with a total energy given by

$$E = \sum_{s=1}^{N} \frac{p_s^2}{2m}. \tag{16.54}$$

This N-particle wave function does not satisfy the symmetry properties for identical particles, however, which can be achieved by symmetrizing or anti-symmetrizing, for example, through a Slater determinant for the fermions:

$$\phi^a_{\mathbf{p}_1,\ldots,\mathbf{p}_N}(\mathbf{r}_1,\ldots,\mathbf{r}_N) = \frac{1}{\sqrt{N!}}\begin{vmatrix} \phi_{\mathbf{p}_1}(\mathbf{r}_1) & \phi_{\mathbf{p}_1}(\mathbf{r}_2) & \cdots & \phi_{\mathbf{p}_1}(\mathbf{r}_N) \\ \phi_{\mathbf{p}_2}(\mathbf{r}_1) & \phi_{\mathbf{p}_2}(\mathbf{r}_2) & \cdots & \phi_{\mathbf{p}_2}(\mathbf{r}_N) \\ \vdots & \vdots & \ddots & \vdots \\ \phi_{\mathbf{p}_N}(\mathbf{r}_1) & \phi_{\mathbf{p}_N}(\mathbf{r}_2) & \cdots & \phi_{\mathbf{p}_N}(\mathbf{r}_N) \end{vmatrix}$$

$$= \frac{1}{\sqrt{N!}} \sum_{\{\mathbf{p}'_1,\mathbf{p}'_2,\ldots,\mathbf{p}'_N\}} (-1)^\sigma \phi_{\mathbf{p}'_1}(\mathbf{r}_1)\phi_{\mathbf{p}'_2}(\mathbf{r}_2)\cdots\phi_{\mathbf{p}'_N}(\mathbf{r}_N), \tag{16.55}$$

where $\{\mathbf{p}'_1, \mathbf{p}'_2, \ldots, \mathbf{p}'_N\}$ denotes any permutation of the indices $\{\mathbf{p}_1, \ldots, \mathbf{p}_N\}$ and σ is the parity of the permutation. The Slater determinant form of the wave function of (16.55) also conforms to the Pauli exclusion principle, which states that no two fermions can occupy the same energy level. This is clear as the determinant will vanish if two momentum indices for any two rows are the same (namely, the corresponding energy levels hold two fermions).

For the bosons, a similar symmetrization procedure can be defined by dropping the sign factor $(-1)^\sigma$, i.e.,

$$\phi^s_{\mathbf{p}_1,\ldots,\mathbf{p}_N}(\mathbf{r}_1,\ldots,\mathbf{r}_N)$$
$$= \frac{1}{\sqrt{N!n_1!n_2!\cdots}} \sum_{\{\mathbf{p}'_1,\mathbf{p}'_2,\ldots,\mathbf{p}'_N\}} \phi_{\mathbf{p}'_1}(\mathbf{r}_1)\phi_{\mathbf{p}'_2}(\mathbf{r}_2)\cdots\phi_{\mathbf{p}'_N}(\mathbf{r}_N), \tag{16.56}$$

where a different normalization factor is used for the bosons due to the fact that more than one boson can occupy the same energy level, and $n_s, s = 1, 2, \ldots,$ is the number of bosons occupying the s-energy level. The factor $\sqrt{n_s!}$ accounts for the number of possible permutations among those n_s bosons within the same energy level, which gives the same product in the sum of (16.56) and a non-zero contribution to the computation of the norm of the wave function $\phi^s_{\mathbf{p}_1,\ldots,\mathbf{p}_N}$ [4, p. 289].

In both cases, the total energy of the non-interacting particles is given by

$$E_{\mathbf{n}_s} = \sum_{s=1}^{\infty} n_s \frac{p_s^2}{2m}. \tag{16.57}$$

Moreover, we have, for $1 \le s < \infty$,

- bosons:

$$n_s = 0, 1, 2, 3, \ldots, N; \tag{16.58}$$

- fermions:

$$n_s = 0, 1. \tag{16.59}$$

We now discuss the partition functions for various ensembles for the quantum system. As each of the microstates identified by $\mathbf{n}_s$ is in fact an eigenstate of the non-interacting system (16.2) with $U = 0$ with energy (16.57), the partition functions can be defined similarly as in (16.39) or (16.47).

Canonical Ensemble

For an NVT ensemble, the partition function simply sums over all occupation numbers $\mathbf{n}_s$ subject to the constraint $\sum_{s=1}^{\infty} n_s = N$ indicated by the notation $\sum'_{\mathbf{n}_s}$:

$$Z_N(T) = \sum_{\mathbf{n}_s}{}' \mathrm{e}^{-\beta E_{\mathbf{n}_s}} = \sum_{\mathbf{n}_s}{}' \mathrm{e}^{-\beta \sum_{s=1}^{\infty} n_s \varepsilon_s}. \tag{16.60}$$

It should be noted that the sum in (16.60) could involve degenerate energy levels, namely, that eigenstates represented by different $\mathbf{n}_s$ can have the same energy level $E_{\mathbf{n}_s}$. When degeneracy arises, the probability P_r for $E_r = E_{\mathbf{n}_s}$ in (16.36) will be given a different formula to reflect the degeneracy [1, p. 55].

Due to the constraint on $\mathbf{n}_s$, no further simplifications on the partition function will in general be carried out.

Grand Canonical Ensemble

In this case, we have

$$\begin{aligned} Z_G(\mu, T) &= \sum_{N_r=1}^{\infty} \sum_{\mathbf{n}_s}{}' \mathrm{e}^{-\alpha N_r - \beta E_{\mathbf{n}_s}} = \sum_{N_r=1}^{\infty} \mathrm{e}^{-\alpha N_r} \sum_{\mathbf{n}_s}{}' \mathrm{e}^{-\beta E_{\mathbf{n}_s}} \\ &= \sum_{N_r=1}^{\infty} \mathrm{e}^{-\alpha N_r} Z_{N_r}(T), \end{aligned} \tag{16.61}$$

with the understanding again that degenerate energy levels could be involved in this sum.

The average particle number for the system can be calculated using

$$\overline{N} = \frac{\sum_{N_r} N_r \mathrm{e}^{-\alpha N_r} Z_{N_r}(T)}{Z_G(\mu, T)} = -\left(\frac{\partial \ln Z_G}{\partial \alpha}\right)_{\beta, V} = \frac{1}{\beta}\left(\frac{\partial \ln Z_G}{\partial \mu}\right)_{\beta, V}. \tag{16.62}$$

Meanwhile, the constrained sum over $\mathbf{n}_s$ subject to $\sum_{s=1}^{\infty} n_s = N_r$ can be turned into an unconstrained sum by considering the fact that $1 \le N_r < \infty$, namely, by using (16.49) and (16.50) we have

$$\begin{aligned} Z_G(\mu, T) &= \sum_{\mathbf{n}_s=(0,0,\ldots)}^{\mathbf{n}_s^{\infty}} \mathrm{e}^{-\beta \sum_{s=1}^{\infty} n_s(\varepsilon_s - \mu)} \\ &= \sum_{\mathbf{n}_s=(0,0,\ldots)}^{\mathbf{n}_s^{\infty}} (\mathrm{e}^{-\beta(\varepsilon_1-\mu)})^{n_1} (\mathrm{e}^{-\beta(\varepsilon_2-\mu)})^{n_2} \cdots = \prod_{s=1}^{\infty} \sum_{n_s=0}^{n_s^{\infty}} \mathrm{e}^{-\beta n_s(\varepsilon_s-\mu)}, \end{aligned} \tag{16.63}$$

where $\mathbf{n}_s^{\infty} = (\infty, \infty, \ldots)$ for bosons and $\mathbf{n}_s^{\infty} = (1, 1, \ldots)$ for fermions.

It should be noted that the single-particle energy level ε_s for a many-particle system in general is unknown, especially if electron–electron or electron–phonon interactions are involved.

Bose Statistics ($n_s = 0, 1, 2, \ldots$)

For bosons, we have $\mathbf{n}_s^\infty = (\infty, \infty, \ldots)$; thus

$$Z_G(\beta, T) = \prod_{s=1}^{\infty} \sum_{n_s=0}^{\infty} \mathrm{e}^{-\beta n_s(\varepsilon_s - \mu)} = \prod_{s=1}^{\infty} \frac{1}{1 - \mathrm{e}^{-\beta(\varepsilon_s - \mu)}}, \tag{16.64}$$

and then

$$\ln Z_G(\beta, T) = -\sum_{s=1}^{\infty} \ln\left[1 - \mathrm{e}^{-\beta(\varepsilon_s - \mu)}\right]. \tag{16.65}$$

We have, from (16.62):

$$\overline{N} = \frac{1}{\beta}\left(\frac{\partial \ln Z_G}{\partial \mu}\right)_{\beta, V} = \sum_{s=1}^{\infty} \frac{1}{\mathrm{e}^{\beta(\varepsilon_s - \mu)} - 1} = \sum_{s=1}^{\infty} \overline{n}_s, \tag{16.66}$$

where

$$\overline{n}_s \equiv \frac{1}{\mathrm{e}^{\beta(\varepsilon_s - \mu)} - 1}$$

is the average number of particles at the energy ε_s, also identified as the Bose–Einstein distribution for bosons at the chemical potential μ:

$$f_{\mathrm{BE}}(\varepsilon_s - \mu) = \frac{1}{\mathrm{e}^{\beta(\varepsilon_s - \mu)} - 1}. \tag{16.67}$$

Fermi Statistics ($n_s = 0, 1$)
For fermions, we have instead $\mathbf{n}_s^\infty = (1, 1, \ldots)$; therefore

$$Z_G(\mu, T) = \prod_{s=1}^{\infty} \sum_{n_s=0}^{1} \mathrm{e}^{-\beta n_s(\varepsilon_s - \mu)} = \prod_{s=1}^{\infty} \left[1 + \mathrm{e}^{-\beta(\varepsilon_s - \mu)}\right].$$

Now

$$\ln Z_G = \sum_{s=1}^{\infty} \ln\left[1 + \mathrm{e}^{-\beta(\varepsilon_s - \mu)}\right]. \tag{16.68}$$

Then

$$\overline{N} = \frac{1}{\beta}\left(\frac{\partial \ln Z_G}{\partial \mu}\right)_{\beta,V} = \frac{1}{\beta}\sum_{s=1}^{\infty}\frac{\beta e^{-\beta(\varepsilon_s-\mu)}}{1+e^{-\beta(\varepsilon_s-\mu)}}$$
$$= \sum_{s=1}^{\infty}\frac{1}{1+e^{\beta(\varepsilon_s-\mu)}} = \sum_{s=1}^{\infty}\overline{n}_s, \qquad (16.69)$$

where

$$\overline{n}_s \equiv \frac{1}{1+e^{\beta(\varepsilon_s-\mu)}}$$

is the average number of particles at energy ε_s, also identified as the Fermi–Dirac distribution for fermions at the chemical potential μ:

$$f_{\mathrm{FD}}(\varepsilon_s-\mu) = \frac{1}{1+e^{\beta(\varepsilon_s-\mu)}}. \qquad (16.70)$$

16.2 Density Operator $\hat{\rho}$ for Quantum Systems

From the discussion in Sect. 16.1, the probability that a quantum fermion system at thermal equilibrium with its environment can be found in a microstate with energy E_r is given by the Fermi–Dirac distribution in (16.70), and the average energy of the system can be calculated using this distribution. A dynamical variable in a quantum system is represented by a Hamiltonian operator and an observable of the dynamical variable as quantum average with respect to the microstate wave function. For systems with an ensemble distribution as given in Sect. 16.1, the concept of the density operator $\hat{\rho}$ introduced by von Neumann [5, 6] embodies the statistical distributional characteristics of the ensemble. With this density operator, the trace of the product of the dynamic operator with $\hat{\rho}$ will accomplish both the quantum and the statistical averages. Moreover, the coordinate representation of the density operator, the *density matrix*, can be used to define the quantum mechanics analog of the Boltzmann distribution in phase space, the Wigner distribution, and its kinetic theory for non-equilibrium systems [7].

For a given N-particle quantum system represented by a wave function $|\psi\rangle$, we introduce a coordinate representation as

$$\psi(\mathbf{x}) \equiv \langle\mathbf{x}|\psi\rangle, \qquad (16.71)$$

where $|\mathbf{x}\rangle = (x_1, \ldots, x_N)^{\mathrm{T}}$.

If the state of the quantum system is prepared such that it is described by one microstate $|\psi\rangle$, namely, the system is in a *pure state*, we have a complete knowledge of the system within the context of quantum mechanics. Then, the average of any physical dynamical variable $\hat{A}$ can be calculated by considering the quantum average with respect to the wave function:

$$\left\langle \hat{A} \right\rangle = \left\langle \psi \left| \hat{A} \right| \psi \right\rangle . \tag{16.72}$$

This average can also be calculated via a density operator $\hat{\rho}$ defined as the projection operator onto the 1-D subspace spanned by $|\psi\rangle$:

$$\hat{\rho} = P|_{\psi} = |\psi\rangle \langle\psi| , \tag{16.73}$$

through a trace operation, i.e.,

$$\mathrm{tr}(\hat{\rho}\hat{A}) = \mathrm{tr}(|\psi\rangle \langle\psi| \hat{A}) = \mathrm{tr}\left(\left\langle \psi \left| \hat{A} \right| \psi \right\rangle\right) = \left\langle \hat{A} \right\rangle . \tag{16.74}$$

In general, the trace operation $\mathrm{tr}(\hat{O})$ on an operator is defined as

$$\mathrm{tr}(\hat{O}) = \sum_{i=1}^{\infty} O_{ii}, \tag{16.75}$$

where $O_{ii} = \left\langle \phi_i | \hat{O} | \phi_i \right\rangle$ is the diagonal element of the operator $\hat{O}$ under any complete basis $\{|\phi_n\rangle\}|_{n=1}^{\infty}$ in the space where $\hat{O}$ is defined. Equation (16.74) can also be shown by setting $|\phi_1\rangle = |\psi\rangle$ in an orthogonal augmentation for the complete basis.

In practice, only partial information about the state of a system is available. As a result, a statistical description of the state of the system is required due to the uncertain effects arising from the variables outside the given (partial) system. In other words, no single wave function can completely define this system. Instead, a probability distribution will be used to describe the chance of the system in one of the many microstates $\{|\psi_i\rangle\}$, with some probability ω_i; such a quantum system is said to be in a *mixed state*. The probability ω_i is exactly the frequency of the wave function $|\psi_i\rangle$ that appears in the ensemble collection of $\mathcal{N}$ copies of microstates of the system $\left\{ |\psi^{(k)}\rangle \right\}_{k=1}^{\mathcal{N}}$ of (16.13) for the system under study. Namely,

$$\omega_i = \frac{\#\text{ of } |\psi^{(k)}\rangle = |\psi_i\rangle}{\mathcal{N}}, \quad \mathcal{N} \to \infty, \tag{16.76}$$

where

$$\omega_i \geq 0, \quad \sum_i \omega_i = 1.$$

Therefore, the average of the dynamical variable $\hat{A}$ can be simply defined as in (16.17):

$$\left\langle \hat{A} \right\rangle \equiv \frac{1}{\mathcal{N}} \sum_{k=1}^{\mathcal{N}} \left\langle \psi^{(k)} \left| \hat{A} \right| \psi^{(k)} \right\rangle$$
$$= \sum_i \omega_i \left\langle \psi_i \left| \hat{A} \right| \psi_i \right\rangle = \sum_i \omega_i \mathrm{tr}(P|_{\psi_i} \hat{A}) = \mathrm{tr}(\hat{\rho}\hat{A}), \tag{16.77}$$

where the density operator has been identified to take the following equivalent forms [8]:

$$\hat{\rho} = \frac{1}{\mathcal{N}} \sum_{k=1}^{\mathcal{N}} |\psi^{(k)}\rangle \langle \psi^{(k)}| = \sum_i \omega_i P|_{\psi_i} = \sum_i \omega_i \; |\psi_i\rangle \langle \psi_i| \,. \tag{16.78}$$

In (16.77), two types of averages are involved: the statistical average, with probability ω_i due to the statistical description of the system, and the quantum average with respect to each microstate $|\psi_i\rangle$.

The density operator can have different matrix representations, giving different forms of a density matrix, under various bases such as the physical coordinates or energy eigenstates of the Hamiltonian of the system. In the physical coordinates, the density matrix for a canonical ensemble of an N-particle system is defined as

$$\rho(\mathbf{x}, \mathbf{x}') = \langle \mathbf{x}|\hat{\rho}|\mathbf{x}' \rangle = \sum_i \omega_i \psi_i^*(\mathbf{x}') \psi_i(\mathbf{x}), \tag{16.79}$$

where $|\mathbf{x}\rangle$ is identified as a complete basis for the physical coordinate space.

To get a representation of the density operator under any other complete basis $\{|\phi_n\rangle\}|_{n=1}^{\infty}$, we express each of the microstates in the ensemble as [1, 6]

$$|\psi^{(k)}\rangle = \sum_{n=1}^{\infty} a_n^{(k)} \; |\phi_n\rangle \,, \; k = 1, 2, \ldots . \tag{16.80}$$

Therefore

$$\hat{\rho} = \frac{1}{\mathcal{N}} \sum_{k=1}^{\mathcal{N}} \sum_{m=1}^{\infty} \sum_{n=1}^{\infty} a_n^{(k)*} a_m^{(k)} \; |\phi_m\rangle \langle \phi_n|$$
$$= \sum_{m=1}^{\infty} \sum_{n=1}^{\infty} \frac{1}{\mathcal{N}} \left(\sum_{k=1}^{\mathcal{N}} a_n^{(k)*} a_m^{(k)} \right) |\phi_m\rangle \langle \phi_n|$$
$$= \sum_{m=1}^{\infty} \sum_{n=1}^{\infty} \rho_{mn} \; |\phi_m\rangle \langle \phi_n| \,, \tag{16.81}$$

where

$$\rho_{mn} = \frac{1}{\mathcal{N}} \left(\sum_{k=1}^{\mathcal{N}} a_n^{(k)*} a_m^{(k)} \right) \tag{16.82}$$

is the entry in the matrix representation of the density operator under the basis $\{|\phi_n\rangle\}|_{n=1}^{\infty}$.

Finally, we want to appreciate better the source of statistical uncertainty embodied in the statistical density operator, arising from the lack of complete knowledge of the outside environment for the system under study (whose variables are denoted by $\mathbf{y}$, while $\mathbf{x}$ denotes the coordinates for the subsystem). If $\psi(\mathbf{y}, \mathbf{x})$ is a microstate wave function for the combined system, we can expand the wave function under an orthogonal basis $\{|\phi_n\rangle\}$ above with $\mathbf{y}$-dependence coefficients, i.e.,

$$\psi(\mathbf{y}, \mathbf{x}) = \sum_{n=1}^{\infty} C_n(\mathbf{y})\phi_n(\mathbf{x}), \tag{16.83}$$

where

$$C_n(\mathbf{y}) = \langle \phi_n | \psi \rangle . \tag{16.84}$$

Then, the quantum average of the dynamical variable of the subsystem $\hat{A} = \hat{A}(\mathbf{x})$ is given by

$$\begin{aligned} \left\langle \hat{A} \right\rangle = \left\langle \psi | \hat{A} | \psi \right\rangle &= \sum_{n=1}^{\infty} \sum_{n'=1}^{\infty} \langle C_n(\mathbf{y}) | C_{n'}(\mathbf{y}) \rangle \left\langle \phi_n \left| \hat{A} \right| \phi_{n'} \right\rangle \\ &= \sum_{n=1}^{\infty} \sum_{n'=1}^{\infty} \rho_{nn'} A_{n'n} = \operatorname{tr}(\hat{\rho}\hat{A}), \end{aligned} \tag{16.85}$$

where the density matrix element $\rho_{nn'}$ can be identified with

$$\rho_{nn'} = \langle C_n(\mathbf{y}) | C_{n'}(\mathbf{y}) \rangle , \tag{16.86}$$

indicating clearly the dependence on the outside $\mathbf{y}$-variable [4, 9].

16.2.1 One-Particle Density Matrix $\rho(x, x')$

The density matrix defined for the many-particle system in (16.79) is difficult to compute in practice as the wave functions $|\psi_i\rangle$ are impossible to obtain for a many-particle Hamiltonian. Fortunately, for electron transport, the one-particle Schrödinger description of the quantum device in Sect. 15.3.1 is a very good approximation, with good selections of the effective mean field potential to model the many-electron interactions (Hartree and correlations) [10, Sects. 3.1–3.2] as in the density functional theory [11], discussed in Chap. 17. For the rest of Part III of this book (except in Chap. 17), we will use the one-particle Schrödinger equation, for which the density matrix is simply

$$\rho(x, x') = \left\langle x | \hat{\rho} | x' \right\rangle . \tag{16.87}$$

If $\{|\phi_\alpha\rangle\}_\alpha$ are the one-particle energy eigenstates for the Schrödinger equation, the density operator is given by

$$\hat{\rho} = \sum_\alpha \omega_\alpha |\phi_\alpha\rangle \langle\phi_\alpha| , \tag{16.88}$$

with a given probability ω_α. Then, the density matrix representation of the density operator becomes

$$\rho(x, x') = \langle x| \sum_\alpha \omega_\alpha |\phi_\alpha\rangle \langle\phi_\alpha| x'\rangle = \sum_\alpha \omega_\alpha \phi_\alpha(x)\phi_\alpha^*(x'). \tag{16.89}$$

From (16.70), the occupation for electrons at energy ε_α obeys the Fermi–Dirac distribution; thus we have

$$\omega_\alpha = f_{\mathrm{FD}}(\varepsilon_\alpha - \mu). \tag{16.90}$$

The diagonal element of the density matrix gives the density distribution of the electron:

$$\hat{\rho}(x, x) = \sum_\alpha f_{\mathrm{FD}}(\varepsilon_\alpha - \mu)|\phi_\alpha(x)|^2. \tag{16.91}$$

As $|\phi_\alpha\rangle$ are the eigenstates of the Hamiltonian of the quantum system, we have

$$f_{\mathrm{FD}}(\hat{H} - \mu I) |\phi_\alpha\rangle = f_{\mathrm{FD}}(\varepsilon_\alpha - \mu) |\phi_\alpha\rangle , \tag{16.92}$$

which implies that

$$\begin{aligned} \hat{\rho} &= \sum_\alpha f_{\mathrm{FD}}(\varepsilon_\alpha - \mu) |\phi_\alpha\rangle \langle\phi_\alpha| = \sum_\alpha f_{\mathrm{FD}}(\hat{H} - \mu I) |\phi_\alpha\rangle \langle\phi_\alpha| \\ &= f_{\mathrm{FD}}(\hat{H} - \mu I) \sum_\alpha |\phi_\alpha\rangle \langle\phi_\alpha| , \end{aligned} \tag{16.93}$$

namely

$$\hat{\rho} = f_{\mathrm{FD}}(\hat{H} - \mu I), \tag{16.94}$$

due to the completeness and the orthogonality of the basis $|\phi_\alpha\rangle$.

16.3 Wigner Transport Equations and Wigner–Moyal Expansions

In order to describe the dynamics of a quantum system in phase space, like the Boltzmann distribution for classical mechanics system, Wigner introduced the concept of the Wigner distribution [12, 13]. Although the Wigner distribution is not a real probability distribution function due to its possible negative values, the Wigner function can serve as a distribution [14–16], for example, in calculating number densities, current densities, etc. Using the Wigner equation to investigate quantum transport has become more relevant [17, 18] when the quantum behavior of semiconductor devices cannot be neglected as the size is down to nano-scale.

The Wigner distribution function $f(x, k, t)$ for a pure state $\psi(x)$ in the phase space $(x, k) \in \mathbb{R}^2$ for position x and wave number k is defined through a Fourier transform [7, 12]

$$f_\psi(x, k, t) = \frac{1}{2\pi} \int_{-\infty}^{\infty} \exp(-iky)\psi\left(x + \frac{y}{2}\right)\psi^*\left(x - \frac{y}{2}\right) \mathrm{d}y. \tag{16.95}$$

It is easy to show that the Wigner distribution function defined above is a real quantity; however, it is not necessarily positive. Meanwhile, the density operator for the pure state defined in (16.73) and (16.87) is given by a projection operator, i.e.,

$$\hat{\rho}_\psi \equiv |\psi\rangle \langle\psi| \,,$$

whose matrix representation is simply

$$\rho_\psi(x, x') = \langle x|\hat{\rho}_\psi|x'\rangle = \psi(x)\,\psi^*(x').$$

Therefore, the Wigner function for the pure state in terms of the density matrix is

$$f_\psi(x, k, t) = \frac{1}{2\pi} \int_{-\infty}^{\infty} \exp(-iky)\rho_\psi\left(x + \frac{y}{2}, x - \frac{y}{2}, t\right) \mathrm{d}y. \tag{16.96}$$

The density matrix entry $\rho_\psi\,(x + y/2, x - y/2)$ reflects the correlation of electrons at locations $x \pm y/2$, with x identified as the center of mass of the two electrons and y as the distance in between. The former can be considered as the slow variable and the latter as the fast variable; the Wigner distribution $f_\psi(x, k, t)$ via the Fourier transform contains the frequency information of the density matrix in the fast variable.

For a general quantum system in a mixed state involving microstates $|\psi_\alpha\rangle$, the density operator is a linear superposition of the density operators associated with each microstate (refer to (16.78)), i.e.,

$$\hat{\rho} = \sum_\alpha \omega_\alpha\, |\psi_\alpha\rangle \langle\psi_\alpha| = \sum_\alpha \omega_\alpha \hat{\rho}_{\psi_\alpha}, \tag{16.97}$$

whose matrix representation is

$$\rho(x, x') = \sum_\alpha \omega_\alpha \psi_\alpha(x)\,\psi^*_\alpha(x') = \sum_\alpha \omega_\alpha \rho_{\psi_\alpha}(x, x'). \tag{16.98}$$

Equation (16.98) indicates a linear superposition property of the density matrices. Therefore, the Wigner distribution function can also be defined as a linear superposition of the corresponding Wigner distributions of the density matrices as follows:

$$\begin{aligned} f(x,k,t) &\equiv \sum_{\alpha} \omega_{\alpha} f_{\psi_{\alpha}}(x,k,t) \\ &= \sum_{\alpha} \omega_{\alpha} \frac{1}{2\pi} \int_{-\infty}^{\infty} \exp(-iky) \rho_{\psi_{\alpha}} \left(x + \frac{y}{2}, x - \frac{y}{2}, t\right) \mathrm{d}y \\ &= \frac{1}{2\pi} \int_{-\infty}^{\infty} \exp(-iky) \sum_{\alpha} \omega_{\alpha} \rho_{\psi_{\alpha}} \left(x + \frac{y}{2}, x - \frac{y}{2}, t\right) \mathrm{d}y, \end{aligned} \tag{16.99}$$

resulting in the following definition of the Wigner distribution for a general quantum system:

$$f(x,k,t) = \frac{1}{2\pi} \int_{-\infty}^{\infty} \exp(-iky) \rho \left(x + \frac{y}{2}, x - \frac{y}{2}\right) \mathrm{d}y. \tag{16.100}$$

The dynamic equations for the Wigner distribution can be directly obtained from the Schrödinger equations governing the time evolution of a pure state or the microstate involved in the definition of the density operator (16.98). Due to the principle of linear superposition in the definition of the Wigner distribution function for the mixed states (16.99), we only have to derive the dynamic equation for the case of a pure state, i.e., $\rho(x, x') = \psi(x)\psi^*(x')$, where the wave function is assumed to satisfy the following Schrödinger equation:

$$\frac{\partial}{\partial t}\psi(x,t) = \frac{\mathrm{i}\hbar}{2m} \frac{\partial^2}{\partial x^2}\psi(x,t) + \frac{q}{\mathrm{i}\hbar} V(x)\psi(x,t). \tag{16.101}$$

From (16.95), we have

$$\begin{aligned} &\frac{\partial}{\partial t} f(x,k,t) \\ &= \frac{1}{2\pi} \int_{-\infty}^{\infty} \mathrm{e}^{-iky} \left[\frac{\partial}{\partial t}\psi^* \left(x - \frac{y}{2}, t\right) \psi \left(x + \frac{y}{2}, t\right) \right. \\ &\quad \left. + \psi^* \left(x - \frac{y}{2}, t\right) \frac{\partial}{\partial t}\psi \left(x + \frac{y}{2}, t\right) \right] \mathrm{d}y \end{aligned}$$

$$
\begin{aligned}
&= \frac{i\hbar}{2m}\frac{1}{2\pi}\int_{-\infty}^{\infty} e^{-iky}\left[-\frac{\partial^2}{\partial x^2}\psi^*\left(x-\frac{y}{2},t\right)\psi\left(x+\frac{y}{2},t\right)\right.\\
&\left.+\,\psi^*\left(x-\frac{y}{2},t\right)\frac{\partial^2}{\partial x^2}\psi\left(x+\frac{y}{2},t\right)\right]dy\\
&+\frac{q}{i\hbar}\frac{1}{2\pi}\int_{-\infty}^{\infty} e^{-iky}\left[V\left(x+\frac{y}{2}\right)-V\left(x-\frac{y}{2}\right)\right]\psi^*\left(x-\frac{y}{2},t\right)\psi\left(x+\frac{y}{2},t\right)dy\\
&= I + II.
\end{aligned}
\tag{16.102}
$$

We will simplify both terms I and II individually. By using the identities

$$
\frac{\partial}{\partial y}\psi^*\left(x+\frac{y}{2}\right)=\frac{1}{2}\frac{\partial}{\partial x}\psi^*\left(x+\frac{y}{2}\right), \tag{16.103}
$$

$$
\frac{\partial}{\partial y}\psi\left(x-\frac{y}{2}\right)=-\frac{1}{2}\frac{\partial}{\partial x}\psi\left(x-\frac{y}{2}\right), \tag{16.104}
$$

the first term can be rewritten as

$$
\begin{aligned}
I &= \frac{4i\hbar}{2m}\frac{1}{2\pi}\int_{-\infty}^{\infty} e^{-iky}\left[-\frac{\partial^2}{\partial y^2}\psi^*\left(x-\frac{y}{2},t\right)\psi\left(x+\frac{y}{2},t\right)\right.\\
&\left.+\,\psi^*\left(x-\frac{y}{2},t\right)\frac{\partial^2}{\partial y^2}\psi\left(x+\frac{y}{2},t\right)\right]dy\\
&= \frac{2i\hbar}{m}\frac{1}{2\pi}\int_{-\infty}^{\infty} e^{-iky}\frac{\partial}{\partial y}\left[-\frac{\partial}{\partial y}\psi^*\left(x-\frac{y}{2},t\right)\psi\left(x+\frac{y}{2},t\right)\right.\\
&\left.+\,\psi^*\left(x-\frac{y}{2},t\right)\frac{\partial}{\partial y}\psi\left(x+\frac{y}{2},t\right)\right]dy.
\end{aligned}
\tag{16.105}
$$

Using integration by parts, assuming the vanishing boundary condition of the wave functions at infinity, and using the identities (16.103) and (16.104) again, I can be further simplified as follows:

$$
\begin{aligned}
I &= -\frac{2i\hbar}{m}(-ik)\frac{1}{2\pi}\int_{-\infty}^{\infty} e^{-iky}\left[-\frac{\partial}{\partial y}\psi^*\left(x-\frac{y}{2},t\right)\psi\left(x+\frac{y}{2},t\right)\right.\\
&\left.+\,\psi^*\left(x-\frac{y}{2},t\right)\frac{\partial}{\partial y}\psi\left(x+\frac{y}{2},t\right)\right]dy\\
&= -\frac{k\hbar}{m}\frac{1}{2\pi}\int_{-\infty}^{\infty} e^{-iky}\left[\frac{\partial}{\partial x}\psi^*\left(x-\frac{y}{2},t\right)\psi\left(x+\frac{y}{2},t\right)\right.
\end{aligned}
$$

$$+\,\psi^*\left(x-\frac{y}{2},t\right)\frac{\partial}{\partial x}\psi\left(x+\frac{y}{2},t\right)\Big]\mathrm{d}y$$

$$=-\frac{k\hbar}{m}\frac{\partial}{\partial x}\frac{1}{2\pi}\int\limits_{-\infty}^{\infty}\mathrm{e}^{-\mathrm{i}ky}\psi\left(x+\frac{y}{2},t\right)\psi^*\left(x-\frac{y}{2},t\right)\mathrm{d}y$$

$$=-\frac{k\hbar}{m}\frac{\partial}{\partial x}f(x,k,t). \tag{16.106}$$

Next, to simplify II, we will express the potential difference $V(x+y/2)-V(x-y/2)$ in terms of its Fourier transform, the Wigner potential $V_w(x,k)$ as follows:

$$\begin{aligned}V_w(x,k)&=\mathrm{i}\int\limits_{-\infty}^{\infty}\mathrm{e}^{-\mathrm{i}ky}\left[V\left(x+\frac{y}{2}\right)-V\left(x-\frac{y}{2}\right)\right]\mathrm{d}y\\&=2\int\limits_{0}^{\infty}\left[V\left(x+\frac{y}{2}\right)-V\left(x-\frac{y}{2}\right)\right]\sin ky\ \mathrm{d}y\\&=4\int\limits_{0}^{\infty}[V(x+y)-V(x-y)]\sin 2ky\ \mathrm{d}y,\end{aligned} \tag{16.107}$$

namely

$$V\left(x+\frac{y}{2}\right)-V\left(x-\frac{y}{2}\right)=\frac{1}{2\pi\mathrm{i}}\int\limits_{-\infty}^{\infty}\exp(\mathrm{i}ky)V_w(x,k)\mathrm{d}k. \tag{16.108}$$

Using (16.108), II becomes

$$\begin{aligned}II&=\frac{q}{\mathrm{i}\hbar}\frac{1}{2\pi}\int\limits_{-\infty}^{\infty}\mathrm{e}^{-\mathrm{i}ky}\left[V\left(x+\frac{y}{2}\right)-V\left(x-\frac{y}{2}\right)\right]\psi^*\left(x-\frac{y}{2},t\right)\psi\left(x+\frac{y}{2},t\right)\mathrm{d}y\\&=-\frac{q}{2\pi\hbar}\frac{1}{2\pi}\int\limits_{-\infty}^{\infty}\mathrm{e}^{-\mathrm{i}ky}\int\limits_{-\infty}^{+\infty}\mathrm{e}^{\mathrm{i}k'y}V_w(x,k')\mathrm{d}k'\psi^*\left(x-\frac{y}{2},t\right)\psi\left(x+\frac{y}{2},t\right)\mathrm{d}y\\&=-\frac{q}{2\pi\hbar}\frac{1}{2\pi}\int\limits_{-\infty}^{\infty}V_w(x,k')\mathrm{d}k'\int\limits_{-\infty}^{\infty}\mathrm{e}^{-\mathrm{i}(k-k')y}\psi^*\left(x-\frac{y}{2},t\right)\psi\left(x+\frac{y}{2},t\right)\mathrm{d}y\\&=-\frac{q}{2\pi\hbar}\int\limits_{-\infty}^{\infty}V_w(x,k')f(x,k-k',t)\mathrm{d}k'.\end{aligned} \tag{16.109}$$

Finally, a dynamic equation for the Wigner distribution can be obtained:

$$\frac{\partial}{\partial t} f(x,k,t) + \frac{\hbar k}{m}\frac{\partial}{\partial x} f(x,k,t) + \frac{q}{2\pi\hbar}\int_{-\infty}^{\infty} V_w(x,k - k') f(x,k',t)\mathrm{d}k' = 0. \tag{16.110}$$

An alternative form of the Wigner equation can be obtained by using a Taylor expansion for the potential function, if it exists,

$$V\left(x+\frac{y}{2}\right) - V\left(x-\frac{y}{2}\right) = 2\sum_{s=0}^{\infty}\frac{V^{(2s+1)}(x)}{(2s+1)!}\left(\frac{y}{2}\right)^{2s+1}, \tag{16.111}$$

and the identity

$$\frac{\partial^n}{\partial k^n} f(x,k,t) = \frac{1}{2\pi}\int_{-\infty}^{\infty} (-\mathrm{i}y)^n \mathrm{e}^{-\mathrm{i}ky}\psi\left(x+\frac{y}{2}\right)\psi^*\left(x-\frac{y}{2}\right)\mathrm{d}y. \tag{16.112}$$

Then the second term II becomes

$$\begin{aligned} II &= \frac{q}{\mathrm{i}\hbar}\frac{1}{2\pi}\int_{-\infty}^{\infty} \mathrm{e}^{-\mathrm{i}ky}\left[V\left(x+\frac{y}{2}\right) - V\left(x-\frac{y}{2}\right)\right]\psi^*\left(x-\frac{y}{2},t\right)\psi\left(x+\frac{y}{2},t\right)\mathrm{d}y \\ &= \frac{2q}{\mathrm{i}\hbar}\frac{1}{2\pi}\int_{-\infty}^{\infty} \mathrm{e}^{-\mathrm{i}ky}\sum_{s=0}^{\infty}\frac{V^{(2s+1)}(x)}{(2s+1)!}\left(\frac{y}{2}\right)^{2s+1}\psi^*\left(x-\frac{y}{2},t\right)\psi\left(x+\frac{y}{2},t\right)\mathrm{d}y \\ &= \frac{2q}{\mathrm{i}\hbar}\sum_{s=0}^{\infty}\frac{1}{(2s+1)!}V^{(2s+1)}(x)\left(\frac{1}{-2\mathrm{i}}\right)^{2s+1}\frac{\partial^{2s+1}}{\partial k^{2s+1}} f(x,k,t) \\ &= \frac{q}{\hbar}\sum_{s=0}^{\infty}\frac{(-1)^s}{4^s(2s+1)!}V^{(2s+1)}(x)\frac{\partial^{2s+1}}{\partial k^{2s+1}} f(x,k,t), \end{aligned} \tag{16.113}$$

resulting in the following Wigner–Moyal expansion [7].

- Wigner–Moyal expansion of the Wigner equation:

$$\frac{\partial}{\partial t} f(x,k,t) + \frac{\hbar k}{m}\frac{\partial}{\partial x} f(x,k,t) = \frac{q}{\hbar}\sum_{s=0}^{\infty}\frac{(-1)^s\, V^{(2s+1)}(x)}{4^s(2s+1)!}\frac{\partial^{2s+1}}{\partial k^{2s+1}} f(x,k,t). \tag{16.114}$$

Furthermore, we can rewrite (16.114) using the pseudo-differential operator notation as follows by continuing from the third line of (16.113):

$$
\begin{aligned}
II &= \frac{2q}{\mathrm{i}\hbar}\sum_{s=0}^{\infty}\frac{V^{(2s+1)}(x)}{(2s+1)!}\left(\frac{1}{-2\mathrm{i}}\right)^{2s+1}\frac{\partial^{2s+1}}{\partial k^{2s+1}}f(x,k,t)\\
&= \frac{2q}{\mathrm{i}\hbar}\left[V\left(x+\frac{1}{-2\mathrm{i}}\frac{\partial}{\partial k}\right)-V\left(x-\frac{1}{-2\mathrm{i}}\frac{\partial}{\partial k}\right)\right]f(x,k,t)\\
&= \frac{2q}{\mathrm{i}\hbar}\left[V\left(x+\frac{\mathrm{i}}{2}\frac{\partial}{\partial k}\right)-V\left(x-\frac{\mathrm{i}}{2}\frac{\partial}{\partial k}\right)\right]f(x,k,t),
\end{aligned}
\tag{16.115}
$$

and the Wigner equation becomes [19]

$$
\frac{\partial}{\partial t}f(x,k,t)+\frac{\hbar k}{m}\frac{\partial}{\partial x}f(x,k,t)+\frac{2q}{\mathrm{i}\hbar}\left[V\left(x-\frac{\mathrm{i}}{2}\frac{\partial}{\partial k}\right)-V\left(x+\frac{\mathrm{i}}{2}\frac{\partial}{\partial k}\right)\right]f(x,k,t)=0.
\tag{16.116}
$$

Even though the Wigner distribution function is not positive, and thus cannot be treated as a joint probability in the phase space as the Boltzmann distribution would be, it can be shown [7] that for a pure state density operator $\rho(x,x')=\psi(x)\psi^*(x')$, using the plane wave representation of the Dirac δ function

$$
\delta(x)=\frac{1}{2\pi}\int_{-\infty}^{\infty}\mathrm{e}^{\mathrm{i}kx}\mathrm{d}k,
\tag{16.117}
$$

the integration over the momentum will give a coordinate probability density,

$$
\int_{-\infty}^{\infty}f(x,k,t)\mathrm{d}k=|\psi(x,t)|^2=\mathrm{n}(x,t),
\tag{16.118}
$$

while the integration over the spatial coordinate gives a momentum probability density,

$$
\int_{-\infty}^{\infty}f(x,k,t)\mathrm{d}x=2\pi|\hat{\psi}(k,t)|^2,
\tag{16.119}
$$

where $\hat{\psi}(k,t)$ is the Fourier transform of the wave function $\psi(x,t)$, i.e.,

$$
\hat{\psi}(k)=\frac{1}{2\pi}\int_{-\infty}^{\infty}\mathrm{e}^{-ikx}\psi(x)\mathrm{d}k.
\tag{16.120}
$$

Most importantly, like the Boltzmann probability distribution, the Wigner distribution can be used to compute the average of any quantum variable $\hat{A}$ [7, p. 354]

$$\left\langle \hat{A} \right\rangle = \mathrm{tr}(\hat{\rho}\hat{A}) = \int_{-\infty}^{\infty}\int_{-\infty}^{\infty} A(x,k) f(x,k)\, \mathrm{d}x\, \mathrm{d}k, \tag{16.121}$$

provided that the dynamical function $A(x, k)$ is defined through the Weyl correspondence [7, Eq. (2.55)]

$$A(x,k) = 2\int_{-\infty}^{\infty} \mathrm{e}^{2iky} \left\langle x + y \left| \hat{A} \right| x - y \right\rangle \mathrm{d}y. \tag{16.122}$$

Moreover, the Wigner function can be used to calculate the electron density $\mathrm{n}(x, t)$ as in (16.118) and the current density $\mathrm{j}(x, t)$ by

$$\mathrm{j}(x,t) = \frac{\hbar}{m}\int_{-\infty}^{\infty} k f(x,k,t)\mathrm{d}k, \tag{16.123}$$

and, by integrating the Wigner equation (16.110) over k, the following continuity equation for the electron is obtained:

$$\frac{\partial}{\partial t}\mathrm{n}(x,t) + \frac{\partial}{\partial x}\mathrm{j}(x,t) = -\mathrm{p}(x,t) = 0, \tag{16.124}$$

where

$$\mathrm{p}(x,t) = \frac{q}{2\pi\hbar}\int_{-\infty}^{\infty} \mathrm{d}k \int_{-\infty}^{\infty} \mathrm{d}k' V_w(x, k - k') f(x, k', t) = 0, \tag{16.125}$$

due to the anti-symmetry of $V_w(x, k)$ in k. This continuity equation corresponds to the conservation of the zeroth moment, i.e., the charge conservation.

16.4 Quantum Wave Transmission and Landauer Current Formula

Due to the wave nature of electrons in nano-devices, the transport of the electrons should be described by transmission coefficients as the waves pass through potential barriers created by either impurities or hetero-junctions. The major difference between a classical particle and a quantum electron is the tunneling phenomena through potential barriers shown by the latter. The Landauer theory treats the transmission coefficient as the probability of an electron passing through a conductor.

The Landauer current formula was used for the current–voltage in the early 1930s to describe tunneling junctions [20] and later was derived by Tsu and Esaki to calculate the I-V characteristics of the resonant tunneling diode, which they co-invented [21, 22].

16.4.1 Transmission Coefficient $T(E)$

Transmission Over a Semi-infinite Potential Step

Consider the transmission of a free electron $\psi^{\mathrm{inc}}(x) = \mathrm{e}^{\mathrm{i}k_1 x}$, $x < 0$, over a semi-infinite potential step:

$$V(x) = \begin{cases} 0, & \text{if } x < 0, \\ V_0, & \text{if } x > 0. \end{cases} \tag{16.126}$$

The solution to the Schrödinger equation,

$$-\frac{\hbar^2}{2m}\psi'' + V(x)\psi = E\psi, \tag{16.127}$$

is composed of the reflected ($x < 0$, region 1) and the transmitted waves ($x > 0$, region 2):

$$\psi(x) = \begin{cases} A\mathrm{e}^{\mathrm{i}k_1 x} + B\mathrm{e}^{-\mathrm{i}k_1 x}, & \text{if } x < 0, \\ C\mathrm{e}^{\mathrm{i}k_2 x} + D\mathrm{e}^{-\mathrm{i}k_2 x}, & \text{if } x > 0, \end{cases} \tag{16.128}$$

where

$$A = 1,\ k_1^2 = \frac{2mE}{\hbar^2},\ k_2^2 = \frac{2m(E - V_0)}{\hbar^2}. \tag{16.129}$$

The continuity of the wave function and its derivative (due to the uniform effective electron mass m) at $x = 0$ relates the coefficients A and B to C and D by a transmission matrix T^{21}:

$$\begin{bmatrix} C \\ D \end{bmatrix} = T^{21} \begin{bmatrix} A \\ B \end{bmatrix}, \tag{16.130}$$

where

$$T^{21} \equiv T(k_1, k_2) = \frac{1}{2k_2} \begin{bmatrix} k_2 + k_1 & k_2 - k_1 \\ k_2 - k_1 & k_2 + k_1 \end{bmatrix}. \tag{16.131}$$

For the incident free electron defined by a plane wave transmitting over a semi-infinite potential step, we have $A = 1$ and $D = 0$; meanwhile, we denote $B = r$ and $C = t$, where

$$t = \frac{2k_1}{k_1 + k_2},\ r = \frac{k_1 - k_2}{k_1 + k_2}. \tag{16.132}$$

Define the flux transmission coefficient T and the reflection coefficient R as the ratio of the incident wave current (as defined in (15.125)) over the transmitted current and the reflected current, respectively. We have

$$T = \frac{\hbar k_2/m}{\hbar k_1/m}|t|^2 = \frac{k_2}{k_1}|t|^2 = \frac{4k_1k_2}{(k_1+k_2)^2}, \tag{16.133}$$

$$R = \frac{\hbar k_1/m}{\hbar k_1/m}|r|^2 = |r|^2 = \frac{(k_1-k_2)^2}{(k_1+k_2)^2}. \tag{16.134}$$

Remark 16.1 If $E < V_0$, then

$$t = \frac{2k_1}{k_1 + \mathrm{i}\kappa_2}, \quad r = \frac{k_1 - \mathrm{i}\kappa_2}{k_1 + \mathrm{i}\kappa_2}, \tag{16.135}$$

$\psi = t\mathrm{e}^{-k_2 x}$. Thus, the transmission current is zero, $\mathbf{J} = \mathbf{0}$, and the transmission coefficient will vanish, i.e., $T = 0$, and as a result the reflection coefficient $R = 1$. Thus, there is no transmitted current, and we have a case of total reflection.

Transmission Over a Single Barrier

A single finite barrier with width a is given by the potential function

$$V(x) = \begin{cases} 0, & \text{if } |x| > a/2, \\ V_0, & \text{if } |x| < a/2. \end{cases} \tag{16.136}$$

The transmission matrix over a single finite barrier can be obtained by the composition of the transmission matrix defined in (16.131) by [23]

$$\begin{aligned} T^{31} &= \begin{bmatrix} \mathrm{e}^{-\mathrm{i}k_1 a/2} & 0 \\ 0 & \mathrm{e}^{\mathrm{i}k_1 a/2} \end{bmatrix} T(k_1, k_2) \begin{bmatrix} \mathrm{e}^{\mathrm{i}k_2 a/2} & 0 \\ 0 & \mathrm{e}^{-\mathrm{i}k_2 a/2} \end{bmatrix} \\ &\quad \cdot \begin{bmatrix} \mathrm{e}^{\mathrm{i}k_2 a/2} & 0 \\ 0 & \mathrm{e}^{-\mathrm{i}k_2 a/2} \end{bmatrix} T(k_2, k_1) \begin{bmatrix} \mathrm{e}^{-\mathrm{i}k_1 a/2} & 0 \\ 0 & \mathrm{e}^{\mathrm{i}k_1 a/2} \end{bmatrix} \\ &= \begin{bmatrix} T^{31}_{11} & T^{31}_{12} \\ T^{31}_{21} & T^{31}_{22} \end{bmatrix}, \end{aligned} \tag{16.137}$$

where $k_1 = \sqrt{2mE/\hbar^2}$, $k_2 = \sqrt{2m(E - V_0)/\hbar^2}$, and

$$T^{31}_{21} = \frac{\mathrm{i}(k_1^2 - k_2^2)\sin k_2 a}{2k_1k_2}, \tag{16.138}$$

$$T^{31}_{22} = \frac{2k_1k_2\cos k_2 a - \mathrm{i}(k_1^2 + k_2^2)\sin k_2 a}{2k_1k_2}\mathrm{e}^{\mathrm{i}k_1 a}, \tag{16.139}$$

$$T_{11} = T_{22}^*, \quad T_{12} = T_{21}^*. \tag{16.140}$$

For the transmitted wave magnitude behind the barrier, we have

$$t = \frac{T_{11}T_{22}^* - T_{12}T_{21}^*}{T_{22}} = \frac{1}{T_{22}} = \frac{2k_1k_2 e^{-ik_1a}}{2k_1k_2 \cos k_2a - i(k_1^2 + k_2^2)\sin k_2a}; \tag{16.141}$$

the transmission flux coefficient is given by

$$T = |t|^2 = \frac{1}{1 + \frac{V_0^2}{4E(E-V_0)}\sin^2 k_2a},$$

and the reflection flux coefficient is

$$R = 1 - T.$$

Remark 16.2 Again for $E < V_0$, we have

$$k_2 = i\kappa_2, \quad \kappa_2 = \sqrt{\frac{2m(V_0 - E)}{\hbar^2}}, \quad T = \frac{1}{1 + \frac{V_0^2}{4E(E-V_0)}\sinh^2 k_2a}. \tag{16.142}$$

Figure 16.1 shows the quantum transmission coefficient $T(E)$ for one barrier, together with the classical transmission coefficient, which shows no transmission when $E < V_0$ and total transmission otherwise.

16.4.2 Current Formula Through Barriers Via $T(E)$

Density of States in Contacts

To describe the current through potential barriers, we first need to know the density of states of the electrons for the contacts in various physical dimensions in terms of the momentum vector k. This can be achieved simply by counting the states in k space for one to three dimensions.

- **1-D device contact**

For electrons in a 1-D system $[0, L]$ with the periodic boundary condition $\psi(0) = \psi(L)$ and $\psi'(0) = \psi'(L)$, the quantum states are given by

$$\psi_n(x, t) = A_n e^{-i\frac{\varepsilon_n}{\hbar}t} e^{ik_nx}, \tag{16.143}$$

with discrete energy levels

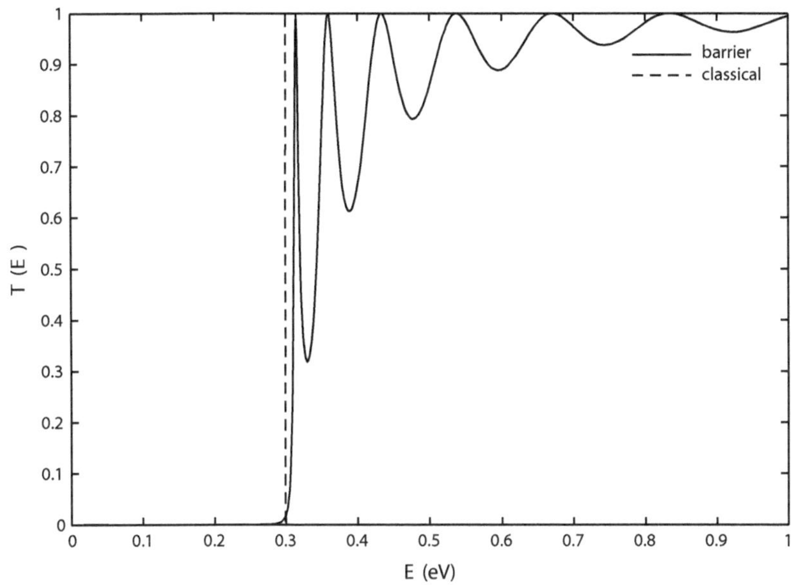

Fig. 16.1 Transmission coefficient through one barrier: quantum transmission (solid line), and classical transmission (dashed line)

$$\varepsilon_n = \frac{\hbar^2 k_n^2}{2m}, \; k_n = n\frac{2\pi}{L}, \qquad n = 1, 2, 3, \ldots. \tag{16.144}$$

Denote by $n(k)$ the number of quantum states per unit k and per unit length. Then $N(k) = n(k) * L$ is the number of quantum states per unit k for the whole device, and

$$N(k)\delta k = 2(\text{spins})\frac{\delta k}{2\pi/L} = \frac{L}{\pi}\delta k \;\Rightarrow N(k) = \frac{L}{\pi}. \tag{16.145}$$

Thus

$$n(k) = \frac{1}{\pi}. \tag{16.146}$$

We could also consider the number of states $n(E)$ per unit energy by relating the momentum k to the energy E using the dispersion relation for the electrons:

$$E = E(k) = \frac{\hbar^2 k^2}{2m}. \tag{16.147}$$

Next, using the fact that $E(-k) = E(k)$ and

$$n(E)\delta E = 2n(k)\delta k = \frac{2}{\pi}\delta k, \tag{16.148}$$

we have

$$n(E) = \frac{2}{\pi}\left(\frac{\mathrm{d}E}{\mathrm{d}k}\right)^{-1} = \frac{2}{\pi\hbar v_g(E)}, \tag{16.149}$$

where, using the relation $\omega = E/\hbar$, the group velocity is given by

$$v_g = \frac{\mathrm{d}\omega}{\mathrm{d}k} = \frac{1}{\hbar}\frac{\mathrm{d}E}{\mathrm{d}k} = \frac{\hbar k}{m}, \tag{16.150}$$

and

$$n_{1\mathrm{D}}(E) = \frac{2m}{\pi\hbar^2 k} = \frac{1}{\pi\hbar}\sqrt{\frac{2m}{E}}. \tag{16.151}$$

Similarly, it can be shown [23] that the density state per unit energy per unit volume of 2-D and 3-D devices are as given in the following:

- **2-D device contact**

$$n_{2\mathrm{D}}(E) = \frac{m}{\pi\hbar^2}; \tag{16.152}$$

- **3-D device contact**

$$n_{3\mathrm{D}}(E) = \frac{mk}{\pi^2\hbar^2} = \frac{m}{\pi^2\hbar^3}\sqrt{2mE}. \tag{16.153}$$

Current Through 1-D Devices and Tsu–Esaki Formula

In the Landauer theory, the electron current is defined with charge q, velocity v, and electron density n by

$$J = nqv, \tag{16.154}$$

which is applied in the calculation of the right-going current through a 1-D device by integrating the electron distribution from the contact over the whole range of the positive momentum $k > 0$ vector:

$$\begin{aligned} I_L &= q\int_0^\infty f_{\mathrm{FD}}(\varepsilon(k), \mu_L)v(k)T(k)n_{1\mathrm{D}}(k)\mathrm{d}k \\ &= q\int_0^\infty f_{\mathrm{FD}}(\varepsilon(k), \mu_L)v(k)T(k)\frac{\mathrm{d}k}{\pi}, \end{aligned} \tag{16.155}$$

where $f_{\mathrm{FD}}(\varepsilon(k), \mu_L)$ is the Fermi–Dirac distribution probability, $v(k)$ is the velocity of the electron, $T(k)$ is the transmission coefficient, $n_{1\mathrm{D}}(k)\mathrm{d}k = \mathrm{d}k/\pi$ is the density of states in the range of $\mathrm{d}k$ as shown in (16.146), and the transmission coefficient

$T(k)$ is the probability of a given electron with momentum k going through the device barrier.

Now, using the dispersion relation again in (16.152) to convert the integration over the momentum k to that of the energy E via the group velocity (the velocity of the electron $v(k) = v(E) = v_g(E)$), we have

$$I_L = q \int_{E_c^L}^{\infty} f_{\text{FD}}(E, \mu_L) v(E) T(E) \frac{\mathrm{d}E}{\pi \hbar v_g}$$
$$= \frac{2q}{h} \int_{E_c^L}^{\infty} f_{\text{FD}}(E, \mu_L) T(E) \mathrm{d}E, \tag{16.156}$$

where $h = 2\pi\hbar$.

Similarly, we can calculate the left-going current from the other contact using

$$I_R = -\frac{2q}{h} \int_{E_c^R}^{\infty} f_{\text{FD}}(E, \mu_R) T(E) \mathrm{d}E. \tag{16.157}$$

By combining (16.156) and (16.157), we obtain the total current through the device as

$$I = I_L + I_R = \frac{2q}{h} \int_{E_c^L}^{\infty} [f_{\text{FD}}(E, \mu_L) - f_{\text{FD}}(E, \mu_R)] T(E) \mathrm{d}E, \tag{16.158}$$

after assuming that no electrons exist in $[E_c^R, E_c^L]$. Equation (16.158) is the Tsu–Esaki formula [21, 22].

At low temperature $T \to 0$, the Fermi–Dirac distribution tends to be a step function, i.e., for $T = 0$,

$$I = \frac{2q}{h} \int_{\mu_c^R}^{\mu_L} T(E) \mathrm{d}E. \tag{16.159}$$

For $T > 0$,

$$\mu_L = \mu + \frac{qV}{2}, \quad \mu_R = \mu - \frac{qV}{2},$$

we have the following approximation:

$$f_{\rm FD}(E,\mu_L) - f_{\rm FD}(E,\mu_R) \approx qV\frac{\partial f_{\rm FD}(E,\mu)}{\partial \mu} = -qV\frac{\partial f_{\rm FD}}{\partial E};$$

then, the current formula (16.158) simplifies to

$$I = \frac{2q^2V}{h}\int_{E_c^L}^{\infty}\left(-\frac{\partial f_{\rm FD}}{\partial E}\right)T(E)\mathrm{d}E. \tag{16.160}$$

Therefore, the conductance is given by

$$G = \frac{I}{V} = \frac{2q^2}{h}\int_{E_c^L}^{\infty}\left(-\frac{\partial f_{\rm FD}}{\partial E}\right)T(E)\mathrm{d}E;$$

as $T \to 0$, the Fermi–Dirac distribution function $f_{\rm FD}$ becomes a step function; thus

$$-\frac{\partial f_{\rm FD}}{\partial E} = \delta(E-\mu),$$

and we obtain the well-known Landauer formula [10]:

$$G = \frac{2q^2}{h}T(\mu). \tag{16.161}$$

Current Through Quasi-2-D Devices

Consider a potential as a function of only the longitudinal variable z, i.e., $V(x, y, z) = V(z)$, such as quantum wells with infinite size in transverse dimensions, and denote $\mathbf{r} = (x, y)$, $\mathbf{k} = (\mathbf{k}_t, k_z)$, where $\mathbf{k}_t = (k_x, k_y)$ is the transverse momentum vector. Then the free electrons in the contact are given by the following plane waveform:

$$\Psi_{\mathbf{k}_t,k_z} = \mathrm{e}^{\mathrm{i}\mathbf{k}_t\cdot\mathbf{r}}u_{\mathbf{k}_t}(z), \tag{16.162}$$

$$\varepsilon(k) = E_c^L + \frac{\hbar^2|\mathbf{k}_t|^2}{2m} + \frac{\hbar^2k_z^2}{2m}, \tag{16.163}$$

and, using the density of states for the 2-D electrons, the right-going current from the left contact will be

$$
\begin{aligned}
J_L &= 2q \int \frac{\mathrm{d}\mathbf{k}_t}{(2\pi)^2} \int_0^\infty \frac{\mathrm{d}k_z}{2\pi} f_{\mathrm{FD}}(\varepsilon(k), \mu_L) v_z(k) T(k_z) \\
&= q \int_0^\infty \frac{\mathrm{d}k_z}{2\pi} \frac{\hbar k_z}{m} T(k_z) \left[2 \int \frac{\mathrm{d}\mathbf{k}_t}{(2\pi)^2} f_{\mathrm{FD}} \left(E_c^L + \frac{\hbar^2 |\mathbf{k}_t|^2}{2m} + \frac{\hbar^2 k_z^2}{2m}, \mu_L \right) \right] \\
&= q \int_0^\infty \frac{\mathrm{d}k_z}{2\pi} \frac{\hbar k_z}{m} T(k_z) f_{\mathrm{2D}} \left(\mu_L - E_c^L - \frac{\hbar^2 {k_z}^2}{2m} \right),
\end{aligned}
\tag{16.164}
$$

where

$$
f_{\mathrm{2D}}(\mu) = \frac{m k_B T}{\pi \hbar^2} \ln \left(1 + \mathrm{e}^{\frac{\mu}{k_B T}} \right) = \int_0^\infty \frac{m}{\pi \hbar^2} \frac{1}{1 + \mathrm{e}^{\frac{E-\mu}{k_B T}}} \, \mathrm{d}E. \tag{16.165}
$$

Thus,

$$
J_L = \frac{q}{h} \int_{E_c^L}^\infty f_{\mathrm{2D}}(\mu_L - E) T(E) \mathrm{d}E. \tag{16.166}
$$

Similarly, we have, for the right contact,

$$
J_R = -\frac{q}{h} \int_{E_c^R}^\infty f_{\mathrm{2D}}(\mu_R - E) T(E) \mathrm{d}E. \tag{16.167}
$$

The total current is then given by

$$
J = J_L + J_R = \frac{q}{h} \int_{E_c^L}^\infty [f_{\mathrm{2D}}(\mu_L - E) - f_{\mathrm{2D}}(\mu_R - E)] T(E) \mathrm{d}E. \tag{16.168}
$$

16.5 Non-Equilibrium Green's Function (NEGF) and Transport Current

In this section, we will define the NEGF and show how to use it to compute the transmission coefficient $T(E)$ in the Tsu–Esaki formula (16.158) or the Landauer current formula (16.161) for general nano-devices. First, we rewrite the density operator using the identity from the Sokhatsky–Weierstrass theorem [24]:

$$2\pi\delta(E-\varepsilon_\alpha) = \mathrm{i}\left(\frac{1}{E-\varepsilon_\alpha+\mathrm{i}0^+} - \frac{1}{E-\varepsilon_\alpha-\mathrm{i}0^+}\right). \tag{16.169}$$

Under a complete orthogonal basis $\{|\phi_\alpha\rangle\}$, we have

$$\begin{aligned}\hat{\rho}_{\alpha\beta} &= \langle\phi_\beta|f_{\mathrm{FD}}(\varepsilon_\alpha-\mu)|\phi_\alpha\rangle \\ &= \int_{-\infty}^{\infty}\langle\phi_\beta|f_{\mathrm{FD}}(E-\mu)\delta(E-\varepsilon_\alpha)|\phi_\alpha\rangle\mathrm{d}E.\end{aligned} \tag{16.170}$$

On defining non-equilibrium Green's functions of the device as

$$\begin{aligned}\mathbf{G}(E) &= [(E+\mathrm{i}0^+)\mathbf{I}-\mathbf{H}_d]^{-1}, \\ \mathbf{G}^+(E) &= [(E-\mathrm{i}0^+)\mathbf{I}-\mathbf{H}_d]^{-1}\end{aligned} \tag{16.171}$$

in the $\{|\phi_\alpha\rangle\}$ basis, which is formed with the eigen-functions of the Hamiltonian $\mathbf{H}_d$ of the device (to be specified in Sect. 16.5.1 and Chap. 18), we have the following diagonal matrix representation, with the α-diagonal entry given by

$$\begin{aligned}\mathbf{G}(E) &\longleftrightarrow \frac{1}{(E+0\mathrm{i})-\varepsilon_\alpha}, \\ \mathbf{G}^+(E) &\longleftrightarrow \frac{1}{(E-0\mathrm{i})-\varepsilon_\alpha}.\end{aligned} \tag{16.172}$$

Then

$$\hat{\rho}_{\alpha\beta} = \int_{-\infty}^{\infty} f_{\mathrm{FD}}(E-\mu)\frac{\mathrm{i}}{2\pi}[\mathbf{G}(E)-\mathbf{G}^+(E)]\mathrm{d}E. \tag{16.173}$$

Denoting the spectral operator

$$\mathbf{A}(E) = \mathrm{i}(\mathbf{G}-\mathbf{G}^+) \tag{16.174}$$

in the energy E-space, which indicates the available states for the quantum system, we have

$$\hat{\rho} = \frac{1}{2\pi}\int_{-\infty}^{\infty} f_{\mathrm{FD}}(E-\mu)\mathbf{A}(E)\mathrm{d}E. \tag{16.175}$$

Most of the time, a quantum device is considered as an open system consisting of a local device region Ω and contacts which extend to the outside environment. If we are only interested in the local device region, we can define the device density operator $\hat{\rho}$ similarly. We follow the presentation of [10, Sect. 9.2] closely in the following.

16.5.1 Quantum Devices with One Contact

Let us denote

$$[0, L] - \text{device}, \quad (-\infty, 0) - \text{contact}, \tag{16.176}$$

and decompose the wave function for the contact–device system according to the physical partition

$$\psi = \begin{pmatrix} \psi^c \\ \psi^d \end{pmatrix}, \tag{16.177}$$

where ψ^c is for the contact and ψ^d is for the device, and $\psi^c(0) = \psi^d(0)$.

The electrons in a contact at chemical potential μ are distributed according to the equilibrium Fermi–Dirac distribution. Once the contact is attached to the nano-device, each incident electron wave function ψ^c_{inc} will induce a wave function ψ^d inside the device, while the latter will produce a reflecting wave χ into the contact. Therefore, the wave function for the combined contact–device system can be decomposed into the following so-called scattering states [10, 14, 16]:

$$\psi^c = \psi_{\text{inc}} + \chi. \tag{16.178}$$

The Hamiltonian for the contact is given by

$$H = U_c - \frac{\hbar^2}{2m_e}\frac{\mathrm{d}^2}{\mathrm{d}x^2}, \tag{16.179}$$

and the electrons are assumed to occupy the eigenstates $|\phi_\alpha\rangle$ over the contact $[-R, 0]$ with $|R| \gg 1$:

$$\phi_\alpha(x) = \phi_{k_\alpha}(x) = \frac{1}{\sqrt{R}} \exp\left(\mathrm{i}k_\alpha x\right), \tag{16.180}$$

with discrete energy levels given by

$$\varepsilon_\alpha = U_c + \frac{\hbar^2 k_\alpha^2}{2m}, \; k_\alpha = \pm\frac{2\alpha\pi}{R}, \qquad \alpha = 0, 1, 2, 3, \ldots. \tag{16.181}$$

The contact system can be described by its density matrix $\rho_c(x, x')$ given in (16.89):

$$\rho_c(x, x') = \sum_\alpha \phi_\alpha(x) f_{\text{FD}}(\varepsilon_\alpha - \mu)\phi^*_\alpha(x'). \tag{16.182}$$

Let us consider a finite difference discretized version of the Hamiltonian on a mesh $x_i = ia$, where $x_0 = 0$ is considered to be the left-most point of the device and $x_{-1} = -a$ is the right-most point of the contact, and

$$H \simeq U_n - \tau(\Delta_+ + \Delta_- - 2), \tag{16.183}$$

where $U_n = U(x_n)$, Δ_+ and Δ_- are the $+1$ and -1 shift operator on the index, respectively, and

$$\tau = \frac{\hbar^2}{2m_e a^2}. \tag{16.184}$$

We define the Hamiltonian for the contact with M mesh points $\{-\infty < x_{-M} < \cdots < x_{-1} < 0\}$ as

$$\mathbf{H}_c = \begin{pmatrix} U_{-M}+2\tau & -\tau & 0 & \cdots & 0 \\ -\tau & U_{-M+1}+2\tau & -\tau & \cdots & 0 \\ \vdots & \ddots & \ddots & \ddots & \vdots \\ 0 & 0 & -\tau & U_{-2}+2\tau & -\tau \\ 0 & 0 & 0 & -\tau & U_{-1}+2\tau \end{pmatrix}_{M\times M}, \tag{16.185}$$

and the Hamiltonian for the device with N mesh points as

$$\mathbf{H}_d = \begin{pmatrix} U_0+2\tau & -\tau & 0 & \cdots & 0 \\ -\tau & U_1+2\tau & -\tau & \cdots & 0 \\ \vdots & \ddots & \ddots & \ddots & \vdots \\ 0 & 0 & -\tau & U_{N-2}+2\tau & -\tau \\ 0 & 0 & 0 & -\tau & U_{N-1}+2\tau \end{pmatrix}_{N\times N}. \tag{16.186}$$

The reflective wave in (16.178) can be considered as having been caused by a source term resulting from the incident wave ψ_{inc} [10]:

$$\mathbf{s}_{\text{inc}} = (EI - \mathbf{H}_c)\psi_{\text{inc}}, \tag{16.187}$$

where

$$\psi_{\text{inc}} = (\psi_{\text{inc}}(x_{-M}), \ldots, \psi_{\text{inc}}(x_{-1}))^{\text{T}}. \tag{16.188}$$

It was then proposed [10, Eq. (9.2.1)] that the wave function for the contact–device system satisfies the following inhomogeneous Schrödinger equation:

$$\begin{bmatrix} E\mathbf{I} - \mathbf{H}_c + \mathrm{i}\eta & -\mathbf{C}^+ \\ -\mathbf{C} & E\mathbf{I} - \mathbf{H}_d \end{bmatrix} \begin{bmatrix} \psi_{\text{inc}} + \chi \\ \psi^d \end{bmatrix} = \begin{bmatrix} \mathbf{s}_{\text{inc}} \\ \mathbf{0} \end{bmatrix}, \tag{16.189}$$

where $\eta > 0$ is a small perturbation and C is the coupling matrix in the Hamiltonian for the combined contact–device system, and

$$\psi^d = (\psi^d(x_0), \ldots, \psi^d(x_{N-1}))^{\text{T}}. \tag{16.190}$$

The first equation in (16.189), together with the definition of $\mathbf{s}_{\text{inc}}$ in (16.187), gives

$$[E\mathbf{I} - \mathbf{H}_c + \mathrm{i}\eta]\chi - \mathbf{C}^+\psi^d = \mathbf{0}, \tag{16.191}$$

implying that the reflection wave is given by

$$\chi = \mathbf{G}_c \mathbf{C}^+ \psi^d, \tag{16.192}$$

where the device Green's function is defined as

$$\mathbf{G}_c = (E\mathbf{I} - \mathbf{H}_c + i\eta)^{-1}. \tag{16.193}$$

If $g(x, x')$ is the analytical form for the device Green's function, then

$$G^c_{i,j} = g(x_i, x_j; E), \ -M \le i, j \le -1. \tag{16.194}$$

Meanwhile, the second equation in (16.189) reads

$$(E\mathbf{I} - \mathbf{H}_d)\psi^d - \mathbf{C}\chi = \boldsymbol{C}\psi_{\text{inc}}. \tag{16.195}$$

Substituting (16.192) into (16.195), we have

$$(E\mathbf{I} - \mathbf{H}_d)\psi^d - \mathbf{C}\mathbf{G}_c\mathbf{C}^+\psi^d = C\psi_{\text{inc}}, \tag{16.196}$$

namely

$$[E\mathbf{I} - \mathbf{H}_d - \boldsymbol{\Sigma}]\psi^d = \mathbf{s}, \tag{16.197}$$

where

$$\boldsymbol{\Sigma} = \mathbf{C}\mathbf{G}_c\mathbf{C}^+, \tag{16.198}$$

$$\mathbf{s} = C\psi_{\text{inc}}. \tag{16.199}$$

So we have

$$\psi^d = \boldsymbol{G}\boldsymbol{s} = \boldsymbol{G}\boldsymbol{C}\psi_{\text{inc}}, \tag{16.200}$$

where the device Green's function is defined as

$$\mathbf{G} = (E\mathbf{I} - \mathbf{H}_d - \boldsymbol{\Sigma})^{-1}. \tag{16.201}$$

Now, from (16.199) and (16.200), for each of the eigenstates ϕ_α as the incident excitation electron wave, we have $\mathbf{C}\psi^c_{\text{inc}} = \mathbf{C}\phi_\alpha$, and the corresponding device wave function is

$$\psi^d_\alpha = \mathbf{G}\mathbf{C}\phi_\alpha.$$

Assuming a non-correlation among the states of $\{|\psi^d_\alpha\rangle\}$, the density operator $\hat{\rho}_d$ for the device can be written as

$$
\begin{aligned}
\hat{\rho}_d &= \sum_{k_\alpha} f_{\mathrm{FD}}(\varepsilon_\alpha - \mu)\, |\psi_\alpha^d\rangle \langle\psi_\alpha^d| \\
&= \int \mathrm{d}E f_{\mathrm{FD}}(E-\mu) \sum_{k_\alpha} \delta(E-\varepsilon_\alpha)\, |\psi_\alpha\rangle \langle\psi_\alpha| \\
&= \int \mathrm{d}E f_{\mathrm{FD}}(E-\mu) \mathbf{G}\mathbf{C} \sum_{k_\alpha} \delta(E-\varepsilon_\alpha)\, |\phi_\alpha\rangle \langle\phi_\alpha|\, \mathbf{C}^{+}\mathbf{G}^{+} \\
&= \int \mathrm{d}E f_{\mathrm{FD}}(E-\mu) \mathbf{G}\mathbf{C}\mathbf{A}_c(E)\mathbf{C}^{+}\mathbf{G}^{+},
\end{aligned}
\tag{16.202}
$$

where a spectral function $\mathbf{A}_c(E)$ for the contact is defined as

$$
\mathbf{A}_c(E) = 2\pi \sum_{k_\alpha} \delta(E-\varepsilon_\alpha)\, |\phi_\alpha\rangle \langle\phi_\alpha| \,. \tag{16.203}
$$

It should be noted that for devices with one contact only, the summation over k_α in (16.202) is limited over $k_\alpha > 0$ as only right-going electron waves are involved. However, in a real device, at least two contacts are involved as input and output; therefore, in general all values of k_α will be used as in the derivation above.

Now define the dissipation $\mathbf{\Gamma}$ for the contact as

$$
\mathbf{\Gamma} = \boldsymbol{C}\boldsymbol{A}_c(E)\mathbf{C}^{+}, \tag{16.204}
$$

a spectral function for the device

$$
\mathbf{A}(E) = \boldsymbol{G}\mathbf{\Gamma}\boldsymbol{G}^{+}, \tag{16.205}
$$

and an electron correlation in the device

$$
\mathbf{G}^n(E) = f_{\mathrm{FD}}(E-\mu)\mathbf{A}(E). \tag{16.206}
$$

Then, similar to (16.175), the density matrix for the device can be written as

$$
\hat{\rho}_d = \frac{1}{2\pi} \int f_{\mathrm{FD}}(E-\mu)\mathbf{A}(E)\mathrm{d}E = \int \frac{\mathrm{d}E}{2\pi} \mathbf{G}^n(E). \tag{16.207}
$$

Next, we will show that the device spectral function $\mathbf{A}(E)$ can in fact be obtained directly from the device Green's function, instead of using the contact spectral function $\mathbf{A}_c(E)$.

The device Green's function $g(x, x')$ can be represented through the eigenfunction expansion following the completeness of the eigenfunctions as follows:

$$
g(x, x', E) = \sum_{k_\alpha} \frac{\phi_\alpha(x)\phi_\alpha^*(x')}{E - \varepsilon_\alpha + \mathrm{i}0^+}, \quad -R \le x, x' \le 0, \tag{16.208}
$$

which, by using the identity (16.169), implies that the contact spectral function $A_c(x, x', E)$ in the matrix notation is given by

$$A_c(x, x', E) = \mathrm{i}\left[g(x, x', E) - g^*(x, x', E)\right] \tag{16.209}$$

or

$$A_c(E)_{i,j} = \mathrm{i}\left[G^c_{i,j} - G^{c+}_{i,j}\right].$$

Then, using (16.198), we have

$$\begin{aligned}\mathbf{\Gamma} &= \mathbf{C}\mathbf{A}_c(E)\mathbf{C}^+ = \mathbf{i}\mathbf{C}(\mathbf{G}_c - \mathbf{G}_c^+)\mathbf{C}^+ \\ &= \mathrm{i}(\mathbf{\Sigma} - \mathbf{\Sigma}^+) = -2\,\mathrm{Im}(\mathbf{\Sigma}).\end{aligned} \tag{16.210}$$

Moreover, from the definition of the device Green's function in (16.201), we can see that

$$\mathbf{\Gamma} = \mathrm{i}(\mathbf{\Sigma} - \mathbf{\Sigma}^+) = -\mathrm{i}\left[\mathbf{G}^{-1} - (\mathbf{G}^+)^{-1}\right]. \tag{16.211}$$

Finally, substituting (16.211) into (16.205), we have the spectral function for the device only in terms of the device Green's function:

$$\mathbf{A}(E) = \boldsymbol{G}\mathbf{\Gamma}\boldsymbol{G}^+ = \mathrm{i}(\mathbf{G} - \mathbf{G}^+) = -2\,\mathrm{Im}(\mathbf{G}). \tag{16.212}$$

16.5.2 Quantum Devices with Two Contacts

A nano-device $[0, L]$ usually has at least two contacts, such as the source and the drain in a metal-oxide-semiconductor field-effect transistor (MOSFET), so we will need the density matrix for a device with two contacts. In general, the chemical potentials for different contacts are not the same, so the contact–device system cannot be treated as a thermal equilibrium system in the strictest sense. However, we will follow the same procedure as before to obtain the results [10] for the density operator:

$$\hat{\rho}_d = \int \frac{\mathrm{d}E}{2\pi}\mathbf{G}^n(E), \tag{16.213}$$

where

$$\mathbf{G}^n(E) = \mathbf{A}^{(1)}(E) f_{\mathrm{FD}}(E - \mu_1) + \mathbf{A}^{(2)}(E) f_{\mathrm{FD}}(E - \mu_2), \tag{16.214}$$

and the device partial spectral functions $\mathbf{A}^{(1)}(E)$ and $\mathbf{A}^{(2)}(E)$ associated with the contacts are defined as

$$\mathbf{A}^{(1)}(E) = G\Gamma_1\mathbf{G}^+, \quad \mathbf{A}^{(2)}(E) = G\Gamma_2\mathbf{G}^+. \tag{16.215}$$

Here, the device Green's function $\mathbf{G}$ is given in (16.201); however, the self-energy $\boldsymbol{\Sigma}$ will now consist of the self-energies from the two contacts, i.e.,

$$\Sigma = \Sigma^{(1)} + \Sigma^{(2)}, \tag{16.216}$$

where

$$\Sigma^{(i)} = \mathbf{C}_i\mathbf{G}_{c,i}\mathbf{C}_i^+ \tag{16.217}$$

and $\mathbf{C}_1$ and $\mathbf{C}_2$ are the coupling matrices for the two contacts, and $\mathbf{G}_{c,i}$ is the contact Green's function,

$$\mathbf{G}_{c,i} = [E\mathbf{I} - \mathbf{H}_{ci} + \mathrm{i}\eta]^{-1}, \; i = 1, 2. \tag{16.218}$$

Similar to (16.204), for the two contacts we define their dissipations as

$$\Gamma_1 = \mathbf{C}_1\mathbf{A}_c^{(1)}(E)\mathbf{C}_1^+, \quad \Gamma_2 = \mathbf{C}_2\mathbf{A}_c^{(2)}(E)\mathbf{C}_2^+, \tag{16.219}$$

and the spectral functions for the two contacts are defined through their respective Green's functions as follows:

$$\mathbf{A}_c^{(i)}(E) = \mathrm{i}(\mathbf{G}_{c,i} - \mathbf{G}_{c,i}^+) = -2\,\mathrm{Im}(\mathbf{G}_{c,i}). \tag{16.220}$$

From (16.219), (16.220), and (16.217), we can see that the dissipation Γ_i is again related to the imaginary part of the self-energy $\Sigma^{(i)}$ as

$$\Gamma_i = -2\,\mathrm{Im}(\Sigma^{(i)}), \quad i = 1, 2. \tag{16.221}$$

In Chap. 18, details of how to obtain the self-energies $\Sigma^{(i)}$ for the two contacts will be given in terms of the boundary treatment for the device region in the framework of both finite difference and finite element discretization of the Hamiltonians.

16.5.3 Green's Function and Transport Current Formula

The device NEGF introduced in this section can be shown [25] to be related to the transmission coefficients for the quantum device. Specifically, the transmission coefficient $T^{1\text{-}2}$ from a source input (1) to a drain output (2) can be expressed in terms of the device Green's function (16.201) and the imaginary part of the self-energy (dissipation) (16.221) as follows:

$$T^{1\text{-}2}(E) = \mathrm{tr}\left(\Gamma_1\mathbf{G}\Gamma_2\mathbf{G}^+\right) = \mathrm{tr}\left(\Gamma_2\mathbf{G}\Gamma_1\mathbf{G}^+\right). \tag{16.222}$$

The electron current between contacts 1 and 2 is given for a ballistic transport by the Tsu-Esaki formula (16.158):

$$I = \frac{q}{\pi\hbar}\int_{-\infty}^{+\infty} T^{1\text{-}2}(E)(f_{\mathrm{FD}}(E-\mu_1) - f_{\mathrm{FD}}(E-\mu_2))\mathrm{d}E. \tag{16.223}$$

16.6 Summary

The averaged physical quantities in a quantum system, such as the electron density, can be calculated using the density operator or its matrix representation (i.e., the density matrix). The electron transport can be described by the quantum analog of the Boltzmann kinetic theory, namely, the kinetic equation of the Wigner distribution in the phase space, which is the Fourier transform of the density operator. Alternatively, the transport current can be described by the Landauer formula using the transmission coefficient $T(E)$ of an electron at energy level E as the probability of the electron propagating from an input to an output. Finally, the non-equilibrium Green's function (NEGF) method, which originates from the quantum field theory of many-particle quantum systems [26, 27], allows the numerical calculation of the transmission coefficient $T(E)$ for a device of general shape [25, 28]. Chapters 18 and 19 describe the numerical methods used for solving the NEGF and the Wigner distributions, respectively.

References

1. R.K. Pathria, *Statistical Mechanics*, 2nd edn. (Butterworth-Heinemann, Oxford, 1996)
2. J.W. Gibbs, *Elementary Principles in Statistical Mechanics* (Yale University Press, New Haven, 1902)
3. F. Bloch, J.D. Walecka, *Fundamentals of Statistical Mechanics: Manuscript and Notes of Felix Bloch* (World Scientific, Singapore, 2001)
4. W. Greiner, L. Neise, H. Stöcker, *Thermodynamics and Statistical Mechanics* (Springer, Berlin, 1995)
5. J. von Neumann, Wahrscheinlichkeitstheoretischer aufbau der quantenmechanik. Nachr. Ges. Wiss. Gottingen **1**, 245–272 (1927)
6. R.C. Tolman, *The Principles of Statistical Mechanics* (Oxford University Press, Oxford, 1950)
7. R.L. Liboff, *Kinetic Theory, Classical, Quantum, and Relativistic Descriptions*, 3rd edn. (Springer, Berlin, 2003)
8. C. Cohen-Tannoudji, B. Diu, F. Laloe, *Quantum Mechanics*, vol. I (Wiley-Interscience, New York, 2006)
9. R. Feynman, *Statistical Mechanics* (Westview Press, Boulder, 1972)
10. S. Datta, *Quantum Transport: Atom to Transistor* (Cambridge University Press, Cambridge, 2005)
11. R.G. Parr, W.T. Yang, *Density Functional Theory of Atoms and Molecules* (Oxford University Press, Oxford, 1989)

12. E. Wigner, On the quantum correction for thermodynamic equilibrium. Phys. Rev. **40**, 749–759 (1932)
13. V.I. Tatarskiĭ, The Wigner representation of quantum mechanics. Sov. Phys. Usp. **26**, 311–327 (1983)
14. C. Jacoboni, R. Brunetti, P. Bordone, A. Bertoni, Quantum transport and its simulation with the Wigner function approach. Int. J. High Speed Electron. Syst. **11**, 387–423 (2001)
15. P.A. Markowich, C.A. Ringhofer, C. Schmeiser, *Semiconductor Equations* (Springer, Berlin, 2002)
16. C. Jacoboni, P. Bordone, The Wigner-function approach to non-equilibrium electron transport. Rep. Prog. Phys. **67**, 1033–1071 (2004)
17. W.R. Frensley, Wigner-function model of a resonant-tunneling semiconductor device. Phys. Rev. B **36**, 1570–1580 (1987)
18. N.C. Kluksdahl, A.M. Kriman, D.K. Ferry, C. Ringhofer, Self-consistent study of the resonant-tunneling diode. Phys. Rev. B **39**, 7720–7735 (1989)
19. P.A. Markowich, C.A. Ringhofer, An analysis of the quantum Liouville equation. ZAMM - J. Appl. Math. Mech. **69**, 121–127 (1989)
20. C.B. Duke, *Tunneling Phenomena in Solids* (Academic Press, New York, 1969)
21. L. Esaki, R. Tsu, Superlattice and negative differential conductivity in semiconductors. IBM J. Res. Develop. **14**, 61–65 (1970)
22. R. Tsu, L. Esaki, Tunneling in a finite superlattice. Appl. Phys. Lett. **22**, 562–564 (1973)
23. J.H. Davis, *Physics of Low Dimensional Semiconductors: An Introduction* (Cambridge University Press, Cambridge, 1997)
24. P. Blanchard, E. Brüning, *Mathematical Methods in Physics* (Birkhauser, Cambridge, 2002). Ex. 3.3.1 4
25. S. Datta, *Electronic Transport in Mesoscopic Systems* (Cambridge University Press, Cambridge, 1997)
26. A.L. Fetter, J.D. Walecka, *Quantum Theory of Many Particle Systems* (McGraw-Hill, New York, 1971)
27. A.A. Abrikosov, L.P. Gorkov, I.E. Dzyaloshinski, *Methods of Quantum Field Theory in Statistical Physics* (Dover, New York, 1975)
28. H. Haug, A.P. Jauho, *Quantum Kinetics in Transport and Optics of Semiconductors*, 2nd edn. (Springer, Berlin, 2007)

Chapter 17
Density Functional Theory Computation of Electronic Structure

Understanding the electronic structure of an N-electron system requires solving an eigenvalue problem of a $3N$-dimensional Schrödinger equation, which suffers from the curse of dimensionality in computational costs. For small systems, quantum Monte Carlo methods [1] have been used quite effectively. For large systems and bulk material studies, the density functional theory (DFT) has become the main computational method for various chemistry, biology, and material science applications. The DFT transforms a high-dimensional linear eigenvalue value problem to a nonlinear 3-D eigenvalue problem for an electron density-dependent Hermitian operator.

In this chapter, we will first review the Kohn–Sham DFT theory, where one-particle orbitals are used to compute the DFT solutions. Then, frequency domain plane wave methods and real-space finite element methods will be introduced to solve the Kohn–Sham orbital equations. Finally, to compute the ground state of a large system, an $O(N)$ linear scaling density matrix minimization (DMM) method will be presented, where a discontinuous Galerkin space is used to represent the density matrix of the system using a hybrid basis of local atomic orbitals and piecewise polynomials for efficient discretization.

17.1 Kohn–Sham (KS) DFT: From a 3N-D Linear Eigenvalue Problem to a 3-D Nonlinear One

The state of an interacting N-electron $\{\mathbf{r}_i\}_{i=1}^N$ and M-nuclei quantum system with nuclei positions $\mathbf{R} = \{\mathbf{R}_I\}_{I=1}^M$ and charges $\{Z_I\}_{I=1}^M$ in $\mathbb{R}^3$ is governed by the following Hamiltonian (with the electron mass, unit charge, and Planck constant in (16.2) all set to be 1 in atomic units, i.e., $m = e = \hbar = 1$),

W. Cai, *Deterministic, Stochastic, and Deep Learning Methods for Computational Electromagnetics*, https://doi.org/10.1007/978-981-96-0100-4_17

$$\widehat{H} = \widehat{H}(\mathbf{r}_1, \cdots, \mathbf{r}_N) = \sum_{i=1}^{N} -\frac{1}{2}\nabla_{\mathbf{r}_i}^2 + \sum_{i<j}^{N} \frac{1}{|\mathbf{r}_i - \mathbf{r}_j|}$$
$$+ \sum_{i=1}^{N} V_{\text{ext}}(\mathbf{r}_i; \{\mathbf{R}_I\}) + \sum_{I<J}^{M} \frac{Z_I Z_J}{|\mathbf{R}_I - \mathbf{R}_J|}$$
$$\equiv \widehat{T} + \widehat{V}_{\text{ee}} + \widehat{V}_{\text{en}} + \widehat{E}_{\text{nn}}, \tag{17.1}$$

where the second equation line shows the operators for electron kinetic energy, electron–electron Coulombic interaction energy, external potential energy from electron–nucleus interaction, and nucleus–nucleus interaction energy, respectively, i.e.,

$$\widehat{T} = \sum_{i=1}^{N} -\frac{1}{2}\nabla_{\mathbf{r}_i}^2, \quad \widehat{V}_{\text{ee}} = \sum_{i<j}^{N} \frac{1}{|\mathbf{r}_i - \mathbf{r}_j|}, \tag{17.2}$$

$$\widehat{V}_{\text{en}} = \sum_{i=1}^{N} \widehat{V}_{\text{ext}}(\mathbf{r}_i; \{\mathbf{R}_I\}), \; \widehat{V}_{\text{ext}}(\mathbf{r}; \{\mathbf{R}_I\}) \equiv \widehat{V}_{\text{ext}}(\mathbf{r}) = \sum_{I=1}^{M} \frac{Z_I}{|\mathbf{r} - \mathbf{R}_I|}, \tag{17.3}$$

$$\widehat{E}_{\text{nn}} = \sum_{I<J}^{M} \frac{Z_I Z_J}{|\mathbf{R}_I - \mathbf{R}_J|}. \tag{17.4}$$

In a Born-Oppenheimer approximation, the heavier nuclei are considered stationary relative to the electrons; thus, the nucleus–nucleus interaction energy for a given atomic configuration $\{\mathbf{R}_I\}$ will be considered a constant, i.e., $\widehat{E}_{\text{nn}} = E_{\text{nn}}$, when the state of the electrons is considered.

In reality, each electron will also have a spin degree of freedom $\sigma \in \{\uparrow, \downarrow\}$ for up and down spins, but we ignore the spin degree of freedom in our discussion, and the electrons will be considered "spinless" for simplicity of presentation. The eigenfunction $\Psi(\mathbf{r}_1, \cdots, \mathbf{r}_N)$ for the ground state of the electron system is given by the following eigenvalue problem for the smallest eigenvalue E,

$$\widehat{H}\Psi = E\Psi, \tag{17.5}$$

where the eigenfunction for the electron system also satisfies the anti-symmetric property and the Pauli exclusion principle

$$\Psi(\mathbf{r}_1, \cdots, \mathbf{r}_i, \cdots, \mathbf{r}_j, \cdots, \mathbf{r}_N) = -\Psi(\mathbf{r}_1, \cdots, \mathbf{r}_j, \cdots, \mathbf{r}_i, \cdots, \mathbf{r}_N), \; i \neq j. \tag{17.6}$$

As shown in (16.55), the Slater determinant from the N single particle wavefunction will satisfy this anti-symmetric property as well as the Pauli exclusion principle.

Using the Courant–Fisher minimax theorem [2], the eigenvalue for the ground state eigenvalue is given by the following minimization problem over an anti-symmetric wavefunction Hilbert space $\mathcal{A}_N$,

$$E = \inf_{|\Psi>\in\mathcal{A}_N, <\Psi|\Psi>=1} <\Psi|H|\Psi>, \tag{17.7}$$

where the bra-ket $<\Psi|\Psi>$ of the wavefunction $|\Psi>$ corresponds to the conventional inner product of the Hilbert space, and the normalization of the wavefunction is given by

$$<\Psi|\Psi>:=\int \Psi^*(\mathbf{r}_1,\cdots,\mathbf{r}_N)\Psi(\mathbf{r}_1,\cdots,\mathbf{r}_N)d\mathbf{r}_1\cdots d\mathbf{r}_N = 1. \tag{17.8}$$

The computation of the linear eigenvalue problem (17.5) suffers from the curse of dimensionality if traditional numerical methods such as spectral and finite element methods are used because the number of unknowns will be $m^d, d = 3N$ if m degrees of freedom are used for each spatial variable, and grows exponentially with the dimension of the problem d. A breakthrough was made by Hohenberg and Kohn [3], who developed the famous density functional theory, which states that the ground state energy can in fact be uniquely determined by the electron density $\rho(\mathbf{r})$ of the system,

$$\rho(\mathbf{r}) = N\int |\Psi(\mathbf{r},\mathbf{r}_2\cdots,\mathbf{r}_N)|^2 d\mathbf{r}_2\cdots d\mathbf{r}_N. \tag{17.9}$$

The DFT reformulates the minimization problem (17.7) over the wavefunction space $\mathcal{A}_N$, into an equivalent two-step minimization, the first one over a space of 3-dimensional density function $\rho(\mathbf{r})$, $\mathbf{r}\in\mathbb{R}^3$, and the second one over wavefunctions with a given density $\rho(\mathbf{r})$, i.e.,

$$E = \inf_{|\Psi>\in\mathcal{A}_N, <\Psi|\Psi>=1} <\Psi|\widehat{H}|\Psi> = \inf_{\rho\in\mathcal{D}_N}\left\{\inf_{\Psi\in\mathcal{A}_N, \Psi\longmapsto\rho} <\Psi|\widehat{H}|\Psi>\right\}, \tag{17.10}$$

where the density function space $\mathcal{D}_N$ is defined as

$$\mathcal{D}_N = \{\rho\geq 0: \int\rho(\mathbf{r})d\mathbf{r} = N, \nabla\sqrt{\rho}\in L^2(\mathbb{R}^3)\}, \tag{17.11}$$

and the nonemptiness of the set of functions Ψ for any given $\rho\in\mathcal{D}_N$ is proven in [4].

Using the definition of the Hamiltonian $\widehat{H}$ in (17.1), we can show that

$$\begin{aligned} E &= \inf_{\rho\in\mathcal{D}_N}\left\{\inf_{\Psi\in\mathcal{A}_N, \Psi\longmapsto\rho} <\Psi|\widehat{T}+\widehat{V}_{\text{ee}}|\Psi> + \int\rho(\mathbf{r})V_{\text{ext}}(\mathbf{r})d\mathbf{r}\right\} + E_{\text{nn}} \\ &\equiv \inf_{\rho\in\mathcal{D}_N}\left\{F[\rho] + \int\rho(\mathbf{r})V_{\text{ext}}(\mathbf{r})d\mathbf{r}\right\} + E_{\text{nn}}, \end{aligned} \tag{17.12}$$

where a universal functional $F[\rho]$ is defined for the electron kinetic energy and Coulomb electron-electron energy, depending only on the distribution of the electron density of the system, not on the external potential V_{ext}, i.e.,

$$F[\rho] = \inf_{\Psi \in \mathcal{A}_N, \Psi \longmapsto \rho} < \Psi | \widehat{T} + \widehat{V}_{\text{ee}} | \Psi > . \tag{17.13}$$

Equation (17.12) indicates that the ground state energy can be obtained by a minimizer over the density function space and there exists a universal functional $F[\rho]$.

Kohn–Sham orbitals A significant development in the DFT is the introduction of single particle orbitals in computing the functional $F[\rho]$ by Kohn and Sham (1965) [5]. Given N one-particle orthonormal orbitals $\{\psi_i\}_{i=1}^N$,

$$\int \psi_i^*(\mathbf{r})\psi_j(\mathbf{r})d\mathbf{r} = \delta_{ij}, \quad i, j = 1, \cdots, N, \tag{17.14}$$

from which we can construct an anti-symmetrized N-particle wavefunction (similar to a noninteracting N-electron system, i.e., $\widehat{V}_{\text{ee}} = 0$ in (17.2)),

$$\Psi(\mathbf{r}_1, \cdots, \mathbf{r}_N) = \{\psi_i\}^{\text{as}}, \tag{17.15}$$

using the Slater determinant as in (16.55) with $\phi_{\mathbf{p}_i}(\mathbf{r})$ replaced by $\psi_i(\mathbf{r})$. A density operator $\widehat{\rho}$ for the one-electron system with N-electrons distributed over the orbitals can be defined as

$$\widehat{\rho} = \sum_{i=1}^N |\psi_i >< \psi_i|, \tag{17.16}$$

whose spatial representation gives the density matrix

$$\rho(\mathbf{r}, \mathbf{r}') = \sum_{i=1}^N \psi_i^*(\mathbf{r}')\psi_i(\mathbf{r}), \tag{17.17}$$

which can be used to compute the electron density as

$$\rho(\mathbf{r}) = \rho(\mathbf{r}, \mathbf{r}) = \sum_{i=1}^N |\psi_i(\mathbf{r})|^2. \tag{17.18}$$

Next, the energy functional $F[\rho]$ in (17.13) will be split into three terms as follows:

$$F[\rho] = \inf_{\Psi \in \mathcal{A}_N^{\text{as}}, \Psi \longmapsto \rho} < \Psi | \widehat{T} | \Psi > + \frac{1}{2} \int \frac{\rho(\mathbf{r})\rho(\mathbf{r}')}{|\mathbf{r} - \mathbf{r}'|} d\mathbf{r} d\mathbf{r}' + E_{\text{xc}}(\rho), \tag{17.19}$$

where $\mathcal{A}_N^{\text{as}}$ is the space of the Slater determinants of N orbitals in (17.15), and the first term is the kinetic energy of N noninteracting electrons occupying those orbitals

$$T(\{\psi_i\}) = < \Psi|\widehat{T}|\Psi >= \text{Tr}(\widehat{\rho}\widehat{T}) = \frac{1}{2}\sum_{i=1}^{N} \int |\nabla\psi_i(\mathbf{r})|^2 d\mathbf{r}, \tag{17.20}$$

and the second term as the Hartree approximation of the Coulomb repulsion energy experienced by one electron in a mean-field electron density approach

$$E_{\text{H}}(\rho) = \frac{1}{2}\int\int \frac{\rho(\mathbf{r})\rho(\mathbf{r}')}{|\mathbf{r}-\mathbf{r}'|} d\mathbf{r}d\mathbf{r}' \ \text{(e-e Hartree interaction)}, \tag{17.21}$$

and the last term the exchange–correlation energy $E_{\text{xc}}(\rho)$ of the N-electron system (accounting for what is left from the first two terms, and the latter may make up 95% of the total energy [5]). Thus, in the Kohn–Sham DFT approach, the ground state energy E is now given by

$$E^{\text{KS}} = \inf_{\Psi\in\mathcal{A}_{N,}^{\text{as}}\,\Psi\longmapsto\rho} E^{\text{KS}}(\{\psi_i\}, \mathbf{R}), \tag{17.22}$$

where a KS total energy functional is defined as

$$E^{\text{KS}}(\Psi, \mathbf{R}) = E^{\text{KS}}(\{\psi_i\}, \mathbf{R}) = T(\Psi) + E_{\text{H}}(\rho) + E_{\text{xc}}(\text{æ}) + E_{\text{ext}}(\rho, \mathbf{R}) + E_{\text{nn}}(\mathbf{R}), \tag{17.23}$$

where $\Psi =: \{\psi_i\}$, and the external electronic potential

$$E_{\text{ext}}(\rho, \mathbf{R}) = -\sum_{J=1}^{M} \int \rho(\mathbf{r}) \frac{Z_J}{|\mathbf{r}-\mathbf{R}_J|} d\mathbf{r} \quad \text{(e-n interaction)}, \tag{17.24}$$

and the constant nucleus–nucleus interaction energy $E_{\text{nn}}(\mathbf{R})$ in (17.4).

We can then consider the following minimization of the KS energy (17.23), subject to orthogonality constraints (17.14) enforced by Lagrangian multipliers, λ_{ij}, i.e.,

$$\inf_{\Psi\in\mathcal{A}_{N,}^{\text{as}}\,\Psi\longmapsto\rho,\lambda_{ij}} \left\{ E^{\text{KS}}(\Psi, \mathbf{R}) + \sum_{i,j=1}^{N} \lambda_{ij} \left(\int \psi_i^*(\mathbf{r})\psi_j(\mathbf{r})d\mathbf{r} - \delta_{ij} \right) \right\}, \tag{17.25}$$

where the variation of the energy functional is

$$\frac{\delta E^{\text{KS}}(\Psi, \mathbf{R})}{\delta\psi_i^*(\mathbf{x})} = \left(-\frac{1}{2}\nabla^2 + V_{\text{H}}[\rho] + V_{\text{xc}}[\rho] + V_{\text{ext}} \right) \psi_i \triangleq \hat{H}^{\text{KS}}[\rho]\psi_i,$$

and the stationary condition of (17.25) gives the Euler–Lagrangian equation for the orbitals

$$\widehat{H}^{\mathrm{KS}}[\rho]\psi_i = \left(-\frac{1}{2}\nabla^2 + V_{\mathrm{H}}[\rho](\mathbf{r}) + V_{\mathrm{xc}}[\rho] + V_{\mathrm{ext}}(\mathbf{r})\right)\psi_i = \varepsilon_i\psi_i(\mathbf{x}), \quad (17.26)$$

where the energy level ε_i is the eigenvalue of the Lagrangian multiplier matrix (λ_{ij}) [6] and the one-particle Kohn–Sham Hamiltonian $\widehat{H}^{\mathrm{KS}}[\rho]$ is defined for a self-consistent potential depending on the density ρ in (17.18), and the Hartree potential

$$V_{\mathrm{H}}[\rho](\mathbf{r}) = \int \frac{\rho(\mathbf{r}')}{|\mathbf{r}-\mathbf{r}'|}d\mathbf{r}', \quad (17.27)$$

and the exchange–correlation potential

$$V_{\mathrm{xc}}[\rho] = \frac{\delta E_{\mathrm{xc}}(\rho)}{\delta\rho}. \quad (17.28)$$

Due to the Hermitian nature of $\widehat{H}^{\mathrm{KS}}[\rho]$, the orthogonality of the orbital functions given by (17.26) is assured.

- Approximation of the exchange–correlation energy $E_{\mathrm{xc}}[\rho]$.

A key component of the DFT is the exchange–correlation functional E_{xc}, which is approximated by a so-called Jacob-ladder of functionals [7, 8] with increasing complexity and accuracy, including both local and nonlocal functionals, the latter requiring special computational treatment [6]. The first level of approximation to the functional is the local density approximation (LDA), introduced by Kohn and Sham in 1965 [9], which assumes that

$$E_{\mathrm{xc}}[\rho] = \int \epsilon_{\mathrm{xc}}(\rho(\mathbf{r}))\rho(\mathbf{r})d\mathbf{r}, \quad (17.29)$$

and, thus, the corresponding potential (17.28)

$$V_{\mathrm{xc}}[\rho(\mathbf{r})] = \frac{\partial\left(\rho\epsilon_{\mathrm{xc}}\right)}{\partial\rho}. \quad (17.30)$$

The next approximation is the general gradient approximation (GGA) [10–12], which assumes that

$$E_{\mathrm{xc}}[\rho] = \int \epsilon_{\mathrm{xc}}(\rho(\mathbf{r}), |\nabla\rho(\mathbf{r})|^2)d\mathbf{r} \quad (17.31)$$

and, the corresponding potential (17.28) [6],

$$V_{\mathrm{xc}}[\rho(\mathbf{r})] = \frac{\partial\epsilon_{\mathrm{xc}}}{\partial\rho} - 2\nabla\cdot\left(\frac{\partial\epsilon_{\mathrm{xc}}}{\partial\nabla\rho(\mathbf{r})}\nabla\rho\right). \quad (17.32)$$

17.2 De-Singularizing $V_{\text{ext}}(\mathbf{r})$ in the KS Equations with an Effective Pseudo-Potential

The singular bare ionic potential $\hat{V}(\mathbf{r}) = -\frac{Ze}{|\mathbf{r}|}$ from the $\hat{V}_{\text{ext}}(\mathbf{r})$ of (17.3) causes rapid oscillations for wavefunctions near the ions, resulting in a large range of Fourier frequencies if a plane wave basis is used to represent the KS orbitals, or an extremely fine grid near the ions for a real-space finite element method. Fortunately, the filled shells of the core electrons of atoms are not affected by the crystalline environment of the extended (bulk) material or neighboring atoms in a cluster of molecules. In fact, most chemical bonding between atoms is caused by the interaction between the outer layer valence electrons from open shells, especially in metals and semiconductors. In view of these facts, a natural idea is to introduce an effective potential, i.e., pseudo-potential (PS), which removes the singularity at the ion center, but the effects of core electrons are accounted for by its design, thus, producing the same effect beyond some cutoff region to the other atoms.

To illustrate the general principle of the pseudo-potential for one atom, let us consider the Hamiltonian $\widehat{H}$ for an atom with Z nuclei charges and N-electrons and denote the wavefunction solution for the Schrödinger equation for the inner core electrons by $|\psi_c>$ and those for the valance electrons by $|\psi_v>$, respectively. Therefore, we have

$$\widehat{H}|\psi_\varkappa >= E_\varkappa|\psi_\varkappa >, \quad \varkappa = c, v. \tag{17.33}$$

We decompose the valence electron wavefunction $|\psi_v>$ into a smooth part $|\varphi_v>$ of some prescribed shape near the ion without high oscillation, and a localized part near the ion with all of its oscillations (see Fig. 17.1). Namely, for all n_v valence electrons in the outermost layer of the orbitals, we have

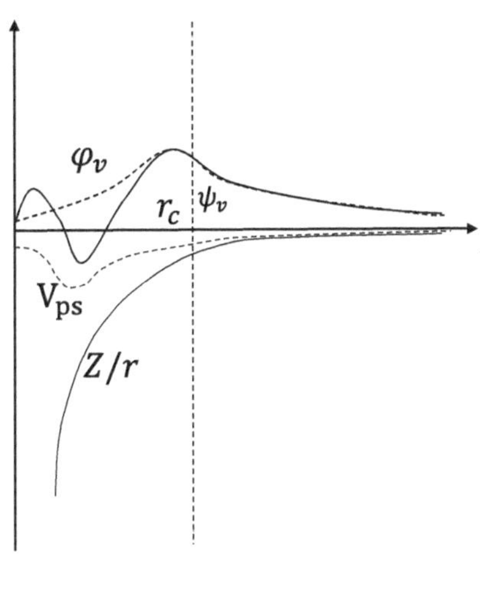

Fig. 17.1 Smooth pseudo-potential V_{ps} and non-oscillatory pseudo-wave function φ_v vs bare ionic potential Z/r and oscillatory valence wavefunction ψ_{v}

$$|\psi_v>=|\varphi_v> + \sum_c \alpha_{vc}|\psi_c>, \ v = 1, \cdots, n_v, \tag{17.34}$$

and naturally $\alpha_{vc} = -<\psi_c|\varphi_v>$ due to the orthogonality of the electron wavefunctions.

The Schrödinger equation for the valence pseudo-wavefunction $|\varphi_v>$ will be

$$\widehat{H}|\varphi_v>=E_v|\varphi_v> - \sum_c (E_v - E_c)|\psi_c><\psi_c|\varphi_v>, \tag{17.35}$$

which can be rewritten as

$$\widehat{H}^{\text{ps}}|\varphi_v>=E_v|\varphi_v>, \tag{17.36}$$

where the pseudo-potential Hamiltonian is [13]

$$\widehat{H}^{\text{ps}} = \widehat{H}^{\text{ps}}(E_v) = \widehat{H} - \sum_c (E_c - E_v)|\psi_c><\psi_c|. \tag{17.37}$$

The PS Hamiltonian $\widehat{H}^{\text{ps}}$ and the valence PS wavefunction $|\varphi_v>$ are not unique, but a general criterion is given as the Hamann-Schluter–Chiang (HSC) conditions [14] for a norm-conserved pseudo-potential: (1) wavefunctions $|\varphi_v>$ and $|\psi_v>$ share the same eigenvalues E_v with their corresponding Hamiltonian, (2) the wavefunctions agree beyond a cutoff region $r \geq r_c$, and (3) the pseudo-potential wavefunction satisfies $<\varphi_v|\varphi_v>=1$ (normal conservation). Various proposals of norm-conserved PSs have been proposed, including the Troullier–Martins (TM) pseudo-potential [15], the Hartwigsen–Goedecker–Hutter (HGH) pseudo-potential [16], and the optimized norm-conserving Vanderbilt (ONCV) pseudo-potential [17]. The procedure usually starts with a prescribed form for the PS wavefunction $|\varphi_v>$ inside the cutoff region, applying the DFT theory for the one-atom system, solving for the pseudo-potential wavefunctions, and then extract the corresponding effective pseudo-potential by inverting the Schrödinger equation using the PS wavefunctions. More details on the procedure can be found in [6, 13].

For instance, a class of pseudo-potential V_{ps} in the Kleinman-Bylander separable and nonlocal form [18] can be defined as follows

$$\begin{aligned} V_{\text{ps}}(\mathbf{r}, \mathbf{r}') &= V_{\text{loc}}(\mathbf{r})\delta_{\mathbf{r},\mathbf{r}'} + \sum_{lm} \gamma_l b_{lm}(\mathbf{r}) b_{lm}^*(\mathbf{r}') \\ &= V_{\text{loc}}(\mathbf{r})\delta_{\mathbf{r},\mathbf{r}'} + V_{\text{nl}}(\mathbf{r}, \mathbf{r}'), \end{aligned} \tag{17.38}$$

where the sum over the magnetic number m is $-l \leq m \leq l$, and $V_{\text{ps}}(\mathbf{r}, \mathbf{r}')$ acts as the kernel of an integral operator, i.e.,

$$\begin{aligned}\hat{V}_{\text{ps}}(\mathbf{r},\mathbf{r}')\psi_i(\mathbf{r}) &\equiv \int V_{\text{ps}}(\mathbf{r},\mathbf{r}')\psi_i(\mathbf{r}')d\mathbf{r}' \\ &= V_{\text{loc}}(\mathbf{r})\psi_i(\mathbf{r}) + \int V_{\text{nl}}(\mathbf{r},\mathbf{r}')\psi_i(\mathbf{r}')d\mathbf{r}',\end{aligned} \tag{17.39}$$

where

$$\int V_{\text{nl}}(\mathbf{r},\mathbf{r}')\psi_i(\mathbf{r}')d\mathbf{r}' = \sum_{lm} d_{lm}\gamma_l b_{lm}(\mathbf{r}), \tag{17.40}$$

$$d_{lm} = \int b_{lm}^*(\mathbf{r}')\psi_i(\mathbf{r}')d\mathbf{r}'.$$

The derivation of the local potential $V_{\text{loc}}(\mathbf{r})$ and "projector" $b_{lm}(\mathbf{r})$ can be found in [6].

Under the pseudo-potential V_{ps}, the electrostatic potential energy is

$$\begin{aligned}E_{\text{ext}}^{\text{ps}}(\rho,\mathbf{R}) = \text{Tr}(\widehat{\rho}\widehat{V}_{\text{ps}}) &= \sum_{i=1}^{N}\int\int f_i\psi_i^*(\mathbf{r})V_{\text{ps}}(\mathbf{r},\mathbf{r}')\psi_i(\mathbf{r}')d\mathbf{r}d\mathbf{r}' \\ &= \int V_{\text{loc}}(\mathbf{r})\rho(\mathbf{r})d\mathbf{r} + \sum_{i=1}^{N}\int\int f_i\psi_i^*(\mathbf{r})V_{\text{nl}}(\mathbf{r},\mathbf{r}')\psi_i(\mathbf{r}')d\mathbf{r}d\mathbf{r}' \\ &= \int V_{\text{loc}}(\mathbf{r})\rho(\mathbf{r})d\mathbf{r} + \sum_{i=1}^{N}\int f_i\psi_i^*(\mathbf{r})\widehat{V}_{\text{nl}}\psi_i(\mathbf{r})d\mathbf{r},\end{aligned} \tag{17.41}$$

where the occupation number f_i of an electron on the i-th orbital is $f_i = 1$ for the zero temperature ground state energy considered here. Meanwhile, for a finite temperature case, it is given by the Fermi-Dirac distribution $f_i = f_{\text{FD}}(\varepsilon_i - \mu)$ in (16.70).

Finally, the KS total energy with a pseudo-potential is

$$E_{\text{ps}}^{\text{KS}}(\{\psi_i\},\mathbf{R}) = T(\{\psi_i\}) + E_{\text{H}}(\rho) + E_{\text{xc}}(\text{æ}) + E_{\text{ext}}^{\text{ps}}(\rho,\mathbf{R}) + E_{\text{nn}}(\mathbf{R}), \tag{17.42}$$

where the kinetic energy $T(\{\psi_i\})$ is given in (17.20); the Hartree energy $E_{\text{H}}(\rho)$ in (17.21); the exchange–correlation energy $E_{\text{xc}}(\text{æ})$ in (17.29) or (17.31); and the PS potential energy $E_{\text{ext}}^{\text{ps}}(\rho,\mathbf{R})$ in (17.41).

The Euler–Lagrangian equation from the stationary condition of the KS total energy $E_{\text{ps}}^{\text{KS}}$ leads to a PS Hamiltonian eigenvalue problem for the KS (valence) orbitals $|\psi_i>$ for a N-electron system (excluding the core electrons accounted for already by the PS potential)

$$\widehat{H}^{\text{ps}}|\psi_i> \equiv \left(-\frac{1}{2}\nabla^2 + V_{\text{H}}[\rho] + V_{\text{ps}} + V_{\text{xc}}[\rho]\right)|\psi_i> = \varepsilon_i|\psi_i>,\ i = 1,\cdots,N, \tag{17.43}$$

and the Hartree potential $V_{\mathrm{H}}(\mathbf{r})$ satisfies the Poisson equation,

$$-\nabla^2 V_{\mathrm{H}}(\mathbf{r}) = 4\pi\rho(\mathbf{r}), \quad \mathbf{r} \in \mathbb{R}^3. \tag{17.44}$$

For convenience, an effective KS-potential can be defined as

$$V^{\mathrm{KS}}(\rho) = V_{\mathrm{H}}[\rho] + V_{\mathrm{ps}} + V_{\mathrm{xc}}. \tag{17.45}$$

17.3 Self-Consistent Solution of the KS Nonlinear Eigenvalue Problem

The Kohn–Sham equations (17.43) and (17.44) is a nonlinear eigenvalue problem that must be solved self-consistently through an iterative procedure to ensure the convergence of the density $\rho(\mathbf{r})$ as follows:

- Step 0. Given an error tolerance ε, and making an initial guess for the electron density $\rho^{(0)}(\mathbf{r})$. Start with $\rho^{(n)}(\mathbf{r})$ for $n = 0, 1, \cdots,$
- Step 1. Compute the KS potential $V^{\mathrm{KS}}(\mathbf{r})$ (17.45) with electron density $\rho^{(n)}(\mathbf{r})$, while the Hartree potential $V_{\mathrm{H}}(\mathbf{r})$ can be obtained by solving the Poisson equation (17.44) by various numerical methods (finite element methods with a multigrid acceleration or FMM accelerated integral equation methods).
- Step 2. Discretize the KS equations (17.43), and solve numerically with some eigensolvers for at least N eigenfunctions (KS orbitals) $\psi_i(\mathbf{r})$ and eigenvalues.
- Step 3. Update the density $\tilde{\rho}^{(n+1)}(\mathbf{r}) = \sum_{i=1}^{N} |\psi_i(\mathbf{r})|^2$.
- Step 4. Mix density $\beta\tilde{\rho}^{(n+1)}(\mathbf{r}) + (1-\beta)\rho^{(n)}(\mathbf{r}) = \rho^{(n+1)}(\mathbf{r})$, $\beta = 0.3$, to ensure convergence.
- Step 5. Check convergence if $|\rho^{(n+1)}(\mathbf{r}) - \rho^{(n)}(\mathbf{r})| \leq \varepsilon$, stop, otherwise, return to step 2.

Various eigensolvers can be used in Step 2 including: (a) direct solvers—LAPACK [19] and ScaLAPCK [20], (b) Krylov subspace iterative methods—Davidson–Liu method [21, 22] and LOBPCG method [23], and (c) filtering method based on Chebyshev polynomial maximization [24]. Meanwhile, other more sophisticated mixing schemes using more previous density history, such as Anderson and Broyden mixing, can be used in Step 4 [25, 26].

In the following, we will present two common discretization methods for the Kohn–Sham equations (17.43): a plane wave method in the frequency space and a finite element method in real physical space.

17.3.1 Plane Wave Solutions for the KS Equations

Let us consider a crystalline material with $\mathbf{a}_i$ $i = 1, 2, 3$ as its lattice vector and $\mathbf{b}_j$, $j = 1, 2, 3$ as its reciprocal vectors as defined in (7.2) in a unit cell Λ and the atoms inside the unit cell at $\mathbf{R}_J$, $J = 1, \cdots, M$. Then, the PS potential within the unit cell is given by

$$V_{\rm ps}(\mathbf{r},\mathbf{r}') = \sum_{J=1}^{M} V_{\rm ps}^{J}(\mathbf{r}-\mathbf{R}_J, \mathbf{r}'-\mathbf{R}_J) \equiv V_{\rm loc}(\mathbf{r}) + V_{\rm nl}(\mathbf{r},\mathbf{r}'), \quad \mathbf{r},\mathbf{r}' \in \Lambda, \quad (17.46)$$

where $V_{{\rm ps},J}(\mathbf{r},\mathbf{r}')$ is the semi-local PS potential at atom $\mathbf{R}_J$ as given in (17.38) and

$$V_{\rm loc}(\mathbf{r}) = \sum_{J=1}^{M} V_{\rm loc}^{J}(\mathbf{r}-\mathbf{R}_J)\delta_{\mathbf{r}-\mathbf{R_J},\mathbf{r}'-\mathbf{R_J}}, \quad (17.47)$$

and

$$V_{\rm nl}(\mathbf{r},\mathbf{r}') = \sum_{J=1}^{M} V_{\rm nl}^{J}(\mathbf{r}-\mathbf{R}_J, \mathbf{r}'-\mathbf{R}_J) = \sum_{J=1}^{M}\sum_{lm} \gamma_l^J b_{lm}^J(\mathbf{r}-\mathbf{R}_J) b_{lm}^{J*}(\mathbf{r}'-\mathbf{R}_J), \quad (17.48)$$

which defines a non-local, but separable operator as

$$\widehat{V}_{\rm nl}\psi_i(\mathbf{r}) := \int_{\Lambda} V_{\rm nl}(\mathbf{r},\mathbf{r}')\psi_i(\mathbf{r}')d\mathbf{r}' = \sum_{J=1}^{M}\sum_{lm} d_{lm}^J \gamma_l^J b_{lm}(\mathbf{r}-\mathbf{R}_J), \quad (17.49)$$

$$d_{lm}^J = \int_{\Lambda} b_{lm}^*(\mathbf{r}'-\mathbf{R}_J)\psi_i(\mathbf{r}')d\mathbf{r}',$$

and the subscript J indicates possible different local and nonlocal behaviors for each atom at $\mathbf{R}_J$, and all pseudo-potentials are extended periodically outside Λ.

The plane wave solutions [27, 28] for the KS equation are based on the Bloch wave solution for the periodic structure of (14.4), the PS valence wavefunction is given by

$$\psi_{\mathbf{k},n}(\mathbf{r}) = e^{i\mathbf{k}\cdot\mathbf{r}} \sum_{\mathbf{G}} c_{\mathbf{k},n}(\mathbf{G}) e^{i\mathbf{G}\cdot\mathbf{r}}, \quad (17.50)$$

where the wave vector $\mathbf{k}$ is in the first Brillouin zone $\mathbf{B}$ of (14.27) and the $n = 1, 2, \cdots$ indicates the band index (i.e., the ordering of the eigenvalue of KS equations), and the infinite sum of the $\mathbf{G}$-vector is truncated for the numerical solution for the unknown coefficients $c_{\mathbf{k},n}(\mathbf{G})$. If an isolated system (a cluster of molecules), the periodic image of the unit cell will be virtual and may introduce some artifact due to long range charge interactions; thus, the unit cell will be taken large enough to avoid such artificial effects, and the wavenumber is also set to $\mathbf{k} = \mathbf{0}$.

As in Sect. 14.1.4, the KS equations (17.43) will become in the frequency domain as [13]

$$\sum_{G'} \widehat{H}_{\mathbf{G},\mathbf{G}'}(\mathbf{k}) c_{\mathbf{k},n}(\mathbf{G}') = \varepsilon_{\mathbf{k},n} c_{\mathbf{k},n}(\mathbf{G}), \tag{17.51}$$

where the frequency domain form of the Hamiltonian is

$$\begin{aligned} \widehat{H}_{\mathbf{G},\mathbf{G}'}(\mathbf{k}) = & \frac{1}{2}|\mathbf{k}+\mathbf{G}|^2 \delta_{\mathbf{G},\mathbf{G}'} + V_{\mathrm{ps}}(\mathbf{k}+\mathbf{G}, \mathbf{k}+\mathbf{G}') \\ & + V_{\mathrm{H}}(\mathbf{G}-\mathbf{G}') + V_{\mathrm{xc}}(\mathbf{G}-\mathbf{G}'), \end{aligned} \tag{17.52}$$

and the frequency domain form for the PS is

$$V_{\mathrm{ps}}(\mathbf{k}+\mathbf{G}, \mathbf{k}+\mathbf{G}') = \sum_{J=1}^{M} V_{\mathrm{ps}}^{J}(\mathbf{k}+\mathbf{G}, \mathbf{k}+\mathbf{G}') e^{i(\mathbf{G}-\mathbf{G}')\cdot \mathbf{R}_J}, \tag{17.53}$$

$$V_{\mathrm{ps}}^{J}(\mathbf{G}, \mathbf{G}') = V_{\mathrm{loc}}^{J}(\mathbf{G}-\mathbf{G}') + V_{\mathrm{nl}}^{J}(\mathbf{G}, \mathbf{G}'), \tag{17.54}$$

$$V_{\mathrm{nl}}^{J}(\mathbf{G}, \mathbf{G}') = \sum_{lm} \gamma_l b_{lm}^{J}(\mathbf{G}) b_{lm}^{J*}(\mathbf{G}'), \tag{17.55}$$

where the Fourier coefficients $b_{lm}^{J}(\mathbf{G})$ for the projector function $b_{lm}^{J}(\mathbf{r})$ will be needed at a cost of $O(N_{\mathrm{pw}})$ for each magnetic number l and m, N_{pw} is the total number of plane waves used in (17.50).

The Hartree and exchange–correlation potentials depend on the electron density $\rho(\mathbf{r})$ which involves a nonlinear quadratic function evaluation of the KS wavefunction $\psi_{\mathbf{k},n}(\mathbf{r})$. The evaluation is implemented in a pseudo-spectral manner by using inverse FFT (going between the frequency $\mathbf{k}$-space and real $\mathbf{r}$-space (see Sect. 2.2.3), namely,

$$\begin{aligned} c_{\mathbf{k},n}(\mathbf{G}) \overset{\mathrm{FFT}^{-1}}{\rightarrow} \psi_{\mathbf{k},n}(\mathbf{r}) \rightarrow \rho(\mathbf{r}) = & \sum_{\mathbf{k},n=1}^{n_v} f(\varepsilon_{\mathbf{k},n})|\psi_{\mathbf{k},n}(\mathbf{r})|^2 \\ \rightarrow V_{\mathrm{H}}(\rho(\mathbf{r})), V_{\mathrm{xc}}(\rho(\mathbf{r})) & \overset{\mathrm{FFT}}{\rightarrow} V_{\mathrm{H}}(\mathbf{G}), V_{\mathrm{xc}}(\mathbf{G}), \end{aligned} \tag{17.56}$$

the sum (i.e., integration) over $\mathbf{k} \in B$ is computed by a quadrature rule.

Numerical linear algebra algorithms will then be used to diagonalize the matrix equation (17.51) [29]. The band structure of the electron system from the computed eigenvalues will look similar to Fig. 14.1.

17.3.2 Finite Element Methods for the KS Equations

Finite element methods [30–35] as a real space method provide a systematically improvable adaptive discretization method with the help of posterior estimate [36] for solving the KS equations, compared with orbital and Gaussian function-based discretization.

Using the PS potential approximation for the external potential, the KS orbitals in the real space $\psi = \{\psi_i(\mathbf{r})\}_{i=1}^N$ for a system with N-electrons and M nuclei $\mathbf{R} = \{\mathbf{R}_1, \mathbf{R}_2, \cdots, \mathbf{R}_M\}$ satisfy the equations (17.43) and (17.44) as a result of a variational problem of the KS energy functional (17.42) [37, 38].

Let us consider a triangulation $\{T_k\}$ of the computational domain Ω,

$$\Omega = \bigcup_{k=1}^{K} T_k, \tag{17.57}$$

with T_k being the usual tetrahedral and cube elements. The wavefunctions $\psi_i(\mathbf{r})$, $1 \le i \le N$ and the Hartree potential $V_{\mathrm{H}}(\mathbf{r})$ are expressed in terms of finite element basis functions $\{N_k(\mathbf{r})\}_{k=1}^{n_K}$ over the triangulation $\{T_k\}$

$$\psi_i(\mathbf{r}) = \sum_{k=1}^{n_K} \psi_i^k N_k(\mathbf{r}), \quad V_{\mathrm{H}}(\mathbf{r}) = \sum_{k=1}^{n_K} V_{\mathrm{H}}^k N_k(\mathbf{r}), \quad \mathbf{r} \in \Omega. \tag{17.58}$$

Then, multiplying (17.43) and (17.44) by a test function $N_j(\mathbf{r})$ and integrating by parts, we can arrive at the following discretized finite element system

$$H\Psi_i = \varepsilon_i S \Psi_i, \tag{17.59}$$

$$A\mathbf{V}_{\mathrm{H}} = \mathbf{b}, \tag{17.60}$$

where the unknown vectors $\Psi_i = (\psi_i^1, \psi_i^2, \cdots, \psi_i^{n_K})^\top$, $\mathbf{V}_{\mathrm{H}} = (V_{\mathrm{H}}^1, V_{\mathrm{H}}^2, \cdots, V_{\mathrm{H}}^{n_K})^\top$, and the discretized Hamiltonian matrix $H = (H_{jk})_{n_K \times n_K}$

$$H_{jk} = H_{jk}^{\mathrm{kin}} + H_{jk}^{\mathrm{loc}} + H_{jk}^{\mathrm{nl}}, \tag{17.61}$$

and

$$H_{jk}^{\mathrm{kin}} = \frac{1}{2} \int_{\Omega} \nabla N_j(\mathbf{r}) \cdot \nabla N_k(\mathbf{r}) d\mathbf{r}\,, \tag{17.62}$$

and

$$H_{jk}^{\mathrm{loc}} = \int_{\Omega} V_{\mathrm{loc}}^{\mathrm{total}}(\mathbf{r}) N_j(\mathbf{r}) N_k(\mathbf{r}) d\mathbf{r}, \quad V_{\mathrm{loc}}^{\mathrm{total}}(\mathbf{r}) = V_{\mathrm{H}}(\mathbf{r}) + V_{\mathrm{xc}}(\mathbf{r}) + V_{\mathrm{loc}}(\mathbf{r}), \tag{17.63}$$

and the nonlocal part (17.48) of the discretized Hamiltonian matrix is given by

$$H_{jk}^{\mathrm{nl}} = \sum_{J=1}^{M} \sum_{lm} \gamma_l^J c_{lm,j}^J c_{lm,k}^{J*}, \tag{17.64}$$

$$c_{lm,k}^{J*} = \int_{\Omega} b_{lm}^{J*}(\mathbf{r}-\mathbf{R}_J) N_k(\mathbf{r}) d\mathbf{r}, \quad c_{lm,j}^{J} = \int_{\Omega} b_{lm}^{J}(\mathbf{r}-\mathbf{R}_J) N_j(\mathbf{r}) d\mathbf{r}, \tag{17.65}$$

and $S = \left(s_{jk}\right)_{n_K \times n_K}$ is the mass (overlap) matrix

$$s_{jk} = \int_{\Omega} N_j(\mathbf{r}) N_k(\mathbf{r}) d\mathbf{r}, \tag{17.66}$$

and the stiffness matrix for the Poisson equation $A = (a_{jk})_{n_K \times n_K}, a_{jk} = 2H_{jk}^{\mathrm{kin}}$, and the right hand side vector $\mathbf{b} = \{b_j\} \in \mathbb{R}^{n_K}$,

$$b_j = 4\pi \int_{\Omega} \rho(\mathbf{r}) N_j(\mathbf{r}) d\mathbf{r}. \tag{17.67}$$

17.4 A Linear Scaling Discontinuous Galerkin Density Matrix Minimization (DG-DMM) Method

Due to the required orthogonality of the KS orbitals, the plane wave and finite element methods in the previous section have an $O(N^3)$ complexity for an N-electron system. For a large system, linear scaling $O(N)$ complexity is needed for practical applications. Many linear scaling methods have been developed, including the Fermi operator expansion method [39], the Fermi operator projection method [40], the divide-and-conquer method [41], the density-matrix minimization approach [42], the orbital minimization approach [43], and the optimal basis density-matrix minimization scheme [44], pole expansion and selected inversion (PEXSI) method [45]. The important characteristics of $O(N)$ methods is that the calculation of energy and forces (for MD type applications) does not require the calculation of the eigenvalues and eigenvectors of the KS Hamiltonian.

In this section, we present the DG-DMM [46], which combines the discontinuous Galerkin (DG) discretization and the density matrix minimization method (DDM) [47]. In the DMM approach, the idempotency constraint on the density operator, i.e. $\hat{\rho} = \hat{\rho}^2$, is not strictly enforced, and instead, a weaker condition is used when minimizing E^{KS}, although the constraint of N-electrons equal to the trace of the density operator is included. The discontinuous Galerkin method [48] allows for the optimal discretization of the Hamiltonian energy operator by combining physics-based local orbitals around atoms inside a given element K, and traditional piecewise polynomial basis for systematically improvable numerical convergence. The DG discretization also provides easy parallelism of the computation, further improving the linear scaling performance of the DMM approach.

17.4.1 Enforcing Idempotency of Density Matrix $\rho(\mathbf{r}, \mathbf{r}')$

The ground state will be calculated by minimizing the total energy E^{KS} (17.42) with respect to the density matrix $\rho(\mathbf{r}, \mathbf{r}')$ [44], which is required to satisfy the idempotency condition

$$\hat{\rho}^2 = \hat{\rho}, \tag{17.68}$$

namely,

$$\rho(\mathbf{r}, \mathbf{r}') = \int_{\Omega} d\mathbf{r}'' \rho(\mathbf{r}, \mathbf{r}'')\rho(\mathbf{r}'', \mathbf{r}'), \quad \mathbf{r}, \mathbf{r}' \in \mathbb{R}^3, \tag{17.69}$$

where Ω is the domain containing the system.

The total number of electrons is given by the trace of the density operator,

$$N = \mathrm{Tr}(\hat{\rho}) = \int_{\Omega} \rho(\mathbf{r}, \mathbf{r}) d\mathbf{r}. \tag{17.70}$$

Idempotency condition The idempotency constraint (17.68) can be satisfied at the convergence of a purification procedure proposed in McWeeny [49], by casting ρ as

$$\rho = 3\vartheta * \vartheta - 2\vartheta * \vartheta * \vartheta, \tag{17.71}$$

for some auxiliary two-point function $\vartheta(\mathbf{r}, \mathbf{r}')$. Here, the asterisk represents the continuum analog of matrix multiplication, i.e.,

$$\vartheta * \vartheta(\mathbf{r}, \mathbf{r}') \equiv \int_{\Omega} d\mathbf{r}'' \vartheta(\mathbf{r}, \mathbf{r}'')\vartheta(\mathbf{r}'', \mathbf{r}'). \tag{17.72}$$

The rationale behind (17.71) is that if λ is an eigenvalue of ϑ, then the corresponding eigenvalue of ρ is $f(\lambda) = 3\lambda^2 - 2\lambda^3$, which is a map guaranteeing that if ϑ is close to being idempotent, and ρ will be closer to being idempotent as $\{0, 1\}$ are the limits of the iterations given by the map. During the process of minimizing E^{KS}, using ρ in the form of (17.71) has the effect of driving the eigenvalues toward zero or unity, namely, ρ is driven toward idempotency. For more details of this procedure, refer to Li et al. [42].

Linear scaling via local support functions for $\vartheta(\mathbf{r}, \mathbf{r}')$ The density matrix embodies the well-known "nearsightedness" of electrons [50, 51], namely, $\rho(\mathbf{r}, \mathbf{r}') \to 0$ as $|\mathbf{r} - \mathbf{r}'| \to \infty$. This fact provides the basis for developing $O(N)$ linear scaling algorithms for finding the ground state energy. To encode the decay condition of far off-diagonal entries of the density matrix, the support functions $\phi_\alpha(\mathbf{r})$ are introduced to express the KS orbitals as

$$\psi_i(\mathbf{r}) = \sum_\alpha c_{i\alpha}\phi_\alpha(\mathbf{r}), \quad 1 \le i \le N. \tag{17.73}$$

Thus, the density matrix can be rewritten as

$$\rho(\mathbf{r}, \mathbf{r}') = \sum_{\alpha,\beta} \phi_\alpha(\mathbf{r}) K_{\alpha\beta}\phi_\beta(\mathbf{r}'), \tag{17.74}$$

where $K_{\alpha\beta} = \sum_i c_{i\alpha}c^*_{i\beta}$. Moreover, kinetic energy through the trace operation becomes

$$\mathrm{Tr}[\widehat{\rho}\widehat{T}] = \sum_{\alpha,\beta} \int_\Omega d\mathbf{r}\phi_\beta(\mathbf{r}) K_{\alpha\beta}\left(-\frac{1}{2}\nabla^2\right)\phi_\alpha(\mathbf{r}), \tag{17.75}$$

and similarly, the total energy

$$E^{\mathrm{KS}} = \mathrm{Tr}[\widehat{\rho}\widehat{H}^{\mathrm{ps}}] = \sum_{\alpha,\beta} K_{\alpha\beta}H_{\beta\alpha}, \tag{17.76}$$

$$H_{\beta\alpha} = \int_\Omega \phi_\alpha(\mathbf{r})\widehat{H}^{\mathrm{ps}}\phi_\beta(\mathbf{r})d\mathbf{r}, \tag{17.77}$$

where $\widehat{H}^{\mathrm{ps}}$ is defined in (17.43); finally, the electron number (17.70) becomes

$$N = \int_\Omega \rho(\mathbf{r}, \mathbf{r})d\mathbf{r} = \mathrm{Tr}(KS), \tag{17.78}$$

where $S_{\alpha\beta}$ is the overlap matrix of the support functions,

$$S_{\alpha\beta} = \int_\Omega d\mathbf{r}\phi_\alpha(\mathbf{r})\phi_\beta(\mathbf{r}) = \sum_k \int_{T_k} d\mathbf{r}\phi_\alpha(\mathbf{r})\phi_\beta(\mathbf{r}) \equiv \sum_k S^k_{\alpha\beta}, \tag{17.79}$$

where the finite element triangulation (17.57) of the domain Ω is used.

We can also express the matrix $\vartheta(\mathbf{r}, \mathbf{r}')$ in terms of the support functions $\phi_\alpha(\mathbf{r})$ as well

$$\vartheta(\mathbf{r}, \mathbf{r}') = \sum_{\alpha,\beta} \phi_\alpha(\mathbf{r}) L_{\alpha\beta} \phi_\beta(\mathbf{r}'). \tag{17.80}$$

Using the relation given in (17.71), we can see that the coefficients $L = (L_{\alpha\beta})$ are related to $K = (K_{\alpha\beta})$ as follows

$$K = 3LSL - 2LSLSL. \tag{17.81}$$

The support functions $\phi_\alpha(\mathbf{r})$ are selected to be nonzero only within a localized spatial region, referred to as *support regions*, and the matrix elements $L_{\alpha\beta}$ will be nonzero only if the corresponding regions are separated by less than a chosen cutoff distance R_{cut}. The strategy of density matrix minimization (DMM) algorithm now is to minimize the total energy both with respect to the support functions $\phi_\alpha(\mathbf{r})$ and the $L_{\alpha\beta}$ coefficients (parameters in the density matrix), subject to the condition that the number of electrons remains at N.

17.4.2 Support Functions in Discontinuous Galerkin Space

The support regions for $\phi_\alpha(x)$ are chosen to be spherical with a radius R_{reg} and are centered on the atoms. The support functions will be approximated by piecewise polynomials enriched with local orbitals over the triangulation (17.57),

$$\phi_\alpha(\mathbf{r}) = u_\alpha^{\text{loc}}(\mathbf{r}) + u_\alpha(\mathbf{r}) = u_\alpha^{\text{loc}}(\mathbf{r}) + \sum_j u_{\alpha,j}^k \varphi_j^k(\mathbf{r}), \ \mathbf{r} \in T_{\mathbf{k}} \subset \Omega, \tag{17.82}$$

using a hybrid basis, i.e., $u_\alpha^{\text{loc}}(\mathbf{r})$ is an analytical local orbital function, possibly localized by a truncation, related to the one-atom Hamiltonian and selected to capture the dominant feature of the support function $\phi_\alpha(\mathbf{r})$. Meanwhile, $u_{\alpha,j}^k$ $0 \le j \le p$ are the coefficients of the p-th order polynomial basis functions $\{\varphi_j^k(\mathbf{r})\}_{j=0}^p$ over element T_k. The support function $\phi_\alpha(\mathbf{r})$ is optimized to improve the approximation to the eigenvalues of the system.

17.4.3 DG-DMM Method

The ground state energy will be sought by minimizing the total energy E^{KS} over the density matrix coefficients K (i.e., L and the coefficient $u_{\alpha,j}^k$ for optimal support functions), an implementation of conjugate gradient methods such as the Polak-Ribiere variant of the conjugate gradient method [52] can be used, which requires the computations of the following gradients.

Gradient with respect to $L_{\alpha\beta}$. Adding an additional term to enforce the total electron number condition (17.78) with a penalty μ, the following quantity will be minimized to approximate the E^{KS}

$$E^{\text{KS}} = 2\text{Tr}\left((3LSL - 2LSLSL)H\right) + \mu\left[N - 2\text{Tr}\left((3LSL - 2LSLSL)S\right)\right]^2, \tag{17.83}$$

where the factor of 2 in front of Tr reflects the 2-spins of electrons being considered here, the Hamiltonian matrix H is given in (17.77), the matrix K has been replaced with $(3LSL - 2LSLS)$ to carry out the McWeeny's purification, and the total electron number constraint is handled by a penalty term μ [53]. The analytical expressions for the derivatives $\partial E^{\text{KS}}/\partial L_{\alpha\beta}$ are given by

$$\frac{\partial E^{\text{KS}}}{\partial L_{\alpha\beta}} = 6(SL\tilde{H} + \tilde{H}LS)_{\beta\alpha} - 4\left(SLSL\tilde{H} + SL\tilde{H}LS + \tilde{H}LSLS\right)_{\beta\alpha}. \tag{17.84}$$

where $\tilde{H} = H - 2\mu\left[N - 2\text{tr}\left((3LSL - 2LSLSL)S\right)\right] S$.

Gradient with respect to coefficient $u^k_{\alpha,j}$ on element T_k We will also include an optimization of the support function $\phi_\alpha(\mathbf{r})$ by minimizing the total energy E^{KS} with respect to its expansion coefficients $u^k_{\alpha,j}$ under the DG basis functions (17.82). The chain rule can be used to evaluate $\frac{\partial E^{\text{KS}}}{\partial u^k_{\gamma j}}$ as follows

$$\frac{\partial E^{\text{KS}}}{\partial u^k_{\gamma j}} = \sum_{\alpha\beta} \frac{\partial E^{\text{KS}}}{\partial S_{\alpha\beta}} \frac{\partial S_{\alpha\beta}}{\partial u^k_{\gamma j}} + \sum_{\alpha\beta} \frac{\partial E^{\text{KS}}}{\partial H_{\alpha\beta}} \frac{\partial H_{\alpha\beta}}{\partial u^k_{\gamma j}}, \tag{17.85}$$

where

$$\frac{\partial S_{\alpha\beta}}{\partial u^l_{\gamma i}} = \sum_k \frac{\partial S^k_{\alpha\beta}}{\partial u^l_{\gamma i}}, \tag{17.86}$$

and

$$\frac{\partial S^k_{\alpha\beta}}{\partial u^l_{\gamma i}} = \delta_{kl}\left[\delta_{\alpha\gamma} S^k u^k_\beta + \delta_{\beta\gamma} S^k u^k_\alpha\right]_i. \tag{17.87}$$

Moreover, by observing (17.83), we have

$$\begin{aligned}\frac{\partial E^{\text{KS}}}{\partial S_{\alpha\beta}} &= \left[6(LHL) - 4(LSLHL + LHLSL)\right]_{\beta\alpha} \\ &\quad + 2\mu\left[2\text{tr}\left((3LSL - 2LSLSL)S\right) - N\right](6LSL - 6LSLSL)_{\beta\alpha},\end{aligned}$$

$$\frac{\partial E^{\text{KS}}}{\partial H_{\alpha\beta}} = 2\,(3LSL - 2LSLSL)_{\beta\alpha}. \tag{17.88}$$

Computing Hamiltonian in the discontinuous Galerkin method. Next let us consider the Hamiltonian matrix $H_{\beta\alpha} = \int_{\Omega} \phi_\alpha(\mathbf{r})\hat{H}^{\text{ps}}\phi_\beta(\mathbf{r})d\mathbf{r}$. For the Hamiltonian operator given in (17.100), we have

$$H_{\beta\alpha} = H^{\text{kinetic}} + H^{\text{potential}}, \tag{17.89}$$

where the kinetic energy

$$H^{\text{kinetic}}_{\beta\alpha} = -\frac{1}{2}\int_{\Omega} \phi_\alpha(\mathbf{r})\nabla^2\phi_\beta(\mathbf{r})d\mathbf{r}, \tag{17.90}$$

and the potential energy

$$H^{\text{potential}}_{\beta\alpha} = \int_{\Omega} V^{\text{KS}}(\mathbf{r})\phi_\alpha(\mathbf{r})\phi_\beta(\mathbf{r})d\mathbf{r}, \tag{17.91}$$

where the KS potential V^{KS} is given in (17.45) and the computation of $H^{\text{potential}}_{\beta\alpha}$ can be done similar to (17.63) and (17.65) for the finite element methods.

To compute the matrix elements of the kinetic energy in the discontinuous Galerkin method, where the basis functions are discontinuous across element interfaces for parallelism and adaptivity, we first rewrite (17.90) as

$$H^{\text{kinetic}}_{\beta\alpha} = -\frac{1}{2}\int_{\Omega} \phi_\alpha(\mathbf{r})\nabla\cdot\boldsymbol{\sigma}_\beta(\mathbf{r})d\mathbf{r}, \tag{17.92}$$

where a new gradient variable

$$\boldsymbol{\sigma}_\alpha(\mathbf{r}) = \nabla\phi_\alpha(\mathbf{r}), \tag{17.93}$$

is introduced, and $\phi_\alpha(\mathbf{r})$ and $\phi_\beta(\mathbf{r})$ are expressed using the DG basis functions

$$\phi_\alpha(\mathbf{r}) = \sum_j u^k_{\alpha,j}\varphi^k_j(\mathbf{r}), \quad \phi_\beta(\mathbf{r}) = \sum_j u^k_{\beta,j}\varphi^k_j(\mathbf{r}), \quad \mathbf{r} \in T_k, \tag{17.94}$$

and, similarly, $\sigma_\alpha(\mathbf{r})$ and $\sigma_\beta(\mathbf{r})$

$$\boldsymbol{\sigma}_\alpha(\mathbf{r}) = \sum_{,j} \boldsymbol{\sigma}^k_{\alpha,j}\varphi^k_j(\mathbf{r}), \quad \boldsymbol{\sigma}_\beta(\mathbf{r}) = \sum_j \boldsymbol{\sigma}^k_{\beta,j}\varphi^k_j(\mathbf{r}), \quad \mathbf{r} \in T_k. \tag{17.95}$$

Then, (17.92) is calculated by integration by parts and introducing appropriate numerical fluxes for $\sigma_\beta(\mathbf{r})$ at the boundary of each element T_k, namely,

$$H^{\text{kinetic}}_{\beta\alpha} = \sum_{T_k} H^{\text{kinetic},k}_{\beta\alpha}, \tag{17.96}$$

where

$$
\begin{aligned}
H_{\beta\alpha}^{\text{kinetic},k} &= -\frac{1}{2}\int\limits_{T_k} \phi_\alpha(\mathbf{r})\nabla\sigma_\beta(\mathbf{r})dx \\
&= -\frac{1}{2}\left[\int\limits_{\partial T_k} \phi_\alpha(\mathbf{r}\,)h_\sigma(\sigma_\beta^-, \sigma_\beta^+)\cdot\mathbf{n}ds - \int\limits_{T_k} \nabla\phi_\alpha(\mathbf{r})\sigma_\beta(\mathbf{r})d\mathbf{r}\right] \\
&= -\frac{1}{2}\left[\int\limits_{\partial T_k} \phi_\alpha(\mathbf{r}\,)h_\sigma(\sigma_\beta^-, \sigma_\beta^+)\cdot\mathbf{n}ds - \int\limits_{T_k} \sigma_\alpha(\mathbf{r})\sigma_\beta(\mathbf{r})d\mathbf{r}\right],
\end{aligned}
\tag{17.97}
$$

where σ_β^- and σ_β^+ denote the values of σ_β inside and outside of T_k on the boundary ∂T_k, and the numerical flux $h_\sigma(\sigma_\beta^-, \sigma_\beta^+)$ can be defined using an upwinding approach or a central average of values on both sides of the interface [54] as follows:

$$
h_\sigma(\sigma_\beta^-, \sigma_\beta^+)(\mathbf{r}) = \frac{1}{2}(\sigma_\beta^- + \sigma_\beta^+). \tag{17.98}
$$

To illustrate the linear scaling of the DG-DMM method, we calculate the ground state energy of a simple lattice model

$$
\hat{H}\psi_i = \varepsilon_i\psi_i, \tag{17.99}
$$

where the Hamiltonian for an N-lattice 1-D system is given by

$$
\hat{H} = -\frac{1}{2}\nabla^2 + V^{\text{KS}}(x). \tag{17.100}
$$

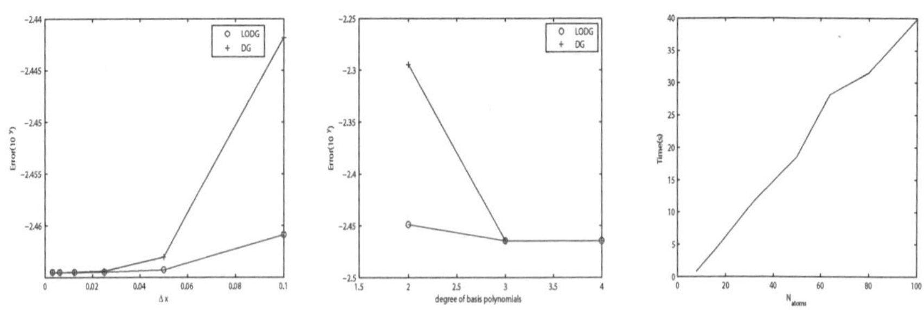

Fig. 17.2 (Left): Error of the total energy with different mesh sizes by LODG and DG methods in h-refinement. (Middle): Error of total energy with different degrees of basis polynomials by LODG and DG in p-refinement. (Right): CPU time vs the number of atoms in the lattice system. Reprinted from [46], copyright (2013) by Global Science, Ltd.

The effective potential V^{KS} is taken to be simply the sum of Gaussian wells located at the atom sites

$$V^{\mathrm{KS}}(x) = -\sum_{i \in \mathbb{Z}} \frac{a}{\sqrt{2\pi\sigma^2}} \exp\left[-\left(x - X_i\right)^2 / 2\sigma^2\right]. \tag{17.101}$$

This model has two parameters—a, which characterizes the depth of the wells, and σ, which characterizes its width. We will consider the case corresponding to a semiconductor [55], i.e., $a = 10$ and $\sigma = 0.3$,

The DG-DMM method is used to compute the ground state energy with a quadratic polynomial basis enriched with local orbitals (LODG) and without (DG). Fig. 17.2 shows the mesh and p-order convergence for the ground state energy by both DG and LODG (left and middle plots) and the CPU time linear scaling in terms of the number of atoms in the system.

17.5 Summary

Density functional theory, recognized by the 1998 Nobel Prize awarded to Walter Kohn, is now the main computational tool, with many developed software packages (such as ABINIT, Quantum EXPRESSO, SIESTA, VASP, and NWChem) [56], which have been used to investigate chemical, optical, magnetic, and mechanical properties of molecules, proteins, and bulk materials. Not discussed in this chapter is the time-dependent DFT theory [57], which was developed to study the transient and optical properties of quantum systems, especially for plasmonic systems arising from nanophotonics. Moreover, beyond the one-particle density approach of DFT, other many-body perturbation theory (MBPT) approaches such as the non-equilibrium Green's function (NEGF) and the GW approximations [58] are active research areas addressing the shortcomings of DFT approaches. Finally, deep neural network learning has opened a new avenue to leverage the DFT computations by providing training data to build molecular interaction potentials for molecular dynamic simulation of millions of molecules with quantum accuracy [59].

References

1. R.M. Martin, L. Reining, D.M, Ceperley, *Interacting Electrons* (Cambridge University Press, 2016)
2. R. Courant, D. Hilbert, *Method of Mathematical Physics, Vol. I* (Wiley-Interscience, 1989)
3. P. Hohenberg, W. Kohn, Inhomogeneous electron gas. Phys. Rev. B **136**, 864–871 (1964)
4. E.H. Lieb, Density functionals for Coulomb systems. Int. J. Quantum Chem. **24**, 243–277 (1983)
5. W. Kohn, L. Sham, Self-consistent equations including exchange and correlation effects. Phys. Rev. A **140**, 1133–1138 (1965)

6. L. Lin, J. Lu, L. Ying, Numerical methods for Kohn-Sham density functional theory. Acta Numerica **28**, 405–539 (2019)
7. J.P. Perdew, K. Schmidt, Jacob's ladder of density functional approximations for the exchange-correlation energy. AIP Conference Proceedings **577**, 1–20 (2001)
8. J.P. Perdew, Climbing the ladder of density functional approximations. MRS Bull. **38**, 743–750 (2013)
9. D. Bagayoko, Understanding density functional theory (DFT) and completing it in practice. AIP Adv. **4**(12) (2014)
10. C. Lee, W. Yang, R.G. Parr, Development of the Colle-Salvetti correlation-energy formula into a functional of the electron density. Phys. Rev. B **37**, 785–789 (1988)
11. A.D. Becke, Density-functional exchange-energy approximation with correct asymptotic behavior. Phys. Rev. A **38**, 3098–3100 (1988)
12. J.P. Perdew, K. Burke, M. Ernzerhof, Generalized gradient approximation made simple. Phys. Rev. Lett. **77**, 3865–3868 (1996)
13. F. Nogueira, A. Castro, M.A. Marques, A tutorial on density functional theory, in *A Primer in Density Functional Theory* (2003), pp. 218–256
14. D.R. Hamann, M. Schluter, C. Chiang, Norm-conserving pseudopotentials. Phys. Rev. Lett. **43**, 1494–1497 (1979)
15. N. Troullier, J.L. Martins, Efficient pseudopotentials for plane-wave calculations. Phys. Rev. B **43**, 1993–2006 (1991)
16. C. Hartwigsen, S. Goedecker, J. Hutter, Relativistic separable dual space Gaussian pseudopotentials from H to R^n. Phys. Rev. B **58**, 3641–3662 (1998)
17. D.R. Hamann, Optimized norm-conserving Vanderbilt pseudopotentials. Phys. Rev. B **88**, 085117 (2013)
18. L. Kleinman, D.M. Bylander, Efficacious form for model pseudopotentials. Phys. Rev. Lett. **48**, 1425–1428 (1982)
19. E. Anderson, Z. Bai, C. Bischof, S. Blackford, J. Demmel, J. Dongarra, J. Du Croz, A. Greenbaum, S. Hammarling, A. McKenney, D. Sorensen, *em LAPACK Users' Guide*, 3ed edn. (SIAM, 1999)
20. L.S. Blackford, J. Choi, A. Cleary, E. D'Azevedo, J. Demmel, I. Dhillon, S. Hammarling, G. Henry, A. Petitet, K. Stanley, D. Walker, R.C. Whaley, *ScaLAPACK Users' Guide* (SIAM, 1997)
21. E.R. Davidson, The iterative calculation of a few of the lowest eigenvalues and corresponding eigenvectors of large real symmetric matrices. J. Comput. Phys. **17**, 87–94 (1975)
22. B. Liu, *The simultaneous expansion method for the iterative solution of several of the lowest eigenvalues and corresponding eigenvectors of large realsymmetric matrices, in Report LBL-8158, Lawrence Berkeley Laboratory* (University of California, Berkeley, 1978)
23. A.V. Knyazev, Toward the optimal preconditioned eigensolver: Locally optimal block preconditioned conjugate gradient method. SIAM J. Sci. Comput. **23**, 517–541 (2001)
24. Y. Zhou, Y. Saad, M.L. Tiago, J.R. Chelikowsky, Self-consistent- field calculations using Chebyshev- ltered subspace iteration. J. Comput. Phys. **219**, 172–184 (2006)
25. D.G. Anderson, Iterative procedures for nonlinear integral equations. J. ACM (JACM) **12**(4), 547–60 (1965)
26. C.G. Broyden, A class of methods for solving nonlinear simultaneous equations. Math. Comput. **19**(92), 577–93 (1965)
27. M.C. Payne, M.P. Teter, D.C. Allen, T.A. Arias, J.D. Joannopoulos, Iterative minimization techniques for ab initio total energy calculation: molecular dynamics and conjugate gradients. Rev. Mod. Phys. **64**, 1045–1097 (1992)
28. G. Kresse, J. Furthmuller, Efficient iterative schemes for ab initio total energy calculations using a plane-wave basis set. Phys. Rev. B **54**, 11169–11186 (1996)
29. Y. Saad, J.R. Chelikowsky, S.M. Shontz, Numerical methods for electronic structure calculations of materials. SIAM Rev. **52**(1), 3–54 (2010)
30. E. Tsuchida, Augmented orbital minimization method for linear scaling electronic structure calculations. J. Phys. Soc. Japan **76**, 034708 (2007)

31. E. Tsuchida, M. Tsukada, Electronic-structure calculations based on the nite-element method. Phys. Rev. B **52**, 5573–5578 (1995)
32. P. Suryanarayana, V. Gavani, T. Blesgen, K. Bhattacharya, M. Ortiz, Non-periodic finite-element formulation of Kohn-Sham density functional theory. J. Mech. Phys. Solids **58**, 258–280 (2010)
33. G. Bao, G. Hu, D. Liu, An h-adaptive nite element solver for the calculations of the electronic structures. J. Comput. Phys. **231**, 4967–4979 (2012)
34. L. Lin, J. Lu, L. Ying, E. Weinan Adaptive local basis set for Kohn-Sham density functional theory in a discontinuous Galerkin framework, I. J. Comput. Phys. **231**, 2140–2154 (2012)
35. L. Lin, J. Lu, L. Ying, E. Weinan, Optimized local basis function for Kohn-Sham density functional theory. J. Comput. Phys. **231**, 4515–4529 (2012)
36. H. Chen, X. Dai, X. Gong, L. He, A. Zhou, Adaptive finite element approximations for Kohn-Sham models. Multiscale Model. Simul. **12**, 1828–1869 (2014)
37. P. Motamarri, M.R. Nowak, K. Leiter, J. Knap, V. Gavini, Higher-order adaptive finite-element methods for Kohn-Sham density functional theory. J. Comput. Phys. **253**, 308–43 (2013)
38. P. Motamarri, S. Das, S. Rudraraju, K. Ghosh, D. Davydov, V. GaviniV, DFT-FE-A massively parallel adaptive finite-element code for large-scale density functional theory calculations. Comput. Phys. Commun. **246**, 106853 (2020)
39. S. Goedecker, L. Colombo, Efficient linear scaling algorithm for tight-binding molecular dynamics. Phys. Rev. Lett. **73**, 122–125 (Jul1994)
40. S. Goedecker, Low complexity algorithms for electronic structure calculations. J. Comput. Phys. **118**, 261–268 (1995)
41. W. Yang, Direct calculation of electron density in density-functional theory. Phys. Rev. Lett. **66**, 1438–1441 (1991)
42. X.-P. Li, R.W. Nunes, D. Vanderbilt, Density-matrix electronic-structure method with linear system-size scaling. Phys. Rev. B **47**(16), 10891–10894 (1993)
43. F. Mauri, G. Galli, R. Car, Orbital formulation for electronic-structure calculations with linear system-size scaling. Phys. Rev. B **47**(15), 9973–9976 (1993)
44. E. Hernández, M.J. Gillan, Self-consistent first-principles technique with linear scaling. Phys. Rev. B **51**(15), 10157–10160 (1995)
45. Y. Lin, A. Baumketner, S. Deng et al., An image-based reaction field method for electrostatic interactions in molecular dynamics simulations of aqueous solutions. J. Chem. Phys. **131**, 154103 (2009)
46. T. Lu, W. Cai, J. Xin, Y. Guo, Linear scaling discontinuous Galerkin density matrix minimization method with local orbital enriched finite element basis: 1-D lattice model system. Commun. Comput. Phys. **14**(2), 276–300 (2013)
47. G. Galli, Linear scaling methods for electronic structure calculations and quantum molecular dynamics simulations. Current Opinion Solid State Mater. Sci. **1**(6), 864–874 (1996)
48. B. Cockburn, C.-W. Shu, The local discontinuous Galerkin method for time-dependent convection-diffusion systems. SIAM J. Numer. Anal. **35**, 2440–2463 (1998)
49. R. McWeeny, The density matrix in self-conssitent field theory. I. iterative construction of the density matrix. Proc. Roy. Soc. Lond. A **235**, 496–509 (1956)
50. W. Kohn, Density functional and density matrix method scaling linearly with the number of atoms. Phys. Rev. Lett. **76**, 3168–3171 (1996)
51. E. Prodan, W. Kohn, Nearsightedness of electronic matter. Proc. Nat. Acad. Sci. **102**, 11635–11638 (2005)
52. R. Klessig, E. Polak, Efficient implementations of the Polak-Ribière conjugate gradient algorithm. SIAM J. Control **10**(3), 524–49 (1972)
53. D. Luenberger, Y. Ye. *Linear and Nonlinear Programming*, 3rd edn. (Springer, 2008)
54. T. Lu, W. Cai, P. Zhang, Conservative local discontinuous Galerkin methods for time dependent Schrödinger equation. Int. J. Numer. Anal. Model. **2**(1), 75–84 (2005)
55. C.J. García-Cervera, J. Lu, Y. Xuan, E. Weinan, Linear-scaling subspace-iteration algorithm with optimally localized nonorthogonal wave functions for kohn-sham density functional theory. Phys. Rev. B **79**, 115110 (2009)

56. C.D. Sherrill, D.E. Manolopoulos, T.J. Martínez, A. Michaelides, Electronic structure software. J. Chem. Phys. **153**(7) (2020)
57. C.A. Ullrich, *Time-Dependent Density-Functional Theory: Concepts and Applications* (Oxford University Press, 2011)
58. G. Onida, L. Reining, A. Rubio, Electronic excitations: Densityfunctional versus many-body Green's-function approaches. Rev. Mod. Phys. **74**, 601–659 (2002)
59. H. Wang, L. Zhang, J. Han, W.E. DeePMD-kit, A deep learning package for many-body potential energy representation and molecular dynamics. Comput. Phys. Commun. **228**, 178–84 (2018)

Chapter 18
Non-equilibrium Green's Function (NEGF) Methods for Transport

In this chapter, we will present the quantum transport method based on non-equilibrium Green's functions (NEGFs) for computing electron currents in an open quantum system. The system is usually connected to surrounding environments at different chemical potentials, such as nano-electronics connected to input contacts at different electric potentials. The NEGF allows the calculation of electron transmission coefficients through general devices; the latter are then used for current calculations with the Landauer transport theory discussed in Sect. 16.4.

18.1 NEGFs for 1-D Devices

18.1.1 1-D Device Boundary Conditions for Green's Functions

For a 1-D ultra-small device with two large contacts as shown in Fig. 18.1, the electric potential is of the form

$$V(x) = \begin{cases} v^{(1)}, & \text{if } -\infty < x < x_1, \\ v(x), & \text{if } x_1 \le x \le x_N, \\ v^{(2)}, & \text{if } x_N < x < +\infty, \end{cases} \tag{18.1}$$

where $v^{(\alpha)}$ is the constant potential in the contact α ($\alpha = 1, 2$). A Green's function is defined by

$$\left(E - V(x) + \frac{\hbar^2}{2} \frac{\partial}{\partial x} \frac{1}{m} \frac{\partial}{\partial x} \right) G(x, x') = \delta(x - x'), \quad x, x' \in (-\infty, +\infty). \tag{18.2}$$

W. Cai, *Deterministic, Stochastic, and Deep Learning Methods for Computational Electromagnetics*, https://doi.org/10.1007/978-981-96-0100-4_18

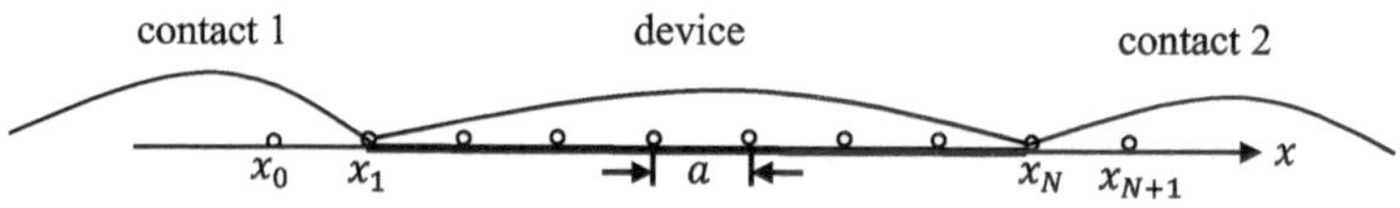

Fig. 18.1 A 1-D uniform mesh with grid size a for the computational domain $\Omega_D = [x_1, x_N]$ (the device region)

To compute the electron current through the device region, taken to be the interval $[x_1, x_N]$, we only need Green's function within the device. Therefore, boundary conditions at $x = x_1$ and $x = x_N$ are needed to determine Green's function G inside the device. To illustrate, let us consider the left boundary point $x = x_1$; $x = x_N$ can be handled similarly. First, an auxiliary Green's function $g(x, x_e')$ is defined in the contact-1 domain $\Omega_1 = (-\infty, x_1)$ by

$$\left(E - v^{(1)} + \frac{\hbar^2}{2m^{(1)}}\frac{\partial^2}{\partial x^2}\right) g(x, x_e') = \delta(x - x_e'), \quad x, x_e' \in (-\infty, x_1), \tag{18.3}$$

which can be viewed as the restriction of (18.2) onto the semi-infinite region Ω_1 plus a yet to be determined boundary condition on $x = x_1$. Here, the subscript e denotes the exterior of the device domain Ω_D. Subtracting the product of (18.2) and $g(x, x_e')$ from the product of (18.3) and $G(x, x')$ with $x' \in \Omega_D$, integrating with respect to x on Ω_1, and then using Green's formula, we have

$$\begin{aligned} G(x_e', x') &= \int_{\Omega_1} G(x, x')\delta(x - x_e')\mathrm{d}x - \int_{\Omega_1} g(x, x_e')\delta(x - x')\mathrm{d}x \\ &= \int_{\Omega_1} \frac{\hbar^2}{2m^{(1)}}\left(\frac{\partial^2}{\partial x^2} g(x, x_e')G(x, x') - \frac{\partial^2}{\partial x^2} G(x, x')g(x, x_e')\right)\mathrm{d}x \\ &= \frac{\hbar^2}{2m^{(1)}}\left(\frac{\partial g(x, x_e')}{\partial x}G(x, x') - \frac{\partial G(x, x')}{\partial x}g(x, x_e')\right)\Bigg|_{x=-\infty}^{x_1} \\ &= \frac{\hbar^2}{2m^{(1)}}\left(\frac{\partial g(x_1, x_e')}{\partial x}G(x_1, x') - \frac{\partial G(x_1, x')}{\partial x}g(x_1, x_e')\right), \end{aligned} \tag{18.4}$$

where $m^{(1)}$ is the effective mass in contact 1. Here we have used the facts that $m^{(1)}$ is independent of position inside the contact region Ω_1, and that both $G(x, x')$ and $g(x, x_e')$ satisfy the Sommerfeld radiation condition as $x \to -\infty$.

According to (18.4), by assuming different boundary conditions for the auxiliary Green's function $g(x, x_e')$, we can arrive at different boundary conditions for Green's function $G(x, x')$.

1. **Homogeneous Dirichlet condition**, i.e., $g(x, x_e')|_{x=x_1} = 0$. It can be shown using the method of images that $g(x, x_e')$ is just

$$g(x, x_e') = g_0(x, x_e') - g_0(2x_1 - x, x_e'), \tag{18.5}$$

where $g_0(x, x_e')$ is a retarded Green's function in the infinite 1-D wire which satisfies

$$\left(E - v^{(1)} + \frac{\hbar^2}{2m^{(1)}}\frac{\partial^2}{\partial x^2}\right) g_0(x, x_e') = \delta(x - x_e'),\ x \in (-\infty, \infty), \tag{18.6}$$

namely

$$g_0(x, x') = \frac{m^{(1)}}{\mathrm{i}k^{(1)}\hbar^2} \exp(\mathrm{i}k^{(1)}|x - x_e'|), \tag{18.7}$$

where $k^{(\alpha)} = \sqrt{2m^{(\alpha)}(E - v^{(\alpha)})/\hbar^2}$ for $\alpha = 1, 2$.
Using this homogeneous auxiliary Green's function in (18.4), we obtain the following boundary condition for the device Green's function:

$$G(x_e', x') = \frac{\hbar^2}{2m^{(1)}}\frac{\partial g(x_1, x_e')}{\partial x} G(x_1, x'),\ x_e' \in (-\infty, x_1), x' \in [x_1, x_N]. \tag{18.8}$$

This boundary condition relates the value of Green's function $G(x_e', x')$ at a location x_e' outside the device region to the value at the left boundary of the device $x = x_1$. Such a condition will be useful for the finite difference discretization of (18.2) at the boundaries of the device, as solution unknowns at "ghost" points outside the device region will be needed.

2. **Homogeneous Neumann condition**, i.e., $\frac{\partial g(x, x_e')}{\partial x}|_{x=x_1} = 0$. Then

$$g(x, x_e') = g_0(x, x_e') + g_0(2x_1 - x, x_e'), \tag{18.9}$$

and, after plugging (18.9) into (18.4), we obtain the following boundary condition for the device Green's function:

$$G(x_e', x') = -\frac{\hbar^2}{2m^{(1)}}\frac{\partial G(x_1, x')}{\partial x} g(x_1, x_e'),\quad x_e' \in (-\infty, x_1], x' \in [x_1, x_N], \tag{18.10}$$

and, in particular, for $x_e' = x_1$,

$$G(x_1, x') = -\frac{\hbar^2}{2m^{(1)}}\frac{\partial G(x_1, x')}{\partial x} g(x_1, x_1^-) = -\frac{1}{\mathrm{i}k^{(1)}}\frac{\partial G(x_1, x')}{\partial x},\quad x' \in [x_1, x_N]. \tag{18.11}$$

The boundary condition (18.11) relates the derivative of Green's function $\partial G(x_1, x')/\partial x$ at the boundary to the boundary value of Green's function $G(x_1, x')$, which in fact defines the so-called Dirichlet-to-Neumann (DtN) mapping. This boundary condition will be useful when a finite element method is used to discretize (18.2).

18.1.2 Finite Difference Methods for 1-D Device NEGFs

Equation (18.2) over the device region $[x_1, x_N]$ can be discretized by a second-order central difference scheme over the mesh $x_1 < x_2 < \cdots < x_N$, $x_i = x_1 + (i-1)a$, using the following difference formula:

$$\frac{\partial}{\partial x}\left(\frac{1}{m}\frac{\partial u}{\partial x}\right)_{x=x_i} \approx \frac{1}{a^2}\left(\frac{u_{i+1}-u_i}{m_{i+1/2}} - \frac{u_i - u_{i-1}}{m_{i-1/2}}\right), \tag{18.12}$$

where $u_i = u(x_i)$ and $m_{i\pm 1/2} = m((x_i + x_{i\pm 1})/2)$, $i = 1, 2, \ldots, N$. For $i = 1$, the finite difference scheme becomes

$$\frac{\partial}{\partial x}\left(\frac{1}{m}\frac{\partial u}{\partial x}\right)_{x=x_1} \approx \frac{1}{a^2}\left(\frac{u_2-u_1}{m_{3/2}} - \frac{u_1 - u_0}{m_{1/2}}\right), \tag{18.13}$$

which implies that we need to specify the nodal value $u_0 \equiv G_{0,j} = G(x_0, x_j)$ in terms of the nodal values $u_i \equiv G_{i,j}, i = 1, 2, \ldots, N$, for any given $j \in \{1, 2, \ldots, N\}$. This can be achieved by using the boundary condition (18.8). Setting $x_e' = x_0$ in (18.8), we have

$$u_0 = G(x_0, x') = \frac{\hbar^2}{2m^{(1)}}\frac{\partial g(x_1, x_0)}{\partial x} G(x_1, x') \equiv \omega_1^{(1)} u_1, \tag{18.14}$$

where

$$\omega_q^{(1)} = \frac{\hbar^2}{2m^{(1)}}\frac{\partial g(x_1, x_0)}{\partial x}\delta_{q,1} = \exp(\mathrm{i}k^{(1)}a)\delta_{q,1},\ 1 \le q \le N. \tag{18.15}$$

Similarly, at the right end boundary $x = x_N$, we have the following boundary condition:

$$u_{N+1} = G(x_{N+1}, x') = \omega_N^{(2)} u_N, \tag{18.16}$$

where

$$\omega_q^{(2)} = \exp(\mathrm{i}k^{(2)}a)\delta_{q,N}, \quad 1 \le q \le N. \tag{18.17}$$

Applying the discretization (18.12) in (18.2) at x_i, $i = 1, 2, \ldots, N$, together with the boundary conditions (18.14) and (18.16), we arrive at the linear system

$$(\mathcal{E} - \mathbf{H}_d - \Sigma)\overline{\mathbf{G}} = \overline{\mathbf{I}}, \tag{18.18}$$

where $\mathcal{E} = E\overline{\mathbf{I}}$, the jth column of the matrix $\overline{\mathbf{G}}$ contains $u_i \equiv G_{i,j} = G(x_i, x_j)$, $i = 1, 2, \ldots, N$, the matrix $\mathbf{H}_d$ is the device (discrete) Hamiltonian

$$\mathbf{H}_d = \mathbf{V} + \frac{\hbar^2}{2a^2} D_N, \tag{18.19}$$

where $\mathbf{V} = \text{diag}(V(x_1), V(x_2), \ldots, V(x_N))$, and the difference derivative matrix

$$D_N = \begin{bmatrix} -\frac{1}{m_{1/2}} - \frac{1}{m_{3/2}} & \frac{1}{m_{3/2}} & \cdots & \cdots & 0 \\ \frac{1}{m_{3/2}} & -\frac{1}{m_{3/2}} - \frac{1}{m_{5/2}} & \ddots & \ddots & \vdots \\ 0 & \frac{1}{m_{5/2}} & \ddots & \ddots & \vdots \\ \vdots & \ddots & \ddots & \ddots & 0 \\ \vdots & \ddots & \ddots & -\frac{1}{m_{N-3/2}} - \frac{1}{m_{N-1/2}} & \frac{1}{m_{N-1/2}} \\ 0 & \cdots & 0 & \frac{1}{m_{N-1/2}} & -\frac{1}{m_{N-1/2}} - \frac{1}{m_{N+1/2}} \end{bmatrix}. \tag{18.20}$$

Moreover, the matrix $\Sigma = (\Sigma_{p,q})$ contains the coefficients in the boundary conditions (18.14) and (18.16). Specifically,

$$\Sigma_{p,q} = \Sigma^{(1)}_{p,q} + \Sigma^{(2)}_{p,q}, \tag{18.21}$$

where

$$\Sigma^{(1)}_{p,q} = -\frac{\hbar^2}{2m_{1/2}a^2}\omega^{(1)}_q \delta_{p,1} = -\frac{\hbar^2}{2m^{(1)}a^2} \exp(ik^{(1)}a)\delta_{q,1}\delta_{p,1}, \tag{18.22}$$

$$\Sigma^{(2)}_{p,q} = -\frac{\hbar^2}{2m_{N+1/2}a^2}\omega^{(2)}_q \delta_{p,N} = -\frac{\hbar^2}{2m^{(2)}a^2} \exp(ik^{(2)}a)\delta_{q,N}\delta_{p,N}. \tag{18.23}$$

Here, the complex quantities $\Sigma^{(1)}$ and $\Sigma^{(2)}$ are called the self-energies [1] of contacts 1 and 2, respectively, which represent the influence of the contacts on the current transport through the device.

18.1.3 Finite Element Methods for 1-D Device NEGFs

Denote by $\varphi_i(x)$ the nodal shape function of a finite element space, namely

$$\varphi_i(x_j) = \delta_{i,j}, \tag{18.24}$$

and assume that Green's function over the device region $\Omega_D = [x_1, x_N]$ is approximated by

$$G_h(x, x_j) = \sum_{i=1}^{N} G_{i,j}\varphi_i(x), \ x, x_j \in [x_1, x_N]. \tag{18.25}$$

The weak form of (18.2) for the device region implies that, for any test function $\varphi(x)$, we have

$$\begin{aligned} E\int_{\Omega_D} G_h\varphi\, \mathrm{d}x - \frac{\hbar^2}{2}\int_{\Omega_D}\frac{1}{m}\frac{\partial G_h}{\partial x}\frac{\partial \varphi}{\partial x}\mathrm{d}x - \int_{\Omega_D} VG_h\varphi \mathrm{d}x \\ -\frac{\hbar^2}{2}\frac{1}{m}\frac{\partial G_h}{\partial x}\varphi\bigg|_{x=x_1} + \frac{\hbar^2}{2}\frac{1}{m}\frac{\partial G_h}{\partial x}\varphi\bigg|_{x=x_N} = \varphi(x_j), \end{aligned} \tag{18.26}$$

where we have set the source $x' = x_j$ in (18.2).

The derivative of Green's function $\partial G_h/\partial x$ at the boundaries will be replaced by the DtN boundary condition in (18.11), which can be rewritten as

$$\frac{\partial G_h(x_1, x_j)}{\partial x} = \sum_{i=1}^{N}\omega_i^{(1)}G_{i,j}, \quad \frac{\partial G_h(x_N, x_j)}{\partial x} = \sum_{i=1}^{N}\omega_i^{(2)}G_{i,j}, \tag{18.27}$$

where

$$\omega_i^{(1)} = -\mathrm{i}k^{(1)}\delta_{i,1}, \ \omega_i^{(2)} = -\mathrm{i}k^{(2)}\delta_{i,N}. \tag{18.28}$$

Therefore, plugging (18.27) into (18.26), we obtain the following linear system:

$$(\mathcal{E} - \mathbf{H}_d - \Sigma)\mathbf{G} = \bar{\mathbf{I}}, \tag{18.29}$$

where $\mathcal{E} = E\mathbf{S}$, in which $\mathbf{S} = (S_{p,q})$, with

$$S_{p,q} = \int_{\Omega_D}\varphi_q\varphi_p\, \mathrm{d}x, \tag{18.30}$$

and

$$(\mathbf{H}_d)_{p,q} = \frac{\hbar^2}{2}\int_{\Omega_D}\frac{1}{m}\frac{\partial \varphi_q}{\partial x}\frac{\partial \varphi_p}{\partial x}\, \mathrm{d}x + \int_{\Omega_D} V\varphi_q\varphi_p\, \mathrm{d}x, \tag{18.31}$$

and again the self-energies $\Sigma^{(1)}$ and $\Sigma^{(2)}$ contain the coefficients resulting from the boundary conditions:

$$\Sigma_{p,q}^{(1)} = \frac{\hbar^2}{2m_1}\omega_q^{(1)}\delta_{p,1}, \ \Sigma_{p,q}^{(2)} = -\frac{\hbar^2}{2m_N}\omega_q^{(2)}\delta_{p,N}. \tag{18.32}$$

Finally, by inverting the matrix system (18.18) or (18.29), we obtain the device Green's function, as defined in (16.201), which will play a key role in the NEGF method for electron transport:

$$\mathbf{G}(E) = (\mathcal{E}(E) - \mathbf{H}_d(E) - \Sigma(E))^{-1}, \tag{18.33}$$

where the self-energy is given by

$$\Sigma(E) = \Sigma^{(1)}(E) + \Sigma^{(2)}(E). \tag{18.34}$$

18.2 NEGFs for 2-D Devices

In this section, we will extend the preceding results on 1-D devices to 2-D devices [2]; as a matter of fact, all derivations are directly applicable to 3-D devices. Green's function in two dimensions is defined on the domain $\Omega = \Omega_D \cup (\sum_\alpha \Omega_\alpha)$, which consists of the device and the contacts, $\Gamma = \partial\Omega$, as depicted in Fig. 18.2. Here, Ω_D is the device region, Ω_α is the area of the contact α which extends to infinity, and $\Gamma_\alpha = \Gamma_D \cap \partial\Omega_\alpha$ with $\Gamma_D = \partial\Omega_D$. For a given energy E, Green's function $G(\mathbf{r}, \mathbf{r}')$ is defined by

$$(E - H)G(\mathbf{r}, \mathbf{r}') = \delta(\mathbf{r} - \mathbf{r}'), \quad \mathbf{r}, \mathbf{r}' \in \Omega, \tag{18.35}$$

where

$$H = -\frac{\hbar^2}{2}\nabla \cdot \left(\frac{1}{m(\mathbf{r})}\nabla\right) + V(\mathbf{r}) \tag{18.36}$$

is the Hamiltonian of the infinite system with an effective mass $m(\mathbf{r})$, Planck constant is given by $2\pi\hbar$, and $V(\mathbf{r})$ is the potential energy. The Green's function $G(\mathbf{r}, \mathbf{r}')$ vanishes on the boundary Γ and satisfies the Sommerfeld radiation condition at infinity. Again, we only need to compute Green's function on the device region Ω_D, without the details of Green's function in the remaining infinite exterior domain.

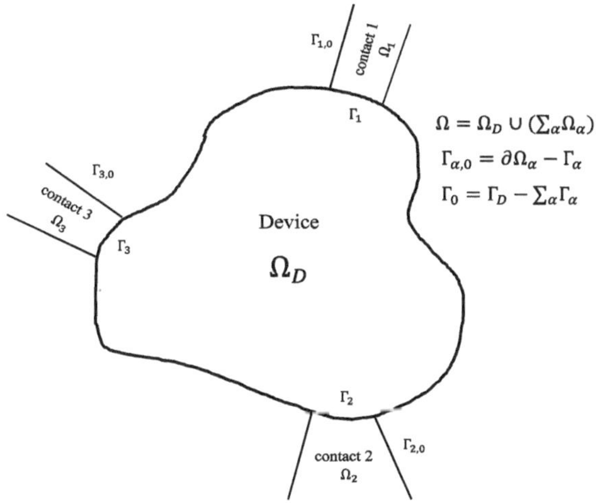

Fig. 18.2 A device Ω_D and its contacts: Ω_D (the bold curve), and Ω_α—the contact α with $\alpha = 1, 2, \ldots$. Γ_α is the boundary between Ω_D and Ω_α while the rest of $\partial\Omega_\alpha$ is $\Gamma_{\alpha,0}$; $\Gamma_D = \partial\Omega_D$

18.2.1 2-D Device Boundary Conditions for Green's Functions

We assume that, in the contact region, the potential $V(\mathbf{r})$ is invariant by translation along the transport direction, and also that the effective mass $m(\mathbf{r})$ is independent of position. As before, an auxiliary Green's function $g(\mathbf{r}, \mathbf{r}_e')$ is defined [3] by

$$(E - H)g(\mathbf{r}, \mathbf{r}_e') = \delta(\mathbf{r} - \mathbf{r}_e'), \quad \mathbf{r}, \mathbf{r}_e' \in \Omega_\alpha, \tag{18.37}$$

which can be viewed as the restriction of (18.36) onto the semi-infinite region Ω_α plus a yet to be determined boundary condition on Γ_α. Here, the subscript e denotes the exterior of the device Ω_D. Subtracting the product of (18.35) and $g(\mathbf{r}, \mathbf{r}_e')$ from the product of (18.37) and $G(\mathbf{r}, \mathbf{r}')$ with $\mathbf{r}' \in \Omega_D$, integrating with respect to $\mathbf{r}$ on Ω_α, and using Green's formula, we have

$$\begin{aligned}
G(\mathbf{r}_e', \mathbf{r}') &= \int_{\Omega_\alpha} G(\mathbf{r}, \mathbf{r}')\delta(\mathbf{r} - \mathbf{r}_e')\mathrm{d}\mathbf{r} - \int_{\Omega_\alpha} g(\mathbf{r}, \mathbf{r}_e')\delta(\mathbf{r} - \mathbf{r}')\mathrm{d}\mathbf{r} \\
&= \int_{\Omega_\alpha} \frac{\hbar^2}{2m^{(\alpha)}} \left(\nabla^2 g(\mathbf{r}, \mathbf{r}_e')G(\mathbf{r}, \mathbf{r}') - \nabla^2 G(\mathbf{r}, \mathbf{r}')g(\mathbf{r}, \mathbf{r}_e')\right) \mathrm{d}\mathbf{r} \\
&= \int_{\partial\Omega_\alpha} \frac{\hbar^2}{2m^{(\alpha)}} \left(\frac{\partial g(\mathbf{r}, \mathbf{r}_e')}{\partial \mathbf{n}} G(\mathbf{r}, \mathbf{r}') - \frac{\partial G(\mathbf{r}, \mathbf{r}')}{\partial \mathbf{n}} g(\mathbf{r}, \mathbf{r}_e')\right) \mathrm{d}s \\
&= \int_{\Gamma_\alpha} \frac{\hbar^2}{2m^{(\alpha)}} \left(\frac{\partial g(\mathbf{r}, \mathbf{r}_e')}{\partial \mathbf{n}} G(\mathbf{r}, \mathbf{r}') - \frac{\partial G(\mathbf{r}, \mathbf{r}')}{\partial \mathbf{n}} g(\mathbf{r}, \mathbf{r}_e')\right) \mathrm{d}s,
\end{aligned} \tag{18.38}$$

where $m^{(\alpha)}$ is the effective mass in the contact α and $\mathbf{n}$ is the normal vector exterior to the boundary $\partial\Omega_\alpha$. Here we have used the facts that $m^{(\alpha)}$ is independent of position and that both $G(\mathbf{r}, \mathbf{r}')$ and $g(\mathbf{r}, \mathbf{r}_e')$ satisfy the Sommerfeld radiation condition at infinity and the homogeneous Dirichlet conditions on $\Gamma_{\alpha,0}$.

1. **Homogeneous Dirichlet condition,** $g(\mathbf{r}, \mathbf{r}_e') = 0$ for $\mathbf{r} \in \Gamma_\alpha$; then

$$G(\mathbf{r}_e', \mathbf{r}') = \int_{\Gamma_\alpha} \frac{\hbar^2}{2m^{(\alpha)}} \frac{\partial g(\mathbf{r}, \mathbf{r}_e')}{\partial \mathbf{n}} G(\mathbf{r}, \mathbf{r}')\mathrm{d}s, \quad \mathbf{r}_e' \in \Omega_\alpha, \mathbf{r}' \in \Omega_D. \tag{18.39}$$

2. **Homogeneous Neumann condition,** $\frac{\partial g(\mathbf{r}, \mathbf{r}_e')}{\partial \mathbf{n}} = 0$ for $\mathbf{r} \in \Gamma_\alpha$; then

$$G(\mathbf{r}_e', \mathbf{r}') = -\int_{\Gamma_\alpha} \frac{\hbar^2}{2m^{(\alpha)}} \frac{\partial G(\mathbf{r}, \mathbf{r}')}{\partial \mathbf{n}} g(\mathbf{r}, \mathbf{r}_e')\mathrm{d}s, \quad \mathbf{r}_e' \in \Omega_\alpha, \mathbf{r}' \in \Omega_D. \tag{18.40}$$

The auxiliary Green's function $g(\mathbf{r}, \mathbf{r}_e')$ for both boundary conditions above are given in Sect. 18.4 for a strip-shaped contact.

We note that Green's function satisfies the following continuity conditions for $\mathbf{r} \in \Gamma_\alpha$ and $\mathbf{r}' \in \Omega_D$:

$$\begin{cases} G(\mathbf{r}_-, \mathbf{r}') = G(\mathbf{r}_+, \mathbf{r}'), \\ \frac{1}{m(\mathbf{r}_-)}\frac{\partial}{\partial \mathbf{n}}G(\mathbf{r}_-, \mathbf{r}') = \frac{1}{m(\mathbf{r}_+)}\frac{\partial}{\partial \mathbf{n}}G(\mathbf{r}_+, \mathbf{r}'), \end{cases} \tag{18.41}$$

where − (+) denotes the limit from the exterior (interior) of Ω_D.

Remark 18.1 The continuity equation (18.41) needs some delicate interpretation when both the source point $\mathbf{r}'$ and the field point $\mathbf{r}$ are on the device boundary Γ_α in deriving the device Green's function. For this case, we will consider the source point $\mathbf{r}'$ by a limiting process from inside the device toward the device boundary, and in this way the continuity conditions (18.41) for the device Green's function can be used on the device boundary. This continuity is necessary to connect the values of the device Green's function from both sides of the device boundary and obtain the self-energies $\Sigma^{(\alpha)}$ for the contacts.

Equations (18.39) and (18.40) yield boundary conditions for $G(\mathbf{r}, \mathbf{r}')$ provided that $g(\mathbf{r}, \mathbf{r}'_e)$ is known. As shown in the case of 1-D devices, these boundary conditions will define the self-energy $\Sigma^{(\alpha)}$ corresponding to the contact α. Equation (18.39) can be used in a finite difference method to eliminate the unknowns at "ghost" points $\mathbf{r}'_e$ in Ω_α outside the computational domain Ω_D in terms of the solutions at the boundary points $\mathbf{r}$. Equation (18.40) is the so-called Neumann-to-Dirichlet (NtD) mapping on Γ_α by letting $\mathbf{r}'_e \to \mathbf{r}_\alpha$ with $\mathbf{r}_\alpha \in \Gamma_\alpha$, and can be used in the finite element method to connect the solution and its normal derivative. In practice, it is more convenient to use a Dirichlet-to-Neumann (DtN) mapping, which is the inverse of (18.40). We could get the DtN mapping from (18.39). Differentiating (18.39) with respect to $\mathbf{r}'_e$, letting $\mathbf{r}'_e \to \mathbf{r}_\alpha$, and taking the normal derivative at $\mathbf{r}_\alpha$, we obtain

$$\frac{\partial G(\mathbf{r}_\alpha-, \mathbf{r}')}{\partial \mathbf{n}_\alpha} = \frac{\hbar^2}{2m^{(\alpha)}}\int_{\Gamma_\alpha} \frac{\partial^2 g(\mathbf{r}, \mathbf{r}_\alpha-)}{\partial \mathbf{n}_\alpha \partial \mathbf{n}} G(\mathbf{r}, \mathbf{r}')ds, \tag{18.42}$$

where $\mathbf{n}_\alpha$ denotes the outward normal of Ω_D at $\mathbf{r}_\alpha$. Equation (18.42) is to be treated as the Hadamard finite part when $\mathbf{r}'$ is on the boundary.

We will use (18.38), (18.39), and (18.42) to derive the self-energies $\Sigma^{(\alpha)}$ for all contacts and then calculate Green's function.

18.2.2 Finite Difference Methods for 2-D Device NEGFs

If a 2-D quantum device is wide in the y-direction, we can assume that $G(\mathbf{r}, \mathbf{r}')$ is independent of y, i.e., it is a function of (x, z) only. We consider an ultra-small MOSFET simulation in the strip region Ω (see Fig. 18.3), which consists of three sub-domains: the contact-1 area Ω_1, the device area Ω_D, and the contact-2 area Ω_2.

As in the 1-D case, the computational domain is denoted as Ω_D. Let L be the thickness of the silicon layer, or the combined thickness of the silicon layer and the oxide layers if tunneling effects are to be included. When electron tunneling into the

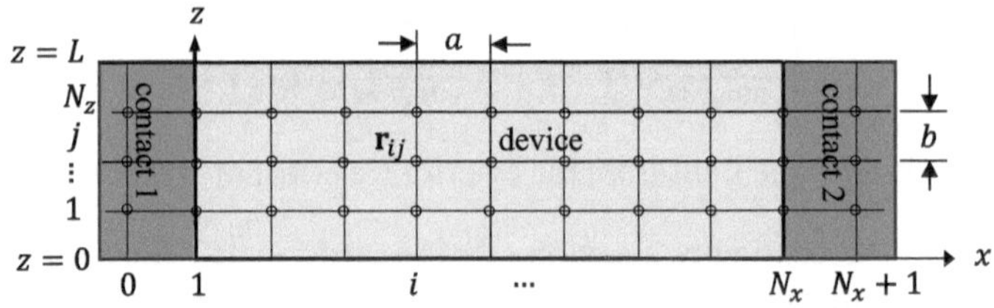

Fig. 18.3 A 2-D uniform mesh (gray lines) with grid space a in the x-direction and b in the z-direction. The computational domain Ω_D is indicated by the light gray region. The width in the z-direction is L. The homogeneous Dirichlet conditions on the top and bottom boundaries are used

oxide regions is neglected, the homogeneous Dirichlet conditions can be used on the top and bottom boundaries. We also assume that the band structure of the contact is independent of x; thus, we have

$$V(\mathbf{r}) = \begin{cases} v^{(1)}(z), & \text{if } \mathbf{r} \in \Omega_1, \\ v(x, z), & \text{if } \mathbf{r} \in \Omega_D, \\ v^{(2)}(z), & \text{if } \mathbf{r} \in \Omega_2, \end{cases} \tag{18.43}$$

where $\mathbf{r} = (x, z) \in \Omega$. The relevant Green's function is defined by

$$\left[E - V(\mathbf{r}) + \frac{\hbar^2}{2}\nabla \cdot \left(\frac{1}{m}\nabla\right)\right] G(\mathbf{r}, \mathbf{r}') = \delta(\mathbf{r} - \mathbf{r}'), \quad \mathbf{r}, \mathbf{r}' \in \Omega, \tag{18.44}$$

where $\nabla = (\partial/\partial x, \partial/\partial z)$. In order to obtain a finite discrete system, we need to set suitable boundary conditions on Γ_α, to obtain the self-energies $\Sigma^{(\alpha)}$ ($\alpha = 1, 2$) as in the 1-D case. Let us deal only with Γ_1 as an example, i.e., the computation of $\Sigma^{(1)}$. The calculation of $\Sigma^{(2)}$ is similar. Again, we define an auxiliary Green's function $g(\mathbf{r}, \mathbf{r}_e')$, which satisfies in Ω_1 the following:

$$\left(E - v^{(1)}(z) + \frac{\hbar^2}{2m^{(1)}}\nabla^2\right) g(\mathbf{r}, \mathbf{r}_e') = \delta(\mathbf{r} - \mathbf{r}_e'), \quad \mathbf{r}, \mathbf{r}_e' \in \Omega_1. \tag{18.45}$$

We will consider the finite difference method first for Green's function G. The unknowns are at the nodes $\mathbf{r}_q = \mathbf{r}_{i,j}$ (see Fig. 18.3). For a given source $\mathbf{r}' = \mathbf{r}_{q'} = \mathbf{r}_{i',j'}$, denoting $G_{q,q'} = G(\mathbf{r}_q, \mathbf{r}_{q'})$ and $\mathbf{G} = [G_{q,q'}]_{N \times N}$, the unknown vector is the q'th column of $\mathbf{G}$, with $N = N_x N_z$ being the number of unknowns. The relation between the indices q and (i, j) is based on the dictionary order of the mesh points inside the device (Fig. 18.3), i.e.,

$$q = j + (i - 1)N_z. \tag{18.46}$$

The notations $G_{i,j,i',j'}$, $G_{i,j,q'}$, and $G_{q,q'}$ will be used interchangeably. While using the second-order central difference scheme, we need boundary conditions when computing the unknowns at Γ_1. To compute $G_{1,j,q'}$ (q' is fixed), $j = 1, \ldots, N_z$, we need to know the "ghost values" $G_{0,j,q'}$ using $G_{q,q'}$, $q = 1, \ldots, N$. For this purpose, we insert the expression of $g(\mathbf{r}, \mathbf{r}'_e)$ from Sect. 18.4 into (18.39) to obtain

$$G(x'_e, z'_e, x', z') = \int_0^L G(x_1, z, x', z') \sum_l \chi_l^{(1)}(z)\chi_l^{(1)}(z'_e) \exp\left(-\mathrm{i}k_l^{(1)}(x'_e - x_1)\right) \mathrm{d}z, \tag{18.47}$$

with $\chi_l^{(\alpha)}(z)$ and $k_l^{(\alpha)}$ defined in Sect. 18.4. From (18.47) with a N_z-point trapezoidal rule for the integration along Γ_1, we have

$$\begin{aligned} G(x_0, z_j, x', z') &= \sum_l \int_0^L G(x_1, z, x', z')\chi_l^{(1)}(z)\chi_l^{(1)}(z_j) \exp\left(\mathrm{i}k_l^{(1)}a\right) \mathrm{d}z \\ &\approx \sum_l \sum_{j_1=1}^{N_z} bG(x_1, z_{j_1}, x', z')\chi_l^{(1)}(z_{j_1})\chi_l^{(1)}(z_j) \exp\left(\mathrm{i}k_l^{(1)}a\right) \\ &= \sum_{j_1=1}^{N_z} \sum_l bG(x_1, z_{j_1}, x', z')\chi_l^{(1)}(z_{j_1})\chi_l^{(1)}(z_j) \exp\left(\mathrm{i}k_l^{(1)}a\right), \end{aligned} \tag{18.48}$$

which implies that

$$G_{0,j,q'} = \sum_{q=1}^{N} \omega_q^{(1,j)} G_{q,q'}, \tag{18.49}$$

and

$$\omega_q^{(1,j)} = \begin{cases} \sum_l b\chi_l^{(1)}(z_q)\chi_l^{(1)}(z_j) \exp\left(\mathrm{i}k_l^{(1)}a\right), & \text{if } q \in \{1, \ldots, N_z\}, \\ 0, & \text{otherwise.} \end{cases} \tag{18.50}$$

As a result, Green's function satisfies a similar equation as in (18.18) with the following Hamiltonian form:

$$\mathbf{H}_d\mathbf{G} = \mathbf{V}\mathbf{G} + \frac{\hbar^2}{2a^2}\mathbf{G} \star D_{N_x} + \frac{\hbar^2}{2b^2} D_{Nz} \star \mathbf{G}, \tag{18.51}$$

where the 2-D finite difference operator $\mathbf{G} \star D_{N_x}$ is applied to each q'-column of $\mathbf{G}$, denoted as a vector $\mathbf{u} = \mathbf{G}(\cdot, q') \in \mathbb{R}^N$, as follows. If the vector $\mathbf{u}$ is arranged into an $N_z \times N_x$ matrix $\mathbf{U}_{N_z \times N_x}$, then we have

$$\mathbf{G} \star D_{N_x} = \mathbf{U}_{N_z \times N_x} \otimes (D_{Nx})^{\mathrm{T}}, \tag{18.52}$$

where $\otimes$ indicates the multiplication of each row of the matrix $\mathbf{U}_{N_z \times N_x}$ by the matrix $(D_{Nx})^{\mathrm{T}}$; $D_{Nz} \star \mathbf{G}$ can be defined similarly.

Meanwhile, the self-energy for the contacts is given by

$$\Sigma^{(1)}_{p,q} = \begin{cases} -\dfrac{\hbar^2}{2m_{1/2,p}a^2}\omega_q^{(1,p)}, & \text{if } p \in \{1, \ldots, N_z\}, \\ 0, & \text{otherwise,} \end{cases} \tag{18.53}$$

where $m_{1/2,p} = m((x_0 + x_1)/2, z_p)$.

Truncating the infinite series for $\omega_q^{(1,j)}$ to a finite order M, we obtain the self-energy as

$$\Sigma^{(1)} = Q\Lambda Q^{\mathrm{T}}, \tag{18.54}$$

where

$$Q = \begin{bmatrix} \chi_1^{(1)}(z_1) & \chi_2^{(1)}(z_1) & \cdots & \chi_M^{(1)}(z_1) \\ \chi_1^{(1)}(z_2) & \chi_2^{(1)}(z_2) & \cdots & \chi_M^{(1)}(z_2) \\ \vdots & \vdots & \ddots & \vdots \\ \chi_1^{(1)}(z_{N_z}) & \chi_2^{(1)}(z_{N_z}) & \cdots & \chi_M^{(1)}(z_{N_z}) \\ 0 & 0 & \cdots & 0 \\ \vdots & \vdots & \ddots & \vdots \\ 0 & 0 & \cdots & 0 \end{bmatrix}_{N \times M} \tag{18.55}$$

and

$$\Lambda = -\frac{\hbar^2 b}{2m^{(1)}a^2}\mathrm{diag}\left(\exp\left(\mathrm{i}k_1^{(1)}a\right), \exp\left(\mathrm{i}k_2^{(1)}a\right), \ldots, \exp\left(\mathrm{i}k_M^{(1)}a\right)\right)_{M \times M}. \tag{18.56}$$

18.2.3 *Finite Element Methods for 2-D Device NEGFs*

The nodal shape function $\varphi_q(\mathbf{r})$, corresponding to the node $\mathbf{r}_q$, satisfies

$$\varphi_q(\mathbf{r}_{q'}) = \delta_{q,q'}. \tag{18.57}$$

The approximate Green's function, for a given source point $\mathbf{r}_{q'}$, can be written as

$$G_h(\mathbf{r}, \mathbf{r}_{q'}) = \sum_{q=1}^{N} G_{q,q'}\varphi_q(\mathbf{r}). \tag{18.58}$$

The weak form of (18.44) in the computational domain Ω_D for any test function $\varphi(\mathbf{r})$ is then given by

$$E\int_{\Omega_D} G_h\varphi\,\mathrm{d}\mathbf{r} - \int_{\Omega_D} VG_h\varphi\,\mathrm{d}\mathbf{r} - \frac{\hbar^2}{2}\int_{\Omega_D}\frac{1}{m}\nabla G_h\cdot\nabla\varphi\,\mathrm{d}\mathbf{r} \\ + \frac{\hbar^2}{2}\int_{\Gamma_D}\frac{1}{m}\frac{\partial G_h}{\partial\mathbf{n}}\varphi\,\mathrm{d}s = \varphi(\mathbf{r}_{q'}), \tag{18.59}$$

where $\mathbf{n}$ is the outward unit normal of Ω_D, and the source is located at $\mathbf{r}_{q'}$. By noting that Ω_D is the rectangular region shown in Fig. 18.3, the surface integral in (18.59) can be rewritten as

$$\frac{\hbar^2}{2}\int_0^L\left(\frac{1}{m}\frac{\partial G_h}{\partial x}\varphi\right)\Bigg|_{x=x_1}^{x=x_{N_x}}\mathrm{d}z + \frac{\hbar^2}{2}\int_{x_1}^{x_{N_x}}\left(\frac{1}{m}\frac{\partial G_h}{\partial z}\varphi\right)\Bigg|_{z=0}^{z=L}\mathrm{d}x. \tag{18.60}$$

The second integral in (18.60) is zero due to the homogeneous Dirichlet conditions, while the first one is as follows:

$$-\frac{\hbar^2}{2}\int_0^L\frac{1}{m(x_1,z)}\frac{\partial G_h(x_1,z,\mathbf{r}_{q'})}{\partial x}\varphi(x_1,z)\mathrm{d}z \\ + \frac{\hbar^2}{2}\int_0^L\frac{1}{m(x_{N_x},z)}\frac{\partial G_h(x_{N_x},z,\mathbf{r}_{q'})}{\partial x}\varphi(x_{N_x},z)\mathrm{d}z. \tag{18.61}$$

Again, the derivatives of Green's function at the boundary will be replaced by the boundary condition (18.42), which becomes after using the analytical expression of $g(\mathbf{r},\mathbf{r}_e')$ from (18.73)

$$\frac{\partial G(x_1^-,z_e',x',z')}{\partial x} = \int_0^L G(x_1,z,x',z')\sum_l \chi_l^{(1)}(z)\chi_l^{(1)}(z_e')(-\mathrm{i}k_l^{(1)})\mathrm{d}z, \tag{18.62}$$

which can be rewritten as

$$\frac{\partial G_h(x_1,z,\mathbf{r}_{q'})}{\partial x} = \hat{\Sigma}^{(1)}\cdot G_h(x_1,\tilde{z},\mathbf{r}_{q'}), \tag{18.63}$$

where

$$\hat{\Sigma}^{(1)}\cdot\varphi_q(x_1,\tilde{z}) = \frac{m(x_1,z)}{m^{(1)}}\int_0^L\varphi_q(x_1,\tilde{z})\sum_l\chi_l^{(1)}(\tilde{z})\chi_l^{(1)}(z)(-\mathrm{i}k_l^{(1)})\mathrm{d}\tilde{z}. \tag{18.64}$$

Using (18.63), the first integral in (18.61) can be written as

$$\begin{aligned}
&\frac{\hbar^2}{2}\int_0^L \frac{1}{m(x_1,z)}\frac{\partial G_h(x_1,z,\mathbf{r}_{q'})}{\partial x}\varphi(x_1,z)\mathrm{d}z \\
&= \frac{\hbar^2}{2}\int_0^L \frac{1}{m(x_1,z)}\left(\hat{\Sigma}^{(1)}\cdot G_h(x_1,\tilde{z},\mathbf{r}_{q'})\right)\varphi(x_1,z)\mathrm{d}z \\
&= \frac{\hbar^2}{2}\int_0^L \frac{1}{m(x_1,z)}\left[\hat{\Sigma}^{(1)}\cdot\left(\sum_{q=1}^{N} G_{q,q'}\varphi_q(x_1,\tilde{z})\right)\right]\varphi(x_1,z)\mathrm{d}z \\
&= \sum_{q=1}^{N} G_{q,q'}\frac{\hbar^2}{2}\int_0^L \frac{1}{m(x_1,z)}(\hat{\Sigma}^{(1)}\cdot\varphi_q(x_1,\tilde{z}))\varphi(x_1,z)\mathrm{d}z,
\end{aligned} \tag{18.65}$$

from which we see that the self-energy $\Sigma^{(1)}$ can be defined as

$$\Sigma^{(1)}_{p,q} = \frac{\hbar^2}{2}\int_0^L \frac{1}{m(x_1,z)}(\hat{\Sigma}^{(1)}\cdot\varphi_q(x_1,\tilde{z}))\varphi_p(x_1,z)\mathrm{d}z, \tag{18.66}$$

namely

$$\begin{aligned}
\Sigma^{(1)}_{p,q} &= \frac{\hbar^2}{2}\int_0^L \frac{1}{m(x_1,z)}\left(\frac{m(x_1,z)}{m^{(1)}}\int_0^L \varphi_q(x_1,\tilde{z})\right. \\
&\quad\times\left.\sum_l \chi_l^{(1)}(\tilde{z})\chi_l^{(1)}(z)(-\mathrm{i}k_l^{(1)})\mathrm{d}\tilde{z}\right)\varphi_p(x_1,z)\mathrm{d}z \\
&= \frac{\hbar^2}{2}\sum_l \frac{-\mathrm{i}k_l^{(1)}}{m^{(1)}}\left[\int_0^L \chi_l^{(1)}(z)\varphi_p(x_1,z)\mathrm{d}z\right]\left[\int_0^L \chi_l^{(1)}(\tilde{z})\varphi_q(x_1,\tilde{z})\mathrm{d}\tilde{z}\right].
\end{aligned} \tag{18.67}$$

In computation, the summation in (18.67) is truncated to a finite order and the device Green's function is defined by the 2-D version of (18.29).

18.3 NEGF Simulation of a 29 nm Double Gate MOSFET

The transport current for the NEGF is computed by the Tsu–Esaki/Landauer formula (16.158), where the self-energies for contacts for finite difference and finite element have been detailed in Sects. 18.1 and 18.2. The electron density $n(\mathbf{r})$ is given by the diagonal elements of the density matrix $\hat{\rho}_d$ in (16.213), which depends on the potential $V(\mathbf{r})$ in the device by the definition of the device Green's function (16.201). To account for the space charge effect, a self-consistent procedure with a Poisson equation for the potential will be needed, where the potential distribution is then determined by

$$-\nabla\cdot(\epsilon(\mathbf{r})\nabla V(\mathbf{r})) = q(-n(\mathbf{r}) + N_d(\mathbf{r})), \tag{18.68}$$

where $N_d(\mathbf{r})$ is the doping density, $\epsilon(\mathbf{r})$ is the dielectric constant, and q is the electron charge.

The geometry of a double gate MOSFET is shown in Fig. 18.4 [4]. The width of the device is assumed to be large, and the potential is invariant along the y-direction. The silicon layer is sandwiched by two symmetric oxide layers. The source and the drain are heavily doped.

Both finite difference and finite element methods are used to solve the 2-D coupled Poisson equation and NEGF. The meshes are shown in Fig. 18.3. The second-order central difference scheme and the linear finite element will be used in finite difference and finite element methods, respectively.

The self-consistent iteration solution is obtained as follows.

1. Start with an initial potential distribution $V(\mathbf{r}) = V_0$; let V_j be the resulting potential of the jth iteration, and we will compute V_{j+1}.
2. For a given energy E, solve Green's function $\mathbf{G}(E)$ and self-energies $\Sigma^{(\alpha)}(E)$ as discussed in Sects. 18.1 and 18.2 based on V_j, and then the spectral function $\mathbf{A}^{(\alpha)}(E)$ as defined in (16.215).
3. Calculate the electron density $n(\mathbf{r})$ by integrating the density matrix $\hat{\rho}_d$ defined in (16.213) with respect to energy E. It is noted that we need to repeat Step 2 for different sampling values of E for such an integration.
4. Insert the electron density $n(\mathbf{r})$ into the Poisson equation (18.68) to obtain a new potential, namely V_{j+1}.
5. Check $|V_{j+1} - V_j| < \varepsilon$ (a given stop accuracy): if yes, stop; otherwise go to Step 2.

Remark 18.2 Direct use of (18.68) leads to slow convergence. Instead, we will solve a nonlinear Poisson equation using a Newton method [4].

The Poisson equation is solved in the rectangular region ABCD including the silicon layer and the oxide layers with the boundary condition

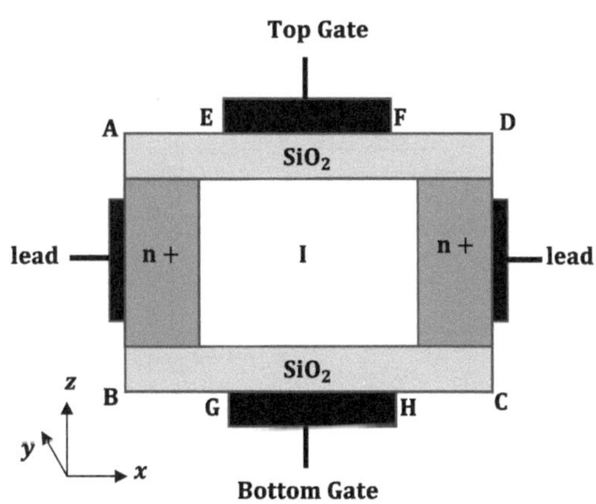

Fig. 18.4 An ultra-thin double gate MOSFET. The rectangle ABCD is the computational domain, which includes the $n+$ doped regions, two SiO_2 dielectric layers, and an intrinsic region (I)

$$\begin{cases} V(\mathbf{r}) = V_g, & \text{if } \mathbf{r} \in \text{EF, GH}, \\ \dfrac{\partial V(\mathbf{r})}{\partial \mathbf{n}} = 0, & \text{if } \mathbf{r} \in \text{AB, BG, HC, CD, DF, EA}, \end{cases} \tag{18.69}$$

where $\mathbf{n}$ is the outward normal of the rectangular region and V_g is the gate voltage. Here, electron penetration into the oxide regions is neglected, so the transport equation is considered only in the silicon layer, and the gate voltage V_g is imposed on gates EF and GH. The floating boundary condition, i.e., a homogeneous Neumann condition, maintains macroscopic space charge neutrality at the source (drain) end despite the biasing condition. The rectangular region is taken to be the computational domain for Green's function.

The potential distribution and the density under the gate bias $V_g = 0.4$ eV and the drain bias $V_{ds} = 0.4$ eV in the silicon layer can be found in [2, Figs. 12, 13].

18.4 Derivation of Green's Function in 2-D Strip-Shaped Contacts

The retarded Green's function in an infinite strip wire satisfies

$$\left(E - v^{(\alpha)}(z) + \frac{\hbar^2}{2m^{(\alpha)}}\nabla^2\right) g_0(\mathbf{r}, \mathbf{r}') = \delta(\mathbf{r} - \mathbf{r}'), \tag{18.70}$$

the solution of which reads, for $\mathbf{r} = (x, z)$ and $\mathbf{r}' = (x', z')$,

$$g_0(\mathbf{r}, \mathbf{r}') = \sum_l \frac{m^{(\alpha)}}{\mathrm{i}k_l^{(\alpha)}\hbar^2}\chi_l^{(\alpha)}(z)\chi_l^{(\alpha)}(z')\exp\left(\mathrm{i}k_l^{(\alpha)}|x - x'|\right), \tag{18.71}$$

where $\chi_l^{(\alpha)}(z)$ satisfies

$$\left(-\frac{\hbar^2}{2m^{(\alpha)}}\frac{\partial^2}{\partial z^2} + v^{(\alpha)}(z)\right)\chi_l^{(\alpha)}(z) = \lambda_l^{(\alpha)}\chi_l^{(\alpha)}(z) \tag{18.72}$$

and $k_l^\alpha = \sqrt{2m^{(\alpha)}(E - \lambda_l^{(\alpha)})/\hbar^2}$. Then, Green's function in a semi-infinite strip wire with a straight line boundary $x = d$ is as follows:

1. if $g(d, z, \mathbf{r}') = 0$, then

$$g(\mathbf{r}, \mathbf{r}') = g_0(x, z, \mathbf{r}') - g_0(2d - x, z, \mathbf{r}'); \tag{18.73}$$

2. if $\partial g(d, z, \mathbf{r}')/\partial x = 0$, then

$$g(\mathbf{r}, \mathbf{r}') = g_0(x, z, \mathbf{r}') + g_0(2d - x, z, \mathbf{r}'). \tag{18.74}$$

18.5 Summary

The transmission coefficient $T(E)$ of a nano-device, the key quantity in the Landauer transport theory for nano-devices, is related to the non-equilibrium Green's function, which can be computed by either finite element methods or finite difference methods. Treatment of the boundary condition at the input and output contacts of the device for both methods can be formulated in terms of a boundary Dirichlet-to-Neumann mapping, which reflects the effect of the contacts on the transport of the electron through the nano-devices. Simulation tools based on the NEGF for nano-electronics have been developed in [4].

References

1. S. Datta, *Electronic Transport in Mesoscopic Systems* (Cambridge University Press, Cambridge, 1997)
2. H. Jiang, S. Shao, W. Cai, P. Zhang, Boundary treatments in the non-equilibrium Green's functions (NEGF) methods for transport in nano-MOSFETs. J. Comput. Phys. **227**, 6553–6573 (2008)
3. P. Havu, V. Havu, M.J. Puska, R.M. Nieminen, Nonequilibrium electron transport in two-dimensional nanostructures modeled using Green's functions and the finite-element method. Phys. Rev. B **69**, 115325 (2004)
4. Z. Ren, R. Venugopal, S. Goasguen, S. Datta, M.S. Lundstrom, NanoMOS 2.5: a two-dimensional simulator for quantum transport in double-gate MOSFETs. IEEE Trans. Electron Dev. **50**, 1914–1925 (2003)

Chapter 19
Numerical Methods for Wigner Quantum Transport

As a kinetic approach, Wigner equations for quantum transport in nano-devices and their numerical solutions will be discussed in this chapter. First, we address the issues of the phase space truncations for the Wigner distributions in computational simulations and the Frensley inflow boundary conditions at the physical boundaries of the devices. Then, a conservative adaptive spectral element method based on cell averages will be given, followed by an upwinding finite difference method. Numerical results on a resonant tunneling diode (RTD) will be presented using both methods.

19.1 Wigner Equations for Quantum Transport

Let us consider the Wigner equation for the Wigner distribution $f(x, k, t)$ of a 1-D quantum device in the phase space $(x, k) \in \mathbb{R}^2$:

$$\frac{\partial}{\partial t} f(x, k, t) + \frac{\hbar k}{m} \frac{\partial}{\partial x} f(x, k, t) = \Theta_V[f](x, k, t), \tag{19.1}$$

where

$$\Theta_V[f](x, k, t) = -\frac{q}{2\pi\hbar} \int_{-\infty}^{\infty} V_w(x, k - k') f(x, k', t) \mathrm{d}k', \tag{19.2}$$

and the non-local Wigner potential $V_w(x, k)$ is defined in (16.107).

W. Cai, *Deterministic, Stochastic, and Deep Learning Methods for Computational Electromagnetics*, https://doi.org/10.1007/978-981-96-0100-4_19

19.1.1 Truncation of Phase Spaces and Charge Conservation

The definition of the Wigner potential $V_w(x,k)$ in (16.107) has no meaning when the Fourier transform of $[V(x+r/2) - V(x-r/2)]$ does not exist. However, as (16.100) implies that $\rho(x+r/2, x-r/2)$ is the inverse Fourier transform of the Wigner function $f(x,k)$, i.e.,

$$\rho\left(x+\frac{r}{2}, x-\frac{r}{2}\right) = \int_{-\infty}^{\infty} f(x,k')\exp(\mathrm{i}k'r)\mathrm{d}k'; \tag{19.3}$$

the right-hand side of (19.1) can be shown to be equivalent to

$$\int_{-\infty}^{\infty}\left[V\left(x+\frac{r}{2}\right) - V\left(x-\frac{r}{2}\right)\right]\rho\left(x+\frac{r}{2}, x-\frac{r}{2}\right)\exp(-\mathrm{i}kr)\mathrm{d}r. \tag{19.4}$$

As in many cases it is reasonable to assume finite coherence length in the density operator, i.e.,

$$\rho\left(x+\frac{r}{2}, x-\frac{r}{2}\right) \to 0, \quad \text{as} \quad r\to\infty,$$

so the Fourier transform in (19.4) is then well defined. For numerical calculations, it is necessary to truncate the infinite integration domain in (19.4) to a finite coherence interval $[-L_{\mathrm{coh}}, L_{\mathrm{coh}}]$, i.e.,

$$\int_{-L_{\mathrm{coh}}}^{L_{\mathrm{coh}}}\left[V\left(x+\frac{r}{2}\right) - V\left(x-\frac{r}{2}\right)\right]\rho\left(x+\frac{r}{2}, x-\frac{r}{2}\right)\exp(-\mathrm{i}kr)\mathrm{d}r, \tag{19.5}$$

which defines an effective Wigner potential for (19.1):

$$\hat{V}_w^{\mathrm{eff}}(x,k) = 2\int_0^{L_{\mathrm{coh}}}\sin(ky)\left[V\left(x+\frac{y}{2}\right) - V\left(x-\frac{y}{2}\right)\right]\mathrm{d}y. \tag{19.6}$$

Moreover, we only need to compute the Wigner distribution for $|k| < L_k/2$ for some large value L_k; therefore, we will zero out the distribution function $f(x,k) = 0$ if $|k| > L_k/2$, and thus (19.1) is only solved for $|k| < L_k/2$ in the following modified form. For $(x,k) \in \mathcal{X}\times\mathcal{K}$,

$$\frac{\partial}{\partial t}f(x,k,t) + \frac{\hbar k}{m}\frac{\partial}{\partial x}f(x,k) + \frac{q}{2\pi\hbar}\int_{-L_k/2}^{L_k/2}\hat{V}_w^{\mathrm{eff}}(x,k-k')f(x,k')\mathrm{d}k' = 0, \tag{19.7}$$

where $\mathcal{X}$ is the computational domain in x-space and $\mathcal{K} = [-L_k/2, L_k/2]$.

Next, the Wigner potential $\hat{V}_w^{\mathrm{eff}}(x,k)$ will be replaced by a discretized version, say a N_{coh} point trapezoidal rule with a spacing h_{coh} as follows:

$$\hat{V}_w^{\text{eff}}(x,k) \approx V_w^h(x,k) = 2h_{\text{coh}} \sum_{\mu=1}^{N_{\text{coh}}} \sin(ky_\mu) \left[V\left(x + y_\mu/2\right) - V\left(x - y_\mu/2\right) \right], \tag{19.8}$$

where

$$N_{\text{coh}} h_{\text{coh}} = L_{\text{coh}}, \quad y_\mu = \mu h_{\text{coh}}, \quad \mu = 1, 2, \ldots, N_{\text{coh}}. \tag{19.9}$$

In order to keep the charge conservation, we require

$$\int_{\mathcal{K}} \mathrm{d}k \int_{\mathcal{K}} \mathrm{d}k' \; V_w^h(x, k-k') f(x, k', t) = 0, \tag{19.10}$$

for $V_w^h(x,k)$ with $k \in \mathcal{K}$. Frensley [1, 2] suggested a sufficient condition

$$|\mathcal{K}| h_{\text{coh}} = 2\pi, \tag{19.11}$$

which guarantees that $V_w^h(x,k)$ is not only odd, but also periodic in k with a period $|\mathcal{K}|$. Equation (19.10) can then be easily verified by noting that, with the condition (19.11),

$$\begin{aligned} &\int_{\mathcal{K}} \sin[(k-k')y_\mu] \mathrm{d}k = \frac{\cos[y_\mu(k_{\min} - k')] - \cos[y_\mu(k_{\max} - k')]}{y_\mu} \\ &= \frac{1}{\mu h_{\text{coh}}} \left(\cos\left[\mu h_{\text{coh}}(k_{\min} - k')\right] - \cos\left[\mu h_{\text{coh}}(k_{\min} - k') + \mu h_{\text{coh}} |\mathcal{K}|\right] \right) = 0. \end{aligned} \tag{19.12}$$

19.1.2 Frensley Inflow Boundary Conditions

The Wigner equation only contains a first-order spatial x-derivative, and therefore we need to specify only one boundary condition in x-space. For instance, we can use the following inflow boundary conditions [1]:

$$\begin{aligned} f(x_{\min}, k, t) &= f_L(k), \quad \text{if} \quad k > 0, \\ f(x_{\max}, k, t) &= f_R(k), \quad \text{if} \quad k < 0. \end{aligned} \tag{19.13}$$

For open systems, as free electrons are supposed to be injected from infinity, their dispersion relation in the left contact $E = E(k)$ is given by

$$E(k_1) = \frac{\hbar^2 k_1^2}{2m} + v_1, \tag{19.14}$$

and a similar dispersion relation holds for the electrons in the right contact.

The left boundary condition for the Wigner function will be

$$f_L(k) = \frac{mk_BT}{\pi\hbar^2}\log\left[1+\exp\left(\frac{\mu_s - \frac{\hbar^2k^2}{2m} - v_1}{k_BT}\right)\right], \qquad k > 0, \tag{19.15}$$

and the right boundary condition is given by

$$f_R(k) = \frac{mk_BT}{\pi\hbar^2}\log\left[1+\exp\left(\frac{\mu_d - \frac{\hbar^2k^2}{2m} - v_2}{k_BT}\right)\right], \qquad k < 0, \tag{19.16}$$

where μ_s and μ_d are the Fermi energies of the left contact and the right contact, and v_1 and v_2 are the external bias potentials of the left contact and the right contact, respectively.

19.2 Adaptive Spectral Element Method (SEM)

We will present an adaptive conservative SEM [3], which uses cell averages centered at Gauss–Chebyshev points in k-space and Gauss–Lobatto collocation points in x-space. The computational domain $\mathcal{X} \times \mathcal{K}$ is divided into $Q \times R$ non-overlapping elements (sub-domains) as follows:

$$\mathcal{X} \times \mathcal{K} = \bigcup_{q=1}^{Q}\bigcup_{r=1}^{R} \mathcal{X}_q \times \mathcal{K}_r, \quad \mathcal{X}_q = [g_{q-1}, g_q], \quad \mathcal{K}_r = [d_{r-1}, d_r], \tag{19.17}$$

with $g_0 = x_{\min}$, $g_Q = x_{\max}$, $d_0 = k_{\min}$, and $d_R = k_{\max}$. In an element $\mathcal{X}_q \times \mathcal{K}_r$ ($q = 1, 2, \ldots, Q$ and $r = 1, 2, \ldots, R$), the collocation points are $\{(x_{i;q,r}, k_{j;q,r})\}$, with $i = 0, 1, \ldots, M_{q,r}$ and $j = 1, \ldots, N_{q,r}$, so we have $(M_{q,r} + 1) \times N_{q,r}$ collocation points. Here, we set $x_{i;q,r}$ to be the Gauss–Lobatto points and $k_{j;q,r}$ to be the Gauss–Chebyshev points in (5.33) (5.34) to use the fast Fourier transforms [4]:

$$\begin{aligned} &\text{(Gauss–Lobatto)} \ x_{i;q,r} = g_{q-1} + \frac{G_q}{2}\left(1 + \cos\frac{i\pi}{M_{q,r}}\right), \\ &\text{(Gauss–Chebyshev)} \ k_{j;q,r} = d_{r-1} + \frac{D_r}{2}\left[1 + \cos\left(j - \frac{1}{2}\right)\frac{\pi}{N_{q,r}}\right], \end{aligned} \tag{19.18}$$

where $G_q = g_q - g_{q-1}$ and $D_r = d_r - d_{r-1}$. Denote by $f_{q,r}(x, k, t)$ the restriction of the Wigner function $f(x, k, t)$ on the element $\mathcal{X}_q \times \mathcal{K}_r$, and $f_{j,i;q,r}(t) := f_{q,r}(x_{i;q,r}, k_{j;q,r}, t)$.

A non-adaptive SEM means choosing the same $(M_{q,r}, N_{q,r})$ for all (q, r)-elements, i.e., $M_{q,r}$ and $N_{q,r}$ are two constants, while an adaptive SEM allows different $(M_{q,r}, N_{q,r})$ in different elements. The total number of unknowns is denoted by $\mathcal{N}$:

$$\mathcal{N} = \sum_{q=1}^{Q} \sum_{r=1}^{R} (M_{q,r} + 1) \times N_{q,r}. \tag{19.19}$$

We will take the element $\mathcal{X}_q \times \mathcal{K}_r$ as an example to illustrate the adaptive conservative cell average SEM, and the subscripts q and r for the (q, r)-element under consideration will be dropped from $x_{i;q,r}$, $k_{j;q,r}$, $M_{q,r}$, $N_{q,r}$, $f_{q,r}(x, k, t)$, and $f_{j,i;q,r}(t)$.

19.2.1 Cell Averages in k-Space

A cell I_j in k-space is given as (see Fig. 19.1)

$$I_j = [k_{j+1/2}, k_{j-1/2}], \tag{19.20}$$

with

$$k_{j\mp 1/2} = d_{r-1} + \frac{D_r}{2}\left[1 + \cos\left(j - \frac{1}{2} \mp \frac{1}{2}\right)\frac{\pi}{N}\right]. \tag{19.21}$$

Obviously, we have $\mathcal{K}_r = \bigcup_{j=1}^{N} I_j$. We define local quantities corresponding to $\mathrm{n}(x, t)$, $\mathrm{j}(x, t)$, and $\mathrm{p}(x, t)$ in each computational cell as follows:

$$\mathrm{n}_j(x, t) = \int_{I_j} f(x, k, t)\mathrm{d}k, \tag{19.22}$$

$$\mathrm{j}_j(x, t) = \frac{\hbar}{m}\int_{I_j} kf(x, k, t)\mathrm{d}k, \tag{19.23}$$

$$\mathrm{p}_j(x, t) = \frac{q}{2\pi\hbar}\int_{I_j}\int_{\mathcal{K}} V_w^h(x, k - k')f(x, k', t)\,\mathrm{d}k'\,\mathrm{d}k. \tag{19.24}$$

Then, a local continuity equation for the cell I_j is defined as

$$\frac{\partial}{\partial t}\mathrm{n}_j(x, t) + \frac{\partial}{\partial x}\mathrm{j}_j(x, t) + \mathrm{p}_j(x, t) = 0. \tag{19.25}$$

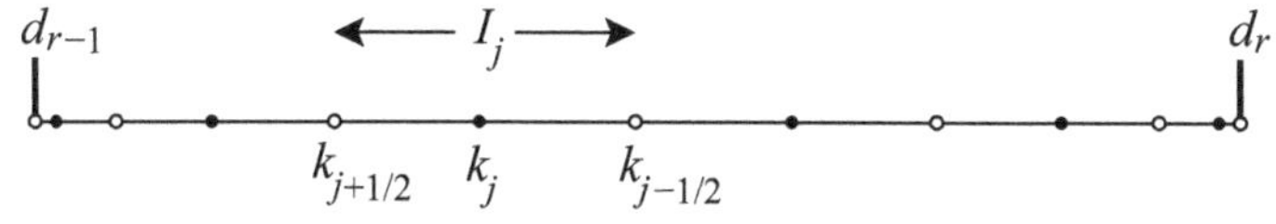

Fig. 19.1 The k-mesh in the sub-domain $\mathcal{K}_r = [d_{r-1}, d_r]$. The black points are the Gauss–Chebyshev collocation points, denoted by k_j. The ends of cells are $k_{j\mp 1/2}$ displayed in small circles. Reprinted from [3], copyright (2011) by Global Science, Ltd.

From (19.22)–(19.24), we can see that both cell averages and point values of the Wigner function $f(x,k,t)$, expressed in terms of the Chebyshev polynomials of the k-variable over each cell, are involved in (19.25). Fortunately, in the Chebyshev polynomial space, there is an analytical relation between the expansions for $\mathrm{n}_j(x,t)$, $\mathrm{j}_j(x,t)$, and $\mathrm{p}_j(x,t)$ as shown below.

For $k \in \mathcal{K}_r$, $\eta \in [-1,1]$, and $\theta \in [0,\pi]$, we use the following transform:

$$k = d_{r-1} + \frac{D_r}{2}(1+\eta), \quad \eta = \cos\theta, \tag{19.26}$$

to define

$$C_l(k) = T_l(\eta) = \cos l\theta, \quad S_l(k) = \sin(l+1)\theta, \quad l = 0,1,\ldots,N-1, \tag{19.27}$$

where $T_l(\eta)$ is the Chebyshev polynomial of the first kind. Then, we have a spectral approximation

$$f(x,k,t) \approx \tilde{f}(x,k,t) = \sum_{l=0}^{N-1} a_l(x,t)C_l(k), \quad k \in \mathcal{K}_r. \tag{19.28}$$

Consequently, plugging the approximation for the Wigner function into (19.22)–(19.24), we have spectral approximations for n_j, j_j, and p_j, which are denoted by $\tilde{\mathrm{n}}_j, \tilde{\mathrm{j}}_j$, and $\tilde{\mathrm{p}}_j$, respectively.

We proceed by substituting the above expansion (19.28) into (19.22) and (19.23), and using the following two identities for $\theta, \theta \pm \Delta\theta/2 \in [0,\pi]$, which relate the cell averages of the first kind of the Chebyshev polynomials to the second kind [5]:

$$\int_{\cos(\theta-\Delta\theta/2)}^{\cos(\theta+\Delta\theta/2)} d\eta\, T_l(\eta) = \begin{cases} -\frac{\sin\Delta\theta}{2}\sin 2\theta, & l=1,\\ \frac{\sin\frac{(l-1)\Delta\theta}{2}}{l-1}\sin(l-1)\theta - \frac{\sin\frac{(l+1)\Delta\theta}{2}}{l+1}\sin(l+1)\theta, & l\neq 1, \end{cases}$$

$$\int_{\cos(\theta-\Delta\theta/2)}^{\cos(\theta+\Delta\theta/2)} d\eta\, \eta T_l(\eta) = \begin{cases} -\frac{\sin 2\Delta\theta}{8}\sin 4\theta, & l=2,\\ \frac{\sin\frac{(l-2)\Delta\theta}{2}}{2(l-2)}\sin(l-2)\theta - \frac{\sin\frac{(l+2)\Delta\theta}{2}}{2(l+2)}\sin(l+2)\theta, & l\neq 2, \end{cases}$$

and with (19.22), (19.23), and (19.27) we obtain

$$\tilde{\mathrm{n}}_j(x,t) = \sum_{l=0}^{N-1} a_l(x,t)\left[\int_{I_j} C_l(k)\mathrm{d}k\right] = -\sum_{l=0}^{N-1} b_l(x,t)S_l(k_j), \tag{19.29}$$

$$\tilde{\mathrm{j}}_j(x,t) = \frac{\hbar}{m}\sum_{l=0}^{N-1} a_l(x,t)\left[\int_{I_j} kC_l(k)\mathrm{d}k\right] = -\frac{\hbar}{m}\sum_{l=0}^{N} c_l(x,t)S_l(k_j). \tag{19.30}$$

Here, the expansion coefficients b_l and c_l can be expressed in terms of a_l recursively as follows:

$$a_l = \begin{cases} -b_l/\sigma_l, & l = N-1, N-2, \\ a_{l+2} - b_l/\sigma_l, & l = N-3, \dots, 1, \\ \frac{1}{2}(a_2 - b_0/\sigma_0), & l = 0, \end{cases} \tag{19.31}$$

$$b_l = \begin{cases} \sigma_0(a_2 - 2a_0), & l = 0, \\ \sigma_l(a_{l+2} - a_l), & l = 1, 2, \dots, N-3, \\ -\sigma_l a_l, & l = N-2, N-1, \end{cases} \tag{19.32}$$

$$c_l = \begin{cases} \bar{d}_r b_0 + \frac{D_r \sigma_0}{4}(a_3 - a_1), & l = 0, \\ \bar{d}_r b_1 + \frac{D_r \sigma_1}{4}(a_4 - 2a_0), & l = 1, \\ \bar{d}_r b_l + \frac{D_r \sigma_l}{4}(a_{l+3} - a_{l-1}), & l = 2, 3, \dots, N-4, \\ \bar{d}_r b_l - \frac{D_r \sigma_l}{4} a_{l-1}, & l = N-3, N-2, N-1, \\ -\frac{D_r \sigma_N}{4} a_{N-1}, & l = N, \end{cases} \tag{19.33}$$

where $\bar{d}_r$ denotes the center of $\mathcal{K}_r$, and

$$\sigma_l = \frac{D_r \sin[(l+1)\frac{\pi}{2N}]}{2(l+1)}, \quad l = 0, 1, \dots, N. \tag{19.34}$$

It is easily seen that we can transform $b_l(x,t)$ to $\tilde{\mathrm{n}}_j(x,t)$ in (19.29) via a fast sine transform (FST), and $c_l(x,t)$ to $\tilde{\mathrm{j}}_j(x,t)$ in (19.30) via another FST.

Meanwhile, substituting (19.28) into (19.24) and carrying out some careful algebraic calculations, we can show that $\mathrm{p}_j(x,t) \approx \tilde{\mathrm{p}}_j(x,t)$, with

$$\begin{aligned} \tilde{\mathrm{p}}_j(x,t) =& \frac{q\Delta y}{\pi^2 \hbar} \sum_{\mu=1}^{N_y} [V(x+y_\mu) - V(x-y_\mu)] \sum_{r'=1}^{R} \int_{\mathcal{K}_{r'}} \mathrm{d}k' \\ & \left\{ \int_{I_j} \sin[2y_\mu(k-k')]\mathrm{d}k \right\} \sum_{l=0}^{N_{q,r'}-1} a_{l;q,r'}(x,t) C_{l;q,r'}(k') \\ =& -\frac{q\Delta y}{\pi^2 \hbar} \sum_{\mu=1}^{N_y} \sin(y_\mu \Delta k_j) \chi(x, y_\mu) \sum_{r'=1}^{R} D_{r'} \\ & \times \left\{ \sin[2y_\mu(\bar{k}_j - \bar{d}_{r'})] \mathrm{Re}[\lambda(x, y_\mu, t, r')] \right. \\ & \left. - \cos[2y_\mu(\bar{k}_j - \bar{d}_{r'})] \mathrm{Im}[\lambda(x, y_\mu, t, r')] \right\}, \end{aligned} \tag{19.35}$$

where

$$\chi(x, y_\mu) = \frac{V(x+y_\mu) - V(x-y_\mu)}{2y_\mu},$$

$$\lambda(x, y, t, r') = \sum_{l=0}^{N_{q,r'}-1} a_{l;q,r'}(x,t) O_l(D_{r'} y), \tag{19.36}$$

$\bar{k}_j$ is the middle point of the cell I_j, and we have used the spectral expansion in the element $\mathcal{X}_q \times \mathcal{K}_{r'}$, i.e.,

$$f_{q,r'}(x,k,t) \approx \sum_{l=0}^{N_{q,r'}-1} a_{l;q,r'}(x,t) C_{l;q,r'}(k). \tag{19.37}$$

Here r' comes from the integral with respect to k' in the sub-domain $\mathcal{K}_{r'}$, $\Delta k_j = k_{j-1/2} - k_{j+1/2}$, and $O_l(z)$ is an oscillatory integral given in (19.66) of Sect. 19.4.

In order to determine the expansion coefficients a_l in (19.28), we solve the approximated local continuity equation

$$\frac{\partial}{\partial t}\tilde{\mathrm{n}}_j(x,t) + \frac{\partial}{\partial x}\tilde{\mathrm{j}}_j(x,t) + \tilde{\mathrm{p}}_j(x,t) = 0, \tag{19.38}$$

with the spectral approximations (19.29), (19.30), and (19.35). It is noted that there are only spectral errors associated with the Chebyshev polynomial expansion of $f(x,k,t)$ in (19.28), since all the integrals in (19.22), (19.23), and (19.24) are calculated analytically.

Remark 19.1 (*Exact charge conservation*) The cell equation (19.38) involves the cell averages of the Wigner function, the cell fluxes, involving the point values of $f(x,k,t)$ over the whole cell, and the local Wigner potential term p_j, involving $f(x,k,t)$ and the Wigner potential, where all integrals are carried out exactly. If our primary unknowns are selected to be the cell averages $\tilde{\mathrm{n}}_j(x,t)$, such an exact calculation is only possible if the distribution function $f(x,k,t)$ is represented by a global (Chebyshev) polynomial in the domain $\mathcal{K}_r$ due to the analytical relation between the cell averages of $\tilde{f}(x,k,t)$ ($\tilde{\mathrm{n}}_j(x,t)$) and the point values without numerical errors. As a result, we can sum all cell equations for $\tilde{f}(x,k,t)$, and the summation of $\tilde{\mathrm{p}}_j(x,t)$ for all elements will be zero, i.e., for all x, we have

$$\Theta(x,t) = \sum_{r=1}^{R}\sum_{j=1}^{N_{q,r}} \tilde{\mathrm{p}}_{j;q,r}(x,t) = \int_{\mathcal{K}} \mathrm{d}k \int_{\mathcal{K}} \mathrm{d}k' \; V_w^h(x,k-k')\tilde{f}(x,k',t) = 0, \tag{19.39}$$

thanks to (19.12). Therefore, we can see that the cell average SEM is capable of maintaining the charge conservation exactly for the spectral solution $\tilde{f}(x,k,t)$ in a non-uniform mesh.

19.2.2 Chebyshev Collocation Methods in *x*-Space

After forming the above conservative cell average SEM in k-space, we will solve the local continuity equation (19.38) to obtain the expansion coefficients in (19.28) by using a traditional collocation SEM with the Gauss–Lobatto points in x-space for easy implementation of boundary conditions and fast cosine transforms.

For $x \in \mathcal{X}_q$, $\eta \in [-1, 1]$, and $\theta \in [0, \pi]$, the transform

$$x = g_{q-1} + \frac{G_q}{2}(1+\eta), \quad \eta = \cos\theta,$$

is used to define

$$\phi_\nu(x) = T_\nu(\eta) = \cos\nu\theta, \quad \nu = 0, 1, \ldots, M. \tag{19.40}$$

Then, we have the spectral expansion for the coefficients in (19.28) as

$$a_l(x,t) \approx \sum_{\nu=0}^{M} \beta_{l,\nu}(t)\phi_\nu(x), \quad x \in \mathcal{X}_q, \tag{19.41}$$

where $\beta_{l,\nu}$ are the expansion coefficients. Based on such an expansion, we can obtain the first derivative by a recurrence [4, 6] with $O(M)$ operations. Namely, if a function is expressed in terms of the Chebyshev polynomials, then its first derivative can be obtained directly as [7, 8]

$$\frac{\partial a_l(x,t)}{\partial x} \approx \sum_{\nu=0}^{M} \tilde{\beta}_{l,\nu}(t)\phi_\nu(x), \quad x \in \mathcal{X}_q, \tag{19.42}$$

where

$$\tilde{\beta}_{l,\nu}(t) = \frac{2}{G_q} \times \begin{cases} 0, & \nu = M, \\ 2M\beta_{l,M}(t), & \nu = M-1, \\ \tilde{\beta}_{l,\nu+2}(t) + 2(\nu+1)\beta_{l,\nu+1}(t), & \nu = M-2, \ldots, 1, \\ \frac{1}{2}\tilde{\beta}_{l,2}(t) + \beta_{l,1}(t), & \nu = 0. \end{cases} \tag{19.43}$$

Hence, we could obtain the x-spatial derivative term via a fast cosine transform (FCT) and a recurrence, and the total cost is $O(M \log M)$.

19.2.3 Time Discretization

For the time discretization, we employ explicit multi-step Runge–Kutta methods. If a system of ordinary differential equations is given in a compact operator form,

$$\frac{\mathrm{d}}{\mathrm{d}t}U(t) = L(U), \tag{19.44}$$

then the fourth-order Runge–Kutta scheme is given by [9]

$$\begin{aligned} U^{(1)} &= U^n + \frac{1}{2}\Delta t L(U^n), \\ U^{(2)} &= U^n + \frac{1}{2}\Delta t L(U^{(1)}), \\ U^{(3)} &= U^n + \Delta t L(U^{(2)}), \\ U^{n+1} &= \frac{1}{3}\left[U^{(1)} + 2U^{(2)} + U^{(3)} - U^n + \frac{1}{2}\Delta t L(U^{(3)})\right]. \end{aligned} \tag{19.45}$$

Let $t^n = n\Delta t$, $n = 0, 1, 2, \ldots$, and $f^n_{j,i} \equiv f_{j,i}(t^n)$. The time step size Δt is restricted by the Courant–Friedrichs–Lewy (CFL) condition as follows:

$$\frac{\Delta t}{\min\limits_i\{\Delta x_i\}} \le \frac{m}{\hbar \max\limits_{k\in\mathcal{K}}\{|k|\}}, \tag{19.46}$$

where $\Delta x_i = |x_{i+1} - x_i|$. After discretization in both k-space and x-space, we have the spectral element approximation at the time step t^n:

$$f^n_{j,i} \approx \tilde{f}^n_{j,i} = \sum_{l=0}^{N-1} a^n_{l,i} C_l(k_j) = \sum_{l=0}^{N-1}\sum_{\nu=0}^{M} \beta^n_{l,\nu}\phi_\nu(x_i) C_l(k_j). \tag{19.47}$$

Once the coefficients $\beta^n_{l,\nu}$ are obtained, we are able to compute the Wigner function at any position (x, k) in the element $\mathcal{X}_q \times \mathcal{K}_r$ at t^n through the global spectral approximation

$$f(x, k, t_n) \approx \sum_{l=0}^{N-1}\sum_{\nu=0}^{M} \beta^n_{l,\nu}\phi_\nu(x) C_l(k). \tag{19.48}$$

When evolving from t^n to t^{n+1}, we need boundary conditions in $\mathcal{X}_q \times \mathcal{K}_r$. These boundary conditions are given according to the inflow rule, from the solution in the adjacent elements at t^n, i.e.,

(a) if $k < 0$, then

$$f_{q,r}(g_q, k, t^{n+1}) = \begin{cases} f_{q+1,r}(g_q, k, t^n), & 1 \le q < Q, \\ f_R(k, t^n), & q = Q; \end{cases} \tag{19.49}$$

(b) if $k > 0$, then

$$f_{q,r}(g_{q-1}, k, t^{n+1}) = \begin{cases} f_{q-1,r}(g_{q-1}, k, t^n), & 1 < q \le Q, \\ f_L(k, t^n), & q = 1. \end{cases} \tag{19.50}$$

Here, we set $k = 0$ to be the end point of an element, so $k = 0$ is not a collocation point.

19.2.4 Adaptive Meshes for Wigner Distributions

Following [10, 11], we simulate the motion of a Gauss wave packet (GWP) to investigate the performance of the SEM [3]. The GWP in free space is given by

$$f(x, k, t) = 2\exp\left[-\frac{(x - x_0 - v_0 t)^2}{2a^2(1 + \beta^2 t^2)}\right] \times \exp\left\{-2a^2(1 + \beta^2 t^2)\left[(k - k_0) - \frac{\beta t(x - x_0 - v_0 t)}{2a^2(1 + \beta^2 t^2)}\right]^2\right\}, \quad (19.51)$$

where x_0 is the center of the GWP at $t = 0$, a is the minimum position spread, $v_0 = \hbar k_0/m$ is the average velocity, and $\beta = \hbar/(2ma^2)$. The kinetic energy of such a GWP is $E_0 = \hbar^2 k_0^2/(2m)$. Actually, the GWP given in (19.51) is the analytical solution to the Wigner equation without a Wigner potential [11, 12]. In the numerical simulations, we take $a = 2.825$ and $m = 0.0665 m_e$.

The distributions and the corresponding meshes at different instants for the case $H = 1.3$ are shown in Fig. 19.2, from which we conclude that the p-adaptive methods capture the movement of the GWP accurately.

19.3 Upwinding Finite Difference Scheme

Next, let us consider a finite difference method for the time-independent Wigner equation in $(x, k) \in [-L/2, L/2] \times [-L_k/2, L_k/2]$. We set h_x as the mesh size of x-space, and let N be the number of mesh points

$$h_x = \frac{L}{N}, \quad x_i = -\frac{L}{2} + i h_x, \quad i = 0, 1, 2, \ldots, N. \quad (19.52)$$

From (19.1), only the boundary condition in x-space is required. An upwinding difference method can be used to approximate the first-order derivative in x. Let N_{coh} denote the mesh points in $[0, L_{\text{coh}}/2]$ with a spacing $h_{\text{coh}} = L_{\text{coh}}/N_{\text{coh}}$. Similarly, L_k is the integration length in (19.7), N_k is the number of mesh points in k-space, and $h_k = L_k/N_k$ is the corresponding mesh spacing. In order to avoid $k = 0$, which would lead to a zero diagonal element in the discretization matrix, we choose the mesh points as $k_j = L_k/2 - (j + 1/2) h_k, \ j = 0, 1, \ldots, N_k - 1$.

Using a middle point formula for the integration with respect to k' in (19.7) and a first-order upwinding finite difference scheme for the spatial derivative, we arrive

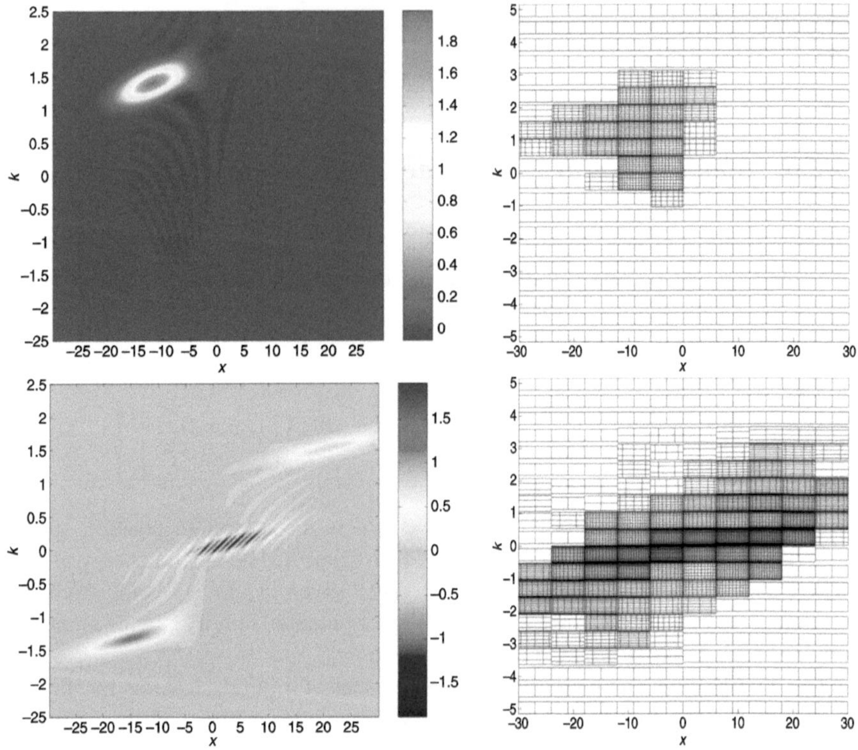

Fig. 19.2 Calculated Wigner functions with corresponding meshes in (x, t) phase space at $t = 7.5$ (top figures) and $t = 20$ (bottom figures). The GWP of energy level $E_0 \approx 1.12$, interacting with a Gauss barrier with a barrier height $H = 1.3$, is separated into two wave packets: one traveling across the barrier while the other being reflected by the barrier. Reprinted from [3], copyright (2011) by Global Science, Ltd.

at the finite difference equation at (x_i, k_j):

$$\frac{\hbar k_j}{m_x}\frac{f(x_i,k_j)-f(x_{i-1},k_j)}{h_x}+\frac{qh_k}{2\pi\hbar}\sum_{j'=0}^{N_k-1}\hat{V}_w(x_i,k_j-k'_{j'})f(x_i,k'_{j'})=0,\ k_j>0,$$
$$\frac{\hbar k_j}{m_x}\frac{f(x_{i+1},k_j)-f(x_i,k_j)}{h_x}+\frac{qh_k}{2\pi\hbar}\sum_{j'=0}^{N_k-1}\hat{V}_w(x_i,k_j-k'_{j'})f(x_i,k'_{j'})=0,\ k_j<0,$$
(19.53)

for $i = 1, 2, \ldots, N-1$, $j = 0, 1, \ldots, N_k - 1$. Here, $\hat{V}_w(x_i, k_j - k'_{j'})$ is calculated in (19.8) by a N_{coh}-point trapezoidal rule with a spacing h_{coh}, which in practice can be taken as $h_{\text{coh}} = 2h_x$.

Meanwhile, the current density can be simply computed as

$$j\left(x+\frac{h_x}{2}\right)=\frac{h_k}{2\pi}\left[\sum_{k_j<0}\frac{\hbar k_j}{m}f(x+h_x,k_j)+\sum_{k_j>0}\frac{\hbar k_j}{m}f(x,k_j)\right]. \tag{19.54}$$

The above definition ensures that the current density calculated for the steady-state solution is independent of x. In Sect. 19.3.1, we will address the issue of how to choose L_k and L_{coh} to ensure the conservation of the electrons and applicability of discrete fast Fourier transforms.

19.3.1 Selections of L_{coh}, N_{coh}, L_k, and N_k

Firstly, for the conservation of electrons in k-space, we have from (19.11) that

$$L_k h_{\text{coh}} = 2\pi. \tag{19.55}$$

Secondly, as L_{coh} is the coherence length in the density operator which defines the effective Wigner potential (19.6), the latter is discretized into a sum (19.8). As the continuous Fourier transform is changed to a discrete one, in order to take advantage of the fast discrete Fourier transform to evaluate the sum (19.8), we like to have the following equality:

$$k_j y_\mu = j\mu\frac{h_k L_{\text{coh}}}{N_{\text{coh}}} = j\mu\frac{2\pi}{N_{\text{coh}}};$$

namely, we require that

$$L_{\text{coh}} h_k = 2\pi. \tag{19.56}$$

In summary, the values of L_k and L_{coh} and the corresponding discretization mesh points N_k and N_{coh} are related by

$$N_k = \frac{L_k}{h_k} = \frac{L_{\text{coh}}}{h_{\text{coh}}} = N_{\text{coh}}. \tag{19.57}$$

It is important to satisfy (19.55), (19.56), and (19.57) in solving the Wigner equation numerically with a finite difference scheme.

19.3.2 Self-consistent Algorithm Through the Poisson Equation

So far, we have introduced two quantum transport models: the NEGF and the Wigner equation. For a 1-D device imposed with a bias potential V, the self-consistent potential energy $v_s(x)$ is related to the static potential $\nu(x)$ by $v_s(x) = q\nu(x)$, which should be determined by a Poisson equation,

$$-\frac{\partial}{\partial x}\left(\epsilon(x)\frac{\partial}{\partial x}\right)\nu(x) = q\left(-n(x) + N_d(x)\right), \tag{19.58}$$

with a Dirichlet boundary condition at X_1 and X_2:

$$\nu(X_1) = 0, \ \nu(X_2) = -V,$$

where $\epsilon(x)$ is the dielectric constant and $N_d(x)$ is the doping density. The intrinsic Fermi energy level is always set to zero as the reference energy.

- **The Wigner equation method with the Poisson equation**

Given an error tolerance $\varepsilon > 0$:

1. guess an initial potential function $\nu(x)$;
2. solve the transport equation with the Wigner equation method (19.7) to calculate the electron density $n(x)$ with the formula (16.118);
3. solve the Poisson equation (19.58) with a Newton iteration method—with the updated potential $\nu(x)$; repeat Steps 2 and 3 until the potential distribution $\nu(x)$ is convergent within the given error tolerance ε;
4. solve the Wigner equation (19.1) with the convergent potential $\nu(x)$, and calculate the current by (19.54).

19.3.3 Currents in RTD by NEGF and Wigner Equations

The RTD structure [13, 14] as shown in Fig. 19.3 is the classical 1-D hetero-structure device with a negative resistance.

Two thin AlGaA layers are sandwiched by GaAs layers to form two energy barriers and one quantum well. For a space-dependent effective mass in the RTD, the 1-D Schrödinger equation should be

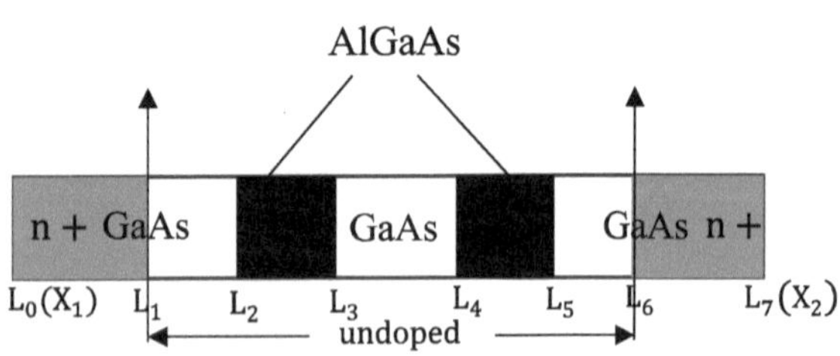

Fig. 19.3 Resonant tunneling diode structure with contacts doped

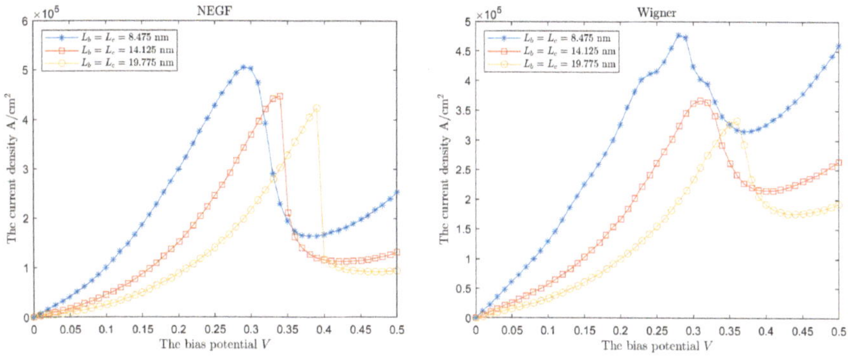

Fig. 19.4 The I–V curves by the NEGF and Wigner function for different contact length L_c and barrier thickness L_b

$$-\frac{\hbar^2}{2}\frac{\partial}{\partial x}\left(\frac{1}{m(x)}\frac{\partial}{\partial x}\right)\psi(x) + v(x)\psi(x) = E\psi(x), \tag{19.59}$$

where the effective mass of GaAs is $m_{\text{GaAs}} = 0.067m_0$, $m_{\text{AlGaAs}} = 0.0919m_0$ for AlGaAs, and m_0 is the electron mass in the vacuum. However, here a constant effective mass $m_x = m_{\text{GaAs}} = 0.067m_0$ is used. The prototype RTD is a symmetric structure, and we state that $L_{ij} = L_j - L_i$, where L_i is the position of the material interface in the hetero-structure. Then, the length of the device is $L = X_2 - X_1 = L_{70}$. The black barrier region is set to $L_{32} = L_{54} = 2.825$ nm, the length of the quantum well is $L_{43} = 4.52$ nm, the length of the contact regions is denoted by $L_c = L_{10} = L_{76}$, depicted as the gray area, and the buffer region $L_b = L_{12} = L_{56}$ is introduced to decrease the scattering of electrons. We set the intrinsic Fermi energy in GaAs as the benchmark of the energy. So the conduction band energy $v_e(x) = 0$ eV in GaAs and $v_e(x) = 0.27$ eV in AlGaAs, where 0.27 eV is the conduction band gap between GaAS and AlGaAs. A coarse mesh size $a = 0.565$ nm is equal to the lattice constant of GaAs. The length of the contacts L_c and the length of the buffer L_b are important parameters which will affect the density and the current. Also we will adjust L_c and L_b to analyze the accuracy of the inflow boundary condition. For self-consistent simulations, the doping profile in the contacts is depicted as Fig. 19.3, where the gray areas are doped with a concentration 0.2×10^{19} cm^{-3} and the white and the black areas are doped intrinsically.

The I–V curves [15] obtained by the NEGF method and by the Wigner function method are plotted in Fig. 19.4, in which the length of the contacts is the same as the size of the buffer layer. The I–V curves obtained by the Wigner function method approach those obtained by the NEGF method for low bias potentials as the lengths of the contacts and the buffer layers increase, which verifies that the boundary condition of the Wigner function method is valid for large enough contact length.

19.4 Calculation of Oscillatory Integrals $O_n(z)$

The oscillatory integrals are defined as

$$O_n(z) = \int_{-1}^{1} e^{izx} T_n(x) dx. \tag{19.60}$$

Firstly, we exploit the expansion of e^{izx} in terms of the spherical Bessel functions of the first kind $j_k(z)$ and the Legendre polynomials $P_k(x)$; see [16] (1972, Eq. (10.1.47)):

$$e^{izx} = \sum_{k=0}^{\infty} (2k+1) i^k j_k(z) P_k(x), \tag{19.61}$$

and then we have

$$\int_{-1}^{1} e^{izx} P_n(x) dx = 2 i^n j_n(z), \tag{19.62}$$

where we have used the orthogonality relation,

$$\int_{-1}^{1} P_n(x) P_k(x) = \frac{2}{2n+1} \delta_{n,k}. \tag{19.63}$$

Secondly, we can express the Chebyshev polynomials $T_n(x)$ using the Legendre polynomials $P_k(x)$ as

$$T_n(x) = \sum_{k=0}^{n} c_{k,n} P_k(x), \tag{19.64}$$

where the coefficient $c_{k,n}$ is defined by

$$c_{k,n} = \frac{2k+1}{2} \int_{-1}^{1} T_n(x) P_k(x) dx, \tag{19.65}$$

and has a recurrence relation

$$c_{0,n} = \begin{cases} 0, & n = 1, \\ \frac{1}{2} \cdot \frac{1+(-1)^n}{1-n^2}, & n \neq 1, \end{cases}$$

$$c_{1,n} = \begin{cases} 0, & n = 2, \\ \frac{3}{2} \cdot \frac{-1+(-1)^n}{n^2-4}, & n \neq 2, \end{cases}$$

$$c_{k+2,n} = \begin{cases} 0, & k = n-3, \\ \frac{2k+5}{2k+1} \cdot \frac{n^2-k^2}{n^2-(k+3)^2} \cdot c_{k,n}, & k = 0, 1, 2, \ldots, n-4, n-2. \end{cases}$$

Therefore, we get the final formula [17]:

$$O_n(z) = \int_{-1}^{1} e^{izx} \left[\sum_{k=0}^{n} c_{k,n} P_k(x) \right] dx = \sum_{k=0}^{n} c_{k,n} \int_{-1}^{1} e^{izx} P_k(x) dx$$
$$= 2 \sum_{k=0}^{n} i^k c_{k,n} j_k(z). \tag{19.66}$$

19.5 Summary

The Wigner transport equation for nano-devices provides a direct link to the classical Boltzmann transport model for micro-devices and allows easy inclusion of scattering effects between electrons and phonons or impurities by using appropriate quantum or even classical collision operators. The Frensley inflow boundary condition for the Wigner distribution assumes that the electrons from the input contact enter into the device active region, without reflection while passing the inflow boundary. Due to the high dimensionality of the phase space, appropriate truncation of the computational domain in the phase space and adaptive meshing are detailed here, as both are important to reduce the overall cost of numerical methods (e.g., an adaptive spectral element and an upwinding finite difference method) so the Wigner transport model can be applied to realistic low-dimensional nano-devices.

References

1. W.R. Frensley, Wigner-function model of a resonant-tunneling semiconductor device. Phys. Rev. B **36**, 1570–1580 (1987)
2. W.R. Frensley, Boundary conditions for open quantum systems driven far from equilibrium. Rev. Mod. Phys. **62**, 745–791 (1990)
3. S. Shao, T. Lu, W. Cai, Adaptive conservative cell average spectral element methods for transient Wigner equation in quantum transport. Commun. Comput. Phys. **9**, 711–739 (2011)
4. J.P. Boyd, *Chebyshev and Fourier Spectral Methods*, 2nd edn. (Dover Publications, New York, 2001)
5. W. Cai, D. Gottlieb, A. Harten, Cell averaging Chebyshev methods for hyperbolic problems. Comput. Math. Appl. **24**, 37–49 (1992)
6. W.H. Press, S.A. Teukolsky, W.T. Vetterling, B.P. Flannery, *Numerical Recipes in FORTRAN: The Art of Scientific Computing*, 2nd edn. (Cambridge University Press, Cambridge, 1992)
7. D. Gottlieb, S. Orszag, *Numerical Analysis of Spectral Methods: Theory and Applications* (SIAM-CBMS, Philadelphia, 1987)
8. J. Shen, T. Tang, L.L. Wang, *Spectral Methods: Algorithms, Analysis and Applications* (Springer Science & Business Media, 2011)
9. S. Gottlieb, C.-W. Shu, Total variation diminishing Runge-Kutta schemes. Math. Comput. **67**, 73–85 (1998)
10. N.C. Kluksdahl, A.M. Kriman, D.K. Ferry, C. Ringhofer, Self-consistent study of the resonant-tunneling diode. Phys. Rev. B **39**, 7720–7735 (1989)
11. B.A. Biegel, *Quantum electronic device simulation*. Unpublished Ph.D. thesis (Stanford University, 1997)
12. B.A. Biegel, *SQUADS Technical Reference* (Unpublished, Stanford University, 1996)

13. L. Esaki, R. Tsu, Superlattice and negative differential conductivity in semiconductors. IBM J. Res. Develop. **14**, 61–65 (1970)
14. R. Tsu, L. Esaki, Tunneling in a finite superlattice. Appl. Phys. Lett. **22**, 562–564 (1973)
15. H. Jiang, W. Cai, R. Tsu, Accuracy of the Frensley inflow boundary condition for Wigner equations in simulating resonant tunneling diodes. J. Comput. Phys. **230**, 2031–2044 (2011)
16. M. Abramowitz, I.A. Stegun, *Handbook of Mathematical Functions with Formulas, Graphs, and Mathematical Tables*, 9th edn. (Dover Publications, New York, 1972)
17. P.K. Kythe, M.R. Schäferkotter, *Handbook of Computational Methods for Integration* (Chapman & Hall/CRC, Boca Raton, 2005)

Chapter 20
Hydrodynamic Electron Transport and Finite Difference Methods

Having discussed quantum transport models in the preceding two chapters, we now turn to the semi-classical Boltzmann descriptions and their moment equations in the hydrodynamic model in electron transport in complex media including semiconductors and plasmas. Then, finite difference methods for solving the hydrodynamic equations for semiconductor devices will be discussed. Because of the high field effect in sub-micron devices, the electron velocity may develop a sharp transition profile resembling shock waves [1], as in high-speed gas dynamics. Therefore, shock capturing schemes developed for gas dynamics [2, 3] can be applied for device simulations. Here, we will present three methods: the traditional Godunov methods, the weighted essentially non-oscillatory (ENO) finite difference methods, and the central differencing methods. It should be mentioned that the discontinuous Galerkin method can also be used to compute the semiconductor hydrodynamic equations [4].

20.1 Semi-classical and Hydrodynamic Models

20.1.1 Semi-classical Boltzmann Equations

A classical Boltzmann equation can be derived from a semi-classical limit of the Wigner–Moyal expansion of the 3-D Wigner equation (16.114) by keeping only the first few terms of the expansion and adding a collision term to account for the scattering of particles such as electrons with other particles or impurities.

W. Cai, *Deterministic, Stochastic, and Deep Learning Methods for Computational Electromagnetics*, https://doi.org/10.1007/978-981-96-0100-4_20

For $\mathbf{r}, \mathbf{k} \in \mathbb{R}^3$, the Wigner–Moyal expansion including collision effects becomes

$$\frac{\partial}{\partial t} f(\mathbf{r}, \mathbf{k}, t) + \frac{\hbar \mathbf{k}}{m} \cdot \nabla_{\mathbf{r}} f(\mathbf{r}, \mathbf{k}, t) = \frac{q}{\hbar} \sum_{s=0}^{\infty} \frac{(-1)^s}{4^s (2s+1)!} \nabla_{\mathbf{r}}^{(2s+1)} V(\mathbf{r}) \cdot \nabla_{\mathbf{k}}^{(2s+1)} f(\mathbf{r}, \mathbf{k}, t) + \left(\frac{\partial f}{\partial t} \right)_{\text{coll}}, \tag{20.1}$$

where the collision $(\partial f / \partial t)_{\text{coll}}$ is defined by the following integral operator:

$$\left(\frac{\partial f}{\partial t} \right)_{\text{coll}} (\mathbf{r}, \mathbf{k}) = \frac{1}{(2\pi)^3} \sum_j \int \left[f(\mathbf{r}, \mathbf{k}', t) S_j(\mathbf{k}', \mathbf{k}) - f(\mathbf{r}, \mathbf{k}, t) S_j(\mathbf{k}, \mathbf{k}') \right] \mathrm{d}\mathbf{k}'. \tag{20.2}$$

Here, the j summation is over all types of scattering events (electron–electron, electron–ion, and electron–impurity, etc.), and the scattering rate $S_j(\mathbf{k}', \mathbf{k})$ is calculated from the Fermi golden rule [5].

Meanwhile, a Poisson equation for the potential is coupled to (20.1) as follows:

$$\nabla^2 V(\mathbf{r}) = \frac{q}{\epsilon} \left[\int f(\mathbf{r}, \mathbf{k}, t) \mathrm{d}\mathbf{k} - p(\mathbf{r}, t) + N_A(\mathbf{r}) - N_D(\mathbf{r}) \right], \tag{20.3}$$

where $p(\mathbf{r}, t)$ is the hole concentration and $N_A(\mathbf{r})$ and $N_D(\mathbf{r})$ are the acceptor and donor concentrations, respectively. The hole concentration satisfies the following drift-diffusion equation [6] (together with (20.3) forming the Poisson–Nernst–Planck (PNP) semiconductor model):

$$\frac{\partial p(\mathbf{r}, t)}{\partial t} + \nabla_{\mathbf{r}} \left[\mu_p p(\mathbf{r}, t) \nabla_{\mathbf{r}} V(\mathbf{r}) + \mu_p V_t \nabla_{\mathbf{r}} p(\mathbf{r}) \right] = R(V, n, p) - G_{ii}(n, p), \tag{20.4}$$

where μ_p is the hole mobility, $V_t = k_B T / q$ denotes the thermal voltage, and $R(V, n, p)$ and $G_{ii}(n, p)$ are the recombination rate and the hole generation rate from impact ionization, respectively.

From (20.1), we can obtain the classical Boltzmann equation for f by keeping only the first term in the expansion (20.1):

$$\frac{\partial f}{\partial t} + \frac{\hbar \mathbf{k}}{m} \cdot \nabla_{\mathbf{r}} f(\mathbf{r}, \mathbf{k}, t) - \frac{q}{\hbar} \nabla_{\mathbf{r}} V(\mathbf{r}) \cdot \nabla_{\mathbf{k}} f(\mathbf{r}, \mathbf{k}, t) = \left(\frac{\partial f}{\partial t} \right)_{\text{coll}}, \tag{20.5}$$

which is also called the semi-classical Boltzmann equation as the scattering rates in the collision term can be calculated using quantum mechanical approaches. The Boltzmann kinetic equation for the one-particle distribution function $f(\mathbf{r}, \mathbf{k}, t)$ can also be derived directly by considering the rates of particles entering and exiting a phase space element due to the collisions between particles [7–9].

The higher terms in the summation in (20.1) will be identified as quantum corrections to the classical description of the electrons. The first quantum correction model will be

$$\frac{\partial f}{\partial t}+\frac{\hbar\mathbf{k}}{m}\cdot\nabla_{\mathbf{r}}f(\mathbf{r},\mathbf{k},t)-\frac{q}{\hbar}\nabla_{\mathbf{r}}V(\mathbf{r})\cdot\nabla_{\mathbf{k}}f(\mathbf{r},\mathbf{k},t)+\frac{q}{24\hbar}\nabla_{\mathbf{r}}^{3}V(\mathbf{r})\cdot\nabla_{\mathbf{k}}^{3}f(\mathbf{r},\mathbf{k},t)$$
$$=\left(\frac{\partial f}{\partial t}\right)_{\text{coll}}. \tag{20.6}$$

20.1.2 Hydrodynamic Equations

The phase space description of the electrons or ions by the Wigner or Boltzmann distributions still pose great challenges for two- and three-dimensional systems as it demands a $(2d+1)$-dimensional ($2d$ dimensions in space and one in time) computational effort for a d-dimensional system. Thus, in many situations, especially for micro-scale devices where the quantum phenomena are not strong, moment equations for (20.5) are used instead. Two popular models can be derived based on the first one or the first three moments of (20.5): drift-diffusion and hydrodynamic models, respectively. The hydrodynamic model resembles the conservation laws for the Newtonian fluid flows and takes into consideration the temperature and the energy gradient of the electrons in the model, while a constant temperature is assumed for the drift-diffusion model.

Using the definition of the momentum $\mathbf{p}=\hbar\mathbf{k}$ in quantum mechanics, we will denote by $f_\alpha(\mathbf{r},\mathbf{v},t)$ the classical Boltzmann distribution of (20.6) for the α-type particles (which could be electrons, ions, or neutral particles) in terms of the position $\mathbf{r}$ and the velocity variable $\mathbf{v}$:

$$\mathbf{v}=\frac{\mathbf{p}}{m}=\frac{\hbar\mathbf{k}}{m}, \tag{20.7}$$

and we have

$$\nabla_{\mathbf{v}}=\frac{m}{\hbar}\nabla_{\mathbf{k}}. \tag{20.8}$$

Introducing the notation

$$\mathbf{a}=\frac{\mathbf{F}}{m}, \tag{20.9}$$

where $\mathbf{F}$ represents the force on a charged particle of charge q, which is the electric force $\mathbf{F}=-q\nabla V(\mathbf{r})$, then the Boltzmann equation (20.5) takes the following form for a classical particle system:

$$\frac{\partial f_\alpha}{\partial t}+\mathbf{v}\cdot\nabla_{\mathbf{r}}f_\alpha(\mathbf{r},\mathbf{v},t)+\mathbf{a}\cdot\nabla_{\mathbf{v}}f_\alpha(\mathbf{r},\mathbf{v},t)=\left(\frac{\partial f_\alpha}{\partial t}\right)_{\text{coll}}. \tag{20.10}$$

In the original Boltzmann kinetic theory of dilute gas, the collisions between particles are assumed to be instantaneous and binary, namely, many-body collisions are not considered and the process is Markovian, such that two particles after a collision will have no memory of their past, and their dynamics will only depend on their states after the collision. In this case, the collision in (20.10) can be expressed in terms of the following Boltzmann collision operator [8] for two particles $(\mathbf{r}, \mathbf{v})$ and $(\mathbf{r}_1, \mathbf{v}_1)$, while $(\mathbf{r}', \mathbf{v}')$ and $(\mathbf{r}_1', \mathbf{v}_1')$ denote their position and velocities after a collision, respectively:

$$\left(\frac{\partial f_\alpha}{\partial t}\right)_{\text{coll}}(\mathbf{r}, \mathbf{v}) = \sum_\beta \int \mathrm{d}\mathbf{v}_1 \left[f_\alpha(\mathbf{r}, \mathbf{v}') f_\beta(\mathbf{r}, \mathbf{v}_1') - f_\alpha(\mathbf{r}, \mathbf{v}) f_\beta(\mathbf{r}, \mathbf{v}_1) \right] g\sigma(\Omega)\mathrm{d}\Omega, \tag{20.11}$$

where $g = |\mathbf{v} - \mathbf{v}_1|$ and $\sigma(\Omega)$ is the scattering cross- section which depends on the specific potential governing the interaction of the two particles. The physical meaning of $\sigma(\Omega)$ is given as follows [8]:

$$I\sigma(\Omega)\mathrm{d}\Omega = \begin{array}{l}\text{number of particles deflected into the element}\\ \text{of solid angle } \mathrm{d}\Omega \text{ about } \Omega \text{ per second}\end{array}$$

for an incident beam of particle of intensity I (number/s $\cdot$ cm^2).

Next, we will derive a general macroscopic equation using moment equations for the Boltzmann equation for various conservative quantities following [10]. Given any conservative quantity $\chi(\mathbf{r}, \mathbf{v})$, we have the following moment equation by integrating over the velocity variable the product of $\chi(\mathbf{r}, \mathbf{v})$ and (20.10):

$$\begin{aligned}\int \chi(\mathbf{r}, \mathbf{v}) \left[\frac{\partial f_\alpha}{\partial t} + \mathbf{v} \cdot \nabla_\mathbf{r} f_\alpha(\mathbf{r}, \mathbf{v}, t) + \mathbf{a} \cdot \nabla_\mathbf{v} f_\alpha(\mathbf{r}, \mathbf{v}, t) \right] \mathrm{d}\mathbf{v} \\ = \int \chi(\mathbf{r}, \mathbf{v}) \left(\frac{\partial f_\alpha}{\partial t}\right)_{\text{coll}} \mathrm{d}\mathbf{v}.\end{aligned} \tag{20.12}$$

To proceed, the number density per unit volume is defined as

$$n_\alpha(\mathbf{r}, t) = \int f_\alpha(\mathbf{r}, \mathbf{v}, t)\, \mathrm{d}\mathbf{v}, \tag{20.13}$$

and the average $\langle \chi \rangle_\alpha$ of the quantity $\chi(\mathbf{r}, \mathbf{v})$ in the phase space is defined as

$$\langle \chi \rangle_\alpha = \frac{\displaystyle\int \chi(\mathbf{r}, \mathbf{v}) f_\alpha\, \mathrm{d}\mathbf{v}}{\displaystyle\int f_\alpha(\mathbf{r}, \mathbf{v}, t)\, \mathrm{d}\mathbf{v}} = \frac{\displaystyle\int \chi(\mathbf{r}, \mathbf{v}) f_\alpha\, \mathrm{d}\mathbf{v}}{n_\alpha(\mathbf{r}, t)}. \tag{20.14}$$

Then, the first term on the left-hand side of (20.12) becomes

$$\int \chi(\mathbf{r}, \mathbf{v}) \frac{\partial f_\alpha}{\partial t}\, \mathrm{d}\mathbf{v} = \frac{\partial}{\partial t} \int \chi(\mathbf{r}, \mathbf{v}) f_\alpha\, \mathrm{d}\mathbf{v} = \frac{\partial}{\partial t} \left(n_\alpha \langle \chi \rangle_\alpha \right), \tag{20.15}$$

and the second term, after using the fact that $\nabla_{\mathbf{r}} \cdot \mathbf{v} = \mathbf{0}$, becomes

$$\int \chi(\mathbf{r}, \mathbf{v})\, \mathbf{v} \cdot \nabla_{\mathbf{r}} f_\alpha(\mathbf{r}, \mathbf{v}, t)\, \mathrm{d}\mathbf{v} = \nabla_{\mathbf{r}} \cdot \int \chi(\mathbf{r}, \mathbf{v})\, \mathbf{v} f_\alpha(\mathbf{r}, \mathbf{v}, t)\, \mathrm{d}\mathbf{v} = \nabla_{\mathbf{r}} \cdot \left(n_\alpha \langle \chi \mathbf{v} \rangle_\alpha\right). \tag{20.16}$$

Meanwhile, the third term can be obtained by using integration of parts to become

$$\begin{aligned} \int \chi(\mathbf{r}, \mathbf{v})\mathbf{a} \cdot \nabla_{\mathbf{v}} f_\alpha(\mathbf{r}, \mathbf{v}, t)\mathrm{d}\mathbf{v} &= \int \chi(\mathbf{r}, \mathbf{v})\, \nabla_{\mathbf{v}} \cdot (\mathbf{a} f_\alpha(\mathbf{r}, \mathbf{v}, t))\, \mathrm{d}\mathbf{v} \\ = -\int \mathbf{a} \cdot \nabla_{\mathbf{v}} \chi(\mathbf{r}, \mathbf{v})\; f_\alpha(\mathbf{r}, \mathbf{v}, t)\mathrm{d}\mathbf{v} &= -n_\alpha \langle \mathbf{a} \cdot \nabla_{\mathbf{v}} \chi \rangle_\alpha , \end{aligned} \tag{20.17}$$

provided that

$$\nabla_{\mathbf{v}} \cdot \mathbf{a} = 0, \tag{20.18}$$

which is used in the second equality of (20.17). Condition (20.18) is true for many types of forces $\mathbf{F}$ in (20.9) when they do not depend on the velocity $\mathbf{v}$, for instance the electric force $\mathbf{F}(\mathbf{r}) = -q\nabla V(\mathbf{r})$, which depends only on the position $\mathbf{r}$.

Substituting (20.15), (20.16), and (20.17) into (20.12), we obtain the general transport equation:

$$\frac{\partial}{\partial t}\left(n_\alpha \langle \chi \rangle_\alpha\right) + \nabla_{\mathbf{r}} \cdot \left(n_\alpha \langle \chi \mathbf{v} \rangle_\alpha\right) - n_\alpha \langle \mathbf{a} \cdot \nabla_{\mathbf{v}} \chi \rangle_\alpha = \left\langle \left(\frac{\partial f_\alpha}{\partial t}\right)_{\mathrm{coll}} \right\rangle, \tag{20.19}$$

where $n_\alpha \langle \chi \mathbf{v} \rangle_\alpha$ in the second term is identified as the conserved quantity flux $\mathbf{J}_\chi$:

$$\mathbf{J}_\chi = n_\alpha \langle \chi \mathbf{v} \rangle_\alpha . \tag{20.20}$$

By setting $\chi = m_\alpha$, the mass of the α-type particle, we have the continuity equation

$$\frac{\partial}{\partial t}\rho_\alpha + \nabla_{\mathbf{r}} \cdot (\rho_\alpha \mathbf{u}_\alpha) = S_\alpha, \tag{20.21}$$

where the density of the α-type particle ρ_α is defined as

$$\rho_\alpha = m_\alpha n_\alpha, \tag{20.22}$$

the average velocity of the α-type particle at the location $\mathbf{r}$ is given by

$$\mathbf{u}_\alpha(\mathbf{r}, t) = \langle \mathbf{v} \rangle_\alpha , \tag{20.23}$$

and the source term S_α results from the collision experienced by the α-type particle:

$$S_\alpha = m_\alpha \int \left[\frac{\partial f_\alpha}{\partial t}\right]_{\mathrm{coll}} \mathrm{d}\mathbf{v}; \tag{20.24}$$

for mass conservation,

$$\sum_{\alpha} S_{\alpha} = 0. \tag{20.25}$$

Next, setting $\chi = m_{\alpha}\mathbf{v}$, (20.19) becomes

$$\frac{\partial}{\partial t}\left(n_{\alpha}\left\langle m_{\alpha}\mathbf{v}\right\rangle_{\alpha}\right) + \nabla_{\mathbf{r}}\cdot\left(n_{\alpha}\left\langle m_{\alpha}\mathbf{v}\mathbf{v}\right\rangle_{\alpha}\right) - n_{\alpha}\left\langle \mathbf{a}\cdot\nabla_{\mathbf{v}} m_{\alpha}\mathbf{v}\right\rangle_{\alpha} = \left\langle\left(\frac{\partial\left(m_{\alpha}\mathbf{v}\right)}{\partial t}\right)_{\text{coll}}\right\rangle. \tag{20.26}$$

The first term in (20.26) is

$$\frac{\partial}{\partial t}\left(n_{\alpha}\left\langle m_{\alpha}\mathbf{v}\right\rangle_{\alpha}\right) = \frac{\partial}{\partial t}\rho_{\alpha}\left\langle \mathbf{v}\right\rangle_{\alpha} = \frac{\partial}{\partial t}\rho_{\alpha}\mathbf{u}_{\alpha}, \tag{20.27}$$

and the second term in (20.26) is

$$\nabla_{\mathbf{r}}\cdot\left(n_{\alpha}\left\langle m_{\alpha}\mathbf{v}\mathbf{v}\right\rangle_{\alpha}\right) = \nabla_{\mathbf{r}}\left(\rho_{\alpha}\left\langle \mathbf{v}\mathbf{v}\right\rangle_{\alpha}\right), \tag{20.28}$$

which involves the second moment $\mathbf{vv}$. Consider the velocity fluctuation with respect to the average velocity $\mathbf{u}_{\alpha}$ of the α-type particle:

$$\mathbf{v} = \mathbf{u}_{\alpha} + \mathbf{v}', \tag{20.29}$$

and

$$\left\langle \mathbf{v}'\right\rangle_{\alpha} = 0. \tag{20.30}$$

It follows from (20.30) that the term following $\nabla_{\mathbf{r}}$ in (20.28) becomes

$$\begin{aligned}\rho_{\alpha}\left\langle \mathbf{v}\mathbf{v}\right\rangle_{\alpha} &= \rho_{\alpha}\left\langle\left(\mathbf{u}_{\alpha}+\mathbf{v}'\right)\left(\mathbf{u}_{\alpha}+\mathbf{v}'\right)\right\rangle_{\alpha}\\ &= \rho_{\alpha}\mathbf{u}_{\alpha}\mathbf{u}_{\alpha} + \rho_{\alpha}\left\langle \mathbf{v}'\mathbf{v}'\right\rangle_{\alpha} = \rho_{\alpha}\mathbf{u}_{\alpha}\mathbf{u}_{\alpha} + P_{\alpha},\end{aligned} \tag{20.31}$$

where the pressure dyadic $\mathbf{P}_{\alpha}$ is defined as

$$\mathbf{P}_{\alpha} = \rho_{\alpha}\left\langle \mathbf{v}'\mathbf{v}'\right\rangle_{\alpha}, \tag{20.32}$$

i.e.,

$$\left(P_{\alpha}\right)_{ij} = \rho_{\alpha}\left\langle v_i' v_j'\right\rangle_{\alpha}. \tag{20.33}$$

So the second term in (20.26) becomes

$$\nabla_{\mathbf{r}}\cdot\left(n_{\alpha}\left\langle m_{\alpha}\mathbf{v}\mathbf{v}\right\rangle_{\alpha}\right) = \nabla_{\mathbf{r}}\cdot\left(\rho_{\alpha}\mathbf{u}_{\alpha}\mathbf{u}_{\alpha} + \mathbf{P}_{\alpha}\right), \tag{20.34}$$

where the divergence operator $\nabla_{\mathbf{r}}$ applies to each column of the dyadics $\mathbf{u}_{\alpha}\mathbf{u}_{\alpha}$ and $\mathbf{P}_{\alpha}$.

Next, the third term in (20.26) is

$$n_\alpha \langle \mathbf{a} \cdot \nabla_\mathbf{v} m_\alpha \mathbf{v} \rangle_\alpha = n_\alpha \langle (\mathbf{F} \cdot \nabla_\mathbf{v}) \mathbf{v} \rangle_\alpha = n_\alpha \langle \mathbf{F} \rangle_\alpha . \tag{20.35}$$

Putting (20.27), (20.34), and (20.35) into (20.26), we have the momentum equation for the α-type particle:

$$\frac{\partial \rho_\alpha \mathbf{u}_\alpha}{\partial t} = -\nabla \cdot (\rho_\alpha \mathbf{u}_\alpha \mathbf{u}_\alpha) + n_\alpha \mathbf{F}_\alpha - \nabla \mathbf{P}_\alpha + \mathbf{A}_\alpha , \tag{20.36}$$

where

$$\mathbf{F}_\alpha = \langle \mathbf{F} \rangle_\alpha$$

and

$$\mathbf{A}_\alpha = m_\alpha \int \mathbf{v} \left[\frac{\delta f_\alpha}{\delta t} \right]_{\text{coll}} \mathrm{d}\mathbf{v} ; \tag{20.37}$$

for momentum conservation

$$\sum_\alpha \mathbf{A}_\alpha = \mathbf{0}. \tag{20.38}$$

For later use, we can also define a scalar pressure via the trace of the dyadic $\mathbf{P}_\alpha$ as

$$p_\alpha = \frac{1}{3} \sum_{i=1}^{3} (P_\alpha)_{ii} . \tag{20.39}$$

For an isotropic gas, we have a diagonal dyadic with $\mathbf{P}_\alpha = \mathrm{diag}(p_\alpha, p_\alpha, p_\alpha)$ and

$$p_\alpha = \frac{\rho_\alpha}{3} \left\langle v'^2 \right\rangle_\alpha , \tag{20.40}$$

where $v'^2 \equiv |\mathbf{v}'|^2$.

Meanwhile, the temperature T_α for the α-type particle is defined as

$$\begin{aligned} k_B T_\alpha &= \frac{m_\alpha}{3} \left\langle v'^2 \right\rangle_\alpha = \frac{m_\alpha}{3 n_\alpha} \int v'^2 f_\alpha(\mathbf{r}, \mathbf{v}) \mathrm{d}\mathbf{v} \\ &= \frac{m_\alpha}{3 n_\alpha} \int (v - u_\alpha)^2 f_\alpha(\mathbf{r}, \mathbf{v}) \mathrm{d}\mathbf{v}, \end{aligned} \tag{20.41}$$

and, comparing (20.40) and (20.41), we have the *equation of state* for an ideal gas:

$$p_\alpha = n_\alpha k_B T_\alpha . \tag{20.42}$$

To derive the energy equation for the α-type particle, we take $\chi = m_\alpha v^2/2$. The general transport equation (20.19) now becomes

$$\frac{\partial}{\partial t}\frac{1}{2}\rho_\alpha \left\langle v^2\right\rangle_\alpha + \nabla_{\mathbf{r}} \cdot \frac{\rho_\alpha}{2}\left\langle v^2 \mathbf{v}\right\rangle_\alpha - n_\alpha \left\langle \mathbf{F}\cdot\mathbf{v}\right\rangle_\alpha = 0. \tag{20.43}$$

The first term on the left-hand side of (20.43) is

$$\frac{1}{2}\rho_\alpha \left\langle v^2\right\rangle_\alpha = \frac{1}{2}\rho_\alpha u_\alpha^2 + \frac{1}{2}\rho_\alpha \left\langle v'^2\right\rangle_\alpha + \rho_\alpha \mathbf{u}_\alpha \cdot \left\langle \mathbf{v}'\right\rangle_\alpha = \frac{1}{2}\rho_\alpha u_\alpha^2 + \frac{3}{2}p_\alpha. \tag{20.44}$$

Next, we examine the term inside the second term in (20.43):

$$\begin{aligned}\left\langle v^2\mathbf{v}\right\rangle_\alpha &= \left\langle (u_\alpha^2 + 2\mathbf{u}_\alpha\cdot\mathbf{v}' + v'^2)(\mathbf{u}_\alpha + \mathbf{v}')\right\rangle_\alpha \\ &= \left[u_\alpha^2\mathbf{u}_\alpha + 2\mathbf{u}_\alpha \cdot \left\langle\mathbf{v}'\right\rangle_\alpha \mathbf{u}_\alpha + \left\langle v'^2\right\rangle_\alpha \mathbf{u}_\alpha\right] \\ &\quad + \left[u_\alpha^2\left\langle\mathbf{v}'\right\rangle_\alpha + 2\mathbf{u}_\alpha\left\langle\mathbf{v}'\mathbf{v}'\right\rangle_\alpha + \left\langle v'^2\mathbf{v}'\right\rangle_\alpha\right] \\ &= u_\alpha^2\mathbf{u}_\alpha + \left\langle v'^2\right\rangle_\alpha \mathbf{u}_\alpha + 2\mathbf{u}_\alpha\left\langle\mathbf{v}'\mathbf{v}'\right\rangle_\alpha + \left\langle v'^2 v'\right\rangle_\alpha.\end{aligned} \tag{20.45}$$

Therefore, the second term in (20.43) (inside the divergence operator $\nabla_{\mathbf{r}}$) becomes

$$\begin{aligned}\frac{\rho_\alpha}{2}\left\langle v^2\mathbf{v}\right\rangle_\alpha &= \frac{\rho_\alpha}{2}u_\alpha^2\mathbf{u}_\alpha + \frac{\rho_\alpha}{2}\left\langle v'^2\right\rangle_\alpha \mathbf{u}_\alpha \\ &\quad + \rho_\alpha\mathbf{u}_\alpha\left\langle\mathbf{v}'\mathbf{v}'\right\rangle_\alpha + \frac{\rho_\alpha}{2}\left\langle v'^2\mathbf{v}'\right\rangle_\alpha \\ &= \frac{\rho_\alpha}{2}u_\alpha^2\mathbf{u}_\alpha + \frac{3}{2}p_\alpha\mathbf{u}_\alpha + P_\alpha\mathbf{u}_\alpha + \mathbf{q}_\alpha,\end{aligned} \tag{20.46}$$

where a heat flux $\mathbf{q}_\alpha$ is defined as

$$\mathbf{q}_\alpha = \frac{1}{2}\rho_\alpha\left\langle v'^2\mathbf{v}'\right\rangle_\alpha. \tag{20.47}$$

The third term in (20.43) is

$$n_\alpha\left\langle \mathbf{F}\cdot\mathbf{v}\right\rangle_\alpha = n_\alpha\int f_\alpha \mathbf{F}\cdot\mathbf{v}d\mathbf{v} = n_\alpha\mathbf{F}\cdot\int f_\alpha\cdot\mathbf{v}d\mathbf{v} = n_\alpha\mathbf{F}\cdot\mathbf{u}_\alpha. \tag{20.48}$$

Substituting (20.44), (20.46), and (20.48) into (20.43), we obtain the energy equation for the α-type particle as follows:

$$\frac{\partial W_\alpha}{\partial t} + \nabla\cdot\left[\left(\frac{\rho_\alpha}{2}u_\alpha^2 + \frac{3}{2}p_\alpha + P_\alpha\right)\mathbf{u}_\alpha + \mathbf{q}_\alpha\right] = n_\alpha\mathbf{F}\cdot\mathbf{u}_\alpha + M_\alpha, \tag{20.49}$$

where the energy density for the α-type particle W_α is defined as

$$W_\alpha = \frac{1}{2}\rho_\alpha u_\alpha^2 + \frac{3}{2}p_\alpha = \frac{1}{2}\rho_\alpha u_\alpha^2 + \frac{3}{2}n_\alpha k_B T_\alpha, \tag{20.50}$$

and the energy source term from the collision is given by

$$M_\alpha = \frac{m_\alpha}{2}\int v^2 \left[\frac{\partial f_\alpha}{\partial t}\right]_{\text{coll}} d\mathbf{v}, \tag{20.51}$$

and the energy conservation

$$\sum_\alpha M_\alpha = 0. \tag{20.52}$$

Finally, the heat flux in semiconductor devices can usually be approximated by the gradient of the temperature as follows [11]:

$$\mathbf{q} = -\kappa \nabla T, \tag{20.53}$$

with κ being the thermal conductivity of the semiconductor material of the device, approximated by [12]

$$\kappa = \frac{5k_B^2 n_\alpha T_\alpha}{2m_\alpha v_p(w)}. \tag{20.54}$$

Then, we can obtain the energy flux as

$$\begin{aligned}\mathbf{J}_w &= \left(\frac{\rho_\alpha}{2}u_\alpha^2 + \frac{3}{2}p_\alpha + P_\alpha\right)\mathbf{u}_\alpha + \mathbf{q}_\alpha = \mathbf{u}_\alpha\left(W_\alpha + P_\alpha\right) + \mathbf{q}_\alpha \\ &= \mathbf{u}_\alpha W_\alpha + \left(n\mathbf{u}_\alpha - \frac{\kappa}{k_B}\nabla\right) k_B T_\alpha. \end{aligned} \tag{20.55}$$

Using (20.50) to eliminate the temperature $k_B T_\alpha$, we finally obtain the energy equation:

$$\frac{\partial W_\alpha}{\partial t} = -\nabla\cdot\left[\mathbf{u}_\alpha W_\alpha + \frac{2}{3}\left(n_\alpha \mathbf{u}_\alpha - \frac{\kappa}{k_B}\nabla\right)\left(\frac{W_\alpha}{n_\alpha} - \frac{m_\alpha \mathbf{u}_\alpha^2}{2}\right)\right] + n\mathbf{F}_\alpha\cdot\mathbf{u}_\alpha + M_\alpha. \tag{20.56}$$

In summary, (20.21), (20.36), and (20.56) form the hydrodynamic model for a multiple-species particle system.

In the following, we will consider a system of electrons only without the heat flux term (dropping the index α) whose hydrodynamic equations are written in a conservative divergence form:

$$\mathbf{U}_t + \nabla\bar{\mathbf{f}}(\mathbf{U}) = \mathbf{s}(\mathbf{U}), \tag{20.57}$$

and

$$\mathbf{U} = (\rho, \rho\mathbf{u}, W)^{\mathrm{T}}, \tag{20.58}$$

$$\bar{\mathbf{f}}(\mathbf{q}) = (\rho\mathbf{v}, \rho\mathbf{u}^2 + P, \mathbf{u}(W+P))^{\mathrm{T}}, \tag{20.59}$$

$$W = \frac{1}{2}\rho u^2 + \frac{3}{2}nk_B T, \tag{20.60}$$

$$P = nk_B T = (\gamma - 1)\left(W - \frac{1}{2}\rho u^2\right), \tag{20.61}$$

where $\gamma = 5/3$, and the source term contains the collision relaxation:

$$\mathbf{s}(\mathbf{U}) = (0, qn\mathbf{F} + \mathbf{A}, qn\mathbf{F}\cdot\mathbf{u} + M)^{\mathrm{T}}. \tag{20.62}$$

In many cases, the collision term can be modeled with the relaxation times τ_ρ, τ_p, and τ_w for the mass density, the momentum, and the energy, respectively, i.e.,

$$\mathbf{s}(\mathbf{U}) = \left(\frac{\rho - \rho_0}{\tau_\rho}, qn\mathbf{F} + \frac{\rho\mathbf{u}}{\tau_p}, qn\mathbf{F}\cdot\mathbf{u} + \frac{W - \frac{3}{2}nk_B T_0}{\tau_w}\right)^{\mathrm{T}}, \tag{20.63}$$

where m, ρ, $\mathbf{u}$, P, and W are the mass, mass density, velocity, pressure, and total energy, respectively. Note that ρ_0 and T_0 are the mass density and temperature, respectively, in an equilibrium state of the system [13]. Moreover, the electric field $\mathbf{F} = -\nabla\phi$ and the potential energy $V = -q\phi$; the latter is governed by a Poisson equation,

$$\nabla\cdot\epsilon\nabla V = q^2(N - n), \tag{20.64}$$

where N is a given doping density for the semiconductors.

20.2 High-Resolution Finite Difference Methods of Godunov Type

Consider the 1-D version of the nonlinear conservation law (20.57) with $\mathbf{s} = \mathbf{0}$:

$$\mathbf{U}_t + \mathbf{f}_x(\mathbf{U}) = \mathbf{0}. \tag{20.65}$$

The Godunov method is based on a finite volume formulation of (20.65) by integrating between any two points $a < b$ to arrive at

$$\frac{\partial}{\partial t}\overline{\mathbf{U}} = \frac{\partial}{\partial t}\frac{1}{b-a}\int_a^b \mathbf{U}(x,t)\mathrm{d}x = -\frac{1}{b-a}\left[\mathbf{f}(\mathbf{U}(b)) - \mathbf{f}(\mathbf{U}(a))\right], \tag{20.66}$$

where the overbar indicates the cell-averaged values of the related quantity.

The cell-averaged value $\overline{\mathbf{U}}_j$ over the interval $[x_{j-1/2}, x_{j+1/2}]$ is defined as

$$\overline{\mathbf{U}}_j(t) = \frac{1}{x_{j+\frac{1}{2}} - x_{j-\frac{1}{2}}}\int_{x_{j-\frac{1}{2}}}^{x_{j+\frac{1}{2}}} \mathbf{U}(x,t)\mathrm{d}x, \tag{20.67}$$

for which the time evolution is

$$\frac{\partial}{\partial t}\overline{\mathbf{U}}_j + \frac{1}{\Delta x}\left[\mathbf{f}(\mathbf{U}_{j+\frac{1}{2}}) - \mathbf{f}(\mathbf{U}_{j-\frac{1}{2}})\right] = \mathbf{0}. \tag{20.68}$$

It should be noted that both the point values $\mathbf{U}_{j+1/2}$ and the cell-averaged value $\overline{\mathbf{U}}_j$ of the physical quantity are involved in (20.68), while the time evolution is given for the cell-averaged values. Once we know the cell averages at a given time instant, the point values at the cell boundaries must be established in order to evaluate the flux terms (which allows the cell averages to be predicted at a new time); this is called the "reconstruction step" in the Godunov-type scheme [14].

A semi-discretization scheme for the conservation law can be constructed based on (20.68) for the cell averages on a given mesh:

$$\frac{\partial}{\partial t}\overline{\mathbf{U}}_j + \frac{1}{\Delta x}\left(\hat{\mathbf{f}}_{j+\frac{1}{2}} - \hat{\mathbf{f}}_{j-\frac{1}{2}}\right) = \mathbf{0}, \tag{20.69}$$

where the numerical flux $\hat{\mathbf{f}}_{j+1/2}$ is expected to approximate the analytical flux $f(\mathbf{U}_{j+1/2})$ if the numerically computed cell averages $\overline{\mathbf{U}}_j$ approximate those of the exact solution. Therefore, the key step in the semi-discretization scheme (20.69) is how to define the numerical fluxes once the cell averages $\{\overline{\mathbf{U}}_j\}_{j=1}^N$ at the current time t are given. In the original Godunov scheme [15], this numerical flux is obtained by solving an initial value problem for (20.65) around the cell interface $x_{j+1/2}$ with the initial data at time t made of two constants $\overline{\mathbf{U}}_j$ and $\overline{\mathbf{U}}_{j+1}$ to the left and right sides of $x_{j+1/2}$, respectively. This solution is called a Riemann problem, whose solution is self-similar in time for the hydrodynamic equations, and is denoted as

$$\mathbf{U}^{(R)}\left(x - x_{j+\frac{1}{2}}, t+\tau; \overline{\mathbf{U}}_j, \overline{\mathbf{U}}_{j+1}\right) = \mathbf{U}^{(R)}\left(\xi; \overline{\mathbf{U}}_j, \overline{\mathbf{U}}_{j+1}\right), \tag{20.70}$$

where $\xi = (x - x_{j+1/2})/\tau$. Then, the *Godunov numerical flux* is defined simply as

$$\hat{\mathbf{f}}_{j+\frac{1}{2}} = f\left(u^{(R)}\left(\xi = 0; \overline{\mathbf{U}}_j, \overline{\mathbf{U}}_{j+1}\right)\right). \tag{20.71}$$

High-resolution schemes of Godunov type can be similarly defined if the initial data for the above Riemann problem are replaced with piecewise polynomials of first, or second, or even higher orders. This step thus requires a reconstruction of the point values of the solution from its cell averages using higher-order piecewise polynomials such as the MUSCL (monotone upstream scheme for conservation laws) introduced by van Leer [16–18]. In the case of piecewise linear polynomials, the reconstruction can be defined for each component as follows:

$$\mathbf{U}^L_{j+\frac{1}{2}} = \overline{\mathbf{U}}_j + \frac{1}{2}\Phi(r_{j+\frac{1}{2}})\left(\overline{\mathbf{U}}_j - \overline{\mathbf{U}}_{j-1}\right), \tag{20.72}$$

$$\mathbf{U}^R_{j+\frac{1}{2}} = \overline{\mathbf{U}}_{j+1} - \frac{1}{2}\Phi\left(\frac{1}{r_{j+\frac{3}{2}}}\right)\left(\overline{\mathbf{U}}_{j+2} - \overline{\mathbf{U}}_{j+1}\right), \tag{20.73}$$

where $r_{j+1/2}$ is the ratio of the kth component of the differences of the cell averages,

$$r_{j+\frac{1}{2}} = \frac{\left(\overline{\mathbf{U}}_{j+1} - \overline{\mathbf{U}}_j\right)_k}{\left(\overline{\mathbf{U}}_j - \overline{\mathbf{U}}_{j-1}\right)_k}, \quad k = 1, 2, 3,$$

and a "limiter" $\Phi(r)$ is introduced to limit the slope of the linear interpolation such that the overall piecewise polynomial maintains a monotonic profile if the given data $\{\overline{\mathbf{U}}_j\}_{j=1}^N$ are monotone componentwise. Many limiter functions have been studied (for a review refer to [19]); for example, the so-called "superbee" limiter proposed by [20] is given by

$$\Phi(r) = \max[0, \min(2r, 1), \min(r, 2)]. \tag{20.74}$$

The numerical flux for the higher-order spatial resolution approximation is given by

$$\hat{\mathbf{f}}_{j+\frac{1}{2}} = f\left(\mathbf{U}^{(R)}\left(\xi = 0; \mathbf{U}^L_{j+\frac{1}{2}}, \mathbf{U}^R_{j+\frac{1}{2}}\right)\right), \tag{20.75}$$

where $\mathbf{U}^L_{j+\frac{1}{2}}$ and $\mathbf{U}^R_{j+\frac{1}{2}}$ are the point values at the left and right side of $x_{j+\frac{1}{2}}$, respectively

The remaining issue is to get the Riemann solution $\mathbf{U}^{(R)}(\cdot; \mathbf{U}^L_{j+1/2}, \mathbf{U}^R_{j+1/2})$ for general piecewise polynomial reconstructed initial data (point values). We should note that the Godunov numerical flux, using the Riemann solutions, is in essence a generalization of an upwinding finite difference method, which can be seen for a linear wave equation

$$f(u) = au, \tag{20.76}$$

so, the solution to the Riemann problem in this case is a simple convection of the initial data to the left or right according to the sign of the wave speed a, and we have

$$u^{(R)}(0; \overline{u}_j, \overline{u}_{j+1}) = \begin{cases} \overline{u}_j, & \text{if } a > 0, \\ \overline{u}_{j+1}, & \text{if } a < 0, \end{cases} \tag{20.77}$$

resulting in a numerical flux for the positive speed a,

$$\hat{f}_{j+\frac{1}{2}} = a\overline{u}_j, \tag{20.78}$$

and a downwind finite difference approximation to the scalar version of (20.65) with (20.76),

$$\frac{\partial}{\partial t}\overline{u}_j + \frac{a}{\Delta x}\left(\overline{u}_j - \overline{u}_{j-1}\right) = 0. \tag{20.79}$$

We now consider a system of linear equations (20.65) with the flux defined by a constant matrix A:

$$\mathbf{f}(\mathbf{U}) = A\mathbf{U}, \tag{20.80}$$

where A could have both positive and negative eigenvalues, and we could split the matrix A into A^+ and A^- with positive and negative eigenvalues, respectively, i.e.,

$$A^+ = P\Lambda^+P^{-1}, \quad A^- = P\Lambda^-P^{-1}, \tag{20.81}$$

where P contains the right eigenvectors of the matrix A as its columns and

$$\begin{aligned}
A &= P\Lambda P^{-1}, \quad \Lambda = \text{diag}(\lambda_1, \ldots, \lambda_n),\\
\Lambda^+ &= \frac{1}{2}\text{diag}(\lambda_1 + |\lambda_1|, \ldots, \lambda_n + |\lambda_n|),\\
\Lambda^- &= \frac{1}{2}\text{diag}(\lambda_1 - |\lambda_1|, \ldots, \lambda_n - |\lambda_n|),\\
|\Lambda| &= \Lambda^+ - \Lambda^-, \quad |A| = A^+ - A^-.
\end{aligned} \tag{20.82}$$

Then, the upwinding scheme applied to (20.80) will be

$$\frac{\partial}{\partial t}\overline{\mathbf{U}}_j + \frac{A^+}{\Delta x}\left(\overline{\mathbf{U}}_j - \overline{\mathbf{U}}_{j-1}\right) + \frac{A^-}{\Delta x}\left(\overline{\mathbf{U}}_{j+1} - \overline{\mathbf{U}}_j\right) = 0, \tag{20.83}$$

which can be rewritten in the conservative form of (20.69) with the following numerical flux:

$$\hat{\mathbf{f}}_{j+\frac{1}{2}} = \frac{1}{2}(\mathbf{f}_j + \mathbf{f}_{j+1}) - \frac{1}{2}|A|(\overline{\mathbf{U}}_{j+1} - \overline{\mathbf{U}}_j). \tag{20.84}$$

The Riemann problem and the definition of the numerical flux for the hydrodynamic system, a nonlinear system of conservation laws, are much more difficult. However, approximate Riemann solvers can be found; one of the most popular is the Roe–Riemann solver, which in principle linearizes the flux function at a reference state, called the Roe-averaged state $\hat{\mathbf{U}}$.

Consider the eigenvalues of the Jacobian matrix of the flux function $\mathbf{f}(\mathbf{U})$:

$$A(\mathbf{U}) = \frac{\partial \mathbf{f}(\mathbf{U})}{\partial \mathbf{U}}, \tag{20.85}$$

which are, for a 1-D problem [14],

$$\lambda_1(\mathbf{U}) = u - c, \ \lambda_2(\mathbf{U}) = u, \ \lambda_3(\mathbf{U}) = u + c, \tag{20.86}$$

where the speed of sound is $c = (\gamma P/\rho)^{1/2}$ and $\rho = mn$ is the mass density. The corresponding right eigenvectors are

$$\mathbf{r}_1(\mathbf{U}) = \begin{pmatrix} 1 \\ u - c \\ H - uc \end{pmatrix}, \quad \mathbf{r}_2(\mathbf{U}) = \begin{pmatrix} 1 \\ u \\ \frac{1}{2}u^2 \end{pmatrix}, \quad \mathbf{r}_3(\mathbf{U}) = \begin{pmatrix} 1 \\ u + c \\ H + uc \end{pmatrix}, \tag{20.87}$$

where $H = (W + P)/\rho = c^2/(\gamma - 1) + u^2/2$ is the enthalpy.

The corresponding left eigenvectors $\{\mathbf{l}_k(\mathbf{U})\}$, which are bi-orthonormal to $\{\mathbf{r}_k(\mathbf{U})\}$, are given by

$$\begin{aligned} \mathbf{l}_1(\mathbf{U}) &= \frac{1}{2}(b_2 + u/c, -b_1 u - 1/c, b_1), \\ \mathbf{l}_2(\mathbf{U}) &= \frac{1}{2}(1 - b_2, b_1 u, -b_1), \\ \mathbf{l}_3(\mathbf{U}) &= \frac{1}{2}(b_2 - u/c, -b_1 u + 1/c, b_1), \end{aligned} \tag{20.88}$$

where

$$b_1 = \frac{\gamma - 1}{c^2}, \quad b_2 = \frac{1}{2}u^2 b_1.$$

Then, the Roe averages are defined as follows [21]:

$$\begin{aligned} \hat{u} &= \langle \sqrt{\rho} u \rangle / \langle \sqrt{\rho} \rangle, \\ \hat{H} &= \langle \sqrt{\rho} H \rangle / \langle \sqrt{\rho} \rangle, \\ \hat{c} &= \sqrt{\gamma - 1} \sqrt{\hat{H} - \frac{1}{2}\hat{u}^2}, \end{aligned} \tag{20.89}$$

where the angled brackets denote the arithmetic average of two states:

$$\langle b \rangle = \frac{1}{2}\left[b\left(\mathbf{U}^L_{j+\frac{1}{2}}\right) + b\left(\mathbf{U}^R_{j+\frac{1}{2}}\right)\right]. \tag{20.90}$$

The Jacobian matrix $A(\mathbf{U})$ evaluated at the Roe-averaged state will be denoted as

$$\hat{A}_{j+\frac{1}{2}} = A\left(\hat{\mathbf{U}}_{j+\frac{1}{2}}\right), \tag{20.91}$$

and we have

$$\hat{A}_{j+\frac{1}{2}} = P \Lambda P^{-1}, \tag{20.92}$$

where

$$P = \left(\mathbf{r}_1(\hat{\mathbf{U}}), \mathbf{r}_2(\hat{\mathbf{U}}), \mathbf{r}_3(\hat{\mathbf{U}})\right). \tag{20.93}$$

Finally, the Godunov–Roe flux is given in a form similar to (20.84) by

$$\hat{\mathbf{f}}^{(\mathrm{Roe})}_{j+\frac{1}{2}} = \frac{1}{2}\left[\mathbf{f}\left(\mathbf{U}^L_{j+\frac{1}{2}}\right) + \mathbf{f}\left(\mathbf{U}^R_{j+\frac{1}{2}}\right)\right] - \frac{1}{2}|\hat{A}_{j+\frac{1}{2}}|\left(\mathbf{U}^R_{j+\frac{1}{2}} - \mathbf{U}^L_{j+\frac{1}{2}}\right). \tag{20.94}$$

The Godunov–Roe flux needs an entropy fix to avoid the generation of non-physical expansion shock waves. It was given in [22] by modifying the eigenvalue $|\widehat{\lambda}_{j+\frac{1}{2}}|$ of $|\hat{A}_{j+\frac{1}{2}}|$ as follows:

$$|\widehat{\lambda}_{j+1/2}|_{\mathrm{mod}} = \begin{cases} |\widehat{\lambda}_{j+1/2}|, & \text{if } \; |\widehat{\lambda}_{j+1/2}| \geq \varepsilon, \\ \frac{1}{2}\left(\frac{\widehat{\lambda}^2_{j+1/2}}{\varepsilon} + \varepsilon\right), & \text{if } \; |\widehat{\lambda}_{j+1/2}| < \varepsilon, \end{cases} \tag{20.95}$$

where $\varepsilon = \max\left[0, (\widehat{\lambda}_{j+\frac{1}{2}} - \lambda_j), (\lambda_{j+1} - \widehat{\lambda}_{j+\frac{1}{2}})\right]$.

20.3 Weighted Essentially Non-Oscillatory (WENO) Finite Difference Methods

We illustrate the idea of the WENO finite difference method for a 1-D scalar conservation law:

$$u_t + f_x(u) = 0. \tag{20.96}$$

Extensions to the system of conservation laws will be discussed later. The time derivative and the spatial derivative will be discretized separately. For illustration purposes, we use the Euler forward difference method for the time derivative, and the numerical scheme will be written in the following conservative form:

$$u^{n+1}_j = u^n_j - \lambda\left(\hat{f}_{j+\frac{1}{2}} - \hat{f}_{j-\frac{1}{2}}\right), \tag{20.97}$$

where $u^n_j \approx u(x_j, t^n)$, $x_j = j\Delta x, t^n = n\Delta t$, Δx and Δt are the spatial mesh size and the time step size, respectively,

$$\lambda = \frac{\Delta t}{\Delta x}, \tag{20.98}$$

and $\hat{f}_{j\pm 1/2}$ are the numerical fluxes which should approximate the analytical fluxes $f(u(x_{j\pm 1/2}))$. In order to construct an accurate numerical flux, we consider a function $h(x)$ which is related to $f(u)$ in the following cell-averaging operation [23]:

$$f(u(x)) = \frac{1}{\Delta x}\int_{x-\frac{\Delta x}{2}}^{x+\frac{\Delta x}{2}} h(\xi)\mathrm{d}\xi. \tag{20.99}$$

Equation (20.99) implies that the derivative $f_x(u)$ can be expressed as

$$f_x(u(x_j)) = \frac{h\left(x_j + \frac{\Delta x}{2}\right) - h\left(x_j - \frac{\Delta x}{2}\right)}{\Delta x}, \tag{20.100}$$

which suggests that the numerical fluxes $\hat{f}_{j+1/2}$ should approximate $h(x_j + \Delta x/2)$ as $\Delta x \to 0$.

Therefore, our goal is to find the function $h(x)$ from knowledge of the solution $u(x)$, which will be achieved as follows. Define the primitive function $H(x)$ of $h(x)$ as

$$H(x) = \int_{-\frac{\Delta x}{2}}^{x} h(\xi)\mathrm{d}\xi. \tag{20.101}$$

Assuming that (20.99) holds, then, for $0 \le j \le N$, we have

$$H(x_{j+\frac{1}{2}}) = \int_{-\frac{\Delta x}{2}}^{x_{j+\frac{1}{2}}} h(\xi)\mathrm{d}\xi = \sum_{k=0}^{j} \int_{x_{k-\frac{1}{2}}}^{x_{k+\frac{1}{2}}} h(\xi)\mathrm{d}\xi = \Delta x \sum_{k=0}^{j} f(u(x_k)). \tag{20.102}$$

Now let $q_r(x)$ be an rth-order polynomial over the interval $[x_j, x_{j+1}]$, which interpolates $H(x_{j+1/2})$ over $r+1$ mesh points including x_j and x_{j+1}. There are, however, r choices of the stencil for this purpose, as we can use each of the following stencils S_k, $k = 0, 1, \ldots, r-1$:

$$S_k = \{x_{j+k-r+1}, x_{j+k-r+2}, \ldots, x_{j+k}\}, \tag{20.103}$$

and we have

$$q_r(x_j) = H(x_{j+\frac{1}{2}}), \qquad \text{for } j+k-r+1 \le i \le j+k. \tag{20.104}$$

Then, the numerical flux can be defined as

$$\hat{f}_{j+\frac{1}{2}} = \frac{\mathrm{d}}{\mathrm{d}x} q_r(x_{j+\frac{1}{2}}). \tag{20.105}$$

It should be noted that the numerical scheme thus obtained is based on a direct finite difference type approximation to the differential equation (20.96), different from the finite volume type approximation in the Godunov scheme in Sect. 20.2. However, the approach in this section can be extended to multi-dimensional problems without much more cost, while the reconstruction step in the Godunov-type method involves finding the point values of a function from its multi-dimensional cell averages, which is not an easy task.

Table 20.1 Coefficients $a^r_{k,l}$

r	k	$l=0$	$l=1$	$l=2$
2	0	$-1/2$	$3/2$	
	1	$1/2$	$1/2$	
3	0	$1/3$	$-7/6$	$11/6$
	1	$-1/6$	$5/6$	$1/3$
	2	$1/3$	$5/6$	$-1/6$

As the primitive function of $h(x)$ is related to $q_r(x_{j+1/2})$ through a derivative operation in (20.102), Eq. (20.105) implies that the numerical flux will be a linear combination of the values $f(u(x_i))$, $j+k-r+1 \le i \le j+k$,, i.e.,

$$\hat{f}_{j+\frac{1}{2}} = L_{r,k}(f_{j+k-r+1}, f_{j+k-r}, \ldots, f_{j+k}), \tag{20.106}$$

where $L_{r,k}$ is a linear operator of the following form:

$$L_{r,k}(g_0, \ldots, g_{r-1}) = \sum_{l=0}^{r-1} a^r_{k,l} g_l. \tag{20.107}$$

Here, the coefficients $a^r_{k,l}$ can be obtained from the interpolating polynomial $q_r(x)$ in (20.105). Table 20.1 contains the coefficients $a^r_{k,l}$ for $r=2$ and 3 [24].

The original essentially non-oscillatory (ENO) idea selected a stencil S_k that had the smoothest numerical data based on comparing the relative magnitude of undivided differences for the numerical solution in a recursive manner (namely, starting from the first undivided difference (20.113) up to the rth undivided difference (20.114)) [14, 23]. In contrast, the weighted ENO will use the result of (20.106) on all stencils S_k in a weighted convex combination as follows [25]:

$$\begin{aligned}\hat{f}_{j+\frac{1}{2}} &= L_{2r-1,r-1}(f_{j-r+1}, \ldots, f_{j+r-1}) \\ &+ \sum_{k=0}^{r-1} \left(\omega_k - C^r_k\right) L_{r,k}(f_{j+k-r+1}, f_{j+k-r}, \ldots, f_{j+k}),\end{aligned} \tag{20.108}$$

where coefficients C^r_k are defined by the following conversion formula:

$$L_{2r-1,r-1}(f_{j-r+1}, \ldots, f_{j+r-1}) = \sum_{k=0}^{r-1} C^r_k L_{r,k}(f_{j+k-r+1}, f_{j+k-r}, \ldots, f_{j+k}). \tag{20.109}$$

Some optimal choices for C_k^r are $C_0^2 = 1/3$, $C_1^2 = 2/3$; $C_0^3 = 1/10$, $C_1^3 = 6/10$, and $C_2^3 = 3/10$ [24]. The weights ω_k are defined as

$$\omega_k = \frac{\alpha_k}{\alpha_0 + \cdots + \alpha_{r-1}}, \quad k = 0, 1, \ldots, r-1, \tag{20.110}$$

where

$$\alpha_k = \frac{C_k^r}{(\varepsilon + IS_k)^p}. \tag{20.111}$$

Here, IS_k is a smoothness indicator of the solution, i.e.,

$$IS_k = \sum_{l=1}^{r-1}\sum_{i=1}^{r-l} \frac{(f[j+k+i-r, l])^2}{r-l}, \tag{20.112}$$

where $f[\cdot,\cdot]$ is the lth undivided difference:

$$f[j,0] = f_j, \tag{20.113}$$

$$f[j,l] = f[j+1, l-1] - f[j, l-1], \quad l = 1, 2, \ldots, r-1. \tag{20.114}$$

- **WENO finite difference for hydrodynamic equations of electron transport**

In order to apply the WENO scheme for the hydrodynamic equations (20.57), we apply the WENO formula (20.108) to each of the s-characteristic fields, where the characteristic directions are defined by the left eigenvectors $\mathbf{l}_s(\overline{\mathbf{u}}_{j+1/2})$, $s = 1, 2, 3$, where the overbar indicates the Roe-averaged state between $\mathbf{u}_j$ and $\mathbf{u}_{j+1}$, (20.89).

We define the flux for the system as

$$\widetilde{f}_{j+\frac{1}{2},s} = \sum_{k=0}^{r-1} w_{k,s} L_{r,k}(\mathbf{l}_s \cdot \mathbf{f}_{j+k-r+1}, \ldots, \mathbf{l}_s \cdot \mathbf{f}_{j+k}), \tag{20.115}$$

where $w_{k,s}$ are the weights to be used for the sth characteristic field:

$$w_{k,s} = \omega_k(\mathbf{l}_s \cdot \mathbf{f}_{j-r+1}, \ldots, \mathbf{l}_s \cdot \mathbf{f}_{j+r}), \tag{20.116}$$

defined in (20.110)–(20.112).

Finally, the numerical flux is computed by combining the fluxes for all s-components:

$$\hat{f}_{j+\frac{1}{2}} = \sum_{k=1}^{3} \widetilde{f}_{j+\frac{1}{2},k}\mathbf{r}_k, \tag{20.117}$$

where $\mathbf{r}_k = \mathbf{r}_k(\overline{\mathbf{u}}_{j+1/2})$.

20.4 Central Differencing Schemes with Staggered Grids

The NT central differencing scheme of Nessyahu and Tadmor [26] is based on the cell-average form of the conservation law of (20.96). Consider a cell average centered at x:

$$\overline{u}(t) = \frac{1}{\Delta x}\int_{x-\frac{\Delta x}{2}}^{x+\frac{\Delta x}{2}} u(\xi, t)\mathrm{d}\xi. \tag{20.118}$$

In particular, we could define the cell average $\overline{u}_{j+1/2}(t)$ centered at $x_{j+1/2}$ as

$$\overline{u}_{j+\frac{1}{2}}(t) = \frac{1}{\Delta x}\int_{x_{j+\frac{1}{2}}-\frac{\Delta x}{2}}^{x_{j+\frac{1}{2}}+\frac{\Delta x}{2}} u(x, t)\mathrm{d}x, \tag{20.119}$$

or the cell average $\overline{u}_j(t)$ centered at x_j as

$$\overline{u}_j(t) = \frac{1}{\Delta x}\int_{x_j-\frac{\Delta x}{2}}^{x_j+\frac{\Delta x}{2}} u(x, t)\mathrm{d}x. \tag{20.120}$$

These two sets of cell averages are defined on a staggered grid, and the NT central differencing scheme will provide the procedure for obtaining one set of the two cell averages at t_{n+1} from the other set at t_n, for instance

$$\left\{\overline{u}_j^n\right\} \to \left\{\overline{u}_{j+1/2}^{n+1}\right\}.$$

To illustrate this procedure, let us assume that the cell averages $\overline{u}_j^n$ over the interval $[x_{j-1/2}, x_{j+1/2}]$ are given at t_n. To find the cell averages $\overline{u}_{j+1/2}^{n+1}$ over the staggered interval $[x_j, x_{j+1}]$ at t_{n+1}, we integrate the conservation law in the space–time volume $[x_j, x_{j+1}] \times [t_n, t_n + \Delta t]$ as indicated by the shaded region in Fig. 20.1, and we obtain the evolution equation for the cell averages $\overline{u}_{j+1/2}(t)$:

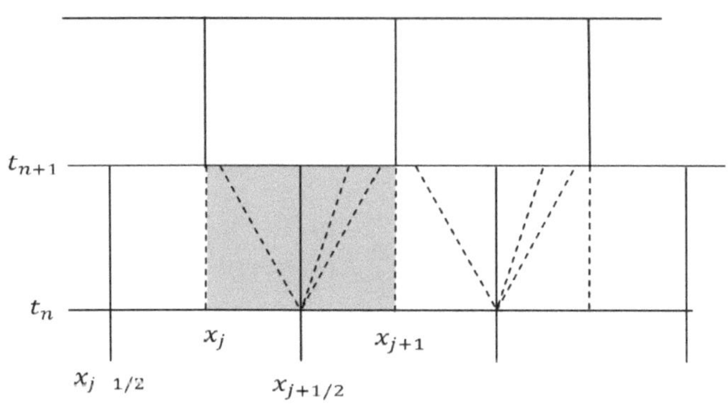

Fig. 20.1 Staggered grids for the NT central differencing scheme

$$\overline{u}_{j+\frac{1}{2}}(t_n + \Delta t) = \overline{u}_{j+\frac{1}{2}}(t_n) - \frac{1}{\Delta x}\left[\int_{t_n}^{t_n+\Delta t} f(u(x_{j+1}, \tau))\mathrm{d}\tau - \int_{t_n}^{t_n+\Delta t} f(u(x_j, \tau))\mathrm{d}\tau\right]. \tag{20.121}$$

The evolution equation for $\overline{u}_{j+1/2}(t_n + \Delta t)$ is exact provided that the flux function $f(u(x_j, t))$ is known for all $t \in [t_n, t_n + \Delta t]$ at all mesh points x_j. The second-order NT scheme is based on a middle-point approximation of the integrals on the right-hand side of (20.121), whose second-order accuracy will be ensured if the function $f(u(x_j, t))$ is smooth along the time interval. The latter smoothness condition is made possible if we assume the solution at t_n is represented by a piecewise polynomial with breaking points at $x_{j+1/2}$ and is due to the finite speed of propagation of the discontinuity, for sufficiently small Δt, such that

$$\frac{\Delta t}{\Delta x}\rho(A(u)) < \frac{1}{2}, \tag{20.122}$$

where $A(u) = \partial f(u)/\partial u$ is the Jacobian matrix of the flux function and $\rho(A)$ denotes the spectral radius of the matrix A.

The time integral of the flux function is calculated using the middle-point rule

$$\int_{t_n}^{t_n+\Delta t} f(u(x_j, \tau))\mathrm{d}\tau = \Delta t f\left(u\left(x_j, t_n + \frac{\Delta t}{2}\right)\right) + O(\Delta t^3). \tag{20.123}$$

Next, the quantity $u(x_j, t_n + \Delta t/2)$ will be approximated by a Taylor expansion in time and a Lax–Wendroff technique, which replaces the time derivative by the spatial derivative with the help of the partial differential equations:

$$\begin{aligned} u\left(x_j, t_n + \frac{\Delta t}{2}\right) &= u(x_j, t_n) + \frac{\Delta t}{2} u_t(x_j, t_n) + O(\Delta t^2) \\ &= u(x_j, t_n) - \frac{\Delta t}{2} f'(u(x_j, t_n)) + O(\Delta t^2). \end{aligned} \tag{20.124}$$

Then, (20.121), (20.123), and (20.124) give the following approximate scheme for $\overline{u}^{n+1}_{j+1/2}$:

$$\begin{aligned} \overline{u}^{n+1}_{j+\frac{1}{2}} = \overline{u}^{n}_{j+\frac{1}{2}} - \frac{\Delta t}{\Delta x}\Bigg[& f\left(u(x_{j+1}, t_n) - \frac{\Delta t}{2} f'(u(x_{j+1}, t_n))\right) \\ & - f\left(u(x_j, t_n) - \frac{\Delta t}{2} f'(u(x_j, t_n))\right)\Bigg]. \end{aligned} \tag{20.125}$$

Next, we will try to approximate cell average $\overline{u}^n_{j+1/2}$ from the given cell average $\overline{u}^n_j$. The numerical solution at each time t_n can be assumed to be a piecewise polynomial over the interval $[x_{j-1/2}, x_{j+1/2}]$. For instance, for second-order accuracy,

$$u(x,t) = L_j(x,t) \equiv \overline{u}_j(t) + (x - x_j)\frac{1}{\Delta x}u'_j, \quad x \in [x_{j-\frac{1}{2}}, x_{j+\frac{1}{2}}], \tag{20.126}$$

and the cell average of the solution coincides with the point value at the center of the cell:

$$\overline{u}_j(t) = u_j(t). \tag{20.127}$$

To ensure the second-order accuracy of the approximation in (20.126), we require the approximate derivative u'_j to satisfy

$$\frac{1}{\Delta x}u'_j = \frac{\partial}{\partial x}u(x_j, t) + O(\Delta x). \tag{20.128}$$

The ENO idea could be used for this purpose [27–29]:

$$u'_j = MM\left(d_{j-\frac{1}{2}}u + \frac{1}{2}MM(D_{j-1}u, D_j u), d_{j+\frac{1}{2}}u - \frac{1}{2}MM(D_j u, D_{j+1}u)\right), \tag{20.129}$$

where

$$d_{j+\frac{1}{2}}u = u_{j+1} - u_j, \tag{20.130}$$

$$D_j u = u_{j+1} - 2u_j + u_{j-1}, \tag{20.131}$$

and the min-mod function

$$MM(x,y) = \begin{cases} \text{sign}(x) \cdot \min(|x|, |y|), & \text{if } \text{sign}(x) = \text{sign}(y), \\ 0, & \text{otherwise.} \end{cases} \tag{20.132}$$

The staggered cell average $\overline{u}_{j+1/2}(t_n)$ can be computed from the piecewise polynomial representation (20.126) of the solution at $t = t_n$, i.e.,

$$\begin{aligned} \overline{u}_{j+\frac{1}{2}}(t) &= \frac{1}{\Delta x}\left[\int_{x_j}^{x_{j+\frac{1}{2}}} L_{j-1}(x,t)\mathrm{d}x + \int_{x_{j+\frac{1}{2}}}^{x_{j+1}} L_j(x,t)\mathrm{d}x\right] \\ &= \frac{1}{2}[\overline{u}_j(t) + \overline{u}_{j+1}(t)] + \frac{1}{8}\left(u'_j - u'_{j+1}\right). \end{aligned} \tag{20.133}$$

Finally, substituting (20.133) into the approximate evolution formula (20.125), we obtain the following relation:

$$\overline{u}_{j+\frac{1}{2}}(t_n + \Delta t) = \frac{1}{2}[\overline{u}_j(t) + \overline{u}_{j+1}(t)] + \frac{1}{8}\left(u'_j - u'_{j+1}\right)$$

$$
\begin{aligned}
&- \lambda \left[f \left(u(x_{j+1}, t_n) - \frac{\Delta t}{2} f'(u(x_{j+1}, t_n)) \right) \right. \\
&\left. - f \left(u(x_j, t_n) - \frac{\Delta t}{2} f'(u(x_j, t_n)) \right) \right] \\
&+ O(\Delta t^3), \qquad (20.134)
\end{aligned}
$$

which gives the following evolution of the cell averages between two staggered grids from t_n to t_{n+1}.

Second-order NT central differencing scheme over the staggered mesh: $\{\overline{u}_j^n\} \rightarrow \{\overline{u}_{j+1/2}^{n+1}\}$

(1) Predictor step

Given the cell averages over the cell $[x_{j-1/2,} x_{j+1/2}]$ at t_n, namely, $\overline{u}_j^n$, calculate

$$u_j^{n+\frac{1}{2}} = \overline{u}_j^n - \frac{\Delta t}{2} f'(u_j^n), \qquad (20.135)$$

where the derivative $f'(u_j^n)$ can be computed through the Jacobian matrix:

$$f'(u_j^n) = A(u_j^n) u_j'^n, \qquad (20.136)$$

with $u_j'^n$ computed using (20.129)–(20.131) and(20.127):

$$u_j'^n = MM \left(d_{j-\frac{1}{2}} \overline{u}^n + \frac{1}{2} MM(D_{j-1}\overline{u}^n, D_j \overline{u}^n), d_{j+\frac{1}{2}} \overline{u}^n - \frac{1}{2} MM(D_j \overline{u}^n, D_{j+1}\overline{u}^n) \right). \qquad (20.137)$$

(2) Corrector step

$$\overline{u}_{j+\frac{1}{2}}^{n+1} = \frac{1}{2} \left(\overline{u}_j^n + \overline{u}_{j+1}^n \right) - \lambda \left(g_{j+1} - g_j \right), \qquad (20.138)$$

where

$$g_j = f \left(u_j^{n+\frac{1}{2}} \right) + \frac{1}{8\lambda} u_j'^n. \qquad (20.139)$$

- **Numerical results of a 1-D n^+–n–n^+ diode**

In the following, we present some numerical results for a 1-D GaAs diode, as shown in Fig. 20.2, which consists of 0.25 μm source and drain sections, highly doped with electron donors n^+, and a 0.25 μm channel section of n-type lightly doped material. In the n^+ doped region, the doping density $N = 5 \times 10^{17}$ cm^{-3}, while the channel region dope density $N = 2 \times 10^{15}$ cm^{-3}. The effective mass of the electron for GaAs is $m = 0.063 m_e$ at temperature $T = 300$ K, where m_e is the free electron mass, and the relative dielectric constant for GaAs $\epsilon = 12.9$. The relaxation times

Fig. 20.2 A n^+–n–n^+ diode

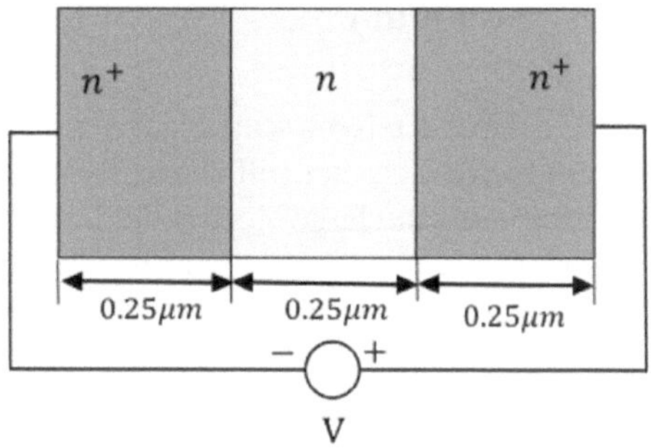

are $\tau_p = \tau_w = 0.2$ ps. At the left and right boundaries of the diode, the constant number density $n = N$ is used, and $T = T_0$ is given at the left boundary; finally, the potential $\phi = 0$ at the left and $\phi = 1$ V at the right boundary. Figure 20.3 [30] shows the velocity, density, temperature, and electric field across the diode. It shows the formation of a shock wave at $x = 0.2$ μm into the channel section and a high temperature increase as the electron leaves the channel to go into the drain section to the right, and also the space charge profile in the device. The solid lines are results obtained by the Godunov-type scheme implemented in the software CLAWPACK [2], and the dotted lines are those by the central differencing scheme presented in this section.

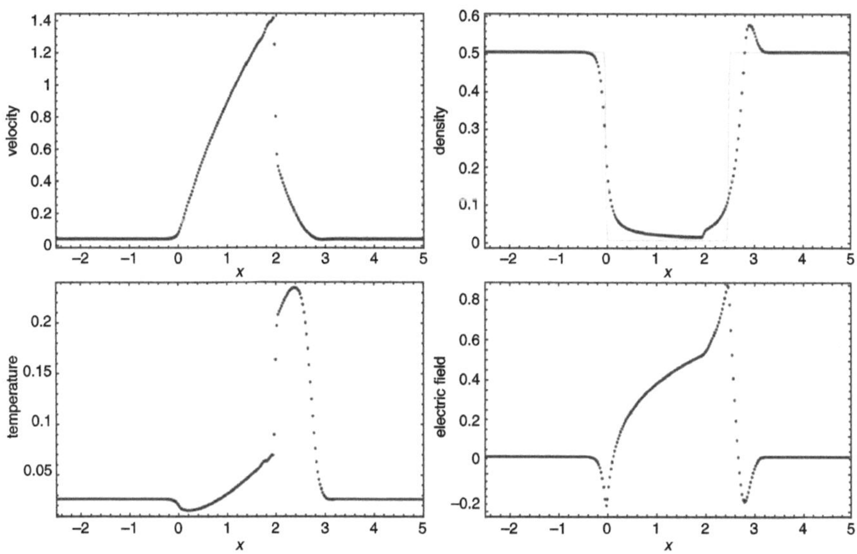

Fig. 20.3 Results for a n^+–n–n^+ diode. Top left: electron velocity; top right: electron density; bottom left: electron temperature; bottom right: electric field. The solid line depicts a Godunov method and, the dotted line depicts the central differencing scheme. The x-unit is 0.1 μm. Reprinted from [30], copyright (2002) by Hindawi.com

20.5 Summary

While devices at micron scales can be simulated by drift-diffusion models, for devices of sub-micron sizes but still above the scales of quantum transport regimes, temperature variance and large electric field will occur (as shown in Fig. 20.3). Therefore, hydrodynamic equations should be used to describe the transport of the so-called hot electrons, i.e., electrons that have gained energy from the large electric field and obtained a velocity distribution far away from that of an equilibrium Maxwellian. The shock profile of the electron velocity and density requires the non-oscillatory approximation of the numerical solutions, which is achieved by limiters in the reconstruction steps in Godunov-type methods or schemes using the weighted essentially non-oscillatory (WENO) interpolation techniques. More sophisticated positivity-preserving schemes can be used to maintain the physical properties of density and pressure [31, 32].

References

1. C.L. Gardner, Numerical simulation of a steady-state electron shock wave in a submicrometer semiconductor device. IEEE Trans. Electron Dev. **38**, 392–398 (1991)
2. R.J. LeVeque, *Finite Volume Methods for Hyperbolic Problems* (Cambridge University Press, Cambridge, 2002)
3. C. Hirsch, *Numerical Computation of Internal and External Flows: The Fundamentals of Computational Fluid Dynamics*, vol. 2, 2nd edn. (Butterworth-Heinemann, Oxford, 2011)
4. Y. Liu, C.-W. Shu, Local discontinuous Galerkin methods for moment models in device simulations: performance assessment and two-dimensional results. Appl. Numer. Math. **57**, 629–645 (2007)
5. S. Datta, *Quantum Phenomena* (Addison-Wesley, Reading, 1989)
6. M. Lundstrom, *Fundamentals of Carrier Transport*, 2nd edn. (Cambridge University Press, Cambridge, 2000)
7. K. Huang, *Statistical Mechanics*, 2nd edn. (Wiley, New York, 1987)
8. R.L. Liboff, *Kinetic Theory, Classical, Quantum, and Relativistic Descriptions*, 3rd edn. (Springer, Berlin, 2003)
9. T.T. Li, T.H. Qin, *Physics and Partial Differential Equations*, vol. 1 (SIAM, Philadelphia, 2012)
10. J.A. Bittencourt, *Fundamentals of Plasma Physics*, 3rd edn. (Springer, Berlin, 2004)
11. K. Bløtekjaer, Transport equations for electrons in two-valley semiconductors. IEEE Trans. Electron Dev. **17**, 38–47 (1970)
12. G. Baccarani, M.R. Wordeman, An investigation of steady-state velocity overshoot in silicon. Solid State Electron. **28**, 407–416 (1985)
13. K. Tomizawa, *Numerical Simulation of Submicron Semiconductor Devices* (Artech House Publishers, Norwood, 1993)
14. A. Harten, B. Engquist, S. Osher, S.R. Chakravarthy, Uniformly high order accurate essentially non-oscillatory schemes. III. J. Comput. Phys. **131**, 3–47 (1997)
15. S.K. Godunov, A finite-difference method for the numerical computation and discontinuous solutions of the equations of fluid dynamics. Mat. Sb. **47**, 271–306 (1959)
16. B. van Leer, Towards the ultimate conservative difference scheme. IV. A new approach to numerical convection. J. Comput. Phys. **23**, 276–299 (1977)
17. B. van Leer, Towards the ultimate conservative difference scheme. V. A second-order sequel to Godunov's method. J. Comput. Phys. **32**, 101–136 (1979)

18. P. Colella, P.R. Woodward, The piecewise-parabolic method (PPM) for gas-dynamical simulations. J. Comput. Phys. **54**, 174–201 (1984)
19. P.K. Sweby, High resolution schemes using flux limiters for hyperbolic conservation laws. SIAM J. Numer. Anal. **21**, 995–1011 (1984)
20. P.L. Roe, Upwind schemes using various formulations of the Euler equations, in *Numerical Methods for the Euler Equations of Fluid Dynamics*. ed. by F. Angrand, A. Dervieux, R. Glowinski, J.A. Desideri (SIAM, Philadelphia, 1985), pp.14–31
21. P.L. Roe, Approximate Riemann solvers, parameter vectors, and difference schemes. J. Comput. Phys. **43**, 357–372 (1981)
22. A. Harten, J.M. Hyman, Self adjusting grid methods for one-dimensional hyperbolic conservation laws. J. Comput. Phys. **50**, 235–269 (1983)
23. C.-W. Shu, S. Osher, Efficient implementation of essentially non-oscillatory shock capturing schemes. II. J. Comput. Phys. **83**, 32–78 (1989)
24. G.-S. Jiang, C.-W. Shu, Efficient implementation of weighted ENO scheme. J. Comput. Phys. **126**, 202–228 (1996)
25. X.-D. Liu, S. Osher, T. Chan, Weighted essentially non-oscillatory schemes. J. Comput. Phys. **115**, 200–212 (1994)
26. H. Nessyahu, E. Tadmor, Non-oscillatory central differencing for hyperbolic conservation laws. J. Comput. Phys. **87**, 408–463 (1990)
27. X.-D. Liu, E. Tadmor, Third order non-oscillatory central scheme for hyperbolic conservation laws. Numer. Math. **79**, 397–425 (1998)
28. F. Bianco, G. Puppo, G. Russo, High-order central schemes for hyperbolic systems of conservation laws. SIAM J. Sci. Comput. **21**, 294–322 (1999)
29. V. Romano, G. Russo, Numerical solution for hydrodynamical models of semiconductors. Math. Model Meth. Appl. Sci. **10**, 1099–1120 (2000)
30. C.L. Gardner, A. Gelb, J. Hernandez, A comparison of modern hyperbolic methods for semiconductor device simulation: NTK central schemes vs. CLAWPACK. VLSI Des. **15**, 721–728 (2002)
31. X. Zhang, C.W. Shu, Positivity-preserving high order discontinuous Galerkin schemes for compressible Euler equations with source terms. J. Comput. Phys. **230**(4), 1238–1248 (2011)
32. K. Wu, C.W. Shu, Geometric quasilinearization framework for analysis and design of bound-preserving schemes. SIAM Rev. **65**(4), 1031–1073 (2023)

Chapter 21
Transport Models in Plasma Media and Numerical Methods

In this, the final chapter of the book, we study the transport phenomena in plasma due to electrons and ions. These phenomena have wide applications in astrophysics, confined nuclear thermal reactions, and in high-density laser plasma interactions, etc. Plasma, considered as the fourth state of matter in the universe, differs from solids, liquids, and gases, in that there is a much weakened bond strength between its constituent particles. The plasma state of a medium is created via high-temperature external heating, which results in an increase in both the thermal energy and the number of atomic ionizations (i.e., an electron in an outer shell of an atom escapes from its nuclear force confinement once it obtains enough external energy from photon excitation or collision), which produces free-moving electrons and ions. The primary force in a plasma medium is described by the long-range Coulomb forces of electrostatics. The motions of the electrons and the ions can be described by a kinetic theory with special treatment of the collision under the long-range electric potential, which includes the Boltzmann–Fokker–Planck equations or the Balescu–Lenard equations. A macroscopic description of the electron/ion density can also be obtained through the moments of the kinetic equation as the magneto-hydrodynamic (MHD) equations. Three types of numerical methods based on kinetic and hydrodynamic models will be discussed in this chapter. The first type is the Boltzmann–Fokker–Planck solver in phase space, and the second is the particle-in-cell method, which tracks the dynamics of individual particles under the Lorentz force of the electromagnetic fields; the latter is also coupled to the charge distributions of the moving particles. Finally, the third type is a constrained transport method of finite difference type for the MHD equations, which observes the divergence-free constraint on the magnetic field.

W. Cai, *Deterministic, Stochastic, and Deep Learning Methods for Computational Electromagnetics*, https://doi.org/10.1007/978-981-96-0100-4_21

21.1 Kinetic and Macroscopic Magneto-Hydrodynamic (MHD) Theories

21.1.1 Vlasov–Fokker–Planck Equations

The kinetic Boltzmann–Vlasov equation given in (20.10) will be used to describe electron and ion transport in a plasma medium. However, due to the long-range interaction of the Coulomb force, the binary collision mechanism (20.2) in the Boltzmann collision (Sect. 20.1.1) will not apply here as each electron will interact with many electrons or ions over a large distance, i.e., they will experience multiple Coulombic collisions, as well as interact with neutral particles in the system though through a different interaction potential. However, if the multiple Coulombic collisions can be considered as a sequence of binary collisions at grazing angles, the binary Boltzmann collision operator can be used to arrive at the Fokker–Planck equation [1].

Let $f_\alpha(\mathbf{r}, \mathbf{v}, t)$ denote the distribution function for the α-type particles (such as electrons, ions, and neutral particles). The Boltzmann collision operator for a conserved quantity χ (20.11) can be rewritten as

$$\left[\frac{\delta n_\alpha \langle\chi\rangle}{\delta t}\right]_{\text{coll}} = \sum_\beta \int_{\mathbf{v}_1}\int_{\mathbf{v}}\int_\Omega f_\alpha f_{\beta_1}(\chi' - \chi) g\sigma(\Omega) \mathrm{d}\Omega\, \mathrm{d}\mathbf{v}\, \mathrm{d}\mathbf{v}_1, \tag{21.1}$$

by using the identity

$$\sum_\beta \int_{\mathbf{v}_1}\int_{\mathbf{v}}\int_\Omega f'_\alpha f'_{\beta_1} \chi g\sigma(\Omega)\mathrm{d}\Omega \mathrm{d}\mathbf{v}\mathrm{d}\mathbf{v}_1 = \sum_\beta \int_{\mathbf{v}_1}\int_{\mathbf{v}}\int_\Omega f_\alpha f_{\beta_1} \chi' g\sigma(\Omega)\mathrm{d}\Omega \mathrm{d}\mathbf{v}\mathrm{d}\mathbf{v}_1, \tag{21.2}$$

where $f_\alpha = f_\alpha(\mathbf{r}, \mathbf{v}, t)$, $f_{\beta_1} = f_{\beta_1}(\mathbf{r}, \mathbf{v}_1, t)$, $f'_\alpha = f_\alpha(\mathbf{r}, \mathbf{v}', t)$, and $f'_{\beta_1} = f_{\beta_1}(\mathbf{r}, \mathbf{v}'_1, t)$.

Assuming that only the grazing angle collision is involved in (21.1), namely

$$\mathbf{v}' = \mathbf{v} + \Delta\mathbf{v}, \tag{21.3}$$

where $|\Delta\mathbf{v}| \ll 1$, we can apply the Taylor expansion to $\chi' = \chi(\mathbf{v}')$:

$$\chi' = \chi(\mathbf{v} + \Delta\mathbf{v}) = \chi(\mathbf{v}) + \sum_i \frac{\partial\chi}{\partial v_i}\Delta v_i + \frac{1}{2}\sum_{i,j}\frac{\partial^2\chi}{\partial v_i \partial v_j}\Delta v_i \Delta v_j + \cdots. \tag{21.4}$$

Then, the time rate of change of the quantity χ due to the collision between the α- and β-type particles is given by

$$\int_{\mathbf{v}} \chi(\mathbf{v}) \left[\frac{\delta f_\alpha}{\delta t}\right] \mathrm{d}\mathbf{v}$$
$$= \sum_\beta \int_{\mathbf{v}_1} \int_{\mathbf{v}} \int_\Omega f_\alpha f_{\beta_1} \left[\sum_i \frac{\partial \chi}{\partial v_i} \Delta v_i + \frac{1}{2} \sum_{i,j} \frac{\partial^2 \chi}{\partial v_i \partial v_j} \Delta v_i \Delta v_j\right] g\sigma(\Omega) \mathrm{d}\Omega \, \mathrm{d}\mathbf{v} \, \mathrm{d}\mathbf{v}_1. \tag{21.5}$$

Applying integration by parts for terms with $\partial \chi / \partial v_i$ and $\partial^2 \chi / \partial v_i \partial v_j$, and using the fact that f_α and $\partial f_\alpha / \partial v_i$ vanish as $|\mathbf{v}| \to \infty$, we obtain

$$\int_{\mathbf{v}} \chi(\mathbf{v}) \left[\frac{\delta f_\alpha}{\delta t}\right]_{\text{coll}} \mathrm{d}\mathbf{v}$$
$$= \sum_\beta \int_{\mathbf{v}_1} \int_{\mathbf{v}} \int_\Omega \chi(v) f_{\beta_1} \left[-\sum_i \frac{\partial}{\partial v_i} \Delta v_i f_\alpha g\sigma(\Omega) \mathrm{d}\Omega\right.$$
$$\left. + \frac{1}{2} \sum_{i,j} \frac{\partial^2}{\partial v_i \partial v_j} f_\alpha \Delta v_i \Delta v_j g\sigma(\Omega) \mathrm{d}\Omega \right] \mathrm{d}\mathbf{v} \, \mathrm{d}\mathbf{v}_1$$
$$= \int_{\mathbf{v}} \chi(\mathbf{v}) \left(-\sum_i \frac{\partial}{\partial v_i} f_\alpha \langle \Delta v_i \rangle + \frac{1}{2} \sum_{i,j} \frac{\partial^2}{\partial v_i \partial v_j} f_\alpha \langle \Delta v_i \Delta v_j \rangle \right) \mathrm{d}\mathbf{v}, \tag{21.6}$$

where the averaging operators $\langle \Delta v_i \rangle$ and $\langle \Delta v_i \Delta v_j \rangle$ are defined as

$$\langle \Delta v_i \rangle = \sum_\beta \int_{\mathbf{v}_1} \int_\Omega f_\beta(\mathbf{r}, \mathbf{v}_1, t)(v_i' - v_i) g\sigma(\Omega) \mathrm{d}\Omega \, \mathrm{d}\mathbf{v}_1, \tag{21.7}$$

and

$$\langle \Delta v_i \Delta v_j \rangle = \sum_\beta \int_{\mathbf{v}_1} \int_\Omega f_\beta(\mathbf{r}, \mathbf{v}_1, t)(v_i' - v_i)(v_j' - v_j) g\sigma(\Omega) \mathrm{d}\Omega \, \mathrm{d}\mathbf{v}_1, \tag{21.8}$$

respectively.

It follows from (21.6) that the collision term for the plasma takes the following form:

$$\left[\frac{\partial f_\alpha}{\partial t}\right]_{\text{coll}} = -\sum_i \frac{\partial}{\partial v_i} (c_i f_\alpha) + \frac{1}{2} \sum_{i,j} \frac{\partial^2}{\partial v_i \partial v_j} (\mu_{ij} f_\alpha) \equiv L_{\text{fp}}(f_\alpha), \tag{21.9}$$

which resembles a Fokker–Planck differential operator with friction and diffusion coefficients defined as

$$c_i = \langle \Delta v_i \rangle, \quad \mu_{ij} = \langle \Delta v_i \Delta v_j \rangle. \tag{21.10}$$

The Boltzmann–Vlasov equation in (20.10), together with (21.9), makes the Vlasov–Fokker–Planck kinetic equation for the plasma:

$$\frac{\partial f_\alpha}{\partial t} + \mathbf{v} \cdot \nabla f_\alpha + \mathbf{a} \cdot \nabla_{\mathbf{v}} f_\alpha = L_{\text{fp}}(f_\alpha). \tag{21.11}$$

21.1.2 MHD Equations for Plasma as a Conducting Fluid

As the kinetic equation is difficult to solve due to its six-dimensional phase space variables, macroscopic equations can be derived from its moment equation. For plasma, the force on the electrons or ions will be the Lorentz force:

$$\mathbf{F} = q(\mathbf{E} + \mathbf{v} \times \mathbf{B}). \tag{21.12}$$

It is easy to verify that

$$\nabla_{\mathbf{v}} \cdot \mathbf{F} = 0, \tag{21.13}$$

as the ith component of $\mathbf{F}$ will not involve the ith component of the velocity vector $\mathbf{v}$ from the definition of the vector cross product. Therefore, the condition in (20.18) is satisfied, and we can obtain the macroscopic equations for the plasma using the general moment equation (20.19) in Sect. 20.1.2.

MHD for plasma as a single conducting fluid

For many situations, we can use a single-fluid model where we can define an average mass density ρ,

$$\rho = \sum_\alpha \rho_\alpha, \quad \rho_\alpha = m_\alpha n_\alpha, \tag{21.14}$$

and an average charge density ρ_e,

$$\rho_e = \sum_\alpha n_\alpha q_\alpha. \tag{21.15}$$

An average velocity $\mathbf{u}$ is defined by

$$\rho \mathbf{u} = \sum_\alpha \rho_\alpha \mathbf{u}_\alpha, \tag{21.16}$$

namely through a mass density weighted average of the velocity of each type of particle.

The velocity fluctuation $\mathbf{v}'$ with respect to the average velocity $\mathbf{u}_\alpha$ defined in (20.23) for the α-type particle has been defined previously and is repeated here:

$$\mathbf{v} = \mathbf{u}_\alpha + \mathbf{v}', \tag{21.17}$$

and

$$\left\langle \mathbf{v}' \right\rangle_\alpha = 0. \tag{21.18}$$

In reference to the average velocity $\mathbf{u}$ of the single-fluid model, a new velocity fluctuation $\mathbf{v}^*$ can be defined as

$$\mathbf{v} = \mathbf{u} + \mathbf{v}^*, \tag{21.19}$$

with which we define a different pressure dyadic $\mathbf{P}_\alpha^*$ for each α-type particle:

$$\left(P_\alpha^*\right)_{ij} = \rho_\alpha \left\langle v_i^* v_j^* \right\rangle_\alpha . \tag{21.20}$$

For the single plasma fluid, an overall pressure dyadic $\mathbf{P}$ is defined by

$$\mathbf{P} = \sum_\alpha \rho_\alpha \left\langle \mathbf{v}^* \mathbf{v}^* \right\rangle_\alpha , \tag{21.21}$$

and the global scalar pressure p is then

$$p = \frac{1}{3} \sum_{i=1}^{3} P_{ii} = \frac{1}{3} \sum_\alpha \rho_a \left\langle v^{*2} \right\rangle_\alpha . \tag{21.22}$$

The mass density averaged pressure dyadic in (21.21) is in general different from the linear sum of the pressure p_α of (20.39) for each individual type of particle. In fact, by subtracting (21.17) from (21.19), we can see that the fluctuation with respect to the average velocity $\mathbf{u}_\alpha$ of the α-type particle and the global mean velocity of the one-fluid plasma $\mathbf{u}$ are related by

$$\mathbf{v}^* = \mathbf{v}' + (\mathbf{u}_\alpha - \mathbf{u}) = \mathbf{v}' - \mathbf{w}_\alpha \tag{21.23}$$

and

$$\mathbf{w}_\alpha = \mathbf{u} - \mathbf{u}_\alpha , \tag{21.24}$$

where $\mathbf{w}_\alpha$ is considered as the diffusive velocity of the α-type particle with respect to the overall plasma velocity, and we also have the following identity using (21.18):

$$\left\langle \mathbf{v}^* \right\rangle_\alpha = -\mathbf{w}_\alpha . \tag{21.25}$$

Now, substituting (21.23) into (21.21), we have

$$\mathbf{P} = \sum_\alpha \mathbf{P}_\alpha + \sum_\alpha \rho_\alpha \, \mathbf{w}_\alpha \mathbf{w}_\alpha . \tag{21.26}$$

Thus, the total scalar pressure p (trace of the dyadic) can be shown to be related to the individual scalar pressure p_α of (20.39) as follows:

$$p = \sum_\alpha p_\alpha + \frac{1}{3}\sum_\alpha \rho_\alpha\, w_\alpha^2. \tag{21.27}$$

Next, the hydrodynamic equations for the single-fluid plasma can be obtained from (20.21), (20.36), and (20.49). Firstly, by summing the continuity equation (20.21) over α and using the definition (21.16) and the conservation identity (20.25), we have the *continuity equation for the plasma fluid*:

$$\frac{\partial \rho}{\partial t} + \nabla \cdot (\rho \mathbf{u}) = 0. \tag{21.28}$$

Secondly, summing the momentum equation (20.36) over α, after using (21.16) and (20.38), we have

$$\frac{\partial \rho \mathbf{u}}{\partial t} = \frac{\partial}{\partial t}\sum_\alpha \rho_\alpha \mathbf{u}_\alpha = -\nabla \sum_\alpha \rho_\alpha \mathbf{u}_\alpha \mathbf{u}_\alpha - \sum_\alpha n_\alpha \mathbf{F}_\alpha - \nabla \sum_\alpha P_\alpha. \tag{21.29}$$

Using the decomposition (21.24), the sum in the first term on the right-hand side of (21.29) becomes

$$\begin{aligned}\sum_\alpha \rho_\alpha \mathbf{u}_\alpha \mathbf{u}_\alpha &= \sum_\alpha \rho_\alpha \mathbf{u}_\alpha (\mathbf{u} - \mathbf{w}_\alpha) = \sum_\alpha \rho_\alpha \mathbf{u}_\alpha \mathbf{u} - \sum_\alpha \rho_\alpha \mathbf{u}_\alpha \mathbf{w}_\alpha \\ &= \rho \mathbf{u}\mathbf{u} + \left(-2\sum_\alpha \rho_\alpha \mathbf{u}\mathbf{w}_\alpha + \sum_\alpha \rho_\alpha \mathbf{w}_\alpha \mathbf{w}_\alpha\right) \\ &= \rho \mathbf{u}\mathbf{u} + \sum_\alpha \rho_\alpha \mathbf{w}_\alpha \mathbf{w}_\alpha. \end{aligned} \tag{21.30}$$

The last equation in (21.30) holds due to the fact that

$$\begin{aligned}\sum_\alpha \rho_\alpha \mathbf{w}_\alpha &= -\sum_\alpha \rho_\alpha (\mathbf{u}_\alpha - \mathbf{u}) = -\sum_\alpha \rho_\alpha \mathbf{u}_\alpha + \sum_\alpha \rho_\alpha \mathbf{u} \\ &= -\rho \mathbf{u} + \rho \mathbf{u} = \mathbf{0}. \end{aligned} \tag{21.31}$$

The second term on the right-hand side of (21.29) is

$$\begin{aligned}\sum_\alpha n_\alpha \mathbf{F}_\alpha &= \sum_\alpha n_\alpha q_\alpha (\mathbf{E} + \mathbf{u}_\alpha \times \mathbf{B}) = \rho_e \mathbf{E} + \left(\sum_\alpha n_\alpha q_\alpha \mathbf{u}_\alpha\right) \times \mathbf{B} \\ &= \rho_e \mathbf{E} + \mathbf{J} \times \mathbf{B}, \end{aligned} \tag{21.32}$$

where the current $\mathbf{J}$ is defined as

$$\mathbf{J} = \sum_\alpha n_\alpha q_\alpha \mathbf{u}_\alpha. \tag{21.33}$$

Substituting (21.30) and (21.32) into (21.29), and using the definition of the global pressure dyadic in (21.26), we have the *momentum equation for the plasma fluid*:

$$\frac{\partial \rho \mathbf{u}}{\partial t} = -\nabla(\rho \mathbf{u}\mathbf{u}) - (\rho_e \mathbf{E} + \mathbf{J} \times \mathbf{B}) - \nabla P. \tag{21.34}$$

Finally, to arrive at an energy equation for the single plasma fluid, we use the energy equation for the α-type particle with the general macroscopic equation (20.19):

$$\frac{\partial}{\partial t} n_\alpha \langle \chi \rangle_\alpha + \nabla \cdot (n_\alpha \langle \chi \mathbf{v} \rangle_\alpha) - n_\alpha \langle \mathbf{a}_\alpha \cdot \nabla_{\mathbf{v}} \chi \rangle_\alpha = \frac{\delta}{\delta t} \left[n_\alpha \langle \chi \rangle_\alpha \right]_{\text{coll}}, \tag{21.35}$$

with $\chi = m_\alpha v^2/2$, and the facts that

$$\mathbf{a}_\alpha = \frac{1}{m_\alpha} \mathbf{F}_\alpha, \; \mathbf{F}_\alpha = q_\alpha (\mathbf{E} + \mathbf{v} \times \mathbf{B}), \;\; \text{and} \quad \nabla_{\mathbf{v}} \chi = \frac{1}{2} m_\alpha \mathbf{v}. \tag{21.36}$$

After a summation over α (the energy loss terms cancel due to conservation), we have

$$\frac{\partial}{\partial t} \frac{1}{2} \sum_\alpha \rho_\alpha \left\langle v^2 \right\rangle_\alpha + \nabla \cdot \sum_\alpha \frac{\rho_\alpha}{2} \left\langle v^2 \mathbf{v} \right\rangle_\alpha - \sum_\alpha n_\alpha \langle \mathbf{F} \cdot \mathbf{v} \rangle_\alpha = 0. \tag{21.37}$$

Now we use the velocity decomposition with respect to the global velocity $\mathbf{u}$ in (21.19) to evaluate each of the terms in (21.37). To simplify the first term on the left-hand side of (21.37), we consider

$$\begin{aligned} \frac{1}{2} \sum_\alpha \rho_\alpha \left\langle v^2 \right\rangle_\alpha &= \frac{1}{2} \sum_\alpha \rho_\alpha u^2 + \frac{1}{2} \sum_\alpha \rho_\alpha \left\langle v^{*2} \right\rangle_\alpha \\ &+ \sum_\alpha \rho_\alpha \left\langle \mathbf{u}\mathbf{v}^* \right\rangle_\alpha = \frac{1}{2} \rho u^2 + \frac{3}{2} p. \end{aligned} \tag{21.38}$$

The last equality is due to the fact that

$$\sum_\alpha \rho_\alpha \left\langle \mathbf{u}\mathbf{v}^* \right\rangle_\alpha = \mathbf{u} \sum_\alpha \rho_\alpha \left\langle \mathbf{v}^* \right\rangle_\alpha = -\mathbf{u} \sum_\alpha \rho_\alpha \mathbf{w}_\alpha = 0, \tag{21.39}$$

where the identity (21.25) is used for the second equality and the final equality is due to (21.31).

Here, the energy density W of the plasma is defined as

$$W = \frac{3}{2} p + \frac{1}{2} \rho u^2. \tag{21.40}$$

Next, we examine the term inside the second summation in (21.37):

$$\begin{aligned}\langle v^2\mathbf{v}\rangle_\alpha &= \langle (u^2 + 2\mathbf{u}\cdot\mathbf{v}^* + v^{*2})(\mathbf{u}+\mathbf{v}^*)\rangle_\alpha \\ &= \left[u^2\mathbf{u} + 2\mathbf{u}\cdot\langle\mathbf{v}^*\rangle_\alpha\mathbf{u} + \langle v^{*2}\rangle_\alpha\mathbf{u}\right] \\ &\quad + \left[u^2\langle\mathbf{v}^*\rangle_\alpha + 2\mathbf{u}\langle\mathbf{v}^*\mathbf{v}^*\rangle_\alpha + \langle v^{*2}\mathbf{v}^*\rangle_\alpha\right] \\ &= \left[u^2\mathbf{u} - 2\mathbf{u}\cdot\mathbf{w}_\alpha\mathbf{u} + \langle v^{*2}\rangle_\alpha\mathbf{u}\right] \\ &\quad + \left[-u^2\mathbf{w}_\alpha + 2\mathbf{u}\langle\mathbf{v}^*\mathbf{v}^*\rangle_\alpha + \langle v^{*2}\mathbf{v}^*\rangle_\alpha\right].\end{aligned} \tag{21.41}$$

Therefore, using (21.31), the second term in (21.37) becomes

$$\begin{aligned}\sum_\alpha \frac{\rho_\alpha}{2}\langle v^2\mathbf{v}\rangle_\alpha &= \sum_\alpha \frac{\rho_\alpha}{2}u^2\mathbf{u} + \sum_\alpha\frac{\rho_\alpha}{2}\langle v^{*2}\rangle_\alpha\mathbf{u} \\ &\quad + \sum_\alpha\rho_\alpha\mathbf{u}\langle\mathbf{v}^*\mathbf{v}^*\rangle_\alpha + \sum_\alpha\frac{\rho_\alpha}{2}\langle v^{*2}\mathbf{v}^*\rangle_\alpha \\ &= \frac{\rho}{2}u^2\mathbf{u} + \frac{3}{2}p\mathbf{u} + P\mathbf{u} + \mathbf{q},\end{aligned} \tag{21.42}$$

where the heat flux $\mathbf{q}$ is defined as

$$\mathbf{q} = \frac{1}{2}\sum_\alpha \rho_\alpha\langle v^{*2}\mathbf{v}^*\rangle_\alpha. \tag{21.43}$$

The third term in (21.37) is

$$\begin{aligned}\sum_\alpha n_\alpha\langle\mathbf{F}\cdot\mathbf{v}\rangle_\alpha &= \sum_\alpha n_\alpha\int_{\mathbf{v}} f_\alpha q_\alpha(\mathbf{E}+\mathbf{v}\times\mathbf{B})\cdot\mathbf{v}\,\mathrm{d}\mathbf{v} \\ &= \sum_\alpha n_\alpha\int_{\mathbf{v}} f_\alpha q_\alpha\mathbf{E}\cdot\mathbf{v}\,\mathrm{d}\mathbf{v} = \left(\sum_\alpha n_\alpha q_\alpha\mathbf{u}_\alpha\right)\cdot\mathbf{E} \\ &= \mathbf{J}\cdot\mathbf{E}.\end{aligned} \tag{21.44}$$

Substituting (21.38), (21.42), and (21.44) into (21.37), we obtain the *energy equation for the plasma fluid*:

$$\frac{\partial W}{\partial t} + \nabla\cdot(W+P)\mathbf{u} + \nabla\cdot\mathbf{q} = \mathbf{J}\cdot\mathbf{E}. \tag{21.45}$$

Also, a single global temperature for the plasma fluid can be defined through the relation

$$p = nk_BT, \tag{21.46}$$

where $n = \sum_\alpha n_\alpha$.

In many situations concerning plasmas, the time dependence of the electric field (displacement current) in Maxwell's equations is ignored. The magnetic flux is then related to the current by the simplified static Ampère's law,

$$\nabla \times \mathbf{B} = \mu_0 \mathbf{J}, \tag{21.47}$$

where the current $\mathbf{J}$ is given by a generalized Ohm's law [2] for the plasma conducting fluid:

$$\mathbf{J} = \sigma(\mathbf{E} + \mathbf{u} \times \mathbf{B}). \tag{21.48}$$

Using Faraday's law, and (21.48) to replace $\mathbf{E}$, we have

$$\begin{aligned} \frac{\partial \mathbf{B}}{\partial t} &= -\nabla \times \mathbf{E} = -\nabla \times \left(\frac{1}{\sigma}\mathbf{J} - \mathbf{u} \times \mathbf{B} \right) \\ &= -\nabla \times \left(\frac{1}{\sigma \mu_0} \nabla \times \mathbf{B} - \mathbf{u} \times \mathbf{B} \right), \end{aligned}$$

where (21.47) is used to obtain the final equation. Next, using the zero divergence of the magnetic induction $\mathbf{B}$, we finally have the dynamic equation

$$\frac{\partial \mathbf{B}}{\partial t} = \nabla \times (\mathbf{u} \times \mathbf{B}) + \frac{1}{\sigma \mu_0} \nabla^2 \mathbf{B}. \tag{21.49}$$

The complete ideal MHD equations ($\sigma = \infty$), consisting of (21.28), (21.34), (21.45), and (21.49), can be written in the following system of conservation laws:

$$\frac{\partial \mathbf{U}}{\partial t} + \frac{\partial \mathbf{f}}{\partial x} + \frac{\partial \mathbf{g}}{\partial y} + \frac{\partial \mathbf{h}}{\partial z} = \mathbf{s}, \tag{21.50}$$

where the unknown $\mathbf{U}$ consists of the conserved density, momentum, total energy $\mathcal{E} = W + |\mathbf{B}|^2/2 = \rho u^2/2 + p/(\gamma - 1) + |\mathbf{B}|^2/2$, and the magnetic induction in (21.49)

$$\mathbf{U} = (\rho, \rho \mathbf{u}, \mathcal{E}, \mathbf{B})^{\mathrm{T}}. \tag{21.51}$$

The fluxes are defined as

$$\mathbf{f} = \begin{pmatrix} \rho u_x \\ \rho u_x^2 + p + \frac{1}{2}|\mathbf{B}|^2 - B_x^2 \\ \rho u_x u_y - B_x B_y \\ \rho u_x u_z - B_x B_z \\ u_x \left(\mathcal{E} + p + \frac{1}{2}|\mathbf{B}|^2 \right) - B_x (\mathbf{u} \cdot \mathbf{B}) \\ 0 \\ u_x B_y - u_y B_x \\ -(u_z B_x - u_x B_z) \end{pmatrix}, \tag{21.52}$$

$$\mathbf{g} = \begin{pmatrix} \rho u_y \\ \rho u_x u_y - B_x B_y \\ \rho u_y^2 + p + \frac{1}{2}|\mathbf{B}|^2 - B_y^2 \\ \rho u_y u_z - B_y B_z \\ u_y \left(\mathcal{E} + p + \frac{1}{2}|\mathbf{B}|^2\right) - B_y(\mathbf{u}\cdot\mathbf{B}) \\ -\left(u_x B_y - u_y B_x\right) \\ 0 \\ u_y B_z - u_z B_y \end{pmatrix}, \tag{21.53}$$

and

$$\mathbf{h} = \begin{pmatrix} \rho u_z \\ \rho u_x u_z - B_x B_z \\ \rho u_y u_z - B_y B_z \\ \rho u_z^2 + p + \frac{1}{2}|\mathbf{B}|^2 - B_z^2 \\ u_z \left(\mathcal{E} + p + \frac{1}{2}|\mathbf{B}|^2\right) - B_z(\mathbf{u}\cdot\mathbf{B}) \\ u_z B_x - u_x B_z \\ -\left(u_y B_z - u_z B_y\right) \\ 0 \end{pmatrix}, \tag{21.54}$$

where the equation for the energy $\mathcal{E}$ is obtained by combining (21.45) and the conservation of the magnetic second moment $|\mathbf{B}|^2$ based on (21.49), and the source term $\mathbf{s}$ contains the remaining undifferentiated terms.

21.2 Vlasov–Fokker–Planck (VFP) Schemes

A fundamental microscopic approach to model electron transport and plasma-laser interaction is to solve the Boltzmann equation in (20.10) directly within the Landau approximation of the collision operator in the form of nonlinear Fokker–Planck operator [3–5]. This kinetic approach is particularly required when there is a large temperature gradient in the plasma medium so that the heat flux approximation in the hydrodynamic model fails and the heat flux will have a non-local dependence on temperature and density [6]. When there is a strong non-local effect due to the comparable scale of the electron mean free path (mfp) and the plasma [7], or a strong collision heating by the laser field [8], the electron distribution function will be non-Maxwellian, which was used for moment closure for the hydrodynamic model.

In the VFP scheme [5], the electron distribution function $f(\mathbf{r}, \mathbf{p}, t)$ is approximated by the spherical harmonics as follows:

$$f(\mathbf{r}, \mathbf{p}, t) = \sum_{l=0}^{\infty} \sum_{m=-l}^{l} f_l^m(r, |\mathbf{p}|, t) P_l^{|m|}(\cos\theta) e^{im\varphi}, \;\; f_l^{-m} = \left(f_l^m\right)^*. \tag{21.55}$$

The Boltzmann equation (20.10) can be rewritten in terms of the expansion coefficients as

$$\frac{\partial f_l^m}{\partial t} - \mathcal{A}_{l,x}^m - \mathcal{A}_{l,y}^m - \mathcal{A}_{l,z}^m - \mathcal{E}_{l,x}^m - \mathcal{E}_{l,y}^m - \mathcal{E}_{l,z}^m - \mathcal{B}_l^m = C_{l,i}^m + C_{l,e}^m + S, \quad (21.56)$$

where various terms in (21.56) are projections under the spherical harmonics basis for the spatial advection, electric and magnetic forces, and collisions, and S is the source term from possible laser fields. Note that $C_{l,i}^m$ and $C_{l,e}^m$ are the angular scattering contributions of electrons off ions and between electrons, respectively [4]. The spatial advection terms are given by

$$\begin{aligned}
\mathcal{A}_{l,x}^{m\geq 0} &= -v\partial_x \left[\left(\frac{l-m}{2l-1}\right) f_{l-1}^m + \frac{l+m+1}{2l+3} f_{l+1}^m\right], \\
\mathcal{A}_{l,y}^{m>0} + \mathcal{A}_{l,z}^{m>0} &= -\frac{v}{2}\left[\frac{\partial_y - \mathrm{i}\partial_z}{2l-1} f_{l-1}^{m-1} - \frac{\partial_y + \mathrm{i}\partial_z}{2l-1}(l-m)(l-m-1) f_{l-1}^{m+1}\right. \\
&\quad \left. - \frac{\partial_y - \mathrm{i}\partial_z}{2l+3} f_{l+1}^{m-1} + \frac{\partial_y + \mathrm{i}\partial_z}{2l+3}(l+m+1)(l+m+2) f_{l+1}^{m+1}\right], \\
\mathcal{A}_{l,y}^0 + \mathcal{A}_{l,z}^0 &= \mathrm{Re}\left\{-v(\partial_y + \mathrm{i}\partial_z)\left[-\frac{l(l-1)}{2l-1} f_{l-1}^1 + \frac{(l+1)(l+2)}{2l+3} f_{l+1}^1\right]\right\}.
\end{aligned} \quad (21.57)$$

The electric field terms are given by

$$\begin{aligned}
\mathcal{E}_{l,x}^{m\geq 0} &= E_x \left[\frac{l-m}{2l-1} G_{l-1}^m + \frac{l+m+1}{2l+3} H_{l+1}^m\right], \\
\mathcal{E}_{l,y}^{m>0} + \mathcal{E}_{l,z}^{m>0} &= \frac{1}{2}\left[\frac{E_y - \mathrm{i}E_z}{2l-1} G_{l-1}^{m-1} - \frac{E_y + \mathrm{i}E_z}{2l-1}(l-m)(l-m-1) G_{l-1}^{m+1}\right. \\
&\quad \left. - \frac{E_y - \mathrm{i}E_z}{2l+3} H_{l+1}^{m-1} + \frac{E_y + \mathrm{i}E_z}{2l+3}(l+m+1)(l+m+2) H_{l+1}^{m+1}\right], \\
\mathcal{E}_{l,y}^0 + \mathcal{E}_{l,z}^0 &= \mathrm{Re}\left\{(E_y + \mathrm{i}E_z)\left[-\frac{l(l-1)}{2l-1} G_{l-1}^1 + \frac{(l+1)(l+2)}{2l+3} H_{l+1}^1\right]\right\},
\end{aligned} \quad (21.58)$$

where

$$G_l^m = p^l \partial_p (p^{-l} f_l^m), \quad H_l^m = p^{-l-1}\partial_p (p^{l+1} f_l^m),$$

and the magnetic field terms are

$$\begin{aligned}
B_l^{m>0} &= -\mathrm{i}B_x m f_l^m \\
&\quad + \frac{1}{2}\left[(l-m)(l+m+1)(B_z - \mathrm{i}B_y) f_l^{m+1} - (B_z + \mathrm{i}B_y) f_l^{m-1}\right], \quad (21.59) \\
B_l^0 &= l(l+1)\mathrm{Re}\left\{(B_z - \mathrm{i}B_y) f_l^1\right\}. \quad (21.60)
\end{aligned}$$

The periodic or reflection boundary condition can be used for the distribution; for the latter case, we have

$$f_{Rx}(x_b \pm x, p_x) = f(x_b \mp x, -p_x) = \sum_{l=0}^{\infty}\sum_{m=-l}^{l}(-1)^{l+m} f_l^m P_l^{|m|}(\cos\theta)\mathrm{e}^{\mathrm{i}m\varphi},$$

$$f_{Ry}(y_b \pm y, p_y) = f(y_b \mp y, -p_y) = \sum_{l=0}^{\infty}\sum_{m=-l}^{l}(-1)^{l+m} f_l^{-m} P_l^{|m|}(\cos\theta)\mathrm{e}^{\mathrm{i}m\varphi}. \tag{21.61}$$

Next, we investigate the collision in the Boltzmann equation (20.10) between a particle of mass m and charge $q = ze$ (e is the electron charge here) corresponding to the distribution f and a scattering center (electron or ion) of mass $M = \mu m$ and charge $Q = Ze$, described by a distribution F. The Fokker–Planck collision operator [9], which is given by

$$\frac{1}{\Gamma_{zZ}}\left(\frac{\delta f}{\delta t}\right) = \frac{4\pi}{\mu}Ff + \frac{\mu-1}{\mu+1}\nabla\mathbb{H}(F)\cdot\nabla f + \frac{\nabla\nabla\mathbb{G}(F)\cdot\nabla\nabla f}{2}, \tag{21.62}$$

where ∇ is the gradient operator in the velocity space,

$$\Gamma_{zZ} = 4\pi(zZe^2)^2\ln\Lambda/m^2, \quad \Lambda = \frac{3T}{Ze^2}\lambda_D. \tag{21.63}$$

Note that λ_D is the Debye length in the plasma medium ($\lambda_D = \kappa^{-1}$, where κ is defined in (2.17)), and $\mathbb{H}(F)$ and $\mathbb{G}(F)$ are the Rosenbluth potentials [10] in the form of integral operators for F:

$$\mathbb{H}(F) = \frac{1+\mu}{\mu}\sum_{l=0}^{\infty}\sum_{m=-l}^{l}\frac{1/v}{2l+1}\left[I_l(F_l^m) + J_{-1-l}(F_l^m)\right]P_l^{|m|}(\cos\theta)\mathrm{e}^{\mathrm{i}m\varphi}, \tag{21.64}$$

$$\mathbb{G}(F) = \sum_{l=0}^{\infty}\sum_{m=-l}^{l} v\left[\frac{I_{l+2}(F_l^m) + J_{-1-l}(F_l^m)}{(2l+1)(2l+3)} - \frac{I_l(F_l^m) + J_{1-l}(F_l^m)}{(2l-1)(2l+1)}\right]P_l^{|m|}(\cos\theta)\mathrm{e}^{\mathrm{i}m\varphi}, \tag{21.65}$$

where

$$I_j = \frac{4\pi}{v^j}\int_0^v F_l^m(u)u^{j+2}\,\mathrm{d}u, \tag{21.66}$$

$$J_j = \frac{4\pi}{v^j}\int_v^\infty F_l^m(u)u^{j+2}\,\mathrm{d}u. \tag{21.67}$$

Appropriate finite difference discretization for the spatial derivatives can be used in (21.57), and numerical quadratures will be used for the integrations in (21.66) and (21.67) [5].

21.3 Particle-In-Cell (PIC) Schemes

The motion of charges follows Newton's law according to the Lorentz force of the electromagnetic fields **E** and **B**, which are coupled self-consistently to the charge distribution and current in Maxwell's equations. The particle-in-cell (PIC) schemes are designed to simulate the motion of particles in space, while a fixed mesh is used to define the electromagnetic fields [11, 12]. Consider N_p charged particles located at $\mathbf{x}_i(t)$ with velocity $\mathbf{v}_i(t)$ and mass m, $1 \leq i \leq N_p$, which are moved by the Lorentz force according to the following set of equations:

$$\begin{aligned} \frac{\mathrm{d}\mathbf{x}_i}{\mathrm{d}t} &= \mathbf{v}_i, \\ \frac{\mathrm{d}m\mathbf{v}_i}{\mathrm{d}t} &= q\left[\mathbf{E}(\mathbf{x}_i) + \mathbf{v}_i \times \mathbf{B}(\mathbf{x}_i)\right]. \end{aligned} \tag{21.68}$$

The electromagnetic fields **E** and **B** are defined on a fixed mesh in the space, which could be the Yee mesh in the finite difference method or given as piecewise polynomials in a discontinuous Galerkin (DG) method. In Maxwell's equations, the current $\mathbf{J}(\mathbf{x})$ in the Ampère–Maxwell equation can be computed by interpolating the individual point-charge currents

$$\mathbf{J}(\mathbf{x}) = \sum_{i=1}^{N_p} q_i \mathbf{v}_i S(|\mathbf{x}_i - \mathbf{x}|), \tag{21.69}$$

where the weight function $S(|\mathbf{x}_i - \mathbf{x}|)$ interpolates the point-charge current $q_i\mathbf{v}_i$ to the location $\mathbf{x}$, and the space charge density $\rho(\mathbf{x})$ can be similarly defined:

$$\rho(\mathbf{x}) = \sum_{i=1}^{N_p} q_i S(|\mathbf{x}_i - \mathbf{x}|). \tag{21.70}$$

Various weight functions S can be used; the simplest one can be of a polynomial type:

$$S(r) = \begin{cases} \dfrac{n+1}{\pi a^2}\left[1 - \left(\dfrac{r}{a}\right)^2\right]^n, & \text{if } r \leq a, \\ 0, & \text{if } r > a, \end{cases} \tag{21.71}$$

where n is an integer.

With the current so defined in (21.69), the electromagnetic fields **E** and **B** can be found at a new time step by using either the finite difference method [12] or the DG method [13]. Then, the particle position and velocity can be advanced by (21.68). However, as the field quantities at the particle locations $\mathbf{x}_i$ are required, which are not at the lattice points of a finite difference mesh, an interpolation of the field quantities to the particle location will also be needed. In the case of finite element approximations (including DG) for the field quantities, such an interpolation is not required as the finite element solutions are provided at all spatial locations by the basis function representation of the solutions.

The self-consistent procedure above, however, does not take the Gauss equation for the electric field into consideration, so in general the electric field **E** thus computed does not satisfy Gauss's law for the given charge density defined in (21.70), which implies that the continuity equation is violated. Various methods have been proposed to enforce Gauss's law in the PIC schemes [14–16]. A global projection method [12] (similar to the projection methods for enforcing the divergence-free condition of the velocity field in incompressible fluids [17, 18]) will make corrections to the electric field based on a Helmholtz decomposition for the provisional value of the electric field at a current time step, denoted by $\mathbf{E}^*$, i.e.,

$$\mathbf{E}^* = \mathbf{E} + \nabla\phi, \tag{21.72}$$

where **E** will assume the given divergence, namely it satisfies the Gauss equation

$$\epsilon_0 \nabla \cdot \mathbf{E} = \rho(\mathbf{x}), \tag{21.73}$$

and the correction potential will satisfy the following Poisson equation with a zero boundary condition on the computational domain Ω:

$$\begin{cases} \epsilon_0 \nabla^2 \phi = \epsilon_0 \nabla \cdot \mathbf{E}^* - \rho(\mathbf{x}), \\ \phi|_{\partial\Omega} = 0. \end{cases} \tag{21.74}$$

21.4 $\nabla \cdot \mathbf{B} = 0$ Constrained Transport Methods for MHD Equations

The MHD equations describe the dynamics of a charged system under an interaction with a magnetic field and the conservation of the mass, momentum, and energy for the plasma system. The dynamics is considered constrained, as the magnetic field of the system develops with the constraint of zero divergence, namely $\nabla \cdot \mathbf{B} = 0$. Numerical modeling of plasmas has shown that the observance of the zero divergence of the magnetic field plays an important role in reproducing the correct physics in the plasma fluid [19]. Various numerical techniques have been devised to ensure that the computed magnetic field remains divergence-free [20]. In the original work of

Brackbill and Barnes [19] a projection approach as discussed in Sect. 21.3 was used to correct the magnetic field to ensure zero divergence.

A more natural way of satisfying this constraint is through a class of so-called constrained transport (CT) numerical methods based on the ideas in [21]. As noted in [22], a piecewise $H(\mathrm{div})$ vector field on a finite element triangulation of a spatial domain can be a global $H(\mathrm{div})$ field if and only if the normal components on the interface of adjacent elements are continuous. Thus, in most of the CT algorithms for the MHD equations, the surface-averaged magnetic flux over the surfaces of a 3-D element will be used to represent the magnetic field, while the volume-averaged quantities are used for mass, momentum, and energy. In this section, we briefly illustrate the key ideas of the CT numerical methods using a second-order Godunov scheme [23]. Various higher-order methods have been developed to increase the accuracy under the same framework of the CT methods.

Consider a cubic cell C_{ijk} whose center is indexed with (i, j, k). The unknowns in a Godunov scheme will be the cell-averaged values for the solution of the conserved hydrodynamic variables in (21.50):

$$\overline{\mathbf{U}}_{i,j,k} \approx \frac{1}{|C_{ijk}|} \int_{C_{ijk}} \mathbf{U}(\mathbf{x}, t) d\mathbf{x}, \tag{21.75}$$

and also the cell-surface-averaged magnetic flux variables defined over the surfaces of each cell C_{ijk} (refer to Fig. 13.3):

$$B_{x,i+\frac{1}{2},j,k}, \quad B_{y,i,j+\frac{1}{2},k}, \quad B_{z,i,j,k+\frac{1}{2}}, \tag{21.76}$$

which correspond to the cell-surface-averaged magnetic flux on the front, the right, and the top surfaces of the cube (refer to Fig. 13.3). These surface-averaged magnetic fluxes will be the primary variables for the magnetic field, whose evolution will be based on the integral form of the induction equation (13.65), as in the case of the Yee scheme.

As shown in the Yee scheme, it can be directly verified by using (13.80)–(13.82) that a discrete version of divergence satisfies

$$\left(\nabla_h \mathbf{B}^{n+1}\right)_{ijk} = \left(\nabla_h \mathbf{B}^{n}\right)_{ijk}, \tag{21.77}$$

where

$$\begin{aligned} \left(\nabla_h \mathbf{B}^{n+1}\right)_{ijk} \equiv & \frac{1}{\Delta x}\left(B^{n+1}_{x,i+\frac{1}{2},j,k} - B^{n+1}_{x,i-\frac{1}{2},j,k}\right) + \frac{1}{\Delta y}\left(B^{n+1}_{y,i,j+\frac{1}{2},k} - B^{n+1}_{y,i,j-\frac{1}{2},k}\right) \\ & + \frac{1}{\Delta z}\left(B^{n+1}_{z,i,j,k+\frac{1}{2}} - B^{n+1}_{z,i,j,k-\frac{1}{2}}\right); \end{aligned} \tag{21.78}$$

thus if initially $\nabla_h \mathbf{B}^0 = 0$, then $\nabla_h \mathbf{B}^n = 0$ for all later times.

In (13.80)–(13.82), the electric field $\mathbf{E}$ along the edge of the cell is needed at $t^{n+1/2}$. Using the fact that in an ideal plasma fluid the electric field is related to the magnetic field by

$$\mathbf{E} = -\mathbf{u} \times \mathbf{B}, \tag{21.79}$$

and also noting [23] that the components of the right-hand side of (21.79) appear in the last three components of the fluxes in (21.52), (21.53), and (21.54), we can use the fluxes to compute the required electric field quantities once the former are defined from the Riemann solver of the underlying Godunov scheme at the same time level. Thus, as in [23], the following formulae can be used:

$$E^{n+\frac{1}{2}}_{x,i,j+\frac{1}{2},k+\frac{1}{2}} = \frac{1}{4}\left(\hat{h}^{n+\frac{1}{2}}_{7,i,j,k+\frac{1}{2}} + \hat{h}^{n+\frac{1}{2}}_{7,i,j+1,k+\frac{1}{2}} - \hat{g}^{n+\frac{1}{2}}_{8,i,j+\frac{1}{2},k} - \hat{g}^{n+\frac{1}{2}}_{8,i,j+\frac{1}{2},k+1}\right), \tag{21.80}$$

$$E^{n+\frac{1}{2}}_{y,i+\frac{1}{2},j,k+\frac{1}{2}} = \frac{1}{4}\left(\hat{f}^{n+\frac{1}{2}}_{8,i+\frac{1}{2},j,k} + \hat{f}^{n+\frac{1}{2}}_{8,i+\frac{1}{2},j,k+1} - \hat{h}^{n+\frac{1}{2}}_{6,i,j,k+\frac{1}{2}} - \hat{h}^{n+\frac{1}{2}}_{6,i+1,j,k+\frac{1}{2}}\right), \tag{21.81}$$

$$E^{n+\frac{1}{2}}_{z,i+\frac{1}{2},j+\frac{1}{2},k} = \frac{1}{4}\left(\hat{g}^{n+\frac{1}{2}}_{6,i,j+\frac{1}{2},k} + \hat{g}^{n+\frac{1}{2}}_{6,i+1,j+\frac{1}{2},k} - \hat{f}^{n+\frac{1}{2}}_{7,i+\frac{1}{2},j,k} - \hat{f}^{n+\frac{1}{2}}_{7,i+\frac{1}{2},j+1,k}\right), \tag{21.82}$$

where $\hat{\boldsymbol{f}}$, $\hat{\boldsymbol{h}}$, and $\hat{\boldsymbol{g}}$ are the numerical fluxes of a Godunov scheme for (21.50).

The enforcement of the zero divergence for the magnetic field in higher-order schemes can be ensured for the Godunov type [24, 25], central differencing [26], and WENO schemes [27]. A key step towards generating the higher-order divergence-free discretizations of the MHD equations is to obtain a divergence-free magnetic field **B** (required to compute with high accuracy the energy density and the pressure in the momentum equation) composed of piecewise polynomials and consistent with the primary-face-averaged magnetic flux unknowns on the element faces. Such a high-order reconstruction can be found in [24, 27].

Finally, we present the eigensystems for the MHD fluxes which are needed in the characteristic projection during the WENO interpolation in Sect. 20.3 or the Roe–Riemann solver for the definition of numerical fluxes in the Godunov scheme of Sect. 20.2. Let us consider the 1-D version of the MHD system (21.50), i.e., all physical quantities are functions of one variable only, say x. Define the primitive variables

$$\mathbf{V} = (\rho, u_x, u_y, u_z, B_y, B_z, p)^{\mathrm{T}}, \tag{21.83}$$

where the magnetic field B_x is a constant due to the zero divergence of the magnetic induction flux **B**. The MHD system can be rewritten as the following hyperbolic system in a non-conservative form:

$$\frac{\partial \mathbf{V}}{\partial t} + A(\mathbf{V})\frac{\partial \mathbf{V}}{\partial x} = 0, \tag{21.84}$$

where the coefficient matrix $A(\mathbf{V})$ is related to the Jacobian matrix of the numerical flux $\mathbf{f}(\mathbf{U})$ by

$$A(\mathbf{V}) = U_v \frac{\partial \mathbf{f}}{\partial \mathbf{U}} U_v^{-1}, \ U_v = \frac{\partial \mathbf{U}}{\partial \mathbf{V}}, \tag{21.85}$$

and

$$A(\mathbf{V}) = \begin{pmatrix} u_x & \rho & 0 & 0 & 0 & 0 & 0 \\ 0 & u_x & 0 & 0 & \frac{B_y}{\rho} & \frac{B_z}{\rho} & \frac{1}{\rho} \\ 0 & 0 & u_x & 0 & -\frac{B_x}{\rho} & 0 & 0 \\ 0 & 0 & 0 & u_x & 0 & -\frac{B_x}{\rho} & 0 \\ 0 & B_y & -B_x & 0 & u_x & 0 & 0 \\ 0 & B_z & 0 & -B_x & 0 & u_x & 0 \\ 0 & \rho c^2 & 0 & 0 & 0 & 0 & u_x \end{pmatrix}, \tag{21.86}$$

where c is the speed of sound in the plasma

$$c^2 = \gamma p/\rho. \tag{21.87}$$

The matrix $A(\mathbf{V})$ has seven eigenvalues which define seven plane wave structures of the hyperbolic system with corresponding left eigenvectors $\mathbf{l}_k$ and right eigenvectors $\mathbf{r}_k$ listed as follows [28, 29].

- One entropy wave:

$$\begin{aligned} \lambda_e &= u_x, \\ \mathbf{l}_e &= \left(1, 0, 0, 0, 0, 0, -\frac{1}{c^2}\right), \\ \mathbf{r}_e &= (1, 0, 0, 0, 0, 0, 0)^{\mathrm{T}}. \end{aligned} \tag{21.88}$$

- Two Alfvén waves:

$$\begin{aligned} \lambda_a &= u_x \pm \frac{B_x}{\sqrt{\rho}}, \\ \mathbf{l}_a &= \left(0, 0, -B_z, B_y, \pm\frac{B_z}{\sqrt{\rho}}, \mp\frac{B_y}{\sqrt{\rho}}, 0\right), \\ \mathbf{r}_a &= (0, 0, -B_z, B_y, \pm\sqrt{\rho}B_z, \mp\sqrt{\rho}B_y, 0)^{\mathrm{T}}. \end{aligned} \tag{21.89}$$

- Four magneto-acoustic waves:

$$\lambda_{f,s} = u_x \pm c_{f,s},$$

$$\mathbf{l}_{f,s} = \left(0, \pm\rho c_{f,s}, \mp\frac{B_x B_y \rho c_{f,s}}{\rho c_{f,s}^2 - B_x^2}, \mp\frac{B_x B_z \rho c_{f,s}}{\rho c_{f,s}^2 - B_x^2}, \frac{B_y \rho c_{f,s}^2}{\rho c_{f,s}^2 - B_x^2}, \frac{B_z \rho c_{f,s}^2}{\rho c_{f,s}^2 - B_x^2}, 1\right),$$

$$\mathbf{r}_{f,s} = \left(\rho, \pm c_{f,s}, \mp \frac{B_x B_y \rho c_{f,s}}{\rho c_{f,s}^2 - B_x^2}, \mp \frac{B_x B_z \rho c_{f,s}}{\rho c_{f,s}^2 - B_x^2}, \frac{B_y \rho c_{f,s}^2}{\rho c_{f,s}^2 - B_x^2}, \frac{B_z \rho c_{f,s}^2}{\rho c_{f,s}^2 - B_x^2}, \gamma p \right)^{\mathrm{T}}, \tag{21.90}$$

where c_f and c_s are the fast and slow magneto-acoustic wave speeds, respectively, i.e.,

$$c_{f,s}^2 = \frac{1}{2}\left[c^2 + \frac{|\mathbf{B}|^2}{\rho} \pm \sqrt{\left(c^2 + \frac{|\mathbf{B}|^2}{\rho}\right) - 4\frac{c^2 B_x^2}{\rho}} \right]. \tag{21.91}$$

We can also label all the eigenvalues in the increasing order as λ_k, i.e.,

$$\lambda_1 < \lambda_2 < \cdots < \lambda_7.$$

Then, we have

$$\mathbf{l}_k A = \lambda_k \mathbf{l}_k, \quad A\mathbf{r}_k = \lambda_k \mathbf{r}_k, \tag{21.92}$$

and

$$\mathbf{l}_i \mathbf{r}_j = 0, \quad i \neq j. \tag{21.93}$$

Next, a numerical upwinding flux based on the Roe–Riemann solver can be defined for any two states on both sides of $x = 0$ as

$$\widetilde{\mathbf{f}}(\mathbf{U}_L, \mathbf{U}_R) = \frac{1}{2}\left[\mathbf{f}(\mathbf{U}_L) + \mathbf{f}(\mathbf{U}_R)\right] - \frac{1}{2}\sum_{k=1}^{7} \alpha_k |\lambda_k| \mathbf{R}_k, \tag{21.94}$$

where $\mathbf{R}_k = U_v(\overline{\mathbf{U}})\mathbf{r}_k(\overline{\mathbf{U}})$, $\overline{\mathbf{U}}$ is the Roe-averaged state, and α_k is the projection of the difference vector $\mathbf{U}_L - \mathbf{U}_R$ along the $\mathbf{R}_k$ direction.

21.5 Summary

The foundation of the transport theory in plasma is based on the Boltzmann–Vlasov–Landau kinetic equation for the statistical distribution function $f(\mathbf{r}, \mathbf{v}, t)$ in the 6-D $(\mathbf{r}, \mathbf{v})$ phase space, self-consistently coupled with Maxwell's equations. The kinetic theory is based on the critical assumption that two-particle (electron or ion) correlation in the system will go to zero as the intra-particle distance grows, similar to Boltzmann's molecular chaos assumption. Considering the Debye screening effect and small-angle collision assumptions, the Landau or Fokker–Planck collision operator is obtained, based on which the VFP algorithm is developed here. In order to avoid the cost from the 6-D phase space, the particle-in-cell (PIC) method can be used, where the distribution function in the phase space can be used for sampling the initial "super" particle; then, the self-consistent evolution of the particles under the Lorentz force can be materialized, just as in a typical molecular dynamics simulation. Even so, the charge conservation in a PIC scheme needs careful attention when observing the ever-important Gauss's law in a charged system. Finally, for a faster and more macroscopic simulation of the electron transport in the plasma, the MHD equations for the plasma treated as a conducting fluid can be derived from the moment equations of the Boltzmann–Vlasov equation and a generalized Ohm's law relating the current and the electric field in the plasma; finite difference methods can then be used, with special attention paid to the divergence-free constraint of the magnetic field, i.e., $\nabla \cdot \mathbf{B} = 0$.

References

1. L. Landau, The transport equation in the case of Coulomb interactions. Phys. Z. Sowj. Union, (Reprinted in D. TerHaar (ed.) (1965), Collected Papers of L. D. Landau, vol. 10 (Pergamon Press, Oxford, 1936), pp. 154–164
2. J.A. Bittencourt, *Fundamentals of Plasma Physics*, 3rd edn. (Springer, Berlin, 2004)
3. F. Alouani-Bibi, M.M. Shoucri, J.-P. Matte, Different Fokker-Planck approaches to simulate electron transport in plasmas. Comput. Phys. Commun. **164**, 60–66 (2004)
4. A.R. Bell, A.P.L. Robinson, M. Sherlock, R.J. Kingham, W. Rozmus, Fast electron transport in laser-produced plasmas and the KALOS code for solution of the Vlasov-Fokker-Planck equation. Plasma Phys. Control Fusion **48**, R37–R57 (2006)
5. M. Tzoufras, A.R. Bell, P.A. Norreys, F.S. Tsung, A Vlasov–Fokker–Planck code for high energy density physics. J. Comput. Phys. **230**, 6475–6494 (2011)
6. L. Spitzer Jr., R. Härm, Transport phenomena in a completely ionized gas. Phys. Rev. **89**, 977–981 (1953)
7. F. Alouani-Bibi, J.-P. Matte, Influence of the electron distribution function shape on nonlocal electron heat transport in laser-heated plasmas. Phys. Rev. E **66**, 066414 (2002)
8. A.B. Langdon, Nonlinear inverse Bremsstrahlung and heated-electron distributions. Phys. Rev. Lett. **44**, 575–579 (1980)
9. I.P. Shkarofsky, T.W. Johnston, M.P. Bachynski, *The Particle Kinetics of Plasmas* (Addison-Wesley, Reading, 1966)
10. M.N. Rosenbluth, W.M. MacDonald, D.L. Judd, Fokker-Planck equation for an inverse-square force. Phys. Rev. **107**, 1–6 (1957)
11. R.W. Hockney, J.W. Eastwood, *Computer Simulation Using Particles* (Taylor & Francis, New York, 1981)
12. C.K. Birdsall, A.B. Langdon, *Plasma Physics via Computer Simulation* (Taylor & Francis, New York, 2004)
13. G.B. Jacobs, J.S. Hesthaven, High-order nodal discontinuous Galerkin particle-in-cell method on unstructured grids. J. Comput. Phys. **214**, 96–121 (2006)
14. J. Villasenor, O. Buneman, Rigorous charge conservation for local electromagnetic field solvers. Comput. Phys. Commun. **69**, 306–316 (1992)
15. TZh. Esirkepov, Exact charge conservation scheme for particle-in-cell simulation with an arbitrary form-factor. Comput. Phys. Commun. **135**, 144–153 (2001)
16. T. Umeda, Y. Omura, T. Tominaga, H. Matsumoto, A new charge conservation method in electromagnetic particle-in-cell simulations. Comput. Phys. Commun. **156**, 73–85 (2003)
17. A.J. Chorin, Numerical solution of the Navier-Stokes equations. Math. Comput. **22**, 745–762 (1968)
18. R. Temam, Une méthode d'approximation des solutions des équations Navier-Stokes. Bull. Soc. Math. France **98**, 115–152 (1968)

19. J.U. Brackbill, D.C. Barnes, The effect of nonzero $\nabla \cdot \mathbf{B}$ on the numerical solution of the magnetohydrodynamic equations. J. Comput. Phys. **35**, 426–430 (1980)
20. G. Tóth, The $\nabla \cdot \mathbf{B} = 0$ constraint in shock-capturing magnetohydrodynamics codes. J. Comput. Phys. **161**, 605–652 (2000)
21. C.R. Evans, J.F. Hawley, Simulation of magnetohydrodynamic flows: a constrained transport method. Astrophys. J. **332**, 659–677 (1988)
22. P. Monk, *Finite Element Methods for Maxwell's Equations* (Oxford University Press, Oxford, 2003)
23. D.S. Balsara, Divergence-free adaptive mesh refinement for magnetohydrodynamics. J. Comput. Phys. **174**, 614–648 (2001)
24. D.S. Balsara, Second order accurate schemes for magnetohydrodynamics with divergence-free reconstruction. Astrophys. J. Suppl. S. **151**, 149–184 (2004)
25. T.A. Gardiner, J.M. Stone, An unsplit Godunov method for ideal MHD via constrained transport in three dimensions. J. Comput. Phys. **227**, 4123–4141 (2008)
26. S. Li, A fourth-order divergence-free method for MHD flows. J. Comput. Phys. **229**, 7893–7910 (2010)
27. D.S. Balsara, Divergence-free reconstruction of magnetic fields and WENO schemes for magnetohydrodynamics. J. Comput. Phys. **228**, 5040–5056 (2009)
28. K.G. Powell, *An Approximate Riemann Solver for Magnetohydrodynamics (That Works in More Than One Dimension)*. Technical Report 94–24, ICASE, Langley, VA, 1994
29. P.L. Roe, D.S. Balsara, Notes on the eigensystem of magnetohydrodynamics. SIAM J. Appl. Math. **56**, 57–67 (1996)

Index

A
Absorbing boundary conditions
  Bayliss–Turkel for Maxwell's equations, in auxiliary variables, 240
  Bayliss–Turkel for scalar wave, in auxiliary variables, 238
  Engquist–Majda one-way wave, 236
  Higdon boundary condition, 238
  uniaxial PML (UPML), 242
Auxiliary differential equation
  Debye materials, 424
  Drude materials, 430

B
Band gap calculations
  finite element, frequency domain, 402
  finite element, time domain, 407
  plane wave methods, 396
  transmission spectra, 398
Band gap, photonic structures, 395
Beam propagation, discontinuous Galerkin method, 449
Bloch theory for periodic structure, 392
Bloch wave expansions, 394
Boltzmann equations, 570
  moment equations, 572
Bosons, 477
Boundary conditions
  dielectric interfaces, 228
  Leontovich impedance, 230
Boundary element method (BEM)
  Nyström hyper-singular method for Maxwell's equations, 320
  Nyström weak singular integral method for Maxwell's equations, 324
  Poisson–Boltzmann equation, 86
Boundary integral equation and WOS (BIE-WOS), 134
  iterative BIE-WOS for mixed problems, 134
Bragg reflection coefficients, 399
Bragg transmission coefficients, 399
Bravais lattice vectors, 187
Brownian motions, 114
  local time, 127
  reflecting, 127

C
Cauchy principal integrals, 74
  direct computation, 91
Cavity field, 16
Cavity resonance, 315
Central difference scheme, 587
Charges, bound, 6
Charging, 41
Charging energy, 41
Charging process, 41
Clausius–Mossotti formula, 11
Collision operators
  Boltzmann, 572
  Fokker–Planck, 596
Combined integral equations and resonance, 315
Constrained transport methods for MHD equations, 608

D
Debye–Hückel inverse length, 35
Debye–Hückel theory, 34
Deep neural network (DNN), 161
  activation functions, 162
  multiscale DNN (MscaleDNN), 177

W. Cai, *Deterministic, Stochastic, and Deep Learning Methods for Computational Electromagnetics*, https://doi.org/10.1007/978-981-96-0100-4

phase shift DNN (PhaseDNN), 175
reduction of spectral bias, 179
spectral bias, 175
stochastic gradient descent (SGD), 163
Density functional theory (DFT), 512
DFT exchange-correlation energy, 514
DG-density matrix minimization (DG-DMM) method, 522
finite element solution, 521
Kohn-Sham (KS) DFT, 512
KS DFT orbital equations, 513
linear scaling methods, 522
plane wave solution, 519
pseudopotential, 516
self-consistent solution, 518
Density matrix, 484
device with contacts, 499
Density of states, electron in contacts, 494
Density operator, 482
DFT exchange-correlation energy, 514
general gradient approximation (GGA), 514
local density approximation (LDA), 514
Dielectric fluctuation formula, 24
total dipole moment, 18
Dielectric formula
Clausius–Mossotti, 11
Onsager formula for dipolar liquid, 13
Diode, n^+–n–n^+, 590
Dipole, 3
Directing field, 13
Discontinuous Galerkin methods for Maxwell's equations, 371
Displacement flux **D**, 7
DNN methods for PDEs, 164
DeepBSDE method, 168
DeepRitz method, 164
diffusion Monte Carlo (DMC) method, 174
FBSNN method, 169
Galerkin weak form learning method, 167
martingale based learning method, 171
Physics-informed neural network (PINN), 167
Drift-diffusion model, 570
Duffy mapping, 324
Dyadic Green's functions
homogeneous media, 256
layered media, 255
longitudinal components, 260
transverse components, 259
vector potential, 266
Dynkin's formula, 118
with stopping time, 118

E
Electrical impedance tomography (EIT), 129
WOS method for local solution, 130
Electric field integral equation (EFIE), 312
Electron continuity equation, 456
Electron correlation function, 504
Electron density, 511
Ensemble theory
canonical ensemble, 471
grand canonical ensemble, 474
microcanonical ensemble, 470
Ewald summation for charges and dipoles, 187
Exit time, 119
Extinction principle, 306
Extraordinary optical transmission (EOT), 420

F
Fast multipole method (FMM), 197
conversion operators for 3-D electrostatics, 205
for Helmholtz equations, wideband, 207
Helmholtz equations in layered media, 295
interaction list, 202
local expansions, 200
local expansions for free space, 286
local expansions for layered media, 289
local to local translations, 203
multipole expansions, 198
multipole expansions for free space, 286
multipole expansions for layered media, 289
multipole to local translations, 200
multipole to multipole translations, 203
polarization sources for ME/LE in layered media, 291
pseudo-code, 204
recursion for local expansions (downward), 201
recursion for multipole expansions (upward), 203
translation operator for free space, 287
Fermions, 477
Feynman–Kac formula, 120
Dirichlet problem, 120
mixed BVP problem, 122

G

Generalized Born approximations, 42
  Born radius, FFT-based calculations, 45
  Born radius, integral expression, 43
Godunov scheme, 579
  plasma MHD models, 610
  semiconductor hydrodynamic models, 579
Guided modes of optical waveguides, 441

H

Hadamard finite part integrals, 74
  direct computation, 91
Hamiltonian of N-electron systems, 509
Hankel transform, 258
  fast calculation with window-based filtering, 269
Helmholtz double layer, 36
Helmholtz vector decomposition, 226
Hertz dipoles, 253
  horizontally directed, 261
  vertically directed, 261
Hierarchical basis for DG methods, 375
  2-D and 3-D quadrilateral elements, 375
  2-D triangular elements, 377
  3-D tetrahedral elements, 377
Huygens' principle, 306
Hybrid model for electrostatics, 211
  dielectric constant calculated, 214
  molecular dynamics, 214
  reaction field, 211
Hydrodynamic equations, 577
  central difference scheme, 587
  weighted essentially non-oscillatory (WENO) method, 583
Hyper-singular integrals, 75
  direct computation, 91
  regularization method, 87

I

Image charges, 52
  conducting cylinder, 54
  conducting half-space, 52
  conducting sphere, 52
  dielectric cylinder, 53
  dielectric half-space, 53
  dielectric sphere in ionic solvent, 61
  dielectric sphere, Friedman, 56
  dielectric sphere, multiple discrete images, 61
  dielectric sphere, Neumann's line images, 57
  layered ionic solvent, 66
  layered non-ionic solvent, 65
Impedance boundary condition, Leontovich, 230
Integral equations
  combined integral equations and resonance, 315
  electric field integral equation, 312
  Galerkin method, 325
  magnetic field integral equation, 313
  Nyström collocation method, 320
  singular and hyper-singular integral equations, 307
  Stratton–Chu formula, 313
  surface IE for conductors, 314
  surface IE for dielectrics, 315
  surface IE for PEC, 314
  volume integral equations for Maxwell's equations, 416
  volume integral equations for quantum dots, 457
Ionic strength, 35
Itô integral, 114
  martingale property, 114
Itô processes, 114
Itô's formula, 116

K

Karhunen-Loève expansion, 149
Kirkwood expansion, 54
Kohn–Sham DFT orbital equations, 513
Kohn–Sham total energy functional, 513

L

Landauer formula, 498
Laplace–Beltrami operator, 242
Local time of reflecting Brownian motion, 121
Longitudinal field components, 260
Lorentz local field, 12

M

Magnetic field integral equation (MFIE), 313
Magnetization, 222
Magneto-hydrodynamics (MHD), 598
  eigen-systems, 610
Maxwell's equations, 224
  Ampère–Maxwell law, 223
  discontinuous Galerkin methods, 371
  Faraday's law, 221

finite difference Yee scheme, 383
Galerkin integral equation methods, 325
Gauss's law, 224
integral form, 384
magnetization, 222
Nyström method for hyper-singular integral equations, 320
Nyström method for weak singular integral equations, 324
potentials, electric and magnetic, 226
Rankine–Hugoniot conditions, 374
singular and hyper-singular integral equations, 307
time-harmonic, 224
uniaxial PML, 246
vector wave equations, 225
volume integral equations, 416
weak form, 369
MOSFET, double gate, 547

N

Nédélec edge elements, 345
2-D hierarchical basis in a quadrilateral, 356
2-D hierarchical basis in a triangle, 357
2-D reference square, 345
2-D reference triangle, 347
3-D hierarchical basis in a 3-D cube, 360
3-D hierarchical basis in a 3-D tetrahedron, 361
3-D reference cube, 349
3-D reference tetrahedron, 349
Piola transform, 344
Non-equilibrium Green's function (NEGF), 533
1-D device, 534
1-D device, boundary conditions, 535
1-D device, finite difference method, 536
1-D device, finite element method, 537
1-D device, self-energy, 537
2-D device, 540
2-D device, boundary conditions, 540
2-D device, finite difference method, 541
2-D device, finite element method, 544
2-D device, self-energy, 546
transmission coefficients, 507
Numerical DG fluxes for Maxwell's equations, 375
Numerical methods
adaptive spectral element method, 555
BIE-WOS for Dirichlet BVP, 134
BIE-WOS for mixed BVP, 134
central difference scheme, 587
collocation boundary element methods, 86
constrained transport methods for MHD equations, 608
DeepBSDE method, 168
DeepRitz method, 164
DFT linear scaling methods, 522
DG-density matrix minimization (DG-DMM), 522
diffusion Monte Carlo DNN method, 174
discontinuous Galerkin methods for dispersive media, 421
discontinuous Galerkin methods for Maxwell's equations, 371
discontinuous Galerkin methods for Schrödinger equations, 435
Ewald summation, 187
fast multipole method (FMM), 197
fast multipole method (FMM) in layered media, 295
FBSNN method, 169
finite element methods for vector Helmholtz eigenvalues, 402
Galerkin integral equation methods for Maxwell's equations, 325
generalized DG beam propagation method, 449
Godunov scheme, 579
image charge method, 51
Karhunen-Lovè expansion, 149
Kohn-Sham DFT, 513
finite element solution, 521
plane wave solution, 519
martingale based learning method, 171
multiscale DNN (MscaleDNN), 177
Nédélec edge elements for Maxwell's equations, 345
non-equilibrium Green's function (NEGF), 533
Nyström hyper-singular integral method for Maxwell's equations, 320
Nyström weak singular integral method for Maxwell's equations, 324
particle-in-cell (PIC) schemes for plasmas, 607
particle-mesh Ewald (PME) method, 194
phase shift DNN (PhaseDNN), 175

Physics-informed neural network (PINN), 167
plane wave methods for band gaps, 396
Poisson–Boltzmann equation
collocation boundary element methods, 86
finite element methods, 99
immersed interface methods (IIM), 102
polynomial chaos expansion, 147
random batch Ewald (RB-Ewald), 196
spectral methods (deterministic), 139
spectral methods (stochastic), 146
stochastic collocation methods, 153
stochastic Galerkin methods, 150
upwinding finite difference method, 561
Vlasov–Fokker–Planck (VFP) schemes, 604
volume integral equations for Maxwell's equations, 416
volume integral equations for quantum dots, 457
walk-on-spheres (WOS), 125
weak adversarial network (WAN), 167
weighted essentially non-oscillatory (WENO) method, 583
Yee scheme for Maxwell's equations, 383

O
Onsager formula for dipolar liquids, 13

P
Pardoux–Peng theory of nonlinear Feynman–Kac formula, 124
Particle-in-cell (PIC) schemes, 607
Particle-mesh Ewald (PME) method, 194
Perfectly matched layer (PML), uniaxial absorbing media, 242
Planck constant h, 454
Plasma transport models
MHD single-fluid, 598
particle-in-cell, 607
Vlasov–Fokker–Planck kinetic, 596
Poisson–Boltzmann equation (PBE), 35
energy variational, 42
hyper-singular surface integral equations, 85
linearized, 35
modified with steric size effect, 36
surface integral representations, 79
Poisson equation, 32
Poisson–Nernst–Planck (PNP) model, 570
Polarization density, 6
Polarization field, 8
Polarization sources, 292
Polynomial chaos (PC)
generalized PC (gPC), 149
Wiener PC, 147
Potentials of electromagnetic fields, 226
Sommerfeld, 266
transverse, 268
Pseudopotential, 516

Q
Quantum dots, volume integral equation method, 457
Quantum many particles
bosons, 479
fermions, 479
partition functions, 477
Pauli exclusion principle, 477
Slater determinants, 476

R
Radiation conditions
Silver–Müller, 233
Sommerfeld, 233
Random batch Ewald (RB-Ewald), 196
Rayleigh–Bloch waves, 409
Reaction field for Green's function in layered media, 289
Reciprocal lattice vectors **G**, 188
Resonant tunneling diode (RTD), 564
I-V curves by Wigner equation and NEGF, 565
Rigorous coupled wave analysis (RCWA)
transmission spectra, 398
RWG current basis, 328
hierarchical basis, 328
triangular–quadrilateral patches, 336
triangular–triangular patches, 331

S
Schrödinger equations
for electrons with an effective mass, 455
paraxial approximation in waveguides, 449
Singular integral potentials
double-layer, 74
single-layer, 74
Singular sources
charges, 251

currents, 253
Hertz dipoles, 253
Skorokhod problem, 121
Solvation, 31
Solvation energy
Born, 40
charging process, 41
electrostatic, 40
Still's generalized Born approximations, 42
Sommerfeld radiation condition and uniqueness, 234
Sparse grid, 153
Spectral function, 504
Spectral methods
deterministic, 139
Galerkin vs collocation, 146
spectral accuracy (deterministic collocation method), 145
spectral accuracy (deterministic Galerkin method), 142
stochastic, 146
Statistical distribution
bosons, 479
fermions, 480
Stochastic gradient descent (SGD), 163
adaptive moment estimation (ADAM), 163
Stochastic spectral methods
sparse grids and collocation methods, 153
stochastic collocation methods, 153
stochastic Galerkin methods, 150
Stopping time, 119
Stratton–Chu formula, 313
Surface differential operators, 318
Surface plasmons
dispersion relation, 414
localized modes, 414
propagating modes, 411
resonant coupling, 421

T

Thermal equilibrium, 468
Time-harmonic Maxwell's equations, 224
Transmission coefficient, 492
single-barrier, 493
Transport models (classical)
hydrodynamic equations, 577
semi-classical Boltzmann equations, 570
Transport models (quantum)
non-equilibrium Green's function, 507
Wigner equations, 491
Transverse electric (TE) wave, 342
Transverse field components, 259
Transverse magnetic (TM) wave, 425
Tsu–Esaki current formula, 496

U

Uncertainty quantification, 155
capacitance of rough microchip interconnects, 155
Uniformly polarized sphere, 26

V

Vlasov–Fokker–Planck (VFP) equations, 596
Vlasov–Fokker–Planck (VFP) schemes, 604

W

Walk-on-spheres (WOS), 125
Weak adversarial network (WAN), 167
Weak form of Maxwell's equations, 369
Weighted essentially non-oscillatory (WENO) method, 583
Weyl correspondence of quantum operators, 491
Whispering gallery mode (WGM), 380
Wigner distribution, 485
Wigner equation
adaptive spectral element method, 555
continuity equation, 491
current density, 491
derivation, 486
electron density, 490
Frensley inflow boundary condition, 554
mesh selection for finite difference method, 563
truncation in phase space, 552
upwinding finite difference method, 561
Wigner–Moyal expansion, 489

Y

Yee finite difference method for Maxwell's equations, 385